Proceedings of an International Congress on Silvopastoralism and Sustainable Management held in Lugo, Spain, in April 2004

Edited by
M.R. Mosquera-Losada and A. Rigueiro-Rodríguez
Universidad de Santiago de Compostela
Lugo
Spain
And
J. McAdam
Queen's University Belfast
Northern Ireland

CABI Publishing

This book has been funded by the European Union as Accompanying Measure QLAM-2001-00512

CABI Publishing is a division of CAB International

CABI Publishing
CAB International
Wallingford
Oxfordshire OX10 8DE
UK

CABI Publishing
875 Massachusetts Avenue
7th Floor
Cambridge, MA 02139
USA

Tel: +44 (0)1491 832111
Fax: +44 (0)1491 833508
E-mail: cabi@cabi.org
Website: www.cabi-publishing.org

Tel: +1 617 395 4056
Fax: +1 617 354 6875
E-mail: cabi-nao@cabi.org

©CAB International 2005. All rights reserved. No part of this publication may be reproduced in any form or by any means, electronically, mechanically, by photocopying, recording or otherwise, without the prior permission of the copyright owners.

A catalogue record for this book is available from the British Library, London, UK.

Library of Congress Cataloging-in-Publication Data

International Congress on Silvopastoralism and Sustainable Management (2004 : Lugo, Spain)
 Silvopastoralism and sustainable land management : proceedings of an International Congress on Silvopastoralism and Sustainable Management held in Lugo, Spain, in April 2004 / Edited by M.R. Mosquera-Losada, A. Rigueiro-Rodríguez and J. McAdam.
   p. cm.
 Includes bibliographical references and index.
 ISBN 1-84593-001-0 (alk. paper)
1. Agroforestry-Congresses. I Mosquera-Losada, M.R. II. Rigueiro-Rodríguez, A. III. McAdam, J. (Jim) IV. Title.

A464.5.A45I574 2006
634.9'9--dc22

ISBN-10:   1 84593 001 0
ISBN-13: 978 1 84593 001 1

Printed and bound in the UK, from copy supplied by the editors, by Cromwell Press, Trowbridge.

CONTENTS
9. Introduction
10. Acknowledgements and Congress information

12. Session 1. Silvopastoral systems. Main types and designs

Main invited Key note
13. Traditional and improved silvopastoral systems and their importance in sustainability of livestock farms
    M. Ibrahim, C. Villanueva and J. Mora.

Invited Key notes
19. Silvopastoral systems in North-West Europe
    J. H. McAdam.

24. Silvopastoral systems in the Neotropics
    E. Murgueitio.

30. European types of silvopastoral systems in the Mediterranean area: dehesa
    L. Olea, R. J. López-Bellido and M. J. Poblaciones.

36. Mediterranean European silvopastoral systems
    A. San Miguel-Ayanz.

Offered papers
41. Agroforestry for improving farm smallholdings
    M. R. Alam, M. D. Jahiruddin and M. S. Islam.

44. Improved fallows with forage trees for a slash and burn maize system in the Yucatán Peninsula
    S. A. Ayala, V. G. Uribe and G. J. A. Basulto.

46. The influence of climate and soil type on the "montado" production system
    A. Quintas, M. S. Pereira, P. J. Carmona, D. Ramirez-Cruzado and C. C. Belo.

49. Silvopastoralism with native tree species in Uruguay
    L. Gallo.

51. Classification principles and use of grazed forest stands in the Ukraine
    G. B. Gladun.

53. Traditional silvoarable systems and their evolution in Greece
    K. Mantzanas, E. Tsatsiadis, I. Ispikoudis and V. P. Papanastasis.

55. Biodiversity and sustainable development in the silvopastoral systems of the Cantabrian mountains
    M. Mayor López, J. A. Oliveira Prendes and M. Fernández Benito.

58. Goat production system in mountain zones: the case of Peneda's mountain
    L. F. Pacheco, J. Pires, A. Iglesias, J. Cantalapiedra, D. Barreto and J. P. Araújo.

60. Silvopastoralism in Evritania, central Greece
    A. Pantera and A. M. Papadopoulos.

62. The effect of duck grazing on cocoa yields in São Tomé island
    A. Pardini.

64. Types of land use in the montado (dehesa) production system
    M. S. Pereira, A. Quintas, I. S. Coelho and C. C. Belo.

66. Contribution of silvopastoral systems to land upkeep: differences and determining factors in a French upland region
    H. Rapey and R. Lifran.

68. From forest grazing to multipurpose shelterbelts in Hungary
   V. Takács and N. Frank.

71. **Session 2. Productivity, quality and management of silvopastoral systems**

Main invited Key note
72. Silvopasture in south-eastern United States: more than just a new name for an old practice
   P. K. R. Nair, M. E. Bannister, V. D. Nair, J. R. R. Alavalapati, E. A. Ellis, S. Jose and A. J. Long.

Invited Key notes
83. Nutritive value of trees and shrubs for ruminants
   J. C. Ku-Vera.

87. Quality of vegetation in silvopastoral systems
   M. P. González-Hernández.

93. The herbaceous component in temperate silvopastoral systems
   M. R. Mosquera-Losada, M. Pinto-Tobalina and A. Rigueiro-Rodríguez.

Offered papers
101. The effect of pruning intensity on acorn production in a holm-oak "dehesa"
   R. Alejano, J. Alaejos, E. Torres, J. A. Forero, R. Tapias and M. Fernández.

103. The effect of tree density on pasture yield
   H. Arzani.

106. An alternative to improve goat production in the mountain areas of north-west Tunisia
   A. Boubaker, C. Kayouli and A. Buldgen.

108. Seasonal fodder biomass production of palo dulce and tepozan
   D. Camacho-Morfín, C. Sandoval-Castro, A. Ayala-Burgos and L. Morfín-Loyden.

110. Pruning influence on acorn yield in cork-oak open woodland
   I. Cañellas, S. Roig and G. Montero.

112. Effects of grazing on fuel biomass in mountain pastures of the south-eastern Pyrenees
   C. Dal Zennaro, J. M. Arenas, G. Argenti, A. Pardini and T. M. Sebastia.

114. Seed production and influence of heat on the germination of *Genista scorpius* (L.) DC
   I. Delgado, M. J. Ochoa and F. Múñoz.

117. Lithium chloride and ipecacuanha syrup to induce sheep aversion to the intake of coffee plants
   C. A. Durantes, J. A. Torres, P. A. Martínez, R. Castro, A. Arroyo and J. G. Cruz.

119. Dispersed trees in pasturelands of cattle farms in a dry ecosystem of Costa Rica
   H. Esquivel, M. Ibrahim, C. A. Harvey, C. Villanueva, F. L. Sinclair and T. Benjamín.

121. Micropropagation of three clones of *Morus alba* L. selected for fodder use
   J. L. Fernández-Lorenzo, V. Pérez, S. Liñayo, M. R. Mosquera-Losada and A. Rigueiro-Rodríguez.

124. Estimating diet selection of goats and sheep grazing on gorse-heathland vegetation with areas of improved pasture
   L. M. M. Ferreira, M. Oliván, M. A. M. Rodrigues, U. García and K. Osoro.

127. Validation of the alkane technique to estimate complex diets in sheep and goats fed on heathland vegetation
   L. M. M. Ferreira, M. Oliván, M. A. M. Rodrigues, U. García and K. Osoro.

129. Effects of grazing on *Quercus faginea* Lam. forests in Navarra (Spain)
   V. Ferrer, C. Ferrer, A. Broca and M. Maestro.

132. Effect of breed and stocking rate on live weight changes of goats grazing heath-gorse vegetation community
    U. García, R. Celaya, B. M. Jáuregui, A. Martínez and K. Osoro.

134. Response of sheep fed with tropical tree legume foliages and Taiwan grass (*Pennisetum purpureum*)
    D. Grande, F. Reyes, H. Losada, J. Nahed, N. Romero, G. Valdivieso and F. Pérez-Gil.

137. Nutritional characterization of some Mediterranean forestry resources
    A. Hajer, S. Lopez and A. Chermiti.

140. Behaviour at grazing of calves with Galician Blonde suckler cows
    A. Iglesias, J. A. Carballo, A. López and L. Monserrat.

142. Vegetation dynamics of burned heath-gorse communities grazed by sheep or goats
    B. M. Jáuregui, R. Celaya, U. García and K. Osoro.

145. Adaptation of herbaceous plant species in the understorey of *Pinus brutia*
    Z. Koukoura and A. Kyriazopoulos.

148. Transition of an abandoned Dutch agrosilvopastoral landscape to 'new wilderness' by extensive grazing with free-ranging cattle and horses
    A. T. Kuiters.

150. Live fences as a means of sustainable integration of livestock, trees and crops in Ségou, Mali
    V. Levasseur, A. Niang and A. Olivier.

152. Comparison of terrestrial and aerial oversowing of ski lanes grazed by sheep in the northern Apennines (central Italy)
    F. Longhi, A. Pardini, L. Ghiselli and R. Tallarico.

154. Tree growth and pasture production under sewage sludge fertilization
    M. L. López-Díaz, M. R. Mosquera-Losada and A. Rigueiro-Rodríguez.

157. Utilization of gorse and heather communities by cattle in mountain grazing
    N. Mandaluniz, L. M. Oregui and A. Aldezabal.

160. Importance of holm-oak (*Quercus ilex*) as a woody food resource in Spanish ibex (*Capra pyrenaica*) diet in Mediterranean forest
    T. Martínez.

163. Woodland grazing in northern Ireland: effects on botanical diversity and tree regeneration
    P. McEvoy and J. H. McAdam.

165. Application of limed sewage sludge to a young *Pinus radiata* plantation on acid soil
    B. Omil-Ignacio, M. R. Mosquera-Losada, A. Rigueiro-Rodríguez and A. Merino-García.

168. Effect of season on the ingestive behavior of cattle grazing *Leucaena leucocephala*
    R. L. Ortega, H. J. Castillo and P. F. Rivas.

170. Horse grazing on a mixture of *Trifolium brachycalycinum* and *Cynodon dactylon* in firebreaks of Tuscan Maremma (central Italy)
    A. Pardini and F. Natali.

172. Dry matter production and nutritive value of cocksfoot (*Dactylis glomerata* L.) grown under different light regimes
    P. L. Peri, R. J. Lucas and D. J. Moot.

174. Evaluation of the production of acorns of the oak (*Quercus ilex* Lam. ssp. *ballota*) from south-west of Extremadura (Spain)
    M. J. Poblaciones, R. López-Bellido, L. Olea and C. Benito.

177. Effect of different tree and shrub densities on the form, agronomic performance and quality of *Panicum maximum* in the Chaqueño Mountains, Bolivia
   P. P. del Pozo, R. Rizzo and E. Fernández.

180. Effects of management on acorn production and viability in holm-oak dehesas
   F. J. Pulido, E. García, J. J. Obrador and M. J. Montero.

182. Effect of fertilization, tree species, plantation density and sowing mixture on pasture production and soil characteristics in silvopastoral systems
   A. Rigueiro-Rodríguez, E. Fernández-Núñez and M. R. Mosquera-Losada.

185. Structure characterization of *Quercus pyrenaica* Willd open woodlands in the Spanish Central Mountain.: implications for silvopastoral management
   S. Roig, M. Río, I. Cañellas, A. Bravo and G. Montero.

188. Pasture production under different tree species and spacing in an Atlantic silvopastoral system
   M. J. Rozados, M. P. González-Hernández and F. J. Silva-Pando.

192. Effects of dairy sludge application on pasture production and arbuscular mycorrhizae in a sown meadow established in a hill soil
   M. J. Sainz, X. A. Alonso, A. Vilariño and M. E. López-Mosquera.

195. Forage yield and botanical composition of a hill sown meadow fertilized with dried pelletized broiler litter
   M. J. Sainz, M. J. Bande, F. Cabaleiro and M. E. López-Mosquera.

197. Establishment of a bovine/*Quercus* silvopastoral experiment in lowland Ireland
   I. Short, J. McAdam, N. Culleton and G. Douglas.

200. Integration of horses to orange tree plantations
   L. Simón, M. D. Sánchez, M. Hernández, S. Sánchez and C. Mendoza.

202. Facilitation of tree regeneration in pasture woodlands
   C. Smit, D. Béguin, A. Buttler and H. Müller-Schärer.

204. Effect of sheep grazing on coffee quality
   J. A. Torres.

206. The influence of spacing and coppice height on herbage mass and other growth characteristics of *Robinia pseudoacacia* in a south-eastern USA silvopastoral system
   L. J. Unruh Snyder, J. P. Mueller, J. M. Luginbuhl and C. Brownie.

209. Session 3. Ecological implications of the silvopastoral systems: biodiversity and sustainable management

Invited Key notes
210. How much carbon can be stored in Canadian agroecosystems using a silvopastoral approach?
   A. M. Gordon, R. P. F. Naresh and V. Thevathasan.

219. Silvopastoral systems in Latin America and their contribution to sustainable development and biodiversity
   I. Hernández and M. D. Sánchez.

223. Compaction and erosion: effects on soil ecology and soil quality
   A. Paz González and E. Vidal Vázquez.

231. Indigenous breeds and silvopastoral systems
   L. Sánchez.

Offered papers
236. Silvopastoral systems to prevent soil losses in sustainable livestock systems
   Z. G. Acosta, G. Reyes and J. L. Montejo.

239. Spatial dependence and seasonal patterns of cattle activity
    A. Buttler, F. Kohler, H. Wagner and F. Gillet.

241. Impacts of cutting and fertilization on pasture systems in the Cantabrian mountain range in León province (NW Spain)
    L. Calvo, A. Fernández, E. Marcos, L. Valbuena, R. Tárrega and E. Luis.

244. Floristic stability of pastures in the Sierra Mágina nature reserve, Andalusia, Spain
    A. Cano-Ortiz, A. García-Fuentes, J. A. Torres, R. Montilla, L. Ruiz, C. Salazar and E. Cano.

246. Effects of breed and stocking rate on vegetation dynamics and biodiversity in heath-gorse communities grazed by goats
    R. Celaya, B. M. Jáuregui, U. García and K. Osoro.

248. Biodiversity and dynamics of traditional sylvopastoral systems in Galicia (north-west Spain): *Cytisus* scrubs
    T. Cornide, E. Díaz-Vizcaíno and M. Casal.

251. Root density and soil water relationships of a silvopastoral system of the tropical region of Yucatan, Mexico
    H. G. Delgado, L. A. Ramírez, J. V. Ku, P. M. Velásquez and J. B. Escamilla.

255. Cover crops effects on plant and insect biodiversity in Western Australian vineyards
    A. Dinatale, A. Pardini and G. Argenti.

257. Live weight changes of sheep and goats grazing a burned heath-gorse vegetation community
    U. García, B. M. Jáuregui, R. Celaya, A. Martínez and K. Osoro.

259. Management and ecological implications of silvopastoral systems in the Alps
    A. C. Mayer.

261. Light availability for understorey pasture in holm-oak dehesas
    M. J. Montero and G. Moreno.

263. Consequences of dehesa management on tree-understorey interactions
    G. Moreno, J. Obrador, E. García, E. Cubera, M. J. Montero and F. Pulido.

266. Pasture establishment for extensive systems
    M. R. Mosquera-Losada, S. Rodríguez-Barreira and A. Rigueiro-Rodríguez.

269. Biological diversity in communities of Erica ciliaris and Erica tetralix: analysis of their spatial and temporal variation
    A. Muñoz, X. M. Pesqueira, R. Álvarez, O. Reyes and M. Casal.

272. Silvopasture as an approach to reducing nutrient loading of surface water from farms
    V. D. Nair and R. S. Kalmbacher.

275. Some ecological impacts of *Quercus rotundifolia* trees on the understorey environment in the "montado" agrosilvopastoral system, southern Portugal
    J. Nunes, M. Madeira and L. Gazarini.

278. Soil nutrient status and forage yield at varying distances from trees in four dehesas in Extremadura, Spain
    J. J. Obrador and G. Moreno.

281. Rangeland health assessment in silvopastoral systems of northern Greece
    Ch. I. Pantazopoulos, M. S. Vrahnakis, D. Chouvardas, M. Papadimitriou and V. P. Papanastasis.

283. Post-fire ecological and dynamic characterization of shrubland communities in Galicia using structural variables
    X. M. Pesqueira, A. Muñoz, R. Álvarez, O. Reyes and M. Casal.

285. Extensive livestock systems as tools for environmental management: impact of grazing on the vegetation of a protected mountain area
J. L. Riedel, I. Casasús, A. Sanz, M. Blanco, R. Revilla and A. Bernués.

288. Historical effects of grazing on tree establishment in the Cantabrian lowlands, northern Spain: a dendroecological analysis in two old-growth forests
V. Rozas.

290. Changes in biodiversity after abandonment in dehesa systems in the province of León
R. Tárrega, L. Calvo, C. Diez, E. Luis, L. Valbuena and E. Marcos.

293. Session 4. Economic, social and cultural benefits of the silvopastoral systems

Main invited Key note
294. Economic considerations of silvopastoralism in California oak woodlands
R. B. Standiford, L. Huntsinger, P. Campos-Palacín and A. Caparrós.

Invited Key notes
299. Silvopastoral management in temperate and Mediterranean areas. Stakes, practices and socio-economic constraints
M. Etienne.

312. Conservation "matching funds" from working woodlands in California
L. Huntsinger, A. Sulak, R. Standiford and P. Campos-Palacín.

319. Cultural aspects of silvopastoral systems
I. Ispikoudis and K. M. Sioliou.

324. Comparative analysis of the EAA/EAF and AAS agroforestry accounting systems: theoretical aspects
P. Campos-Palacín, P. Ovando-Pol and Y. Rodríguez-Luengo.

330. Comparative analysis of the EAA/EAF and AAS agroforestry accounting systems: application to a dehesa estate
Y. Rodríguez-Luengo, P. Campos-Palacín and P. Ovando-Pol.

335. Preliminary analysis of the impact of payment for environmental services on land-use changes: a case study on livestock farms in Costa Rica
J. Mora, M. Ibrahim, J. Cruz, F. Casasola, M. Rosales and V. A. Holguin.

Offered papers
343. Adaptation of an agrosilvopastoral system to land-use dynamics: local-level analysis of strategies and practice changes in north-eastern Portugal
J. Alonso and J. Bento.

346. Agropastoral systems in Cholistan
A. Farooq, F. Gulzar, S. A. Safdar, F. Sameera and A. Zulfiqar.

348. An evaluation of the effects of forest conservation on reservoir capacity: a case study in the "Cuerda del Pozo" reservoir (Soria)
R. García Díaz, F. García Robredo and P. A. Medrano Ceña.

351. Characterization of tree species in silvopastoral systems in the mountain region of Tabasco, Mexico
D. Grande, G. Pérez, H. Losada, M. Maldonado, J. Nahed and F. Pérez-Gil.

355. Non-wood products in Russia
A. V. Griazkin and T. D. Smelkova.

357. Transhumance and silvopastoral dependence in the Great Himalayan National Park Conservation Area – a landscape-level assessment
P. K. Mathur and B. S. Mehra.

359. Economic, social and cultural benefits of silvopastoral systems in Nigeria
I. O. Oladele.

361. Ancient wood-pastures in Scotland
M. A. Smith.

363. Environmental and resource economics: would the Delphi method follow a non-iterative process of surveys?
M. Soliño.

365. The influence of goat grazing on ground vegetation and trees in a forest stand
A. Zingg and P. Kull.

368. Session 5. Future perspectives of silvopastoral systems in a global context

Main invited Key note
369. Silvopastoral systems for rural development on a global perspective
A. Pardini.

Invited Key notes
374. A silvopastoral system for Eastern Europe – based on the example of Poland
K. Boron.

376. From silvopastoral to silvoarable systems in Europe: sharing concepts, unifying policies
C. Dupraz.

380. Silvopastoral systems as a forest fire prevention technique
A. Rigueiro-Rodríguez, M. R. Mosquera-Losada, R. Romero-Franco, M. P. González-Hernández and J. J. Villarino-Urtiaga.

388. Forestry, pastoral systems and multiple use woodland
F. J. Silva-Pando.

Offered papers
395. An assessment of the role of grazing in European habitats
R. G. H. Bunce, M. Pérez-Soba, B. S. Elbersen and W. K. R. E. van Wingerden.

397. Situation and perspectives of silvopastoral systems in Germany
P. Finck, U. Riecken and F. Glaser.

400. Wood-pasture and parkland: overlooked jewels of the English countryside
R. Isted.

402. Local capabilities development and silvopastoral intervention in Chiapas, Mexico
G. Jiménez-Ferrer, L. Soto-Pinto, J. Nahed-Toral, T. Aleman, B. Ferguson, M. Ibrahim and F. Sinclair.

403. Silvopastoralism as a land-use option for sustainable development on grassland farms in Northern Ireland
J. H. McAdam.

406. The potential for agroforestry in the Falkland Islands
J. H. McAdam.

408. Future perspectives for silvopastoral systems in NW Spain
A. Rigueiro-Rodríguez, M. Rois-Díaz, M. Pinto, A. Oliveira and M. R. Mosquera-Losada.

411. Outdoor pig production in the Basque country (Spain)
R. Ruiz, A. Domingo and L. M. Oregui.

413. Technology transfer in silvopastoral agroforestry: a toolbox for UK farmers
A. R. Sibbald.

415. Model and procedures for decision-making in management of diffusion, adoption and improvement process of agroforestry technology
   M. J. Suárez Hernández, G. Hernández Pérez and R. Suárez Mella.

418. Declaration for silvopastoralism
419. Summary - silvopastoral systems conclusions
422. Index

INTRODUCTION

Agroforestry and silvopasture (the main subject of the Congress) are ancient ways of managing forestland that should be encouraged as they increase productivity in the short, medium and long term (in comparison with forestland), biodiversity (in comparison with farmland) and the sustainability of land (multiproduct system). Silvopastoralism will address important problems related to the multiple benefits of forests, fire and erosion risk reduction and countryside conservation.

The importance at a global scale of silvopastoral systems is highlighted in the main international documents related to sustainable use of the resources like **Rio documents** (Convention on Biological Diversity, Convention on Climatic Change, Convention to Combat Desertification and Rio Declaration on Environment Development and Agenda 21), and **Johannesburg** Summits. In this sense, in Agenda 21 where agroforestry practices, and therefore silvopastoral systems, are mentioned as a sustainable way of land management in order to fulfil the objectives of the Chapters 11 (*Combating deforestation*), 12 (*Managing fragile ecosystems: combating desertification and drought*), 13 (*Managing fragile ecosystems: sustainable mountain development*), 14 (*Promoting sustainable agriculture and rural development*) and 15 (*Conservation of biological diversity*) of this important global action plan: Agenda 21.

One of the main conclusions of the congress is that the implementation of silvopastoralism should be based on knowledge of traditional and current silvopastoral systems (which therefore takes into account social and cultural aspects) and on the improvement of those through the knowledge of the specific dynamics of forest ecosystems, relationship between forest operations and ecosystem stability, nutrient cycling, sustainability in forest management and evaluation of the multibenefits of forests, which are important objectives into the global action research on multifunctional management of forest.

The book is structured in the five main sessions of the Congress. The first one includes the characterization of silvopastoral systems in a global context and the second involves the effects of the management tools on the productivity and quality of silvopastoral systems; the third session is related to the ecological implications of the silvopastoral systems, highlighting the biodiversity and sustainable management aspects; the subject of the fourth session is the main economical, social and cultural aspects of silvopastoral systems; finally, the perspectives of these systems in a global and European context will be evaluated in the fifth session. From the general discussion, the main conclusions of the Congress as well as a declaration were delivered and they can be seen at the end of the book.

Acknowledgements and Congress information
The editors would like to thank the European Union for financial assistance (QLAM-2001-00512) and the sponsors FAO, EFI, Xacobeo, Diputación provincial de Lugo, Xunta de Galicia, Concello de Lugo, Junta de Castilla y León, Sargadelos, Ayuntamiento de Salamanca Universidad de Extremadura, Asociación Pura raza cabalo galego, Asociación forestal de Salamanca.

President
Dr A. Rigueiro-Rodríguez
Univ. Santiago de Compostela, Spain

Secretary
Dr M.R. Mosquera-Losada
Univ. Santiago de Compostela, Spain

Organizing Institutions
Crop Production Department (High Polytechnic School of Lugo). Santiago de Compostela University
Agroforestry systems working group of Spanish Society of Forestry Science

Address
Silvopastoralism and Sustainable Management International Congress
Crop Production Department
Escuela Politécnica Superior
Univ. Santiago de Compostela
Campus de Lugo. 27002-LUGO
SPAIN

Fax 00-34-982-285926
E-mail: ssm2004@lugo.usc.es

More Information
http://www.usc.es/ssm2004

Original, Design and Composition:
Divina Vázquez Varela
Esther Fernández Núñez
José Javier Santiago Freijanes
Silvia Rodríguez Barreiro
Teresa Piñeiro López

Translation into English:
Jim McAdam
Dr Scott Laidlaw
Alan Sibbald

Scientific Committee

Dr. A. Rigueiro-Rodríguez, Univ. Santiago, Spain
Dr. M.R. Mosquera-Losada, Univ. Santiago Spain
Dr. M. Sánchez, FAO, Italy
Dr. J. McAdam, Univ. Belfast, United Kingdom
Dr. McEvoy Peter, Queen's University, Northern Ireland.
Dr. E. Murgueitio, CIPAV, Colombia
Dr. A. Pardini, Univ. Florence, Italy
Dr. J. Benavides, BCS, Costa Rica
Dr. V. Papanastasis, Aristotle Univ., Greece
Dr. J. Coelho, Univ. Porto, Portugal
Dr. C. Boron, Univ. de Krakov, Poland
Dr. J. Piñeiro-Andión, CIAM (Consellería Política Agroalimentaria y Desarrollo Rural, Xunta de Galicia), Spain
Dr. Muhammad Ibrahim, CATIE, Costa Rica
Dr. Leonel Simón, Est. Exp. Pastos y Forrajes Indio Hatuey, Mº. de Ed. Superior, Cuba
Dr. P.K. Nair, IFAS, Univ. Florida, USA
D.M. Rois-Díaz, Univ. Santiago, Spain

Dr. J.M. de Miguel, Univ. Comp. Madrid, Spain
Dr. M.P. González-Hernández, Univ. Santiago, Spain
Dr. M. Pinto, NEIKER, Spain
Dr. A. San Miguel, Univ. Politécn. Madrid, Spain
Dr. T. Vidrih, Univ. Ljubljana, Eslovenia
Dr. P. Campos Palacín, CSIC, Spain
Dr. F.J. Silva-Pando, CIFA Lourizán (Consellería de Medio Ambiente, Xunta de Galicia), Spain
Dr. C. Dupraz, INRA, Francia
Dr. L. Olea, Univ. Extremadura, Spain
Dr. V. Nair (Univ. Florida), USA
Dr. S. Roig, INIA, Spain
Dr. I. Cañellas, INIA, Spain
Dr. G. Montero, INIA, Spain
Dr. I. Hernández-Venéreo, Est. Exp. Pastos y Forrajes Indio Hatuey, Mº. de Ed. Superior, Cuba
Dr. González-Rodríguez, CIAM (Consellería Política Agroalimentaria y Desarrollo Rural, Xunta de Galicia), Spain
Dr. A. Sibbald, United Kingdom Agroforestry Forum, United Kingdon
Ilmo. Sr. D. Francisco Maseda Eimil, Director General del Centro de Desarrollo Sostenible (Consellería de Medio Ambiente, Xunta de Galicia), Spain
D.E. Villada, Consellería de Política Agroalimentaria y Desarrollo Rural, Xunta de Galicia, Spain

Organizing committee

Dr. A. Rigueiro-Rodríguez, Univ. Santiago, Spain
Dr. M.R. Mosquera-Losada, Univ. Santiago, Spain
Dr. M. Sánchez, FAO, Italy
Dr. J.M. de Miguel, Univ. Comp. Madrid, Spain
Dr. M.P. González-Hernández, Univ. Santiago, Spain
Dr. R. Romero-Franco, Univ. Santiago, Spain
Dr. J.J. Villarino-Urtiaga, Univ. Santiago, Spain
Dr. M.L. López-Díaz, Univ. Extremadura, Spain
Dr. L. Olea, Univ. Extremadura, Spain
Dr. J.A. Oliveira-Prendes, Univ. Oviedo, Spain
Dr. M. Pinto, NEIKER, Spain
Dr. A. San Miguel, Univ. Politécn. Madrid, Spain
Dr. F.J. Silva-Pando, CIFA de Lourizan (Consellería de Medio Ambiente), Xunta de Galicia, Spain
Dr. M.A. Rodríguez-Guitián, Univ. Santiago, Spain
Dr. J.L. Fernández-Lorenzo, Univ. Santiago, Spain
Dr. B. Pajari, EFI, Filandia
D.M. Rois-Díaz, Univ. Santiago, Spain
D.J.J. Santiago-Freijanes, Univ. Santiago, Spain
D.M.C. Mosquera-Losada, Univ. Santiago, Spain
D.E. Fernández-Núñez, Univ. Santiago, Spain
D.S. Rodríguez-Barreira, Univ. Santiago, Spain
Ilmo. Sr. D. Francisco Maseda Eimil, Director General del Centro de Desarrollo Sostenible (Consellería de Medio Ambiente, Xunta de Galicia), Spain
Ilmo. Sr. D. Tomás Fernández-Couto Juanas (Director General de Montes e Industrias Forestales, Consellería Medio Ambiente, Xunta de Galicia), Spain.

Technical Personnel

D. Divina Vázquez-Varela, Univ. Santiago, Spain
D. Teresa Piñeiro-López, Univ. Santiago, Spain

Session 1.
Silvopastoral systems. Main types and designs

Traditional and improved silvopastoral systems and their importance in sustainability of livestock farms

M. Ibrahim [1], C. Villanueva [2] and J. Mora [3]
Environmental Livestock Management Group, CATIE, Costa Rica, [2] Researcher in Silvopastoral Systems, CATIE, Costa Rica, [3] Coordinator of Livestock, Environment and Development Initiative LEAD/FAO, CATIE, Costa Rica, mibrahim@catie.ac.cr, cvillanu@catie.ac.cr, jmora@catie.ac.cr

Abstract
Conversion of primary forest to pasture is widespread in Central and Latin America and the progressive removal of trees has resulted in environmental degradation and a decline in productivity, simultaneously threatening rural livelihoods and regional biodiversity. This chapter presents data on the ability of traditional and improved silvopastoral systems to enhance livestock productivity and rural livelihoods. We hypothesized that silvopastoral systems are more productive and provide greater social and environmental benefits than traditional cattle production systems in Central America.

Introduction
Conversion of primary forest to pasture is widespread in Central America and the progressive removal of trees has resulted in environmental degradation and a decline in productivity, simultaneously threatening rural livelihoods and regional biodiversity (Pezo and Ibrahim, 1999; Szott et al., 2000). Pasture covers more than 9 million ha of Central America alone, half of which is estimated to be degraded, affecting the livelihoods of 10 million people (Szott et al., 2000). From an environmental point of view, the biophysical conditions (e.g. soil, climate and topography) in the Central American tropical areas are inappropriate for production models based on management of grass monocultures and the use of this pasture leads to land and environmental degradation. Throughout Latin America, livestock farmers are engaged in establishing silvopastoral systems which involve the integration of woody perennials (shrubs and trees) with animal and pasture components and are classified based on the structure and functional role of trees in these systems (Pezo and Ibrahim, 1999). The level of interactions between components in silvopastoral systems depends on the spatial arrangements and/or configuration and diversity of trees in the pastures. Silvopastoral systems have the potential to reduce the impact of livestock systems on the environment in the long term and enhance both livestock productivity and rural livelihoods. Research in Central America in both the highlands and the lowlands has demonstrated that traditional cattle production systems are not labour intensive (2-3 field workers/farm) and have a lower demand for labour compared to other agricultural systems (Ruiz García, 2002; Gobbi and Ibrahim, 2004).

Retaining and managing trees on pasture may improve farm productivity and sustainability by generating tree products and providing ecosystem services (Ibrahim et al., 2001a). Within fragmented forest landscapes, farm trees may also represent critical habitats and corridors for plant and animal species and help maintain local and regional biodiversity (Harvey and Haber, 1999).

Recently, farmers have become interested in managing trees in pastures because they can provide high-quality feed, especially during the dry season, and other benefits such as timber, carbon sequestration and the conservation of biodiversity (Harvey and Haber, 1999; López et al., 1999; Souza de Abreu et al., 2000). Studies have been carried out on traditional silvopastoral systems, fodder bank, live fences, alley pasture systems and grazing in forest plantations in Central America. This chapter presents data on traditional and improved silvopastoral systems and their ability to enhance livestock productivity and rural livelihoods. It is hypothesized that silvopastoral systems are more productive and provide greater social and environmental benefits than traditional cattle production systems in Central America.

Traditional silvopastoral systems
In Latin America there are good examples of traditional silvopastoral systems where trees establish in pastures by natural regeneration. These systems are characterized by multi-strata of trees and shrubs that are grown in isolation and/or in groups (Cajas-Giron and Sinclair, 2001; Villanueva et al., 2004). A study conducted by the FRAGMENT project in Costa Rica and Nicaragua showed that the tree resources present in pastures play an important role in cattle farms: dispersed trees were in more than 90% of the farms, while the live fences varied between 49 and 89% of the farms surveyed in both countries. The use of live fences is associated with a low availability of posts in or near the farms and a reduction in the cost of establishment and maintenance when compared to the cost of purchasing posts (Holmann et al., 1992). The total tree cover in pastures (dispersed trees and live fences) varied between 8 and 29% and total tree density ranged between 19 and 53 trees/ha (Villacis, 2003; Villanueva et al., 2004; Ruiz and Gómez unpublished). In Central America, farm landscapes show richness and density of dispersed trees in pastures that range from 72 to 107 species and 10.36 to 32.31 trees/ha respectively. In these landscapes, the five most abundant species represent more than 55% of the total individuals (Table 1).

Table 1. Species composition and structure of dispersed trees based on trees with diameters bigger than 10 cm in pastures in Cañas, Río Frío, Rivas and Matiguas. Source: Villanueva et al. (2004) (Cañas and Rivas); Villacís (2003) (Río Frío); Ruiz and Gómez (unpublished) (Matiguás). Different letters within the rows show statistically significant difference ($P < 0.05$) according to the Duncan test.

Variable	Cañas, Costa Rica (n = 5896 trees in pastures)	Río Frío, Costa Rica (n = 2482 trees in pastures)	Rivas, Nicaragua (n = 2297 trees in pastures)	Matiguás, Nicaragua (n = 7994 trees in pastures)
Ecological life zone	Tropical dry forest	Tropical wet forest	Tropical dry forest	Transition from tropical dry forest to tropical humid forest
Main cattle production system	Beef	Milk and beef	Dual purpose + agriculture	Dual purpose
Mean density of trees/ha	10.36a ± 1	21.34ab ± 3	16.22a ± 5.11	32.31b ± 5.82
Mean no. of tree species/farm	28.46a ± 3.78	26.60a ± 1.85	24.88a ± 2.59	36.03 ± 2.94
Total no. of tree species in landscape	101	107	72	101
Five most common tree species present (% of all trees surveyed)	Tabebuia rosea (12.8%) Guazuma ulmifolia (12.6%) Cordia alliodora (12%) Acrocomia aculeata (10.2%) Byrsonima crassifolia (7.4%)	Cordia alliodora (25.9%) Psidium guajava (22.5%) Pentaclethra macroloba (4.7%) Citrus sinensis (4.7%) Citrus limon (3.1%)	Cordia alliodora (22.7%) Guazuma ulmifolia (15.2%) Tabebuia rosea (7.1%) Byrsonima crassifolia (6.6%) Gliricidia sepium (6.4%)	Guazuma ulmifolia (35.7%) Cordia alliodora (12.9%) Tabebuia rosea (5.8%) Enterolobium cyclocarpum (5.7%) Samanea saman (4.9%)

The variation in species richness and density depends on factors such as the agroecological zone, density and distribution of seed trees, pasture management, farm size, dependence of family on farm as a source of income and dynamic of land use change from crops to young secondary forest or from crops to livestock. For example, in Rivas, Nicaragua, dispersed tree density in pastures is higher than in the Cañas, Costa Rica; both sites are relatively close (approximately 160 km) and are in a similar life zone. However, there are differences in farmer livelihoods, since in Nicaragua farmers depend economically on their farms and the tree resource is valued for the products (e.g. firewood, posts and timber) it delivers. For example, around 95% of Nicaraguan farmers used firewood compared to 58% in Costa Rica. This forces them to use strategies to maintain more trees in pastures by managing natural regeneration. Likewise this situation is related to a lesser fixed capital per farm in Nicaragua than in Costa Rica (339 vs. 3135 $/ha).

Local knowledge and how farmers make decisions about the retention, removal and planting of trees in pastures are being documented and this information can be used in the participatory design and management of improved silvopastoral systems. The studies in Cañas, Costa Rica, showed that livestock farmers tend to maintain a mixture of tree species in pastures depending on their value and functional roles; highly valued timber trees with small crowns (e.g. *Cordia alliodora, Cedrella odorata, Platymiscium pleiostachyum*) are retained in pasture in combination with species that have larger crowns that provide shade for animals, especially those species that produce forage and fruits (e.g. *Enterolobium cyclocarpum, Pithecellobium saman, Guazuma ulmifolia*) for animal consumption. Tree shaded pastures are associated with improvements in milk production and live weight gains due to the reduction in heat stress of animals (Souza de Abreu, 2002) and increases in voluntary intake (Djimde et al., 1989; Pezo and Ibrahim, 1999) (Table 2). In the humid tropics, Souza de Abreu (2002) found that those cows that had access to tree shade in pasture had lower respiratory rates compared to cows grazing open pastures without trees (65 *vs* 80 times/min).

Table 2. Influence of dispersed trees in pastures on animal production. Tree density was around 18 trees/ha.

Ecosystem	Production system	Tree cover (%)	Animal production	Season	Reference
Sub-humid tropical forest	Dual purpose cattle	Low (0 – 7%)	4.1 l milk /cow per day	Dry	Betancourt et al. (2004)
		High (22 – 30%)	3.1 l milk/cow per day		
Dry tropical forest	Beef cattle	Low (7%)	-104 / 777 g/heifer per day	Dry/Rainy	Restrepo et al. (2004)
		Medium (14%)	-160 / 768 g/heifer per day		
		High (27%)	-93 / 893 g/heifer per day		
Humid tropical forest	Dairy cattle	Medium (10 – 15%)	12.7 / 9.0 l milk/cow per day	Dry/Rainy	Souza de Abreu (2002)
		0 (Without shade)	11.1 / 9.2 l milk/cow per day		

In Central and Latin America trees and shrubs are commonly used as live fence posts to divide pasture and to delimit farm boundaries (Cajas-Giron and Sinclair, 2001; Ibrahim et al., 2001a; Harvey et al., 2005). Studies about the richness and diversity of live fences have been carried out in Central America by the FRAGMENT project. The studies showed that the density of trees in live fences varied between 67 and 242 trees/km and the total species richness 27 to 85 species (Table 3) (Harvey et al., 2005). Most species were native, showing the importance of these live fences in conservation of biodiversity. The integration of timber trees in the live fences may increase the value of these systems and should be evaluated through a participatory approach with farmers.

Table 3. Species composition and structure of live fences based on trees with diameters bigger than 10 cm in Cañas, Río Frío, Rivas and Matiguas. Source: Harvey et al. (2005). Different letters within the rows show statistically significant difference ($P < 0.05$) according to the Duncan test.

Variable	Cañas, Costa Rica (n = 20974 trees in 385 live fences)	Río Frío, Costa Rica (n = 3812 trees in 409 live fences)	Rivas, Nicaragua (n = 1852 trees in 71 live fences)	Matiguás, Nicaragua (n = 3464 trees in 330 live fences)
Total live fence length per site (km)	83.55	35.5	35.6	24.3
Mean density of trees per km	241.87a ± 11.78	88.10b ± 7.46	67.53c ± 6.92	92.1b ± 5.8
Mean no. of tree species in live fences per farm	24.8a ± 2.40	4.8c ± 0.70	17.33b ± 2.50	20.3ab ± 2.0
Total no. of tree species found in live fences	85	27	73	72
Five most common tree species present in live fences (% of all trees surveyed)	Bursera simaruba (54.2%) Pachira quinata (27.6%) Ficus sp (3.8%) Gliricidia sepium (1.9%) Tabebuia rosea (1.9%)	Erythrina costarricense (75.6%) Gliricida sepium (11.1%) Cordia alliodora (2.8%) Bursera simaruba (2.6%) Dracaena fragrans (1.8%)	Guazuma ulmifolia (9.06%) Cordia dentata (8.44%) Acacia collinsii (7.01%) Myrospermum frutescens (6.67%) Simarouba glauca (6.3%)	Bursera simaruba (50.1%) Guazuma ulmifolia (8.7%) Pachira quinata (7.1%) Gliricida sepium (5.5%) Erythrina berteroana (4.4%)

Improved silvopastoral systems
Fodder banks
Management of fodder trees and shrubs at high densities (20,000–40,000 plants/ha) with adequate pruning frequencies (3–6 months) can result in significant fodder production; e.g. 7–14 t DM/ha per year (Benavides, 1994; Ibrahim et al., 1998). Fodder trees and shrubs can maintain higher nutritive value in their foliages (*in vitro* dry matter digestibility (IVDMD) between 54 and 65% and crude protein (CP) > 18%) in the dry season than most tropical grasses (IVDMD < 38% and CP < 7%) (Elena et al., 1997; Hernández et al., 1999; Ibrahim et al., 2001b). In Cuba, dairy cows grazing shrub fodder banks (e.g. *Leucaena leucocephala*) or multi-strata forage systems (*L. leucocephala* + *Stylosanthes guianensis* + *Neonotonia wightii* + *Panicum maximum*) produced between 9 and 11 kg of milk/cow per day without the use of supplementary concentrates. This is an impressive result considering that potential milk yield with tropical grasses is 7 to 8 kg/cow per day (Archibald, 1984). In Costa Rica in the dry season, dual purpose cattle with a diet of *Cratylia argentea*, sugar cane and wheat bran produced 6.1 kg/cow per day, which is similar in amount and quality of milk produced with the supplementation of chicken litter (Ibrahim et al., 2001b). In the same country, fat steers grazing mixed pastures (*Paspalum fasciculatium*, *Axonopus compressus* and *Cynodon nlemfluensis*) and a fodder bank of *Erythrina berteroana* (two hours daily) achieved 21 to 25% more live weight than for animals only grazing pasture (Ibrahim et al., 2000). These data show that trees and shrubs that are found locally in livestock farming systems can be managed to improve livestock production in Latin America.

Alley pasture systems
More complex silvopastoral systems have been designed to improve interactions between the components. Among these, "alley pasture" silvopastoral systems have been evaluated with the objective of integrating trees to enhance nutrient cycling and to increase production of high quality forage (Pezo and Ibrahim, 1999). In these systems multipurpose trees or shrubs are generally planted in rows (4 to 8 m apart) and grasses are sown within rows. Studies on acid soils in Calabacito, Panama, showed that silvopastoral systems with *Acacia mangium* planted 8-10 metres between rows in *Brachiaria humidicola* pastures contributed to improvements in soil P and N in 98 and 38% respectively over pasture without trees. In the silvopastoral systems the grass yield was 40% higher than that of the grass monoculture (Bolivar, 1998). The silvopastoral systems play an important role in the carbon sequestration (in the soil, pasture and

tree components), which could be a potential additional economic benefit for livestock farmers (e.g. environmental services payment or products with added value). Studies show that the total carbon in these systems ranged between 95 and 205 t/ha, with most carbon stored in the soil (López et al., 1999; Avila et al., 2001; Ruiz García, 2002; Villanueva and Ibrahim, 2002).

Multi-strata system

Multi-strata silvopastoral systems have been designed to improve above- and below-ground resource use and to increase the diversity of species in the system. Trees of large stature provided shade and produce timber, trees of medium stature produced fruits or pods, while other species were managed as shrubs producing green leaf fodder. A study conducted in the Caribbean region of Colombia showed that dual purpose cows grazing in diverse multi-strata silvopastoral systems maintained relatively high forage intake (> 12 kg DM/cow per day) and that a large percentage (> 45%) of the diet was selected from forage trees and shrubs (*Leucaena leucocephala*, *Gliricidia sepium* and *Crescentia cujete*) (Cajas-Giron et al., 2001). Growth of timber trees (*Pachira quinata*, *Swietenia macrophylla* and *Tabebuia rosea*) in this multi-strata system was relatively good when compared to their observed growth in plantations. These results are very promising and participatory research with farmers should be conducted to promote the adoption of multi-strata silvopastoral systems for the diversification of livestock production systems.

Why do the silvopastoral systems contribute to enhance the social benefits?

In order for alternatives to extensive grazing to be viable they must fulfil, at least partially, the biophysical and socioeconomic niche occupied by cattle in current farming systems. They should help reduce environmental degradation and poverty at the same time as improving and diversifying farm productivity. In this respect, silvopastoral systems are some of the most promising holistic approaches for sustainable livestock production and may have the greatest probability of early adoption since more than 70% of cattle farmers have some experience of managing trees in their pastures (Souza de Abreu, 2002; Villacis, 2003). Silvopastoral systems can mimic forest ecosystems to some extent and are of significant value for improving farm productivity and generating social and environmental benefits to livestock farmers, through provision of product and services such as fruits, timber, forage, fuel, biodiversity, carbon stored and scenic beauty.

Some studies have showed the benefits of silvopastoral systems. For example, a case study conducted in San Carlos, Costa Rica, showed that dual purpose cattle farmers maintained a higher proportion of timber trees in pastures, which is a good indication that these farmers minimized risk by diversifying production (Table 4). Profitability of small Costa Rican dairy farms was increased, especially when labour costs increased, through diversification with highly valued timber species such as *Cordia alliodora* (2188 vs 1478 US$/farm per year for farms with and without the trees) (Holmann et al., 1992). At the same time, cattle farms with 50% of young secondary forest and 50% of silvopastoral systems with improved grass are an alternative that increases productivity by about 12-fold, compared to traditional farming, and could help to reach a carbon fixation rate of 2.8 t C/year (Ruiz García, 2002).

Table 4. Abundance of trees by cattle production system (La Fortuna San Carlos, 1999) (average number of trees/ha). Source: Souza de Abreu et al. (2000). [1]Values with the same letter in the same row are not significantly different ($P < 0.05$). [2]*Zanthoxilum belizense*, *Terminalia oblonga*, *Pentaclethra macroloba*. [3]*Citrus* sp, *Inga* sp and *Ficus* sp (fruit and shade trees).

Trees	Mixed ($n = 4$)	Specialized in milk production ($n = 3$)	Dual purpose ($n = 3$)
Cordia alliodora	7.33b[1]	10.34b	16.08a
Cedrela odorata	0.63a	1.44a	0.62a
Other timber trees[2]	2.51a	4.33a	1.26a
Non-timber trees[3]	1.99b	6.00a	2.51b
Total	12.46a	22.11b	20.47b

Impact and constraints for adoption

The impact of silvopastoral systems has also resulted in financial benefits, especially in increases in the net incomes of silvopastoral farmers compared to traditional farmers. Farmers already know and are able to use trees for shade for animals in pastures. For example, a case study in Belize showed that 80% of the farmers were familiar with the benefits provided with the use of foliage from trees such as *Guazuma ulmifolia* and *Brosimum alicastrum* for feeding cattle: 63% of the farmers indicated that forage was harvested from multipurpose trees to feed cattle in the dry season (Alonzo and Ibrahim, 2001). Similar results were found from a study conducted in Green Park, Jamaica, which concluded that more than 70% ($n = 45$) of dairy farmers knew about the use of fodder trees and shrubs for feeding cattle (Morrison et al., 1996).

The major limiting factors in the adoption of silvopastoral technologies as reported by the farmers in Belize are risk, capital, markets and poor genetic stock (Alonzo and Ibrahim, 2001). Farmers are reluctant to change from traditional systems to new technologies associated with higher risks. This reaction is mainly due to a lack of knowledge of these systems (Aldy et al., 1998). The capital required for the initial investment to establish these systems was reported as a limiting factor, especially since farmers are reluctant to take loans from credit institutions, mainly due to the high interest rates and collateral requirements (Jansen et al., 1997). Farmers also agree that better, more secure, markets need to be identified for their product as producer prices are too low for them to make a fair profit. In this sense, CATIE is developing in Costa Rica, Nicaragua and Colombia a strategy of paying for environmental services to promote the adoption of silvopastoral systems in cattle farms (Gobbi and Ibrahim, 2004). This project results in an increase in environmental benefits at a global level such as carbon sequestration and biodiversity conservation, but also a change or improvement of the traditional systems to a more intensive one with silvopastoral systems can have an impact of increasing the labour demand (Ruiz García, 2002). In general, the adoption of silvopastoral systems increases the number of day-workers needed for operating the farm and scenarios modelled indicate that the percentages of annual increase in labour requirements on farms with silvopastoral systems range from 34% to 106% in the three countries (Gobbi and Ibrahim, 2004).

Conclusions

Traditional cattle systems are characterized by a diversity of woody perennials which significantly improve animal productivity and conserve natural resources. The selection and management of improved grasses and woody perennials in silvopastoral systems can increase the capacities of soil and livestock and thereby free fragile grazing lands for reafforestation. In addition, silvopastoral systems could contribute to enhancing social benefits and increased environmental diversity.

Acknowledgements

Acknowledgements to the FRAGMENT Project ("Developing methods and models for assessing the impacts of trees on farm productivity and regional biodiversity in fragmented landscapes"). It is funded by the European Community Fifth Framework Programme (INCO-Dev ICA4-CT-2001-10099) confirming the international role of community research. The authors are solely responsible for the material reported here; this publication does not represent the opinion of the Community and the Community is not responsible for any use of the data appearing herein. In addition, we would like to acknowledge Dr Celia Harvey's comments on this article.

References

Aldy, J.E., Hrubovcak, J. and Vasavada, U. (1998) The role of technology in sustaining agriculture and the environment. *Ecological Economics* 26, 81-96.

Alonzo, Y. and Ibrahim, M. (2001) Potential of silvopastoral systems for economic dairy production in Cayo, Belize and constraints for their adoption. In: Ibrahim, M. (ed.) *International Symposium on Silvopastoral Systems and Second Congress on Agroforestry and Livestock Production in Latin America*. CATIE, San Jose, Costa Rica, pp. 465-470.

Archibald, K. (1984) Dairy cattle feeding in the humid tropics or high rainfall tropics. In: Smith, A.J. (ed.) *Milk Production in Developing Countries*. University of Edinburgh, Edinburgh, UK, pp. 110-132.

Avila, G., Jimenez, F., Beer, F., Gómez, M. and Ibrahim, M. (2001) Carbon sequestration and storage and environmental services assessment in agroforestry systems in Costa Rica (Almacenamiento, fijación de carbono y valoración de servicios ambientales en sistemas agroforestales en costa Rica). *Agroforestería en las Américas* 8 (30), 32-35. (In Spanish.)

Benavides, J. (1994) Research on fodder trees (La investigación en árboles forrajeros). In: Benavides, J. (ed.) *Árboles y Arbustos Forrajeros en América Central*. CATIE, Turrialba, Costa Rica, pp. 3-28. (In Spanish.)

Betancourt, K., Ibrahim, M., Harvey, C. and Vargas, B. (2004) Effect of the canopy cover on animal behaviour in double purpose farms in Matiguas, Matagalpa, Nicaragua (Efecto de la cobertura arbórea sobre el comportamiento animal en fincas ganaderas de doble propósito en Matiguás, Matagalpa, Nicaragua). *Agroforesteria en las Américas* 10, 39-40. (In Spanish.)

Bolivar, D. (1998) Contribution of *Acacia mangiun* to the improvement of fodder quality of *Brachiaria humidicola* and fertility in acid soils in humid tropic (Contribución de *Acacia mangiun* al mejoramiento de la calidad forrajera de *Brachiaria humidicola* y la fertilidad de un suelo acido del tropico humedo). MSc thesis. CATIE. Turrialba, Costa Rica. (In Spanish.)

Cajas-Giron, Y. and Sinclair, F. (2001) Characterization of multistrata silvopastoral systems on seasonally dry pastures in the Caribbean region of Colombia. *Agroforestry Systems* 53, 215-225.

Cajas-Giron, Y., Mayes, R. and Sinclair, F. (2001) Estimating feed intake of browse species in biodiverse silvopastoral system. In: Ibrahim, M. (ed.) *International Symposium on Silvopastoral Systems and Second Congress on Agroforestry and Livestock Production in Latin America*. CATIE, San Jose, Costa Rica, pp. 280-284.

Djimde, M., Torres, F. and Migongo-Bake, W. (1989) Climate, animals and agroforestry. Meteorology and agroforestry. In: Reifsnyder, W.S. and Darnhofer, T.O. (eds) *Proceedings of an International Workshop on the Application of Meteorology to Agroforestry Systems Planning and Management, 1987*. International Council for Research in Agroforestry (ICRAF), Nairobi, Kenya, pp. 463-470.

Elena, M., Rodriguez, L., Murgueitio, E., Ines, C., Rosales, M., Hernan, C., Molina, H. and Molina, J. (1997) *Fodder Trees and Shrubs Used for Animal Feed as Protein Source (Árboles y arbustos forrajeros utilizados en alimentación animal como fuente de proteica)*. CIPAV, Cali, Colombia. (In Spanish.)

Gobbi, J. and Ibrahim, M. (2004) Creating win-win situations: The strategy of paying for environmental services to promote adoption of silvopastoral systems. In: Mannetje, L., Ramírez, L., Ibrahim, M., Sandoval, C., Ojeda, N. and Ku, J. (eds) *The Importance of Silvopastoral Systems for Providing Ecosystems Services and Rural Livelihoods*. University of Yucatán, Mérida, Mexico, pp. 98-101.

Harvey, C. and Haber, W. (1999) Remnant trees and the conservation of biodiversity in Costa Rican pastures. *Agroforestry Systems* 44, 37-68.

Harvey, C.A., Villanueva, C., Villacís, J., Chacón, M., Muñoz, D., López, M., Gómez, R., Ibrahim, M., Sinclair, F.L., Taylor, R., Martínez, J., Navas, A., Saenz, J., Sanchez, D., Medina, A., Vilchez, S., Hernández, B., Perez, A. and Lang, I. (2005) Contribution of live fences to the productivity and ecological integrity of agricultural landscapes in Central America. *Agriculture, Ecosystems and Environment* (In press.)

Hernández, D., Mirta, C. and Reyes, F. (1999) Establishment of a multi-associated silvopastoral system. *Pastos y Forrajes* 22, 123-134.

Holmann, F., Romero, F., Montenegro, J., Chana, C., Oviedo, E. and Baños, A. (1992) Profitability of silvopastoral systems of small dairy producers in Costa Rica: first approach (Rentabilidad de sistemas silvopastoriles con pequeños productores de leche en Costa Rica:primera aproximación). *Turrialba Costa Rica* 42 (1) 79-89. (In Spanish.)

Ibrahim, M., Canto, G. and Camero, A. (1998) Establishment and management of fodder banks for livestock feeding in Cayo. In: Ibrahim, M. and Beer, J. (eds) *Agroforestry Prototypes for Belize.* CATIE, Turrialba, Costa Rica, pp. 15-43.

Ibrahim, M., Holmann, F., Hernández, M. and Camero, A. (2000) Contribution of *Erythrina* protein banks and rejected bananas for improving cattle production in the humid tropics. *Agroforestry Systems* 49, 245-254.

Ibrahim, M., Schlonvoigt, A., Camargo, J.C. and Souza, M. (2001a) Multi-strata silvopastoral systems for increasing productivity and conservation of natural resources in Central America. In: Gomide, J.A., Mattos, W.R.S. and da Silva, S.C. (eds) *Proceedings of the XIX International Grassland Congress.* FEALQ, Piracicaba, Brazil, pp. 645-650.

Ibrahim, M., Franco, M., Pezo, D., Camero, A. and Araya, J.L. (2001b) Promoting intake of *Cratylia argentea* as a dry season supplement for cattle grazing *Hyparrhenia rufa* in the sub-humid tropics. *Agroforestry Systems* 51, 167-175.

Jansen, H.G.P., Ibrahim, M., Nieuwenhuyse, A. and Mannetje, L. (1997) The economics of improved pasture and silvipastoral technologies in the Atlantic zone of Costa Rica. *Tropical Grasslands* 6 (31), 588-598.

López, A., Schlönvoigt, A., Ibrahim, M., Kleinn, C. and Kanninen, M. (1999) Accounting of stored carbon in soil in a silvopastoral systems in the Atlantic region of Costa Rica (Cuantificación del carbono almacenado en el suelo de un sistema silvopastoril en la zona Atlántica de Costa Rica). *Agroforestería en las Américas* 6 (23), 51-53. (In Spanish.)

Morrison, B., Gold, M. and Lantagne, D. (1996) Incorporating indigenous knowledge of fodder trees into small-scale silvopastoral systems in Jamaica. *Agroforestry Systems* 34, 101-117.

Pezo, D. and Ibrahim, M. (1999) *Agroforestry formation – Silvopastoral systems* (Modulo de enseñanza agroforestal N° 2 Sistemas silvopastoriles). 2nd ed. CATIE, Turrialba, Costa Rica. (In Spanish.)

Restrepo, C., Ibrahim, M., Harvey, C., Harmmand, M. and Morales, J. (2004) Relationship between the canopy cover in pasture and bovine production in farms in the dry tropic, Cañas, Costa Rica (Relaciones entre la cobertura arbórea en potreros y la producción bovina en fincas ganaderas en el trópico seco, Cañas, Costa Rica). *Agroforesteria en las Américas* 10, 39-40. (In Spanish.)

Ruiz Garcia, A. (2002) Carbon sequestration and storage in silvopastoral systems and economic competitiveness in Maniguas, Nicaragua (Fijación y almacenamiento de carbono en sistemas silvopastoriles y competitividad económica en Matiguás, Nicaragua). MSc thesis. CATIE, Turrialba, Costa Rica. (In Spanish.)

Souza de Abreu, M.H., Ibrahim, M., Harvey, C. and Jiménez, F. (2000) Characterization of tree component in husbandry systems in La Fortuna de San Carlos, Costa Rica (Caracterización del componente arbóreo en los sistemas ganaderos de La Fortuna de San Carlos, Costa Rica). *Agroforestería en las Américas* 7 (26), 53-56. (In Spanish.)

Souza de Abreu, M.H. (2002) Contribution of trees to the control of heat stress in dairy cows and the financial viability of livestock farms in humid tropics. PhD thesis. CATIE, Turrialba, Costa Rica.

Szott, L., Ibrahim, M. and Beer, J. (2000) *The Hamburger Connection Hangover. Cattle Pasture Land Degradation and Alternative Land Use in Central America.* CATIE, Turrialba, Costa Rica.

Villacís, J. (2003) Relationship between the canopy cover and the intensification level in husbandry farms in Rio Frio, Costa Rica (Relaciones entre la cobertura arborea y el nivel de intensificación de las fincas ganaderas en Rio Frio, Costa Rica). MSc thesis. CATIE, Turrialba, Costa Rica. (In Spanish.)

Villanueva, C. and Ibrahim, M. (2002) Assessment of the silvopastoral systems impact on the restoration of degraded pastures and the contribution to carbon sequestration in dairy farms in Costa Rica (Evaluación del impacto de los sistemas silvopastoriles sobre la recuperación de pasturas degradadas y su contribución al secuestro de carbono en lecherías de altura de Costa Rica). *Agroforestería en las Américas* 9 (35-36), 69-74. (In Spanish.)

Villanueva, C., Ibrahim, M., Harvey, C., Sinclair, F., Gómez, R., López, M. and Esquivel, H. (2004) Tree resources on pastureland in cattle production systems in the dry pacific region of Costa Rica and Nicaragua. In: Mannetje, L., Ramírez, L., Ibrahim, M., Sandoval, C., Ojeda, N. and Ku, J. (eds) *The Importance of Silvopastoral Systems for Providing Ecosystems Services and Rural Livehoods.* University of Yucatán, Mérida, Mexico, pp. 183-188.

Silvopastoral systems in North-West Europe

J. H. McAdam
Dept. of Applied Plant Science, Queen's University Belfast and Applied Plant Science Division, Dept. of Agriculture and Rural Development, Newforge Lane, Belfast, BT9 5PX, jim.mcadam@dardni.gov.uk

Abstract
Silvopastoral practices in which trees are grown in grazed pasture in cool temperate regions of Western Europe are reviewed. In many areas there are concerns over problems with grassland intensification and fragmentation of woodland habitats. Objectives for silvopastoral systems are: reduction in agricultural production; increased biodiversity; retention of nutrients; landscape enhancement; improved stock welfare; reduction in timber imports; maintenance of diversified rural economies.

Silvopastoral systems have been researched but little practised although there are examples of woodland grazing, re-spaced forests and wide-spaced, protected trees in pasture. Such systems, where a range of livestock species are incorporated, exist in Britain and other parts of Western Europe where pastoralism is common.

This chapter reviews these systems and shows how silvopastoral systems align with current and proposed EU policy on sustainability and rural development. The factors governing uptake of systems in the light of current pressures on farming and on sectoral support will be considered. It also shows how the decoupling of subsidy from production within a reformed CAP offers substantial opportunities to increase the areas and types of silvopasture.

Key words: grassland intensification, rural development, sustainability, grazing, trees

Definition of silvopastoral systems
To set the topic of silvopastoral types in an NW Europe context it is necessary to define systems, set geographical boundaries, outline the objectives for the systems, their policy context, uptake, support and likely future prospects.

Agroforestry is a collective name for land-use practices where trees are combined with crops and/or animals on the same unit of land and there are significant ecological or economic interactions between the tree and the agricultural components. For the purposes of this chapter the strict definition of silvopastoral systems will be considered as those where trees are grown in grazed pasture in a regular or varied pattern. It is accepted, however, that in Western Europe many forests are grazed to a greater or lesser degree by wild or domesticated livestock. For example, in most Scandinavian countries, where tree cover is high and pressure on agricultural land is low, the most likely integration of stock and forests will be in the form of forest grazing. Wooded pasture still exists to a greater or lesser extent in mountain areas (e.g. the Jura Mountains of Switzerland, Gillet and Garlandat, 1996). Forest grazing has disappeared from most of temperate Europe as a result of population pressure, agricultural intensification, the disappearance of traditional transhumance patterns and forest specialization (Dupraz and Newman, 1997; see Finck et al. in this volume). These forest-grazing systems will not be considered in this chapter.

There are several ways of achieving silvopastoral systems. Two of the most common are (1) re-spacing an established woodland or planted forest by selective thinning and establishing pasture, i.e. silvopasture which has evolved from either a pure forest or forest-grazing scenario, or (2) planting trees into already established pasture. In the latter scenario these trees will generally have to be protected from grazing initially or the pasture cut for silage or hay until the trees are above browsing height. Hence the subject is essentially being covered from an agricultural viewpoint where wooded pasture is considered as a unit of intensive or semi-intensive grazing.

The primary purpose of this review is to consider the types of silvopasture which are practised or could be introduced into grazed pasture (both intensive and extensive) in the relatively high rainfall areas of cool-temperate North-Western Europe.

Objectives for silvopastoral systems
The evolution of land use and objectives for silvopastoralism essentially determine the types of silvopastoral systems practised.
(a) Intensity of land use. The Common Agricultural Policy (CAP) has been the main driver of land-use practice in Europe since the early 1970s. In the context of pastoralism in Western Europe there has been a gradual trend towards intensification fuelled by CAP production-based subsidies. This intensification resulted in: loss of biodiversity on farmland; increase in eutrophication of watercourses and groundwater; deterioration of soil structure and inherent fertility; decline in mixed, multi-output farming; loss in scrub and woodland cover; increase in farm unit size and corresponding decline in the farm workforce. These changes happened to a different extent across western Europe, depending on history (particularly post -1945) and the stability of rural communities and populations. The consequences of intensification were production beyond consumption (food 'mountains' resulting), a distortion in world trade balances and a negative environmental impact. By the mid-1980s it was realized that such a policy was unsustainable

and this, coupled with an upswelling of environmental concern, led to reform of the CAP in 1992 and 2000 and ultimately to decoupling of production support.

(b) Role of silvopasture. Research on silvopastoral systems in western Europe since the 1980s has shown them to: be sustainable (e.g. Sibbald, 1999); create spatial heterogeneity, which enhances biodiversity (Burgess, 1999; McAdam *et al.*, 1999); have the potential to reduce nutrient leakage (Sinclair, 1999); enhance the rural landscape; improve animal welfare criteria (McAdam *et al.*, 1999); have multifunctional outputs (Brann, 1988). Hence the current objectives for silvopasture in North-Western Europe are to: reduce levels of livestock production; improve the physical and biological environment; deliver acceptable animal welfare systems; be sustainable; increase tree cover in the rural landscape; augment rural development plans; reduce timber imports.

Types of silvopastoral systems

Agroforestry systems have been practised for centuries, ever since man started to manage trees, food crops and domestic animals on a particular area of land. However, the scientific study of agroforestry is young, and research has been devoted to tropical systems (Etienne, 1996) and little to silvopastoral systems. It has been said (Nair, 1993) that agroforestry is a new name for a set of old practices and, among this set, silvopasture is probably one of the oldest. Etienne (1996) recognizes three types of silvopastoral systems within a farm:

-Ligniculture on swards where widely spaced trees arc planted into established pasture to exploit the component interactions.
- Grazing in forests following thinning and after reseeding.
- Forestry on livestock farms where trees are planted at high density to diversify production at the farm level.

All three of these systems exist in North-Western Europe. The last of these is common in mainland Europe, the woodland component being utilized when investment in the farm business is required (Swain, 1987). Developments in land use in New Zealand whereby trees were seen as having a functional use to the stock rearing component of farms as well as providing additional revenue (Hawke and Knowles, 1997) led to a re-evaluation of the role trees could play in western Europe. This, coupled with the policy developments listed above, identified a wider role which trees could play in pastoral systems in which there was the recognized need to reduce livestock production and to address environmental concerns over land-use practices. In most of the pastoral areas of mainland Europe the tree base was already present and incorporating it into livestock systems was along the lines of moulding the farm-forest interface through successful models of testing the wider role trees could play in a farming context.

In most instances in the British Isles the tree base did not exist. Much of the woodland had gone through successive clearances and agricultural intensification - cereal production in the east and pastoralism in the north and west prevailed. In Ireland the situation was exacerbated by a high rural population in the 19th century and a land ownership hierarchy which resulted in very low tree cover (6%). In such a scenario where farms are often family owned there is little option but to consider establishing silvopasture from a base of open pasture and to plant protected trees at a range of wide spacings. In the 1950s a UK company (Bryant and May) planted wide-spaced poplars (*Populus* sp) on commercial farms for matchstick production. In this system, alleys between rows of poplars were cropped for 7-8 years and then undersown with pasture and grazed as silvopasture until final felling at about 25 years (Sheldrick and Auclair, 2000). Unfortunately the system became uneconomic in the 1970s but it does provide an interesting model for the establishment of future silvopastoral systems, reducing the need for expensive tree protection in the early stages.

Research Findings

Agroforestry systems involve complex interactions between the individual components (Sinclair, 1999) and research has concentrated on quantifying production with lesser resources directed towards the investigation of ecological interactions. Recent interest in agroforestry started in the 1980s with the publication of a number of models (Sibbald and Sinclair, 1990). It was recognized that the necessary biological research to validate these models would be resource demanding and would require a collaborative approach. Most research effort has been devoted to coordinated national networks of silvopastoral (Sibbald *et al.*, 2001) and silvoarable trials. These trials have described many of the issues of establishment and early growth, tree protection, impact on production, environmental significance, ecological interaction and animal welfare issues.

The research has shown that trees can be successfully established and grown when closely integrated with either livestock or an arable crop. In silvopastoral systems livestock performance has not been reduced up to 12 years after planting and output only marginally reduced in the latter two years after planting ash at 400 stems/ha. Overall compatibility with farming systems is better than predicted from modelling. Silvopastoral agroforestry systems have been shown to be sustainable in the widest sense of the term (Sibbald *et al.*, 2001) and to create spatial heterogeneity, which, enhances biodiversity, particularly of invertebrates, birds and plants (McAdam, 2000). There are also strong indications that agroforestry systems may be less 'nutrient leaky' and have reduced pesticide requirement than their counterpart conventional grassland and arable systems respectively. Research on livestock in silvopastoral systems has shown that stock take advantage of the shade and shelter. Utilizing the range of tree species available with the flexibility of planting density and arrangements that may be used enables agroforestry systems to be designed for positive landscape impact. On the basis of this research, a technology transfer process to implement the findings on commercial farms was initiated (Crowe and McAdam, 1999). In a major development in 1993 this Experiment became the basis for an EU-funded research project on alternative land use with fast growing trees (Auclair, 1998). The biophysical factors interacting within an agroforestry system were modelled (Auclair, 1998) and, when set alongside economic models (Thomas and Willis, 2000), a valuable resource was made available for studying the interactions and the likely impact of silvopastoral types on a range of farms.

The Animal Component

Most silvopastoral systems tested in the UK have used sheep and cattle as the livestock component. However, woodland is the natural habitat of wild pigs, poultry and turkeys and there are examples of domestic varieties of these animals being incorporated into silvopastoral systems (Brownlow *et al.*, 2000). In specific situations, non-ruminant livestock can be integrated with trees. In the current search for novel systems of land use incorporating the concept of multiple outputs and with increasing concerns over animal welfare issues, especially to cross-comply with EU subsidy decoupling, there is undoubtedly a place for such systems in the future. Management of such systems is not always as straightforward as for ruminants, but there are potential benefits, namely in terms of marketable animal products, financial returns, improved animal welfare and benefit to the environment (Brownlow *et al.*, 2000).

Policy alignment

Agriculture in the UK is undergoing a period of rapid change and is under extreme pressure for survival in certain instances. Many aspects of the 1992 Reform of the CAP are now strongly influencing these changes and Agenda 2000 will continue to do so. Many commentators on agricultural structural change predict that the industry will develop in two (very broad) groups of agricultural production

(a) 'Competitive pillar' - a relatively intensive agriculture industry competing on world markets in a strictly business-orientated method of raw material/food production with state/EU subsidy directed at increasing competitiveness and efficiency.

(b) 'Recreational pillar' - with state subsidy aimed at producing CARE (conservation, amenity, recreation and environment) goods through funding to farmers/landowners.

The 'bridge' between these two pillars is rural development policy, which can provide benefits in both areas. Agroforestry offers a strategic policy option to realize some of these goals although it is considered that silvopasture is unlikely to feature in the 'competitive pillar' scenario due to the need to compete in relatively low-margin markets on a national or global scale.

The provision of CARE goods through silvopasture is realistic and has been demonstrated in both research and initial technology transfer. The main question here is therefore whether silvopasture provides a significantly better means of delivering CARE goods than more conventional systems of land use. However, while the wider socioeconomic implication(s) of silvopasture appear positive (Doyle and Thomas, 2000; Thomas and Willis, 2000), the extensive adoption of such systems in the UK will depend on whether farmers perceive silvopasture as relatively profitable and ultimately the decision to invest in agroforestry for both 'pillar' categories and for flexible use of land.

Factors governing uptake of systems

It is appropriate to consider the needs of farmers and landowners rather than concentrating on what silvopasture may offer.

Commercial farms have both short/medium-term and long-term needs:

(i) Short/medium-term needs can be realized from systems which have a reasonable cash flow profile, give a quick return on capital, will enhance the value of the property and land, give good return on investment, can exploit currently available skills and expertise on the farm, deliver products for direct sale of processing on the farm, create enough activity to generate employment and help keep people on the land, help gain access to Capital Improvement Grants, and allow access to EU funding through participation in agri-environment schemes.

(ii) Long-term needs can be realized from production systems which can be seen as being viable and sustainable, yield capital, while enhancing assets such as land value, providing security of investment, granting tax considerations for succession, and creating a general enhancement of the landscape and the living environment of the family farm.

UK agriculture has come through a series of crises recently and farm incomes are currently severely depressed. In difficult times farmers generally concentrate on short-term goals and needs, longer-term needs being much less attractive. This tends to severely limit the opportunity for innovative long-term planning. The needs which can be justifiably met by planting silvopasture tend to fall into the 'longer term' category and currently farmers are, by necessity, concentrating on the short- to medium-term goals. Although this fact has always been recognized as a major drawback to farmer investment in woodland-related enterprises, it would appear this limitation is particularly strong at present. The position of silvopasture becomes even more difficult as it is viewed as an unproven technology in a range of existing woodland options which are already considered as limited in achieving short- to medium-term goals. It is reasonable to assume that farmers with no interest in planting woodland will have limited interest in planting agroforestry and that those with woodland interest are likely to be candidates for agroforestry. Speculating on the potential for agroforestry planting, given the current state of the industry, it is likely that agroforestry will substitute for, rather than complement, current or proposed woodland planting. If this is the case, then, dependent on the level of substitution, the area of silvopasture planted would be extremely low. Hence, substitution of candidate land for farm woodland is clearly not the route to follow to encourage adoption of silvopasture.

Current silvopastoral uptake is probably low on UK farms, because
(i) An adequate and dedicated technology transfer mechanism is not in place.
(ii) It is viewed as an unproven technology.
(iii) While it is a system which satisfies long-term goals and aspirations, it is only one of a range of woodland options which are already seen as limited in achieving short- to medium-term goals.

Conclusions and prospects for silvopasture

Silvopastoral systems can be managed for multiple output and offer flexibility as a land-use option. EU policy is to decrease levels of livestock production, tighten nutrient management on farms, increase tree cover, stabilize rural communities and enhance biodiversity through a more sustainable land use. Agroforestry satisfies a strategic policy to realize some of these goals. Currently there is little sectoral support from mainstream forestry or agriculture but the goals of agroforestry align with objectives for agri-environment measures and are likely to be supported by conservation bodies and community development groups.

Because in difficult times, such as at present, commercial farmers generally concentrate on short-term goals, the immediate future for conventional silvopasture is poor. The recent slow uptake of agroforestry systems coincides with a background interest in woodlands and to substitute current conventional woodland planting would make little impact on uptake of silvopasture. The strategic outlook for development of silvopastoral systems is likely to be targeted towards the environmental side of current farming activity, conservation and community groups.

While current levels of support for agroforestry are low, the messages emanating from the research programmes are encouraging and provide a reasonable basis to underpin any future implementation. The whole debate for a more radical review of land use in western Europe has created the conditions and highlighted options which could well favour agroforestry. There is a need for land-use systems to address the short-term goals of commercial producers by promoting some of the non-timber benefits of agroforestry such as fruit, flowers, fodder, foliage, fibre while ensuring close alignment with agri-environment measures. If such an agroforestry strategy is underpinned by an efficient and relevant applied research and technology transfer base, then the realistic prospects for silvopasture in western Europe must be seen as favourable. The most radical reform of the CAP is currently under way (2004) with subsidy payouts being decoupled from production. This will create the opportunity for farmers to change their land-use patterns while retaining the economic security of a fixed annual payment. In particular, the opportunity will exist to reduce livestock numbers and allow more novel land-use practices to be introduced. Such a move will allow the introduction of silvopastoral systems, delivering environmental and landscape benefits while still allowing a level of production and subsidy income. In addition, the expansion of the EU to include countries where food production is likely to be cheaper than in western Europe will further intensify the pressure on farmers to diversify. This should favour the uptake of silvopasture systems on western European farms.

References

Auclair, D. (1998) *Alternative Agricultural Land Use with Fast Growing Trees*. Final technical report, EC DG VI Contract AIR-CT 290134. Brussels, Belgium.

Brann, G. (1988) Farm-scale agroforestry in the Eastern Bay of Plenty. In: Maclaren, P. (ed.) *Agroforestry Symposium Proceedings*. Forest Research Institute Bulletin No. 139, Rotorua, New Zealand.

Brownlow, M., Carruthers, P. and Dorward, P. (2000) Alternatives to grazing livestock. In: Hislop, M. and Claridge, J. (eds) *Agroforestry in the UK*. Forestry Commission Bulletin 122, London, United Kingdom, pp. 58-70.

Burgess, P.J. (1999) Effects of agroforestry on farm biodiversity in the United Kingdom. *Scottish Forestry* 53 (1), 24-27.

Crowe, S.R. and McAdam, J.H. (1999) Silvopastoral practice-on farm agroforestry in N. Ireland. *Scottish Forestry* 53 (1), 33-36.

Doyle, C.J. and Thomas, T. (2000) The social implications of agroforestry. In: Hislop, M. and Claridge, J. (eds) *Agroforestry in the United Kingdom*. Forestry Commission Bulletin 122, Edinburgh, United Kingdom, pp. 99-106.

Dupraz, C. and Newman, S.M. (1997) Temperate agroforestry: the European way In: Gordon, A.M. and Newman, S.M. (eds) *Temperate Agroforestry Systems*. CAB International, Wallingford, United Kingdom, pp. 181-236.

Etienne, M. (1996) Research on temperate and tropical silvopastoral systems: a review. In: Etienne, M. (ed.) *Western European Silvopastoral Systems*. INRA, Paris, France, pp. 5-22.

Gillet, F. and Garlandat, J.D. (1996) Wooded pastures of the Jura Mountains. In: Etienne, M. (ed.) *Western European Silvopastoral Systems*. INRA, Paris, France, pp. 39-55.

Hawke, M.F. and Knowles, R.L. (1997) Temperate agroforestry systems in New Zealand. In: Gordon, A.M. and Newman, S.M. (eds) *Temperate Agroforestry Systems*. CAB International, Wallingford, United Kingdom, pp. 85-118.

McAdam, J.H. (2000) Environmental impacts. In: Hislop, M. and Claridge, J. (eds) *Agroforestry in the United Kingdom*. Forestry Commission Bulletin 122, Edinburgh, United Kingdom, pp. 83-90.

McAdam, J.H., Crowe, S.R. and Sibbald, A.R. (1999) Agroforestry as a sustainable land use option. In: Burgess, P.J., Brierley, E.D.R., Morris, J. and Evans, J. (eds) *Farm Woodlands for the Future*. Bios Scientific Publishers, Oxford, United Kingdom, pp. 127-137.

Nair, P.K.R. (1993) *An Introduction to Agroforestry.* Kluwer, Dordrecht.

Sheldrick, R. and Auclair, D. (2000) Origins of agroforestry and recent history in the United Kingdom. In: Hislop, M. and Claridge, J. (eds) *Agroforestry in the United Kingdom.* Forestry Commission Bulletin 122, Edinburgh, United Kingdom, pp. 7-16.

Sibbald, A.R. and Sinclair, F.L. (1990) A review of agroforestry research and progress in the United Kingdom. *Agroforestry Abstracts* 3, 149-163.

Sibbald, A.R. (1999) Agroforestry principles-sustainable productivity? *Scottish Forestry* 53 (1), 18-23.

Sibbald, A.R., Eason, W.R., McAdam, J.H. and Hislop, A.M. (2001) The establishment phase of a silvopastoral national network experiment in the United Kingdom. *Agroforestry Systems* 39, 39-53.

Sinclair, F.L. (1999) The agroforestry concept - managing complexity. *Scottish Forestry* 53 (1) 12-17.

Swain, P.J. (1987) *A Study of Aspects of Farm Forestry in New Zealand, Canada, Denmark, Sweden, Finland and California.* ADAS/WOAD, Aberystwyth, United Kingdom.

Thomas, T.H. and Willis, R.W. (2000) The economics of agroforestry in the United Kingdom In: Hislop, M. and Claridge, J. (eds) *Agroforestry in the United Kingdom.* Forestry Commission Bulletin 122, Edinburgh, United Kingdom, pp. 107-125.

Silvopastoral systems in the Neotropics

E. Murgueitio
Fundación CIPAV, Cra. 2 Oeste No. 11-54, Cali, Colombia. enriquem@cipav.org.co

Abstract
In the tropical Americas, pasturelands occupy the largest proportion (60% to 80%) of the area designated as agroecosystems in some countries. The expansion of the area of grazed pasture has implied the loss or extreme alteration of natural ecosystems, mainly lowland and montane tropical forests and, to a lesser extent, wetlands. In spite of the increased area of pasturelands, meat and milk production of bovine cattle (the most widely used species in the region) have only increased marginally. Stocking rates and animal production indexes (milk l/ha or meat kg/ha) are low and make only a limited contribution to capitalization and rural employment. Within vast geographical areas cattle production is carried on unsuitable soils. This promotes environmental degradation in the lowland humid tropics (Amazon rainforests and others) and montane areas (Northern Andes and Central American mountains). It is important for the social and environmental development of the region that this trend of cattle production is reversed. Even though the reduction of extensive grazing resulting from careful land-use planning would be the most desirable approach, not one country has made significant advances in this direction. This is because of the complex situation resulting from economic, political and institutional crises, together with the absence of viable alternatives in a sociopolitical context. In the meantime, intensification of cattle production could significantly increase its economic and social contribution. Silvopastoral systems are fundamental to this process of change. The main silvopastoral systems (SPS), including those already investigated and those empirically implemented by farmers in the region, are: scattered trees in pasturelands; SPS based on managed secondary succession; live fences; high tree density SPS; cut-and-carry systems; fodder tree banks and cattle grazing in tree plantations. Of less significance are windbreaks and pastures within tree alleys.

Key words: livestock production, fodder trees and shrubs

Introduction
Cattle ranching began in the 1500s in Latin America, taking advantage of natural savannah ecosystems in different areas within the Caribbean, Orinoco and the Pampa. Its gradual advance was accompanied by clearing of forests in dry and moist ecosystems, mountains and high plains or altiplanos. In time, cattle became fundamental for the consolidation of political and economic land control (Murgueitio, 2003).

The use of fire became a strategy to control plant succession and a tool to transform native forests into millions of hectares of pasturelands. African pasture species such as *Hyparrhenia rufa*, *Melinis minutiflora* and *Panicum maximum* were chosen for their aggressive growth, fire tolerance and fecundity (Parsons, 1972).

Livestock production based on grazing has caused major changes in rural landscapes at a continental scale and thus must be acknowledged as a system with enormous environmental and social repercussions (Bennett and Hoffmann, 1992). Pastures are the main land-use category in the tropical Americas, representing 60-80% of the total area. The expansion of this activity has implied the loss of natural ecosystems such as tropical and montane forests and, to a lesser degree, wetlands.

In Colombia natural forests and similar land-use categories declined from 94.6 to 72.4 million hectares between 1960 and 1995 and area under cattle grazing increased from 14.6 to 35.5 million hectares (Alexander von Humboldt Instituto, 1998). This figure could rise to 40 million hectares (Ministerio de Agricultura y Desarrollo Rural, 2001). Pasturelands occupy 46% of the total land area in Central America (18.4 million hectares) (Szott *et al.*, 2000).

However, even though the expansion of pasture areas in the tropical Americas continues, cattle meat and milk production have increased little. Stocking densities and low production indexes per animal (l of milk or kg of meat per ha respectively) are low and contributions to capitalization and rural employment reduced.

The social and environmental reconversion of livestock production is urgent and a priority for the region, (Murgueitio, 2000); cattle production could significantly increase its economic and social contributions and silvopastoral systems are fundamental in the process of change.

Silvopastoral systems in the tropical Americas
Silvopastoral systems are a type of agroforestry which combines fodder plants such as grasses and leguminous herbs with shrubs and trees for animal nutrition and complementary uses.

Over recent years agroforestry for animal production (AAP) (which includes silvopastoral systems) has aroused interest among researchers, development planners and farmers throughout Latin America, particularly in the Neotropics, as a result of the multiplicity of temporal and spatial arrangements and output variations. These advances have been recently reviewed by the Agroforestry for Animal Production Network led by FAO (Sánchez *et al.*, 2003).

The main silvopastoral systems, either researched or implemented empirically, include scattered trees in pasturelands (SPS), with managed plant succession, live fences, high tree density SPS, cut-and-carry systems, fodder tree banks and livestock grazing within tree plantations. Windbreaks and pastures between tree alleys are of lesser importance.

1. Scattered trees in pasturelands

A considerable proportion of cattle farms have scattered trees that offer shade and food for animals and generate income through timber and fruit. Some of these are remnant trees from the original forests, others have been planted, but the majority appear spontaneously as a result of dispersal by cattle and wildlife.

Recent studies in cattle areas in Central America showed that farmers retained between 88 and 100% of scattered trees in pasturelands (Harvey and Haber, 1999; Souza de Abreu *et al.*, 2000) with high species richness (Esquivel *et al.*, 2004).

In Costa Rica, Nicaragua, Colombia and Venezuela, wide-crown species such as *Samanea* (*Pithecellobium*) *saman*, laurel, *Cordia alliodora* and guanacaste, *Enterlobium cyclocarpum*, are accepted by farmers as shade trees, and also produce fruits, avidly consumed by cattle during the dry season (Roncallo *et al.*, 1996; Viera and Barrios, 1997; Ibrahim and Camargo, 2001; Esquivel *et al.*, 2004; Escalante, unpublished data). This silvopastoral practice is likely to be adopted by many farmers in the near future, given that it requires a lower financial investment than any other option.

2. SPS with plant succession management

Management of plant succession within SPS is accomplished through the elimination of indiscriminate weed control practices such as herbicide spraying, fire and mechanical weeding. After this first step, native vegetation and animals efficiently consolidate the system (some plants are dispersed by cattle). Selective pruning and timber extraction can be practised to encourage pasture growth.

Guava, *Psidium guajava*, one of the tropical fruits with highest vitamin C and mineral contents, has a wide geographic distribution and is directly associated with cattle grazing areas. Even though it was already abundant in tropical America at the time of the Spanish conquest (Patiño, 2002), cattle ranching facilitated its expansion to different ecosystems on the continent as it is readily consumed by livestock and wildlife, especially birds. Guava trees are spread into grazing areas mainly through seeds in cattle manure. Different researchers have proposed management strategies for this association with densities of 10 to 350 guava trees/ha, in order to maximize fruit and timber production and other local products with no detriment to livestock production. The beneficial effects of guava trees as catalysts of succession in grazing areas have also been shown (Somarriba, 1985a,b; Moreno and Latorre, 1999; Calle, 2003).

Other important multipurpose trees in Neotropical silvopastoral systems based on managed plant succession are different species of *Crescentia* (Bignoniaceae). Six species are known, of which the most widely distributed are *C. cujete* in South America and *C. alata* in Central America (Gentry, 1980). Given its drought tolerance, *Crescentia* is considered especially important in the dry tropics and regions which might easily become deserts, though the genus is also well adapted to moist conditions. Fruits are the main useful product of calabash trees. These fruits vary from round to elongated, and are 4 to 25 cm diameter (Arango, 2004) and between 423 g and 1500 g weight. Annual fecundity varies between 27 and 92 fruits or 16.2 and 81.2 kg per tree (Roncallo *et al.*, 1996). Seeds and pulp, with average crude protein content, sugars and minerals contents, are widely used as feedstuffs for cattle, goats, sheep, poultry and fish. Silvopastoral systems based on *C. alata* and *C. cujete* are increasingly successful and widespread in the Pacific lowlands of Nicaragua, the Caribbean and the Andes of Colombia and Venezuela.

Numerous native trees and shrubs found in grazing areas with managed succession are being researched and used by farmers in different countries within the region. Encouraging results in productivity and soil conservation have been found in south eastern Brazil (Minas Gerais) with trees such as *Zeyhera tuberculosa* and *Myracrodruon urundeuva* compared to the currently dominant *Brachiaria brizantha* monoculture (Viana *et al.*, 2002).

3. Live fences

In tropical America timber demand for livestock fencing is one of the greatest threats to existing forest patches in rural landscapes. Fences with barbed wire require 250 to 500 wooden posts per kilometre (each post 1.8 to 2.5 m long), the durability (and thus replacement time) of which depend on climate, soil and type of timber employed. Live fences are those in which trees and shrubs adaptable to short planting distances and frequent pruning are used to replace posts. Originating in the need to delimit properties, this agroforestry practice has major economic and ecological importance because it results in a 54% cost reduction compared to conventional fencing, removes pressure on forests as sources of posts and firewood. At the same time it allows trees to be introduced to grazing areas (Galindo and Murgueitio, 2003).

In Panama, Colombia, Venezuela and Ecuador, options for live fences and barriers include hedges with *Trichanthera gigantea*, *Tithonia diversifolia*, *Malvaviscus penduliflorus* or *Gliricidia sepium*; timber trees such as *Cordia alliodora* and *Bombacopsis quinatum* in the lowlands, and *Eucalyptus globulus* in mountain areas.

As linear elements in the landscape, live fences provide habitat and shelter for wildlife, and can be transformed into wildlife corridors when different native species are used to connect isolated forest fragments throughout the grazing areas. The contribution of hedges and live fences to the conservation of local biodiversity will depend on attributes such as width, structural complexity and species diversity (Murgueitio and Calle, 1999). In the Orinoco, a study of live fences formed through managed succession, with ages between 3 and > 30 years, found 39 bird species compared to four in the adjacent *Brachiaria decumbens* pastureland (the complete bird community contained 106 species); and six reptile species compared to zero in pasturelands (Molano *et al.*, 2003).

4. High tree density SPS

This includes modern intensive SPS developed through scientific research. Fodder shrubs are planted at high densities (1000 to > 10,000 individuals/ha) together with improved pastures with high biomass production, under intensive rotational grazing with electric fencing. They can include timber or fruit trees and leguminous herbs. *Leucaena leucocephala* has been the most successful species in intensive SPS in the tropics and subtropics, given its nutritional quality, nitrogen fixation, fast growth, moderate tolerance to drought and adaptability to browsing (Shelton, 1996). Animal stocking densities above four milk cows/ha with a milk production of 17,026 l/ha have been reported at Valle del Cauca, Colombia, in high density rotational grazing systems without artificial fertilizers (Molina and Uribe, 2002).

The advantages of the intensive *Leucaena leucocephala–Cynodon plectostachyus* SPS compared to a monoculture of the same pasture with nitrogen fertilization in deep fertile soil have been shown (Table 1).

Table 1. Production and quality of fodder biomass in an intensive SPS compared to a pasture monoculture with nitrogen fertilization (per cent increment in right column). Source: Adapted from Molina and Uribe, 2002.

Parameter	*Cynodon plectostachus* monoculture + 184 kg N_2/ha per yr	Intensive *Leucaena-C. plectostachyus* SPS (10,000 trees/ha), no N_2
Biomass (t DM/ha per year)	23.2	29.5 (+27.15%)
Crude protein (t DM/ha per year)	2.5	4.1 (+64.0%)
Metabolizable energy (Mcal/ha per year)	56,876	70,222 (+23.46%)
Calcium (kg/ha)	83.2	142.32 (+71.05%)
Phosphorus (kg/ha)	74.0	88.81 (+20.01%)

In addition, higher production has been achieved at a lower cost as a result of nitrogen fixation and the more efficient photosynthesis. Bird species richness in the system is higher (46 compared to 19 species in nearby sugar cane plantations) and water consumption is 25% lower. This has allowed farmers to obtain a certification for clean production (Molina and Uribe, 2002).

5. Cut-and-carry systems: woody fodder shrub banks

Cut-and-carry systems are widespread throughout the developing countries, especially in small farms, mountainous regions, agricultural lands and densely populated areas. In different Latin American countries, research, technology transfer and development work with farmers and local communities have focused on cut-and-carry systems for cattle and other domestic animals such as pigs, horses, goats, sheep, water buffaloes, rabbits, guinea pigs, poultry, fish and silkworms.

Species more traditionally used are *Gliricidia sepium, Trichanthera gigantea, Morus* sp, *Erythrina poeppigiana, E. berteroana, E. edulis, E. fusca* and *Boehmeria nivea*. Species more recently researched include *Tithonia diversifolia, Cratylia argentea, Malvaviscus penduliflorus, Spondias purpurea, Cajanus cajan* and *Urera caracasana* (Benavides, 1994; Escobar *et al.*, 1996; Gómez *et al.*, 1997; Argel and Lascano, 1999; Ríos, 1999; Murgueitio and Ibrahim, 2001).

The highest known densities in fodder shrub banks (up to one million plants per hectare) are reported with marango, *Moringa oleifera*, a tree native to the Himalayas introduced to the Pacific lowlands of Nicaragua (Foidl *et al.*, 1999).

Cut-and-carry polycultures were developed following the tradition of combining crops.

6. Grazing within tree plantations

In Brazil and other countries large plantations are often established on native or introduced degraded pasturelands, second growth areas and native forests. Different large scale projects (hundreds of thousands of hectares) have been established in the Atlantic and Amazon regions of Brazil. Introduced species from Australia and the Pacific islands dominate these plantations. These include eucalypts *E. tereticornis, E. grandis × E. urophylla* (*E. urograndis*), *E. camaldulensis, E. deglupta* and *E. saligna; Gmelina arborea*, teak, *Tectona grandis*, and different acacias such as *Acacia mangium, A. auriculiformis* and *A. mearnsii* (Schwengber, 1996; Bastos da Veiga *et al.*, 2001; Mochiutti and De Lima Meirelles, 2001; Silva da Silva *et al.*, 2001; Beer *et al.*, 2003). One of the most successful autochthonous species in large scale plantations is the Caribbean pine *Pinus caribea* (Gasca, 1996).

Some reforestation enterprises need a short term source of income to sustain investment during the establishment phase, reduce the cost of weeding competing gramineous and fodder species and minimize the risk of fire during drought periods. In the lowlands, the growth of pastures within timber plantations increases maintenance costs in such a way that the complete business of planting timber becomes unviable. Thus, cattle grazing within these plantations becomes a profitable option (Gasca, 1996; Londoño, 1996). In most cases, young animals graze within the tree plantations. Sheep production for wool could play an important role in different areas within the dry and sub-humid tropics.

7. Windbreaks

Windbreaks are a type of agroforestry system established to protect crop areas against winds, cold air currents and extreme climatic events such as hurricanes. They are formed by various parallel tree lines, sometimes up to 10 m apart. Windbreaks are considered SPS when the protected areas are dedicated to livestock production. In these cases, they help improve small animals' health, reduce heat stress and enhance fodder production through the enrichment of soil and reduction of water loss. They are not widely practised.

8. Pastures in tree and shrub alleys

Pastures in alleys involve planting herbaceous fodder species between lines of trees and shrubs. Arrangements for acid soils in the humid tropics include alleys of *Acacia mangium* or *Eucalyptus deglupta* with *Brachiaria brizantha, B. decumbens* and *Panicum maximum*. Although not widely planted they have shown important benefits for soil and fodder quality (Ibrahim and Camargo, 2001).

9. Expansion of SPS

Silvopastoral systems must be studied and promoted for their technical and ecological contribution to agroecosystems. Recent development of environmental services at the international level in densely populated areas of Latin America (carbon sequestration, soil protection, water regulation, scenic beauty and prevention of natural disasters) is generating new opportunities for transforming cattle production, which currently occupies a significant proportion of agroecosystems (Beer *et al.*, 2003; Murgueitio *et al.*, 2003). The development of paying schemes for these services could act positively to increase the area of silvopastoral systems.

References

Arango, A.J. (2003) *Use of the Diversity of the Totumo (Crescentia cujete L): a Multipurpose Tree for Colombia and Tropical America* (Uso de la diversidad del Totumo (*Crescentia cujete* L): un árbol multipropósito para Colombia y América Tropical). http://www.reuna.edu.co/temporales/memorias/especies/SeminarioMedellin.htm (accessed 5 May 2004) (In Spanish.)

Argel, P. and Lascano, C. (1999) *Cratylia argentea:* a new leguminous species for acid soils in tropical sub-humid areas (*Cratylia argentea*: una nueva leguminosa arbustiva para suelos ácidos en zonas subhúmedas tropicales). In: Sánchez, M. and Rosales, M. (eds) *Agroforestería para la Producción Animal en Latinoamérica.* FAO, Rome, Italy, pp. 259-275. (In Spanish.)

Bastos da Veiga, J., Alves, C.P., Tavares Marques, L.C. and Da Veiga, D.F. (2001) Silvopastoral systems in Eastern Amazon (Sistemas silvipastoris na Amazonia Oriental). In: Carvalho, M., Alvim, M. and Da Costa, C.J. (eds) *Sistemas Agroflorestais Pecuarios. Opcoes de Sustentabilidade para Áreas Tropicais e Subtropicais.* EMBRAPA–FAO–Ministerio da Agricultura Pecuaria e Abastecimento de Brasil. Brazil, pp. 41-76. (In Portuguese.)

Beer, J., Harvey, C., Ibrahim, M., Harmand, J.M., Somarriba, E. and Jiménez, F. (2003) Environmental services of agroforestry systems (Servicios Ambientales de los Sistemas Agroforestales). *Agroforestería de las Américas* 10, 37-38.(In Spanish.)

Benavides, J. (1994) Research on fodder trees (La investigación en Árboles Forrajeros). In: *Árboles y Arbustos Forrajeros en América Central.* CATIE, Turrialba, Costa Rica, pp. 113-120. (In Spanish.)

Bennett, D. and Hoffmann, R. (1992) Husbandry in the new world (La ganadería en el nuevo mundo). In: Viola, H. and Margolis, C. (eds) *Semillas de Cambio.* Smithsonian Institution, Washington and London, pp. 90-110. (In Spanish.)

Calle, Z. (2003) Silvopastoral systems with guayaba trees (Sistemas Silvopastoriles con Árboles de Guayaba). In: *Restauración de Suelos y Vegetación Nativa: Ideas para una Ganadería Andina Sostenible.* CIPAV, Cali, Colombia, pp. 61-66. (In Spanish.)

Escobar, A., Romero, E. and Ojeda, A. (1996) *Gliricidia sepium, a Multipurpose tree* (*Gliricidia sepium.* El matarratón, árbol multipropósito). Fundación Polar, Universidad Central de Venezuela, Caracas, Venezuela. (In Spanish.)

Esquivel, H., Ibrahim, M., Harvey, C. and Villanueva, C. (2004) Dispersed trees in pastures in farms in dry ecosystems in Costa Rica (Árboles dispersos en potreros de fincas ganaderas en un ecosistema seco de Costa Rica). *Agroforestería de las Américas*. (In press.) (In Spanish.)

Foidl, N., Mayorga, L. and Vásquez, W. (1999) Use of the Marango (*Moringa oleifera*) as fresh fodder for livestock (Utilización del Marango (*Moringa oleifera*) como forraje fresco para ganado). In: Sánchez, M. and Rosales, M. (eds) *Agroforestería para la Producción Animal en Latinoamérica.* FAO, Rome, Italty, pp. 341-350. (In Spanish.)

Galindo, W. and Murgueitio, E. (2003) Tools for sustainable management of Andine husbandry (Herramientas de Manejo Sostenible para la Ganadería Andina). In: *Manejo Sostenible de los Sistemas Ganaderos Andinos.* CIPAV, Cali, Colombia. pp. 19-88. (In Spanish.)

Gasca, G. (1996) Suitable species for silvopastoral systems in the Eastern Llanos (Especies promisorias para sistemas silvopastoriles en los Llanos Orientales). In: Álvaro Uribe C. (ed.) *Silvopastoreo: Alternativa para Mejorar la Sostenibilidad y Competitividad de la Ganadería Colombiana. Compilación de las Memorias de dos Seminarios Internacionales sobre Sistemas Silvopastoriles.* Corpoica, Bogotá, Colombia, pp. 265-276. (In Spanish.)

Gentry, A.H. (1980) *Neotropical Flora (Flora neotrópica).* Bignoiniaceae Part I. The New York Botanical Garden, New York, USA.

Gómez, M.E., Rodríguez, L., Murgueitio, E., Ríos, C., Rosales, M., Molina, C.H., Molina, E., Molina, C.H. and Molina, J.P. (1997) *Fodder Trees and Shrubs Used in Animal Feeding as Protein Source* (Árboles y arbustos forrajeros utilizados en alimentación animal como fuente proteica), 2nd edn. CIPAV, Cali, Colombia. (In Spanish.)

Harvey, C.A. and Haber, W.A. (1999) Remnant trees and the conservation of biodiversity in Costa Rican pastures. *Agroforestry Systems* 44, 37-68.

Ibrahim, M. and Camargo, J.C. (2001) Productivity and environmental services of silvopastoral systems: case studies from CATIE (Productividade e servicos ambientais de sistemas silvipastoris: experiencias do CATIE). In: Carvalho, M., Alvim, M. and Da Costa, C.J. (eds) *Sistemas Agroflorestais Pecuarios. Opcoes de Sustentabilidade para Áreas Tropicais e Subtropicais.* EMBRAPA–FAO, Brazil, pp. 331-347. (In Portuguese.)

Instituto Alexander von Humboldt, Ministerio de Medio Ambiente, DNP, PNUMA (1998) *Colombia: Biodiversity in XXI Century* (Colombia: Biodiversidad en el siglo XXI). Santafé de Bogotá, Colombia. (In Spanish.)

Londoño, G. (1996) Silvopastoral programme: Project La Gloria (Programa Silvopastoril: proyecto La Gloria). In: Álvaro Uribe C. (ed.) *Silvopastoreo: Alternativa para Mejorar la Sostenibilidad y Competitividad de la Ganadería Colombiana. Compilación de las Memorias de Dos Seminarios Internacionales sobre Sistemas Silvopastoriles.* Corpoica, Bogotá, Colombia, pp. 253-263. (In Spanish.)

Ministerio de Agricultura y Desarrollo Rural (2001) *Policy for an Environmentally Friendly Agrarian Development* (Política para el Desarrollo Agropecuario Ambientalmente Sostenible). Versión para la presentación ante el Consejo Nacional Ambiental, Bogotá, Colombia. (In Spanish.)

Mochiutti, S. and De Lima Meirelles, P.R. (2001) Silvopastoral systems in the Amapá: current situation and future perspectives (Sistemas silvipastoris no Amapá: situacao atual e perspectivas). In: Carvalho, M., Alvim, M. and Da Costa, C.J. (eds) *Sistemas Agroflorestais Pecuarios. Opcoes de Sustentabilidade para Áreas Tropicais e Subtropicais.* EMBRAPA, FAO, Brazil, pp. 77-99. (In Portuguese.)

Molano, J.G., Quiceno, M.P. and Roa, C. (2003) The role of living fences in an agrarian system in Piedemonte (El papel de las cercas vivas en un sistema agropecuario en el Piedemonte Llanero). In: Sánchez, M. and Rosales, M. (eds) *Agroforestería para la Producción Animal en Latinoamérica.* FAO, Rome, Italy, pp. 45-61. (In Spanish.)

Molina, C.H. and Uribe, F. (2002) Experiences of clean production in grazing husbandry (Experiencias en producción limpia de Ganaderías en Pastoreo). In: *Competitividad en Carne y Leche. Memorias del III Seminario Internacional. Cooperativa Lechera de Antioquia.* COLANTA, Medellín, Colombia, pp. 333-354. (In Spanish.)

Moreno, V.M. and Latorre, S. (1999) Assessment of a silvopastoral system with guayaba trees in Hoya (Evaluación del Sistema Silvopastoril Guayaba-Grama Natural en la Hoya del río Suárez). In: Corpoica. *Memorias del VI Seminario Internacional sobre Sistemas Agropecuarios Sostenibles,* CIPAV (www.cipav.org.co), Cali, Colombia. (In Spanish.)

Murgueitio, E. and Calle, Z. (1999) Biological diversity in bovine husbandry systems in Colombia (Diversidad biológica en sistemas de ganadería bovina en Colombia). In: Sánchez, M. and Rosales, M. (eds) *Agroforestería para la Producción Animal en Latinoamérica.* FAO, Rome, Italy, pp. 53-88. (In Spanish.)

Murgueitio, E. (2000) Agroforestry systems for husbandry production in Colombia (Sistemas Agroforestales para la Producción Ganadera en Colombia). In: Pomareda, C. and Steinfeld, H. (eds) *Intensificación de la Ganadería en Centroamérica - Beneficios Económicos y Ambientales.* CATIE, FAO and SIDE, San José, Costa Rica, pp. 219-242. (In Spanish.)

Murgueitio, E. and Ibrahim M. (2001) Agroforestry for husbandry development in Latinamerica (Agroforestería pecuaria para la reconversión de la ganadería en Latinoamérica). *Livestock Research for Rural Development* 13, 3. http://www.cipav.org.co/lrrd/lrrd13/3/ly133.htm CIPAV, Cali, Colombia. (In Spanish.)

Murgueitio, E., Ibrahim, M., Ramírez, E., Zapata, A., Mejía, C. and Casasola, F. (2003) *Land Use in Husbandry Farms* (Usos de la tierra en fincas ganaderas). Guide of Environmental Payments in the Integrated Silvopastoral Focussed Project for the Ecosystem Management (Guía para el pago de servicios ambientales en el proyecto enfoques silvopastoriles integrados para el manejo de ecosistemas). CIPAV, Cali, Colombia. (In Spanish.)

Murgueitio, E. (2003) *Environmental Impact of Dairy Husbandry in Colombia and Solution Alternatives* (Impacto ambiental de la ganadería de leche en Colombia y alternativas de solución). CIPAV, Cali, Colombia. (In Spanish.)

Parsons, J.J. (1972) Spreading of African pastures in the American tropics (Difusión de los pastos africanos en los trópicos americanos). In: Molano, J. and Fondo, B. (eds) *Las Regiones Tropicales Americanas: Visión Geográfica.* FEN Colombia. Bogotá, Colombia, pp. 355-370. (In Spanish.)

Patiño, V.M. (2002) *History and Distribution of the Native Fruit Trees in the Neotropics* (Historia y distribución de los Frutales Nativos del Geotrópico). Centro Internacional de Agricultura Tropical (CIAT), Asohofrucol y Fondo Nacional de Fomento Hortofrutícola. Publicación CIAT 326, Cali, Colombia. (In Spanish.)

Ríos, C.I. (1999) *Tithonia diversifolia* Hemsl. Gray, a fodder species for sustainable systems (El botón de oro *Tithonia diversifolia* Hemsl. Gray, otra especie forrajera para sistemas sostenibles). In: Sánchez, M. and Rosales, M. (eds) *Agroforestería para la Producción Animal en Latinoamérica.* FAO, Rome, Italy, pp. 311-326. (In Spanish.)

Roncallo, B., Navas, A and Garibella, A. (1996) Potential of the fruits of native plants for the ruminants (Potencial de los frutos de plantas nativas en la alimentación de rumiantes). In: *Sistemas Silvopastoriles: Alternativa para una Ganadería Moderna y Competitiva. Memorias II Seminario Internacional.* Ministerio de Agricultura–CONIF, Santafé de Bogotá, Colombia, pp. 81-92. (In Spanish.)

Sánchez, M., Rosales, M. and Murgueitio, E. (2003) Cattle agroforestry in Latin America (Agroforestería Pecuaria en América Latina). In: Sánchez, M.D. and Rosales, M. (eds) *Agroforestería para la Producción Animal en América Latina – II.* FAO, Rome, Italy, pp. 1-10. (In Spanish.)

Schwengber, D.R. (1996) Silvopastoral systems in Brazil, with focus on the Amazon (Sistemas Silvopastoriles en Brasil, con énfasis en la Amazonia). In: Álvaro Uribe, C. (ed.) *Silvopastoreo: Alternativa para Mejorar la Sostenibilidad y Competitividad de la Ganadería Colombiana.* Bogotá, Colombia, pp. 33-42. (In Spanish.)

Shelton, M. (1996) *Leucaena* genus and its potencial for the tropics (El género *Leucaena* y su potencial para los trópicos). In: Tyrone Clavero (ed.) *Leguminosas Forrajeras Arbóreas en la Agricultura Tropical.* Fundación Polar, Universidad del Zulia, Maracaibo, Venezuela, pp. 17-28. (In Spanish.)

Silva da Silva, J.L., De Saibro, J.C. and De Souza Castilhos, Z.M. (2001) Research situation and use of silvopastoral systems in Rio Grande do Sul (Situacao da pesquisa e utilisacao de sistemas silvopastoris no Rio Grande do Sul). In: Carvalho, M., Alvim, M.

and Da Costa, C.J. (eds) *Sistemas Agroflorestais Pecuarios. Opcoes de Sustentabilidade para Áreas Tropicais e Subtropicais.* EMBRAPA–FAO, Brazil, pp. 257-283. (In Portuguese.)

Somarriba, E. (1985a) *Guayaba trees (Psidium guajava L.) in pasture lands. I Fruit Production and Potential of Seed Dispersal* (Árboles de guayaba (*Psidium guajava* L.) en pastizales. I Producción de Fruta y Potencial de dispersión de semillas). *Turrialba* 35 (3), 289-295.

Somarriba, E. (1985b) *Guayaba Trees (Psidium guajava L.) in Pasturelands.* II Fruit consumption and seed dispersal (Árboles de guayaba (*Psidium guajava* L.) en pastizales. II Consumo de Fruta y dispersión de semillas). *Turrialba* 35 (4), 329-332. (In Spanish.)

Souza de Abreu, M.H., Ibrahim, M., Harvey, C. and Jiménez, F. (2000) Characterization of the tree component in the husbandry systems in La Fortuna de San Carlos, Costa Rica (Caracterización del componente arbóreo en los sistemas ganaderos de La Fortuna de San Carlos, Costa Rica). *Agroforestería de las Américas* 7 (26), 53-56. (In Spanish.)

Szott, L., Ibrahim, M. and Beer, J. (2000) *The Hamburger Connection Hangover: Cattle, Pasture Land Degradation and Alternative Land Use in Central America.* CATIE-DANIDA-GTZ, Turrialba, Costa Rica.

Viana, V., Mauricio, R.M., Matta-Machado, R. and Pimenta, I. (2002) Management of the natural regeneration of native tree species for silvopastoral systems in dry forests in South-Eastern Brazil (Manejo de la regeneración natural de especies arbóreas nativas para la formación de sistemas silvopastoriles en bosques secos del sureste de Brasil). *Agroforestería de las Américas* 9 (33-34), 48-52. (In Spanish.)

Viera, C. and Barrios, C. (1997) Wood production inventory on pastures of Esparza: species, management and wood component management (Exploración sumaria de la producción de maderas en porteros de la zona ganadera de Esparza: especies, manejo, y dinámica de componentes maderables*).* Segundo curso de manejo de forestal, Turrialba, Costa Rica.

European types of silvopastoral systems in the Mediterranean area: dehesa

L. Olea, R. J. López-Bellido and M. J. Poblaciones
Dept. de Biología y Producción de Vegetales, Universidad de Extremadura, Ctra. Cáceres s/n, C.P. 06071, Badajoz, Spain. lolea@unex.es

Abstract
The ecosystem dehesa, which exceeds 3.5 million hectares in the South-West of the Iberian peninsula, is a peculiar silvopastoral system and generally well preserved. It is greatly extensive and has a high environmental value due to the low impact of human activity on the Mediterranean forest.

The dehesa is mainly located in semi-arid areas over acid or neutral soils with a restricted production potential. Results of this investigation on the quality and production of the herbaceous pastures (1500 to 2700 kg DM/ha), improvement of systems (introduction of species and fertilization of sown and natural pastures), management (extensive grazing), fruit production (670 kg DM/ha acorn) and distribution so it can be used by pigs, erosion control and so on are reported and discussed. The extensive set stocking rate is very dependent on the ecosystem (ruminant and monogastric animals). The dehesa must be considered as an ecosystem resulting from extensive stocking, taking into account that the best way to keep it is by means of production.

Key words: extensive pastures, subterranean shamrock, semiarid climate

Introduction
There are many historical events (first reference in 924) which have influenced the formation and expansion of the dehesa. Some of these important events include the role of Mesta, the system of property exchange in the Middle Ages with the different phases of the Reconquista and the afterwards the disaggregation of heritages, especially those of the nobility and religious orders.

Alfonso X (the Wise Man) established the first Fueros de la Mesta, through which some balance was achieved between agriculture and stockbreeding interests. This balance was maintained until the reign of the Catholic kings, which helped to push the stockbreeding.

Dehesa has been defined from biological, socio-economic and extensive considerations nevertheless we choose the more complete one which may be the following: 'Dehesa is a semidry ecosystem where grasses, bushes and tree-shaped species live together and contribute to the feeding of either domestic or wild animals in an extensive and sustainable production'.

Environmental and physical characteristics
Its climatic, soil and geomorphological characteristics give rise to a poor environment for agriculture. The climate in the dehesa is (L'Houerou, 1975) moderately semidry with annual rainfall ranging between 450 and 900 mm with dry and hot summers (lasting 3-5 months) and cold and rainy winters. As a result of this, two different strategies can be observed among the vegetation.
1. Esclerophyllia: Includes the bushy and tree-shaped species. They are provided with powerful root systems that enable the plant to resist long dry periods.
2. Terofitos with an annual cycle: This group is able to produce seeds at the end of the wet season and spend the dry season as seeds, until the autumn rain comes when the seminal lethargy frees them from climatic and handling adversities. The subsoils over which the dehesa is located are composed of materials of Palaeozoic origin (slate, granites, …) or any other material that derives from its erosion (coarse and less coarse sands), resulting in a generally acid or neutral soil in terms of pH, poor in nutrients (P, N and Ca above all) and with a limited depth.
Its relief is undulating with restricted slopes. The interaction of all these factors makes this territory heterogeneous in relation to environments and resources. The dehesa is located in the south-western part of the Iberian peninsula in Spain and Portugal, extending to around 3.5 million hectares. The greatest area is in Extremadura, i.e. 1.25 million ha, followed by that in Andalusia (around 700,000 ha) and in Alentejo (region in the south-west of Portugal about 800,000 ha).

Pastures
The main characteristic of the pasture in the dehesa is its restricted and uneven (irregular) production, highly influenced by the climate. Variation in rainfall throughout the year and between years and in soils over short distances produces big differences and contrasts in production. These differences can be up to 200% as an average of 5 years in different

trials (Olea *et al.,* 1989). The annual distribution of yield is irregular and it has been observed that 70% of its production takes place in spring, 30% in autumn with no growth in summer (Figure 1).

Table 1. Quality of natural pastures. OMD, Organic matter digestibility.

Net protein			OMD			% of Legumes		
Max	Min	Average	Max	Min	Average	Max	Min	Average
14.8	8.5	10.3	63.3	49.0	55.2	24.0	4.0	8.5

The average annual production of pastures is around 1440 kg DM/ha. The quality of these pastures is limited and it is shown in Table 1.

Figure 1. Annual production of natural pastures.

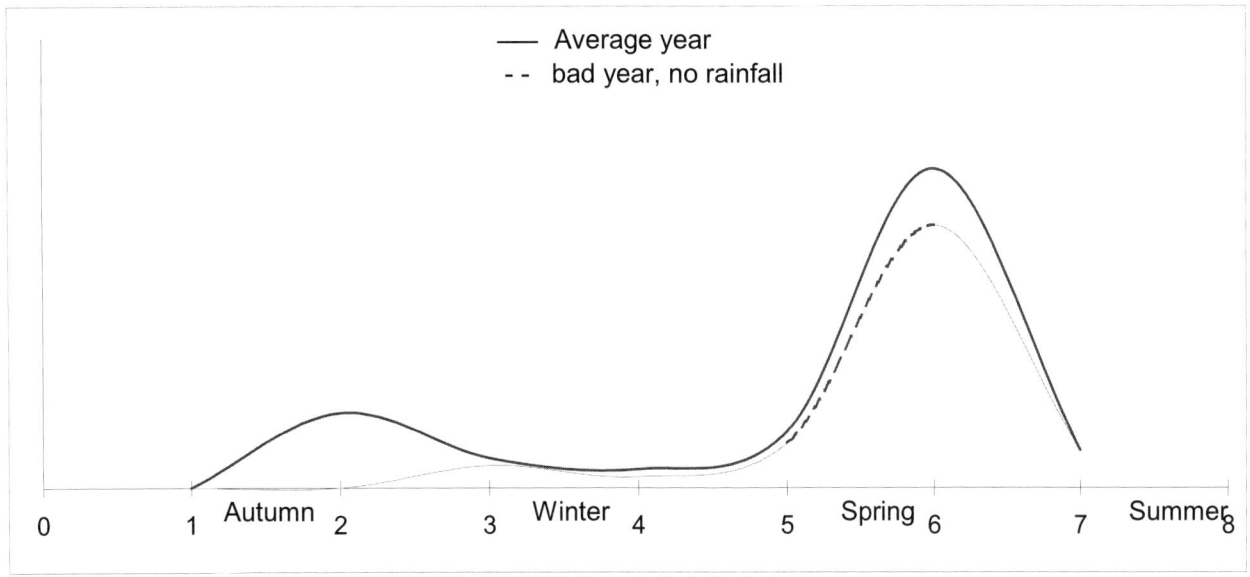

This low quality makes management difficult and makes the use of resources of trees and bushes especially important at certain times throughout the year. Improvement in pasture production in the dehesa is needed and should include a package of integrated actions to achieve the greatest production (quality and quantity) and the thickest plant cover.

The improvement methods to be used can be either fertilization of the natural pastures with proper management or through the introduction of species and varieties, fertilization and correct management.

The method to be used should be selected according to production potential and floristic composition.

Improvement of dehesa pastures: natural pasture fertilization and correct management

Fertilization and management of natural pastures is very helpful as far as the flora is appropriate. The influence of nutritional factors determines the nature and profitability of the pastures (Jiménez Mozo and Martínez Agrella, 1982). The low quantity of organic material and phosphate are the most obvious characteristics of these soils. The most commonly applied fertilizer is usually phosphate, with residual effects. These fertilizers range from 25 to 35 kg P_2O_5/ha during the first year and 18 to 25 thereafter. These are the recommended average fertilization levels.

The phosphate level in soil needs to be raised to 8 or 12 ppm P-Olsen (Granda *et al.,* 1991; Moreno *et al.,* 1994). Details should be known about the method and timing of phosphate application. The traditional method commonly used is based on surface application in autumn. However, recent research (Maldonado *et al.,* 2004) casts doubt on the effectiveness of this system. On the other hand, Moreno *et al.* (1993) make clear that superphosphate lime 18%, when inserted into the soil, reaches the plant sooner and more effectively. However, this effect hasn't yet been confirmed (Maldonado *et al.,* 2004) and creates a greater erosion risk. Research was carried out to replace the usual form of fertilizer, superphosphate lime 18%, with superphosphate rock (ecological product), and the same results were obtained (Maldonado *et al.,* 2004).

The need for potassium in these types of pastures is higher in granite soils. Jiménez Mozo and Martínez Agrella (1982) have estimated that 25 to 35 kg K_2O/ha per year is enough to satisfy needs under grazing conditions.

Variable results on the amount of the oligoelements and secondary elements that should be supplied to these pastures have been obtained (Moreno *et al.,* 1994).

Dehesa pasture improvement: introduction of species and varieties, fertilization and management

This improvement method consists of introducing pasture species and varieties chosen to persist and improve the production along with adequate fertilization and management.

Among these species, the most important components are annual legumes; the use of grassy species is of secondary importance due to low soil fertility, and therefore soils need to recover first with the use of legumes. As pointed out when discussing the previous method, higher phosphate fertilization is advisable (35 to 40 kg P_2O_5/ha) during the first year and 25 to 30 kg P_2O_5/ha for the following years.

The most common species of legumes used in the improvement method are: *Trifolium subterraneum*: ssp: *subterraneum, yanninicum* and *brachycalcynum*). Varieties from SIA in Extremadura, Spain (González López, 1994), are especially relevant. *T. glomeratum, Ornithopus compressus* L., *Medicago polimorfa* and *M. murex, M. truncatula, M. rugosa, M. tribuloides, T. resupinatum, T. michelianum, T. hirtum, T. incarnatum, T. balanza, T. striatum, Hedysarum coronarium* (*zulla*), *Biserula pelicinum*, etc.

Different combinations of these species are used to improve specific areas:
- Combination of 3 to 5 different varieties of *T. subterraneum* (sowing rate: 15 to 20 kg/ha).
- Combination of 3 to 5 species with 2 or 3 varieties of each one (sowing rate: 18 to 25 kg/ha).
- Biodiversity combination: 12 to 16 species and varieties (sowing rate: 20 to 30 kg/ha).

Production of improved pastures (quality and quantity)

Improved pastures from dehesa are more productive and have higher quality than the natural unimproved ones (Table 2).

Table 2. Production (quality and quantity) of pastures.

	Production		Quality (%)		
	kg DM/ha	Average response	Net protein	OMD	Legumes
Natural	1440	-	10.3	52.0	8.5
Natural fert.	2238	55%	11.0	58.9	18.0
Introd. fert.	2670	86%	13.6	62.5	30.0

Some differences are maintained between the three different types. The net protein in improved pastures during autumn and winter varies from 19 to 21%. Two remarkable differences are obvious (Figure 2): net protein levels are better in improved pastures, and introduced pastures have between 2 and 3% more net protein than the control in spring and summer.

The trend in the organic matter digestibility (OMD) of the two types of improved pasture follows that for net protein through the year (Table 3).

Table 3. Results of cover measurements in different times of the year.

Treatment	November	February	March	April
Ploughed	0	0	0	0
Introduced and fertilized pasture (1 year)	18	76	83	99
Burnt pasture	25	68	79	79
Natural fertilized pasture	82	95	97	99
Natural pasture (unimproved)	70	81	84	90

Both pastures are more digestible during the whole cycle than the unimproved control (Figure 3). As dry pastures they achieve levels close to 48%, similar to those obtained by López Gallego (1988) for fertilized pastures in these areas.

Figure 2. Evolution of the net protein with time.

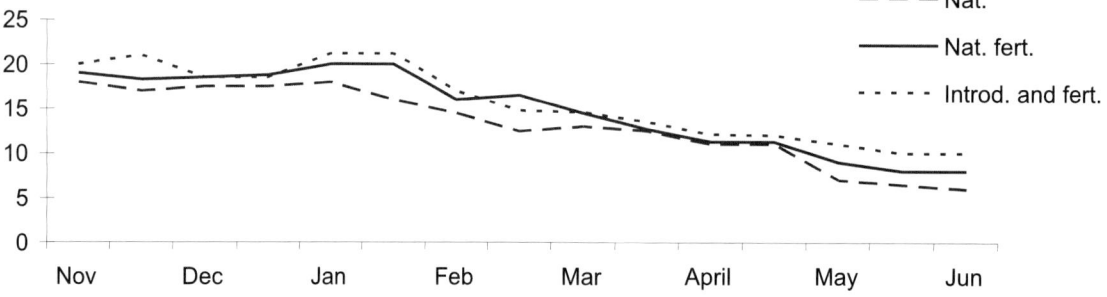

This improved dry pasture quality (over 9% net protein and 48% OMD), along with the seed (average 85 kg/ha) that may be available to animals when there is an increase in the level of legumes, results in less rejection and reduces the amount of pasture needed to be on offer in summer for the animal to survive (Olea *et al.,* 1989).

Figure 3. Evolution of the OMD with time.

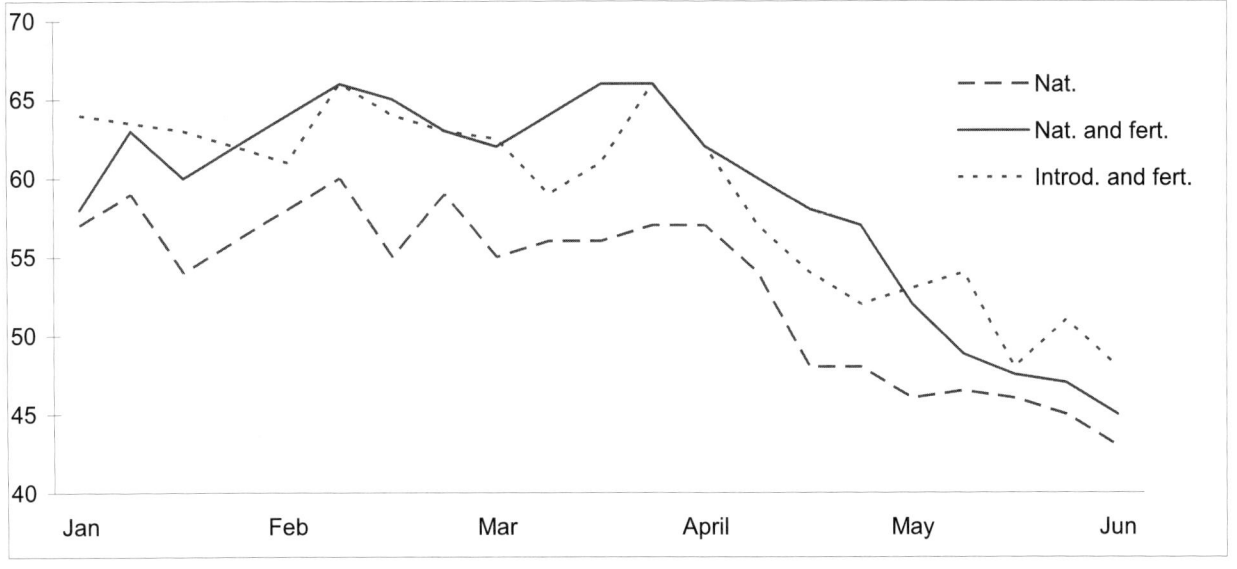

Dehesa conservation

In general terms, the dehesa is the best preserved ecosystem in the Mediterranean. Only in certain areas and under specific conditions are degradation and erosion risks found. Although the correlation between cover and erosion is obvious (Stocking, 1988), the complexity of the relationship presents aspects that are nevertheless still unknown. It is of great interest to know within this ecosystem and under certain climatic conditions how the plant-erosion relation is established throughout the year and what the structural layout is in connection with physical processes that take place.

There is a critical threshold between 50% and 60% of plant cover, below which the loss of soil is a relevant issue, as is shown in Figure 4 (Stocking's curve adapted to the dehesa).

In dehesas the highest risk of erosion is after long drought periods. The maximum percentage loss occurs in April with 60% cover, but these covers are less thick in autumn and winter.

Investigations (research) carried out by the Geography Faculty of Extremadura, the SIA Extremadura and by the Agronomist Engineer School of the Extremadura University (Maldonado *et al.,* 2004) are revealing that, in general terms, plant cover in the dehesa is adequate and sufficiently plentiful, especially when the pastures are improved. The factors that cause the highest risk of erosion are those due to hydric erosion, and this risk reaches its highest level at the beginning of autumn. Any technology or action that results in an increase in the vegetation cover will cause a decrease in the risk of erosion.

Figure 4. Relation between percentage vegetal coverage and loss of soil.

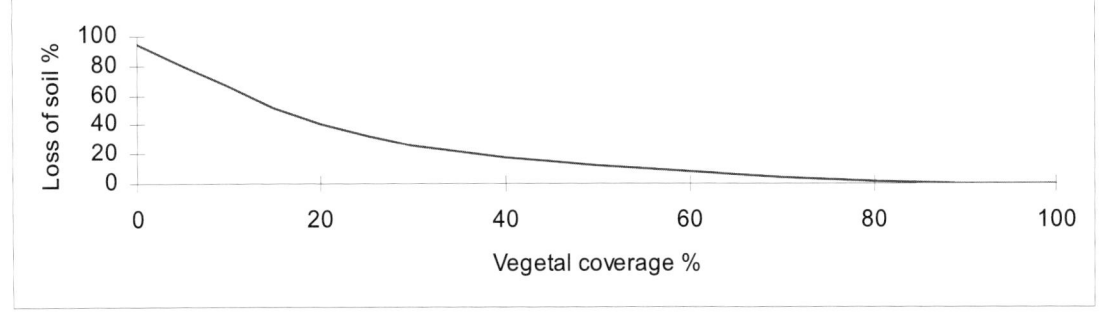

Environmental factors such as the slope, soil texture, present of bushy or tree-shaped vegetation, etc., are reflected in these values.

Research which has already been mentioned has given results as indicated in Figure 5 and Figure 6 for different technologies and for rainy and dry years.

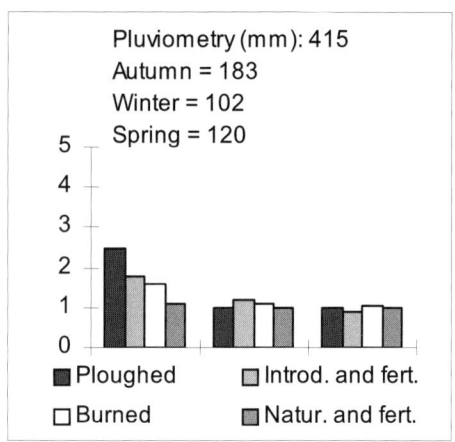

Figure 5. Dragging coefficient in an average rainy year.

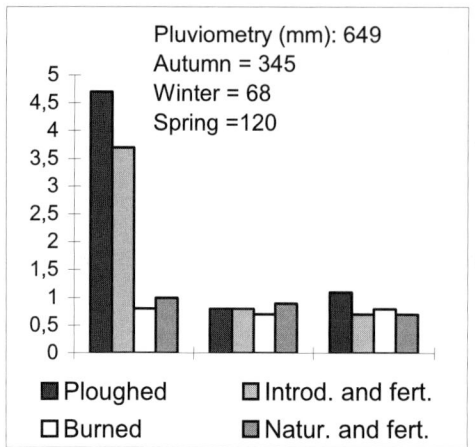

Figure 6. Dragging coefficient in a rainy year.

A high risk of erosion in autumn and winter is shown, especially in rainy years, as is the necessity of decreasing the risk caused by some agricultural activities, among others. These include the introduction of pastures in the first year by the traditional system, especially in shallow soils, with slopes and so on.

When pastures are improved, the erosion risk decreases and besides higher production in terms of both quantity and quality is obtained. Having a thick vegetation cover within the dehesa is a very important factor to take into account to produce more and achieve better preservation, always with the aim of creating sustainable systems.

Management and use of pastures

The use and management of pastures are highly important and they affect production and preservation of the pasture. Management from the grazing point of view must achieve maximum economic and technical efficiency in the use of pasture as well as maintain environmental preservation.

Australian studies (Quinlivan, 1981; Ewing, 1982) and our own research and experience in the management of improved (developed) pastures indicate that it is highly recommended that, during grazing, high grazing pressure should always be avoided during spring (bloom period) and autumn.

Animals will remain grazing for as long as possible. The grazing must be controlled in winter so that enough pasture is left for it to bloom and be productive.

The seasons during which production is highest and when demand is highest should coincide. Biomass surplus in spring can be used to feed animals in summer and autumn if quality is sufficiently high. This will minimize dependence on resources outside the system. If we obtained pastures of adequate quality (18% net protein and 63% OMD) we can offer herbage at around 500 kg DM per sheep per year, whereas using low quality pastures (12% net protein and 45% OMD) it is necessary to offer more than 830 kg of MS per sheep per year (Olea and Viguera., 1998). With this management the vegetation will always be more productive without losing its biodiversity. A high, well-rooted and dense vegetation cover establishes a harmonic unit along with trees and bushes (if they exist) that help in the seasonal feeding of wild and domestic animals.

Dehesas usually have a problem, because this pasture harmonic unit with trees and bushes can produce a restricted efficiency in critical periods of the year (summer, autumn and winter), a problem that it is necessary to solve. This can be achieved by following some practices:

- Grazing shrubs: They can be placed in small areas where there are reservoirs of green forage in drought seasons.
- Trees: Cork trees and *Q. ilex* especially contribute leaves, stems and fruits to the diet of ruminants and monogastric animals (extensive pig-farming).

Research carried out in dehesas in the south-west of the Iberian peninsula by Olea *et al.* (2004) has determined average values of production and quality of montaneras (Table 4).

Table 4. Production and quality (N, Ca, P, metabolic energy and OMD) of the montaneras of the SW of Spain.

Dry matter (kg/ha)	Net protein (%)	Ca (%)	P (%)	Metabolic energy (cal/kg)	OMD (%)
674	4.70	0.22	0.09	3017.42	85.11

Dehesa perspectives

The dehesa is one of the best-preserved ecosystems all over the world, especially within the Mediterranean semi-arid regions, but it is necessary to establish more strongly the double role that must be assumed by the farmer:

- As producer.
- As landscape and environmental guardian.

This can only be done with highly extensive systems and the dehesa plays and must play a major role. To sum up, the best way to preserve the dehesa is to make it reasonably productive using appropriate techniques.

References

Ewing, B.M. (1982) Pasture and crops legumes… their place in woathelt rotations. *Journal of Agriculture W.A.* 2. Perth, Australia.

González López, F. (1994) *Spanish Varieties of Subterranean Clover: Origin, Identification and Use Recommendations* (Variedades españolas de trébol subterráneo: Origen, identificación y recomendaciones para su uso). Junta de Extremadura, Badajoz, Spain. (In Spanish.)

Granda, M., Moreno, V. and Prieto, P.M. (1991*) Improvement and Use of Natural Pastures of the Dehesa* (Mejora and utilización de pastos naturales de dehesa). MAPA, Madrid, Spain. (In Spanish.)

L'Houerou, H.N. (1975) *Bioclimatology of the Mediterranean Region* (Bioclimatología de la región mediterránea). Seminario. La Orden, CRIDA 08, Badajoz, Spain. (In Spanish.)

López Gallego, F. (1988) Lamb systems in pastures in Extremadure (Sistemas de acabado de corderos en pastos de Extremadura). *XXXIX Congreso de la Federación Europea de Zootecnia.* Lisboa, Portugal. (In Spanish.)

Jiménez Mozo, J. and Martínez Agrella, T. (1982) *Pastures Fertilization: I Nutrient Needs of Macroelements* (Fertilización de pastos: I Necesidades nutritivas referentes a macroelementos). Publicaciones SEA-UNEX, Badajoz, Spain. (In Spanish.)

Maldonado, A., Olea, L., Viguera, J. and Poblaciones, M.J. (2004) Effect of applying different sources or phosphorus fertilization on slate soils in dehesas and pasture lands in south-western Spain (Efecto de la aplicación de diferentes fuentes de fertilización fosfórica sobre suelos de pizarra en dehesas and pastizales del S.O. de Spain). *XLVI Reunión Científica de la Sociedad Española para el Estudio de los Pastos (SEEP)*. Salamanca, Spain. (In Spanish.)

Moreno, V., Bueno, C. and Santos, A. (1993) Response to different doses of calcium superphosphate in brown soils in dehesas in Extremadura (Respuesta a distintas dosis de superfosfato de cal en suelos pardos meridionales de la dehesa extremeña). *Actas XXXIII Reunión Científica de la Sociedad Española para el Estudio de los Pastos (SEEP)*. Ciudad Real, Spain, pp. 234-243. (In Spanish.)

Moreno, V., González, F. and Olea, L. (1994) Annual legumes improvement for pastures. *Melhoramento* 33 (1), 230-240.

Olea, L. and Viguera, F.J. (1998) Pastures and crops (Pastizales and cultivos). In: *La Dehesa. Aprovechamiento Sostenible de los Recursos Naturales*. Agrícola Española S.A., Madrid, Spain. (In Spanish.)

Olea, L., Paredes, J. and Verdasco, P. (1989) Productive characteristics of the dehesa pastures in south-western Iberian Peninsula (Características productivas de los pastos de la dehesa del S.O. de la Península Ibérica). *II Reunión Ibérica de la Sociedad Española para el Estudio de los Pastos (SEEP)*. Badajoz, Spain, pp. 194-230. (In Spanish.)

Olea, L., Poblaciones, M.J., Viguera, J. and Olea, B. (2004) Availability distribution of holm-oak (*Quercus ilex lam.* ssp. *ballota*) acorns (quantity and quality) in 'montanera' in South-western Extremadure (Distribución de la "oferta" de bellota (cantidad and calidad) de encina (*Quercus ilex lam.* sp *ballota*) en "montanera" en dehesas del S.O. de Extremadura). *XLIV Reunión Científica de la Sociedad Española para el Estudio de los Pastos (SEEP)*. Salamanca, Spain. (In Spanish.)

Quinlivan, B.J. (1981) *Assignment Terminal Report*. INIA/UNDP/FAO, SPA 71/517, Badajoz, Spain.

Stocking, M.A.(1988) Assessing vegetative cover and management effects. In: Lal, R. (ed.) *Soil Erosion Research Methods*. Soil and Water Conservation Society, Ankeny, Iowa, pp. 163-187.

Mediterranean European silvopastoral systems

A. San Miguel-Ayanz
Dept. Silvopascicultura, E.T.S. Ingenieros de Montes, Universidad Politécnica de Madrid, Ciudad Universitaria s/n, E-28040 Madrid, Spain. asanmiguel@montes.upm.es

Abstract
Differences between Mediterranean European and other silvopastoral systems are described regarding both their human and natural environments. The long history of human utilization of the Mediterranean European environment has resulted in a close relationship between biodiversity, nature conservation and rational management, especially under the current European high socio-economic situation. Therefore, silvopastoral methods should not be regarded as only productive activities but also, and mainly, as major conservation tools. Services (protection, stability, landscape, structural heterogeneity) and non-timber products (cork, fungus, fodder, fruits, honey) from the tree layer are usually much more important than timber products, including fuelwood, the demand for which has almost disappeared due to the availability of other sources of energy. Other consequences of the European high socio-economic situation are the decrease of shepherding and transhumance, the partial substitution of extensive sheep herds by cattle and the excessive use of concentrate feed, which results in an unbalanced distribution of stocking levels and an uneven utilization of grazing and browsing resources. These changes result in local shrub encroachment, increased fire hazard and loss of biodiversity. Another major feature of European silvopastoral systems is the increasing hunting of big and small game as a new and very profitable type of wild livestock.

Finally, a classification of Mediterranean European silvopastoral systems is presented, regarding the nature and distribution of their major components: tree or shrub layer, grass sward and animals.

Key words: agroforestry, silvopastoralism, extensive grazing, biodiversity, forest fires

Major features of Mediterranean European silvopastoral systems
Silvopastoral systems are the result of the historical co-evolution of human communities and their, usually difficult, environments. Therefore, the main features of their components are never casual; they have suffered, and overcome, the double and extremely demanding processes of natural and human selection, the latter changing through time, as human demands change with history and with the socio-economic situation. Therefore, these features will be used to describe and classify Mediterranean European silvopastoral systems.

Many authors (Nair, 1993; Etienne, 1996; Papanastasis, 1996; San Miguel, 2003) have pointed out that silvopastoral systems have at least four components: man, trees (woody vegetation), sward and animals. Therefore, we will describe the current situation and particularities of Mediterranean European silvopastoral systems by describing the most important features of their components.

Human demands and activities
A long history of human utilization is one of the most relevant features of Mediterranean silvopastoral systems (L'Houerou, 1981). There is evidence of human presence (*Homo antecessor*) around the Mediterranean basin for at least 780,000 years (Bermúdez de Castro *et al.*, 1997). The human utilization of fire dates from around 400,000 years, and agriculture and livestock came with the Neolithic revolution, some 10,000-12,000 years ago. As a consequence of this long and intense human activity, almost no primary forests remain around the Mediterranean basin: the whole territory may be considered as a huge agroforestry system. Traditional silvopastoral systems and their natural environment are well adapted to each other. That is why most Mediterranean landscapes and a high percentage of their biodiversity are closely related to human activities and depend upon them. Therefore, within that region, traditional silvopastoral treatments must be regarded as both productive activities and major conservation tools, the latter becoming increasingly important as the European socio-economic situation improves (Bland and Auclair, 1996; Redecker *et al.*, 2002; San Miguel, 2003) and as urban people demand (and pay for) more environmental quality and more preservation of traditional landscapes and culture. The European Nature 2000 network represents living and promising evidence of these demands.

Another necessary way of preserving rural areas is preserving their people. Sustainable rural development is nowadays one of the basic priorities of the Common Agricultural Policy (CAP). Thus, since traditional and current silvopastoral systems can contribute to the achievement of that objective, they should also be considered as socio-economic conservation tools for rural areas.

Socio-economic improvement, which has been dramatic in Europe since the 1960s, has resulted in deep changes in human demands and activities, and consequently in the management and structure of silvopastoral systems. For example, fuelwood, which is an essential resource in the southern Mediterranean, is no longer important in Europe. As a result, most coppices are not subject to silvicultural treatments and show problems of stability (Serrada *et al.*, 1992;

Amorini, E. 2004 unpublished). Shepherding and transhumance, which remained commonplace in Mediterranean Europe until the early 1960s (Beaufoy, 1994), are seldom used today. Local shrub encroachment, increase in fuel biomass (Beaufoy, 1994; Bland and Auclair, 1996), reduction in the flow of genetic material between regions and changes (usually reduction) in biodiversity (Malo et al., 2000) are some results of that situation. On the other hand, the density of the tree layer and its regeneration are benefiting from these changes, with the exception of the dehesa system, where sedentary cattle have substituted transhumant sheep herds and restrict natural tree regeneration (Montero et al., 2000).

Trees and other woody vegetation

As might be expected, the nature, distribution and role of trees and other woody vegetation of Mediterranean silvopastoral systems depend greatly upon the Mediterranean climate. Summer drought and, usually, winter cold impose severe restrictions on grass and browse growth and thus on animal feeding behaviour. They also regulate the major features of leaves, wood and fruits of woody vegetation and therefore determine its role and distribution within silvopastoral systems.

Due to their sclerophyllous evergreen nature, many Mediterranean trees and shrubs become real fodder reserves for livestock and wildlife, especially ungulates but also rabbits. Their nutritive value is neither high nor too low, being lower than green but higher than senescent grass. However, their browse is always available, so livestock and wildlife feeding behaviour is believed to consist of selecting food with the aim of optimizing nutrient uptake while maximizing forage consumption (Perevolotsky, 1996; Danell and Bergström, 2002). The result is that sclerophyllous trees and shrubs mitigate hunger periods (summer and winter) and usually increase wildlife carrying capacity as compared to temperate systems (e.g. 20-30 red deer individuals/km^2 in Mediterranean silvopastoral systems vs 4-10 in temperate forests). That is why oak coppices and shrublands (maquis, garrigue and many others) are so important for wildlife and extensive livestock in Mediterranean Europe (Papanastasis, 1999; San Miguel et al., 1999). However, their ability to resist browsing is limited, so special care must be taken to guarantee their preservation and, especially, their natural regeneration (San Miguel et al., 1999, San Miguel, 2003). Another consequence of the climate on sclerophyllous trees is their low productivity of heavy wood, which is excellent as fuel but not for industrial purposes. Therefore non-timber products (cork, fungus, fodder, fruits, honey) coming from the tree layer are usually much more important than timber products. Finally, fruits, and especially acorns, become a strategic and very valuable complement for herbage and browse resources.

From an ecological point of view, the specific effect of scattered trees in Mediterranean silvopastoral systems is a reduction of climatic stress, an improvement of soil moisture levels (Joffre et al., 1991, 1999) and, therefore, a parallel improvement of herbage production, both in quantity and quality, and also in the length of the vegetative period. That is why, seeking for a maximization of the positive tree-sward relationship, their distribution is usually a uniform one, as opposed to that of many temperate systems, where trees or shrubs are located in lines or clumps with the aim of reducing their competition with the herb layer.

Mediterranean pines are not interesting for livestock or wildlife browsing. However, grazing plays a fundamental role in reducing fuelwood and fire hazard in pine plantations, increases their grazing possibilities and may even improve their wood quality (Perevolotsky and Haimov, 1991; Etienne, 1996, 2000, 2002). Nevertheless, some native junipers (*J. thurifera, J. oxycedrus* and others) show an acceptable browsing value under a continental Mediterranean climate.

A final comment on the role woody vegetation plays in Mediterranean European silvopastoral systems: as a consequence of the variability of the Mediterranean climate, the increasing levels of erosion and desertification and the improvement of the European socio-economic situation, the so-called services and environmental rent (Campos et al., 2001) produced by trees and shrubs is more and more important each day and its value often surpasses that of direct products (Pardini et al., 2003; San Miguel, 2003).

Sward

Mediterranean swards are usually constituted by therophytes and perennial, summer-senescing herbs, the growth of which usually stops or reduces in winter due to cold. Therefore, most Mediterranean European silvopastoral systems show two periods of lack or shortage of green herbage: summer and winter. One response of livestock and wildlife to that situation is migrating: transhumance or shorter seasonal herd movements. Another response is seeking better sources of nutrients, which usually come from woody perennials, as stated above. As a consequence, as spring herbage production is usually very high, only a moderate percentage of total herbage production (usually between 40 and 60%) is effectively used by non-migrating livestock; the rest senesces and is no longer used due to its low nutritional quality. However, the current need to sustain higher stocking rates without transhumance imposes the additional utilization of agricultural by-products, supplements and even (more and more each day) agricultural crops used by grazers, both domestic and wild. With such high quality supplementary food, even senescent grass could be fully used.

Legumes are the most important component of Mediterranean silvopastoral systems swards due to the low nutritional quality of senescent grass and their usually low soil fertility. Senescent legumes usually show an acceptable energy and protein content and provide protein-rich fruits and seeds in late spring and summer. Besides, they contribute to soil enrichment in nitrogen. That is why many pastoral treatments within those systems are aimed at increasing legume abundance in natural and artificial swards.

Another particularity of Mediterranean European swards is their high levels of biodiversity (Pineda *et al.*, 1991; Beaufoy, 1994), especially within silvopastoral systems, which are the result of their long history of low-intensity and diversified utilization by domestic and wild herbivores, and also of transhumance (movement of plant genetic material and fertility by animals). The higher diversity levels shown by silvopastoral systems are the result of an additional source of structural and micro-ecological diversity: the nature, size and distribution of woody perennials. Similar Californian Mediterranean pastures show high levels of biodiversity (Rice, 1989), although these are lower than in Mediterranean Europe. Furthermore, a high percentage of that biodiversity comes from seeds in the soil, which are essential for the persistence of herbage swards under the highly variable Mediterranean climate. That is another difference between Mediterranean and temperate swards.

Animals

The variability of environmental situations and the nature and diversity of feeding resources of Mediterranean European silvopastoral systems have promoted the utilization of a high diversity of livestock species and breeds and their seasonal movement (transhumance). However, dramatic changes in the European socio-economic situation since the 1960s have resulted in the following changes in the role of their livestock and wildlife:

- Partial substitution of extensive, low-intensity grazing for semi-intensive management regimes.
- Dramatic decrease of transhumance and shorter seasonal herd movements.
- Reduction of livestock variety: species and breeds.
- Partial substitution of sheep by cattle, as a result of the lack of shepherds and CAP subsidies.
- Partial substitution of traditional breeds by industrial crossing, which results in a loss of adaptation to the natural environment and native fodder supply.
- Excessive use of concentrate feed.
- Uneven distribution of stocking rates, which results in shrub encroachment in some areas (usually those located far away from villages and roads) and overgrazing in some others (usually those located within the close surroundings of villages and roads).

The most important consequence of these changes is the decoupling of the natural environment and its fodder productivity, on the one hand, and livestock raising practices, on the other. Social, economic and policy aspects are becoming more and more important each day for the design and management of Mediterranean European silvopastoral systems.

As a consequence of shrub encroachment and fire hazard increase, livestock and especially browsers or opportunist herbivores, e.g. goats, are considered suitable (and productive) tools for reducing fuel from coppices, high forests and fuel breaks and are, therefore, regarded as forest allies (Perevolotsky and Haimov, 1991; Etienne, 1996, 2002; Gutman *et al.*, 2001). Pasture improvement is an appropriate treatment to increase stocking rates and browsing.

Another major feature of European silvopastoral systems is the increasing consideration of big game as a new and very profitable type of wild livestock (San Miguel *et al.*, 1999; Pardini *et al.*, 2003). Hunting is currently one of the most important sources of profitability for Mediterranean silvopastoral landowners. That situation could be considered as favourable, since profitability is essential for the preservation of those systems. However, new problems of sustainability are arising for these new types of silvopastoral systems, especially for those including wild ungulates (San Miguel *et al.*, 1999).

Even some endangered wildlife species should be considered products (or services) of silvopastoral systems, since those systems constitute their habitat and since society is concerned about their conservation (which requires suitable silvopastoral management) and pays for it (San Miguel, 2003). Two paradigmatic cases are the Iberian lynx (*Lynx pardinus*), the most endangered feline in the world, and the Iberian imperial eagle (*Aquila adalberti*), one of the most endangered raptors in the world. Three LIFE-Nature projects are being carried out in Spain with the aim of recovering those species, where silvopastoral treatments, aimed at preserving their habitat and at increasing rabbit numbers (their basic prey), are major developments (San Miguel, 2003).

Table 1. Classification of Mediterranean European silvopastoral systems[1].

Grazing and browsing in shrublands or forests[2]	Shrublands	Natural shrublands		• Maquis • Garrigue • Legume shrublands • Xerophytic shrublands • Saline and nitrophyllous shrublands • Other
		Fodder shrub plantations		• *Atriplex* sp • *Medicago arborea* • *Opuntia* sp • Other
	Forests	Coppices		• Usually *Quercus* sp
		High forests	Conifers	• Some wild junipers Provide forage for livestock *and* wildlife
			Broadleaved	• Usually provide browse and sometimes fruits for livestock and wildlife
Scattered trees (or shrubs) on swards	Wild trees	Fodder trees (browse, fruits)		• Dehesas *and* montados • Other
		Non-fodder trees		• Xerophytic conifer Woodlands
	Plantations (usually agricultural trees)			• Olive (*Olea europaea*) • Almond (*Prunus dulcis*) • *Pinus pinea* • Leguminous fodder trees (*Ceratonia, Robinia, Gleditsia,…*) • Other
Mosaic of different land uses within one management unit	Two or more of the following land uses: forests, woodland, shrubland, rangeland, cropland			

1. Regarding plant components. Each one may be used by livestock or game species, and often by both.
2. The separation between forests and shrublands, on one hand, and scattered trees on swards, on the other, lies in the herb layer. That of forests and shrublands is composed mainly of their characteristic species. However, that of scattered trees on swards is mainly composed of species characteristic of early successional stages, strongly influenced by grazing.

Classification of Mediterranean European silvopastoral systems

Many attempts have been made to classify agroforestry systems, and particularly silvopastoral systems (Nair, 1993; Etienne, 1996). We have adapted Etienne's (1996) basic classification, which takes account of the distribution of woody and herbaceous vegetation and animals, and have elaborated a classification of Mediterranean European silvopastoral systems, which is shown in Table 1.

References

Beaufoy, G. (1994) *The Nature of Farming. Low Intensity Farming Systems in Nine European Countries.* Institute for European Environmental Policy, London, United Kingdom.

Bermúdez de Castro, J.M., Arsuaga, J.L., Carbonell, E., Rosa, A., Martínez, I. and Mosquera, M. (1997) A hominid from the Lower Pleistocene of Atapuerca: possible ancestor of Neandertals and modern humans. *Science* 276, 1392-1395.

Bland, F. and Auclair, D. (1996) Silvopastoral aspects of Mediterranean forest management. In: Etienne, M. (ed.) *Western European Silvopastoral Systems.* INRA, París, France, pp. 125-141.

Campos, P., Rodríguez, Y. and Caparrós, A. (2001) Towards the Dehesa total income accounting: theory and operative Monfragüe study cases. *Investigación Agraria: Sistemas and Recursos Forestales fuera de serie* 1, 45-69.

Danell, K. and Bergström, R. (2002) Mammalian herbivory in terrestrial environments. In: Herrera, C.M. and Pellmyr, O. (eds) *Plant-Animal Interactions. An Evolutionary Approach.* Blackwell Publishing, Oxford, United Kingdom, pp. 107-131.

Etienne, M. (1996) Research on temperate and tropical silvopastoral systems: a review. In: Etienne, M. (ed.) *Western European Silvopastoral Systems.* INRA, París, France, pp. 5-19.

Etienne, M. (2000) Pine agroforestry in the West Mediterranean Basin. In: Neeman, G. and Trabaud, L. (eds) *Ecology, Biogeography and Management of Pinus halepensis and P. brutia Forest Ecosystems in the Mediterranean Basin.* Bachuys Publishers, Leiden, The Netherlands, pp. 355-368.

Etienne, M. (2002) Management of Mediterranean forests against forest fires and for biodiversity (Aménagement de la forêt méditerranéenne contre les incendies and biodiversité). *Revue Forestière Française* 53, 121-126. (In French.)

Gutman, M., Perevoltsky, A., Yonatan, R. and Gutman, R. (2001) Grazing as a management tool against fire in open areas. The future of the green Mediterranean. *Euro-Mediterranean Conference.* Alghero, Sardinia, Italy.

Joffre, R., Hubert, B. and Meuret, M. (1991) *Mediterranean Agrosilvopastoral Systems: Stake and Reflexions for a Rational Management* (Les systèmes agro-sylvo-pastoraux méditerranéennes: enjeux et réflexions pour une gestion raisonée). Dossier MAB 10, UNESCO, Paris, France. (In French.)

Joffre, R., Rambal, S. and Ratte, J.P. (1999) The dehesa system in southern Spain and Portugal as a natural ecosystem mimic. *Agroforestry Systems* 45, 57-79.

L'Houerou, N. (1981) Impact of man and his animals on Mediterranean vegetation. In: Di Castri, F., Goodall, D.W. and Specht, R. (eds) *Ecosystems of the World 11. Mediterranean Type Shrublands.* Elsevier, Amsterdam, The Netherlands, pp. 479-522.

Malo, J.E., Jimenez, B. and Suarez, F. (2000) Spatial patterns of herbivore dung and endozoochorous seed input to a Mediterranean grazing system. *Journal of Range Management* 53, 322-328.

Montero, G., San Miguel, A. and Cañellas, I. (2000) *Systems of Mediterranean Silviculture. "La Dehesa".* Grafistaff S.L., Madrid, Spain.

Nair, P. (1993) *An introduction to Agroforestry.* Kluwer A.P., Dordrecht, The Netherlands.

Papanastasis, V.P. (1996) Silvopastoral systems and range management in the Mediterranean region. In: Etienne, M. (ed.) *Western European Silvopastoral Systems.* INRA, Paris, France, pp. 143-156.

Papanastasis, V.P. (1999) Grassland and woody plants in Europe with special reference to Greece. *Grassland Science in Europe* 4, 15-24.

Pardini, A., Longhi, F., Orlandini, S. and Dalla Marta, A. (2003) Integration of pastoral communities in the global econonomy. In: *Procedings Conference Reinventing Regions in a Global Economy.* Pisa, Italy, 18 pp.

Perevolotsky, A. (1996) Factors affecting diet preference of goats grazing on dry Mediterranean scrubland in Israel. In: Etienne, M. (ed.) *Western European Silvopastoral Systems.* INRA, París, France, pp. 103-110.

Perevolotsky, A.S. and Haimov, Y. (1991) The effect of thinning and goat browsing on the structure and development of Mediterranean woodland in Israel. *Forest Ecology and Management* 49, 61-74.

Pineda, F.D., Casado, M.A., Miguel, J.M. and Montalvo, J. (1991) *Biological Diversity.* Fund. Areces, WWF-Adena, Scope, Madrid, Spain.

Redecker, B., Finck, P., Härdtle, W., Riecken, U. and Schröder, E. (2002) *Pasture Landscapes and Nature Conservation.* Springer-Verlag, Berlin.

Rice, K.J. (1989) Competitive interactions in California Annual Grasslands. In: Huenneke, L.F. and Mooney, H.A. (eds) *Grassland Structure and Function.* Kluwer Academic Publishers, Dordrecht, The Netherlands, pp. 59-71.

San Miguel, A. (2003) Silvopastoral management and conservation of species and protected areas (Gestión silvopastoral and conservación de especies and espacios protegidos). In: Robles, A.B., Ramos, M.E., Morales, M.C., Simón, E., González-Rebollar, J.L. and Boza, J. (eds) *Pastos, Desarrollo and Conservación.* Junta de Andalucía, Granada, Spain, pp. 409-422. (In Spanish.)

San Miguel, A., Pérez-Carral, C. and Roig, S. (1999) Deer and traditional agrosilvopastoral systems of Mediterranean Spain. A new problem of sustainability for a new concept of land use. *Cahiers Options Méditerranéennes* 39, 261-264.

Serrada, R., Allué, M. and San Miguel, A. (1992) The coppice system in Spain. Current situation, state of art and major areas to be investigated. *Annali dell'Istituto Sperimentale del la Selvicoltura* 23, 266-275.

Agroforestry for improving farm smallholdings

M. R. Alam [1], M. D. Jahiruddin [2] and M. S. Islam [3]

[1]Animal Science Dept., [2]Soil Science Dept., [3]Agricultural Economics Dept., Bangladesh Agricultural University, Mymensingh. mralam@royalten.net.b, soilbau@mymensingh.net, islamms@royalten.net.bd

Abstract

The potential contribution of multipurpose trees and shrubs in an agroforestry system for improving crop-livestock farming was investigated in Bangladesh. The tree *Leucaena leucocephala* (*Leucaena*) and shrub *Sesbania acculeata* (*Sesbania*) were planted around homesteads for use as forage, fuelwood and live fence and for improving soil fertility. *Leucaena* was lopped at 1 metre height 4 times a year and *Sesbania* was harvested twice. Estimated annual yields of leaf and stem were 7.5 t/ha and 7.0 t/ha, respectively. Soil in plantation sites maintained a high pH and organic matter (OM) suitable for cultivation of crops. Phosphorus (P) and sulphur (S) in soil were increased by 71% by leaves and roots. Supplementation of fresh foliages to lactating cows at the rate of 25% of feed dry matter (DM) requirement/day increased milk yield of cows and live weight of calves by 32% (0.607 l/day) and 12% (0.236 kg/day), respectively. A gross income of US$ 135 per household was earned from increased milk yield, meat production, fodder and fuelwood. In addition, the plantation created an extra 6 h work/month per farm and provided a live fence for protection of crops. Agroforestry systems with fodder tree plantation may improve livestock rearing, soil fertility and income on the farm without affecting the current farming system.

Key words: silvopasture, fodder, livestock, soil, income

Introduction

Livestock are reared in smallholdings at subsistence level as an integral part of farming systems to supplement farm income. Feeding of livestock is mostly dependent on rice straw and extensive use of available vegetation, which provides 87% of feed. In a situation of increasing human population in the country and scarcity of land, tree foliages can be grown in silvopastoral systems in homestead areas and used as protein banks for strategic supplementation to livestock (Alam, 1998). The current study was undertaken to assess the potentials of agroforestry based forage-livestock production, soil fertility and income of the farmers.

Materials and Methods

The study was conducted on-farm with 24 farmers in Boira, Sutiakhali and Dorikathal and on the research farm (control) at Bangladesh Agricultural University under Mymensingh district. On average 80 *Leucaena* were planted 1 metre apart in one to two strata systems on each farm and *Sesbania* was grown as a field crop in between the summer crop. Two year old *Leucaena* trees were lopped at 1 metre height four times a year and 60 days old *Sesbania* shrub were harvested twice every two months from January 2001 to December 2002.

Twenty-four indigenous cows belonging to farmers during their second to third parity were divided into three groups and fed with basal diets of 5 kg rice straw, 5 kg natural grass and 0.5 kg concentrate mixture (wheat bran, rice polish and mustard oil cake). Of these, two groups were supplemented with either *Leucaena* or *Sesbania* while another group remained as control. Foliages were offered at the rate of 25% of feed DM requirement per day, by mixing with basal roughage immediately after calving till 45 days of lactation. During the trial, daily feed intake, milk yield (after complete hand milking), initial and final weight of cows and weekly live weight of calves were recorded. During the study period composite samples of foliages at 3 months interval and soil at 0-15 cm depth from plantation sites were collected at 6 months interval. Foliages were analysed for organic matter (OM), crude protein (CP), neutral detergent fibre (NDF), acid detergent fibre (ADF) (AOAC, 1984), total tannins (TT), condensed tannin (CT) (Makkar, 2000) and soil for pH, OM, total nitrogen (N), P, S, respectively (Page et al., 1982). Data on economic returns were collected by direct interview with the farmers through a pre-tested questionnaire.

Results and Discussion

Plantation soil was maintained at pH 6.9, considered suitable for growing fodder plants (Table 1). OM and N contents in soil remained the same or improved a little due to leaf fall, root exudation, decay of roots and nodules and *Rhizobium* bacteria in roots. Hence the C:N ratio remained constant. After plantation there was a trend of increasing P and S contents in soil ($P < 0.01$), which were contributed by leaf and root (Smucker et al., 1995). The results indicated that soil fertility either remained unchanged or was a little improved by tree plantation. The yield of leaf and stem was found to be higher ($P < 0.01$) in *Leucaena* plants compared to *Sesbania* and both of the plants produced high amounts of leaves for use as fodder (Table 2). Higher biomass yields suggest that agroforestry practices with legume fodder plants can benefit the farmers by supplying forage, fuelwood and maintaining improved soil health.

Table 1. Composition (%) of forages and soil in farm (on-farm) and research farm (on-station). Means with different letters within same parameter differ significantly ($P < 0.05$ or $P < 0.01$).

	Leucaena	Sesbania		On-farm Initial	On-farm Final	On-station Initial	On-station Final
OM	93.0	90.6	pH	6.4	6.8	7.1	7.5
CP	22.4	21.1	OM	1.1	1.4	1.0	1.1
NDF	24.6	27.0	N	0.04	0.07	0.05	0.06
ADF	17.8	20.5	C:N	15.2	14.8	11.1	11.4
TT	4.7	4.2	P (ppm)	27.7a	62.3b	21.1a	31.8b
CT	0.7	0.5	S (ppm)	13.7a	21.5b	9.9a	14.8a

Leaves of *Leucaena* contain higher CP and lower NDF, ADF, TT and CT than *Sesbania* (Table 1). High content of CP in foliages grown in all sites may provide fermentable N for enhancing microbial digestion of low quality roughage in the rumen (Van Soest, 1994). It has been recognized that presence of a high level of TT (5-9%) and CT (> 2%) depresses rumen ammonia through decrease of protein and fibre degradation (Barry, 1983; Leng, 1997). In these foliages, low concentration of tannins suggests that they would not cause any adverse effect and enhance utilization of a low quality basal diet (Leng, 1997).

Table 2. Yield of forages (t/ha per year), milk (l/day) and live weight (kg) of cows and calves. Means with different letters within same parameter differ significantly ($P < 0.05$ or $P < 0.01$).

	Leucaena	Sesbania			Leucaena	Sesbania	Control
On-farm							
Leaf	6.3	6.1	Milk	Initial	1.7	1.7	1.6
Stem	9.3a	3.9b		Final	2.3a	2.2a	1.7b
Total	15.6a	10.0b	Cow wt.	Initial	138.3	137.8	138.1
On-station				Final	139.7	139.2	138.8
Leaf	10.1a	7.6b	Calf wt.	Initial	11.6	11.2	10.4
Stem	10.2a	4.6b		Final	21.7a	21.1a	19.1b
Total	20.3a	12.2b					

Calcium and phosphorus contents in *Sesbania* and *Leucaena* were found to be 140 and 37 g/kg and 156 and 21 g/kg, respectively (R. Alam, 2004, unpublished results) and are also an important source of these minerals for lactating and growing animals. Daily milk yield was increased significantly ($P < 0.01$) by 25% and 32.8% in cows fed with *Sesbania* and *Leucaena*, respectively, compared to no supplementation (Table 2). Lactating cows only gained 1% in weight but their calves in supplemented groups had 10.7% higher live weight gain ($P < 0.05$) compared to control group. This implies that *Sesbania* and *Leucaena* foliages added higher digestible nutrients and increased basal feed utilization, which may have promoted milk yield (Khan et al., 1991) and weight gain of their sucklers. Present findings reaffirmed the trend of increasing milk yield by provision of better nutrition through the foliage (Muinga et al., 1995) and promoting utilization of basal diet and milk synthesis. Based on theoretical estimation of CP and metabolizable energy (ME) requirements for maintenance and production (307 g and 33.4 MJ/day, respectively) and supply from foliages (185 g and 10.9 MJ/day, respectively), it appears that both the foliages supplied 61 and 33% of total requirements of CP and ME, respectively. Therefore, twofold increase of intake would meet these essential nutrients almost at no cost. Based on present market price, the estimated economic benefit per household from fodder, fuel wood, milk yield and weight gain in terms of variable cost, estimated gross income and gross margin were Taka 8568 (US$ 145), 626 (US$ 11) and 7942 (US$ 135), respectively. Additionally, women on the farm put in 6 h/month additional labour following plantation and the cost of fencing the kitchen garden was reduced by 40%.

Conclusions

In a resource constrained and subsistence livestock farming system, cultivation of leguminous trees and shrubs may provide an opportunity to improve livestock production, soil fertility, supply fuel wood, generate income and employment for improving livelihood and agroecology of the farm. All the activities require very small investment and farm labour and would not affect the present crop-livestock farming system.

References

Alam, M.R. (1998) Potential use of legume tree leaves as forage in Bangladesh. In: Daniel, J.N. and Roshtko, J.M. (eds) *Nitrogen Fixing Trees for Fodder Production.* Winrock International, Little Rock, AR, USA, pp. 205-211.
AOAC (1984) *Official Methods of Analysis*, 13th edn. Association of Official Analytical Chemistry, Washington, DC.
Barry, T.M. (1983) The role of condensed tannins in the nutritional value of *Lotus pendunculatus* for sheep. 3. Rates of body and wool growth. *British Journal of Nutrition* 54, 211-217.
Khan, M.A.S., Jabbar, M.A., Akbar, M.A. and Topps, J.H. (1991) *Leucaena leucocephala* as an alternative protein supplement to fishmeal for lactating cows in Bangladesh. *Animal Production* 52, 594.

Leng, R.A. (1997) *Tree Foliage in Ruminant Nutrition*. FAO Animal Production and Health Paper 139, Rome, Italy, pp. 44-87.

Makkar, H.P.S. (2000) *Quantification of Tannins in Tree Foliages*. FAO/IAEA Working Document, IAEA, Vienna, Austria.

Muinga, R.W., Topps, J.H., Rooke, J.A. and Thorpe, W. (1995) The effect of supplementation of *Leucaena leucocephala* and maize bran on voluntary food intake, digestibility, live weight and milk yield of *Bos indicus* × *Bos taurus* dairy cows and rumen fermentation in steers offered dairy *Pennisetum purpureum* ad libitum in semi-humid tropics. *Animal Science* 60, 13-23.

Page, A.L., Miller, R.H. and Keeney, D.R. (1982) *Methods of Soil Analysis, Part 2*. American Society of Agronomy, Madison, Wisconsin, USA.

Smucker, A.J.M., Ellis, B.G. and Kang, B.T. (1995) Root, nutrient and water dynamics in alley cropping on an Alfisol in a forest savanna transition zone. In: Kang, B.T., Larbi, A. and Osiname, A.O. (eds) *AFNETA*. IITA, Ibadan, Nigeria, pp. 103-121.

Van Soest, P.J. (1994) *Nutritional Ecology of the Ruminant*, 2nd edn. Cornell University Press, Ithaca.

Improved fallows with forage trees for a slash and burn maize system in the Yucatán Peninsula

S. A. Ayala, V. G. Uribe and G. J. A. Basulto
Agroforestry Researchers National Research Institute of Forestry Agriculture and Livestock PO Box 50-4, PC 97000, Mérida, Yucatán, Mexico. ayalasa@hotmail.com, guribe@hotmail.com, jbasulto@hotmail.com

Abstract
With the aim of improving soil productivity a shift in secondary natural vegetation was induced with cropped plants in a collaborative project utilizing improved fallows for maize production under a slash and burn system. A survey of 150 families, two experimentation peasant groups and three formal experiments was conducted. Family welfare depends on forage trees which support soil productivity and fertility. Within 4 years peasants working with improved fallows system found best results with *Leucaena leucocephala*, *Caesalpina yucatanenses* and *Piscidis piscipula*. Maize production was improved with *Mucuna pruriens* plus 100 kg/ha of P_2O_5. Two and 4 year-old improved fallows of *M. pruriens* and *L. leucocephala* did not improve maize production but weed competition was reduced.

Introduction
More than 250,000 ha per year of secondary vegetation are being slashed and burnt in the Yucatán Península to grow maize for 3 years and then immediately put the land to fallow for at least 10 years. This system is called "milpa" and actually is in decline because short fallow periods interact with silvopastoral systems. The soil is not productive enough to produce enough food and forestry products for 70,000 families. Natural degradation of resources is a problem in this region because fallow periods are short and cannot support maize and forage production. Peasants are aware of the need to improve the system through tree cropping to enhance fertility, restore soil cover, reduce soil moisture loss, weed competition, and produce different forest products for the welfare of families. Researchers of INIFAP (National Research Institute of Forestry Agriculture and Livestock) and ICRAF (International Centre for Research in Agroforestry) developed a joint project about participative research in agroforestry which encompasses family expectancy, participation in the agroforestry technology and suitable design for traditional systems.

Materials and Methods
In July 1994, a survey of 150 peasants was carried out to investigate "milpa" problems, interest in agrosilvopastoral technologies and family expectations. Between 1995 and 1998 two groups of participative research were integrated in different villages, where land, manpower, ideas and knowledge were shared to design and establish improved fallows. Methods involved peasants, experimentation, diagnosis workshops, semi-structured survey and interchange of experiences. Under rainfed conditions three experiments were conducted in a field station. In one of them 18 forage trees species - *Acacia gaumeri*, *A. glomerosa*, *Bauhinia divaricata*, *B. ungulata*, *Caesalpinea gaumeri*, *C. yucatanensis*, *Gliricidia sepium*, *Leucaena leucocephala*, *Lonchocarpus rugosos*, *L. yucatanensis*, *Mimosa bahamensis*, *Piscidia piscipula*, *Pithecellobium albicans*, *P. dulce*, *P. leucospermun*, *Senna octucifolia*, *S. racemosa*, *Sesbania sesban* - were planted in a randomized complete block design and monitored for 6 years. Establishment, development, biomass production, soil cover and improvement and yield of maize after the fourth year were estimated. In the second trial 14 treatments included two consecutive maize controls with and without fertilizers using 40-100-00 NPK as a treatment in a complete randomized block design for 5 years. Four treatments were - associations of *Mucuna pruriens* and *Canavalia ensiformis* with maize with and without 100 kg/ha of P_2O_5 and four others were of improved fallow for 1 year and four treatments of 4 years, all of them followed by maize consecutively with and without 100 kg/ha of P_2O_5. In a third experiment established in 1996 and still ongoing, a split plot block design arrangement with seven treatments in main plots with improved fallows for 2 and 4 years for *L. leucocephala*, *M. pruriens* and natural vegetation as well as continuous cropping of maize was set up. Three levels of subplots with continuous maize and after improved fallows were fertilized with three dosages: 0, 20-50-00 and 40-100-00 NPK, respectively. Analysis of variance was applied to data using Statistical Analysis System (1993) version 6.12 and mean separation was done by Duncan's test.

Results and Discussion
Most of the peasants believe that the "milpa" system is for self consumption and cannot cover daily requirements; it is therefore necessary to intensify production. Natural fallow is a source of building materials; however, continuous degradation reduces material available from this system. The yield of maize is low due to many factors such as drought, weeds, pests and low soil fertility; thus it is necessary to improve the system. Farmers believe that using agrochemicals and farm machinery is the way to increase yield of maize, while 21% of the peasants regard improved fallows as a useful technology for silvopastoral systems. The 20 participants (40 ha) used to crop 5 ha yearly, put 20 ha in fallow for the "milpa" system and at least half of them used to induce forage areas. Farms were continuously cropped for 2 to 6

years, and fallow from 3 to 15. There is a high level of knowledge but farmers do not save resources and there are few conservation practices. They are aware of low soil fertility but they do not use any of the 40 well-known species to improve soil fertility. When they used improved fallows with *Leucaena* and *Mucuna*, yields of maize were low due to rain scarcity, weed abundance, low fertility and lack of interest. On the other hand, team achievements were organization, diagnosis, cooperation and improved fallow design.

Among 18 forest trees evaluated for 4 years as improved fallows, *L. leucocephala* gave the best canopy cover with 65%, was the highest 7.2 m and weed competition was reduced by 38%. *Piscidis piscipula*, *Caesalpina yucatanensis* and *Gliricidia sepium* still had 100% cover after 4 years. They promoted plant development and maize yield (815 kg/ha per year) when fallow cropping was over.

Evaluation of 14 treatments of *M. pruriens and C. ensiformis* in association with maize or as improved fallow for 1 or 2 years with and without phosphorus showed a lower yield of maize than in control plots (530 *vs* 1351 kg/ha per year). However, yield in control plots declined. Maize and legume associations with and without phosphorus initially reduced maize yield but it became higher by the third year (Buckles *et al.*, 1998). No benefit in maize yield was found with 1 year of fallow. Half yield was achieved (1724 kg/ha) with 2 years of fallow compared to the association. Legumes benefited maize 3 years after the system was established. Fallow improved for 2 and 4 years with *Leucaena*, *Mucuna* and natural forest did not increase maize yield even though *Mucuna* kept the soil moist and reduced weed competition and *Leucaena* and *Mucuna* performed better than the natural forest. The average yield over 6 years of continuous maize cropping was 2162 kg/ha compared to cropped fallow 1922 kg/ha in 2001.

Conclusions

Peasants were encouraged to participate and prioritize their future work even though there were few field results. Three forest trees and an associated legume can increase knowledge for acceptance of the potential for "milpa" and as a component of silvopastoral systems.

References

Buckles, D., Triomphe, B. and Sain, G. (1998) *Cover Crops in Hillside Agriculture. Farmer Innovation with Mucuna*. International Development Centre, International Maize and Wheat Improvement Center, Ottawa, Canada.
Statistical Analysis System (1993) *User Guide. Statistical Methods System*. Cary, North Carolina, USA.

The influence of climate and soil type on the "montado" production system

A. Quintas [1], M. S. Pereira [1], P. J. Carmona [2], D. Ramirez-Cruzado [1] and C. C. Belo [1]

[1] STPA. Estação Zootécnica Nacional-INIAP, Fonte Boa, 2000-763 Vale de Santarém, Portugal. stpa.ezn@mail.telepac.pt, [2] Dept. de Geografia y Ordenacion del Territorio, Universidade da Extremadura. Spain.

Abstract

In an analysis of agricultural activities in Portugal and according to the Agricultural General Census of 1999, farms were allocated into three real evapotranspiration zones: between 400 and 450 mm/year (EZ1); between 450 and 500 mm/year (EZ2), and greater than 500 mm/year (EZ3). The "montado" system is an agrosilvopastoral production system in which cork- and holm-oaks are characteristic elements. "Montado" was found to be important in all zones, and most of these coexisted with crops. Most of the "montado" area was used for permanent pastures rather than for temporary crops (grains, forages and temporary rangelands). Permanent pastures were mostly found in EZ3 rather than in EZ1 and EZ2, whereas the percentage of surface occupied by temporary crops was mostly found in EZ1. In general, temporary crops did not exceed 30% of the total land used for agricultural purposes (LUAP). The area for grain growing decreased and temporary prairies and forage crops increased with growing evapotranspiration. Most of the farms (80%) had livestock. Mixed stocking was frequent, particularly with sheep and cattle. In terms of livestock units (LU), the stocking rate decreased with permanent pastures and increased with the area for temporary crops and fallow land. Grains were grown mostly on schist soils, while temporary prairies and forage crops were grown mostly on sandstone soils. Data obtained from this 1999 census were compared to those obtained from the same census in 1989.

Key words: real evapotranspiration, permanent pastures, crop patterns, mixed grazing

Introduction

The "montado" is a good example of a Mediterranean agrarian system because of the multiple usage of available resources, namely cork-oak and holm-oak trees, which are predominantly used for extensive animal production.

In Portugal, the climate for "montado" regions is characterized by high temperatures and lack of precipitation during long periods of time. This type of semi-arid/arid climate is an indicator of desertification and is observed in large areas (desertification indicators for continental Portugal).

The evaluation of the relative importance of agricultural crops according to evapotranspiration and soil type will make it possible to learn how farmers deal with these challenges and what alternatives are possible. Research has been playing a vital role in developing these alternatives to be applied to this production system.

Materials and Methods

In order to define real evapotranspiration and soil type zones, lithology and real evapotranspiration charts from the Environmental Atlas (scale of 1:1,000,000) and the Administrative Chart of Portugal (scale of 1:250,000) referring to councils, districts and parishes geographically referenced in "Datum Lisboa-military coordinates" in arcview format were used. Corine Land Cover chart (scale of 1:100,000), geographically referenced in the "Datum 73 - rectangular coordinates" and then transformed in "Datum Lisboa - military coordinates" in coverage arcinfo format, was also gathered, as was the chart for forest stands from the Portuguese Forest Services (scale of 1:1,000,000), which is geographically referenced in "Datum Lisboa-military coordinates" and also presented in arcview format. The above mentioned charts made possible the development of a system for geographic information to help select sampling areas (Environmental Atlas, Official Administrative Map of Portugal, Corine Land Cover map of Portugal, map of forest populations).

The study was based on data obtained from the Agricultural General Census (7) regarding 932 farms from Alentejo region (Portugal). All forms of evaluation for the types of agricultural activities refer to farms located in areas defined by three real evapotranspiration zones - from 400 to 450 mm (EZ1), from 450 to 500 mm (EZ2) and greater than 500 mm (EZ3) and two major groups of soil types; sand/sandstone soils and schist soils (Environmental Atlas).

Results and Discussion

"Montado" was found to be important in all zones, and most coexisted with crops. Most of the land was used for permanent pastures–greater than 60% rather than for temporary crops (Figure 1). On farms with less than 400 ha, the area of permanent pastures was greater in EZ3, while a greater percentage of temporary crops was mostly found in EZ1 (Figure 1).

In general, the area with temporary crops was lower than 30% of the total LUAP and only approached this value in zones of lower real evapotranspiration. With increasing real evapotranspiration, the area for growing cereals decreased while the area for temporary pastures and forage crops increased.

Figure 1. "Montado" agricultural activities according to real evapotranspiration and area classes.

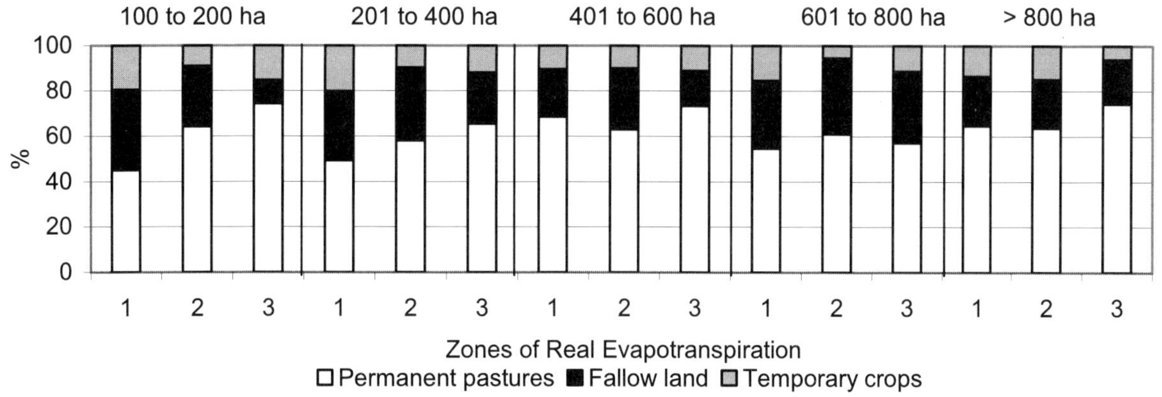

In order to show the type of animal husbandry, the distribution of cattle, sheep and goats was analysed. Information was not gathered on whether or not pigs were grazing. The three animal species were found in all ACs (Agricultural Census) and EZs and mixed stocking was frequent, particularly with sheep and cattle. Working with "livestock standard units" (LSU) shows the predominance of cattle in all EZs and ACs, with the exception of EZ1, in farms with an area of less than 600 ha (Figure 2). If the feeding area is considered, the stocking rate decreased to values below 0.40 LSU, especially for EZ3 (Figure 2). The greater stocking rate values were found for EZ1 and are related to the area of fallow land (Figure 1).

For all classes of area, cereal grains were mostly grown on schist soils, while temporary pastures and forages were mostly grown on sandstone soils. Stocking rate (Figure 2) was also greater on schist soils for all classes of area, reaching values close to 0.60 LSU and showing the importance of the area of fallow land for ruminant production.

When comparing 1989 and 1999 agricultural censuses for the "montado" area, the following differences were observed:

- Permanent pastures, especially natural pastures, increased 112% for cleared land and 92.6% for under cork-oak and holm-oak trees.
- Temporary pastures decreased 55.4% and annual species forages decreased by 51.4%. The only forage which increased over the period (1989/1999) was sorghum (64.8%).
- Cattle was the only animal species that increased (50.8%). The number of goats decreased by 16.5% and sheep by 2.3%.
- Besides the great increase in the number of cattle, the stocking rate decreased in general due to the large increase in the area used for permanent pastures.

Figure 2. Stocking rate according to real evapotranspiration and area classes.

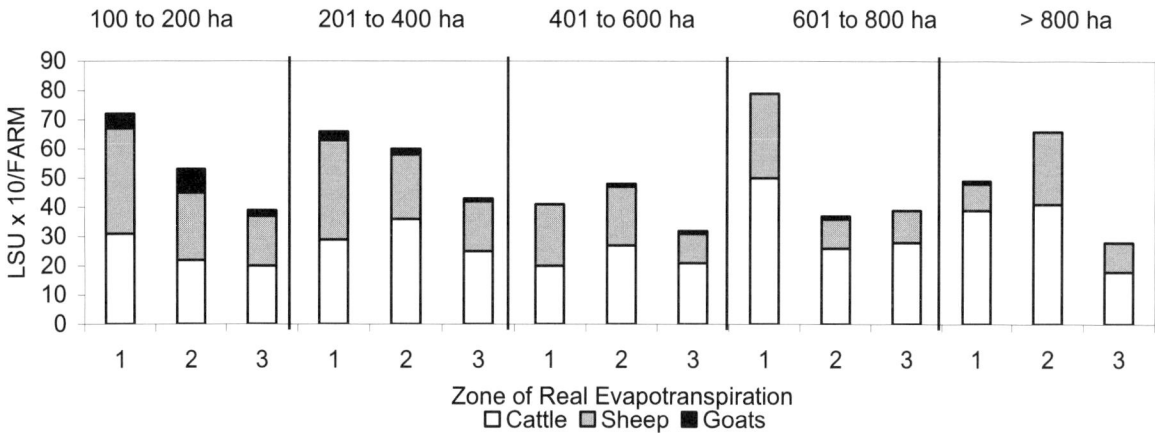

Conclusions

The area for permanent pastures and forages increased with real evapotranspiration, while the area for cereal crops decreased. The stocking rate was greater in areas with lower real evapotranspiration, thus emphasizing the feeding importance of fallow land in these agrarian systems. Cereal grains were mostly grown on schist soils, while temporary pastures and forages were observed on sandstone soils. Between the 1989 and 1999 census, the area for permanent pastures and forages increased, although the stocking rate decreased in most of the farms. During this period, the number of cattle increased by 50%, while that of other ruminants decreased.

References

DGA (Direcção Geral do Ambiente) (1998) Environmental Atlas (Atlas do Ambiente) http://195.22.0.189/atlas/index.html (accessed April 2004).

IGP (Instituto Geográfico Portugés) (2004) Corine Land Cover map of Portugal (Cartas Corine Land Cover de Portugal) Map of forest populations (Carta de Povoamentos Florestais) http://www.igeo.pt/IGEO/portugues/Frameset-produtos.htm (accessed April 2004).

IGP (Instituto Geográfico Portugés) Official Administrative Documento of Portugal (Carta Administrativa Oficial de Portugal (CAOP)) www.igeo.pt (accessed May 1999).

INE (Instituto Nacional de Estadistica) (1989) Agricultural Statistics (Recenseamento Geral da Agricultura) Portugal http://www.ine.pt:8080/biblioteca/search.do (accessed April 2004).

INE (Instituto Nacional de Estadistica) (1999) Agricultural Statistics (Recenseamento Geral da Agricultura) Portugal http://www.ine.pt:8080/biblioteca/search.do (accessed April 2004).

Lúcio Rosário (2004) Desertification Indicators for Continental Portugal (Indicadores de Desertificação para Portugal Continental) Edição da Direcção Geral dos Recursos Florestais. http://biblioteca.dgf.min-agricultura.pt/index1.htm (accessed April 2004).

Silvopastoralism with native tree species in Uruguay

L. Gallo
Dept. of Forestry and Wood Technology, Faculty of Agronomy, University of the Republic, Garzón 780, Montevideo, Uruguay. lgallo@fagro.edu.uy

Abstract
Most farmers in Uruguay do not appreciate the benefits offered by native tree species and do not consider trees, pastures and livestock as a system. Since forests are renewable natural resources, sustainable use may be attained by the application of principles that ensure the perpetuation of resources through time. National information for the adoption, adaptation or technological development of silvopastoral systems is scarce. Technological proposals involving the native forests, livestock management and grassland do not exist. Native tree species (*Acacia caven*, *Prosopis* sp) may be preserved, replanted and managed in traditional pastoral systems, as a valuable complement for sustainable animal husbandry.

Key words: silvopastoral systems, native forests, pastures, livestock

Introduction
Uruguay has introduced sustainable development and the reduction of subsidies in agricultural policies. Organic meat production has developed recently. This chapter aims to contribute to the current policy and shows the potential of native woodland species for silvopastoral systems. The integration of native forest species in animal husbandry systems is a strategy that contributes to counteracting the negative environmental impact of traditional systems. Since the introduction of cattle in 1614, livestock and native forests have been interacting. However, forestry is not deemed to be or actually managed as a silvopastoral system component, but rather regarded as a competing factor, displacing other activities, and one which adds to the complexity of management (driving livestock, loss of animals, etc.). Cattle farmers usually clean the range, clear-felling or uprooting native trees.

Current and potential situation
The natural indigenous range (as opposed to cultivated fields or artificial pastures) extends through approximately 14 million hectares in Uruguay (Del Puerto, 1987). Natural pastures are plant communities where grasses and legumes dominate, together with several other botanical families, which appear in lower frequencies. Since cattle husbandry was introduced in the range, changes have taken place to an extent in which pastures have not yet reached a new balance and can therefore be regarded as disclimax communities. This pastoral disclimax originates mainly from indigenous shrub species. Later on, clear-cutting and burning of native trees, grazing and cultivation of extended areas have determined the evolution towards current pasture communities. However, rapid regression to climax shrub vegetation has been observed following exclusion of grazing. Shrub species do not appear alone; under their shade, the frequency of
high-quality winter herbaceous species increases. These species were formerly much more abundant under forest canopies than at full exposure to sunlight (*Bromus unioloides, B. brachyanthera, B. auleticus, Agropyron cabrifolium, Stipa megapotamica, S. hyalina*, etc.). Shade is also amenable to other summer vegetative cycle species, owing to a more favourable summer water balance (*Paspalum dilatatum, Axonopus compressus, Setaria argentinensis, Stenotaphrum*, etc.). This results in a significant increase in biodiversity and sustainability of productive systems. Native pasture communities are therefore capable of taking advantage of sunlight restrictions imposed by silvopastoral systems, with the resulting increase in winter productivity. Silvopastoral environments can be less stressful for grassland species than traditional range systems during critical periods for pasture communities and may also favour livestock because of lower temperatures during the summer and higher temperatures in winter (Millot, 1999). Native park forest types can be defined as an intermediate condition between a wooded range, where density may be as low as one tree per hectare or less, and a dense forest stand. Park forest types are known as *algarrobal* or *espinillar* (Brussa, 1989), depending on the relative abundance of main species, *algarrobo* (*Prosopis* sp), *espinillo* (*Acacia caven*) or *chañar* (*Geoffroea decorticans*). Under *Prosopis* park types with densities ranging from 400 to 1200 trees per hectare, a dense grass pasture usually develops, extending up to the base of the trees (Rolfo, 1970; Millot, 1999). Sheep usually graze in open, low-density forests rather than in dense stands. Cattle spread throughout the woodland, avoiding overgrazing. Horses are more selective and browsing damage may be negligible, but trampling is often significant. In dense forests growing on fertile sites, stand management should be aimed at increasing livestock grazing capacity, facilitating animal husbandry management, controlling stand density and increasing clean bole length through pruning. Tree species in this forest type regenerate naturally and, in some cases, the passage of seed through the alimentary canal of cattle has been shown to improve its germination. Livestock do not usually browse on *Prosopis* seedlings; some breakage may be observed on young trees, but this is offset by the high regeneration capacity of the species. *Prosopis* sp is a good soil-improving species, through its nitrogen-fixing ability, its nutrient-cycling capacity and its role in erosion

control and sandy soil stabilization (Pezo and Ibrahim, 1998). If a rotational grazing system in small paddocks is adopted, forage availability is higher and browsing damage on tree seedlings is minimized. After 3 or 4 years, regeneration should have grown to a height out of reach of browsing livestock. Forage grass species (*Bromus* sp) and species of lower nutritive value (*Cynodon dactylon*) have exhibited excellent growth under managed forest canopy on sandstone-formed soils (Millot, 1999).

Espinillares are park forest types where *espinillo* (*Acacia caven*) is prevalent, often mixed with *algarrobos* (*Prosopis* sp). Positive effects of *A. caven* on soil properties may be significant, if expressed either as organic matter level or as total available nitrogen and potassium (Ovalle *et al.*, 1990). In regional cattle husbandry systems, promising associations between trees and several high-quality grasses, which benefit from partial shade (*Lolium multiflorum, Festuca, Dactylis, Panicum maximum, etc.*), have been established (Millot, 1999).

Discussion

Animal production levels under current conditions are influenced by the interaction between native woody species and herbaceous vegetation. Establishment of pasture species in native forest areas may be achieved through thinning. The establishment of trees on pasture communities may take place naturally, either as a result of previous litter and soil seed storage or seed dispersal by livestock. Another option which should be considered is trees planted on grasslands.

Conclusions

The adoption of silvopastoral systems involving native grassland and tree species is a viable alternative to traditional husbandry systems in Uruguay, if the preservation of both native pasture and indigenous forest biodiversity has been defined as part of sustainable development policies.

References

Brussa, C. (1989) Characteristics of the indigenous forest (Características del monte indígena). In: *Jornada de Conservación del Monte Indígena*. Facultad de Agronomía, Montevideo, Uruguay, pp. 1-2. (In Spanish.)

Del Puerto, O. (1987) *Uruguay Vegetation* (Vegetación del Uruguay). Facultad de Agronomía, Universidad de la República, Montevideo, Uruguay. (In Spanish.)

Millot, J.C. (1999) Silvopastoral systems (Sistemas Silvopastoriles). In: *Expoforesta 99*, Tacuarembó, Uruguay. (In Spanish.)

Ovalle, C., Aronson, J., Del Pozzo, A. and Avendano, J. (1990) The espinal: agroforestry system of the Mediterranean type climate region of Chile. *Agroforestry Systems* 10, 213-239.

Pezo, D. and Ibrahim, M. (1998) *Silvopastoral Systems* (Sistemas silvopastoriles). Proyecto Agroforestal CATIE/GTZ, Turrialba, Costa Rica. (In Spanish.)

Rolfo. M. (1970) Study of *Prosopis* genus in Uruguay (Estudio del género *Prosopis* en el Uruguay). MSc thesis, Facultad de Agronomía Montevideo, Uruguay. (In Spanish.)

Classification principles and use of grazed forest stands in the Ukraine

G.B. Gladun
Dept. of Landscape Ecology and Agroforestry, Ukrainian Research Institute of Forestry and Forest Melioration named after G.M.Vysotskij, Pushkins'ka st. 86, Kharkiv-24, Ukraine. gladun@uriffm.com.ua

Abstract
The classification of grazed forest stands is based on their role, designation and application. Pastures protected by forest stands are classified into: permanent pastures (pasture-protecting forest belts, pasture-improving feed stands, silvopasture, farm protective stands, green belts); temporary pastures (shelterbelts, shrub rows and belts on sands, areas used in agriculture, green umbrellas, protective stands near water sources, grazed valleys and slopes); natural stands on seasonally grazed pastures (kolki forests, bairak forests, small forests, shrubberies, etc.). Distribution of trees and shrubs and their locations were related to the natural climatic regions of the country. They covered three zones: Polissya, forest-steppe and steppe zone (distinguished as temperate steppe and dry steppe) and usage was based on the aims of grazing in forest stands. Planting of shrubs with a forage value counteracts site degradation. Spatial arrangement of trees was adapted to local conditions, particularly to provide the most favourable conditions for animal feed and rest. The category of landscape relief (upland or water-meadows, wash sites, slopes of different exposure, foothills, etc) determines the permitted grazing level.

Key words: temporary pastures, seasonal grazing, shrub cover, shelterbelts, permanent pasture, zoomeliorative forests

Introduction
The climate of the Ukraine gives rise to the applicaton of functionally distinctive forest stands. All forest stands which protect livestock and pastures are grouped under "zoomeliorative forest stands". This term not only specifies the appropriate protective function of these stands, but also determines their location. Hence, the stands are formed with a definite spatial structure, composed of primary and secondary tree species and shrubs, density, crown density, etc. The objectives of this chapter are to develop classification principles for these zoomelioration stands on the basis of available information and consider their use as a component of farmed landscapes in the Ukraine. Only the main artificial zoomeliorative forest stands are discussed in the chapter.

Materials and Methods
The classification of zoomeliorative stands and their spatial arrangement within farmed landscapes allows harmonized integration with the environment, aids land management decisions, improves livestock conditions and is the basis for sustainable development of large land areas. Livestock are protected from adverse natural phenomena by zoomeliorative forests. They also help in managing natural resources: soil conservation, regulation of surface and soil water regime, health and social functions, especially in forest-poor regions of country. To develop a zoomelioration lands classification, data on target protective stands in arid regions were used. Pastures protected by forest stands were classified into: permanent pastures (pasture-protecting forest belts, pasture-improving feed stands, silvopasture, farm protective stands, green belts); temporary pastures (shelterbelts, shrub rows and belts on sands, areas used in agriculture, green umbrellas in animals resting-places and cattle-movement roads, protective stands near water sources, grazed valleys and slopes); natural stands on seasonally grazed pastures (kolki forests, bairak forests, small forests, shrubberies, etc.). The classification is based on the hypothesis that different zoomelioration stands can be used in complex arrangements to protect and enhance farmed landscapes.

Results and Discussion
Pasture-protecting forest belts, green (tree) umbrellas, stands near farms and cattle breeding complexes, green stands near industrial, residential and communal buildings are established on permanent or temporary pastures in agricultural areas.

Pasture-protecting forest belts are established to shelter pasture and for animal protection from adverse winds, hot winds and snowstorms, extremes of air temperature; for improvement of microclimate and livestock and herbage productivity. The design of protective stands depends on the topography and locally limiting factors to create the best environment for pasture growth and livestock production (Gladun, 1988, 2003a).

The main belts are up to 15 m and 3-4 rows of trees wide, established transversely to the prevalent wind direction for pastures likely to suffer wind damage. Their width is determined by tree row and shrub row spacing and increases with deteriorating site conditions. Depending on the likelihood of the soil to blow, the width between the belts may vary from 50 to 200 m. The transverse belts are placed at a distance apart of not more than 1000 m. Green umbrella are round-shaped forest stands which are formed for animals to rest in and be protected against adverse climatic factors. These are found mainly in steppe and dry steppe zones and are established in relatively good pasture. In contrast to pasture-protective stands, umbrellas are established in different patterns. Green umbrella areas generally cover 0.5-1.2 ha of fodder area (Gladun, 2003b) and are composed of micro-umbrellas (600-900 m^2), which are systematically located on pasture and have up to 40 trees each. Spaces (10-20 m width) allow fodder crops or natural grasses to grow between the micro-umbrellas. The number of animals per umbrella depends on the minimal area necessary for each species (3 m^2 for sheep, 12 m^2 for cow, 0.3 m^2 for bird). The stocking rate will depend on the productivity capacity of the area.

"Comfort stands" are established on pastures at a distance 3-5 km from one another. They are 2-4 ha in size and are designed to protect stock against selected regional climatic peculiarities. These stands are mainly square with one side left open to allow livestock to move in and out. Sides are 85-110 m, and the width of the stand is 30-32 m. The outside rows of trees in the stand are planted more diversely than the inside rows.

Clumps of fodder trees and bushes are used to increase livestock breeding efficiency. Tree and bush species with high nutrient content are eaten by animals and their use for fodder is increasing. Species composition is determined by the region and seasonal grazing pressure. Ameliorative fodder stands are established mainly as 15-40 m wide belts. The width of belts and inter-belt distances depend on the existing herbage condition and possibilities of its improvement by reseeding (Gladun, 2003b). Ameliorative fodder stands take about 2-3 years after planting to be effective. Over this period, grazing loads are strictly regulated. This also follows for bush ameliorative fodder stands. These stands are more stable under grazing and browsing. Typically, ameliorative fodder stands with trees and bushes increase fodder potential of pastures by 1 t DM/ha and play a phytosanitary role, acting as a mechanical obstruction to wind-transferred pathogens and preventing direct contact between animals.

Conclusions

Forest stands integrated with animal production systems are important elements of modern livestock breeding, improve the appearance of steppe landscapes and increase the bioclimatic potential and total productivity in the Ukraine. Their functions are protective, ecological and social, and they enhance biodiversity in farmed landscapes.

References

Gladun, G.B. (1988) *Multipurpose Utilization of Forest Belts. Handbook of Agroforestry*, Urojay, Kyiv. pp. 30-33. (In Ukrainian.)
Gladun, G.B. (2003a) *Forest Meliorations of Agrolandscapes*, Nove slovo, Kharkiv. 164 p. (In Ukrainian.)
Gladun, G.B. (2003b) The Role of Forest Stands in the Sustainable Functioning of Agrarian Landscapes. *Proceedings of XII World Forestry Congress*, Canada. http://www.fao.org/docrep/article/wfc/xii/0130-b1.htm

Traditional silvoarable systems and their evolution in Greece

K. Mantzanas, E. Tsatsiadis, I. Ispikoudis and V. P. Papanastasis
Laboratory of Rangeland Ecology, Aristotle University, PO Box 286, 54124 Thessaloniki, Greece. konman@for.auth.gr

Abstract
Traditional silvoarable systems occur in several parts of Greece and play an important role in the local economy. They also play a very important ecological role because they prevent soil erosion and surface runoff. Several combinations of trees and crops were identified in a survey conducted between 2002 and 2004. The main tree species were oaks, walnuts, poplars and fruit trees. The understorey crops consisted mainly of cereals (wheat, barley, maize), lucerne, tobacco and vegetables. Crops were used for livestock feeding indirectly (grain, hay) or directly (grazing). In the latter case, animals used these areas in the critical periods of the year such as the summer and early autumn (after the crop harvest). Over the last decades, these systems have been substantially reduced due to reasons which include both extensification and intensification. This evolution is thoroughly discussed and recommendations are made for their conservation and management.

Key words: survey, grazing, conservation, silvopasture, crops

Introduction
Agrosilvicultural systems involving trees and crops on the same unit of land are one of the three types of agroforestry systems, the other two being silvopastoral and agrosilvopastoral (Nair, 1991). These systems, also known as silvoarable, constitute a very old practice in the Mediterranean region. They are characterized by a dynamic equilibrium between the tree species and crops in a small area that can ensure sustainable land use. Most of them traditionally function as agrosilvopastoral systems as well, because they are grazed by livestock after the crop harvest, or tree fruits are used as feed (e.g. acorns) by domestic animals (sheep, goats and pigs) in the autumn and winter. The trees may be shredded at the end of summer for the leaves to be used as fodder for the animals during the winter period (Papanastasis, 2003).

Due to the rapid socio-economic changes over the last few decades involving extensification of human activities or intensification of agricultural practices (Papanastasis, 2003), traditional silvoarable systems are in danger of being either abandoned or converted to intensive monocultures. Such an evolution would lead to their extinction with a concomitant loss of biodiversity, landscape stability and accumulated cultural knowledge. This can be prevented by preserving the still viable and functioning systems and restoring the abandoned ones, after adapting them to the current socio-economic conditions.

In Greece there are several traditional agroforestry systems. The most important of these have been already inventoried and described by Shultz et al. (1987). In this study, a detailed survey of the traditional silvoarable systems found in mountainous areas was undertaken as a first attempt to identify the still viable ones and suggest measures of preserving them to meet both ecological and economic objectives.

Materials and Methods
The research was carried out at the municipality of Askio (21° 29′14′′ N and 40° 20′32′′ E), north-western Greece, at an altitude of about 700 m asl in 2002 and 2003. Individual farms (plots) holding silvoarable systems were located in an area of about 20,000 ha and the owners of these farms-plots were identified. A special inventory sheet was prepared and the information describing each plot was collected directly from the farmers-owners of the plots as well as from municipality officers. The inventoried silvoarable systems were grouped according to tree species and crops.

Results and Discussion
Thirty-two different privately owned systems were recorded and the majority of these were rented by the owners to farmers. The main characteristics of these systems were: the small area of plots, the relatively high number of tree species present on the same plot, the presence of trees around the plots without any spacing and the variety of crops grown in a small area with different management practices. Oak species and fruit trees occurred in most systems (Table 1). Oaks and poplars had the tallest height and the biggest diameter compared with the other tree species. Only poplars and fruit trees (including walnuts and almond trees) were planted, the rest being wild. The main output of all these systems was the crop yield while the trees were used only for fuelwood and fruits.

Table 1. Main traditional silvoarable systems in the municipality of Askio. Oak (O); Walnut (W); Apple (A); Almond (Al); Wild pear (Wp); Poplar (P); Elm (E); Cherry (C); Plum (Pl); Hazel (H); Nettle (N); Black locust (Bl); Willow (Wi).

Crops	O	W	A	Al	Wp	P	E	C	Pl	H	N	Bl	Wi	Total
Cereals	1	1	1	1	1	1	1				1	1	1	10
Maize	1	1												2
Lucerne	1	1		1	1	1		1	1					7
Grapes		1	1	1						1				4
Tobacco	1				1		1							3
Vegetables	1	1	1			1								4
Dry beans		1	1											2
Total	5	5	5	3	3	3	2	1	1	1	1	1	1	32

The plots with the combination of trees and cereals, maize and lucerne were grazed after the crop harvest, while animals used the acorns and leaves of oak trees from the cut branches during the autumn and winter. Table 2 shows that 21 of the recorded plots functioned as agrosilvopastoral systems because they were also used by grazing animals.

Table 2. Silvoarable systems used by grazing animals in the municipality of Askio. Oak (O); Walnut (W); Apple (A); Almond (Al); Wild pear (Wp); Poplar (P); Elm (E); Cherry (C); Plum (Pl); Nettle (N); Black locust (Bl); Willow (Wi).

Crops	O	W	A	Al	Wp	P	E	C	Pl	N	Bl	Wi	Total
Cereals	1	1	1	1	1	1	1			1	1	1	10
Maize	1	1											2
Lucerne	1	1		1	1	1		1	1				7
Other crops	2												2
Total	5	2	2	2	2	2	1	1	1	1	1	1	21

The surveyed systems had in general low economic output because they were used extensively. Most of them are maintained due to crop subsidies by the EU, and the farmers, mainly tenants, cared only for the crops and not for the trees. In the few cases where the owners themselves were involved in the cultivation of systems, the inputs in terms of fertilizers, pesticides, etc. were limited. In the areas where the land was consolidated and irrigation became available, the owners and farmers eliminated the trees in order to increase the crop yield and facilitate the use of agricultural machines. The attitude of farmers to trees is negative and all these systems are in danger of been phased out in the near future. The only solution is to provide special incentives to owners to keep the trees, plant new ones and take care of these valuable systems. Similar concern exists for traditional agroforestry systems in other European countries, too (Dupraz, 1999; Mary *et al.*, 1999; Paris *et al.*, 2001).

Conclusions and recommendations

In north-west Greece traditional silvoarable systems are used also as silvopastoral systems but they are in danger of being phased out due to their low economic output.

Special incentives in the form of subsidies should be given to farmers in order to maintain these systems and improve them by planting new trees with high growth and quality timber so that they become economically viable.

Acknowledgements

This research is part of the European project SAFE (Silvoarable Agroforestry For Europe) (European Research Contract QLK5-CT-2001-00560).

References

Dupraz, C. (1999) Adequate design of control treatments in long term agroforestry experiments with multiple objectives. *Agroforestry Systems* 43 (1-3), 35-48.

Mary, F., Dupraz, C., Delannoy, E. and Liagre, F. (1999) Incorporating agroforestry practices in the management of walnut plantations in Dauphine, France: an analysis of farmers' motivation. *Agroforestry Systems* 43, 243-256.

Nair, P.K.R. (1991) State-of the-art of agroforestry sytems. In: Jarvis, P.G. (ed.) *Agroforestry: Principles and Practices.* Elsevier, Amsterdam, The Netherlands, pp. 5-29.

Papanastasis, V.P. (2003) Vegetation degradation and land use changes in agrosilvopastoral systems. *Agrosilvopastoral Symposium* Cáceres, Spain. (in press).

Paris, P., Pisanelli, A., Musicanti, A. and Cannata, F. (2001) Agroforestry in Italy: tradition of the practice and research indications on new models. *Proceedings of the Sino-Italian Workshop. Forestry and Agroforestry for Environmental Protection and Rural Development.* Beijing, China, pp. 2-3.

Schultz, A.M., Papanastasis, V.P., Katelman, T., Tsiouvaras, C., Kandrelis, S. and Nastis, A. (1987) *Agroforestry in Greece. Working Document.* Thessaloniki, Greece.

Biodiversity and sustainable development in the silvopastoral systems of the Cantabrian mountains

M. Mayor López [1], J. A. Oliveira Prendes [2] and M. Fernández Benito [1]

[1]Dept. de Biología de Organismos y Sistemas, Área de Botánica, Universidad de Oviedo, C/Catedrático Rodrigo Uría s/n 33031 Oviedo, Spain; [2]Dept. de Biología de Organismos y Sistemas, Área de Producción Vegetal, Campus de Mieres, Universidad de Oviedo, 33600 Mieres, Spain. mmayor@uniovi.es, oliveira@uniovi.es

Abstract

During 2003 a study was carried out in the beech woods in the mountain passes of Tarna, Pontón, Panderuedas and Somiedo, where cattle graze during the months of July, August and September. In this chapter we present the results of 16 phytosociological inventories on four pasture communities (dry *Nardus* pasture, pasture of *Cynosurus cristatus*, wet *Nardus* pasture and hygrophilous pasture) according to the method of Braun-Blanquet. On dry *Nardus* pastures, *Agrostis tenuis*, *Nardus stricta* and *Avena lodunensis* had the highest cover values. An increase in soil moisture is shown by the appearance of *Juncus squarrosus* and *Nardus stricta* on hygrophilous and wet *Nardus* pastures. *Cynosurus cristatus* and *Trifolium repens* were the most frequent species on pasture of *Cynosurus cristatus*. GPS was used to locate the habitat of each plant community so that their evolution could be followed during the following 5 years.

Key words: forest grazing, pastures, phytosociology

Introduction

According to San Miguel (1995) three types of silvopastoral systems can be distinguished: pure, in lines and in mosaic. Pure systems are systems where grazing is within forests. If forests or shrubs are situated in single or multiple rows along the edge of pasture areas these are silvopastoral systems in lines. Finally, silvopastoral systems in mosaic are when pasture areas are separated from the forests.

We attempt to analyse in this chapter the biodiversity of the third type, that of the mosaics of the Cantabrian beech woods.

In the Cantabrian beech woods we distinguish three vegetation series with their respective mosaics (Díaz Gónzalez and Fernández Prieto, 1994): mesophytic basiphilous beech forest (*Carici silvaticae-Fagetum sylvaticae*), mesophytic acidophilous beech forest (*Blecno spicanti-Fagetum sylvaticae*) and xerophilous and basiphilous beech forest (*Epipactido helleborines-Fagetum sylvaticae*). In our work, we only considered the first two mosaics.

The selection of these four beech woods resulted from the extensive grazing with cattle and horses practised over many years; that is why they maintain great floristic and habitat diversity.

The mosaics of the four beech woods are similar: in the first place there is an arboreal layer with *Fagus sylvatica*, used, in general, by cattle looking for shade. Secondly there is a layer of scrub, broom formations of *Genista obtusirramea*, endemic to the Cantabrian mountains, used by the cattle as protection in the hours of strong heat. Thirdly are the different types of pastures, differing in their floristic composition by the greater or lesser extent of soil moisture, slope and intensity of grazing (Fernández Fernández et al., 2000; Gunnemann et al., 2001; Mayor López, 2002).

The type of extensive grazing of the four beech woods studied is quite common in the Cantabrian mountains. Cattle remain in these woods without the constant presence of a shepherd. The shepherd visits at the end of every day by car to verify the condition of the cattle. Usually, cattle do not receive any other food than grazing from the pasture.

In this chapter we present the initial results using the transect technique and the Braun-Blanquet method to establish the floristic composition of four pasture communities in four mountain passes representative of the Cantabrian beech woods.

Materials and Methods

In 2003 the beech woods of the Tarna (1452 m asl), Pontón (1290 m asl), Panderuedas (1465 m asl) and Somiedo (1,486 m asl) mountain passes were visited. In these zones, the cattle graze during the months of July, August and September. In this chapter we present the results of 16 phytosociological inventories of four different pasture communities: (1) dry *Nardus* pastures, (2) pastures of *Cynosurus cristatus*, (3) wet *Nardus* pastures and (4) hygrophilous pastures. Cover/abundance codes were noted for each species. Six cover/abundance codes (Braun-Blanquet classes): +: few individuals, insignificant cover, 1: scattered individuals (1-5% cover), 2: 5-25% cover, 3: 25-50% cover, 4: 50-75% cover, 5: 75-100% cover (Mueller-Dombois and Ellemberg, 1974). Using GPS, we located the habitats of the different communities with the purpose of following their evolution during the next 5 years.

Results and Discussion

The biodiversity of the pasture communities of the four mountain passes can be observed in the following table (Table 1).

In the analysis of the silvopastoral systems of four beech woods we have differentiated four habitats: dry *Nardus* pasture, pasture of *Cynosurus cristatus*, wet *Nardus* pasture and hygrophilous pasture.

Dry *Nardus* pastures were located in the dry, steepest slopes, usually occupying the zones that abandon broom and heath gorse formations. The contact of these pastures with the heath gorse formations is noted by the presence of small shrubs: *Vaccinium myrtillus* and *Calluna vulgaris*. *Agrostis tenuis*, *Nardus stricta* and *Avena lodunensis* had the highest cover values. However *Trifolium* sp and *Lotus corniculatus* were not present.

Pastures of *Cynosurus cristatus* were located on less steep slopes with moderate soil moisture. These pastures showed the greater degree of floristic diversity, shown by the number of species found. *Cynosurus cristatus* and *Trifolium repens* were the most frequent species. *Cynosurus cristatus* is not preferred by cattle; nevertheless, it is a good ecological indicator of this type of mountain pastures of great agricultural interest.

The increase in soil moisture was shown by the appearance of *Juncus squarrosus*, *Carum verticillatum* and *Nardus stricta* on hygrophilous and wet *Nardus* pastures. These pastures have little nutritional value for the cattle.

Hygrophilous pastures were situated in the depressions of the land where the greatest soil moisture is accumulated, *Juncus conglomeratus* and *Juncus squarrosus* being the most frequent species. They are habitats visited by the cattle in search of water.

Conclusions

In general, these types of pasture communities had a similar floristic composition in the four mountain passes, except in the Panderueda mountain pass, where the wet *Nardus* and the hygrophilous pastures were not present. The differences in floristic composition were more important between the pasture communities than between mountain passes. The knowledge of these ecosystems and their traditional uses will offer the best guarantee for their conservation.

References

Díaz González, T.E. and Fernández Prieto, J.A. (1994) Vegetation in Asturias (La vegetación de Asturias). *Itineraria Geobotánica* 8, 243-258. (In Spanish.)

Fernández Fernández, A., Mayor López, M., Carlón Ruiz, L. and Martínez González, J. (2000) Ecological indicators in beech stands in the Cantabrian Range (Indicadores ecológicos de los hayedos de la Cordillera Cantábrica). *Boletín Ciencias Naturaleza* R.I.D.E.A. (Real Instituto de Estudios Asturianos) 46, 145-165. (In Spanish.)

Gunnemann, H., Mayor López, M., Fernández Fernández, A. and Carlón Ruiz, L. (2001) Ecological and corological aspects in the Cantabrian beech stands (Aspectos ecológicos y corológicos de los hayedos cantábricos). *Boletín Ciencias Nauraleza.* R.I.D.E.A. (Real Instituto de Estudios Asturianos) 47, 7-52. (In Spanish.)

Mayor López, M. (2002) Landscapes of northern Spain and pastoral systems. In: Redecker, B., Fink, P., Härdtle, W., Riecken, U. and Schröder, E. (eds) *Pasture Landscapes and Nature Conservation*. Springer, Berlin, Heidelberg, New York, USA, pp. 67-86.

Mueller-Dombois, D. and Ellemberg, H. (1974) *Aims and Methods of Vegetation Ecology*. Wiley, New York, USA.

San Miguel, A. (1995) Silvopastoral planning (Ordenaciones silvopastorales). *Cuadernos de la Sociedad Española de Ciencias Forestales* 1, 23-30. (In Spanish.)

Table 1. Maximum cover class observed in four pasture communities within the four mountain passes. Six cover/abundance codes (Braun-Blanquet classes): +: few individuals, insignificant cover, 1: scattered individuals (1-5% cover), 2: 5-25% cover, 3: 25-50% cover, 4: 50-75% cover, 5: 75-100% cover.

	Tarna	Pontón	Panderuedas	Somiedo
Dry Nardus pasture				
Agrostis tenuis	3	3	3	1
Avenula lodunensis	2	1	2	2
Calluna vulgaris	1			+
Galium saxatile	+	+	2	+
Gentiana lutea			1	
Nardus stricta	3	1	2	1
Plantago alpina	+			1
Potentilla erecta	1	+	1	1
Trifolium repens	1	+	1	1
Vaccinum myrtillus	1			1
Pasture of Cynosurus cristatus				
Achillea millefolium	1		+	1
Agrostis tenuis	3	4		1
Anthoxantum odoratum		1		1
Cerastium fontanum		1		
Cynosurus cristatus	3	2	2	3
Festuca rubra		1	2	2
Lotus corniculatus		+	1	2
Plantago lanceolata		1		
Plantago media	2		4	2
Prunela grandiflora			1	
Sanguisorba minor	1			1
Trifolium pratense		2	1	2
Trifolium repens	3	2	3	2
Wet Nardus pasture				
Calluna vulgaris		+		
Carum verticillatum	2	1		2
Cynosurus cristatus	1	+		1
Festuca rubra	2	1		3
Juncus squarrosus	+	+		2
Lotus corniculatus	2	2		2
Nardus stricta	3	3		3
Potentilla erecta	2	2		1
Trifolium repens	1	2		1
Hygrophilous pasture				
Agrostis tenuis	1	1		1
Juncus conglomeratus	3			2
Juncus squarrosus	2	3		3
Nardus stricta	1	2		2
Potentilla erecta	+	+		+

Goat production system in mountain zones: the case of Peneda's Mountain

L. F. Pacheco [1], J. Pires [2], A. Iglesias [3], J. Cantalapiedra [3], D. Barreto [2] and J. P. Araújo [2]

[1]DRAEDM, Divisão de Produção Animal, Rua Franca, 534, 4800-875, S. Torcato, Portugal. [2]Escola Superior Agrária de Ponte de Lima - Inst. Politécnico de Viana do Castelo (ESAPL-IPVC), Convento de Refóios – 4990-706 Ponte de Lima, Portugal. [3]Dept. de Anatomía y Producción Animal, Universidad de Santiago de Compostela, 27002 Lugo, Galicia, Spain. filipepacheco@oninet.pt, anigbe@lugo.usc.es, pedropi@esa.ipvc.pt

Abstract

Animal production plays a major social and economic role in Peneda's Mountain (NW of Portugal). In recent decades, population decrease has reached a peak but still continues, leading to a dismantling of social organization and fundamental changes in land management. The aim of this chapter is, through farmer cooperation, to assess how the goat production systems operate, to identify the main problems and to define options for supporting their development. Village flocks were surveyed, farmer's strategies evaluated and the main land-use activities reported. Body condition scoring (BCS) of the goats in winter, along with a high kid mortality, demonstrated a need for improved goat feeding. Supplying goats with concentrates to stimulate the intake of fibrous feeds and a more local agrarian and environmental operation are some of the proposals to improve system efficiency.

Key words: common land, management, sustainability

Introduction

Peneda's Mountain covers three parishes with a total area of approx 90 km^2. A large part of this territory is partially included in the Peneda-Gerês National Park. The common land, more than 50 km^2, is jointly used by the inhabitants. In the village studied, agriculture, carried out in small fields, is the main socio-economic activity (Pires, 2000). Field crops are mainly used to feed the animals and also for human consumption. The animals (cows, goats and horses) are raised to provide income through sales and subsidies. All the families owning indigenous goats also have cows. Goat keeping also improves feed resources located in places inaccessible to other herbivores and has a role in maintaining social structures. Management farmers have been able to carry out has led to a diversified income and yielded an overall profit for the land area. As a consequence, it has allowed the preservation of the resources and landscape of this fragile place. Production and preservation were possible due to a high population density, a strong social organization and common goals. However, in the last decades the rate of population decrease reached a peak in 1991 and is continuing to decline, leading to the dismantling of the social organization and major changes in land management (Machado, 2000). There has been a reduction in the number of goats, whereas cattle numbers have remained stable, mainly because of a lack of family labour. Also the isolation and the difficulty of goat keeping lead to the cessation of this activity. Hence the landscape loses its diversity, the grazing areas decline and there are fewer species for hunting. It is felt that the preservation of the natural resources and landscape of such a fragile territory, in this case the whole mountain, involves maintaining people on the land and engaged in farming activities which can be reorganized with other activities (Graça and Carvalho, 2000). The main objective was to assess, in cooperation with the farmers, the operation of the goat system in order to identify the main problems and to define options for supporting its future development.

Materials and Methods

The village studied had four herds of goats, with between 80 and 250 animals per family and with herds of cows between 1 and 45 animals (Pacheco et al., 2000). The village flock was surveyed (Pacheco, 2001) to identify the shepherds' routines, understand the farmers' strategies and the main events in their work calendars. Body condition scoring (BCS) of the goats throughout a production cycle was assessed according to the method proposed by Russel et al. (1969) and further adapted for goats by Morand-Fehr et al. (1988). A randomized assessment of 10 to 15 male and female goats, on every individual flock in the village, was carried out by three independent assessors during the mating period (July/August), at the end of pregnancy and during suckling.

Results and Discussion

The animals feed exclusively on the semi-natural vegetation on the mountain (*baldio*). Goats must be herded because there are private fields near the *baldio* and risks of wolf attack. In May, goats are taken from villages to the summering places. Grazing follows a daily system: the village flock (700-1000 animals), which is made up of the individual flocks, is kept by two shepherds. This collective practice is labour saving but its benefits are nowadays reduced due to the decrease in the number of goats owned per family. In September 1997, each family was responsible for keeping the flock twice a week (occasionally three times). The same task was performed in 2000 every two days. During winter the animals have higher nutritional needs due to the end of pregnancy and the suckling of kids. In addition, weather conditions (rainfall over 2000 mm per year and snow) and day length make it difficult to meet nutritional requirements by grazing. Furthermore, there is a decrease in the natural vegetation available on the mountain in some places. Both males and females reach their highest BCS in July-August, when the majority of the flock are mating (Table 1).

Table 1. Body condition score (BCS) of male and female goats grazing mountain pastures.

	Males	Females	
	July/ August	July/ August	March
BCS	2.3 ± 0.33	2.3 ± 0.44	0.9 ± 0.41
Number	5	66	43

These data support the shepherds' views that the goats are too weak during winter and their performance is very much reduced. This led us to question the compatibility between feed intake and the actual nutritional requirements of the goats in wintertime, which leads to an increase in the mortality rate. It appears that the kids are starving until March, yet farmers claim the goats' lack of milk to feed the kids is inevitable because:

1. It is well known by all farmers that, in winter, cows need more feed supply than the goats and as there is not enough hay the farmers prefer to supply the cows rather than the goats.
2. The goat pens have not enough space or feed supply systems. So the supply of any feed will result in struggles between animals which may cause abortions and deaths.

Mortality rate of kids in pens is also high because of the lack of space and light and crushing. Many others, born during the night in the crags near the village, are killed by rain, cold or by predators. The small pens require much more labour and make them difficult to manage. It is hard to find appropriate places to herd the goats, because of the difficult topography, grazing and access paths and also due to legal problems, e.g. the strict rules of the Peneda-Gerês National Park and local government.

Conclusions

Because of the lack of labour available, new techniques and routines to manage goats are required. Goat keeping could be simplified using mobile fences or trained dogs. New pens will improve housed space management and improve animal performance. To improve feeding, goats should be fed with concentrates to stimulate the intake of fibrous feeds (Simiane and Meuret, 1994). This will optimize the short time available for goats grazing. Clearing some dense bushes would increase feed available in winter, and improve access for people and animals. The challenge is to make the system economically sustainable.

References

Graça, L.L. and Carvalho, A. (2000) Participatory research and rural development (Investigação participada e desenvolvimento rural). In: Graça, L.L. and Santos, H. (eds) *Cadernos da Montanha, Peneda – 1*. DRAEDM, Braga, Portugal, pp. 8–17. (In Portuguese.)

Machado, C. (2000) Population dynamics and structure (Dinâmica e estrutura populacional). In: Graça, L.L. and Santos, H. (eds) *Cadernos da Montanha, Peneda – 1*. DRAEDM, Braga, Portugal, pp. 40-47. (In Portuguese.)

Morand-Fehr, P., Hervieu, J., Santucci, P. and Branca, A. (1988) Monitoring of the body conditions state during the year (Suivi de l'état corporel au cours de l'année). *CR Réunion du groupe Agrimed (CEE) sur l'état corporel*. FAO-CIHEAM, Zaragoza, Spain. (In French.)

Pacheco, F. (2001) Caprine livestock breeding in the Peneda range: identification, assessment and future perspectives (Práticas da capricultura na serra da Peneda: identificação, avaliação e perspectivas de futuro). MSc thesis. The University of Tras Os Montes, Vila Real, Portugal. (In Portuguese.)

Pacheco, F., Santos, J.C.R. and Ferreiro, R. (2000) Management routines in goat systems on Peneda's Mountain: evaluation and perspectives. In: Gagnaux, D. and Poffet, J.R. (eds) *Livestock Farming Systems: Integrating Animal Science Advances into the Search for Sustainability*. EAAP, Wageningen, The Netherlands, pp. 391-395.

Pires, C.B. (2000) Structure of the agrarian population in Cabreiro, Gavieira and Sistelo (Estrutura das populações agrícolas de Cabreiro, Gavieira e Sistelo). In: Graça, L.L. and Santos, H. (eds) *Cadernos da Montanha, Peneda–1*. DRAEDM, Braga, Portugal, pp. 48-57. (In Portuguese.)

Russel, A.F., Doney, J. and Gunn, R.G. (1969) Subjective assessment of body fat in sheep. *Journal of Agriculture Science* 72, 451-454.

Simiane, M. and Meuret, M. (1994) Un complément spécial parcours. *La Chèvre* 202, 26-29.

Silvopastoralism in Evritania, central Greece

A. Pantera [1] and A. M. Papadopoulos [2]
Technological Education Institute of Lamia, Forestry Department at Karpenissi, Karpenissi, 36100 Greece.
[1]pantera@teilam.gr, [2]ampapadopoulos@teilam.gr

Abstract
Evritania is a mountainous prefecture in central Greece with 54% of its area at altitudes over 1000 m. The total area of Evritania is 186,940 ha, 9060 of which are agricultural land, 82,220 ha forests and 76,000 ha rangelands. Generally, Evritania is a livestock husbandry prefecture due to the physiography of the area and the existing rangelands. Silvopastoralism is of major economic interest as in many areas the additional income from the land use may contribute to the low income of the population. The main silvopastoral systems, their function and environmental adaptability are described in this chapter. Finally, suggestions are proposed on the future of silvopastoralism in the prefecture.

Key words: land use, socio-economic, environment, livestock

Introduction
Silvopastoral systems are complex entities that involve trees, pasture, animal and man (Papanastasis, 1996). Silvopastoralism has been practised from the earliest time of herding and domestication of livestock in most areas of Greece (Schultz et al., 1987) and specifically in mountainous prefectures such as Evritania.

Area description
Evritania belongs to the Greek mountainous highlands (geographical latitude $38°40'-39°17'$ and longitude $21°22'-21°00'$) covering an area of 186,940 ha. Almost half of this area is covered by forests (82,220 ha), only 9060 ha are agricultural land, and 76,000 ha are rangelands (National Statistical Service of Greece, 1995). Relics found in the area indicate human presence since the ancient times despite the difficulty in reaching the highlands. The population's main occupation is animal husbandry even though, lately, Evritania is becoming one of Greece's most popular tourist destinations. The climate can be characterized as mountainous Mediterranean with mean annual temperature ranging from 9.2°C to 15.4°C and mean annual precipitation from 1039 mm to 2035 mm (Papadopoulos, 1999a,b). The vegetation may be classified into the types: subalpine, fir forest of *Abies borisii regis* and *A. cephalonica*, oak stands of *Quercus frainetto* and *Q. pubescens*, evergreen broadleaved, and phrygana with the dominant species *Phlomis fruticosa*.

Main silvopastoral and grazing systems
There are two silvopastoral systems in Evritania, the **open forests** with tree canopy less than 40% and **grazable forests** with tree canopy more than 40% (Papanastasis, 1996). Grazing can be identified as in: (a) **the village or flock system** where goats, sheep and cattle are fed in barns from November to April and allowed to graze outside for the rest of the year, (b) **the intra-transhumant system** where goats and sheep are kept in barns and moved to grazing lands from February to May or May to June and possibly again in November, (c) **the inter-transhumant system** where mainly sheep and a few goats graze in Evritania from the end of May until the middle of October and for the remaining time are moved to lower altitudes (Thanopoulos et al., 1999).

Animal capital and agricultural land
Only a fraction of the total number of cattle and goats (Table 1) are subsidized, based on the requirements posed to farmers by the Greek Ministry of Agriculture. On the contrary, an additional 22,540 sheep of those recorded by the National Statistical Service of Greece (National Statistical Service of Greece, 2001) are subsidized. This is due to the fact that, the number of sheep in subsidies (Table 1) includes the inter-transhumant animals from other prefectures. The arable land in Evritania has been reduced by 31.1% in the last decade (National Statistical Service of Greece, 1995, 2001). A similar decrease was noticed for agricultural land used for vegetables. Pantera et al. (2002), based on previous reports (Papanastasis and Giannakopoulos, 1989; Tassos, 1992), pointed out that this decrease commenced during the decades following the severe depopulation of the prefecture. On the contrary, fallows have been increased by 22.6%.

Table 1. Estimation of the livestock capital (animal units) of the Evritania prefecture. +Source: National Statistical Service of Greece, Census of year 2000. *Source: Department of Agriculture and Fisheries, 2000.

	Animal units+	Animal units subsidized*
Cattle	2,485	1,650
Sheep	43,318	65,858
Goats	47,335	27,217
Pigs	1,783	-
Beehives	7,432	-
Poultry	43,244	-

Discussion

The difficulty in reaching Evritania and the low living standards have resulted in a population decrease with a subsequent decrease in arable land. Most of the available funding was used for road construction with little allotted to husbandry development (Pantera et al., 2002). Furthermore, farmers refused to adopt modern practices that would have increased their production (Thanopoulos et al., 1999). Nowadays, primary production of agricultural goods is restricted to domestic use or for the local markets. However, lately there is a growing interest in the market for meat derived from free-grazing animals and especially beef.

The existing livestock graze mostly in forestland, abandoned reforested fields and partly in forests, and may all be characterized as silvopastoral. Grazing is prohibited only in forests under management, to enhance natural regeneration (G. Karageorgos, Ervitania, 2004, personal communication).

Conclusions

Silvopastoralism, as a traditional land-use system in Evritania, is well adapted to the physiography of the area and must be enhanced and promoted.

Farmers must be informed on the advantages of agroforestry, and specifically of silvopastoral systems.

Qualified personnel should be employed in both the Agricultural Service and the Forest Service, who are able to support and inform husbandry farmers on the use of silvopastoral systems.

References

Department of Agriculture and Fisheries (2000) *Data on the Number of Subsidised Animals for the Year 2000*. Department of Agriculture and Fisheries of the Prefecture of Evritania, Athens, Greece.

National Statistical Service of Greece (1995) *Agricultural Statistics of Greece for the Year 1991*. N.S.S. Greek National Statistics Service, Athens, Greece.

National Statistical Service of Greece (2001) *Results from Agricultural and Animal Production Research (Agricultural and Animal Capital Census) 1999/2000*. N.S.S. Greek National Statistics Service, Athens, Greece.

Pantera, A., Karageorgos, G., Loukas, P. and Phakas, I. (2002) Mountainous rangelands of Evritania (in Greek). In: Platis, P.D. and Papachristou, T.G. (eds) *Proceedings of the Third Range Science and Mountainous Development. National Rangeland Congress*. Hellenic Range and Pasture Society Publication No. 10, Kaprenissi, pp. 25-32.

Papadopoulos, A.M. (1999a) Study of the dendroecological relations of *Abies cephalonica* in Timphristos mountain. In: Hellenic Forestry Society (ed.) *Eighth National Forestry Congress*. Hellenic Forestry Society Publication, Alexandroupoli, Greece pp. 218-226.

Papadopoulos, A.M. (1999b) The NATURA 2000 Network in Evritania. In: *Proceedings of the Workshop on the Conservation of the Natural Environment Prerequisite for Sustainable Development*. Greek Geotechnical Chamber, Karpenissi, Greece, p. 16.

Papanastasis, V.P. (1996) Silvopastoral systems and range management in the Mediterranean region. In: Etienne, M. (ed.) *Western European Silvopastoral Systems*. INRA, Paris, France, pp. 143-156.

Papanastasis, V.P. and Giannakopoulos, A. (1989) *Study on Pasture and Animal Husbandry Development of the Agrafa Area of Evritania*. Agricultural Bank of Greece, Athens, Greece. (In Greek.)

Schultz, A.M., Papanastasis, V., Katelman, T., Tsiouvaras, C., Kandrelis, S. and Nastis, A. (1987) *Agroforestry in Greece. Working Document*. Aristotle University of Thessaloniki, Thessaloniki, Greece.

Tassos, G.D. (1992) *Animal Husbandry Enterprises in the Prefecture of Evritania for the Year (1991): Employment and Economic Results*. Agricultural Development Company "Evritania", Greece. (In Greek.)

Thanopoulos, R., Vlahos, G. and Louloudis, L. (1999) Differentiation between mountain livestock production systems: the impact of social and environmental factors In: Williams, S.M. and Wright, I.A. (eds) *Proceedings of Two International Workshops, in Karpenissi, Greece (22-24/1/1998) and in Vienna, Austria (18-20/6/1998)*. ELPEN, Macaulay Land Use Research Institute, Aberdeen, United Kingdom, pp. 73-85.

The effect of duck grazing on cocoa yields in São Tomé island

A. Pardini
DiSAT-University of Florence, Italy. andrea.pardini@unifi.it

Abstract
A decline in quality and productivity has reduced the value of cocoa for trade on the island of São Tomé. However, both seed and fruit of cocoa remain an important food for local people. Plantations used for subsistence are poorly managed, resulting in poor yields, weed diffusion, reduction of soil fertility and enhanced soil erosion. Grazing by ducks, geese and poultry can help maintain soil fertility to eliminate weeds in small plantations and home gardens on the island. These small animals are frequently reared on the same land to supplement diet, give income if sold in village markets, produce organic manure and help control weeds. Three treatments were carried out in a mature cocoa plantation used for self sustenance of local people, near the capital town: grazed by ducks; not grazed with weeds cut; neither cut nor grazed. Dry cocoa production was compared 5 years after treatments were imposed. Results showed that duck grazing or weed mowing contributed to increased productivity. Duck grazing was preferable to mowing because the ducks produced a good fertilizer for the soil. However, after 5 years of treatment imposition, differences were not yet significant.

Key words: soil fertility, weed control, grazing, mowing

Introduction
Cocoa and coffee have been important components of trade in São Tomé in the past for a long time. Nowadays both crops have poor quality, are less important for export and are used mainly for self subsistence (Pardini *et al.,* 1996). Traditional management practices cause weed diffusion, reduction of soil fertility, enhanced soil erosion and, in turn, poor yields, which reduce after a few years of cropping (CIRAD, 1995). Fertilizer cannot be afforded by local people to replace leached nutrients. Weeds are known to significantly reduce cocoa yields. They are partially reduced by the presence of shadow trees; however, weed control should always be applied (Purseglove, 1968). An experiment is reported whereby duck grazing is used to reduce weeds and fertilize the soil with dung, to increase and sustain production.

Materials and Methods
The research was carried out in the north-east of São Tomé island in a plantation established during the Portuguese period (up to the 1970s) and then practically abandoned. The land has not been restored by either the State or privately and is used by local people from a nearby village. The plantation is made up of now selected and unselected types of trees which are from 15 to 25 years old. The weather is frequently cloudy, nonetheless *Erythrina poeppigiana* trees were planted for shade by the Portuguese colonists and they are still present. Cocoa trees are spaced at 3.5 × 3.5 m; however, the spacing is less regular in some parts of the plantation. An area of the plantation was chosen with trees of the same age, size and health and soil and climate were uniform all over the area. Three treatments were established in plots of 12 trees each. Plots were rectangular and comprised three rows of four trees each. Each treatment was repeated twice. Native plants were mainly grasses of the genera *Pennisetum, Heteropogon, Setaria* and *Loudetia*. The following treatments were compared:

- Grazed all through the year by a local group of ten ducks, which also gave organic manure.
- Not grazed, weeds hand-mown each time the sward was higher than 50 cm, not fertilized.
- Not grazed, not mowed, not fertilized.

The ducks were local breeds and they fed on native pasture only. Measurements taken were:

- Cocoa seed yield in 1997 before start of the trial (mean number of fruits per 10 trees on the average of 10, and mean fruit weight of 10).
- Cocoa seed yield in 2001 at the end of the trial (mean number of fruits per 10 trees on the average of 10, and mean fruit weight of 10).
- Herbage height at fruit harvest.
- Biomass of weeds (hand-mown and dry weight of all the herbage in the sample areas).

Results and Discussion

Initial seed production per tree was very low (Table 1) due to poor traditional practices (leaves and fruits disease, weed infestation, low fertility). After treatments had been imposed for 4 years, dry cocoa yield remained low relative to published values; however, it was higher in the two plots with weed control (mean 0.53 kg plant) than in the area without control (mean 0.28 kg per plant). Initial yields were actually doubled by grazing or hand-mowing. Dry cocoa yield differences in grazed or mowed plots were significant at $P = 0.05$; however, it is possible that this yield difference will be enhanced within the next few years due to positive effect of organic manure from the duck.

Table 1. Dry cocoa seed yield before start of the trial (1997) and at its end (2001), herbage height (cm) and biomass (t/ha DM). Numbers with different letters in columns are significantly different at $P = 0.01$ (capitals) and 0.05.

	Dry cocoa yield/kg/ha		Sward height	Weed DM
	Before start	In 2001	(cm)	(t/ha)
Grazed		458 A a	15 B	1.15 B
Not grazed, mowed	222	411 A b	17 B	1.23 B
Not grazed, not mowed		225 B c	97 A	9.86 A

Differences in sward height and biomass of weeds were not significant between grazed and mowed plots, but, as would be expected, both these values were significantly lower than where swards were not controlled.

Conclusions

Both duck grazing and mowing have controlled weeds and this has increased fruit production. Organic manuring by birds would be expected to increase yield more than mowing because of the nutrient impact; however, 4 years of the trial were not sufficient to verify this. If this is proved, duck grazing can be recommended instead of mowing to control understorey vegetation. Moreover, ducks are an additional source of income and they contribute to a more balanced diet for local people. Some tourism is already present on the island (even if, at the moment, visitors are mainly researchers, technicians and officers of international organizations) and meat can be sold at much higher prices in the capital town than anywhere else in the island. The local authorities have plans to develop this economic sector (ECOFAC, 1995).

It is suggested that duck flocks, or indeed any poultry or different small animals, are introduced to cocoa, coconut, banana, pineapple, coffee plantations of the island, in both home gardens and governmental farms, as a management tool and alternative source of food production.

References

CIRAD (1995) *Rehabilitation of the Experimental Research Station of Poto* (Projecto de rehabilitacion da estacao Experimental de Poto). Rapport annuel d'activité 1995, Paris, France. (In Portuguese.)
ECOFAC (1995) *Reforestations and Agro-Afforestations* (Repovoamento florestal e agro-forestamento). ECOFAC, Sao Tomé e Principe. (In Portuguese.)
Pardini, A., Longhi, F. and Zari, A. (1996) Possibility of eco-compatible economic development in the Democratic Republic of Sao Tomé and Principe (Possibilità di sviluppo economico eco-compatibile nella Repubblica Democratica di Sao Tomé e Principe). *Rivista di Agronomia Subtropicale e Tropicale* 90 (4), 497-514. (In Italian.)
Purseglove, J.W. (1968) *Tropical Crops*, Longmans, Vol. 2., pp 719.

Types of land use in the montado (dehesa) production system

M. S. Pereira [1], A. Quintas [2], I. S. Coelho [2] and C. C. Belo [1]

[1]STPA Dept. EZN-INIAP, 2005-048 Vale de Santarém, Portugal. [2]EESA Dept. EAN-INIAP, Quinta do Marquês, 2780 Oeiras, Portugal. stpa.ezn@mail.telepac.pt

Abstract

This study was based on data obtained from the General Census of Agriculture (RGA, 1999) regarding 932 farms from the Alentejo region (Portugal). Farms were classified by percentage of "montado" according to the land used for agricultural purposes (LUAP) ("montado" classes (MC): MCa 0-33%; MCb 33-66%; MCc 66-100%). There are three types of land use closely associated with the proportion of "montado" on farms. The increase from MCa to MCb resulted in decreases of 50% in the items referring to cleared land, while items referring to under-*Quercus* increased 200%. On the other hand, the increase from MCb to MCc resulted in greater decreases in items referring to cleared land (80%), while items referring to under-*Quercus* increased 50% for temporary crops, 132% for fallow land and 73% for permanent pastures. Forage production, animal types and stocking rate were also characterized.

Key words: stocking rate, silvopasture

Introduction

In Portugal, there are about 1 million ha of "montado". These areas have socio-economic importance because they are agrosilvopastoral and can generate both public and private wealth and services, particularly high-quality animal products. Of all the types of land used for agricultural purposes (LUAP), the area of natural permanent pastures has increased more than all the others. The population of ruminants has also increased, indicating a shift towards extensive animal production. In this work, "montado" production systems were identified and characterized as a precursor to future development and economic sustainability.

Materials and Methods

Ten classes from the General Census of Agriculture (RGA, 1999) were selected because they covered all the types of land used for agricultural purposes. These included total areas of temporary crops, permanent pastures and fallow land in two different situations: cleared land and under-*Quercus*. Variables evaluated were expressed as proportion of the LUAP (per cent). The "montado" production system area was divided into three classes (MC), according to the percentage of "under-*Quercus*" in LUAP: MCa - 0-33% (234 farms); MCb - 33-66% (223 farms); MCc - 66-100% (475 farms, 97 of which were 100%). A discriminant analysis by Fisher's method and forward stepwise analysis (Reis, 1997) was carried out for the three MCs. This method seeks linear combinations (discriminant functions) that maximize the distance between group means. It is possible to classify farms with these discriminant functions and is also possible, by Bayes criterion, to give error probabilities to an *a posteriori* classification. Bayes criterion gives the probability of a farm belonging to one of the three MCs. The farms are then classified into an MC by choosing the biggest probability (Reis, 1997). To confirm the discriminant analysis results a matrix must be calculated. This matrix compares the initial with *a posteriori* classifications of the same farms. The correct classification percentage is then calculated.

Results and Discussion

It was possible to discriminate between MCa, MCb and MCc. The types of utilization of the "montado" production system are the variables with the largest contribution towards the discrimination between MCs. Permanent pastures are the most important crop, followed by fallow land and temporary crops. The discriminant analysis resulted in 99.25% correct classifications. The total means for the types of land use were also evaluated using Scheffé's test (Neter *et al.*, 1996) regarding the characterization of the three MCs.

Table 1. Percentage for main types of land use according to "Montado" classes (area/LUAP). TC – temporary crops; FL – fallow land; PP – permanent pastures; sem – standard error of the means; comparisons were made between MCa, MCb and MCc. Means with different letters are significantly different for $P < 0.05$.

MC/Land use		MCA (% ± sem)	MCB (% ± sem)	MCC (% ± sem)
Cleared land	TC	28.8 c ± 1.2	19.6 b ± 0.9	4.4 a ± 0.8
	FL	19.5 c ± 1.2	12.2 b ± 0.7	2.5 a ± 0.7
	PP	26.8 c ± 1.2	11.5 b ± 1.7	1.5 a ± 0.8
Under-*Quercus*	TC	2.3 a ± 0.9	6.9 b ± 0.8	10.4 c ± 0.9
	FL	2.8 a ± 1.7	8.9 b ± 1.8	20.7 c ± 1.1
	PP	10.7 a ± 2.4	33.3 b ± 2.4	57.7 c ± 1.6

All types of land use – 'cleared' land and 'under-*Quercus*' land were statistically different for all MCs (Table 1). In relation to cleared land, the increase in under-*Quercus* land from MCa to MCb led to a decrease of 32% for temporary crops, 38% for fallow land and 58% for permanent pastures. On the other hand, these types of land use increased by about 200% in the under-*Quercus* situation. When considering MCb farms, the area for temporary crops in under-*Quercus* only represents about one-third of the cleared land area, while the fallow land area is slightly lower and permanent pastures tripled. The decrease observed for the temporary crops is mainly due to a decrease in area used by cereal crops for grain, since both fodder crops (oats, annual mixtures, like oats × vetch, and industrial crops (sunflower) maintained their area. In relation to cleared land and when comparing MCb and MCc values, the percentage of LUAP used for temporary crops, fallow land and permanent pastures all decreased by the same proportion, about 80%, when the area of under-*Quercus* increased (Table 1).

On the other hand, in the under-*Quercus* situation, all types of land use increased (50% for temporary crops, 132% for fallow land and 73% for permanent pastures) regardless of the "montado" class. The decrease observed for temporary crops was also due to the decrease in area used by cereal crops for grain, from 14.2% to 5.5% of LUAP. Permanent pastures, especially natural pastures with low feeding value, represent about 75% of the total permanent *Quercus* under-tree pastures in both MCa and MCb farms, and about 84% in MCc farms. The predominant animal species are cattle and sheep. When the three MCs are considered, there were no significant differences between areas for fodder production and the mean stocking rate supported by it (0.3 standardized LU/ha).

Conclusions

The classification functions obtained through the discriminant analysis resulted in farms being correctly allocated to the appropriate MC.

The decrease in the area given to temporary crops is due to the decrease in cereal crops for grain since the area for fodder crops is maintained. The area for fallow land and temporary crops used for fodder, along with the acorns and leaves, determines how silvopastoralism performs and the stocking rate for this production system, especially for greater "montado" areas. There is also potential to increase the ruminant population since the overwhelming majority of pastures have very low feeding value.

References

Neter, J., Kutner, M.H., Nachtsheim, C.J. and Wasserman, W. (1996) Analysis of factor levels effects. In: Neter, J., Kutner, M.H., Nachtsheim, C.J. and Wasserman, W. (eds) *Applied Linear Statistical Models,* 4th edn. McGraw-Hill, New York, USA, pp. 710-755.

Reis, E. (1997) Discriminating analysis (Análise Discriminante). In: Reis, E. (ed.) *Estatística Multivariada Aplicada*. Edições Silabo, Lisbon, Portugal, pp. 201-244. (In Portuguese.)

RGA (Instituto Nacional de Estadistica). (1999) Agricultural Statistics (Recenseamento Geral da Agricultura) Portugal, http://www.ine.pt:8080/biblioteca/search.do (accessed April 2004)

Contribution of silvopastoral systems to land upkeep: differences and determining factors in a French upland region

H. Rapey [1] and R. Lifran [2]

[1] CEMAGREF, Unité de Recherche Dynamiques and Fonctions des Espaces Ruraux, 24 avenue des Landais, B.P. 50085, 63172 Aubière Cedex 1, France. [2] INRA, Economie and Sociologie Rurale, 2 place Viala, 34060 Montpellier Cedex 1, France. helene.rapey@cemagref.fr, lifran@ensam.inra.fr

Abstract
In many wooded regions, the decline of farming and the trend towards increasingly uniform forested landscapes compromise environmental, social and economic objectives and are therefore of concern to managers. Analysis of farming systems maintaining forested and non-forested areas can help to identify the farming conditions that are conducive to management and opening up of these uniform landscapes. With this view, a study was carried out in a French wooded region presenting a wide variety of silvopastoral systems. The sample of 137 farms surveyed presented four silvopastoral systems clusters, each differing in the pastoral and forestry use of its farm area. Each cluster also displayed specific land structures and specific livestock orientations and dynamics. This analysis emphasizes the links between (i) the size and dynamics of the farming systems and (ii) the extension of forest area upkeep by the farming systems.

Key words: livestock farm, land maintenance, upland area, silvopastoralism

Introduction
Upland areas currently constitute Europe's most densely wooded areas with the greatest forest extent (OEFM, 2000). One of the main aims of farming and forestry is to conserve diversity of landscapes in order to preserve their environmental, economic and social functions. In this context, silvopastoral practices and systems have attracted the interest of managers because of their relevance to the long-term maintenance of large woodlands, shrublands and grasslands. For several decades, a diversity of silvopastoral practices has been implemented and developed on tens of thousands of hectares in the Margeride region in central France (Long and Daget, 1965; Nougarède, 1970; Lienard, 1977; INRA, 1983). A recent field survey on woodlands of this region reports on multiform forest maintenance and its use by farmers (Lifran et al., 1997). Our purpose here was to determine the farming conditions that favour the maintenance of forested areas by farms, based on the Margeride silvopastoral experience.

Materials and Methods
All the farms with grazed forests in one of the 18 heavily wooded villages of Margeride were surveyed. These 137 farms made up 5.7% of the farms in this region, and 10% of its total farm area, of which 16% was forested ("more than 10% of the ground covered by trees" in the farmers' appraisal). Each farm surveyed is described as regards financial and technical aspects, wooded and non-wooded areas, present and past. The analysis was carried out in three stages: (i) data analysis (variance analysis, Principal Component Analysis (PCA), Factorial Component Analysis (FCA)) of farm area covers and uses to identify the most important features of the farmland management, (ii) identification of farms clusters based on their woodland covers and uses, (iii) characterization of each cluster with regard to the farms structure and development.

Results and Discussion
Four farms clusters were identified, differing in land size, management and maintenance patterns.

A first cluster of 56 farms globally managed a large area (≈ 4300 ha; 77 ha/farm), predominantly with grasslands, the wooded grazings comprising a small part of their area (24%). Globally these woodlands presented a small number of tree cuts (fewer than three plots with cuts for timber, pulp or fuel per farm since the farm's establishment). Their global 554 ha of wooded land in only some instances supported a combination of agricultural and forestry practices. Their practices and land maintenance were concentrated on non-wooded areas. These small farms (current average of 77 ha, 30 cows including 18 suckler cows), often with a dual-purpose cattle herd (milk-meat), had grown slightly since their establishment (+ 22 ha) and were now in low debt (current annual repayments of 4120 €).

A second cluster of 48 farms globally managed an even larger area than the previous cluster (≈ 5300 ha; 111 ha/farm), predominantly with grasslands, combining more tree cuts than above on their small woodland areas (more then three plots with cuts per farm since the farm's establishment). Their practices and land maintenance were concentrated on non-wooded areas, and limited on wooded areas. The structures were larger than above (111 ha and 36 cows/farm, including 22 suckler cows, 86 ha initially), often with a dual-purpose cattle herd and a moderate debt (current annual repayments of 5920 €).

A small third cluster of 16 farms managed a large area (156 ha per farm; Σ ≈ 2500 ha), predominantly with grazed forest and few tree cuts. Their woodland totalled 653 ha, essentially with agricultural practices. Cattle numbers were high and predominant or exclusive (56 cows/farm including 52 suckler cows). These farms had nearly doubled their herd and area since establishment, and were now in debt (current annual repayments of 8090 €).

The fourth cluster of 17 farms managed the largest area (201 ha per farm; Σ ≈ 3400 ha), predominantly with grazed forest and a high number of tree cuts. Their large herds (61 cows/farm including 47 suckler cows) were dual-purpose cattle (meat-milk) in almost half the cases. These farms had less than doubled their suckler cows and areas, and were heavily indebted (current annual repayments of 9440 €). Their woodland totalled 768 ha, frequently combining agricultural and forestry practices.

Conclusions

In this survey, three main factors contributed to the extension of silvopastoral practices: enlargement of farmland area, financial investment in the farm and meat herd specialization. In contrast, a dual-purpose cattle herd was associated with less spatial and technical association of agricultural and forestry practices; the labour constraint of such mixed livestock farming may limit silvopastoral land maintenance of these farms. Hence, to favour silvopastoral land management, agriculture and forest regulations and grants must, in their definitions, take into account these variables in farm situation and dynamic.

References

INRA (1983) *The Margeride: the Mountain, the People* (La Margeride: la montagne, les hommes). INRA, Paris, France. (In French.)

Lienard, G. (1977) *Structure of the Management, Land Use, Main Production in the Margeride. Quick analysis and RGA, 1970.* (Structure des exploitations, occupation du territoire, principales productions de la Margeride. Analyse rapide du RGA 1970). INRA, Paris, France. (In French.)

Lifran, R., Rapey, H. and Valadier, A. (1997) New approach of the traditional silvopastoralism in Lozère, Agreste (Nouveau regard sur la tradition silvopastorale en Lozère, Agreste). *Les Cahiers* 21, 17-22. (In French.)

Long, G. and Daget, P. (1965) Contribution to the ecological study of the Margeride Massif (Contribution à l'étude écologique du Massif de la Margeride). *Annales Agronomiques* 16, 4. (In French.)

Nougarède, O. (1970) *Preliminary Observations Concerning the Equilibrium of Agrosilvopastoralism and the Rural Population in Margeride.* Document (Premières réflexions concernant les équilibres agro-sylvo-pastoraux and la société rurales en Margeride. Working document de travail). INRA, Paris, France. (In French.)

OEFM (Observatoire Européen des Forêts de Montagne) (2000) White Book 2000 on the mountain forests in Europe (Livre Blanc 2000 sur la forêt de montagne en Europe). http://www.mtnforum.org/resources/library/oefm00b.htm (accessed April 2004) (In French.)

From forest grazing to multipurpose shelterbelts in Hungary

V. Takács [1] and N. Frank [2]

Dept. of Silviculture, Faculty of Forestry, University of West Hungary, Bajcsy-Zs. u. 4. 9400 Sopron, Hungary.
[1]vik@nyme.hu, [2]frank@emk.nyme.hu

Abstract

Livestock keeping in forests has a tradition both in the plains and in mountainous areas of Hungary. This ensured the stability of under-utilized areas and also sustained a diversity of animal stock. Due to ownership changes after 1989 the structure of agriculture became diversified; some private farms re-formed and new agricultural firms were established with some 100-1000 hectares of area. This allowed a revival of interest in silvopastoral systems.

This research review covers trends and possibilities in landscape management for specific locations and countryside. These farming methods offer the opportunity to embrace conservation, landscape protection and ethnic considerations.

Changes in land structure and the connection of Hungary to the EU have increased not only the economic objectives of afforestation, but also shelterbelts, landscape enhancement, settlement and protection. The spread of alternative land-use options is also expected. Surveys on resource condition need to be followed by long-term planning, which aims to increase the protection of forests. Experiences and potentials are given, but optimal exploitation needs further research and collaboration with the affected sectors.

Key words: land-use change, integration, rehabilitation, afforestation, rural balance

Introduction

Currently land development, agriculture and forestry are trying to solve the problems of the rural sector to keep the economical and ecological balance of the countryside. Soil conservation and use are seen as key environmental issues in Hungary. Sustainable development should be accompanied by a recognition of the impact of grazing on sensitive plant communities. Traditional methods of management ensured the economical stability of the countryside until the 1940s.

The outskirts and the inner areas of rural settlements and villages need to be managed and developed in a unified manner, the aim being to promote organic units, which are far more ecologically sustainable through decentralization and promoting regionality with local, independent infrastructure. Increasing developing and/or re-forming silvopastoral systems where farming and tree growing are carried out on the same area creates the opportunity for this to happen. Livestock breeding has been a centuries-old tradition in Hungary (Hegyi, 1978). The benefits of these ancient extensive livestock systems are becoming recognized in relation to management of small-scale farming. The old Hungarian system of keeping cattle, horses and sheep on the plains and pigs in the highlands is well documented. This type of extensive stock management is known as "acorning" (Hegedűs and Szentesi, 2000). However, it must be stressed that pasture-forests (with grazing as one of the subsiduary uses of forest) were not established for timber production but mainly to protect the soil from erosion. After the Second World War, i.e. from 1945 onwards, such semi-natural types of farming almost died out and despite the imposed programme of socialization there was little rehabilitation of these areas. Following the inception of large state farms which concentrated on cattle and pig keeping and the transfer of private forests to public ownership, forest pasturing came to an end. In 2001 there were 16,416 ha of shelterbelt systems established in the 1960s. These and their connecting forest areas might allow the repatriation of multipurpose management: semi-natural livestock keeping combined with wood management on agricultural land. This is in line with current support for "biofarming", the integration of environmentally sustainable and welfare-friendly livestock keeping systems.

The main goal of this research (supported by the Hungarian Scientific Research Fund OTKA, T 043417) is to find acceptable parameters for re-forming silvopastoral farms. Under current forest law in Hungary, livestock keeping in state-owned forests is not allowed. Hence this project focuses on the shelterbelt systems, farm forests and those small forest areas (area < 1500 m^2) which do not fit into the forest land-use category (as defined by the Hungarian 1996. LIV. 'Law about forest and protecting of forest'). The purpose of this research is to demonstrate how silvopastoral systems can contribute to the expansion of the Hungarian forest area and be introduced in harmony with the restructuring of existing or new shelterbelt systems.

Materials and Methods

This research consists of analysing of resources and optimizing outcomes. Almost 20% of Hungary's territory is forest. The strategic goal is to the increase forest cover to 25-27%, which would be reached by planting 15 -18,000 ha/year. Joining the EU will require a change in the current agricultural policy. One of its main elements is the utilization of particular agricultural areas for afforestation. This would potentially involve change of use of ca. 700,000 ha of potential arable land, (Barátossy and Verbay, 2001).

These arable lands are on unfavourable soils and could be used for forestry and grasslands. Based on research led by the Hungarian State Forest Service (Bán *et al.*, 2001), the potential areas recommended for change of land use are: 683,900 ha of arable land, 56,100 ha of pastures and 38,300 ha of meadows. From these areas 174,000 ha of afforestation are suggested for the 2001-2010 period. The first problem is to study how newly established silvopastoral systems will fit into this concept. Only approximately 1% of Hungary's forest area (ca. 1,850,000 ha) is shelterbelt. Windbreaks to protect agricultural areas account for 16,416 ha of that area. These shelterbelts were established in networks during the 1950s and 1960s. They were expected to reach heights of 10-20 metres to fulfil their expected function. Some tree species in these belts will reach the rotation age in 5-10 years (*Prunus avium*), but others (*Quercus* sp, *Fraxinus* sp, *Acer* sp) will deliver their function for 30-50 years as viable tree stands. Shelterbelts containing these species (while these are usually not qualified as forest area) are suitable for pasturing. This is particularly because, when thinned, they have a suitable spatial structure to incorporate livestock production and encourage development of a predominantly herbaceous ground cover vegetation. Due to the wide tree spacing (2-3 m inter-row and wide trunk spacing) the land can be cultivated and grazed. Areas appropriate for forest livestock keeping have to be reviewed on this basis. Components of these should be the existing woody pastures, the potential areas for new plantings, windbreaks and shelterbelts, farm forestations and the interactions between them. Potential areas for silvopastoral systems on a country level can be selected and designated using geographical information systems (GIS), based on digital maps, aerial photos, satellite images, forestry databases, the State Forest Service and Agricultural Institutes. These areas should have a tradition of agriculture, should be suitable for tree growing, should be areas of land that can be utilized for this purpose and adequately fulfil their soil protection function.

This research is currently underway. The categories and conditions are being defined by studying a chosen NW Hungarian sample area on which data and information are available from the past 60 years both in agriculture (livestock, forage, climate) and forestry (shelterbelts).

Concurrently, the identifying and enumeration of several shelterbelt systems by types are being undertaken for the whole country. The net result there will be to identify those shelterbelts and their networks which are suitable for introducing silvopastoral systems, whereby woodlands can be grazed.

Results and Discussion

The data collected enable the silvopastoral capacity of the studied and reviewed areas and of the shelterbelt types to be determined (Takács, 2003). Proposals will also be made for the silvocultural treatments, cultivation, development or restructuring of these areas. This could be achieved by studying and specifying the forest structure required and tree species suitable for these Hungarian sites.

Most important trees are native oak (*Quercus* sp), the heavier shading lime (*Tilia* sp) and maple (*Acer* sp) etc. and fruit-producing species such as *Pyrus* sp and *Prunus* sp. In addition, treatment/management plans and directives need to be defined for removing wind-sensitive species such as *Populus* sp or the suppression of aggressively spreading foreign species such as *Padus* sp. However, some foreign species could be useful as natural fences, e.g. *Maclurapomifera* and *Gleditsia* sp. Based on the information and results the applicable shelterbelt types and pasture tree planting patterns will be defined according to the different site conditions and usages. By rationalizing forestry use and grazing in the same space, the multipurpose uses of the silvopasture are for timber production, forage production, landscape protection, wind protection and revival of traditional farming patterns of individual farms.

From these preliminary studies and plans, a landowner can apply for competing funds from government ministries and other funds for agricultural development and afforestation.

Conclusions

Hungary has more than 10,000 ha of land which have been shown to be suitable for the formation and establishment of silvopastoral systems. These have to be surveyed and, learning from their traditional use and on a scientific basis of integrated land use, knowledge must be gained on how to make best use of the potential. There is a huge need to produce an inventory which would be accepted by the appropriate professionals and sectors of economy as a theoretical and basic document for planning. The research reported here would give the underlying forestry background for this approach.

References

Bán, I., Király, P., Pluzsik, A., Szabó, P., Wisnovszky, K. and Zétényi, Z. (2001) *Information on the Forest Resources of Hungary*. Hungarian State Forest Service, Budapest, Hungary.

Barátossy, G. and Verbay, J. (2001) The place of afforestation in the vision of the agricultural policy. In: *Forestry Forum 2001*. Forest Research, Budapest, Hungary, pp. 7-14.

Hegedűs, A. and Szentesi, Z. (2000) *Subsidiary Use of Forests*. Hungarian State Forest Service and Ministry of Agriculture, Sopron, Hungary.

Hegyi, I. (1978) *The Historical Forms of the Domestic Forest Utilization*. Akadémiai Kiadó, Budapest, Hungary.

Takács, V. (2003) *Survey of the Shelterbelt System of Sopronhorpács, the Possibilities of the Further Development.* Faculty of Forestry, Sopron, Hungary.

Session 2.
Productivity, quality and management of silvopastoral systems

Silvopasture in south-eastern United States: more than just a new name for an old practice

P. K. R. Nair [1], M. E. Bannister [1], V. D. Nair [2], J. R. R. Alavalapati [1], E. A. Ellis [1], S. Jose [1] and A. J. Long [1]

[1] School of Forest Resources and Conservation, [2] Soil and Water Science Dept., Center for Subtropical Agroforestry, Institute of Food and Agricultural Sciences, University of Florida, Gainesville, Florida 32611, USA. Author for correspondence: pknair@ufl.edu

Abstract

In the industrialized world, silvopasture refers to combining trees, forage plants and livestock in an integrated and intensively managed system. Inherent in the concept of modern silvopasture is the emphasis on spatial and temporal domains and integrated management of the system, such that much of the traditional animal grazing in forests and woodland is not considered as silvopasture. Silvopasture is by far the most prevalent agroforestry practice in south-eastern United States, and indeed much of North America, where it is becoming increasingly attractive to non-industrial private forest landowners and livestock operators who want to diversify their enterprise. With increasing public awareness about environmental quality, tree-based systems such as silvopasture are now receiving substantial interest as a new strategy for agroecosystem management. Our recent research results suggest that properly managed silvopasture can enhance water quality by reducing nutrient runoff from heavily fertilized agricultural and dairy enterprises, and improve small farm profitability by providing multiple income sources. Proper procedures need to be developed for valuation of ecosystem services such as water quality enhancement, carbon sequestration and improvement of wildlife habitat provided by silvopasture. Simultaneously, efforts should be taken to institute policy changes that accommodate the value of such benefits. In spite of some location specificity that is inevitable in such studies, our experience on various aspects presented in the chapter will hopefully stimulate silvopastoral research along these lines in other comparable regions as well.

Key words: agroforestry, carbon sequestration, decision support tools, environmental amelioration, environmental economics, water quality

Introduction

Animal grazing in forests and woodlands is an old form of "natural ecosystem management" in North America just as elsewhere in the world. Although these traditional non-intensive and passive forms of land use produced wood products and forages for livestock and wildlife, they are not considered silvopasture or agroforestry. As Garrett *et al.* (2004) state, silvopasture management, like the other forms of agroforestry, is based on our thinking in both spatial and temporal domains, and it demands skills in managing rather than reducing complexity. An understanding of hierarchical relationships within ecosystems and recognition that defined ecosystem-boundaries exist primarily for managerial convenience are essential to this concept (Garrett *et al.*, 1994; Garrett and Buck, 1997). Thus, silvopasture refers to combining trees, forage plants and livestock in an integrated and intensively managed system. In a broader context, silvopasture involves grazing systems and tree-fodder systems (Nair, 1993). But, in the North American context, the emphasis is on the former, and encompasses practices varying from rotational grazing in forests or plantations to intentional grazing under hardwoods and fruit-tree orchards. Silvopasture is becoming increasingly attractive to non-industrial private forest landowners and livestock operators in south-eastern United States who want to diversify their enterprise. With the realization about its environmental amelioration potential and greater appreciation of the societal value of ecosystem benefits such as carbon sequestration and wildlife habitat improvement, the practice is now receiving substantial interest as a new strategy for agroecosystem management. The scope of this chapter is to review briefly the current efforts in understanding and exploiting the ecological and economic potential of this promising agroforestry practice in North America with particular reference to south-eastern United States, and to develop a premise for its wider application.

Silvopasture reseach in south-eastern United States

Silvopasture is the most commonly practised kind of agroforestry in the southern United States (Zinkhan and Mercer, 1997), pine (*Pinus* sp)-based silvopastures being the most common (Williams *et al.*, 1997; Clason and Sharrow, 2000). The history of silvopasture in the region is fairly short. In the 1930s, L.T. Nieland with the Florida State Forest Service developed a programme designed to protect timber stands from fire by installing 30-m (100-foot), wide improved pasture grass grazing lanes. A survey of cattle ranchers on the south-eastern coastal plain in the 1940s indicated that most of them thought that grazing the forest reduced the threat of fires (Biswell and Foster, 1942). However, interest in pine forest grazing declined when serious efforts on cross-breeding of cattle got under way in the 1940s, and consequently better breeds of cattle requiring higher levels of nutrition than available in native vegetation were introduced (Lewis and Pearson, 1987). Although agronomic evaluation of warm-season grasses and legumes under natural stands of longleaf pine (*Pinus palustris*) and slash pine (*P. elliottii*) started in the 1940s (Halls *et al.*, 1957), it was in the 1950s that the "modern-day silvopasture" began when pine trees were first planted in improved pastures as part of the Conservation Reserve Soil Bank Program (Biles *et al.*, 1984). Several studies have since been conducted on the management of the practice of silvopasture in the region. These include: a pine-pasture-cattle integration study comparing the influence of vegetation, cultivation and fertilization on slash pine growth (Lewis *et al.*, 1983); simultaneous planting of pines and pastures, with hay production in the first 4 years until trees were fully resistant to grazing injury (Lewis *et al.*, 1985); evaluation of the productivity of cool-season exotic grasses under established slash and loblolly pines (*Pinus taeda*) in central Louisiana (Pearson, 1975); testing of Pensacola bahiagrass (*Paspalum notatum*) as a low-cost forage for improving forage resources in regenerating southern pines in north Florida (Lewis *et al.*, 1985); establishment of pines and bahiagrass on row-ploughed-disked areas in Florida (Lewis *et al.*, 1985); evaluation of the dry matter yields of perennial cool-season grasses (*Lolium perenne* and *Festuca arundinacea*) under artificial shade in Mississippi (Watson *et al.*, 1984); planting loblolly pine seedlings in open subterranean clover (*Trifolium subterraneum*) pasture in Louisiana (Lewis and Pearson, 1987); and a study on the effect of grazing cattle in newly established slash pine-ryegrass (*Lolium perenne*) pasture (Pearson and Rollins, 1987).

These studies have generated valuable scientific data on silvopastoral management, especially in the configuration of trees and the type of forage grass used, as well as the economics of silvopasture systems (Grado *et al.*, 2001). For example, it has been clearly demonstrated that combining the production of southern pines and beef on improved pastures offers an opportunity for multiple product yields. Integrating forestry with ranching may increase profitability and help buffer year-to-year variability in income through the sale of forest products and increased opportunities for sale of hunting leases brought about by the creation of wildlife habitat. Pensacola bahiagrass has been shown to be the most shade tolerant of the 23 grasses studied (Pearson, 1975; Lewis and Pearson, 1987). Several legume species have shown potential for production under partial shade (McGraw *et al.*, 2001), lespedeza (*Kummerowia* sp) and white clover (*Trifolium repens*) being the promising ones for silvopasture. Double rows of pines at 2.43 m (8 ft) between rows and 1.21 m (4 ft) between trees and 12.16 m (40 ft)-wide alleys produced more forage and as much wood as the single 2.43 × 3.65 m (8 × 12 ft) rows (Lewis *et al.*, 1985), and this remains the most popular spacing for silvopasture across the region today (Clason and Sharrow, 2000).

Silvopastoral research in the south-eastern United States got a boost in the 1990s with the USDA Forest Service/National Research Council (NRC) initiatives in agroforestry and the establishment of the National Agroforestry Center (http://www.unl.edu/nac/). These were further strengthened in 2001 with the establishment of the Center for Subtropical Agroforestry (CSTAF) (http://cstaf.ifas.ufl.edu/) at the University of Florida/Institute of Food and Agricultural Sciences (UF/IFAS). Areas of CSTAF research currently in progress include pine production and beef cattle management in slash pine silvopasture in Central Florida (http://cstaf.ifas.ufl.edu/research5.htm), performance and management of goats and swine in pine and hardwood silvopasture in north Florida (http://cstaf.ifas.ufl.edu/research6.htm), using silvopasture as an approach to reducing nutrient loading of groundwater from heavily fertilized pastures (Nair and Graetz, 2004), production physiology and interspecific interactions between trees and understorey grasses (Jose *et al.*, 2004; Wanvestraut *et al.*, 2004), integrating a landscape-scale approach in the development of decision support systems for silvopasture and other agroforestry planning (Ellis *et al.*, 2003, 2004), and valuation and policy implications of environmental services such as carbon sequestration and wildlife habitat improvement provided through silvopasture (Shrestha and Alavalapati, 2004; Alavalapati *et al.*, 2004). Another CSTAF project has assessed the perceptions of landowners and extension professionals to agroforestry in general, and evaluated the constraints to silvopasture adoption (Workman *et al.*, 2003; Workman and Allen, 2003). CSTAF has also been recently awarded a so-called "Higher Education Challenge Grant, 2004" for developing a web-based distance education course in agroforestry for disseminating silvopasture and other agroforestry knowledge to technicians and natural resource management professionals. In the following section, we will summarize major aspects of CSTAF's current silvopastoral research.

Current research
Environmental amelioration potential

Agricultural production systems in the industrialized nations are known for the application of large amounts of nutrients to the soil. Animal production systems, which often produce large amounts of manure with relatively small land areas to "dispose" of the manure, also belong to this category. In both these, large amounts of nutrients, phosphorus (P), can accumulate in the soil. Although the agronomic benefits of fertilization are well known, it is becoming apparent that the build-up of nutrients in the soil often increases the potential for nutrient loss from the soil (Edmeades, 2003). Animal manures and fertilizers, for example, pose a serious threat to surface and groundwater quality in major river (Suwannee River) and lake (Lake Okeechobee) basins of Florida. Elevated NO_3-N concentrations have been measured in water from both domestic and monitoring wells in the Suwannee River basin (Katz *et al.*, 1999), exceeding the maximum contaminant level of 10 mg/l set by the USA Environmental Protection Agency. Nitrate-N concentrations above this limit can cause methaemoglobinaemia or "blue baby" syndrome in infants and elderly adults (Mueller and Helsel,

1996). These increased nitrate concentrations in these waters have been attributed to dairy wastes (Andrews, 1994), poultry operations (Hatzell, 1995) and fertilizer applications (Katz *et al.*, 1999).

Similarly, loss of P from agricultural soils also can have serious implications for water quality. Dairy farming and beef ranching in the Lake Okeechobee basin in south Florida are major contributors of P to the lake (Fonyo and Flaig, 1995). Evidence is available that P moves through the sandy soil profile of the Spodosols of this region and that the P could eventually reappear in surface and groundwaters (Nair and Graetz, 2002).

Although P loss potential from silvopasture systems can be high because of high rates of application of manure with a high P:N ratio, the loss of P would be dependent on the P retention capacity of the soil. From an environmental standpoint, erosion of P-laden soil particles in surface runoff from agricultural lands has received most of the attention with regard to P and water quality. However, the vertical and lateral transport of P through soil profiles in which drainage water ultimately enters surface water has recently been identified as a potential source of P to surface waters in certain types of landscapes (Sims *et al.*, 1998; Nair and Graetz, 2002).

Cattle ranching is an important agricultural enterprise in Florida. Ranchlands cover over 2.4 million ha and generate over $300 million each year with over 1.8 million cattle (FAS, 2002). More than 60% of cattle production in Florida occurs near Lake Okeechobee and the Everglades (south Florida), which is a large freshwater lake (1891 km^2) and a drainage basin of approximately 12,950 km^2 (5000 square miles) (Harvey and Havens, 1999). During the twentieth century, Lake Okeechobee and the Everglades have been subjected to massive water control projects to drain parts of the Everglades, to protect areas from flooding and to make land available for agriculture (including cattle ranching) and urban development. The resulting land-use changes have had a profoundly adverse impact on environmental quality in south Florida. In particular, the P content of Lake Okeechobee has more than doubled over the past century, causing eutrophication and subsequent damage to its aquatic life (Harvey and Havens, 1999). Since cattle ranching is one of the sources of P runoff, it has long been perceived as an environmental concern in this region.

The effectiveness of some agroforestry practices such as riparian buffers in reducing non-point source pollution and thereby improving water quality is well documented (Schultz *et al.*, 2000, 2004). Little information is available, however, on the environmental implications associated with silvopastoral practices, particularly in regard to their impact on surface and groundwater quality. The deeper and more extensive tree roots would invariably be able to remove more nutrients (especially N) from the soil compared to crops with shallower root systems. Thus, nutrient leaching rates from soils under agroforestry systems where trees are a major component can be hypothesized to be lower than those from other land-use systems. Preliminary research conducted on a silvopastoral site at a research farm at Ona, Florida, on flatwood soils (Spodosols) suggest that silvopastoral sites are less likely to accommodate nutrients within a soil profile compared to an adjacent fertilized pasture with cattle grazing (Nair and Graetz, 2004) (Table 1). This supports the hypothesis that the tree and the pasture components minimize nutrient losses from the soil because of enhanced nutrient uptake by tree and grass roots from varying soil depths, compared to more localized and shallow rooting depths of the regular pasture.

Table 1. Soluble reactive phosphorus (SRP) concentrations with depth at a native flatwoods site, a silvopastoral site and an adjacent regular pasture at Ona, Florida, USA. Source: Nair and Graetz, 2004.[1] Numbers in parentheses are standard deviation values.

Depth (cm)	Native flatwoods	Silvopasture	Pasture
		SRP (mg/kg)	
0-5	1.59 (1.14)[1]	4.23 (1.34)	9.10 (3.26)
5-15	1.28 (1.17)	2.09 (1.36)	2.79 (2.20)
15-30	0.12 (0.02)	0.79 (0.39)	1.52 (0.94)
30-50	0.00 (0.00)	0.17 (0.14)	1.91 (1.25)
50-75	0.00 (0.00)	0.02 (0.02)	0.42 (0.40)
75-100	0.00 (0.00)	0.04 (0.03)	0.66 (0.52)

Carbon sequestration

It has increasingly been recognized that agroforestry practices such as silvopasture have importance as a carbon sequestration strategy because of carbon storage potential in the multiple plant species and soil (Nair and Nair, 2003). This potential, which seems to be substantial, has not even been adequately recognized, let alone exploited (Montagnini and Nair, 2004). Proper design and management of agroforestry practices can make them effective carbon sinks. As in other land-use systems, the extent of C sequestered will depend on the amounts of C in standing biomass, recalcitrant C remaining in the soil and C sequestered in wood products.

Discussing the area under silvopastoral systems, Clason and Sharrow (2000) argue that, given the widespread co-occurrence of grazing and forestry across North America, the joint production of livestock and tree products is by far the most prevalent form of agroforestry found in the United States and Canada, but the area statistics do not deal with multiple product systems such as silvopasture. According to Brooks (1993), forests currently occupy approximately 314 million ha of land in the United States and 436 million ha in Canada, of which only 70% and 60%, respectively, are timber-producing areas. Based on inventories made on an 11-year-old Douglas-fir (*Pseudotsuga menziesii*)/perennial ryegrass/subclover silvopastoral agroforestry system in western Oregon, Sharrow and Ismail (2004) reported that the agroforests (silvopastures) were more efficient in accreting carbon than tree plantations or pasture monocultures. They attributed this to the advantages of silvopastures in terms of higher biomass production and active nutrient cycling patterns of both forest stands and grasslands, compared to those of pastures or timber stands alone. Even close-canopied forests may produce considerable amounts of vegetation following timber harvests or fire, which could remain unexploited if not grazed by ruminant livestock. Based on these considerations, Clason and Sharrow (2000) conclude that about 70 million ha, more than a quarter of all forest land in the United States, is grazed by livestock. At the average estimated total soil C sequestration potential for USA grazing lands of 69.9 Tg C per year, the potential for silvopastoral systems could thus be around 9.0 Tg C (= 9 million tons of C) per year (Follett et al., 2001). Evidently, our knowledge base on this seemingly substantial environmental benefit is too narrow; further studies are warranted.

Environmental economics and policy

The welfare economics concept that the value of a public good is the sum of what the individuals would be willing to pay for it (Varian, 1992) has been extensively used to value environmental services associated with forestry and agricultural practices (Lohr and Park, 1994; Cooper and Keim, 1996). Such studies have been extended to tree-based systems as well: for example, Cameron et al. (2002) estimated the non-market value of tree plantation on public lands, accounting for the benefits of shade, windbreaks and carbon sequestration; Loomis et al. (2000) assessed the value of buffer strips along streams in terms of erosion control, water quality improvement and fish and wildlife habitat; and Shrestha and Alavalapati (2004) assessed the value of environmental services associated with silvopasture, using a stated-preference-based choice-experiment approach.

Studies on silvopasture in south-eastern United States have found it to be an economically viable enterprise (Lundgren et al., 1983; Clason, 1995; Grado et al., 2001; Stainback and Alavalapati, 2004). Stainback et al. (2004), however, found that, although combining slash pine with cattle production was not competitive with conventional ranching in Florida, it could be when environmental costs and benefits were factored into the analysis. Using a dynamic optimization model, they assessed the impact of a tax on phosphorus runoff and a payment for carbon sequestration on the profitability of traditional cattle ranching and silvopasture. In the absence of pollution tax and carbon payments, land values under traditional ranching are shown to be greater than those of silvopasture. As the pollution tax increases, land value decreases for both traditional ranching and silvopasture. For all pollution tax rates with carbon payments of greater than $1 per ton ($10^6$ gram, Mg) land values under silvopasture are greater than those of traditional ranching. As the pollution tax increases, the carbon payment necessary for silvopasture declines. For instance, for a tax of $10 per 0.45 kg (1 lb), a carbon payment of $0.50 per 0.45 kg is not sufficient to make silvopasture competitive. On the other hand, with a pollution tax of $15 per 0.45 kg, a carbon payment of $0.50 per 0.45 kg would make silvopasture financially competitive. Based on these studies, the authors concluded that the profitability of silvopasture will be greater than that of traditional ranching when both costs of pollution and benefits of carbon sequestration are considered. In this case, payment for carbon is shown to be more effective in influencing ranchers to undertake silvopasture practices in Florida.

A related question is: do households care for carbon sequestration, improvement in water quality and biodiversity associated with silvopasture? Will they be willing to pay for these benefits? If so, how much? Revenues from timber and additional hunting can offset the cost of silvopasture only partially (Shrestha and Alavalapati, 2004). Therefore, cattle ranchers may have little or no motivation to adopt silvopasture unless policy incentives are provided. Various federal and state programmes such as Conservation Compliance, Sodbuster, Swampbuster and the Conservation Reserve Program (CRP) in the United States are designed to encourage conservation practices on farmlands (Heimlich et al., 1998; Feather et al., 1999; Ribaudo et al., 1999; Westcott et al., 2002). Under these programmes, conservation practices such as filter strips, riparian buffers, shelterbelts, windbreaks and grass waterways, which are structurally and functionally similar to silvopasture, qualify for incentives.

Several studies have been conducted to assess the responsiveness of landowners to incentive programmes. Cooper and Keim (1996) studied farmers' willingness to adopt water quality protection practices in the face of incentive payments based on contingent valuation (CV) surveys in four watershed areas across the United States. Kingsbury and Boggess (1999) used a contingent valuation approach to study landowners' willingness to participate in a conservation reserve enhancement programme in Oregon. Shrestha and Alavalapati (2004) investigated Florida ranchers' willingness to adopt silvopasture. Applying a logit model for empirical estimation, they assessed the effect of a premium on beef price and a direct payment on the adoption of silvopasture using a dichotomous choice CV approach. The results (Table 2) show that the variable representing the incentive payment offer has a positive and significant effect on the probability of adoption. The impact is shown to increase at a decreasing rate, suggesting a non-linear relationship. The results also indicate that the existence of attributes such as wildlife presence, presence of creeks and/or streams, marshlands and longleaf pine will increase the probability of silvopasture adoption. The variable representing recreational hunting is positive and highly significant in the price support model, suggesting that ranchers are more likely to adopt silvopasture if they currently use their ranches for recreational hunting.

Table 2. Estimates of "marginal willingness to pay" for environmental improvement by landowners in relation to perceived benefits from silvopastoral practice in Florida, USA. Numbers in parentheses are the 95% confidence interval calculated from 1000 draws from the distribution of coefficients in the model. Source: Shrestha and Alavalapati (2004).

Environmental attributes	Value for environmental improvement ($ per household per year for 5 years)	
	Moderate	High
Water quality	30.24 (19.63–40.86)	71.17 (53.41–88.93)
Carbon sequestration	58.05 (43.72–72.37)	62.72 (54.38–71.06)
Wildlife habitat	49.68 (38.08–61.28)	41.06 (35.41–46.72)

Ranchers' mean "willingness to adopt" (WTA) silvopasture has also been investigated by Shrestha and Alavalapati (2004). They found that on average, a price premium of $0.15/lb (1 kg = 2.205 lb) of beef or a direct payment of $9.32/0.4 hectare (acre)/year was required for ranchers to adopt silvopasture practices. These estimates are much lower compared with previous studies on farmers' willingness to participate in conservation programmes in the United States (e.g. Lohr and Park, 1994; Cooper and Keim, 1996). Lant (1991), for example, reported that average annual payment under the Conservation Reserve Program was $48.93/0.4 hectare at national average, while state averages ranged from $37.48/0.4 hectare in Montana to $81.00/0.4 hectare in Iowa. The lower WTA estimates of this study may be due to the high compatibility of tree growing with cattle ranching.

The importance of silvopasture and other agroforestry systems in generating environmental services such as carbon sequestration, improvement in water quality and biodiversity is being recognized increasingly the world over. If these services are not internalized to the benefit of landowners, the landowners (or producers) will have little incentive to produce them at a socially desirable level. Environmental economic analysis provides an effective framework for internalizing environmental services. Recent advancements in environmental valuation methodologies help quantify the potential demand for and supply of environmental services more accurately. Policy makers can use this information as a basis to formulate incentive or tax policies that further the economy and the environment.

Production dynamics and tree-pasture interactions

Although advances in plant and soil sciences have provided the foundation for today's knowledge of silvopastoral systems, process-oriented research on production physiology and tree-grass interactions in silvopastoral systems have been rare. The lessons learned from many of the temperate silvopastoral experiments remain largely site and species specific (Burner and Brauer, 2003). For example, silvopastoral structure has been cited as causing shade and thereby reducing yield of associated grasses in some temperate systems (Brockway et al., 1998; Lin et al., 1999). However, reports of higher quality forage and better cattle performance and production are also available from the silvopastoral literature (Mitlohner et al., 2001; Burner, 2003). The discrepancy among these reports could probably be explained by the differences in species, microsite conditions and management practices that affected the component interactions and production dynamics.

Recent research in temperate alley cropping and silvopastoral systems has shown below-ground competition for water and nutrients (Bendfeldt et al., 2001; Wanvestraut et al., 2004). Lehmkuhler et al. (1999) reported significant reduction in height growth of black walnut (*Juglans nigra*) trees grown in silvopastoral systems compared to monoculture plantations in south-western Missouri, USA. These authors implied that tree-grass interaction for soil resources (water and nitrogen) led to the observed decrease in tree growth. It is possible that competition for water and nutrients by tree roots reduces yields of pasture grass.

Lessons learned about interspecific interactions in alley cropping systems in both the tropical and temperate regions of the world (Rao et al., 1998; Jose et al., 2004) could have applicability in silvopastoral systems too. Since the presence of trees modifies site microclimate in terms of solar radiation, temperature, water vapour content and wind speed it is reasonable to test the hypothesis of light as a limiting factor and to determine the photosynthetic dynamics of tree and crop components in silvopastoral trials. Understanding the temporal patterns of light distribution in alleys and subsequent influence on pasture photosynthesis, production and quality will significantly enhance our ability to design silvopastoral systems to reduce competitive effects so that economic returns can be increased.

Influence of trees on nutrient dynamics in agroforestry systems can also be complementary depending on species, system design and resource availability. Trees can enhance soil fertility in silvopastoral systems by adding organic matter through leaf and root decay, and biological nitrogen fixation by leguminous trees (Palm, 1995; Nair et al., 1999). For temperate systems, however, the trees often planted are non-nitrogen fixers selected for their product value rather than nutrient input. However, these trees can substantially increase the organic matter input into the intercropped system through the turnover of leaves and roots. Further, deep-rooted trees can take up nutrients that are leached down the soil profile, thereby reducing groundwater pollution, which is the essence of the soil-amelioration potential of silvopastoral systems. These nutrients will be returned to the surface soil through litter (leaves, needles, roots) decomposition and mineralization.

Landscape-level decision support tools
Considerable advances have been made in research, planning and development for a variety of agroforestry systems in a wide range of agroecological regions, from tropical to temperate. Prior to 1991, computer use in agroforestry research began with the development of databases as aids in guiding plant selection (Nair, 1998). As the use of agroforestry has broadened to address such issues as climate change and crop growth, carbon sequestration, biodiversity and even green infrastructure, so has the need to simulate agroforestry's longer-term effects across larger scales, further necessitating use of computer-based decision support tools (DSTs). These early DSTs used in agroforestry were generally those already in place in the fields of agriculture and forestry. For instance, the effect of shelterbelts on maize productivity under hypothesized climate change scenarios was examined using the Erosion-Productivity Impact Calculator (EPIC) crop model, an agricultural model originally developed to determine the relationship between soil erosion and crop productivity (Jones *et al.*, 1991; Easterling *et al.*, 1997). Even today, many of these models developed for agriculture or forestry are still a first choice for use in agroforestry exercises. CO2FIX, a user-friendly model for dynamically estimating the carbon sequestration potential of forest management and afforestation projects, is readily adaptable for agroforestry (Masera *et al.*, 2003). Today we have several DSTs developed exclusively for agroforestry applications for the purposes of selecting suitable species, identifying suitable lands, modelling different systems and predicting outcomes of different scenarios.

Given that GIS (geographical information systems) technology is widely available and affordable today and the fact that agroforestry is directly dependent upon spatial characteristics, it is logical to expect to have several agroforestry-specific GIS DSTs; but the reality is that only a few are available. An early GIS application compiled information on 173 species, including their descriptions, soil and climate preferences and management characteristics for Africa (Booth *et al.*, 1989). This application allowed users to query the database and generate maps showing the climatic suitability for different species. At a regional scale, Booth *et al.* (1990) created a similar application for Zimbabwe, demonstrating how GIS applications can be done at many scales. Unruh and Lefebvre (1995) performed a similar GIS application for sub-Saharan Africa to determine areas suitable for different agroforestry systems. Integrating ICRAF's (International Centre for Research in Agroforestry = World Agroforestry Centre, Nairobi, Kenya) agroforestry database with spatial data on geographical regions, climate and land uses in the region, their application was able to map out potential regions for 21 specific types of agroforestry systems.

Most of the past agroforestry GIS applications mentioned above have been research-oriented. The Southeastern Agroforestry Decision Support System (SEADSS), developed recently by the Center for Subtropical Agroforestry (CSTAF) at the University of Florida, brings on-line GIS capabilities directly to extension agents and landowners; it offers county soils, land use and other spatial data for selecting suitable tree and shrub species in a specified location (Ellis *et al.*, 2003). GIS-guided assessments, derived from publicly available datasets, are currently being used to evaluate four key issues of the Western Corn Belt of the United States: biodiversity, soil protection, water quality and agroforestry products. By combining these assessments, information is generated for use in identifying opportunities and constraints on the landscape where multiple benefits from conservation buffers, especially agroforestry plantings, can be achieved (Bentrup *et al.*, 2000). GIS-guided agroforestry suitability analysis will only improve as spatial data and computer resources become more accessible. Many states and countries are already assembling internet-accessible GIS data clearing houses to facilitate the use of spatial information.

Remote sensing and GIS technologies have made it possible to evaluate the ecological and socio-economic characteristics at multiple scales over broad geographical regions. Satellite imagery, digital orthophotos and secondary GIS data available from Florida's water management districts and other sources are now being used to characterize the landscape composition and structure of our study regions. Composition refers to the different natural communities and human created environments present, whereas structure is the pattern of landscape components, known as matrix, patches and corridors (Forman, 1997). Landscape metrics are used to quantify landscape characteristics such as degree of fragmentation and habitat connectivity (Burel and Baudry, 2003).

Environmental, economic and aesthetic improvements in rural landscapes may be achieved by integrating additional components such as hedgerows, shelterbelts, riparian buffers and silvopastoral systems. These tree-based agrotechnologies help increase landscape diversity, reducing fragmentation and improving connectivity between patches. In addition, they serve as biogeochemical barriers that can reduce soil erosion and improve water quality (Ryszkowski and Jankowiak, 2002). By integrating GIS information with silvopastoral data obtained from our research sites, we explore different land-use scenarios in different study areas that integrate different types of silvopastoral systems and evaluate their potential in increasing landscape diversity, creating environmental benefits and improving socio-economic conditions, including aesthetic characteristics, at landscape scales.

The usefulness of such studies is in bringing together both landscape and site scale for the purpose of planning and extension. A decision support system could be developed synthesizing the information obtained into a user-friendly package designed specifically for landowners and extension agents. Users will be able to spatially and visually evaluate silvopastoral options at various geographical scales as well as assess their potential ecological and economic benefits.

Information dissemination
As in most agroforestry innovations, silvopasture adoption has lagged behind the technical advances in the south-eastern United States. On the one hand, the biophysical knowledge is not fully synthesized and validated so that they may not be ready yet for large-scale adoption. Socio-economic and agroforestry needs assessment studies (Zinkhan and Mercer, 1997; Workman *et al.*, 2003) have shown, on the other hand, that few landowners and extension professionals in the south-east have heard of successful silvopasture implementation or are knowledgeable of silvopasture possibilities. Although technical information on species selection, implementation and management is available through Cooperative Extension Service offices and various university internet services, many landowners and education specialists do not feel comfortable making decisions on whether to adopt the practice. Landowners and educators have stated in these studies that they prefer learning via hands-on methods, including visits to demonstration areas and local farm and university research sites. Increased adoption of silvopasture (or other agroforestry) practices will require an increased number of demonstration sites through which potential 'adoptees' can gain the confidence and information necessary to make their decisions.

Along with the demonstration sites, another effective, and requested, mechanism for disseminating the technologies is access to a wide variety of publications, including both technical and more landowner-friendly publications, multimedia tools such as PowerPoint presentations, images and video. CSTAF's ongoing activities include working directly with landowners, county extension agents and foresters to identify additional landowners who would be willing to convert current pastures or timber stands to silvopastoral configurations. CSTAF will provide technical assistance in site design, as well as direct assistance for the operations necessary to establish silvopasture systems. The demonstration sites will represent a range of silvopasture designs that could be established in different land management situations in the south-east, such as: (1) typical wide alley, double tree rows planted on an existing pasture, (2) wide alley, double tree rows created through the thinning of established pine plantations, (3) random clumps of trees with open pasture surrounding these clusters, (4) even, widely spaced trees planted on an existing pasture, (5) multi-species overstorey combinations, and (6) the introduction of livestock and forage under existing nut or fruit tree plantings.

The Southeastern Agroforestry Network of Demonstration Sites (SANDS) will function to provide an on-line database of progress among the demonstration sites. This web-based participatory forum will fill landowner needs to learn and visualize agroforestry on the ground, and will empower landowners, educators, research specialists and product buyers to explore and learn from each other. Making information regarding demonstration sites available will allow stakeholders to access diverse information sources and share their ideas and experiences. CSTAF already manages a website that contains educational materials, decision-making tools and research information. The SANDS module will enhance website usefulness for both landowners and educators by providing interactive maps where site visitors can search a database of the participating landowners. It will be possible to conduct searches by: county, agroforestry practice or land management objective. The network will also include an interactive listserv, through which CSTAF personnel will facilitate discussion sessions and promote enquiry and exploration by participants. The addition of the SANDS network to the CSTAF website will allow learners to access the practical experience of different agroforestry practices, all electronically.

The website will also include an interactive program that will allow landowners to enter specific information about their land and simulate the growth and development of different silvopasture systems on their property. This virtual simulation will provide information on projected yields, economics and other benefits based on current information. It will be fine-tuned with additional research results from projects included in this grant.

To address diffusion of these technologies among the region's professional staff, an educator curriculum guide has been developed that complements intensive, two-day in-service training sessions. This educator curriculum guide includes images, sample PowerPoint presentations, links to useful websites, as well as a thorough compilation of publications, agroforestry site description forms, educational materials including fact sheets and brochures, lesson plans and instructions for providing field tours. The curriculum guide will be made available to all county extension offices in Florida, Georgia and Alabama.

Prospects
Rapidly urbanizing agroecosystems and the people inhabiting them in southern United States are challenged as never before with natural resource management problems that require an integrated approach. Beginning in the mid-20th century, agricultural development has caused significant changes in the region's land use, land cover and socio-economic settings. Much of the upland forests, particularly longleaf pine (*Pinus palustris* Englem.) ecosystems in central and north Florida, for example, have been lost due to agricultural and residential development (Solecki, 2001; Alavalapati *et al.*, 2002; Sprott and Mazzotti, 2001; Brockway and Lewis, 2003); the average forested area per person in the state has declined from 4 acres (1.75 h) to 1 acre (0.4 h) during the period from 1960 to 2000.

The pressure on ecosystems and the services they provide which until now have been taken for granted is enormous. In south Florida, where the majority of cattle ranches and cropland are located, environmental threats to water quality due to nutrient loading and sediment toxicity are a major concern and are having an impact on the natural systems and interfering with their restoration efforts. The land-use and land-cover changes associated with the removal and fragmentation of natural vegetation for establishment of cattle ranches and real-estate development are responsible, to a large extent, for the decline in biodiversity, invasion of exotic species and alterations to nutrient, energy, and water flows, which often result in soil erosion, deterioration of water quality and environmental pollution (Jones et al., 2001; Solecki, 2001; Hawkins and Selman, 2002).

Few, if any, studies have specifically addressed the impacts of these agroecosystem changes at landscape scales, let alone suggested strategies to address the problem. We believe that silvopasture could be a solution to the problem. Although the nature of efforts described above and the results obtained will, to some extent, be site-specific, we believe that these efforts will provide a broad-based foundation and stimulate interest in undertaking silvopastoral research in other regions of similar socio-economic conditions to address local and regional problems, as, for example, is already happening in Australia (www.rirdc.gov.au). Silvopasture of today is not just a new name for an old laissez-faire activity; it is a promising approach to land care and environmental protection.

Acknowledgements

This research is supported in part by a grant from the US Department of Agriculture/CSREES Initiative for Future Agriculture and Food Systems (IFAFS).

References

Alavalapati, J.R.R., Stainback, G.A. and Carter, D.R. (2002) Restoration of the longleaf pine ecosystem on private lands in the US South: an ecological economic analysis. *Ecological Economics* 40, 411-419.

Alavalapati, J.R.R., Shrestha, R.K. and Stainback, A. (2004) Agroforestry development: an environmental economic perspective. In: Nair, P.K.R., Rao, M.R. and Buck, L.E. (eds) *New Vistas in Agroforestry: A Compendium for the 1st World Congress of Agroforestry 2004*. Kluwer, Dordrecht, The Netherlands, pp. 299-311.

Andrews, W.J. (1994). *Nitrate in Groundwater and Spring Water near Four Dairy Farms in North Florida, 1990-93*: USA Geological survey water-resources investigations report,. 94-4162. Florida, USA. 63 p.

Bendfeldt, E.S., Feldhake, C.M. and Burger, J.A. (2001) Establishing trees in an Appalachian silvopasture: reponse to shelters, grass control, mulch and fertilization. *Agroforestry Systems* 53, 291-295.

Bentrup, G., Dosskey, M., Schoeneberger, M., Wells, M., Leininger, T. and Klenke, K. (2000) Planning for multi-purpose riparian management. In: Beschta, R.L. and Wigington, P.J. (eds) *Riparian Ecology and Management in Multi-Land Use Watersheds*. American Water Resources Association, Middleburg, USA, pp. 423-426.

Biles, L.E., Byrd, N. and Brown J. (1984) Agroforestry experiences in the south. In: Linnartz, N.E. and Johnson, M.K. (eds) *Agroforestry in the Southern United States, Proceedings of the 33rd Annual Forestry Symposium*. Louisiana State University, Baton Rouge, USA, pp. 152-158.

Biswell, H.H. and Foster, J.E. (1942) *Forest Grazing and Beef Cattle Production in the Coastal Plain of North Carolina: Results of a Survey of 100 Cattle Producing Farms*. Experiment Station Bulletin No. 334, North Carolina State College of Agriculture and Engineering, Raleigh, USA.

Booth, T.H., Stein, J.A., Nix, H.A. and Hutchinson, M.F. (1989) Mapping regions climatically suitable for particular species: an example using Africa. *Forest Ecology and Management* 28, 19-31.

Booth, T.H., Stein, J.A., Hutchinson, M.F. and Nix, H.A. (1990) Identifying areas within a country climatically suitable for particular tree species: an example using Zimbabwe. *International Tree Crops Journal* 6, 1-16.

Brockway, D.G. and Lewis, C.E. (2003) Influence of deer, cattle grazing and timber harvest on plant species diversity in longleaf pine bluestem ecosystem. *Forest Ecology and Management* 175, 49-69.

Brockway D.G., Wolters G.L., Pearson H.A., Thill R.E. Baldwin V.C. and Martin, A. (1998) Understorey plant response to site preparation and fertilization of loblolly and shortleaf pine forests. *Journal of Range Management* 51, 47-54.

Brooks, D.J. (1993) *US forests in a global context*. USDA Forest Service General Technical Report RM-228, US Forestry Service Rocky Mountain Forest and Range Experimental Station Fort Collins, CO, USA.

Burel, F. and Baudry, J. (2003) *Landscape Ecology: Concepts, Methods and Applications*. Science Publishers Incorporated, Plymouth, United Kingdom.

Burner, D.M. (2003) Effect of alley crop environment on orchardgrass and tall fescue herbage. *Agronomy Journal* 95, 1163-1171.

Burner, D.M. and Brauer, D.K. (2003) Herbage response to spacing of loblolly pine trees in a minimal management silvopasture in southeastern USA. *Agroforestry Systems* 57, 69-77.

Cameron, T.A., Poe, G.L., Ethier, R.G. and Schulze, W.D. (2002) Alternative non-market value-elicitation: Are the underlying preference the same? *Journal Environmental Economy and Management* 44, 391-422.

Clason, T.R. (1995) Economic implications of silvipastures on southern pine plantations. *Agroforestry Systems* 29, 227-238.

Clason, T.R. and Sharrow, S.H. (2000). Silvopastoral practices. In: Garrett, H.E., Rietveld W.J. and Fisher, R.F. (eds) *North American Agroforestry: An Integrated Science and Practice*. American Society of Agronomy, Madison, USA, pp. 119-147.

Cooper, J.C. and Keim, R.W. (1996) Incentive payments to encourage farmer adoption of water quality protection practices. *American Journal of Agricultural Economics* 78, 54-64.

Easterling, W.F., Hays, C.J., Easterling, M.M. and Brandle, J.R. (1997) Modeling the effect of shelterbelts on maize productivity under climate change: an application of the EPIC model. *Agriculture, Ecosystems and Environment* 61, 163–176.

Edmeades, D.C. (2003) The long-term effects of manures and fertilizers on soil productivity and quality: a review. *Nutrient Cycling and Agroecosystems* 66, 165-180.

Ellis, E.A., Nair, P.K.R. and Jeswani, S.D. (2003) The southeastern agroforestry decision support system (SEADSS): a web-based application for agroforestry planning and tree selection. In: Vacik, H., Lexer, M.J., Rauscher, M.H., Reynolds, K.M. and Brooks, R.T. (eds), *Decision Support for Multiple Purpose Forestry: A Transdiciplinary Conference on the Development and*

Application of Decision Support Tools for Forest Management. University of Natural Resources and Applied Life Sciences, Vienna, Austria, CD-ROM, pp. 1-12.

Ellis, E.A., Bentrup, G. and Schoeneberger, M.M. (2004) Computer-based tools for decision support in agroforestry: current state and future needs. In: Nair, P.K.R., Rao, M.R. and Buck, L.E. (eds) *New Vistas in Agroforestry: A Compendium for the 1st World Congress of Agroforestry 2004*. Kluwer, Dordrecht, The Netherlands, pp. 401-422.

FAS (2002) Beef cattle and calf inventory by county: Livestock, dairy, and poultry summary, Florida Agricultural Statistics Service. The USDA National Agricultural Statistics Service. http://www.nass.usda.gov/fl/.

Feather, P., Hellerstein, D. and Hansen, L. (1999) *Economic Valuation of Environmental Benefits and the Targeting of Conservation Programs: The Case of the CRP*. Economic Research Service, Agricultural Economic Report No. 778, Washington, DC, USA.

Follett, R.F., Kimble, J.M. and Lal, R. (2001) (eds) *The Potential of U.S. Grazing Lands to Sequester Carbon and Mitigate the Greenhouse Effect*. Lewis, Boca Raton, FL, USA.

Fonyo, C. and Flaig, E. (1995) Phosphorus budgets for Lake Okeechobee tributary basins. *Ecology Engineering* 5, 209-227.

Forman, R.T. (1997) *Land Mosaic: The Ecology of Landscapes and Regions*. Cambridge University Press, Cambridge, United Kingdom.

Garrett, H.E. and Buck, L. (1997) Agroforestry practice and policy in the United States of America. *Forest Ecology and Management* 91, 5-15.

Garrett, H.E., Buck, L.E., Gold, M.A., Hardesty, L.H., Kurtz, W.B., Lassoie, J.P., Pearson, H.A. and Slusher, J.P. (1994) *Agroforestry: An Integrated Land-Use Management System for Production and Farmland Conservation*. USDA SCS, Washington, DC, USA.

Garrett, H.E., Kerley, M.S., Ladyman, K.P., Walter, W.D., Godsey, L.D., Van Sambeek, J.W. and Brauer, D.K. (2004) Hardwood silvopasture management in North America. In: Nair, P.K.R., Rao, M.R. and Buck, L.E. (eds) N*ew Vistas in Agroforestry: A Compendium for the 1st World Congress of Agroforestry 2004*. Kluwer, Dordrecht, The Netherlands, pp. 21-34.

Grado, S.C., Hovermale, C.H., and St. Louis, D.G. (2001) A financial analysis of a silvopasture system in southern Mississippi. *Agroforestry Systems* 53, 313-322.

Halls, L.K., Burton, G.W. and Southwell, B.L. (1957) *Some Results of Seeding and Fertilization to Improve Southern Forest Ranges*. USDA Forest Service, Southeastern Experiment Station Paper 78, North Caroline, USA.

Harvey, R. and Havens, K. (1999) Lake Okeechobee Action Plan, South Florida Ecosystem Restoration Working Group. http://www.sfwmd.gov/org/wrp/wrp_okee/2_wrp_okee_info/2_actionplan.html.

Hatzell, H.H. (1995) *Effects of Waste-disposal Practices on Ground-water Quality at Five Poultry (Broiler) Farms in North-central Florida, 1992-1993*. USA Geological Survey Water-Resources Investigations Report 95-4064, Florida, USA.

Hawkins, V. and Selman, P. (2002) Landscape scale planning: exploring alternative land use scenarios *Landscape and Urban Planning* 60, 211-224.

Heimlich, R.E., Wiebe, K.D., Claassen, R., Gadsby, D. and House, R.M. (1998) *Wetlands and Agriculture: Private Interests and Public Benefits*. United States Department of Agriculture, Economic Research Service, Agricultural Economic Report No. AER765, Washington, DC, USA.

Jones, B.K., Neale, A.C., Wade, T.G., Wickham, J.D., Cross, C.L., Edmonds, C.M., Loveland, T.R., Nash, M.S., Riitters, K.H. and Smith, E.R. (2001) The consequences of landscape change on ecological resources: an assessment of the United States Mid-Atlantic region, 1973-1993. *Ecosystem Health* 7, 229-242.

Jones, C.A., Dyke, P.T., Williams, J.R., Kiniry, J.R., Benson, V.W. and Griggs, R.H. (1991) EPIC: an operational model for evaluation of agricultural sustainability. *Agricultural Systems* 37, 341–350.

Jose, S., Gillespie, A.R. and Pallardy, S.G. (2004) Interspecific interactions in temperate agroforestry. In: Nair, P.K.R., Rao, M.R. and Buck, L.E. (eds) *New Vistas in Agroforestry*. Kluwer Academic Publishers, Dordrecht, The Netherlands, pp. 237-256.

Katz, B.G., Hornsby, H.D., Bohlke, J.K. and Mokray, M.F. (1999) *Sources and Chronology of Nitrate Contamination in Springs, Suwannee River Basin, Florida*. USA Geological Survey Water-Resources Investigations Report 99-4252, Florida, USA. 54 p.

Kingsbury, L. and Boggess, W. (1999) An economic analysis of riparian landowners' willingness to participate in Oregon's Conservation Reserve Enhancement Program. Selected paper for the annual meeting of the Agricultural Economics Association, 8-11 August, Nashville, Tennessee, USA.

Lant, C.L. (1991) Potential of the Conservation Reserve Program to control agricultural surface water pollution. *Environment Management* 15, 507-518.

Lehmkuhler, J.W., Kerley, M.S., Garrett, H.E., Cutter, B.E. and McGraw, R.L. (1999) Comparison of continuous and rotational silvopastoral systems for established walnut plantations in southwest Missouri, USA. *Agroforestry Systems* 44, 267-279.

Lewis, C.E. and Pearson, H.A. (1987) Agroforestry using tame pastures under planted pines in the southeastern United States. In: Gholz, H. (ed.) *Agroforestry: Realities, Possibilities, and Potentials*. Martinus Nijhoff Publishers, Dordrecht, The Netherlands, pp. 195-212.

Lewis, C.E., Burton, G.W., Monson, W.G. and McCormick, W.C. (1983) Integration of pines, pastures, and cattle in south Georgia, USA *Agroforestry Systems* 1, 277-297.

Lewis, C.E., Tanner, G.W. and Terry, W.S. (1985) Double *vs* single-row pine plantations for wood and forage production. *Southern Journal of Applied Forestry* 9, 55-60.

Lin, C.II., McGraw, R.L., George, M.F. and Garrett, H.E. (1999) Shade effects on forage crops with potential in temperate agroforestry practices. *Agroforestry Systems* 44, 109–119.

Lohr, L. and Park, T. (1994) Discrete/continuous choices in contingent valuation surveys: conservation decisions in Michigan *Review of Agricultural Economics* 16, 1-15.

Loomis, J.B., Kent, P., Strange, L., Fausch, K. and Covich, A. (2000) Measuring the total economic value of restoring ecosystem services in an impaired river basin: results from a contingent valuation survey. *Ecology Economics* 33, 103-117.

Lundgren, G.K., Conner, J.R. and Pearson, H.A. (1983) An economic analysis of forest grazing on four timber management situation. *Southern Journal Applied Forestry* 7, 119-124.

McGraw, R.L, Navarrete-Tindall, N.E. and Van Sambeek, J.W. (2001) Effects of shade on growth and nodulation of six native legumes with potential for use in agroforestry. In: *Program Abstracts, Seventh Biennial Conference on Agroforestry in North America*. Regina, Canada, p. 50.

Masera, O.R., Garza-Caligaris, J.F., Kanninen, M., Karjalainen, T., Liski, J., Nabuurs, G.J., Pussinen, A., de Jong, B.H.J. and Mohren, G.M.J. (2003) Modeling carbon sequestration in afforestation, agroforestry, and forest management projects: the CO2FIX V.2 approach. *Ecological Modelling* 164, 177–199.

Mitlohner, F.M., Morrow, J.L., Dailey, J.W., Wilson, S.C., Galyean, M.L., Miller, M.F. and McGlone, J.J. (2001) Shade and water misting effects on behavior, physiology, performance, and carcass traits of heat-stressed feedlot cattle. *Journal of Animal Science* 79, 2327–2335.

Montagnini, F. and Nair, P.K.R. (2004) Carbon sequestration: an under-exploited environmental benefit of agroforestry systems. *Agroforestry Systems* 61, 281-299.

Mueller, D.K. and Helsel, D.R. (1996) *Nutrients in the Nation's Waters-Too Much of a Good Thing?* USA Geological Survey Circular 1136, USA.

Nair, P.K.R. (1993) *An Introduction to Agroforestry*. Kluwer Academic Publishers, Dordrecht, The Netherlands and ICRAF, Nairobi, Kenya.

Nair, P.K.R. (1998) Directions in tropical agroforestry research: past, present and future. *Agroforestry Systems* 38, 223–245.

Nair, P.K.R., Buresh, R.J., Mugendi, D.N. and Latt, C.R. (1999) Nutrient cycling in tropical agroforestry systems: myths and science. In: Buck, L.E., Lassoie, J.P. and Fernandes, E.C.M. (eds) *Agroforestry in Sustainable Agricultural Systems*. CRC Press, Boca Raton, USA, pp. 1–31.

Nair, V.D. and Graetz, D.A. (2002) Phosphorus saturation in Spodosols impacted by manure. *Journal of Environmental Quality* 31, 1279-1285.

Nair, P.K.R. and Nair, V.D. (2003) Carbon storage in North American agroforestry systems. In: Kimble, J., Heath, L.S., Birdsey, R.A. and Lal, R. (eds) *The Potential of USA Forest Soils to Sequester Carbon and Mitigate the Greenhouse Effect*. CRC Press, Boca Raton, USA, pp. 333-346.

Nair, V.D. and Graetz, D.A. (2004) Agroforestry as an approach to minimizing nutrient loss from heavily fertilized soils: the Florida experience. In: Nair, P.K.R, Rao, M.R. and Buck, L.E. (eds) *New Vistas in Agroforestry: A Compendium for the 1st World Congress of Agroforestry*. Kluwer, Dordrecht, The Netherlands, pp. 269-280.

Palm, C.A. (1995) Contribution of agroforestry trees to nutrient requirements of intercropped plants. *Agroforestry System* 30, 105-124.

Pearson, H.A. (1975) *Exotic Grass Yields Under Southern Pines*. USDA Forest Service Research Note SO-201, Southern Forest Experiment Station, New Orleans, USA.

Pearson, H.A. and Rollins, D.A. (1987) Ryegrass pasture for supplementing southern pine native range. *Rangelands* 9, 19-20.

Rao, M.R., Nair, P.K.R. and Ong, C.K. (1998) Biophysical interactions in tropical agroforestry systems. *Agroforestry Systems* 38, 3-50.

Ribaudo, M.O., Horan, R.D. and Smith, M.E. (1999) *Economics of Water Quality Protection from Non Point Sources*. USDA/Economic Research Service, Washington, DC, USA.

Ryszkowski, L. and Jankowiak, J. (2002). Development of agriculture and its impact on landscape function. The functional approach to agricultural landscape analysis. In: Ryszkowski, L. (ed.) *Landscape Ecology in Agroecosystems Management*. CRC Press, Boca Raton, USA, pp. 9-27.

Schultz, R.C., Colletti, J.P., Isenhart, T.M., Marquez, W.W. and Ball, C.J. (2000) Riparian forest buffer practices. In: Garrett, H.E., Rietveld, W.J. and Fisher, R.F. (eds) *North American Agroforestry: An Integrated Science and Practice*. American Society of Agronomy, Madison, Wisconsin, USA, pp. 189-281.

Schultz, R.C., Isenhart, T.M., Simpkins, W.W. and Colletti, J.P. (2004) Riparian forest buffers in agroecosystems–lessons learned from the Bear Creek watershed, Central Iowa, USA. In: Nair, P.K.R., Rao, M.R. and Buck, L.E. (eds) *New Vistas in Agroforestry: A Compendium for the 1st World Congress of Agroforestry 2004*. Kluwer, Dordrecht, The Netherlands, pp. 35-50.

Sharrow, S.H. and Ismail, S. (2004) Carbon and nitrogen storage in agroforests, tree plantations, and pastures in western Oregon, USA. *Agroforestry Systems* 60, 123-130.

Shrestha, R.K. and Alavalapati, J.R.R. (2004) Valuing environmental benefits of silvopasture practice: a case study of the Lake Okeechobee watershed in Florida. *Ecological Economics* (in press).

Sims, J.T., Simard, R.R. and Joern, B.C. (1998) Phosphorus loss in agricultural drainage: historical perspective and current research. *Journal of Environmental Quality* 27, 227-293.

Solecki, W.D. (2001) The role of global-to-local linkages in land use/land cover change in South Florida. *Ecological Economics* 37, 339-356.

Sprott, P. and Mazzotti, F.J. (2002) *Habitat Loss, Florida's Changing Landscapes: Upland Forests*. WEC 151. Department of Wildlife Ecology and Conservation, Cooperative Extension Service, Institute of Food and Agricultural Sciences, University of Florida, Gainesville, USA.

Stainback, G.A., and Alavalapati, J.R.R. (2004) Modeling catastrophic risk in economic analysis of forest carbon sequestration. *Natural Resource Modeling* (in press).

Stainback, G.A., Alavalapati, J.R.R., Shrestha, R.K., Larkin, S. and Wong, G. (2004) Effect of pollution taxes and carbon payments on the adoption of silvopasture. *Journal of Agricultural and Applied Economics* (in press).

Unruh, J.D. and Lefebvre, P.A. (1995) A spatial database approach for estimating areas suitable for agroforestry in sub-Saharan Africa: aggregation and use of agroforestry case studies. *Agroforestry Systems* 32, 81–96.

Varian, H.R. (1992) *Microeconomics Analysis*, 3rd edn. W.W. Norton and Company, New York, USA.

Wanvestraut, R., Jose, S., Nair, P.K.R. and Brecke, B.J. (2004). Competition for water in a pecan–cotton alley cropping system. *Agroforestry Systems* 60, 167-179.

Watson, V.H., Pearson, H.A., Knight, W.E. and Hagedorn, C. (1984) Cool season forages for use in pine forests. In: Linnartz, N.E. and Johnson, M.K. (eds) *Agroforestry in the Southern United States. Proceedings of the 33rd Annual Forestry Symposium.* Louisiana State University, Baton Rouge, USA, pp. 79-88.

Westcott, P., Young, C.E. and Price, J.M. (2002) *The 2002 Farm Act: Provision and Implications for Commodity Markets.* USDA Economic Research Service, Agricultural Information Bulletin No. AIB778, Washington, DC. USA.

Williams, P.A., Gordon, A.M., Garrett, H.E. and Buck, L.E. (1997) Agroforestry in North America and its role in farming systems. In: Gordon, A.M. and Newman, S.M. (eds) *Temperate Agroforestry Systems.* CABI, New York, USA, pp. 9-84.

Workman, S.W. and Allen, S.C. (2003) The practice and potential of agroforestry in the Southeastern United States. Center for Subtropical Agroforestry, School of Forest Resources and Conservation, University of Florida, http://cstaf.ifas.ufl.edu/whitepaper.htm.

Workman, S.W., Bannister, M.E. and Nair, P.K.R. (2003) Agroforestry potential in the southeastern United States: perceptions of landowners and extension professionals. *Agroforestry Systems* 59, 73-83.

Zinkhan, F.C. and Mercer, D.E. (1997). An assessment of agroforestry systems in the southern USA. *Agroforestry Systems* 35, 303-321.

Nutritive value of trees and shrubs for ruminants

J. C. Ku Vera
Faculty of Veterinary Medicine and Animal Science, University of Yucatan, Mexico. gschan@tunku.uady.mx

Abstract
There are many species of trees and shrubs present in tropical Latin America which have the potential to be incorporated in ruminant production systems. Chemical composition varies between species, but most trees and shrubs contain good contents of crude protein and low concentrations of neutral detergent fibre. Total dry matter intake of low-quality rations is generally increased as the level of incorporation of foliage of trees and shrubs in the ration is raised. After consumption of foliage of trees and shrubs there is an increase in NH_3-N concentration in the rumen as a result of the degradation of crude protein in this organ. The rise in ammonia concentration results in an increase in the synthesis of microbial protein; consequently, the supply of microbial protein to the small intestine of ruminants is increased as the level of incorporation of foliage in the ration is increased. Rate of passage of liquids and solids throughout the reticulo-rumen is increased as a result of the improvement in voluntary intake of dry matter resulting from the incorporation in the ration of foliage of trees and shrubs. Greater integration of trees into ruminant production systems in Latin American countries is gradually leading to the development of sustainable systems of production, less dependent on imported feedstuffs and probably with less deleterious impact on the environment.

Key words: sustainability

Introduction
The rapid growth of the human population, the continuous process of urbanization throughout the developing world and the rise in per capita income are components that together have led to the prediction of an increase in the demand for livestock products during the next 20 years; this demand will arise fundamentally from developing countries. The rise in the number of extensive cattle production farms in Latin American countries since the beginning of last century occurred along with the destruction of huge area of forests, to create open terrain for the monoculture of introduced grasses. This led to the reduction of biodiversity and to the degradation of soils in many countries. There are several alternatives to reduce the destruction of forests in the tropics which result from the expansionist drive of extensive ruminant production; one of them could be the implementation of practices, such as silvopastoralism, which induce greater integration of trees and shrubs with animal production and which result in more sustainable use of natural resources (Toledo, 2004). There is a large diversity of species of trees and shrubs in tropical Latin America which have potential for incorporation in ruminant production systems (Soto Pinto *et al.*, 1997). Trees and shrubs serve several purposes in production systems in addition to that of forage, such as: live fence (e.g. *Gliricidia sepium*, *Jatropha curcas*), shade (e.g. *Enterolobium cyclocarpum*, *Brosimum alicastrum*, etc.), wood, carbon sequestration (Pomareda, 2004) or they are used simply for ornamental purposes. This chapter describes some aspects of the nutritive value of multipurpose trees and shrubs used in ruminant production systems in the tropics.

Chemical composition
The chemical composition of trees and shrubs present in South Mexico is shown in Table 1.

Table 1. Chemical composition (%) of some trees and shrubs available in South Mexico. OM: organic matter; CP: crude protein; NDF: neutral detergent fibre; ADF: acid detergent fibre.[1] Harvested in the State of Yucatan.[2] Harvested in the State of Chiapas.

	OM	CP	NDF	ADF
B. alicastrum[1]	90.4	15.7	37.5	28.5
B. alicastrum[1]	77.0	14.8	40.4	28.9
E. tinifolia[1]	79.6	15.7	65.7	45.8
G. sepium	91.5	19.3	35.7	21.8
L. leucocephala[2]	91.9	18.6	34.6	18.2
E. mexicana[2]	92.2	12.4	50.6	32.4
C. houstoniana[2]	90.2	12.9	48.4	35.6
C. spectabilis[2]	94.5	15.2	41.4	20.7
G. ulmifolia[2]	93.4	9.5	47.0	31.8
G. sepium[2]	95.0	13.5	41.1	20.0

In general, trees and shrubs have a good content of crude protein (CP); thus foliages can be harvested (cut-and-carry systems) or grazed during the dry season as a supplement for the feeding of ruminant livestock (Shelton, 2004). Table 1 shows that CP concentration in the foliage of *B. alicastrum* is about 15% with a low concentration of NDF.

Concentration of CP in leaves of *L. leucocephala* harvested in Chiapas is relatively low (18.6%) compared to the concentration found in species harvested in the Yucatán Peninsula of Mexico (range: 26-30%). The low concentration of CP in *Gliricidia sepium* harvested in Chiapas (13.5%) is also noticeable when compared to that harvested in the State of Yucatán (19.3%), a fact which suggests variation in chemical composition due possibly to type of soil, age of tree, rainfall, etc. Most of the tree and shrub species of South Mexico are relatively low in NDF, with the exception of *Ehretia tinifolia*.

Chemical composition of fruits of tropical trees

Esquivel Mimenza (2004) studied the chemical composition of fallen fruits from four tree species available on cattle farms in Costa Rica (Table 2). Fruits of leguminous trees (*Samanea saman* and *Enterolobium cyclocarpum*) had significantly higher CP contents than those of fruits from non-leguminous trees (*Guazuma ulmifolia* and *Acrocomia vitifolia*). NDF content was generally low for all fruits. These results demonstrate the potential of tree fruits as a source of nutrients for ruminants in the tropical areas of Central America.

Table 2. Chemical composition of fruits of four tree species available in Costa Rica. CP: crude protein; NDF: neutral detergent fibre.

	Acrocomia vitifolia	Guazuma ulmifolia	Enterolobium cyclocarpum	Samanea saman
CP (%)	5.6 a	7.5 b	13.1 c	15.6 d
NDF (%)	42.4 a	36.3 ab	29.3 bc	24.7 c

Palatability of foliage

Cruz Macías (1999) reported the palatability of various cultivars of *Leucaena leucocephala* harvested in Yucatan, Mexico, by Pelibuey sheep, showing that *L. esculenta paniculata* had the highest acceptability (1.93 g DM/min), whereas *L. macrophyllanelsonii* had the lowest (0.13 g DM/min).

Voluntary intake

Voluntary intake is one of the best indicators of nutritive value of a feed for ruminants. Incorporation of graded levels of foliages such as those of *G. sepium* and *B. alicastrum* in the ration increased DM and OM intakes by Pelibuey sheep (Table 3). The increase in intake of DM was not due to an improvement in the rate or extent of degradation in the rumen of the basal ration of tropical grass *Cynodon nlemfuensis* or *Panicum maximum*, as a result of a better rumen environment (NH_3-N), but rather to the high rumen fermentation of the OM of the tree species *per se*, which provoked a faster emptying of the rumen. Aregheore and Perera (2004) showed that goats fed maize stover and supplemented with foliage of trees in Samoa (South Pacific) had a greater intake of DM and a higher digestibility in the gastrointestinal tract.

Table 3. Voluntary intake of DM and OM (g/head per day) by Pelibuey sheep fed increasing levels of *G. sepium* or *B. alicastrum* (Valdivia Salgado and Ku Vera, 1996; Alayón et al., 1998). DM: dry matter; OM: organic matter.

Item	Treatment			
	Stargrass	Stargrass + 10% *Gliricidia*	Stargrass + 20% *Gliricidia*	Stargrass + 30% *Gliricidia*
DM intake	678.3	707.2	950.1	1039.6
OM intake	606.9	627.1	836.1	921.3
	Guinea grass	Guinea grass + 15% *Brosimum*	Guinea grass + 30% *Brosimum*	Guinea grass + 45% *Brosimum*
DM intake	511.0	848.0	1106.0	1313.0
OM intake	464.0	758.0	1032.0	1191.0

Kinetics of rumen fermentation

Some of the results obtained from the State of Yucatan, Mexico, regarding rumen fermentation of foliage of trees and shrubs by the *in situ* (nylon bag) technique are shown below.

Table 4 suggests that the extent of fermentation of CP of *B. alicastrum*, *G. sepium* and *G. ulmifolia* is high, resulting in an increase in the concentration of NH_3-N in the rumen, stimulating the synthesis of microbial protein and consequently the supply of microbial protein to the small intestine. These tree species represent excellent protein sources for ruminants during the dry season. It is clear that *B. alicastrum* supplies highly fermentable OM for the rumen, since the extent of fermentation of foliage of this species was 86.6%. The possibility of increasing voluntary intake and rumen fermentation of tropical grasses by incorporating foliage of trees and shrubs has considerable practical implications for ruminant feeding; however, in the State of Yucatan, Mexico, it was not possible to increase rumen fermentation of low-quality grass. None the less, some authors (Kennedy et al., 2002) have reported an increase in intake (52%) by ruminants of a low-quality basal ration (*Dichanthium aristatum* grass) by supplementation with fallen leaves of *Albizia lebbeck*.

Table 4. Rate and extent of rumen fermentation of organic matter (OM) and crude protein (CP) of forage trees from the State of Yucatan, Mexico. [1]Valdivia Salgado and Ku Vera (1996). [2]Ramírez-Cancino et al. (2000). [3] Alayón et al. (1998).

Species	OM Rate of fermentation (%/h)	OM Extent of fermentation (%)	CP Rate of fermentation (%/h)	CP Extent of fermentation (%)
B. alicastrum[1]	0.105	86.62	0.113	95.19
B. alicastrum[2]	0.049	90.70	0.057	93.30
G. ulmifolia[2]	0.050	69.40	0.045	83.60
E. tinifolia[2]	0.021	27.50	0.140	45.50
G. sepium[3]	0.096	87.77	0.120	94.50

Supply of microbial N to the small intestine

In the tropics, it is important to attempt to maximize the amount of microbial protein which is synthesized per unit (kg) of organic matter fermented in the rumen, in order to be able to supply to the host animal the required amount of protein for maintenance and production. Alayón et al. (1998) and Valdivia Salgado and Ku Vera (1996) supplemented tree foliage and found that, as the level of foliage of the trees was increased, G. sepium (from 10 to 30%) and B. alicastrum (from 15 to 45%), in the basal ration of low-quality grasses (stargrass or guinea grass), the supply of microbial N to the small intestine of sheep was increased (from 4.9 to 9.6 and from 2.2 to 9.7 g/day, respectively), a fact which suggests that the CP of the tree species was being utilized in the rumen for the synthesis of microbial protein.

Rate of passage of liquids and solids

When tree foliage is incorporated in the ration of ruminants, rate of passage of liquids and solids through the rumen is usually increased as the level of incorporation of foliage in the ration is augmented. The faster rates of passage of solids through the rumen allow ruminants to reach a higher intake of DM because of a faster emptying of the rumen. Other workers (Bonsi et al., 1994) have reported an increase in rate of passage in response to supplementation with foliage of Sesbania sesban and Leucaena leucocephala in sheep fed low-quality straw.

Apparent digestibility (*in vivo*) of DM and OM

Table 5 shows that incorporation of foliage of *Gliricidia sepium* and *Brosimum alicastrum* improved the apparent DM digestibility of the total ration offered to Pelibuey sheep. Yerena et al. (1978) reported an apparent DM digestibility for *Brosimum alicastrum* of 67.1%. Pérez Luna (2000) reported a higher daily weight gain (475 g/head) in young cattle supplemented with hay of *Gliricidia sepium* than in cattle grazing *Cynodon nlemfuensis* which was not supplemented (335 g/head) in Villaflores, Mexico. However, no differences in rumen fermentation of ingested DM of Zebu cattle grazing stargrass with or without supplementation with foliage of *Gliricidia sepium* were observed. Kennedy et al. (2002) reported a positive response to supplementation with fallen leaves of *Albizia lebbeck*, in a basal diet of low-quality grass in terms of an improvement in DM intake and digestibility.

Table 5. Apparent digestibility (%) of dry matter (DM) and organic matter (OM) by Pelibuey sheep supplemented with foliages of trees and shrubs in Yucatan, Mexico (Valdivia Salgado and Ku Vera, 1996; Alayón et al., 1998).

Apparent digestibility: (%)	Treatment			
	Stargrass	Stargrass + 10% *Gliricidia*	Stargrass + 20% *Gliricidia*	Stargrass + 30% *Gliricidia*
DM	43.7	45.6	47.9	50.7
OM	48.4	48.8	52.3	54.7
	Guinea grass	Guinea grass + 15% *Brosimum*	Guinea grass + 30% *Brosimum*	Guinea grass + 45% *Brosimum*
DM	35.7	45.1	46.9	49.6
OM	40.5	49.2	49.4	52.0

General comments

The mechanisms of action of foliages of trees and shrubs in ruminants can be described as follows. Tropical trees and shrubs have good chemical composition in terms of crude protein and neutral detergent fibre, supplying fermentable organic matter to the reticulo-rumen. Total voluntary DM and OM intakes in sheep and cattle are increased when foliages of trees and shrubs are introduced in the ration, due to the high rate and extent of rumen fermentation of the foliage *per se*. Due to the increase in DM intake, rate of passage is augmented, causing a greater rate of emptying of digesta, which results in a greater physical space in the rumen, which stimulates the animal to initiate the consumption of food. The increase in rumen kinetics improves the efficiency of microbial protein synthesis in the rumen, which in turn positively affects the supply of microbial N to the small intestine (Dijkstra et al., 2003). Incorporation of the foliage of trees and shrubs improves apparent digestibility of the total ration. The foreseeable future holds opportunities but also challenges for animal production in tropical Latin America. There is a danger that meat production and processing will become further dominated by vertically integrated large-scale commercial operations to the detriment of smallholders, exacerbating rural poverty, rural migration and child malnutrition. Furthermore, there could be major environmental consequences of the uncontrolled growth of intensive animal production operations. Environmental and other services offered by silvopastoral systems, such as carbon sequestration, agro-tourism and eco-tourism, represent alternatives that can be exploited to generate additional income for farmers in developing countries (Harvey *et al.,* 2004), stimulating economic growth.

References

Alayón, J.A., Ramírez-Avilés, L. and Ku Vera, J.C. (1998) Intake, rumen digestion, digestibility and microbial nitrogen supply in sheep fed *Cynodon nlemfuensis* supplemented with *Gliricidia sepium. Agroforestry Systems* 41, 115-126.

Aregheore, E.M. and Perera, D. (2004) Effects of *Erythrina variegata, Gliricidia sepium* and *Leucaena leucocephala* on dry matter intake and nutrient digestibility of maize stover, before and after spraying with molasses. *Animal Feed Science and Technology* 111, 191-201.

Bonsi, M.L.K., Osuji, P.O., Nsahlai, I.V. and Tuah, A.K. (1994) Graded levels of *Sesbania sesban* and *Leucaena leucocephala* as supplements to teff straw given to Ethiopian Menz sheep. *Animal Production* 59, 235-244.

Cruz Macías, W.O. (1999) Agronomic behaviour and nutritive quality of *Leucaena* species in Yucatán (Comportamiento agronómico y calidad nutritiva de especies de *Leucaena* en el Estado de Yucatán). MSc thesis. Universidad Autónoma de Yucatán. Mérida, Mexico. (In Spanish.)

Dijkstra, J., France, J., Tamminga, S. and Mills, J.A.N. (2003) Predicting the yield of nutrients from microbial metabolism in the rumen. In: Mannetje, L., Ramírez-Avilés, L., Sandoval-Castro, C. and Ku-Vera, J.C. (eds) *Proceedings of the Sixth International Symposium on the Nutrition of Herbivores.* Universidad Autónoma de Yucatán, Mérida, Mexico, pp. 101-127.

Esquivel Mimenza, H. (2004) The relationship between tree cover and cattle farm systems. MSc thesis. Centro Agronómico Tropical de Investigación y Enseñanza (CATIE). Turrialba, Costa Rica.

Harvey, C.A., Saenz, J., Montero, J., Medina, A., Sánchez, D., Vilchez, S., Hernández, B., Maes, J.M. and Sinclair, F.L. (2004) Abundance and species richness of trees, birds, bats, butterflies and dung beetles in a silvopastoral systems in the agricultural landscapes of Cañas, Costa Rica and Rivas, Nicaragua. In: Mannetje, L., Ramírez, L., Ibrahim, M., Sandoval, C., Ojeda, N. and Ku, J. (eds) *Proceedings of the Second International Symposium on Silvopastoral Systems.* Universidad Autónoma de Yucatán, Mérida, Mexico, pp. 73-76.

Kennedy, P.M., Lowry, J.B., Coates, D.B. and Oerlemans, J. (2002) Utilization of tropical dry season grass by ruminants is increased by feeding fallen leaf of siris (*Albizia lebbeck*). *Animal Feed Science and Technology* 96, 175-192.

Pérez Luna, E.J. (2000) Use of *Gliricidia sepium* for feeding bovine livestock in the tropics (Uso de la *Gliricidia sepium* en la alimentación de bovinos en el trópico). MSc thesis. Universidad Autónoma de Yucatán. Mérida, Mexico. (In Spanish.)

Pomareda, C. (2004) Policy issues to develop markets for environmental services. In: Mannetje, L., Ramírez, L., Ibrahim, M., Sandoval, C., Ojeda, N. and Ku, J. (eds) *Proceedings of the Second International Symposium on Silvopastoral Systems.* Universidad Autónoma de Yucatán, Mérida, Mexico, pp. 3-9.

Ramírez-Cancino, L., Ramírez-Avilés, L. and Ku Vera, J.C. (2000) Effect of incorporating multipurpose-trees (MPT's) foliage in a basal ration of poor-quality *Pennisetum purpureum* hay fed to Pelibuey sheep. *Journal of Applied Animal Research* 17, 239-251.

Shelton, H.M. (2004) Importance of tree resources for dry season feeding and the impact on productivity of livestock farms. In: Mannetje, L., Ramírez, L., Ibrahim, M., Sandoval, C., Ojeda, N. and Ku, J. (eds) *Proceedings of the Second International Symposium on Silvopastoral Systems.* Universidad Autónoma de Yucatán, Mérida, Mexico, pp. 158-174.

Soto Pinto, M.L., Jiménez Ferrer, G. and de Jong, B.H. (1997) Agroforestry in Chiapas. The case study of Los Altos region (La agroforestería en Chiapas. El caso de la región de Los Altos). In: Parra Vázquez, M. and Hernández, B.M. (eds) *Los Altos de Chiapas: Agricultura y Crisis Rural. Tomo 1. Los Recursos Naturales.* El colegio de la Fronte Sur, San Cristóbal de las Casas, Mexico, pp. 167-186. (In Spanish.)

Toledo, V. (2004) Rural ecology (La ecología rural). *Ciencia y Desarrollo* 30, 36-43. (In Spanish.)

Valdivia Salgado, V. and Ku Vera, J.C. (1996) Effect of *Brosimum alicastrum* in the rumen digestion and flow of microbial protein in ovine livestock (Pelibuey) fed with *Panicum maximum* pastures (Efecto del ramón (*Brosimum alicastrum*) sobre la digestión ruminal y el flujo de proteína microbiana en ovinos Pelibuey alimentados con pasto guinea (*Panicum maximum*)). In: *Reunión Nacional de Investigación Pecuaria Cuernavaca 1996.* Universidad Autónoma de Morelos, Morelos, Mexico, p. 267. (In Spanish.)

Yerena, F., Ferreiro, H.M., Elliott, R., Godoy, R. and Preston, T.R. (1978) Digestibility of *Brosimum alicastrum, Leucaena leucocephala, Cenchrus ciliaris* and *Agave fourcroydes (*Digestibilidad del ramón (*Brosimum alicastrum*), *Leucaena leucocephala,* pasto Buffel (*Cenchrus ciliaris*) y pulpa de bagazo de henequén (*Agave fourcroydes*). *Producción Animal Tropical* 3, 70-73. (In Spanish.)

Quality of vegetation in silvopastoral systems

M. P. González-Hernández
Dept. of Crop Production, University Santiago Compostela, EPS, Campus Universitario, 27002-Lugo, Spain

Abstract
Energy and protein of range forages are usually the most limiting nutritional factors to range animal productivity. Both are measured by parameters such as crude protein, digestibility, carbohydrates (both structural and non-structural) and nitrogen-complexing compounds. These essential nutritional components of vegetation were studied in common forestry communities in Galicia. The significance of these parameters in herbivore nutritional ecology and their variation as a result of the inherent dynamic and complexity of silvopastoral systems are discussed.

Key words: Galician forests, dry matter digestibility, fibre, lignin, digestible protein, crude protein, tannins

Introduction
Knowledge of the nutritive quality of the vegetation is essential to evaluate the resources available for grazing, as well as in developing effective management strategies in silvopastoral systems. Silvopastoral systems or forest grazing, the preferred term for the Forage and Grazing Terminology Committee (FGTC, 1992), is the combined use of forestland or woodland for both wood production and animal production by grazing of the coexisting indigenous forage, or vegetation that is managed like indigenous forage. This multiple use makes silvopastoral systems and the different parameters that define their quality dynamic. The optimal use of nutritional resources available within the silvopastoral system depends on several complex factors that vary with season and grazing time, among others. Considerable research has been reported that adds greatly to the knowledge of the nutritional response of grazers and browsers to grasses, forbs and shrubs, as well as to the knowledge of the response of vegetation to grazing pressure.

This discussion emphasizes the significance of nutritional parameters that are related to energy and protein levels of range forages. These are coexisting indigenous forages, or vegetation that is managed like indigenous forage. Both parameters are usually the most limiting to range animal productivity (Holechek et al., 2004) and are measured by variables such as crude protein, digestibility, carbohydrates (structural: ADF, NDF, lignin, silica; and non-structural) and phenolic compounds (especially nitrogen-complexing compounds). These essential nutritional components of vegetation are analysed and discussed in a general context based on data obtained from studies of nutritional attributes of forest vegetation, both in Galicia (Spain) and in Oregon (USA). Nutritional quality of common forestry communities in Galicia such as oakwoods, conifer stands, heathlands and gorselands is shown, as well as the variation of their nutritional components as a result of the inherent dynamic and complexity within silvopastoral systems.

Comparative nutritional quality of range forages
Plant species composition of forest communities determines the nutritional quality of vegetation since nutritional quality parameters vary greatly among plant species, the portion of the plant consumed, maturity, genetic and environmental factors (Van Soest, 1982; Bailey, 1984). Digestibility and crude protein levels of forage are important components of diet quality while fibre is usually negatively associated with digestibility. Some of the most common oakwoods and shrublands in Galicia were used to analyse the variation of nutritional quality in vegetation (Table 1).

Table 1. Common species in different types of forest vegetation in Galicia.

Type of vegetation	Most common species
Blechno spicanti-Quercetum roboris (oak woodlands)	*Pyrus cordata, Ilex aquifolium, Frangula alnus, Daboecia cantabrica, Erica arborea, Hedera helix, Lonicera periclymenum, Rubus* sp, *Vaccinium myrtillus, Blechnum spicant*, grasses
Rusco aculeati-Quercetum roboris (oak woodlands)	*Castanea sativa, Frangula alnus, Laurus nobilis, H. helix, L. periclymenum, Rubus* sp, *B. spicant*
Myrtillo-Quercetum roboris (oak woodlands)	*I. aquifolium, P. cordata, D. cantabrica, F. alnus, H. helix, L. periclymenum, Rubus* sp, *Vaccinium myrtillus, E. arborea, B. spicant*, grasses
Linariotriornithophorae-Quercetum pyrenaicae (oak woodlands)	*E. arborea, Rubus* sp, *Lonicera peryclimenum, Cytisus scoparius*, grasses
Ulici europaei-Ericetum cinerae (conifer stands, heathlands, gorselands)	*Calluna vulgaris, Erica arborea, E. australis, E. cinerea, E. umbellata, Halimium lasianthum, Ulex* sp, grasses
Gentiano pneumonante-Ericetum mackaianae (conifer stands, heathlands)	*Erica mackaiana, Erica umbellata, Erica ciliaris, Pterospartum tridentatum, Ulex gallii*
Genistello tridentati-Ericetum aragonensis (heathlands)	*Erica australis, Pterospartum tridentatum, Halimium lasianthum, Agrostis curtissii*

Digestibility

Energy availability is generally expressed in terms of vegetation dry matter digestibility (DMD). Forbs and leaves of shrubs usually reach their potential extent of digestion more quickly than grasses. This appears to be partially explained by thin cell walls that facilitate quicker microbial access to soluble cell components (nitrogen, amino acids, non-structural cabohydrates), which are easily broken down by the animal's digestive system and are a readily available source of energy. Substances resistant to microbial digestion located in the cell wall, and detrimentally associated with digestibility, are lignin, cutin and silica. They are arranged differently in the cellulose/hemicellulose matrix in monocots, compared with dicots (Van Soest, 1982). Hence, digestibility varies greatly among plant species, the portion of the plant consumed and its maturity. An important issue in range management is the fact that forages with rapid rates of digestion generally have higher intakes than those with slower rates (Van Soest, 1982).

In Galicia, oakwood understorey provides higher quality forage than conifer understorey or shrublands. Energy availability is higher in oakwoods, shown by the highest values of dry matter digestibility and the lowest of fibre content when compared to other vegetation types (Figure 1). González-Hernández and Silva-Pando (1999) reported for heaths (*Erica* sp, *Calluna vulgaris*) the lowest DMD values in Galician forests, ranging from 16 to 25%. Other shrubs such as gorse (*Ulex* sp), bilberry (*Vaccinium myrtillus*), blackberry (*Rubus* sp) and grasses averaged 40-50%. Digestibility of black dogwood (*Frangula alnus*), ivy (*Hedera helix*), honeysuckle (*Lonicera periclymenum*), all common in oakwoods, ranged from 50 to 64%. Seasonal fluctuation of digestibility is also highest in oakwoods compared with heathlands, gorselands and conifer understorey. Maximum values of digestibility occurred during the growing season, since soluble components of plant cells decreased at senescence (González-Hernández and Silva-Pando, 1999). Although energy requirements vary with age, sex, reproductive status, weather and cover availability, ruminants in general cannot maintain body weight on diets with DMD coefficients less than 50% (Weiner, 1977; Maizeret *et al.*, 1991). Based on this minimum requirement, digestibility of Galician forests can be considered low.

Figure 1. Plant dry matter digestibility (DMD), crude protein (CP), fibre (ADF) and lignin content (%) of most common Galician woodlands.

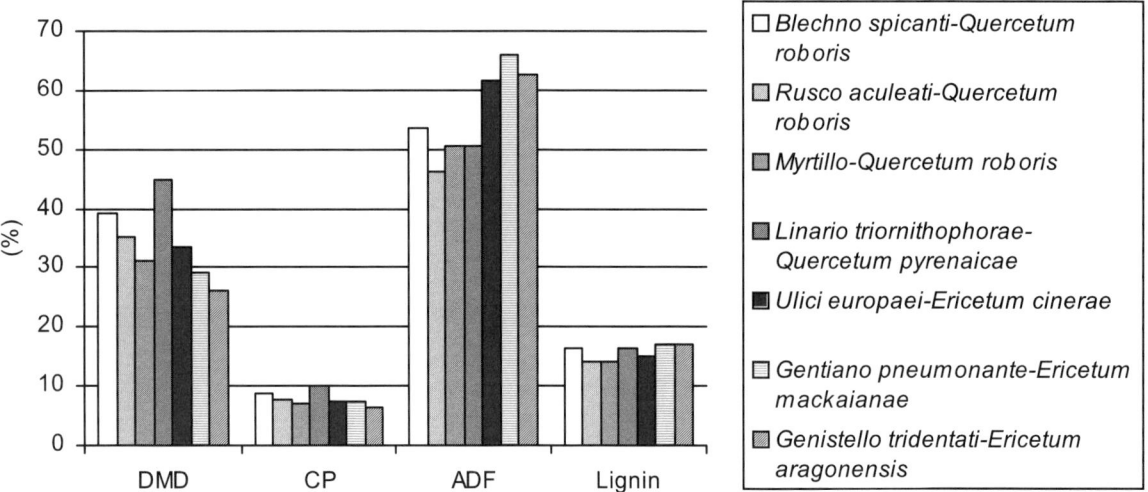

Carbohydrates

Plants have two basic types of carbohydrates, those associated with the cell wall (cellulose, hemicellulose and lignin) and those associated with the cell content (non-structural carbohydrates such as sugars and starches). Digestion of cellulose and hemicellulose is a much slower process than is that of starch and sugar. Lignin cannot be utilized, and reduces quality most in forages. It is the principal constituent of woody material in trees and shrubs.

Non-structural carbohydrates (NC) are a readily available source of energy, and NC levels have commonly been used as a key indicator of plant vigour and an index of the consequences of grazing (Caldwell, 1984). Overharvesting, either by cutting or by grazing, generally reduces NC, and the concept that disappearance or persistence of grazed plants correlates with amount of total NC reserves has been propounded and reviewed many times (Heady and Child, 1994).

In Galicia, unbrowsed plants of *Pterospartum tridentatum*, a legume, contained significantly more NC than browsed plants (Figure 2). NC content also decreased significantly through the grazing period when three different degrees of utilization were monitored (this is represented by the three bars per treatment in Figure 2). In the same study, gorse (*Ulex gallii*) grew better after defoliation than *Pterospartum* (González-Hernández *et al.*, 2002).

Figure 2. Effect of grazing on fibre content (ADF and NDF) and non-structural carbohydrates (NC). *Browsing effect significant ($P < 0.05$). **Significant differences ($P < 0.05$) through the grazing period. Means followed by the same letter are not significantly different.

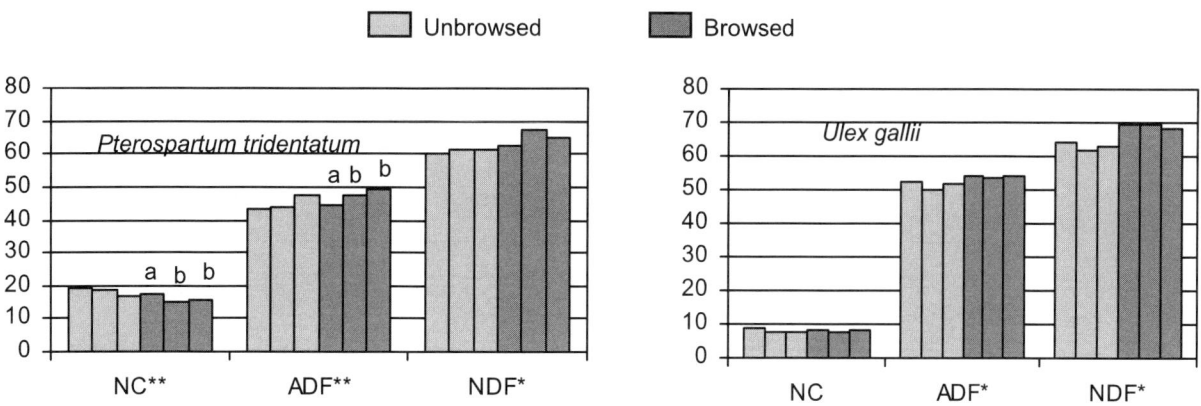

An understanding of the defoliation effect and the potential of plant adaptation is fundamental for the professional rangeland manager. High animal density deteriorates the quality of the vegetation with an increase in lignin content and a decrease in organic matter digestibility (González-Hernández and Silva-Pando, 1996).

Protein

Unlike energy and most minerals, protein cannot be stored by the animal's body, so a continuous supply is required. Proteins are important as enzymes, antibodies against diseases and hormones. Crude protein has been used as an indicator of forage quality. Actively growing plant parts have much higher protein levels than do those that are dormant, and variation in available protein among different vegetation communites will be a function of species composition (González-Hernández and Silva-Pando, 1999). In general, ruminants require a diet containing 5-6% crude protein for maintenance and early pregnancy, increasing to nearly 12% during lactation (Hobbs et al., 1981). Based on this, crude protein of plants in Galician forests would meet ruminants' basic requirements (Figure 1). More recently, it has been reported that crude protein is not a sufficiently good indicator of forage quality to understand the nutritional basis for livestock or ungulate diet selection. Available protein in diets is a function of plant crude protein, the amount of non-digestible fibre-bound protein and the extent of protein precipitation by tannins. Thus, available protein, rather than total protein content, is the physiologically important parameter relative to animal requirements and metabolic capabilities (Robbins et al., 1987).

Nitrogen-complexing compounds

Tannins are phenolic compounds that are widely recognized as digestibility reducers for browsers feeding upon woody plants (Robbins et al., 1987; Happe et al., 1990), and also as compounds that may act as toxins (McArthur et al., 1993). Condensed tannins (i.e. proanthocyanidins) commonly bond to proteins in the digestive tracts and are excreted in faeces, whereas hydrolysable tannins are degraded to low molecular weight phenolics that do not interact with protein (Hagerman et al., 1992), but this may lead to other biological interactions. Digestibility of forage may also be reduced if microbial protein binds with tannin and rumen microflora is negatively influenced by tannins. Both hydrolysable and condensed tannins are present in plants consumed by deer and livestock (Figure 3), with the potential to reduce the availability of protein (Happe et al., 1990; González-Hernández et al., 2003).

To quantify the impact of tannins on nutrient availability for elk and deer, predictive equations were obtained from conducted feeding trials with known quantities of tannins, allowing the prediction of protein digestibility (Robbins et al., 1987): digestible protein (DP) = $-3.87Z + 0.9283X - 11.82Y$, where Z is digestible protein in grams per 100 grams of feed, X is crude protein content as per cent dry matter, and Y is amount of bovine serum albumin precipitated. Solving the equation for the crude protein required for ruminants, assuming lack of astringency, yields dietary requirements of approximately 1% of DP for maintenance and early pregnancy, and about 7% DP for lactation. Hobbs et al. (1981) found that elk diets containing less than 5% crude protein (< 1 per cent DP) would not meet metabolic requirements for protein and also would be likely to result in reduced rates of carbohydrate digestion.

Figure 3. Total hydrolysable and condensed tannins in plants known as components of livestock and deer diets in Galicia. *F. sylvatica* (F.s); *Q. robur* (Q.r); *H. lasianthum* (H.l); *C. vulgaris* (C.v); *E. australis* (E.a); *E. umbellata* (E.u); *V. myrtillus* (V.m); *E. arborea* (E.ab); *Rubus* sp. (R).

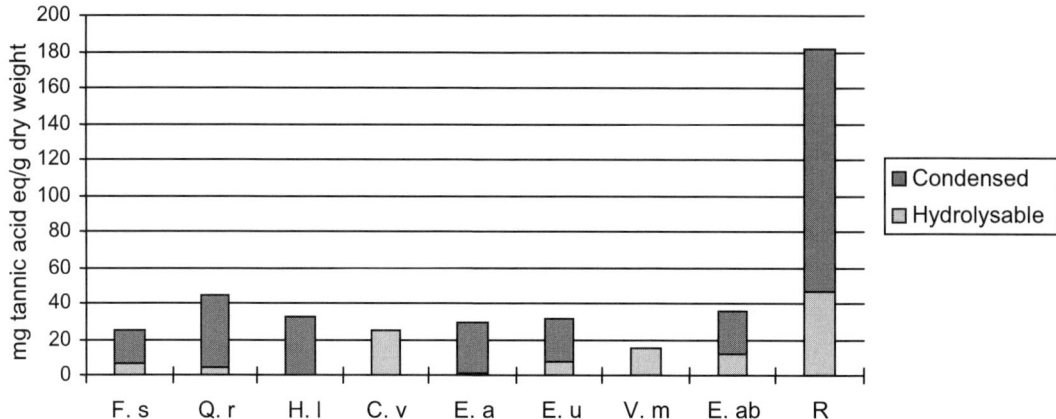

The above equation was solved for some common plants of Galician forests for which crude protein and astringency are known from former research (González-Hernández and Silva-Pando, 1999; González-Hernández *et al.*, 2003). There was a significant reduction in crude protein availability with tannin content (Figure 4).

Figure 4. Effect of tannin astringency on digestible protein of common forages for livestock and deer in Galician forests. *Calluna vulgaris* (calvul), *Halimium lasianthum* (halal), *Erica arborea* (erar), *E. umbellata* (erum), rubus, *Vaccinium myrtillus* (vamy), *Cytisus striatus* (cystr), *Pterospartum tridentatum* (ptri), *Hedera helix* (hehe), *Lonicera periclymenum* (lope).

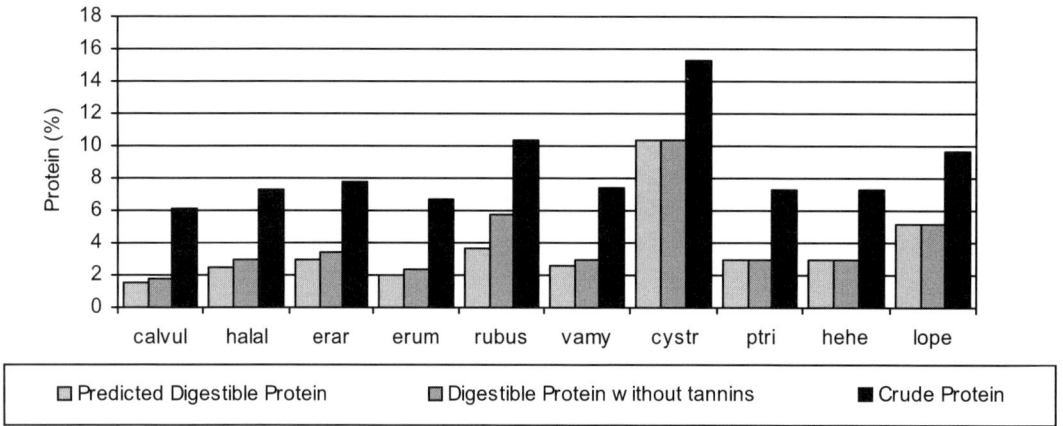

As for other nutritional indicators, tannin content varies with maturity. It has been reported to be the highest in the growing season, and generally decreases from spring through summer, autumn and winter (Happe *et al.*, 1990; Starkey *et al.*, 1999; González-Hernández *et al.*, 2000). The seasonal effect of tannin astringency for *Alnus rubra*, *Erica arborea* and *Rubus* sp., is shown in Figure 5. Changes in concentrations of tannin and non-tannin phenolics, such as phenolic glycosides, are likely to result in changes in taste and may provide olfactory stimuli, respectively. These compounds may be responsible indicators by which herbivores can select forage with relatively greater digestibility of protein (Rowell-Rahier, 1984; McArthur *et al.*, 1993; González-Hernández *et al.*, 2000). An adaptive strategy is reflected in food habits that shift seasonally to take advantage of forages such as grasses or forbs with lower tannin content and/or capacity to precipitate proteins (Starkey *et al.*, 1999).

Figure 5. Effect of tannin astringency on digestible protein after applying the Robbins et al. (1987) equation.

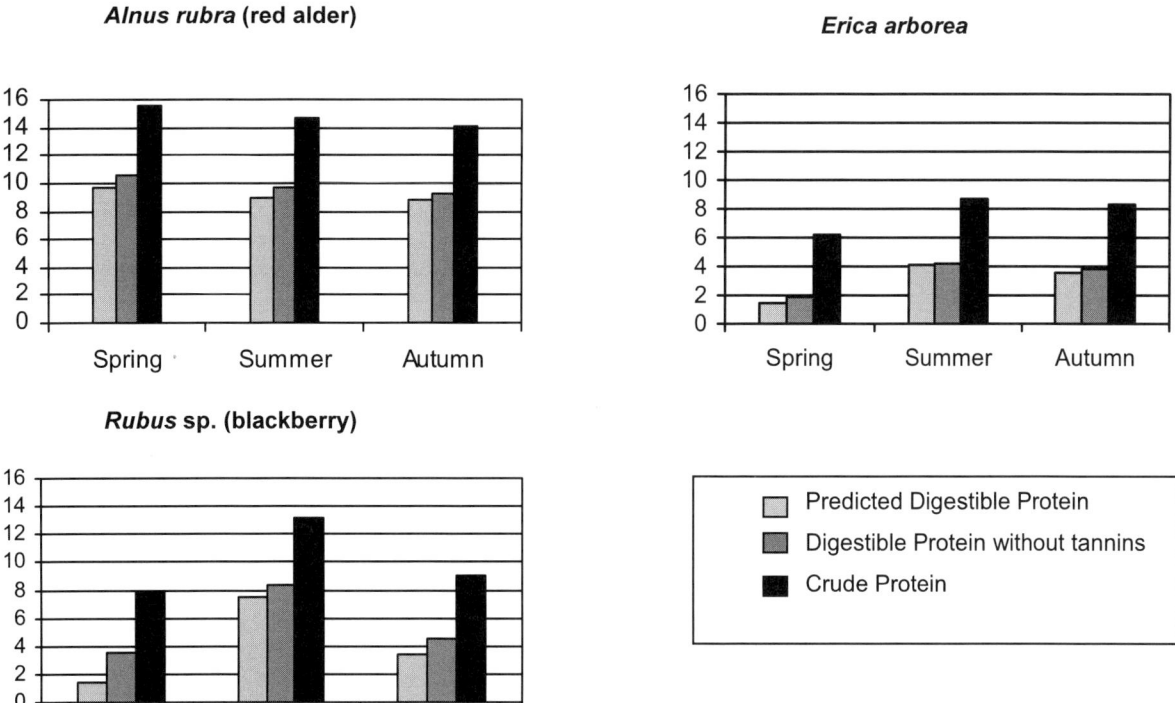

More recently, polyphenols have shown biological effects of nutritional interest, revealing that condensed tannins also form stable complexes with metal ions and are, like other polyphenols, good reducing agents. They can also exert a positive effect, by preventing frothy bloat, or by improving the nutritional utilization of alimentary nitrogen (Jean-Blain, 1998). Although protein protected by tannins and soluble phenolics may improve ruminant performance if consumed at low levels, the complete nutritional significance of the non-specific complexation of proteins is less clear. There are apparently conflicting claims of beneficial and toxic effects caused by hydrolysable tannins in various animal species, including rodents and ruminants (Clifford and Scalbert, 2000), but some hydrolysable tannins may induce severe intoxications in ruminants and in the horse (Jean-Blain, 1998). Not only tannins, but also small non-tannin phenolics are readily absorbed from the gut and may function as metabolic toxins (McArthur et al., 1993). Non-tannin phenolics depressed ruminant in vitro digestibility, some have been reported as compounds with herbivore deterrent properties, and they may be of importance as antinutritional agents for ruminants (Palo, 1984, 1985; González-Hernández et al., 2000).

Conclusions

A pluralistic approach and a holistic overview to the management of grazing ecosystems are necessary. This will involve knowledge of plant herbivore interactions that considers plant nutritional attributes, their phytotoxicity and antifeedant properties as well as the different tolerance of herbivores generally associated with adaptation to tannin-rich environments. The seasonal pattern of plant production and species richness, as well as the seasonality of nutritive value, will determine the temporal character of silvopastoral systems. Changes in the relative proportions of shrubs, herbs and ferns under grazing conditions have implications for sustainable management (González-Hernández et al., 1998).

Many forage species contain significant levels of astringent tannins with the potential to greatly reduce the availability of protein for livestock and deer. Digestible protein is likely to be a significant limiting factor in forest communities containing predominantly shrubs with high tannin content.

References

Bailey, J.A. (1984) *Principles of Wildlife Management*. Wiley and Sons, New York, USA.
Caldwell, M.M. (1984) Plant requirements for prudent grazing. In: National Research Council and National Academic Science (eds) Developing Strategies for Rangeland Management. Westview Press, Boulder, USA. pp. 117-152.
Clifford, M.N. and Scalbert, A. (2000) Ellagitannins - nature, occurrence and dietary burden. *Journal of the Science of Food and Agriculture* 80 (7), 1118-1125.
FGCT (1992) Terminology for grazing lands and grazing animals. *Journal of Production Agriculture* 5, 91-201.

González-Hernández, M.P. and Silva-Pando, F.J. (1996) Grazing effects of ungulates in a Galician oakwood (NW Spain). *Forest Ecology and Management* 88, 65-70.

González-Hernández, M.P. and Silva-Pando, F.J. (1999) Nutritional attributes of understorey plants known as components of deer diets. *Journal of Range Management* 52 (2), 132-138.

González-Hernández, M.P., Silva-Pando, F.J. and Casal Jiménez, M. (1998) Production patterns of understorey layers in several Galician (NW Spain) woodlands. Seasonality, net productivity and renewal rates. *Forest Ecology and Management* 109, 251-259.

González-Hernández, M.P., Starkey, E.E. and Karchesy, J. (2000) Seasonal variation in concentrations of fiber, crude protein and phenolic compounds in leaves of red alder (*Alnus rubra*): nutritional implications for cervids. *Journal of Chemical Ecology* 26 (1), 293-301.

González Hernández, M.P., Castro, P., Rigueiro, A. and Mosquera, M.R. (2002) Effects of grazing pressure in the nutritional quality of 2 understorey species. In: Frochot, H., Collrt, C. and Balandien, P. (eds) *Popular Summaries from the IV International Conference on Forest Management*. INRA, Champenoux, France, pp. 418-420.

González-Hernández, M.P., Starkey, E.E. and Karchesy, J. (2003) Research observation: hydrolysable and condensed tannins in plants of northwest Spain forests. *Journal of Range Management* 56 (5), 461-465.

Hagerman, A.E., Robbins, C.T., Weerasuriya, Y., Wilson, T.C. and McArthur, C. (1992) Tannin chemistry in relation to digestion. *Journal of Range Management* 45 (1), 57-62.

Happe, P.J., Jenkins, K.J., Starkey, E.E. and Sharrow, S.H. (1990) Nutritional quality and tannin astringency of browse in clearcuts and old-growth forests. *Journal of Wildlife Management* 54, 557-566.

Heady, H.F. and Child, R.D. (1994) *Rangeland Ecology and Management*. Westview Press, Boulder, USA.

Hobbs, N.T., Baker, D.L., Ellis, J.E. and Swift, D.M. (1981) Composition and quality of elk winter diets in Colorado. *Journal of Wildlife Management* 45, 156-171.

Holechek, J.L., Pieper, R.D. and Herbel, C.H. (2004) *Range Management. Principles and Practice,* 5th edn. Prentice Hall, Pearson, USA.

Jean-Blain, C. (1998) Nutritional and toxicological aspects of tannins. Aspects nutritionnels et toxicologiques des tannins. *Revue de Médecine Vétérinaire* 149 (10), 911-920. (In French.)

McArthur, C., Robbins C.T, Hagerman, A.E. and Hanley, T.A. (1993) Diet selection by a ruminant generalist browser in relation to plant chemistry. *Canadian Journal of Zoology* 71, 2236-2243.

Maizeret, C., Bidet, F., Boutin, J.M. and Carlino, J.P. (1991) Influence de la composition chimique des végétaux sur les choix alimentaires des chevreuils. *Revue Ecologie* 46, 39-52.

Palo, R.T. (1984) Distribution of birch (*Betula* sp), willow (*Salix* sp), and poplar (*Populus* sp) secondary metabolites and their potential role as chemical defense against herbivores. *Journal of Chemical Ecology* 10 (3), 499-520.

Palo, R.T. (1985) Chemical defense in birch. Inhibition of digestibility in ruminants by phenolics extracts. *Oecologia* 68, 10-14.

Robbins, C.T., Hanley, T.A., Hagerman, A.E., Hjeljord, O., Baker, D.L., Achwartz, C.C. and Mautz, W.W. (1987) Role of tannins in defending plants against ruminants: reduction in protein availability. *Ecology* 68 (1), 98-107.

Rowell-Rahier, M. (1984) The presence or absence of phenolglycosides in *Salix* (Salicaceae) leaves and the level of dietary specialization of some of their herbivorous insects. *Oecologia* 62, 26-30.

Starkey, E.E., Happe, P.J., González-Hernández, M.P., Lange, K. and Karchesy, J. (1999) Tannins as nutritional constraints for elk and deer of the coastal Pacific Northwest. In: Gross, G.G., Hemingway, R.W. and Yoshida, T. (eds) *Plant Polyphenols 2: Chemistry, Biology, Pharmacology, Ecology*. Kluwer Academic/Plenum Publishers, New York, USA, pp. 897-908.

Van Soest, P.J. (1982) *Nutritional Ecology of the Ruminant*. O and B Books, Corvallis, USA.

Weiner, J. (1977) Energy metabolism of the roe deer. *Acta Theriologica* 22, 3-24.

The herbaceous component in temperate silvopastoral systems

M. R. Mosquera-Losada, M. Pinto-Tobalina and A. Rigueiro-Rodríguez
Crop Production Department, University of Santiago de Compostela, 27002-Lugo, Spain. romos@lugo.usc.es

Abstract

The aim of this chapter is to evaluate herbaceous strata development in silvopastoral systems in a western temperate Atlantic-influenced area in Europe. This is a more modern and less well-known system than other classic Mediterranean silvopastoral systems. Pasture production and quality depend on tree type, density, distribution and age. Interaction of responses of pasture and tree productivity to management factors are shown and they depend on soil type and climatic conditions, as well as on tree production aim. There are several different kinds of silvopastoral systems which can be adopted or are currently being used in Europe. In the Atlantic area, higher pasture and tree growth can be expected compared with in drier, Mediterranean areas. Hence it is important to evaluate the interaction between tree development and pasture production and its distribution and quality under different conditions in order to match the social, productive and environmental requirements for woodlands or forests in silvopastoral systems.

Introduction

Silvopastoral systems have been a traditional form of land use throughout Europe. Nowadays, largely as a result of land specialization into either agriculture or forestry, most silvopastoral systems with domestic animals are to be found mainly in mountain regions of Europe. Examples of silvopastoral systems in northern European countries include Norway, Sweden and Finland, where there are almost 800,000 reindeer grazing in forestland (Yrjölä, 2002). Silvopastoral systems are only used for biodiversity enhancement in limited central-western parts of Europe, in The Netherlands (Kuiters, 2004) or in Germany (Finck et al., 2002). The negative effects of shade on pasture production have been described in literature for over 50 years. However, this early work highlights the fact that deciduous *Larix* woodland had a very valuable pasture understorey in the sub-alpine regions of Germany (Klapp, 1971). In southern Europe, the "dehesas" (Spain) and "montado" (Portugal) are among the best examples of European silvopastoral systems. They are mainly associated with *Quercus suber* and *Quercus ilex* subsp. *ilex*, but silvopastoral systems with other Mediterranean broadleaf species such as *Q. pyrenaica*, *Q. faginea*, *Q. ilex* subsp. *ballota*, *Q. humilis* and *Q. coccifera* or Mediterranean coniferous species such as *Juniperus thurifera*, *P. pinea*, *P. halepensis* Mill., *Pinus brutia* and *P. pinaster nigra* are also used (Étienne, 1996; Papanastasis, 1996; Llorente-Gil and Ciria-Ciria, 1998; Merou and Vrahnakis, 1999; Larsson, 2001). In countries such as north-western Spain, western France, Great Britain and Ireland, i.e. the Atlantic areas of Europe, levels of precipitation are adequate and good temperatures are suitable for better tree and pasture growth compared to the previous areas. Trees evaluated for silvopastoral systems include *E. globulus*, *E. nitens*, *P. pinaster*, *P. sylvestris*, *Alnus rubra*, *Acer pseudoplatanus*, *Populus*, *Q. rubra*, *Fraxinus excelsior*, *Betula alba* and *Pinus radiata*. Initially very good results have been obtained from a biodiversity or economic point of view, if adequate management techniques are used (Sibbald, 1996; McAdam et al., 1999; Mosquera-Losada et al., 2001; Teklehaimanot et al., 2002).

Silvopastoral systems can be defined as complex management systems that integrate tree, pasture and animals in a concrete edapho-climatic context (Nair, 1993; Etienne, 1996). Indeed, pasture production in silvopastoral systems depends on the edapho-climatic context, varying in the temperate area of Europe from 0.2 to 2 t DM/ha per year in the Mediterranean area (Gómez Gutierrez and Calabuig, 1992; Papanastasis, 1999) to 4-15 t DM/ha per year in acid soils (Mosquera-Losada et al., 1999; López-Díaz et al., 2001; Peyraud et al., 2004) in maritime areas with low fertilizer inputs.

The productivity of each component of the silvopastoral system can be manipulated by management. Thinning or pruning affects quality timber and tree production (Evans, 1984), fertilization increases pasture production (Mentxaka et al., 1998; Mosquera-Losada et al., 1999) and stocking rate has an impact on animal production (Mosquera-Losada et al., 2000). These management decisions will indirectly definitively change the productivity of the other components. For example, tree thinning or pruning modifies pasture production and therefore animal production, pasture fertilization will change tree-pasture competition (Mosquera-Losada et al., 1999), mainly in young stands, and an increase in the stocking rate can enhance tree growth due to a reduction of competition with pasture or reduce tree production through compaction effects (Sibbald et al., 2001).

This chapter will focus on silvopastoral systems with timber trees, as the need for wood in Europe is important and this product has, at present, a very good market (Xunta de Galicia, 2001). However, trees can also provide non-wood outputs such as animal feed (*Morus alba* L., *Robinia pseudoacacia* L.) during seasonal forage shortages or have both purposes (tree products and forage products: e.g. *Quercus ilex* (acorns and firewood) and *Quercus suber* (cork and acorns)) in the Mediterranean area and *Betula alba* L. (timber and branches for forage) in the Atlantic area as a traditional system.

The aim of this chapter is to evaluate herbaceous strata development in silvopastoral systems in a western temperate Atlantic-influenced area in Europe. This is a more modern and less well-known system than other classic Mediterranean silvopastoral systems.

Tree effects on pasture productivity

The main difference in pasture management and productivity between silvopastoral systems and open sward is the interaction with the tree component. Light transmission to the understorey will modify pasture production and therefore the annual current silvopastoral system profits. When the percentage of light intercepted by a canopy in a forest is higher than 55% then pasture production is seriously limited (Dodd *et al.*, 1972; Sibbald, 1994). Light input to undergrowth vegetation will be mainly determined by the layer of tree cover in each specific climatic context, which will also modify temperature and water availability for pasture (Gómez Gutierrez and Calabuig, 1992; Sibbald, 1999). Generally, this will depend on density but also on tree crown growth capacity, as the tree crown size is larger when it grows at low density and without other tree competition.

Currently the choice of tree species for timber production depends mainly on edapho-climatic conditions, and the layer of tree cover in mature stands in Europe depends on the current aims for forest or woodland production. This has determined the silvicultural techniques used and adopted in practice for each particular species. Hence there is a direct relationship with socio-economic aspects such as property, productivity and rural population density. In those areas where the main output is from the agricultural component, tree density and cover are low, as in dehesa or montado systems. On the other hand current mature dense stands can be as a result of silvicultures aiming for high- or low-quality timber production. When high-quality wood is the main target product from the forestland then tree density is usually high in classic silvicultures, as occurs in most continental oak stands in Germany and France, which can also be used for grazing by animals at adequate stocking rates for conservation purposes, as grazing favours the persistence of herbaceous species which adapt well to grazing and shade as compared to non-grazed areas, which enhance biodiversity (Finck *et al.*, 2002). In this classic oak silviculture, the main silvicultural reason for avoiding heavy thinning to create low density is the appearance of epicormic branches, which will deteriorate the quality of timber (Evans, 1984). On the other hand, when the main objective of trees is to produce wood pulp or chipwood, density is usually also high. This has occurred in the last century in northern Spain, where intense afforestation with fast growing species has been carried out and where in some cases grazing was necessary to reduce the risk of fire.

On the other hand, nowadays, a lower initial planting density is recommended as this allows mechanization in stands and the production of high-quality wood over a short growth period. For example, the use of broadleaved species (such as *Castanea, Quercus* in Spain) has been promoted in order to substitute short rotation species such as *Pinus radiata, Populus, Pinus pinaster, Eucalyptus*, but the economic return from these broadleaved plantations managed as exclusive forest areas is long term and this makes it difficult to convince farmers. In all of these areas, silvopastoralism can be a way of planting new forests at low densities and these make it possible to obtain a short-term return and eventually produce saw wood in a shorter period of time (as individual tree growth is enhanced when density is low).

The effect of the tree layer on pasture production and its seasonal distribution can be positive or negative. To extend the grazing season is a goal for sustainable grazing systems in Europe, as it reduces concentrate use (Delaby *et al.*, 2004). The tree layer can enhance annual pasture production through pasture increase at the start and the end of the growing season (Sibbald *et al.*, 1991) or when it reduces the effects of drought, as happens in the dry summers in Galicia (grazing under trees was a traditional management practice in Galicia during the summer) and in the dehesa system in Spain. This justifies the continuity of this system in the Mediterranean area in spite of the general distinction between forest and agricultural land which occurred in most European countries as a result of intensifying grazing systems. On the other hand, the tree layer can reduce pasture production by competing for available light and water if the density is high (Sibbald, 1996). Although the effect of light input on the undergrowth in silvopastoral systems will depend on the climatic context, tree type, distribution, density and age in the silvopastoral systems will also have an effect. In young silvopastoral systems these tree parameters can change with time and make management of such systems more difficult than in open swards.

Tree type
Tree species modify light input and therefore pasture production due to a crown structure mainly defined by branch distribution and the shape and size of leaves. It is known that *Eucalyptus globulus* Labill. will allow more light input than *Pinus pinaster* Ait. and both more than *Pinus sylvestris*, due to the different light interception of their crowns (Rigueiro-Rodríguez *et al.*, 1998). In this context tree type and its deciduous or evergreen character will affect pasture production differently (Papanastasis, 2004). Deciduous trees will have a substantial effect on soil nutrient cycling and light input, as leaves cover yearly the soil, being a negative or positive effect on pasture production depending on the type of litter and the mineralization rate and which is usually reduced when a high proportion of light is intercepted by the tree canopy (dense mixed stands). It will also significantly affect the soil fauna.

Tree distribution
Tree distribution in silvopastoral systems can be homogeneous (single or grouped trees) or form borderlines (fencing, riparian, shelterbelt) or forest (mosaic landscape) (Figure 1). In comparison with a border distribution, homogeneous distribution minimizes tree competition but tree pasture interaction is maximized, and the intensity of this interaction

depends mainly on density and tree age. From a landscape point of view, homogeneous silvopastoral systems look like new or traditional forests or woodlands and bordering trees can be distributed as fencing, shelterbelts, riparian or mosaic silvopastoral systems.

Trees can also be distributed homogeneously in groups, as occurs with traditional *Ilex aquifolium* silvopastoral systems consisting of groups 18-20 m in diameter associated with propagule spread around existing trees (García-González and Contreras-Olalla, 1998). Animal products and foliage for seasonal use (e.g. at Christmas) are the main outputs of this silvopastoral system. Teklehaimanot *et al.* (2002) concluded that trees in a clumped pattern appear to combine silvicultural benefits for tree growth with the agricultural benefits of maintaining livestock production in young plantations.

Trees surrounding pasture plots (bordering system) in lines can be used as protection from wind or snow (shelterbelts), living fences or for protection of river water quality as riparian buffers. In this case, tree-pasture interaction is minimized and pasture production is maximized in the inner part of the plot (depending on tree and plot size). However, this system will create new habitats (compared to exclusively forest or agricultural land) as it can favour shade-adapted plant growth on the borders and therefore enhance biodiversity for less sensitive grazing species from an ecological point of view (De Miguel, 2002).

Figure 1. Tree distribution in silvopastoral systems with the same density: homogeneous with single trees (a), with groups of trees (b) and bordering (fencing, shelterbelt or riparian) (c) and mosaic landscape silvopastoral systems (d).

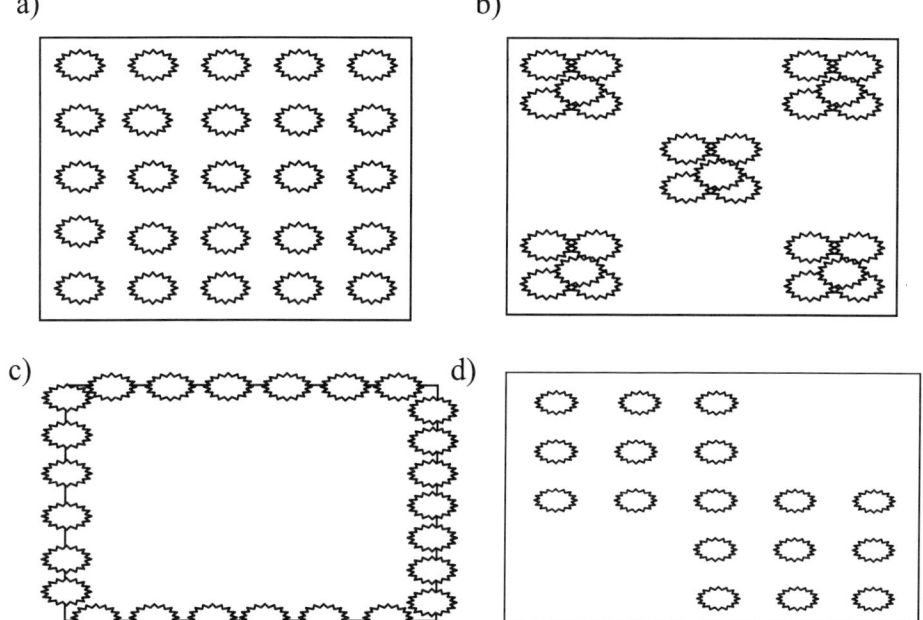

Tree density

Establishment of tree density depends on the objective for the tree component in the silvopasture. Higher densities will enhance tree volume production but will provide lower individual tree sizes than lower densities. In both cases silvopastoral systems can be used to increase short-term land profit from animal production. They can also reduce the cost of scrub clearance, facilitate pruning and thinning and reduce the risk of fire. Examples of this are *Pinus radiata* management at a density of 2500 stems/ha carried out in northern Spain in the mid 20th century in order to produce chipboard or *Eucalyptus globulus*, currently planted at 1700 stems/ha (González-Río *et al.*, 1997). In the former case tree pruning is only made in the first two metres of the stem to make cultivation easier and thinning is carried out to make plantations healthier (Echeverría, 1943; González-Río *et al.*, 1997). There are other kinds of silviculture which use trees at reduced densities. In New Zealand and Australia *Pinus radiata* is used at a low density in a silvopastoral system (Hawke, 1991), and the advantages for these systems for timber production and from a management point of view have been highlighted in Spain (Dans del Valle *et al.*, 1999). This option maximizes pasture production and increases land profit in the long term through tree planting (generally using genetically improved stock) by obtaining high-quality wood. If genetically improved tree plants are not available, initial density should be higher to ensure well-formed trees at final harvest. Initial density can also be higher to produce trees with a large diameter without knots (*Q. robur* and *Q. petraea* in French classical silviculture) and to avoid annual pruning of epicormic buds (Evans, 1984). Therefore, the initial tree density factor will affect pasture production and will have to take into account the wood production objective and tree management costs. For this, in recent decades, trees were planted at wider spacing than in the past to facilitate mechanization and enhance large-size timber production, as required by the EU. This also makes it easier for animals to graze under the trees.

Tree age
Once the initial density has been decided based on the above-mentioned criteria, potential pasture production will be determined by taking into account the age of the plantation and pruning and thinning management (Sibbald *et al.*, 1991). Most of the silvopastoral systems used in Europe have long-lived trees such as those in the dehesas (several centuries) or in northern countries (80-120 years). However, in some new forest areas, the final tree harvest can take place every 15 years or less (*Eucalyptus*, *Populus*) or 25-30 years (*Pinus pinaster*, *Pinus radiata*). Nowadays, some silvopastoral systems, such as the dehesa, have reached maturity, and the effect on pasture production will not change over time. Young plantations usually have little effect on pasture production in the early years, but, as the tree grows, the crown will have a greater effect on pasture development; in some cases it can be negative or positive. For example, with a high-density *Pinus radiata* or *Pinus pinaster* stand, light interception through crown structure can alter the intensity of the negative effect of shade on pasture production as compared to open swards with age. In a young *Pinus pinaster* stand, branches occupy the entire stem, preventing pasture growth under the crown, but, as the trees grow, the lower branches die due to density-dependent lower light input. As the branches fall off, more light reaches the ground, increasing understorey production (generally shrubs); this can also be extended to most of the trees with good natural pruning.

Once the type, age, distribution and density of the tree layer has been defined or adjusted then the optimization of a silvopastoral systems in some areas should take place by establishing a herbaceous layer, as it enhances productivity, quality and reduces the risk of fire.

Establishment of herbaceous strata

Establishment technique will depend on conditions and can be done mechanically or by using animals, the decision being based on where land mechanization is possible. If it is possible, the process will be the same as for open swards: clearance, harrowing, liming, fertilization and sowing. Where mechanization is not possible or is too expensive because of steep slopes or a high density of trees, then animals at an adequate stocking rate can be used, as grazing enhances pasture establishment, if adequate stocking rate used albeit in the long term (Rigueiro *et. al.*, 1998).

If pasture is going to be grazed under young trees, protection will be needed to avoid tree damage (Sibbald and Agnew, 1996; Dupraz *et al.*, 1997). Generally, sowing operations in silvopastoral systems with low tree densities or young trees are similar to those developed in open swards, but the effects of this operation on young tree growth should be evaluated, as at this age they can be affected more by pasture management than when older. However, when trees are older, soils should not be cultivated near the trees to avoid tree root damage, cutting or compaction. In silvopasture, care must be taken when liming, fertilizing and sowing certain pasture species, as this can affect young tree growth and development. Improving soil fertility will enhance pasture production, as usually the tree roots are deeper in the soil than roots of pasture species. When trees are young and their roots are shallow, usually pasture competition reduces tree growth, mostly in tree species with a high growth rate (*Pinus radiata*, *Pinus pinaster*, *Eucalyptus*). This can often be overcome by clearing a bare area around the tree by mulching or with herbicides.

If the soil pH is very acid, and nutrient availability is reduced, pasture production and tree growth are limited and lime should be added, taking into account the saturation aluminium percentage (Al^{3+}/Cation Exchange Capacity). Usually doses of between 1 and 3 t CO_3Ca/ha are enough to encourage pasture production in fertilized acid soils in humid temperate areas. Liming in soils with a pH under 5 and high organic matter improves phosphorus and other nutrient availability, through organic matter mineralization, and reduces the percentage of saturated aluminium in the soil cationic interchange complex for pastures and *Pinus radiata* (Figure 2). The effects of liming and fertilization on pasture and tree growth depend on the soil acidity and type and the kind of fertilizer used, which can be organic or inorganic. The use of inorganic fertilizer or liming alone in very acid soils (under 5.0) enhances tree growth but not pasture production. This is due to the increase in acidity caused by inorganic fertilizer application and the reduced nitrogen availability for pasture when liming alone is applied. In very acid soils the combination of liming and fertilization enhanced pasture production and reduced tree growth (Figure 2). In sandy soils with a low organic matter content and when the saturation percentage of aluminium does not make liming advisable, then the effects of fertilization on tree and pasture growth depend on the type of fertilizer. Inorganic fertilization enhances pasture production but reduces tree growth over no fertilization application. However, when an organic fertilizer is used pasture and tree growth are enhanced compared to no fertilizer or treatment with a mineral fertilizer, respectively. This can be explained by the slow release of the nutrients in the organic fertilizer which is best used when pasture growth is limited by drought, but it enables tree growth to continue in climates with a pronounced summer drought (Rigueiro-Rodríguez *et al.*, 2000). The effects of fertilization on the tree component of silvopastoral systems are more important when species which grow quickly are used. Fertilization had a greater effect on *Pinus radiata* than on *Betula alba* when the same fertilization treatments were applied in fertile soils (pH over 6) as the growth cycle for timber production was around 25 or 50 years, respectively.

Seed sowing should take place when conditions are optimum in order to guarantee profitability. Ryegrasses are the most important sown species used in the more productive systems of temperate regions, due to their high level of productivity and quality (Jung *et al.*, 1996). However, unless tree density is very low when bordering trees are used in the silvopastoral system, their persistence is reduced by tree shade. To combat this, cocksfoot seems to be a more appropriate species for silvopastoral systems in temperate areas as it is shade tolerant and drought resistant (Piñeiro et al., 1988; Koukora, 1999; Peri *et al.,* 2002).

Figure 2. Pasture production (top) and tree growth (bottom) in the first stages of a *Pinus radiata* plantation established two years earlier on fertile abandoned agricultural land (pH 5.5-6) with a low organic matter content (left) and on mountain land with a low pH (4.5-5) and high organic matter content (right). L: organic fertilization as sewage sludge, IN: inorganic fertilization, B: no fertilization. Ial: height increment, Id: diameter increment.

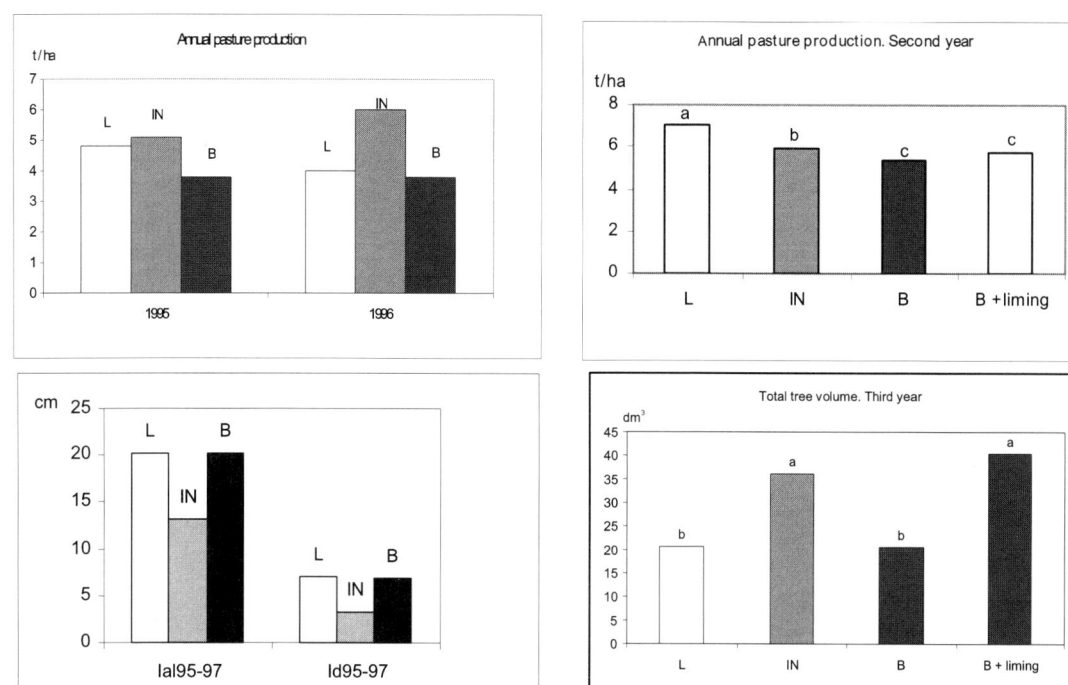

Dactylis glomerata enhanced the quality of *Pinus radiata* tree wood when compared with bare ground (Peri *et al.*, 2002) and the growth of young stands rather than weed species. In an experiment in Galicia, Spain, *Trifolium repens* and *T. pratense* were sown under *Pinus radiata* and *Betula alba* with either perennial ryegrass or cocksfoot. Perennial ryegrass persistence was poor due to its shade intolerance and weed encroachment occurred. This caused a reduction in the soil potassium levels and reduced *Pinus radiata* and *Betula alba* growth (Figure 3). Hence when planting in fertile soils where agricultural land has been abandoned, weed encroachment can cause a reduction in tree growth, even if no fertilizer is applied to increase pasture production. In very acid soils the main type of encroachment comes from shrubs, which usually compete, to a lesser degree, with the trees, due to their slow growth rate. However, the risk of fire is increased.

Figure 3. Height (bars) and diameter (rhombus) of *Pinus radiata* and *Betula alba* after 6 years' growth without fertilization in silvopastoral systems established with *Dactylis glomerata* (Dg) and *Lolium perenne* (Lp).

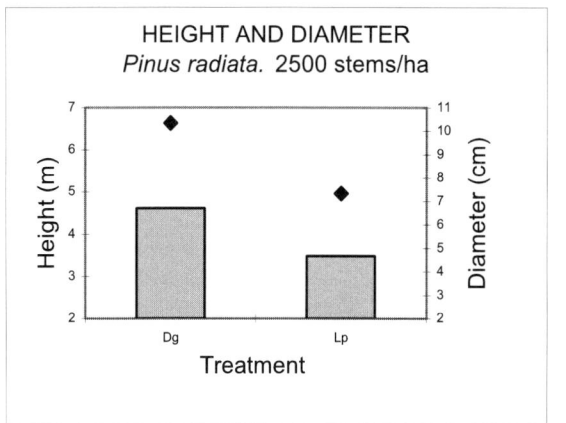
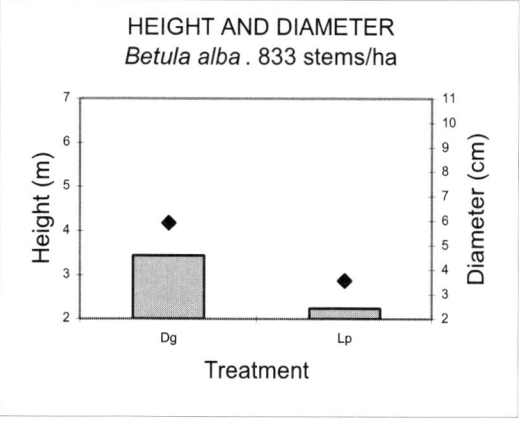

White clover persistence is poor in silvopastoral systems with important tree cover because of its intolerance to shade; hence leguminous species such as *Lotus* which adapt better to shade should be used in Atlantic areas. Other herbaceous species should be evaluated for the highly acidic soils often found in humid Atlantic areas, for example, some of the *Agrostis, Pseudoarrhenaterum, Festuca, Holcus* and *Phleum* genera. Most important attributes will be persistence (especially related to flowering) and production under grazing and shade conditions. Trials have been conducted in Galicia to compare the date of establishment (autumn or spring) of *Lolium multiflorum* Tetilla, *Lolium perenne* cv Brigantia, *Agrostis tenuis* cv. Highland, *Trifolium repens* cv Huia, *Trifolium pratense* cv Marino and *Dactylis glomerata* cv Artabro in open and simulated shaded plots. Establishment, productivity and persistence were better in autumn than spring establishment in the open and better under shade than in the open when sowing was made in spring as the encroachment of annual dicots by shading was limited mainly to *Capsella bursa-pastoris*. However, during autumn establishment, the slow establishment of species of *Dactylis* and *Agrostis* genera allowed encroachment of *Stellaria media* under shade conditions. Both *Agrostis* and *Dactylis* were the species that best adapted to shade conditions (Mosquera-Losada *et al.*, 2001) when annual species disappeared.

It is not advisable to establish pastures by reseeding under mature and dense trees due to the high cost (limited machinery mobility, higher seed rate required) and low success rate of this operation. Some work has been done on establishement of herbs under old trees. Success of this operation depended on the tree type (perennial or deciduous tree) and the crown structure. Generally, herbaceous establishment in autumn sowing under deciduous trees like *Quercus robur* was poor due to the heavy leaf fall. Establishment under *Quercus robur* was better with spring sowing, but, as the light interception of this species is high, pasture productivity is very low for temperate Atlantic area conditions (under 350 kg/ha per year). Species such as *Betula alba* are more suitable for pasture production than *Quercus robur* when the tree crown projects over most of the land surface with a complete canopy. After 6 years of plantation at a high density (2500 trees/ha) pasture production was around 9 t/ha (see Rigueiro-Rodríguez *et al.*, this volume). Light interception by the crown reduces the flowering capacity, which limits the persistence of herbaceous species. On sloping land litter retention is related to herbaceous strata development in soils where there are no shrubs, which limits soil erosion.

It is important to establish a herb layer under trees in areas with a high fire risk. In those areas where short-term income is necessary, this can be obtained through high-quality meat production, tourism woodland use and where there is promotion of the social use of woodlands (Pardini *et al.*, 2002). A herb layer can be established at low cost under heavy shade by grazing alone. Achieving cover will be slower (Rigueiro-Rodríguez *et al.*, this volume).

Conclusions

There are several different kinds of silvopastoral systems which can be adopted or are currently being used in Europe. In an Atlantic area, higher pasture and tree growth can be expected compared with in drier, Mediterranean areas. Hence it is important to evaluate the interaction between tree development and pasture production and its distribution and quality under different conditions in order to match the social, productive and environmental requirements for woodlands or forests in silvopastoral systems.

Acknowledgements

We are grateful to CICYT (Comisión Interministerial de Ciencia y Tecnología), Xunta de Galicia, Gestagua S.A and Agroambprodalt for financial assitance, to Escuela Politécnica Superior for facilities, to Divina Vázquez Varela, José Javier Santiago-Freijanes, Teresa Piñeiro López, Antonio Rodríguez Rigueiro for helping in processing, laboratory and field.

References

Dans del Valle, F., Fernández de Ana-Magán, F.J. and Romero-García, A. (1999) *Pinus radiata Handbook in Galicia* (*Manual de Selvicultura de Pinus Radiata en Galicia*). Proyecto Columela, Universidad de Santiago de Compostela, Santiago de Compostela, Spain. (In Spanish.)

De Miguel, J.M. (2002) Ecology, diversity and sustainable development of agroforesstry systems (Ecología, diversidad y desarrollo sostenible en sistemas agroforestales). *Cuadernos de la Sociedad Española de Ciencias Forestales* 14, 23-31. (In Spanish.)

Dodd, C.J.H. McLean, A. and Brink, V.C. (1972) Grazing values as related to tree-crown covers. *Canadian Journal of Forest Research* 2 (3), 185-189.

Dupraz, C., Bergez, J.E. and Balandier, P. (1997) Improved ventilated tree shelters as a key tool for innovative agroforestry practices in Europe. In: Auclair, D. (eds). *Agroforestry for Sustainable Land use*. Atelier International, Montpellier, France, pp. 275-280.

Echeverría, I. (1943) *Pinus Insignis Treatment. Pruning Intensity, Thinning* (Tratamiento del Pinus insignis. Espesura Podas. Claras). Instituto Forestal de Investigaciones y Experiencias, Madrid, Spain. (In Spanish.)

Etienne, M. (1996) Research on temperatre and tropical silvopastoral systems: a review. In: Etienne, M. (ed). *Western European Silvopastoral Systems*. Institut Nacional de la Recherche Agronomique, Paris, France. 276 pp.

Evans, J. (1984) *Silviculture of Broadleaved Woodland*. Forestry Commission bulletin 62, Her Majesty's Stationery Office, London, United Kingdom. 232 pp.

Finck, P., Riecken, U. and Shröder, E. (2002) Pasture landscapes and nature conservation. New strategies for the preservation of open landscapes in Europe. In: Redecker, B., Finck, P., Härdtle, W., Riecken, U. and Schröder, E. (eds). *Pasture Landscapes and Nature Conservation*, Springer, Bonn, Germany, 2nd edn. pp. 1-13.

García-González, D. and Contreras-Olalla, R. M. (1998) Effect of stocking rate on holy tree regeneration in Garagueta Mountain (*Influencia de la carga ganadera sobre el acebo y su regeneración en el monte Garagueta*). XXXVIII Reunión Científica de la Sociedad Española para el Estudio de los Pastos, 349-352. (In Spanish.)

Gómez Gutiérrez, J.M. and Calabuig, L. (1992) Pasture production (Producción de praderas y pastizales). In: Gómez Gutiérrez, J.M. (ed.) *El Libro de las Dehesas Salmantinas*. Consejería de Medio Ambiente de Castilla y León, Salamanca, Spain, pp. 489-513. (In Spanish.)

González-Río, F., Castellanos, A., Fernández, O. and Gómez, C. (1997) *Eucalypt Selviculture Handbook* (*Manual Técnico de Selvicultura del Eucalipto*). Proyecto Columela, Universidad de Santiago de Compostela, Santiago de Compostela, Spain. (In Spanish.)

Hawke, M.F. (1991) Pasture production and animal performance under pine agroforestry in New Zealand. *Forest Ecology and Management* 458 (1-4), 109-118.

Jung, G.A., Van Wijk, A.J.P., Hunt, W.F. and Watson, C.E. (1996) Ryegrasses. In: Moser, L.E., Buston, D.R., Casler, M.D., Barters, J.M., Peterson, G.A., Baenziger, P.S. and Bigham, J.M. (eds) *Cool-Season Forage Grasses*. Madison, Wicosin, USA. pp. 605-642.

Klapp, (1971) *Prados e pastagems*. Fundacao calouste Gulbenkian, Lisbon, Portugal. (In Portuguese.)

Koukora, Z. and Papanastasis, V. (1996) Establishment and growth of seeded grasses and legumes in pine silvopastoral systems. In: Etienne, M. (ed) *Western Eurpean Silvopastoral Systems*. Institut Nacional de la Recherche Agronomique. Paris, France. pp. 83-92.

Kuiters, A. (2004) *Silvopastoralism and Sustainable Management*. International Congress, Book Abstracs, Lugo, Spain.

Larsson, T.B. (2001) *Biodiversity Evaluation Tools for European Forests*. Ecological Bulletin, 50, Swedish Environmental Protection Agency, Lund, Sweden.

López-Díaz, M.L., Rigueiro-Rodríguez, A. and Mosquera-Losada, M.R. (2001) Liming and organic fertilization effect on *Pinus radiata* D. Don growth and on pasture production in silvopastoral systems (Efecto del encalado y la fertilización orgánica sobre el crecimiento de *Pinus radiata* D. Don y la producción de pasto en sistemas silvopastorales). In: *III Congreso Forestal Español,* Junta de Andalucia, Granada, Spain*,* 5, 171-173. (In Spanish.)

Llorente-Gil., J.A. and Ciria-Ciria, J. (1998) Cervus elephus: an alternative of the silvopastoral system of the Almuerzo Mountain (Soria) (*El ciervo (Cervus elaphus) una alternativa en el aprovechamiento silvopastoral de la sierra del almuerzo (Soria)*). Actas de la XXXVIII Reunión Científica de la Sociedad Española para el Estudio de los Pastos, Soria, Spain. pp. 349-352. (In Spanish.)

McAdam, J.H., Hoppé, G.M., Toal, L. and Whiteide, T. (1999) The use of wide-spaced trees to enhance faunal diversity in managed grasslands. *Grassland Science in Europe* 9, 293-297.

Mentxaka, V., Pinto, M., Rodríguez, M. and Besga, G. (1998) Herbaceous and forestry component interaction in a silvopastoral system (*Interacciones entre el componente herbáceo y el forestal en un sistema silvopastoral*). XXXVIII Reunión Científica de la Sociedad Española para el Estudio de los Pastos Soria, Spain. pp. 357-360. (In Spanish.)

Merou., T. and Vrahnakis, M.S. (1999) Effects of kermes oak (*Quercus coccifera* L.) on understorey plant species diversity and evenness in a Mediterranean-type shrubland. *Grassland Science in Europe* 4, 297-302.

Mosquera-Losada, M.R., Villarino-Urtiaga, J.J. and Rigueiro-Rodríguez, A. (1999) *Ecology and Sward Management* (Ecología y Manejo de praderas). MAPA, Santiago de Compostela, Spain.

Mosquera-Losada, M.R., González-Rodríguez, A. and Rigueiro-Rodríguez, A. (2000) Sward quality affected by different grazing pressures on dairy systems. *Journal of Range Management* 53 (6), 603-611.

Mosquera-Losada, M.R., Rigueiro-Rodriguez, A., Lopez-Díaz, M.L. and Rodríguez-Barreira, S. (2001) Shading and sowing date effect on establishment and production of different sward species (Efecto del sombreado y la época de siembra en el establecimiento y producción de varias especies pratenses). *Investigación Agraria* 16 (2), 169-187. (In Spanish.)

Nair, P.K.R. (1993) *An Introduction to Agroforestry*. Kluwer Academic Publishers, Dordrecht, The Netherlands and ICRAF, Nairobi. Kenya.

Papanastasis, V. (1996) Silvopastoral systems and range management in the Mediterranean Region. In : Etienne, M. (ed) *Western Eurpean Silvopastoral Systems*. Institut Nacional de la Recherche Agronomique, Paris, France. 276 pp.

Papanastasis, V. (1999) Grasslands and woody plants in Europe with special referente to Greece. *Grassland Science in Europe* 4, 15-24.

Papanastasis, V.P. (2004) Vegetation degradation and land cover changes in agrosilvopastoral systems. In: Schnabel, S. and Goncalves, A. (eds) Sustainability of agrosilvopastoral systems-dehesa, montados. *Advances in GeoEcology*. 37, 1-12.

Pardini, A., Mosquera, M.R. and Rigueiro, A. (2002) Land management to develop naturalistic tourism. In: *Proceedings V International IFSA (International Farming Systems Association)*. Florence, Italy, pp. 399-407.

Peyraud, J.L., Mosquera-Losada, M.R. and Delaby, L. (2004) Challenges and tools to develop efficient dairy systems based on grazing: how to meet animal preformase and grazing management. *Grassland science in Europe*, 9, 373-384.

Peri, P.L., McNeil, D.L., Moot, D.J., Varella, A.C. and Lucas, R.J. (2002) Net photosynthetic rate of cocksfoot leaves under continuous and fluctuating shade conditions in the field. *Grass and Forage Science* 57 (2), 157-170.

Piñeiro, J. (1988) Pastures species under pines. In: Proceedings 12 th General Meeting of the European Grassland Federation Dublín, Ireland. pp. 287-291.

Rigueiro, A., Silva, F.J., Rodríguez, R., Castillón, P.A., Álvarez, P., Mosquera, M.R. and González, M.P. (1998) *Handbook of Silvopastoral Systems* (Manual de Sistemas Silvopastorales). Universidad de Santiago de Compostela, Santiago de Compostela, Spain. (In Spanish.)

Rigueiro-Rodríguez, A., Mosquera-Losada, M.R. and López-Díaz, M.L. (1998) Silvopastoral systems in prevention of forest fires in the forests of Galicia (NW Spain). *Agroforestry Forum* 9 (3), 3-8.

Rigueiro-Rodríguez, A., Mosquera-Losada, M.R. and Gatica-Trabanini, E. (2000) Pasture production and tree growth in a young pine plantation fertilized with inorganic fetilizers and milk sewage in northwestern Spain. *Agroforestry Systems* 48, 245-256.

Sibbald, A. (1994) Herbage yield in agroforestry systems as a function of easily measured attributes of the tree canopy. *Forest Ecology and Management* 65, 195-200.

Sibbald, A. (1996) Silvopastoral systems on temperate sown pastures: a personal perspective. In: Étienne, M. (ed.) *Western European Silvopastoral Systems*. INRA, Paris, France, pp. 23-37.

Sibbald, A.R. (1999) Silvopastoral agroforestry: soil-plant-animal interactions in the establishment phase. *Grassland science in Europe* 4, 133-145.

Sibbald, A.R. and Agnew, R.D.M. (1996) United Kingdom silvopastoral network experiment. Annual report, 1995. *Agroforestry Forum* 7 (1), 7-10.

Sibbald A.R., Griffiths, J.H. and Elston, D.A. (1991) The effects of the presence of widely spaced conifers on under-storey herbage production in the United Kingdom. *Forest Ecology and Management* 45 (1-4), 71-77.

Sibbald, A.R., Eason, W.R., McAdam, J.H. and Hislop, A.M. (2001) The establishment phase of a silvopastoral national network experiment in the United Kingdom. *Agroforestry Systems* 53 (1), 39-53.

Teklehaimanot, Z., Jones, M. and Sinclair, F.L. (2002) Tree and livestock productivity in relation to tree planting configuration in a silvopastoral system in North Wales, United Kingdom. *Agroforestry Systems* 56, 47-55.

Xunta de Galicia (2001) *Figures of Galician Forest* (O Monte Galego en Cifras). Dirección Xeral de Montes e Medio Ambiente Natural, Santiago de Compostela, Spain. (In Spanish.)

Yrjölä, T. (2002) Forest management guidelines and practices in Finland, Sweden and Norway. Internal Report of European Forest Institute, 11: http://www.efi.fi/publications/Internal_Reports/IR_11.pdf (Accessed 2 March 2004)

The effect of pruning intensity on acorn production in a holm-oak "dehesa"

R. Alejano, J. Alaejos, E. Torres, J. A. Forero, R. Tapias and M. Fernández

Dept. de Ciencias Agroforestales, University of Huelva, Campus de La Rábida, 21819 Palos de la Frontera, Spain. ralejan@uhu.es

Abstract

In this work the effect of the type of pruning on acorn production in a dehesa of Huelva province (south-west Spain) is evaluated. A plot including 100 holm-oaks *(Quercus ilex* L.) is replanted, and different pruning treatments are carried out (light, moderate, strong and hedge shaped), and there are also trees without pruning. Acorn production is measured in three trees per pruning treatment. Significant differences between acorn production depending on pruning treatment were not found.

Key words: *Quercus ilex*

Introduction

The Iberian pig and cattle raising is one of the most important resources of Mediterranean open woodlands or "dehesas". A better knowledge of acorn production could help to improve its management. Pruning is a silvicultural treatment practised by the greater part of forest owners, who generally accept its positive effect on acorn production (Carbonero *et al.,* 2002). But, because of the scarce scientific results to confirm this hypothesis and the general situation of decline of the holm-oak (*Quercus ilex* L.), carrying out this work was considered very interesting: it contributes scientific results and allows technical criteria to develop for a good open woodland management that does not endanger its conservation.

Materials and Methods

An experimental plot that includes 100 holm-oaks has been replanted at Calañas (Huelva province, south-west Spain) in a grazed dehesa. Four types of pruning were applied in October 2001. We describe the prunings as light, moderate, strong and "hedge-shaped". The plot includes 20 trees per type of pruning and 20 trees without pruning. The trees are 80-100 years old. Three trees for every pruning treatment were selected by stratified sampling to estimate acorn production. The acorn harvesting was carried out by means of containers or traps method. Four containers with 0.45 m diameter at the top were placed under the trees following north, south, east and west directions at 75% of crown radial distance from the base of the tree. Acorn harvesting was carried out during the years 2001 and 2002, every 15 days, from October to January. The acorns were counted and weighed in the laboratory. To test if there were significant differences between acorn production (g/m^2) in 2001 and 2002 depending on pruning treatment a variance analysis has been carried out.

Results and Discussion

Acorn production in 2001 and 2002 depending on pruning treatment is presented in Table 1. We can compare this information with that obtained by Martin *et al.* (1998), who find acorn productions between 11.6 and 285.8 g of g DM/m^2, or with the productions measured by Alvarez *et al.* (2002) with a median of 19 kg/tree (minimum of 0.12 and maximum of 87.85).

Table 1. Acorn production per tree in g DM/m^2 of crown surface and in kg DM/tree acorns in Calañas plot.

Treatment	2001		2002	
	g/m^2	kg/tree	g/m^2	kg/tree
Without pruning	8.91	0.37	91.95	3.08
Light	134.44	7.55	209.25	13.28
Moderate	94.85	9.98	195.86	15.24
Strong	51.51	3.92	121.81	6.60
Hedge-shaped	120.58	4.20	237.39	11.39
Total median	82.06	5.20	171.42	9.92

At the variance analysis we have checked that there are not significant differences in acorn production depending on pruning treatment. In 2001 the least average production came from non-pruned trees and from heavily pruned trees. The greatest production of this year came from lightly pruned and hedge-shaped pruned trees. In 2002 the smallest average production came also from non-pruned and heavily pruned trees. The greatest production of this year came from hedge-shaped trees.

We confirm the importance of masting for acorn production between years. So we can see the difference between average production in 2001 (82.06 g/m^2 and 5.20 kg/tree) and in 2002 (171.42 g/m^2 and 9.92 kg/tree). Those results must be confirmed by researching on the following years. A strong variability in acorn production between individual trees was found.

Conclusions

The kind of pruning does not affect acorn production with statistical signification. As other authors have affirmed, there are factors like masting or the individual capacity of fruiting that hide the effect that pruning could have on acorn production. It will be necessary to continue with the acorn harvesting the following years to obtain more definitive results that can help to manage the "dehesas" in a more sustainable way.

References

Alvarez, S., Morales, R., Bejarano, L. and Durán, A. (2002)Acorn production in the dehesas in Salamanca (Producción de bellota en la dehesa salmantina). In: Actas de la XLII Reunión Científica de la Sociedad Española para el Estudio de los Pastos. SEEP, Salamanca, Spain, pp. 645-650. (In Spanish.)

Carbonero, M.D., Fernández, P. and Navarro, R. (2002) Assessment of acorn production and size of *Quercus ilex* L. subsp. *ballota* (Desf) Samp during a pruning period. Results of the 2001-2002 campaign (Evaluación de la producción and del calibre de bellotas de *Quercus ilex* L. subsp. *ballota* (Desf) Samp a lo largo de un ciclo de poda. Resultados de la campaña 2001-2002). In: *Actas de la XLII Reunión Científica de la Sociedad Española para el Estudio de los Pastos*. SEEP, Lérida, Spain, pp. 633-638. (In Spanish.)

Martín, V., Infante, J. M., García, J., Merino, J. and Fernández, R. (1998) Acorn production in forests and dehesas in south-western Spain (Producción de bellotas en montes and dehesas del suroeste español). *Pastos XXVIII* (2), 237-248. (In Spanish.)

The effect of tree density on pasture yield

H. Arzani
College of Natural Resources, University of Tehran, Karaj, Iran, harzani@ut.ac.ir

Abstract
An experiment was conducted to evaluate the effect of tree density on dry matter yield. Three levels of tree density - high, medium and low - were selected in two different land systems at Manuka and Cultowa in Western Division of New South Wales. At Manuka there were significant differences between tree distance between clumps ($P < 0.05$). No significant difference was found between average tree diameters for medium and high tree density areas, but significant differences were obtained between low, compared with high and medium tree densities ($P < 0.01$). Despite differences in diameter, height and number of trees per transect there were no significant differences between dry matter yield estimated under tree storeys. At Cultowa, there were significant differences ($P < 0.05$) between densities and distance and height between clumps. Significant differences were also observed between average tree diameter and pasture dry matter yield of medium density with others. There were no significant differences between dry matter yield estimates under high and low tree densities.

Key words: overstorey, dry matter, production, silvopasture

Introduction
Due to competition for moisture, pasture production generally is influenced by its overstorey. The effect of trees depends on tree density and their effective canopy cover (Scanlan and Burrows, 1990). Condon (1968) stated that large eucalypts over a canopy diameter of 15-20 m, have a marked influence on pasture growth and mulga (*Acacia aneura*) over a canopy diameter of 4-6 m has a slight influence. Moore and Deiter (1992) established a stand density index (SDI) to predict forage in northern Arizona pine forests. Payke and Zamora (1982) found that canopy coverage of overstorey and sum of tree diameters were the most significantly correlated parameters with forbs production. The objective of the present study was a comparison of dry matter yield under different tree densities.

Materials and Methods
Three levels of tree density (high, medium and low) were selected in two different land systems at Manuka (located 90 km south of Cobar 145°40'E, 32°2'S with average annual rainfall of 364 mm) and Cultowa (175 km west of Cobar, 144°13'E, 31°40'S with average annual rainfall of 248 mm) in Western Division of New South Wales. Within each level two 100 × 10 m belt transects were established, except in the low tree density area at Cultowa, where 150 × 15 m belt transects were used due to the large distances between trees. Tree species were bimble box (*Eucalyptus populnea*), red box (*E. intertexta*), yarran (*Acacia homalophylla*), mallee (*Eucalyptus* sp) and white pine (*Callitris collumellaris*). Coordinates of trees were recorded along X and Y axes of the transects. Height and diameter of each individual tree were measured and four 0.5 m^2 quadrats were located 1, 2, 3 and 4 metres out from the basal trunk area of each tree. Dry matter yield was estimated within each quadrat using the dry-weight-rank method of Mannetje and Haydock (1963). Soil samples were collected from Manuka. Low and medium tree densities were located on the same land class; therefore the same soil samples were used. Physical and chemical soil characteristics were determined according to the procedures described by Melville (1993). Organic carbon was converted to organic matter using 1.9 and 2.5 multipliers (Alison, 1960).

One-way analysis of variance was used to evaluate differences between factors including diameter, distance and height as independent variables and dry matter yields as dependent variables between different levels of tree densities.

Results and Discussion
On average, distance between trees ranged from 4.3 m in the high tree density area to 15.7 m in the low tree density area at Manuka (Table 1). There were significant differences between tree distance between clumps ($P < 0.05$). No significant difference was found between average tree diameters for medium and high tree density areas, but significant differences were obtained between low, compared with high and medium tree densities ($P < 0.01$). Despite differences in diameter, height and number of trees per transect, tree density had no significant differences between dry matter yield estimated under tree storeys during January and September 1992 at Manuka where spear grass (*Stipa variabilis*) and annual medics (*Medicago* sp) were dominant.

At Cultowa there were significant differences ($P < 0.05$) between densities, distance and height between clumps (Table 1). Significant differences were also observed between average tree diameter and pasture dry matter yield of medium tree density with others. *Chenopodium anidiophyllum* and *Atriplex* sp were abundant around and under trees in the medium tree density clump. Dry matter yield was not measured under the medium tree density in September 1992. The highest dry matter yield was recorded for the medium tree density area in December 1991 and the lowest from the high density area in September 1992. There were no significant differences between dry matter yield estimates for high and low tree densities on either occasion.

Table 1. The effect of distance, height, diameter and dry matter yield at Manuka and Cultowa. DMY = dry matter yield, HD = high density, MD = medium density, LD = low density. * = significant difference at $P < 0.05$, ** = significant difference at $P < 0.01$.

Site	Factor	HD	MD	LD
Manuka Jan. 1992	Tree/ha	315.0 **	165.0 **	95.0 **
	Distance (m)	4.3 ± 0.4 *	9.6 ± 2.6 *	15.7 ± 4.4 *
	Diameter (m)	5.3 ± 0.4 ns	6.4 ± 0.5 ns	10.3 ± 0.8 **
	Height (m)	14.6 ± 0.8 ns	15.4 ± 0.9 ns	22.8 ± 1.8 **
	DMY (kg/ha)	192.0 ± 33.5 ns	200.0 ± 54.3 ns	153.0 ± 46.5 ns
Manuka Sep. 1992	DMY (kg/ha)	71 ± 2.0 ns	70.0 ± 3.3 ns	65.5 ± 3.4 ns
Cultowa Dec. 1991	Tree/ha	250.0 **	175.0 **	35.0 **
	Distance (m)	5.2 ± 0.4 ns	9.0 ± 2.8 ns	23.6 ± 8.5 **
	Diameter (m)	5.8 ± 0.4 ns	3.3 ± 0.3 *	5.3 ± 0.7 ns
	Height (m)	15.7 ± 0.8 *	4.8 ± 0.3 *	9.6 ± 1.4 *
	DMY (kg/ha)	173 ± 11.2 ns	501.7 ± 116 **	104.5 ± 15 ns
1 Cultowa Sep. 1992	DMY (kg/ha)	83.3 ± 9.5 ns	Not measured	110.0 ± 18.7 ns

At Manuka soil properties were affected by tree density. Slightly more favourable fertility and lower acidity were found under the high tree density area (Table 2).

Table 2. Soil characteristics under tree storey at Manuka. HD = high density, MD = medium density, LD = low density, SL = sandy loam, CL = clay loam, Text = texture, pH = acidity, EC = electrical conductivity, OC = organic carbon, OM = organic matter, AP = available phosphorus and N = nitrogen.

Location	Layer	Text.	PH	EC (dS/m)	OC%	OM%	AP (mg / kg)	N%
HD	0-18	SL	6.5	0.06	0.99	1.8	0.6	0.08
	18-42	CL	6.4	0.03	0.34	0.85	1.08	0.05
MD and LD	0-18	SL	6.1	0.06	0.75	1.43	0.25	0.06
	18-42	SL	6.3	0.04	0.4	1	1.09	0.04

The level of overstorey effect on herbaceous production varies between and within areas, based on land system characteristics during the year. The reason for higher productivity in high and medium tree density at Manuka is partly due to slightly higher soil fertility, possibly more runoff receipt (because of topographic conditions) and lower evaporation during summer. Although there was no significant difference between dry matter yield estimated under different tree densities, at both Manuka and Cultowa, the relative differences between dry matter yield of low and high tree density during summer was greater than in winter. High evaporation during the summer is a limiting factor in this area and can be reduced by trees. These differences could be also partly related to the influence of shading on vegetation, and partly related to vegetation types beneath each tree storey which has been supported by Frost and McDougald (1989).

Although data from Manuka and Cultowa included different seasons, they were sampled during drought. Therefore it is difficult to confirm or comment on the rating scale generated by Condon (1968). Further investigation is required to create a reliable rating scale, considering other factors which can be used in different vegetation types and land systems.

Conclusions

Level of overstorey effects on herbaceous production varies between areas and during the year according to the land system characteristics. In silvopastoral systems tree density depends on land system, climatic condition, overstorey and understorey species for both wood and forage production.

References

Alison, L.E. (1960) Wet combustion apparatus for organic and inorganic carbon in soil. *Soil Science Society of America Proceedings* 24, 36-40.

Condon, R.W. (1968) Estimation of grazing capacity on arid grazing lands. In: Stewart, G.A. (ed.) *Land Evaluation.* Macmillan of Australia, Brisbane, Australia, pp. 112-124.

Frost, W.E. and McDougald, N.K. (1989) Tree canopy effects on herbaceous production of annual rangeland during drought. *Journal of Range Management* 42, 281-283.

Mannetje, L.T. and Haydock, K.P. (1963) The dry-weight-rank method for the botanical analysis of pasture. Journal of the British Grassland Society 18, 268-275.

Melville, M. (1993) *Soil Laboratory Manual.* University of NSW, Sydney, Australia.

Moore, M.M. and Deiter, D.A. (1992) Stand density index as a predictor of forage production in northern Arizona pine forests. *Journal of Range Management* 45, 267-271.

Payke, D.A. and Zamora, B.A. (1982) Relationships between overstorey structure and understorey production in the Grand Fir/Myrtle boxwood habitat type of Northcentral Idaho. *Journal of Range Management* 35, 769-733.

Scanlan, J.C. and Burrows, W.H. (1990) Woody overstorey impact on herbaceous understorey in *Eucalyptus* sp communities in central Queensland. *Australian Journal of Ecology* 15, 191-197.

An alternative to improve goat production in the mountain areas of north-west Tunisia

A. Boubaker [1], C. Kayouli [1] and A. Buldgen [2]

[1] Laboratoire des Ressources Animales and Alimentaires, INAT, Tunisia ; [2] Faculté des Sciences Agronomiques de Gembloux, Belgium. azizaboubaker@yahoo.com

Abstract

Twenty one local kids (18.75 kg; s.d. ± 1.4) were divided into three equal groups and allowed to browse in shrublands from July to December (dry season). Two groups received feed blocks containing mainly wheat bran and molasses and either included polyethylene glycol (B-PEG) or were PEG-free (B). A control group did not receive any supplement. Kids on B and B-PEG had free access to the blocks in the stalls at night. During the same period, the chemical composition of 12 shrub species from the same browsing area was evaluated in samples collected monthly. Most shrubs are low in crude protein and high in fibre and phenolic compounds. The final body weight of the kids was higher ($P < 0.05$) for B-PEG and B than for the control group. While the control group lost 19 g/day, kids given B and B-PEG grew at rates of 12 and 24 g/day, respectively. It seems that feed block supplementation contributes to a better use of forest resources by kids browsing in the critical period of the year such as the summer. The enrichment with PEG may avoid any risk of tannin toxicity, while reducing pressure on the natural grazing resources and the loss of biodiversity from overgrazing.

Key words: feed blocks, goats, forest

Introduction

The rangelands of Tunisia occupy an area of nearly 5.4% of total land area of the country. Extensive livestock play an important role in the economy of this area. Goats are used primarily for subsistence and frequently as a medium of cash flow. They are reared mainly for meat production under a silvopastoral system dominated by *Quercus suber* and shrubs (mainly *Pistacia lentiscus*, *Myrtus communis* and *Erica arborea*).

To provide adequate nutrition for goats on woody species, Perevolotsky *et al.* (1993) proposed a protein supplement while Gilboa *et al.* (1996) suggested ways to overcome the negative effects of a high level of tannins. Under extensive conditions, providing supplemental nitrogen and/or polyethylene glycol (PEG) to livestock in feed blocks appears to be an interesting alternative to minimize the effects of high tannins level. The manufacture of these blocks is an easy technique and well adapted to small farmer conditions. In this study, the effect of PEG-free and PEG-containing feed blocks on growth of kids browsing during the dry season was investigated.

Materials and Methods

The experiment was carried out in the north-west of Tunisia during the dry season (July-December). The area is a typical Mediterranean shrubland dominated by *Cistus salvifolius*, *Erica arborea*, *Myrtus communis*, *Phillyrea angustifolia*, *Pistacia lentiscus*, *Quercus* sp. Grasses were not available in the period of study. Twenty-one local kids (18.75 kg; s.d. ± 1.4) were used in a 6 months growth trial. All kids browsed together in the shrubland each day for 7 h. At night, animals were separated in three equal groups. Group 1 received blocks PEG-free (B), group two received blocks containing PEG (B-PEG) and the third group was not supplemented (control). Feed blocks were made of wheat bran (54%), cement (10%), molasses (10%), urea (5%), salt (10%), quicklime (5%), bicalcium phosphate (5%) and sulphur (1%). Polyethylene glycol (PEG, mol. wt. 4000) was added to the ingredient mixture before moulding to form 18% of the block. After 2 months of drying in a shady site, blocks were gradually given for experimental animals to avoid any toxicity problems. Feed blocks were offered to animals for 1 hour during the first 3 days and for 4 hours the following 3 days and afterwards they were continuously available (average of three pieces per group). Body weight of kids was recorded every 3 weeks and intake of blocks was controlled daily.

Leaves of 12 woody species seen browsed at varying frequencies were harvested monthly from July to December. Crude protein (AOAC, 1975) and neutral detergent fibre (Goering and Van Soest, 1970) were analysed for each species separately and monthly. Total phenols were determined on samples harvested only in August as described by Scehovic (1990). The effect of supplementation on growth was compared with analysis of variance using the SAS general linear models procedure (SAS, 1985) for a completely randomized design. Duncan multiple range test was used to compare means of treatments.

Results and Discussion

Overall, the woody vegetation was relatively low in crude protein (7.3%), but high in fibre (55.2%) and in phenolic compounds (13.1%). Similar trends have been observed in other Mediterranean shrubs browsed by goats in summer

(Decandia et al., 2000). The CP concentration is typically below the N requirement of actively grazing goats (NRC, 1981). In addition, this type of vegetation may be classified as rich in anti-quality factors which would generally be considered to be potentially detrimental to herbivores.

Kids consumed similar amounts of PEG-free (B) and PEG-containing (B-PEG) blocks (130 g DM/animal per day). During the first 2 months of the trial, the control group lost ($P < 0.05$) more weight (-21%) than B-PEG and B (-6%). The final body weight was higher ($P < 0.05$) for B-PEG and B than for the control group (22.6; 20.8; 15.5 kg, respectively). While the control group lost 19 g/day, kids given B and B-PEG grew at rates of 12 and 24 g/day, respectively. This small difference does not appear to be biologically important, despite its statistical significance, but supplemented kids gained 3.5 kg (B-PEG) and 1.1 kg (B) while the control group lost 2.8 kg during 147 days. Silanikove et al. (1996) reported that, when browses were offered as a sole feed, they did not sustain the maintenance requirement of the animals and weight gain rapidly decreased (100 g/day). The deficiency of essential nutrients (nitrogen, energy, minerals and vitamins) necessary for efficient microbial growth leads to low rates and extent of digestion in the rumen. This reduces voluntary feed intake causing low animal performance. Multi-nutritional blocks have been used to correct nutritional imbalance for rumen bacteria. They are considered to be a good supplement for poor quality diets, as they allow a balanced, synchronous and fractionated supply of nutrients to ruminants. Goats on native shrublands and receiving feed blocks grew at rates significantly higher than those with or without concentrate. In this study, kids receiving B-PEG gained more weight than kids on B. Polyethylene glycol can increase digestibility (Silanikove et al., 1996) and intake of tannin-containing plants (Villalba and Provenza, 2002), thereby enhancing animal performance. Its incorporation in hard blocks is expected to synchronize tannins and PEG consumption and therefore maximizes benefit from PEG use to overcome the negative effects of tannins.

Conclusions

It could be concluded that feed blocks would be a suitable and viable tool to sustain performance of kids reared under a silvopastoral system in the mountain areas of north-west Tunisia. The enrichment of blocks with PEG may avoid any risk of tannin toxicity, reduce the loss of biodiversity from overgrazing and fire risk.

References

Association of Official Analytical Chemists (AOAC) (1975) *Official Methods of Analysis*, 12th edn Association of Official Analytical Chemists, Washington, DC, USA.

Decandia, M., Sitzia, M., Cabiddu, A., Kababya, D. and Molle, G. (2000) The use of polyethylene glycol to reduce the anti-nutritional effects of tannins in goats fed woody species. *Small Ruminant Research* 38, 157-164.

Gilboa, N., Silanikove, N. and Nitsan, Z. (1996) Interactions among supplementation and polyethylene glycol in goat fed tannin rich (*Quercus calliprinos* and *Pistacia lentiscus*) leaves. In: *Recent Advances in Small Ruminant Nutrition.* FAO-CIHEAM Network of Cooperative Research on Sheep and Goats, Rome, Italy, p. 7.

Goering, H.K. and Van Soest, P.J. (1970) *Forage Fibre Analyses (Apparatus, Reagents, Procedures and some Applications).* Agriculture Handbook No. 379, ARS, USDA, Washington, DC, USA.

NRC (1981) *Nutrient Requirements of Domestic Animals, No.15. Nutrient Requirements of Goats.* National Academy of Sciences, Washington, DC, USA.

Perevolotsky, A., Brosh, A., Ehrlich, O., Gutman, M., Henkin, Z. and Holzer, Z. (1993) Nutritional value of common oak (*Quercus calliprinos*) browse as fodder for goats: experimental results in ecological perspective. *Small Ruminant Research* 11, 95-106.

SAS (1985) *SAS User's Guide: Statistics,* 5th edn. SAS Institute, Cary, NC, USA.

Scehovic, J. (1990) Tannins and other phenol polymers of forage plants: identification and their biological activity (Tannins et autres polymères phénoliques dans les plantes de prairies: détermination de leur teneur et de leur activité biologique). *Revue Suisse Agricole* 22, 179-184. (In French.)

Silanikove, N., Gilboa, N., Perevolotsky, A. and Nitsan, Z. (1996) Effect of a daily supplementation of polyethylene glycol on intake and digestion of tannin-containing leaves (*Quercus calliprinos, Pistacia lentiscus* and *Ceratonia siliqua*) by goats. *Journal of Agriculture and Food Chemistry* 44, 199-205.

Villalba, J.J. and Provenza, F.D. (2002) Polyethylene glycol influences selection of foraging location by sheep consuming quebracho tannin. *Journal of Agriculture Science* 80, 1846-1851.

Seasonal fodder biomass production of palo dulce and tepozan

D. Camacho-Morfín [1], C. Sandoval-Castro [2], A. Ayala-Burgos [2] and L. Morfín-Loyden [1]

[1] Bromatology area, UNAM, FES-C, Campo 4, km 2.5 carretera Cuautitlán-Teoloyucan. Cuautitlán Izcalli, Edo. de Méx, Mexico. [2] Faculty of Medicine Veterinary and Zootehnical, University Autonomous of Yucatán, km 15.5 carretera Xmatkuil, Mérida, Yucatán, Mexico. denebcm@correo.unam.mx, ccastro@tunku.uady.mx

Abstract

In order to describe the performance of the seasonal biomass production of palo dulce (*Eysenhardtia polystachya*) (PD) and tepozan (*Buddleia cordata*) (TZ), individuals were cut every 84 days in rotation. It was found that the mortality was higher in TZ (52%) than in PD in the first year and mortality of PD increased in the second year to 23%. Higher forage production was found on PD in summer and biomass production was correlated to the precipitation and minimum temperature. It was concluded that PD and TZ could be combined in order to produce biomass all year long and an 84-day cutting regime is undesirable.

Key word: *Eysenhardtia polystachya*, *Buddleia cordata*, shrubs, forage production

Introduction

The Mezquital Valley is inhabited by the indigenous Hña Hñu. In this area ruminant production is extensive and it has led to overgrazing, soil erosion and compaction. The local government has an interest in natural resources protection and sustainable animal production. It is important to define the use of shrubs and trees. This work is part of the project: "Identification and characterization of woody fodder species from the Mezquital Valley, Hidalgo, Mexico". PD and TZ were selected from 38 fodder species because they showed highest digestibility (52 and 60%, respectively), low condensed tannins contents (2%) and high acceptability by sheep (Camacho-Morfín *et al.*, 1998, 1999). PD and TZ are eaten by ruminants (cattle, goats, sheep and deer) and they are used as fuelwood, medicine, fences and provide wildlife habitat and revegetation in disturbed sites on semiarid areas of Mexico (Granados and Hernández, 1995; Camacho-Morfín *et al.*, 1998). There is no available information on their annual fodder production and the aim of this study was to describe the seasonal biomass production of palo dulce (*Eysenhardtia polystachya*) (PD) and tepozan (*Buddleia cordata*) (TZ), as a precursor to the first step for increasing their use in silvopastoral systems.

Materials and Methods

The experiment was conducted at FES-Cuautitlán, UNAM, in Cuautitlán Izcalli, State of Mexico, Mexico (19° 37' and 19° 45' and 99° 07' and 99° 14'), altitude of 2400 m asl and climate C(Wo)(W), rain in summer and dry winter. The seasonal forage production was evaluated in plots of PD and TZ established with seeds in 1991. Each plot was 6 m × 6 m, plants were transplanted at 1 m × 1 m spacing and cut every year since 1992. Seasonal grass forage production was determined by cuts over two years (periods) (Corrall and Fenlon, 1978). All swards were cut to 0.5 m in August 1997, and the plants were randomly assigned in four groups. Each TZ group had eight plants, each PD group nine plants. All groups were cut every 21 days and the same group was cut every 84. That time range was chosen to represent one full season and estimate seasonal fodder production. The edible DM forage production was determinated on branches of 10 mm diameter. Data on precipitation and temperature were collected from the climatological station of FES-C, UNAM. The interrelations between the seasonal forage production and the precipitation and temperatures were determined by correlation using Graph Pad Prism (1996).

Results and Discussion

The mortality in TZ was so high in the first period that cuts were stopped at the start of the second period (Table 1). The different survival of the species suggests different ways of responding to the stress conditions like cuts, low temperature and drought, because the defoliation involved many stress factors (Raghavendra, 1991; Nilsen and Orcutt, 1996). Forage production in both species was correlated to precipitation and minimum temperature (Table 2).

Table 1. Live and dead plants of palo dulce and tepozan, after two cuts periods. n: number of plants.

	Period		
	First (Sep. 1997-Sep. 1998)		Second (Sep. 1998-Sep. 1999)
	Palo dulce	Tepozan	Palo dulce
Total (n)	35	29	35
Live (%)	94.29	48.28	77.14
Dead (%)	5.71	51.72	22.86

Performance of seasonal fodder biomass production for each species was different in the first year (Figure 1). PD had highest biomass production in summer, but was leafless over winter. TZ produced fodder all the year, more or less at the same quantity every season, except in winter, although this production was lower than PD. Corrall and Fenlon (1978) found that fodder production in grasses was seasonal and species-specific.

Table 2. Correlation among environmental conditions and seasonal forage productions of palo dulce and tepozan.
ns: no significative; *: $P < 0.05$; ** $P < 0.01$; ***: $P < 0.001$.

Species	Precipitation	Days above 0°C	Temperature		
			Maximum	Minimum	Mean
Palo dulce	0.76 ***	-0.38 ns	-0.29 ns	0.63 ***	0.35 ns
Tepozan	0.69 **	-0.52 *	0.14 ns	0.69 **	0.59 *

Annual forage production in PD recorded in the first period (1998) was lower than in the second, probably because of environmental conditions after the homogenization cut. There was an extended and unusually dry, cool period from February to July 1998. Both conditions decreased forage production because the plants were in stress conditions (Nilsen and Orcutt, 1996). TZ produced forage all the year's seasons, but it was lower than PD. Foroughbakhch *et al.* (2001) found a wide variation of foliar biomass production (1092 to 24,072 kg/ha), when they evaluated the foliar biomass of 15 indigenous tree species for fodder, from an arid area in Mexico, 10 years after the plantation establishment. The total annual forage production of PD and TZ were at the lower limit of that interval and near to the *Acacia*'s forage production. PD was near to *A. berlandieri* forage production (2660 kg/ha) and TZ to *A. rigidula* (1410 kg/ha).

Figure 1. Forage production of the two woody species by season. Wi = winter, Sp = spring, Su = summer, Au = autumn.

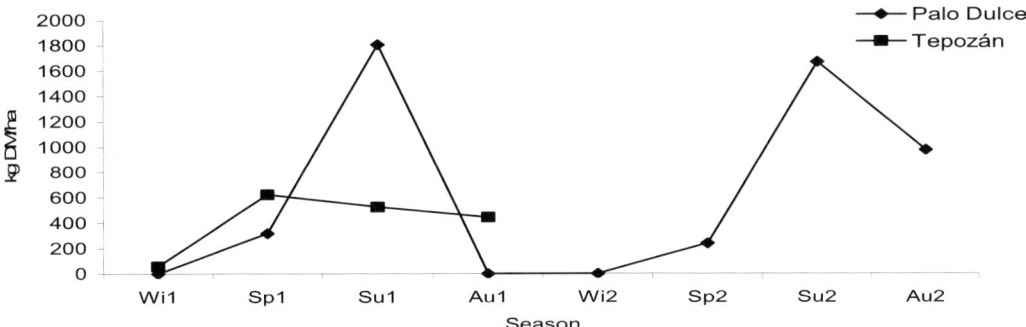

Conclusions

Seasonal biomass production of PD and TZ varied over the 84-day growing period. Management strategy of silvopastoral systems of each species must be different. They could be used on cut and carry systems, for example PD could be harvested and conserved in summer to have fodder for winter.

References

Camacho-Morfín, D., Sandoval, C. and Ayala Burgos, A.J. (1998) Forage trees and shrubs in the Mezquital Valley, Hidalgo, Mexico (Árboles y arbustos forrajeros en el Valle del Mezquital, Hidalgo, Mexico). In: *Memorias del III Taller Internacional "Los árboles y arbustos en la ganadería"*. Estación Experimental de Pastos y Forrajes "Indio Hatuey", Matanzas, Cuba, pp. 11-13. (In Spanish.)

Camacho-Morfín, D., Sandoval, C.A. and Ayala Burgos, A. (1999) Acceptance of four forage woody species (Aceptabilidad del forraje de cuatro especies leñosas). In: *I Reunión Nacional sobre Sistemas Agro y Silvopastoriles*. Huatusco, Veracruz, Mexico. (In Spanish.)

Corrall, A.J. and Fenlon, J.S. (1978) A comparative method for describing the seasonal distribution of production from grasses. *Journal of Agriculture Science* 91, 61-67.

Foroughbakhch, R., Háuad, L.A., Cespedes, A.E., Ponce, E.E. and González, N. (2001) Evaluation of 15 indigenous and introduced species for reforestation and agroforestry in northeastern Mexico. *Agroforestry Systems* 5, 213-221.

Granados, S.D. and Hernández, H.J. (1995) Harvesting system in the Ñahñu community in the Mezquital Valley (Sistema de recolección en una comunidad Ñahñu en el Valle del Mezquital). *Revista Chapingo. Ciencia Forestal* 1, 109-115. (In Spanish.)

Graph Pad Prism 2.01 (1996) *Statistical Package User Manual*. Graph Pad Software San Diego, CA. USA.

Nilsen, E.T. and Orcutt, D.M. (1996) *The Physiology of Plants under Stress. Abiotic Factors*. John Wiley and Sons, New York, USA.

Raghavendra, A.S. (1991) *Physiology of Trees*. John Wiley and Sons, New York, USA.

Pruning influence on acorn yield in cork-oak open woodland

I. Cañellas, S. Roig and G. Montero

Center for Forestry Research (CIFOR-INIA), Ctra. A Coruña km 7.5, 28040 Madrid, Spain. canellas@inia.es

Abstract

Bearing in mind the importance of pruning as a silvicultural treatment in cork-oak (*Quercus suber* L.) open woodland, an experiment was designed to assess the effect of pruning on acorn yield. This work was carried out by comparing annual acorn production from 40 pruned trees and 40 unpruned ones from 1994 till 1999 in kg/m^2 crown cover. By a test of paired comparison of data and analysis of variance, significant differences were found in acorn yield (g/m^2) of crown cover for the pruning year, but not in the following year when comparing pruned and unpruned trees.

Key words: Mediterranean oaks, dehesa, fruit production, silvopastoral system

Introduction

Pruning cork-oak woodland has been a controversial silvicultural practice for a long time, both in theory and in its practical application. It is a silvicultural and economic practice which is part of the traditional silvopastoral system of cork-oak dehesa (open woodland) in Spain. In some zones of the south and north-west of the Iberian peninsula there are cork-oak woodlands at a higher density, where pruning is not common. Although it is not usual in Spanish cork-oak woods, formative pruning is highly valuable in the long term since it rectifies deformed trunks or branch distribution in young trees. This type of pruning seems to increase yield and enhances cork quality by improving cork plank quality, reducing the percentage of cork waste and extraction costs. Light or moderate pruning benefits the tree as a whole. Pruning increases acorn yield, especially on nutrient-poor soils and in areas with deficient summer water supply. This type of pruning does not reduce diameter growth and may even slightly increase it (Hawley and Smith, 1972; Hubert and Courrand, 1989). The economic costs of light or moderate pruning are, however, very high. To compensate for these high costs, intensity of pruning is increased and thereby income is obtained from firewood, charcoal or virgin cork. Intense pruning has another drawback; after the intense pruning there is a period of very low, even no fruit production since the productive surface of the crown has been reduced. Also, the nutrients needed for this production are used in restoring the leaf biomass of the crown. Although acorn yield on the few remaining branches may be abundant, this does not seem to bring increased overall production, and the small remaining crown may be insufficient to maintain diameter growth rhythm (Hawley and Smith, 1972). The reduction of the imbalance between aerial and root biomass is usually through sucker sprouts or epicormic shoots, these can also harm the growth and development of the trees.

In Spain Montero and Curras (1990) developed two experiments to quantify the effect of moderate pruning on firewood and virgin cork production, Cañellas and Montero (2002) and Montero *et al.* (2000) studied the effect of pruning on cork production, Porras (1998) studied the short-term effect of holm-oak pruning on acorn production and Carbonero Muñoz *et al.* (2002, 2003) monitored the evolution of acorn production in a pruning cycle lasting several years.

Materials and Methods

In the El Deheson del Encinar estate in the Castilla-La Mancha region (39°50'N, 5° 05'W, UTM: 30UK23) a mixed holm-oak and cork-oak dehesa at 330 m asl with a continental Mediterranean climate, varying from year to year between arid to sub-humid, was studied. Mean annual rainfall was 573 mm and mean annual temperature 15.2°C with the period liable to frost from October to April. Soil is sandy (> 80% sand) of granitic origin, with pH 5.5 and organic matter < 1%. Pruning was carried out in December 1993 on 40 trees, and these were compared with 40 left unpruned. Cork stripping took place in August 1998, at the end of a 10-year cycle. Trees were selected in pairs according to their circumference over cork at breast height, covering all the diameter classes present in the area. Pruning consisted mainly of eliminating low branches to give a round-shaped crown. Less than 30% of crown volume was cut, this limit only exceeded where the shape of tree or crown made it advisable for the future growth of the tree. Circumference before stripping (CSC), base circumference (CB), mid-point (CM) and the top of the trunk (CS), and height reached in the stripping (HD), were recorded for each tree to assess the effect of pruning. Acorns were collected in one 0.3025 m^2 trap placed randomly under each tree. Material from the traps was gathered monthly from September 1994 to March 1999. Acorns were dried at 80°C for 48 h and weighed to the nearest 0.01 g. Treatments were examined for statistically significant differences using analysis of variance (random block design). Paired comparisons were made using the t-test.

Results and Discussion

The estimated amount of acorn production by crown area units through five annual cycles (1994-1999) as a function of the pruning treatment is presented (Table 1). There were no significant differences in any of the studied variables in relation to pruning treatment, which would seem to indicate that their silvicultural characteristics (tree size and intensity of cork stripping) were similar. There were significant differences in acorn yield between year of pruning but not comparing pruned with unpruned trees in the 3 following years. Other work in Spain on *Q. ilex* (Porras, 1998; Carbonero Muñoz *et al.*, 2002, 2003) concluded that light pruning improves acorn production but only after 3-5 years of pruning. According to Vieira (1950) the benefits obtained in tree growth from rational pruning carried out with biological rather than productivity and economic aims are indisputable. When better conditions for the cork-regeneration layer are created, the branch system is rejuvenated and growth rate and cork yield increased. These results do not seem to confirm this hypothesis since neither the mean cork yield nor yield per unit area of stripping surface seems to be affected by pruning (Cañellas and Montero, 2002). The proportion of acorn yield varied greatly with year but was independent of the treatments, oscillating between 0.7 and to 332.8 g/m^2. Attention should be drawn to the fact that a large fall of acorns in 1994 was followed by a low fall over the next 3 years.

Table 1. Estimated amount of acorn (g/m of crown surface) for five annual cycles (1994-1999). Standard deviation in parentheses.

Years	Not pruned	Pruned
	Annual production period (September-March)	
1994-1995	332.85 (71.01)	137.68 (29.58)
1995-1996	0.74 (0.11)	2.10 (0.32)
1996-1997	58.15 (9.32)	56.76 (9.37)
1997-1998	31.02 (3.58)	32.85 (4.10)
1998-1999	332.57 (43.69)	177.64 (19.86)
Average	155.64 (77.59)	81.21 (32.96)

Conclusions

Moderate pruning (classical pruning developed in cork-oak dehesa systems) had no effect on the quantity of acorns produced per unit area of crown. The mast year has a very severe effect on acorn yield in both pruned and unpruned trees.

References

Cañellas, I. and Montero, G. (2002) The influence of cork-oak pruning on the yield and growth of cork. *Animal and Forage Science* 59, 753-760.

Carbonero Muñoz, M.D., Fernández Rebollo, M.P. and Navarro, R. (2002) Assessment of acorn production and size of *Quercus ilex* L. subsp. *ballota* (Desf) Samp during a pruning period. Results of the 2001-2002 campaign (Evaluación de la producción y del calibre de bellotas de *Quercus ilex* L. subsp. *ballota* (Desf) Samp a lo largo de un ciclo de poda. Resultados de la campaña 2001-2002). In: Chocarro, C., Santiveri, F., Fanlo, R., Bovet, I. and Lloveras J. (eds) *Production of Pastures, Forage and Grasses*. Editions of the University of Lleida, Spain, pp. 633-638. (In Spanish.)

Carbonero Muñoz, M.D., Fernández Rebollo, M.P., Blázquez, A. and Navarro, R. (2003) Assessment of acorn production and size of *Quercus ilex* L. subsp. *ballota* (Desf) Samp during a pruning period. Results of 2001-2002 and 2002-2003 campaigns (Evaluación de la producción y del calibre de bellotas de *Quercus ilex* L. subsp. *ballota* (Desf) Samp a lo largo de un ciclo de poda. Resultados de las campañas 2001-2002 y 2002-2003). In: Robles, A.B., Ramos, E., Morales, M.C., de Simón, E., González Rebollar, J.L. and Boza, J. (eds) *Pastures, Development and Conservation*. Council of Agriculture and Fisheries, Junta de Andalucía, Sevilla, Spain. pp. 463-468. (In Spanish.)

Hawley, R.C. and Smith, D.M. (1972) *Practical Silviculture* (Silvicultura Práctica). Omega, Barcelona, Spain. (In Spanish.)

Hubert, M. and Courrand, R. (1989) *Pruning and Shaping of Forest Trees* (Poda y formación de los árboles forestales). Mundi-Prensa, Madrid, Spain. (In Spanish.)

Montero, G. and Curras, R. (1990) *Pruning of Cork Oak Stands (Quercus suber L.). Quantification of the Products* (La poda del alcornocal (*Quercus suber* L.) Cuantificación de sus productos). Hojas Divulgadoras, MAPA, No. 18-19, Madrid, Spain. (In Spanish.)

Montero, G., Cañellas, I. and Bachiller, A. (2000) The influence of pruning in cork production in dehesas of cork-oak in Extremadura (Spain). In: *Congreso Mundial do Sobreiro e da Cortiça*. Lisbon, Portugal, pp. 19-21.

Porras, C. (1998) Effect of the pruning on hola-oak stands (*Quercus rotundifolia* Lam.) in production and size of acorns (Efecto de la poda de la encina (*Quercus rotundifolia* Lam.) en los aspectos de producción y en el del grosor de las bellotas). In: *Actas de la XXXVIII Reunión Científica de la Sociedad Española para el Estudio de los Pastos*, (SEEP). Soria, Spain, pp. 381-384. (In Spanish.)

Vieira, J. (1950) *Cork Silviculture* (Subericultura). MAPA (Ministry of Agriculture, Fish and Food), Madrid, Spain. (In Spanish.)

Effects of grazing on fuel biomass in mountain pastures of the south-eastern Pyrenees

C. Dal Zennaro [1], J. M. Arenas [2], G. Argenti [1], A. Pardini [1] and T. M. Sebastia [2,3]
[1] DiSAT, University of Florence, Italy; [2] Centre Tecnològic Forestal de Catalunya, Spain; [3] ETSEA, University of Lleida, Spain. andrea.pardini@unifi.it

Abstract
There has been a large decline in the managed pastures of Mediterranean European countries. This has led to decreased biodiversity in the herbaceous layer, shrub encroachment and accumulation of fuel biomass (woody plants). This, in turn, has increased the risk of large forest fires. Consequently pasture management to prevent these effects is nowadays a big issue in many marginal areas. This trend is also present in the Pyrenees (Spain), where research was carried out to analyse the influence of different land use on characteristics related to botanical composition of vegetation and fire hazards. The study was conducted in Vall d'Alinyà (Alt Urgell), where three different pastoral management systems were identified. Botanical composition, biomass production and presence of fuel biomass during summer were assessed. Grazing contributed to reduce shrub encroachment and thus fire risk. Botanical composition and amount of fuel biomass was influenced by actuality management and age of abandoning. However, it emerged that the current animal stocking rate is too low in comparison to the actual carrying capacity and livestock numbers should be increased to limit shrub encroachment and fire risk.

Key words: biomass production, rangeland, management

Introduction
In recent years, utilization and management of pastures of Mediterranean European countries have been greatly reduced. They are characterized, in the last years, by strong reduction of utilization and management. This has led to many negative effects, mainly changes in botanical composition, decreased biodiversity of herbaceous layers, shrub encroachment and increased fuel biomass of woody species. This, in turn, has increased the risk of forest fires, which, since the 1970s, have considerably expanded in number, extent and intensity (Etienne *et al.*, 1990; Papanastasis, 1999; Canals and Sebastia, 2000). Consequently, pasture management is nowadays a big issue in the prevention of negative changes in many marginal areas (Caredda *et al.*, 2002). Accumulation of ungrazed vegetation and presence of shrubs are serious fire hazards as they have a high level of dry biomass. Animal grazing is helpful in reducing fuel biomass in pastures, firebreaks, forests and surrounding areas. Moreover, shrubs are also very dangerous because they transmit fires from herbaceous to woody plants. Thus, the use of animal grazing in forests to reduce shrubs biomass from the storey has been promoted in many Mediterranean countries (Pardini *et al.*, 1993). Results from a survey conducted in an area of Spain on the relationship between management practice and parameters affecting risk of fire in areas with mixed herbaceous and woody plant species are reported in this chapter.

Materials and Methods
The study was conducted in Vall d'Alinyà (Alt Urgell, south-eastern Pyrenees, Spain). Three different pastoral management types were identified: former cropped fields recently abandoned and now managed as pastures, T, mountain cropped fields abandoned a long time ago and now managed as pastures, M and subalpine pastures, S. In each land-use type ten sites were randomly selected and 2 × 2 m areas sampled. The total biomass cut from each sample area was separated into herbaceous and woody parts and also in green or standing dead biomass (personal estimate of tissue flexibility and colour). All components were oven-dried at 60°C and weighted separately. ANOVA was done on all data.

Results and Discussion
The highest biomass was found in M, and this mainly consisted of shrubs. The two other pasture types had lower production than M but had a higher contribution of herbaceous species in the vegetation (Table 1). This difference could be related to grazing, which was intensive in T and S for a large part of the year, and time since abandonment.

Table 1. Green biomass (total and per cent composition) for different pasture types. Values in the same columns with the same letters are not significantly different at $P < 0.05$.

Pasture types	Total green biomass (t/ha)	Vegetation composition (%)	
		Herbage	Woody plants
T	3.1 c	92 a	8 b
M	4.8 a	36 b	64 a
S	4.1 b	91 a	9 b

The lowest value of fuel biomass was found in T, while there was no significant differences between the other types (Table 2). A high presence of woody species did not influence the percentage of dead biomass so there were no significant differences between the studied pastures. This was because shrubs also had green leaves during the summer (period of data collection) when grasses were already dry and inflammable.

Table 2. Fuel biomass (absolute and percentage values) for different pasture types. Values in the same columns with the same letters are not significantly different at $P < 0.05$.

Pasture types	Fuel biomass (t/ha)	Fuel biomass (%)
T	1.5 b	31
M	2.0 a	32
S	2.0 a	33

The total amount of above-ground biomass was highly associated with shrub biomass (Figure 1). This was probably due to the presence of a high proportion of shrubs which influenced total biomass and this was especially true in treatment M. Moreover, shrubs maintain green biomass longer than grass during the hot season.

Conclusions

There were great differences within land-use treatments concerning parameters of fire risk in the studied area. Differences can be related to management (i.e. stocking rate, number of grazing days, animals type) and abandoning of pastoral practices. Grazing has been the most important factor in reducing fire risk while abandonment has had negative consequences in the studied area. However, it is understood that current animal stocking rates are lower than actual carrying capacity and should be increased. Eventually, grazing should be concentrated in some areas and forest reintroduction planned in the less productive pastures. These results are useful for management of areas with mixed pasture and shrub-tree vegetation and where pasture is associated with forest as in silvopastoral systems.

Figure 1. Relationship between shrub biomass and total biomass (herbage plus shrubs).

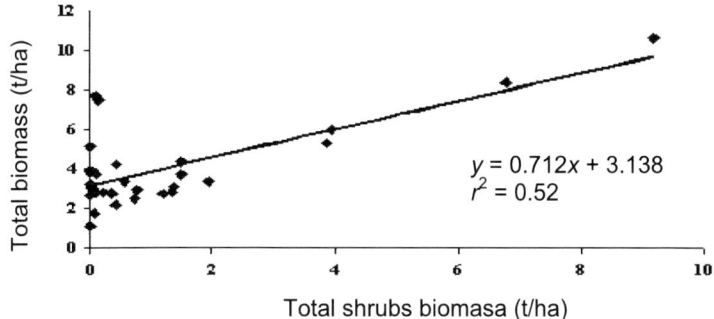

References

Canals, R.M. and Sebastià, M.T. (2000) Analysing mechanisms regulating diversity in rangelands through comparative studies: a case in the south-western Pyrenees. *Biodiversity and Conservation* 9, 965-984.

Caredda, S., Franca, A. and Seddaiu, G. (2002) Firebreaks oversowing: an alternative tool for the wildfire risk reduction in Sardinia. *Grassland Science in Europe* 7, 908-909.

Etienne, M., Hubert, B., Jullian, P., Lecrivain, E., Legrand, C., Meuret, M. and Napoleone, M. (1990) Forest areas, altitude and fires (Espaces forestiers, élevage et incendie). *Revue Forestiére Française* 42, 157-172. (In French.)

Papanastasis, V.P. (1999) Grassland and woody plants in Europe with special reference to Greece. *Grassland Science in Europe* 4, 15-24.

Pardini, A., Piemontese, S. and Argenti, G. (1993) Reduction of forest fires with fire breaks sown for pastures in Toscana (Limitazione degli incendi boschivi con il pascolamento di bande parafuoco inerbite in Toscana). *L'Italia Forestale e Montana* 48, 341-352. (In Italian.)

Seed production and influence of heat on the germination of *Genista scorpius* (L.) DC

I. Delgado, M. J. Ochoa and F. Múñoz
Centro de Investigación y Tecnología Agroalimentaria, Diputación General de Aragón, Apartado 727, 50080 Zaragoza, Spain. idelgado@aragob.es

Abstract
Seed production in north-eastern Spain and the influence of high temperatures (in a range from 50°C to 150°C and exposures from 1 to 15 min) on the germination of *Genista scorpius* (L.) DC. were studied to investigate the possibility of regeneration by seed in silvopastoral systems of Mediterranean calcareous soils. Flowering was sparse, only 71% of plants flowering with a mean of 266.3 flowers/plant, 40.2 pods/plant and 2.9 seeds/pod. Various heat treatments did not increase germination significantly, but temperatures ≥100°C and 5 min of exposure reduced seed hardness and affected the viability of seeds.

Key words: shrubland, flowering, growth, seed hardness, wildfire

Introduction
Genista scorpius (L.) DC is a frequent shrub in silvopastoral systems of the eastern Mediterranean basin with calcareous soils and contributes to the diet of sheep and goats. However, when it spreads too widely, it reduces grass production and increases the risk of wildfires.

As a browsing shrub, cattle utilize it in springtime, selecting the yearly sprouts, flowers and pods which are low in lignin (Muñoz *et al.*, 1996; Delgado *et al.*, 1997, 2002a; Valderrábano and Torrano, 2000).

As an invading shrub, it occupies formerly cultivated fields, covering up to 90% of the surface (Delgado *et al.*, 2002b). Nevertheless, in a study developed to evaluate the spread of *G. scorpius* by seed, it was found that establishment was difficult for the species (Jarauta, 2002). The renewal of shrubs by seed was only 133 plants/ha 4 years after clearing. Wildfire seems to favour the spread by seed of similar species (Añorbe *et al.*, 1990; Tárrega *et al.*, 1992; Herranz *et al.*, 1998), because it combines the elimination of the hard-seededness by the heat and decreases competition from the existing vegetation.

The present work investigates the production of *G. scorpius* seed in silvopastoral systems of north-eastern Spain as well as the effect of heat on the elimination of seed hardness in order to contribute to the knowledge of its spreading by seed.

Materials and Methods
The evaluation of seed production was carried out in five locations of north-eastern Spain: Zuera (41° 57' N, 0° 35' E; 400 m asl; 417 mm yearly rainfall), close to an irrigation ditch (a) and in a rainfed area (b); Las Almunias (42° 16' N, 0° 12' E; 845 m; 636 mm); Bentué (42° 20' N, 0° 4' E; 1062 mm; 957 mm); and San Juan (42° 26' N, 0° 1' E; 1220 m; 1021 mm), during 2002 and 2003. Thirty shrubs/locations were marked at random, where the following data were recorded in springtime: height, major and minor diameter average, number of flowers and pods and number of seeds/pod in 500 pods/location.

The study of reduction in hard-seededness was carried out in July 2003. Seeds collected in 2002 were germinated in plastic bags on two layers of filter paper saturated with water for a period of 28 days. Germinated seeds were discarded and those which did not germinate were considered to be hard seeds and used in experiments. A total of 24 treatments were carried out. The seeds were heated to 50°C, 70°C, 100°C, 130°C and 150°C, with exposure periods of 1, 5, 10 and 15 min in each case. Three additional treatments to reduce the seed hardness in legumes were applied, mechanic scarification of teguments, immersion in hot water for 4 hours and immersion in concentrated sulphuric acid for 4 min. Untreated hard seeds were used as a control.

Seeds were sown in 15 cm diameter Petri dishes, on filter paper saturated with deionized water in darkness at 20°C. There were four replicates of 100 seeds for each treatment. Fungal infection was controlled with a fungicide (Captan-50 at 2.5% concentration). Germinated seeds were discarded every week during 4 weeks. At the end of the germination period, germinated seeds, swollen but not germinated seeds and hard seeds were counted. The results of the final germination were analysed using an ANOVA and the LSD test to detect any significant differences ($P < 0.05$).

Results and Discussion
The average number of flowers ranged between 56.7 and 624.0 (Table 1). Between 7% and 67% of the shrubs did not flower. Pod production was notably lower than flowers, ranging from 1.3 to 62.3 pods/shrub. The average number of

seeds/pod was 2.9. The irregular flowering of *G. scorpius* that had already been found in previous studies was confirmed (Torrano, 2001; Jarauta, 2002).

Table 1. Dimensions of *Genista scorpius* shrubs, dates of initial and final flowering, and number of flowers and pods/shrub at five locations of north-eastern Spain.

Location	Year	Diameter	Height	Flowers	Pods	IF	FF	GP	Veg.	D
San Juan	1	42.1	63.7	438.8	62.3	17/5	13/6	26/7	10	1
	2	47.2	68.5	-	0.7	-	-	23/7	-	0
Bentué	1	29.3	55.4	91.0	11.3	17/5	13/6	26/7	10	0
	2	30.1	54.9	-	2.5	-	-	23/7	-	2
Rodellar	1	37.6	51.2	56.3	1.3	25/4	14/5	3/7	10	0
	2	36.8	52.4	-	17.7	-	-	5/6	-	0
Zuera (a)	1	64.0	84.0	624.0	38.2	12/3	8/4	12/6	2	0
	2	64.8	95.7	235.0	21.2	15/3	9/4	13/6	4	0
Zuera (b)	1	26.0	52.8	80.0	1.3	2/4	30/4	29/5	20	0
	2	27.7	55.4	339.2	83.5	1/4	24/4	6/6	5	1

IF = Initial flowering; FF = final flowering; GP = green pods; Veg = less 10 flowers/plant; D = died plants.

Heat did not improve seed germination significantly ($P > 0.05$) compared to the control, which had 1% germination. No seed germinated in the treatments 100°C for 15'; 130°C for 5', 10' and 15', or at 150°C at all the periods of time. The percentage germination was 64.5% in the scarified seeds, 41.5% in the scalded ones and 34.5% in those immersed in sulphuric acid, the differences between them being highly significant ($P < 0.001$) (Table 2).

Heat favoured the cracking of hard seed coats at temperatures higher than 70°C, causing a significant increase ($P < 0.05$) of swollen but not germinated seeds from 15.5% to 99%, corresponding to the higher percentages at the higher heat treatments (Table 2). However, heat did not produce an increase in germinated seeds after 28 days and seeds that were considered non-viable began to rot.

Table 2. Percentage of germinated seeds, swollen (not germinated) and hard seeds, in different treatments carried out to eliminate hard-seededness in *Genista scorpius* 28 days after beginning of germination.

Treatment	Germinated%	Swollen%	Hard%
50°C 1'	2.00de	8.75i	89.25a
50°C 5'	2.75de	11.75hi	85.50abc
50°C 10'	3.25de	10.5i	86.25abc
50°C 15'	1.00de	11.75hi	87.25abc
70°C 1'	1.00de	20.50gh	78.50cde
70°C 5'	3.25de	25.00efg	71.75def
70°C 10'	0.75de	27.50efg	71.75def
70°C 15'	2.00de	27.50efg	70.50efg
100°C 1'	4.500d	15.50hi	80.00bcd
100°C 5'	2.25de	29.50efg	68.25fg
100°C 10'	2.50de	35.75de	61.75g
100°C 15'	0e	33.00ef	67.00fg
130°C 1'	1.0de	53.00c	46.00h
130°C 5'	0e	86.50b	13.50ij
130°C 10'	0e	85.75b	14.25ij
130°C 15'	0e	81.25b	18.75ij
150°C 1'	0e	97.75a	2.25k
150°C 5'	0e	98.00a	2.00k
150°C 10'	0e	98.50a	1.50k
150°C 15'	0e	99.00a	1.00k
Sulphuric acid 4'	34.25c	53.25c	12.50ij
Scarified	64.50a	35.25de	0.25
Hot water 4 h	41.50b	52.50c	6.00jk
Control	1.0de	10.75i	88.25ab

Conclusions

G. scorpius seed production is limited. Treating seeds with heat reduces the seed hardness, but does not increase germination.

References

Añorbe, M., Gómez, J.M., Pérez, M.A. and Fernández, B. (1990) Influence of the temperature on the germination of *Cytisus multiflorus* Sweet and *Cytisus oromediterraneus* seeds (Influencia de la temperatura sobre la germinación de semillas de *Cytisus multiflorus* (L'Her) Sweet and *Cytisus oromediterraneus*). *Studia Oecologica* 7, 85-100. (In Spanish.)

Delgado, I., Albiol, A., Ochoa, M.J. and Muñoz, F. (1997) Interview with farmers concerning the forage value of the autochthonous flora in Monegros (Aragon) (Encuesta a pastores sobre valoración forrajera de la flora autóctona en Monegros (Aragón)). *ITEA* 18 (1), 245-247. (In Spanish.)

Delgado, I., Ochoa, M.J., Sin, E., Barragán, C., Rodriguez, J. and Nuez, T. (2002a) Pasture restoration by control of *Genista scorpius* DC. In: *Proceedings of 11th Meeting of the FAO-CIHEAM Sub-Network on Mediterranean Pastures and Fodder Crops, Djerba (Tunisia)*.FAO-CIHEAM, Rome, Italy. (In press.)

Delgado, I., Jarauta, E., Andueza, D. and Muñoz, F. (2002b) Implications of the ovine livestock on the expansion of the "aliagar" (Implicación del ganado ovino en la expansión del aliagar). In: Peris, B., Molina, P., Lorente, M. and Garcia, A. (eds) *Producción Ovina y Caprina*. No. XXVII SEOC, University Cardenal Herrera-CEU, Valencia, Venezuela, pp. 831-837.

Herranz, J.M., Ferrandis, P. and Martinez Sanchez, J.J. (1998) Influence of heat on seed germination of seven Mediterranean *Leguminosae* species. *Plant Ecology* 136 (1), 95-103.

Jarauta, E. (2002). Biology and control of the "aliaga" (Biología y control de la aliaga). MSc thesis. Universitaria de Ingeniería Politécnica de Huesca, Spain. (In Spanish.)

Muñoz, F., Andueza, D., Delgado, I. and Ochoa, M.J. (1996) Chemical composition and digestibility of browse plants in a semi-arid region of Spain. In: Gintzgurger, G., Bounejmate, M. and Nefzoui, A. (eds) *Fodder Shrub Development in Arid and Semi-Arid Zones*. ICARDA, Aleppo, Syria, pp. 485-490.

Tárrega, R., Calvo, L. and Trabaud, L. (1992) Effect of high temperatures on seed germination of two woody. *Leguminosae*. *Vegetatio* 102, 139-147.

Torrano, L. (2001) Use of encroached forest areas by caprine livestock and the impact on the understorey vegetation (Utilización por el ganado caprino de espacios forestales invadidos por el matorral y su impacto sobre la vegetación del sotobosque). MSc thesis. University of Zaragoza, Zaragoza, Spain. (In Spanish.)

Valderrábano, J. and Torrano, L. (2000) The potential for using goats to control *Genista scorpius* shrubs in European black pine stands. *Forest Ecology and Management* 126, 377-383.

Lithium chloride and ipecacuanha syrup to induce sheep aversion to the intake of coffee plants

C. A. Durantes [1], J. A. Torres [1], P. A. Martínez [2], R. Castro [1], A. Arroyo [3] and J. G. Cruz [1]

[1]Centro Regional Universitario Oriente, Universidad Autónoma Chapingo, Apdo. 49, CP 94100, Huatusco, Ver., Mexico. tora_sheep@hotmail.com. [2]Dept. de Zootecnia, Universidad Autónoma Chapingo, CP 56230; Texcoco, Edo. Méx., Mexico. pedroarturo@correo.chapingo.mx, [3]Dept. de Agroforestería, Universidad Autónoma Chapingo, CP 56230, Texcoco, Edo. Méx., Mexico. antonioarroyo@taurus1.chapingo.mx

Abstract

Browsing of coffee plants by sheep is one of the major limitations to introducing sheep grazing on coffee plantations. Hence, it is worth finding a mechanism that will prevent animals browsing coffee plants. Emetic agents like lithium chloride (LiCl) have been used successfully in different animal species to induce feed aversion. Ipecacuanha syrup (IS) is an organic emetic agent that has not been tested in ruminants to induce feed aversion. In this study, the ability of both these emetics to induce aversion to coffee foliage intake by ewes was compared. Treatments were in a 3 × 2 factorial arrangement, three emetic agents: LiCl, IS and distilled water (control), and two administration means: cannula to oesophagus and intraruminal injection. The experimental design was completely random with two replicates and the experimental unit was one individually penned ewe. Strength of aversion was measured by the time ewes spent eating coffee foliage after the treatment was applied. Administration path and its interaction with emetic agent was not significant ($P > 0.05$). Strength of aversion varied ($P \leq 0.05$) among emetic agents. LiCl and IS gave similar aversion for the first 10 days after treatments were applied and, from the fifteenth to the nineteenth day after treatments, aversion was decreased for both emetics but LiCl showed a higher strength than IS. Distilled water (control) showed no induced aversion at all. LiCl was more effective than IS at inducing feed aversion. To give long-lasting aversion, more than one application of the induced aversion treatment would need to be used.

Key words: animal behaviour, browse, emetics

Introduction

Sheep grazing in coffee (*Coffea arabica* L.) plantations (traditional and organic) has been implemented by some Mexican coffee producers, principally for weed control and increased profit. However, when these producers see that some sheep or the whole flock start to browse coffee plants, damaging them to the point of bringing coffee yields down, they decide not to continue with the sheep-coffee system. Hence the extent of use and spread of the sheep-coffee system among producers depends on providing them with technologies that will prevent the browsing of coffee plants by sheep. Torres and de Lucas (2002) have shown that this might vary from a few animals to the whole flock depending on season of the year, land condition and animal nutrition. They suggest that the strategic use of emetic agents could become a tool to induce aversion to the browsing of coffee plants by sheep and then reach a more profitable sheep-coffee system. Ralphs *et al.* (2001) pointed out that lithium chloride (LiCl) has been used extensively to induce aversion in lab and farm animals as well as in humans. Pfister (2000) found that LiCl was successful in inducing aversion to toxic plant and pine needle intake by cattle. Extensive on-farm use of LiCl is limited as it demands proper handling, stays within the animal for up to 96 h after dosing and can be excreted in milk (Ralphs and Stegelmeier, 1997). Ipecacuanha (*Cephaelis ipecacuanha* (Brot.) A. Rich.) syrup could be an alternative emetic agent. Spinelli and Reed (1984) indicated that ipecacuanha syrup (IS) has been used in humans, dogs and cats as a vomiting agent. No published information could be found on the use of IS in sheep or any other ruminant species. Hence, the objective of this study was to determine the extent of induced aversion in ewes to the intake of coffee foliage by LiCl and ipecacuanha syrup dosed by intraruminal injection and by cannula inside the oesophagus.

Materials and Methods

The fieldwork was carried out on a commercial farm located in the state of Veracruz, Mexico, 19°13'N, 96°53'W and 1090 m asl. Twelve ewes that were found to browse coffee plants were selected from the flock. These ewes were 2 years old, had an average live weight of 39 kg and were a cross of Pelibuey × Katahdin × St. Croix. Ewes were assigned randomly to one of six treatments, with two ewes per treatment. Treatments were in a 3 × 2 factorial, three emetic agents: LiCl (150 mg/kg LW), ipecacuanha syrup (IS) (25 ml/ewe) and distilled water (25 ml/ewe); and two ways of administration: intraruminal injection and cannula to oesophagus. Ewes were individually penned and given chopped green fodder from the understorey of the coffee plantation, mainly grasses (*Cynodon dactylon* (L.) Pers.; *Panicum maximum* Jacq.; *Pennisetum purpureum* Schumach), *Bidens pilosa* L. and *Sida rhombifolia* L. Once ewes were used to the feeding routine in the pens, feed was withheld for 18 h. After fasting, ewes were exposed to coffee foliage for 1.5 h, at the end of which time the corresponding aversion treatment was applied and feed was withheld up to the next day.

Intermittent offering of coffee foliage (1 hour) and understorey green fodder (2 h) was followed for the next 19 days after aversion treatment. Time ewes spent eating coffee foliage was used to measure the strength of the induced aversion. This time was registered daily from the fifth to the tenth, and from the fifteenth to the nineteenth day after aversion treatment. To determine time spent eating coffee foliage, each ewe was checked every 10 min. Statistical analysis was under a model for a factorial arrangement in a completely random design and with repeated measurements, means were compared by Tukey ($\alpha = 0.05$), and GLM of SAS (1993) was used for statistical calculations.

Results and Discussion

Administration path and its interaction with the emetic agent were not significant ($P > 0.05$) while emetic agent was significant ($P \leq 0.05$). In the first period of measurement (days five to ten after induced aversion) LiCl and IS were equally effective in promoting aversion to coffee intake. In the second period (days 15 to 19 after induced aversion) only LiCl was still effective in promoting some aversion to coffee intake. Over both periods, ewes dosed with distilled water showed no aversion to coffee intake. Strength of induced aversion decreased with time in both LiCl and IS (Figure 1). Burritt and Provenza (1996) also found with sheep that the strength of aversion decreased with time.

Figure 1. Time spent eating coffee foliage by ewes dosed with different emetic agents: lithium chloride (LiCl), ipecacuanha syrup (IS) and distilled water (control).

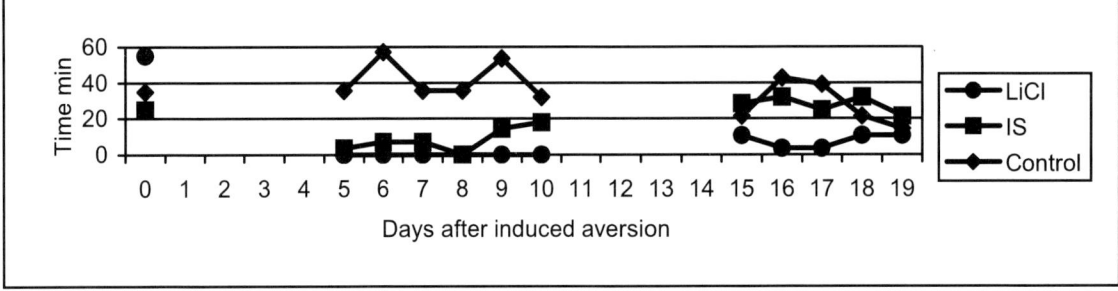

Conclusions

Induced aversion by lithium chloride lasted longer and was stronger than the aversion promoted by ipecacuanha syrup. Time decreased the strength of the induced aversion of both emetic agents. Research still needs to be carried out on dose and times emetics should be applied, sex and age and social interaction to induce aversion to coffee intake by sheep. The strategic use of emetic agents can prevent sheep browsing coffee plants and enhance the functionality of the system. Thus it can help make coffee producers maintain or adopt the sheep-coffee system.

References

Burritt, E.A. and Provenza, F.D. (1996) Amount of experience and prior illness affect the acquisition and persistence of conditioned food aversions in lambs. *Applied Animal Behaviour Science* 48, 73-80.
Pfister, J.A. (2000) Food aversion learning to eliminate cattle consumption of pine needles. *Journal of Range Management* 53, 655-659.
Ralphs, M.H. and Stegelmeier, B.L. (1997) Ability of apomorphine and lithium chloride to create food aversions in cattle. *Applied Animal Behaviour Science* 56, 129-137.
Ralphs, M.H., Provenza, F.D., Pfister, J.A., Graham, D., Duff, G.C. and Greathouse, G. (2001) Conditioned food aversion: From theory to practice. *Rangelands* 23 (2), 14-18.
SAS (1993) SAS/*User's Guide. Version 6,* 4th edn. SAS Institute, Cary, NC, USA.
Spinelli, J.S. and Reed, L. (1984) *Veterinary Pharmacology and Veterinary Therapeutics* (Farmacología y terapeutica veterinaria). Ed. Interamericana, Mexico D.F., Mexico. (In Spanish.)
Torres, J.A. and de Lucas, J. (2002) Hairy ovine in coffee agroecosystems (Los ovinos de pelo en agroecosistemas cafetaleros). In: Pohlan, J. (ed.) *Mexico and the Coffee Crops in Chiapas* (Mexico y la cafeticultura chiapaneca). Shaker Verlag, Aachen, Germany, pp. 269-281. (In Spanish.)

Dispersed trees in pasturelands of cattle farms in a dry ecosystem of Costa Rica

H. Esquivel, M. Ibrahim, C. A. Harvey, C. Villanueva, F. L. Sinclair and T. Benjamín
Dept. CATIE, Costa Rica. hesqui@catie.ac.cr

Abstract

An inventory of all trees dispersed within the pastures of 16 cattle farms in Costa Rica was conducted to characterize and describe richness, composition, abundance and distribution of tree cover. 199 pasture plots were surveyed (835.76 ha) and *Brachiaria brizantha* was the dominant grass species. 5896 trees were found belonging to 99 species and 36 families. The most abundant tree species were *Tabebuia rosea*, *Guazuma ulmifolia*, *Cordia allidora* and *Acrocomia vinifera*. 2941 trees were classified as timber trees, 1571 as forage and 1616 as fruit-bearing trees. Almost half (53.85%) of the total trees were individual trees whereas 46.15% were arranged in clusters. Overall mean crown cover was 6.95%; overall mean tree density was 8.06 trees/ha and overall mean diameter for all trees was 44.8 cm. It is concluded that, although a few species are dominating the landscape, there is a high species richness of trees dispersed in pastures on cattle farms but they are maintained at low densities to reduce interference with pasture productivity.

Key words: richness composition, abundance, silvopastoral systems

Introduction

In Central America, pasturelands are generally established immediately after cutting and burning the forest trees or after 2 to 5 years of shifting cultivation and then sowing with forage grasses (Muchagata and Brown, 2003). These practices have caused large areas of forest to be converted to pastures. The benefits of the conversion of tropical forest to pastures are often temporary due to the rapid depletion of soil nutrient reserves resulting in pasture degradation (Lovejoy, 1985; Nepstad *et al.*, 1990). Maintaining or increasing trees dispersed within the farm represents an option in fragmented landscapes to increase farm productivity and sustainability of livestock farms. The benefits that trees provide for the farms include timber, firewood, medicinal products and food for both people and cattle. Trees also serve as shade and shelter for cattle and provide nesting sites for birds and mammals and conserve biodiversity (Schelhas and Greenberg, 1996; Guevara *et al.*, 1998; Harvey and Haber, 1999; Frankie *et al*, 2001). Many livestock farms maintain trees dispersed in pastures, but their distribution and composition vary considerably across farms and regions (Guevara *et al.*, 1998; Harvey and Haber, 1999; Souza de Abreu *et al.*, 2000). The objective of this research was to characterize and describe the species composition, abundance, richness and the arrangement of trees dispersed within pasturelands of cattle farms in Costa Rica. This will help lead to the design and establishment of better silvopastoral systems.

Materials and Methods

A tree inventory was conducted in Cañas, Guanacaste, in the Pacific zone of Costa Rica from June to December 2002. The area is classified as a tropical dry forest (Holdridge, 1978) with annual rainfall ranging from 1000 to 2500 mm, with most rain falling during May to November (wet season). Mean annual temperature is 27.6°C (23°C-31°C during the year). Soils in the lowlands are of volcanic origin, with an average depth of 1 m. Texture varies from fine to medium and fertility from medium to very high.

Biophysical and socio-economic information available from a semi-structured survey of 35 livestock farms was used to selected a sub-sample of 16 farms based on farmer cooperation. At each farm, a detailed inventory of all trees dispersed within the pasture was carried out from June to December 2002. Riparian trees, live fences and forest patches (group of trees covering > 0.25 ha) were excluded from the pasture tree inventory. All trees bigger than 10 cm in diameter at breast height (dbh) were registered, identified, marked and recorded during the inventory directly in the field with the participation of local farmers. For each tree, the dbh, total height, stem height and crown cover were measured. Tree crown cover area was calculated utilizing the elliptical formula (A = $\pi \times$ Radius 1 $\times$ Radius 2) and total plot crown cover was calculated as the sum of all tree crowns measured in the field for a particular plot divided by the total plot area. In order to characterize the cover of trees dispersed in pastures, trees were classified according to their main uses as timber, forage and fruit trees. Trees were also categorized according to whether they occurred as individuals or in groups (clusters). Descriptive statistics and standard descriptors of vegetation composition (density, abundance, richness) were calculated.

Results and Discussion

Trees were found on all farms and in 86.6% of plots within the farms. A total of 5896 trees belonging to 39 families and 99 species were found dispersed in 835.76 ha of *B. brizantha* pasturelands in 16 surveyed cattle farms. The most frequent tree species were *Tabebuia rosea*, *Guazuma ulmifolia*, *Cordia allidora* and *Acrocomia vinifera*. In contrast, 20 out of 99 tree species were represented by one individual tree. Of the total surveyed trees, 49.80, 26.65 and 27.40% were classified as timber, forage and fruit-bearing trees, respectively. Overall mean crown cover was 6.95% with tree cover ranging from 0.0 to 48.8% while overall mean tree density was 8.06 trees/ha with ranges from 0 to 68.07 trees/ha. The overall mean dbh of all trees surveyed was 44.8 cm with ranges from 10 to 269.7 cm. Most trees (70.9%) had diameters between 20 and 60 cm whereas 573 trees (9.83%) had dbh of 10-20 cm and 221 trees (3.81%) had dbh larger than 1 m. There are various factors which can affect cover of trees dispersed in pastures on cattle farms. On one hand, farmers could have influenced tree cover through their knowledge, preferences, objectives and strategy to leave, cut down or combine tree species with the aim to diversity income, while at the same time securing production of forage and fruits for cattle from forage trees. On the other hand, grazing cattle influence tree cover, causing defoliation and breaking down young trees (< 10 cm dbh). Site condition (climate, soil, rainfall, etc) and competition between grasses and trees could also affect tree cover.

Conclusions

Despite the high species richness of trees at the landscape scale, four tree species dominated the landscape. Tree cover, expressed as either crown cover (m^2/ha) or tree density (trees/ha), was found to be low, irregular and randomly dispersed within the pasturelands, suggesting that trees in cattle farms are maintained at low densities to minimize their effects on pasture production.

References

Frankie, I., de Melo, E., Ferreira, J. and Alexander, V. (2001) Effect of shading by native tree legumes on chemical composition of forage produced by *Penisetum purpureum* in Acre western Brazilia Amazón. In: Ibrahim, M. (ed.) *Silvopastoral Systems for Reforestation of Degraded Tropical Pasture Ecosystems. International Symposium on Silvopastoral Systems and Second Congress on Agroforestry and Livestock Production in Latin America.* CATIE, San José, Costa Rica, pp. 197-202.
Guevara, S., Laborde, J. and Sanchez, G. (1998) Are isolated trees in pastures a fragmented canopy? *Selbyana* 19 (1), 34-43.
Harvey, C.A. and Haber, W.A. (1999) Remnant trees and the conservation of biodiversity in Costa Rican pastures. *Agroforestry Systems* 44, 37-68.
Holdridge, L. (1978) *Ecology Based in Life Zones* (Ecología basada en Zonas de vida). CATIE, San José, Costa Rica. (In Spanish.)
Lovejoy, T.E. (1985). Rehabilitation of degraded tropical forest lands. *Environmentalist* 5, 1-8.
Muchagata, M. and Brown, K. 2003. Cows, colonists and trees: rethinking cattle and environmental degradation in Brazilian Amazonia. *Agricultural Systems* 76, 797-816.
Nepstad, D., Uhl, C. and Serrao, A.S. (1990) Surmounting barriers to forest regeneration in abandoned, highly degraded pastures: a case study from Paragomiinas, Para, Brazil. In: Anderson, A.A. (ed.) *Alternatives to Deforestation: Steps Towards Sustainable Use of the Amazon Rain Forest.* Columbia University Press, New York, USA, pp. 215-229.
Schelhas, J. and Greenberg, R. (1996) *Forest Patches in Tropical Landscapes.* Island Press, Washington, DC, USA.
Souza de Abreu, M.H., Ibrahim, M., Harvey, C. and Jiménez, F. (2000) Characterization of the tree component in the husbandry systems in the Fortuna de San Carlos. Costa Rica (Caracterización del componente árboreo en los sistemas ganaderos de la Fortuna de San Carlos. Costa Rica). *Agroforesteria de las Americas* 7 (26), 53-56. (In Spanish.)

Micropropagation of three clones of *Morus alba* L. selected for fodder use

J. L. Fernández-Lorenzo, V. Pérez, S. Liñayo, M. R. Mosquera-Losada and A. Rigueiro-Rodríguez
Plant Production Department, University of Santiago de Compostela, 27002-Lugo, Spain. juanlufl@lugo.usc.es

Abstract
Morus alba is a forage tree widely used in America and which can be used in some European countries. In order to test the adaptation of American varieties to Europe conditions, *in vitro* culture is needed as a useful tool to produce a large number of plants. Three clones of *Morus alba* varieties selected for fodder use ('Doña Betty', 'Criolla' and 'Acorazonada') were propagated *in vitro* in order to determine their requirements. For the multiplication stage, the MS medium was selected, with the addition of 0.5-1.0 mg/l 6-benzyladenine (BA). 1/2MS or MS, with the addition of 1-3% sucrose and 0-0.1 mg/l 3-indolebutyric acid (IBA) was selected as rooting media. Acclimatization was carried out in a greenhouse on different substrates (sedge peat, 1:1-1:2 peat-moss:perlite), with a survival rate of 53 to 97%.

Introduction
Morus alba is used as fodder in several countries, such as Costa Rica, Cuba or Ethiopia (Benavides, 1999). Some varieties, selected for their digestibility and protein contents, e.g. 'Doña Betty' (DB), 'Acorazonada' (AC) and 'Criolla' (CR), could be of interest in Europe as part of fodder mixtures for direct livestock consumption. Field trials require a substantial amount of plants. Whereas propagation by cuttings is very time-consuming, *in vitro* culture could provide a high number of clonal plants in a relatively short period of time.

Materials and Methods
Plant material: Explants of *Morus alba* 'Doña Betty' (DB), 'Acorazonada' (AC) and 'Criolla' (CR) were obtained from branch segments previously kept in polyethylene bags at 4°C for 2 months.
Forcing of branch segments: Branch segments were thoroughly washed and some of these dipped for 10 min in a fungicide solution (1.2 mg/l benomyl + 0.3 mg/l captan). Once dried, they were stood upright in plastic jars containing tap water, with the bottom 5 cm in water, and stored in a culture room (20-25°C, 16 h/light photoperiod, RFD: $30 \mu mol \times m^{-2} \times s^{-1}$).
In vitro establishment: When 'forced' shoots reached at least 3 cm length, they were excised from the branch, leaves were removed and cut wounds sealed with paraffin. Later they were thoroughly washed, surface-sterilized with 70% ethanol (1 min) + NaClO (0.8% active chloride) (10 min) and rinsed three times in sterile distilled water. Disinfected nodal and apical 10-15 mm long explants were put into test tubes (150 mm × 24 mm) containing MS salts, WPM (Lloyd and McCown, 1981) vitamins, 3% sucrose, 0.7% Difco-Bacto^R agar (DBA) and 1mg/l 6-benzyladenine (BA). pH was adjusted to 5.6 before autoclaving. The tubes were kept in a culture room under the conditions described above. Contamination rate, reactivity and multiplication rate were recorded on day 35.
In vitro multiplication: New shoots obtained from initial cultures were kept under different multiplication conditions, depending on the number of shoots available for cloning. In an initial experiment, two basal media (WPM and MS) were tested, with the addition of BA (0.1, 0.2 or 0.5 mg/l) for clone DB. In a second experiment, different concentrations of DBA in MS medium + 1.0 mg/l BA were tested (0.7%, 0.8% for CR and AC; 0.7%, 0.8% and 0.9% for DB). Multiplication rate, number of shoots per explant and length of longest shoot per explant were recorded after 30-35 days.
In vitro rooting: After six cycles of successive subculturing, rooting experiments were carried out. Three cm long shoot tips taken at the end of the multiplication stage were placed in rooting medium, which consisted of MS or half-strength MS (1/2MS), with the addition of 1% or 3% sucrose and 0, 0.1 or 0.5 mg/l IBA. By day 28 rooting rate, number of roots per rooted shoot and length of the longest root per rooted shoot were recorded.
Acclimatization: Rooted shoots of every clone were transferred to plastic trays (150 cm^3 cavity) and acclimatized under greenhouse conditions. Sedge peat or 1:1-1:2 peat-moss:perlite were used as substrates. Survival rate was recorded after 60 days. Experiments were usually carried out with three sets of 12-15 shoots per treatment in a random design and repeated twice.

Results and Discussion
In vitro establishment: The flushing of buds took place between 17 and 25 days in all clones. Contamination rate was 12.5% using fungicide treatment prior to disinfection and 0% without fungicide treatment.

Table 1. Multiplication rate (MR), number of shoots per reactive explant (N) and length (cm) of the longest shoot per reactive explant (L) for clone DB, depending on the medium and BA concentration. Results are means ± standard deviation. Values followed by different letters differ significantly at level $P < 0.10$, after LSD test.

Medium	mg/l BA	MR	N	L
MS	0.1	1.46 ± 0.35 b	1.43 ± 0.11 a	2.55 ± 0.13 a
	0.2	1.93 ± 0.32 c	1.83 ± 0.05 b	2.48 ± 0.12 a
	0.5	2.46 ± 0.37 d	2.30 ± 0.20 c	1.88 ± 0.35 b
WPM	0.1	1.16 ± 0.11ab	1.86 ± 0.32 b	0.98 ± 0.24 c
	0.2	1.00 ± 0.10 a	1.77 ± 0.32 b	0.96 ± 0.21 c
	0.5	1.06 ± 0.28ab	1.60 ± 0.17ab	1.00 ± 0.16 c

The multiplication rate of the explants after 35 days in initiation medium was 2.2 for AC, 3.7 for DB and 1.8 for CR. The reactivity was 100% in all cases.

Multiplication: The best results for multiplication of clone DB are obtained on MS medium supplied with 0.5 mg/l BA (Table 1). Although a higher concentration (1.0 mg/l BA) yielded better results in the other two clones (results not shown), it caused vitrification in clone DB. With regard to agar concentration, clone DB multiplied better on 0.9% DBA (Table 2). Clones CR and AC grew equally as well as on 0.7 and 0.8% DBA.

Table 2. Multiplication rate (MR), number of shoots per reactive explant (N) and length (cm) of the longest shoot per reactive explant (L) for clones DB, CR and AC, depending on agar (DBA) concentration in medium. Results are means ± standard deviation. Values followed by different letters, for the same clone, differ significantly at level $P < 0.05$, after LSD test.

Clone	%DBA	TDM	N	L
DB	0.7	1.96 ± 0.12 a	1.81 ± 0.16 a	1.79 ± 0.41 a
	0.8	1.77 ± 0.30 a	2.60 ± 0.10 b	2.54 ± 0.23 b
	0.9	2.50 ± 0.26 b	2.43 ± 0.32 b	2.02 ± 0.21ab
CR	0.7	2.27 ± 0.14 a	2.04 ± 0.47 a	2.84 ± 0.14 a
	0.8	2.30 ± 0.20 a	1.95 ± 0.04 a	2.77 ± 0.35 a
AC	0.7	2.24 ± 0.21 a	1.55 ± 0.14 a	3.70 ± 0.51 a
	0.8	2.30 ± 0.22 a	1.45 ± 0.15 a	3.70 ± 0.87 a

Rooting: The results obtained in the rooting experiments (Table 3) show a strong influence of the genotype in response to rooting conditions. Briefly, all clones root well in media without IBA. When there is no IBA in the medium, clone CR roots improve in MS + 3% sucrose, clone AC yields similar results for 1% and 3% sucrose when the rooting medium is half MS and clone DB roots improve in medium supplied with 1% sucrose.

Table 3. Rooting rate (%R), number of roots per rooted shoot (N) and length (cm) of the longest root per rooted shoot (L) of clones 'Criolla', 'Acorazonada' and 'Doña Betty', depending on MS concentration (fs: full strength; ½: half strength), sucrose (S) concentration (10 or 30 g/l) and IBA concentration (0, 0.1 or 0.5 mg/l), by day 28. Values are means ± standard deviation.

			%R			N			L		
Clone	MS	S	0 IBA	0.1 IBA	0.5 IBA	0 IBA	0.1 IBA	0.5 IBA	0 IBA	0.1 IBA	0.5 IBA
CR	fs	10	47.2 ± 40.8	94.5 ± 4.8	75.0 ± 8.3	2.8 ± 2.4	7.8 ± 2.9	6.2 ± 3.5	2.4 ± 1.6	0.9 ± 0.6	0.6 ± 0.4
		30	100 ± 0.0	58.3 ± 8.3	88.9 ± 4.8	7.9 ± 3.3	5.4 ± 3.8	7.4 ± 3.7	4.7 ± 0.7	1.1 ± 0.6	1.0 ± 0.5
	½	10	91.7 ± 8.3	86.1 ± 9.6	861 ± 9.6	5.7 ± 1.8	7.8 ± 3.1	5.7 ± 3.5	2.9 ± 0.9	1.0 ± 0.6	0.6 ± 0.4
		30	88.9 ± 4.8	83.3 ± 8.3	88.9 ± 4.8	6.2 ± 3.0	7.2 ± 3.3	7.4 ± 3.6	3.7 ± 0.7	1.1 ± 0.5	0.6 ± 0.3
AC	fs	10	66.7 ± 8.3	75.0 ± 8.3	25.0 ± 16.7	3.9 ± 1.8	6.6 ± 3.4	2.4 ± 0.9	3.3 ± 1.3	0.3 ± 0.2	0.4 ± 0.3
		30	91.7 ± 8.3	69.4 ±17.4	72.2 ± 17.4	7.0 ± 3.0	5.6 ± 3.6	5.0 ± 3.0	3.8 ± 1.1	0.4 ± 0.2	0.4 ± 0.3
	½	10	97.2 ± 4.8	97.2 ± 4.8	88.9 ± 12.7	6.7 ± 2.5	8.9 ± 2.3	6.0 ± 3.5	3.5 ± 1.1	0.6 ± 0.2	0.4 ± 0.3
		30	94.4 ± 4.8	86.1± 12.7	94.5 ± 4.8	7.6 ± 3.2	7.3 ± 3.3	8.6 ± 2.5	4.5 ± 1.4	1.2 ± 0.7	0.4 ± 0.3
DB	fs	10	88.8 ± 4.8	58.4 ± 28.9	27.8 ± 12.7	6.6 ± 2.6	2.7 ± 1.8	4.9 ± 3.7	3.9 ± 1.7	2.2 ± 1.4	1.0 ± 0.4
		30	36.1 ± 4.8	16.7 ± 8.3	22.2 ± 12.7	7.8 ± 3.3	1.5 ± 0.6	4.6 ± 3.0	4.4 ± 1.4	1.5 ± 0.6	1.9 ± 0.5
	½	10	100 ± 0.0	63.9 ± 9.6	86.1 ± 9.6	6.8 ± 2.1	7.3 ± 2.9	6.8 ± 3.3	3.8 ± 1.9	0.6 ± 0.2	0.6 ± 0.3
		30	83.3 ± 8.3	58.3± 22.1	75.0 ± 0.0	8.4 ± 2.4	6.3 ± 3.3	6.7 ± 2.9	4.4 ± 0.7	0.5 ± 0.3	0.5 ± 0.2

Acclimatization: The survival rate was always over 50% in the conditions assayed (Table 4). A mixture of 1:1 peat-moss:perlite yielded good results for all clones (> 70% survival rate after 2 months). Clone AC reached 96.7% survival rate on a sedge peat substrate.

Table 4. Percentage of survival for clones CR, DB and AC after 2 months on different substrates (Peat moss:perlite - PM:P, 1:1 or 1:2, and sedge peat - SP), under greenhouse conditions.

Clone	PM:P 1:1	PM:P 1:2	SP
CR	70.0 ± 17.3	83.3 ± 15.3	-
DB	70.0 ± 36.1	70.0 ± 10.0	-
AC	73.3 ± 11.6	53.3 ± 20.8	96.7 ± 5.8

Conclusions

The three clones studied show good adaptation to *in vitro* multiplication, with high rates of survival during acclimatization. Their easy rooting on hormone-free media, and in some cases, as in clone DB, with low sucrose concentration (1%), compensates for their relatively high need for BA (up to 1.0 mg/l) during the multiplication stage. Thus, we consider that *in vitro* culture of these clones constitutes a suitable method for obtaining large amounts of plants in the short term in order to undertake field trials for their introduction into Europe as a source of fodder.

References

Benavides, J.E. (1999) Use of *Morus alba* in animal production systems (Utilización de la morera en sistemas de producción animal). In: Sánchez, M.D. and Rosales, M. (eds) *Agroforestería para la Producción Animal en Latinoamérica. Memorias de la Conferencia Electrónica. Estudio FAO Producción y Sanidad Animal 143*. FAO, Rome, Italy, pp. 275-281. (In Spanish.)

Lloyd, G. and McCown, B. (1981). Commercially-feasible micropropagation of mountain laurel, *Kalmia latifolia*, by use of shoot-tip culture. *Combined Proceedings International Plant Propagation Society* 30, 421-427.

Estimating diet selection of goats and sheep grazing on gorse-heathland vegetation with areas of improved pasture

L. M. M. Ferreira [1], M. Oliván [2], M. A. M. Rodrigues [1], U. García [2] and K. Osoro [2]

[1]CECAV, Dept. de Zootecnia, Universidade de Trás-os-Montes e Alto Douro, Apartado 1013, 5001 Vila Real Codex, Portugal. [2]SERIDA, Servicio Regional de Investigación y desarrollo Agroalimentario, Apartado 13, 33300 Villaviciosa, Asturias, Spain. lmf@utad.pt, mrodrigu@utad.pt, mcolivan@serida.org, kosoro@serida.org

Abstract

An experiment was carried out to study the grazing behaviour of goats and sheep managed in mixed grazing on gorse-heathland communities including areas of improved pasture. Equal numbers (42) of ewes and goats were managed in a plot (22 ha) established on natural shrubland dominated by heather and gorse including an area (5 ha) of improved pasture containing rye grass (*Lolium perenne*). Diet composition was estimated from the alkane concentrations in diet and faeces observed on two sampling dates, using least-squares procedures. The faecal recoveries used in the calculations were significantly affected by the proportions of ryegrass and heaths in the diet for both species and sampling days. The diet composition of goats on 3 July (mainly heaths) differed significantly from that calculated for sheep (mainly grass). When the grass availability decreased by 30 July heaths were the main components in the diet of both animal species.

Key words: *n*-alkanes, ruminants, diet composition, natural pastures

Introduction

There are large tracts of heathlands in the less favourable areas of the north-west of the Iberian peninsula. In these areas the accumulation of potentially combustible dry vegetation can cause serious environmental problems due to the increase of fires. The development of sustainable systems of animal production from this vegetation can provide a simple solution to this problem. To do this, a knowledge of grazing behaviour of different herbivore species is important to establish the best grazing strategies for these areas.

The *n*-alkane technique has been used with success to estimate diet selection of grazing ruminants (Dove *et al.*, 1999; Dove and Mayes, 2003). However, most of the studies have been performed on vegetation canopies where herbaceous species were dominant. Therefore there is a lack of information of the accuracy of the *n*-alkane technique to estimate the diet selected by ruminants (goats or sheep) when grazing on heather-gorse communities with intermixed patches of grass and clover. This information would contribute to the improvement of silvopastoral strategies developed on these vegetation communities.

In the present study, the diet composition of goats and sheep managed in a mixed grazing on gorse-heathland communities including an area of improved pasture was estimated at two grazing dates with different availability of the preferred herbaceous species (ryegrass).

Materials and Methods

This study was made on a hill experimental farm (1000 m asl) located in north-western Spain. Forty-two ewes and forty-two goats were managed in a plot (22 ha) established on natural shrubland dominated by heather and gorse with an area (5 ha) of pasture improved by sowing ryegrass. Animals had free access to the whole area. The composition of animal diets was estimated using the *n*-alkane technique on two grazing dates, the beginning (3) and end (30) of July. Between these times availability of the preferred herbaceous species (*Lolium perenne*) decreased from 10.1 cm to 5.6 cm. Faecal and herbage samples of the main vegetal components, heaths (*Calluna vulgaris, Erica umbellata, Erica australis, Erica arborea*), gorse (*Ulex gallii*), and ryegrass (*Lolium perenne*), were collected at each sampling date and the alkane pattern (from C_{22} to C_{36}) was analysed by direct saponification according to the method of Mayes *et al.* (1986) with minor modifications (Oliván and Osoro, 1999). Diet composition was estimated using an optimization procedure which minimizes the sum of squared discrepancies between the actual alkane proportions in faeces (adjusted for the incomplete faecal recoveries) and the calculated proportions (different combinations of diet components). Alkane faecal concentrations (from C_{25} to C_{33}) were corrected with recovery values obtained by the authors in previous validation studies (Ferreira *et al.*, 2003) performed in metabolic pens with animals receiving diets containing different combinations of the main vegetation components present in the field. Goats were fed with diets composed of *Lolium perenne*/*Ulex gallii* (G1), *Lolium perenne*/heather (G2), *Lolium* leaf/stem/spike/heather/*Ulex gallii* (G3) and *Lolium perenne*/*Trifolium repens*/heather (G4), while sheep received diets of *Lolium perenne* (S1) and *Lolium perenne*/*Trifolium repens* (S2). The effect of the application of different faecal recovery values obtained in these validation studies (G1, G2, G3, G4 for goats and S1 and S2 for sheep) on the estimates of diet composition and the effect of animal species (goats or sheep), grazing date (3 or 30 July) and their interaction on diet composition were examined by analysis of variance, using t-tests for comparisons of means.

Results and Discussion

Tables 1 and 2 show diet composition estimates for goats and sheep, respectively, on 3 and 30 July. The results indicate that the faecal recovery values used in the calculations were significantly affected by the estimates of the proportions of *Lolium perenne* and heaths in the diet of both animal species. This had been shown previously by Ferreira *et al.* (2003; see also this volume) in validation studies made on metabolism pens with goats and sheep fed with different proportions of the main vegetation species found in pastures. In contrast, there was no effect on faecal recoveries of the calculated proportion of *Ulex gallii* in the diets, which was always zero.

Table 1. Diet composition (%) of goats on 3 and 30 July, using different faecal recovery values. Faecal recoveries: G1 = diet composed by 70% *Lolium perenne* + 30% *Ulex gallii*; G2 = diet composed by 70% *Lolium perenne* + 30% heather; G3 = diet composed by 28% *Lolium* leaf/stem + 12% *Lolium* spike + 30% heather + 30% *Ulex gallii*; G4 = diet composed by 40% *Lolium perenne* + 30% *Trifolium repens* + 30% heather. Values in the same row with different letters are significantly different ($P < 0.05$). SEM = standard error of means. Sig: significance.

Diet components	Alkane faecal recoveries				SEM	Sig
	G1	G2	G3	G4		
3 July						
Lolium perenne	0.38a	0.11b	0.35a	0.43a	0.034	***
Ulex gallii	0.00a	0.00a	0.00a	0.00a	0.000	ns
Heaths	0.62b	0.89a	0.65b	0.57b	0.034	***
30 July						
Lolium perenne	0.52a	0.42b	0.35c	0.25d	0.023	**
Ulex gallii	0.00a	0.00a	0.00a	0.00a	0.000	ns
Heaths	0.48d	0.58c	0.65b	0.75a	0.023	***

Table 2. Diet composition of sheep on 3 and 30 July, using different faecal recovery values. Faecal recoveries: S1 = diet composed by 100% *Lolium perenne*; S2 = diet composed by 70% *Lolium perenne* + 30% *Trifolium repens*. Values in the same row with different letters are significantly different ($P < 0.05$). SEM = standard error of means. Sig: significance.

Diet components	Alkane faecal recoveries		SEM	Sig
	S1	S2		
3 July				
Lolium perenne	0.73b	0.86a	0.024	**
Ulex gallii	0.00a	0.00a	0.000	ns
Heaths	0.27a	0.14b	0.024	**
30 July				
Lolium perenne	0.52a	0.40b	0.017	**
Ulex gallii	0.00a	0.00a	0.000	ns
Heaths	0.48b	0.60a	0.017	**

The average estimate of diet composition for each animal species is presented in Table 3. At the first sampling date (3 July), when the grass availability was high, there was a significant effect ($P < 0.001$) of animal species on composition of diet, heaths being the main component of the diet in goats (68.3%) while *Lolium perenne* was predominant in the sheep diet (79.2%).

Table 3. Comparison of diet composition of goats and sheep on 3 and 30 July. Values in the same row with different letters are significantly different ($P < 0.05$). SEM = standard error of means.

Diet components	3 July		30 July		SEM	Date × animal species
	Goats	Sheep	Goats	Sheep		
Lolium perenne	0.32c	0.79a	0.39b	0.46b	0.044	***
Heaths	0.68a	0.21c	0.61b	0.54b	0.044	***

On the second collection day (30 July), when the grass availability decreased below 6 cm, animal species had no significant effect on the proportions of the feed components, heaths being predominant in the diet of both goats (61.2%) and sheep (54.2%). These results agree with those reported by Osoro *et al.* (2003) in showing a change in the diet preferences of sheep when grazing on this vegetation community as the availability of the preferred herbaceous species (grass) of the improved areas decreases. Sheep are grass-grazers when the grass availability is high and browsers when the grass amount decreases. However, goats showed preference for heaths on both grazing dates studied.

Conclusions

These results indicate that goats showed browser-grazer behaviour on both dates studied, whereas sheep were grass-grazers when the grass availability was high and browsers when the amount of herbaceous species decreased. The use of different alkane faecal recovery values in the calculations significantly affected the estimates of the composition of the diet of goats and sheep grazing on gorse-heathland communities, despite the fact that the applied recoveries were calculated in metabolic pens with different combinations of the main vegetation components present in the fieldwork.

References

Dove, H. and Mayes, R.W. (2003) Satellite meeting: wild and domestic herbivore diet characterization, In: Ku Vera, J.C. and Rymer, C. (eds) *Proceedings of the Sixth International Symposium on the Nutrition of Herbivore*. Universidad Autónoma de Yucatán, Mérida, Mexico.

Dove, H., Wood, J.T., Simpson, R.J., Leury, B.J., Ciavarella, T.A., Gatford, K.L. and Siever-Kelly, C. (1999) Spray-topping annual grass pasture with glyphosate to delay loss of feeding value during summer. III. Quantitative basis of the alkane-based procedures for estimating diet selection and herbage intake by grazing sheep. *Australian Journal of Agriculture Research* 50, 475-485.

Ferreira, L.M.M., Oliván, M., Rodrigues, M.A.M., Dias-da-Silva, A. and Osoro, K. (2003) The use of alkanes as markers for estimating diet composition in sheep and goats. Wild and domestic herbivore diet characterization. In: Ku Vera, J.C. and Rymer, C. (eds) *Satellite Meeting of the VI International Symposium on the Nutrition of Herbivores*. Universidad Autónoma de Yucatán, Mérida, Mexico, pp. 47-49.

Mayes, R.W., Lamb, C.S. and Colgrove, P.M. (1986) The use of dosed and herbage n-alkanes as markers for the determination of herbage intake. *Journal of Agriculture Science* 107, 161-170.

Oliván, M. and Osoro, K. (1999) Effect of temperature on alkane extraction from faeces and herbage. *Journal of Agricultural Science* 132, 305-312.

Osoro, K., Oliván, M., Martínez, A., García, U. and Celaya, R. (2003) Diet selection and live weight changes in domestic ruminants grazing on heathland vegetation with areas of improved pasture. In: Ku Vera, J.C. and Rymer, C. (eds) *Proceedings of the VI International Symposium on the Nutrition of Herbivores*. Universidad Autónoma de Yucatán, Mérida, Mexico, pp. 491-494.

Validation of the alkane technique to estimate complex diets in sheep and goats fed on heathland vegetation

L. M. M. Ferreira [1], M. Oliván [2], M. A. M. Rodrigues [1], U. García [2] and K. Osoro [2]

[1]CECAV, Dept. de Zootecnia, Universidade de Trás-os-Montes e Alto Douro, Apartado 1013, 5001 Vila Real Codex, Portugal. [2]SERIDA, Servicio Regional de Investigación y desarrollo Agroalimentario, Apartado 13, 33300 Villaviciosa, Asturias, Spain. lmf@utad.pt, mrodrigu@utad.pt, mcolivan@serida.org, koldooo@princast.pt

Abstract

An experiment was carried out to validate the alkane technique for estimation of diets offered to sheep and goats of different combinations of herbaceous (*Lolium perenne* and *Trifolium repens*) and woody species (*Erica* sp and *Ulex gallii*). Two groups of four cross-breed goats (G1 and G2) and four cross-breed sheep (S) were housed in metabolism pens. Diet composition varied among groups: G1 - 28% ryegrass (*Lolium perenne*) leaf/stem, 12% ryegrass seed heads, 30% gorse (*Ulex gallii*) and 30% heather (*Erica* sp); G2 - 40% ryegrass leaf/stem, 30% white clover (*Trifolium repens*) and 30% heather; S - 70% ryegrass leaf/stem and 30% white clover. Diet composition was estimated from the alkane concentrations in the diet and faeces (with or without correction for incomplete faecal recoveries) using the least-squares procedures. There were no significant differences between measured proportions of dietary components and those estimated with alkanes, when applying the faecal recovery corrections. In contrast, the proportions calculated without faecal recovery correction differed significantly ($P < 0.05$) from the actual proportions.

Key words: diet selection, small ruminants, shrubland

Introduction

Knowledge of grazing behaviour, mainly diet selection, of different animal species and breeds on natural vegetation communities is important for the development of sustainable silvopastoral systems. *n*-Alkanes have been used with success as faecal markers to estimate diet composition of herbivores. However, the precision of diet composition estimates depends on the accurate correction of their incomplete faecal recoveries. Ferreira *et al.* (2003) indicate that animal species and diet composition may influence the faecal recovery of alkanes. To date the validation of the method has been mainly confined to two dietary components, generally grass and legume, and little is known on the accuracy of the alkane technique to estimate diets composed by mixtures of herbaceous and woody species, such as those grazed in shrub communities. The objective of this work was to study the accuracy of the alkane technique to estimate the composition of mixed diets offered to sheep and goats composed of different combinations of herbaceous (*Lolium perenne* and *Trifolium repens*) and woody species (*Erica* sp and *Ulex gallii*) and different plant parts of grass (leaf/stem or spike), in order to validate this technique to estimate the diet selection of sheep and goats when grazing on silvopastoral systems composed of gorse-heathland with areas of improved pastures.

Materials and Methods

Twelve animals, divided in three groups of four cross-breed goats (G1: 26 kg live weight), four cross-breed goats (G2: 29 kg live weight) and four cross-breed sheep (S. 30 kg live weight), were housed in metabolism pens to allow total collection of faeces. Animals were offered a daily ration of 0.8 kg DM/100 kg live weight. Diet composition was: G1 - 28% ryegrass leaf/stem, 12% ryegrass seed heads, 30% gorse and 30% heather; G2 - 40% ryegrass leaf/stem, 30% white clover and 30% heather; and S - 70% ryegrass leaf/stem and 30% white clover. Sampling of diet components and total faeces collection (4-day collection period) began after a 7-day adjustment period. Samples of faeces and diet components were freeze-dried and ground prior to alkane analysis. Alkanes (C_{22} to C_{36}) were analysed according to the method of Mayes *et al.* (1986) with minor modifications (Oliván and Osoro, 1999). Alkane faecal recoveries were calculated for each animal as the percentage of ingested marker obtained in faeces. Diet composition was estimated using an optimization procedure which minimizes the sum of squared discrepancies between the actual alkane proportions in faeces (adjusted for the incomplete faecal recoveries) and the calculated proportions (different combinations of diet components). The effect of dietary treatment on faecal recoveries was examined by analysis of variance. Measured and estimated diet compositions were compared using regression analysis and t-tests for paired comparisons.

Results and Discussion

The concentration of alkanes C_{23}, C_{24}, C_{35} and C_{36} were extremely low in most of the vegetation components; therefore they were excluded from the calculations to avoid a bias of estimations (Table 1).

Table 1. Alkane composition of the different dietary components (mg DM/kg). *Trifolium repens* (T.r.); *Lolium* leaf/stem (L. l.); *Lolium* spike (L. s.); *Ulex gallii* (U.g); *Erica* sp (E.).

Alkane	C_{23}	C_{24}	C_{25}	C_{26}	C_{27}	C_{28}	C_{29}	C_{30}	C_{31}	C_{32}	C_{33}	C_{35}	C_{36}
T.r.	2.2	1.9	16	3.8	38	11.3	170	17	206.9	7.2	22.2	2.5	1.6
L. l.	1.6	1.4	10	2.2	22	6.6	124	11	234.9	4.8	38.9	3.2	1.6
L.s.	20	4.1	85	6.9	111	11.7	219	20	337.6	7.8	21.9	1.4	1.8
U.g	0.5	0.9	2	1.5	9	5.5	42.8	12	157.4	5.3	5.3	0.5	1.7
E.	3.4	2.3	11	5.2	49	14.7	278	40	900.4	69	599.4	11.9	1.8

Alkane faecal recoveries showed a tendency to increase with carbon-chain length in all treatments (Table 2). There were significant differences ($P < 0.05$) in alkane recoveries between dietary groups for C_{25}, C_{26}, C_{27} and C_{32}. The same observation was made by Ferreira *et al.* (2003) in a similar experiment with sheep and goats fed on diets of one or two components, although in their study alkane faecal recoveries were lower and showed significant differences between diet treatments for all alkanes from C_{25} to C_{35}.

Table 2. Effect of treatments on alkane faecal recoveries (%). Values in the same line with different letters are significantly different ($P < 0.05$); SEM: standard error of mean; Sig: Significance; *: $P < 0.05$; **: $P < 0.01$; ***: $P < 0.001$; ns: non-significant.

Alkane	C_{25}	C_{26}	C_{27}	C_{28}	C_{29}	C_{30}	C_{31}	C_{32}	C_{33}
G1	70.04b	68.42a	73.11b	76.56a	84.41a	90.54a	95.66a	97.61b	95.61a
G2	95.28a	72.95a	87.25a	78.79a	89.56a	95.33a	97.08a	94.65b	92.34a
S	64.80b	58.76b	70.17b	73.07a	82.68a	96.15a	92.39a	117.01a	98.18a
SEM	4.38	2.882	3.835	3.201	2.763	2.298	2.257	2.488	2.380
Sig	**	*	*	ns	ns	ns	ns	***	ns

The estimation of diet composition with alkanes, when applying faecal recovery corrections, using the mean recoveries for each treatment, was accurate in all treatments (Table 3). However, the proportions calculated without recovery correction differed significantly ($P < 0.05$) from the actual proportions and overestimated *Ulex gallii* and underestimated *Trifolium repens* and *Erica* sp in all diets. *Lolium perenne* was overestimated in groups G2 and S, fed with leaf/stem, and underestimated in group G1, fed with leaf/stem and spikes.

Table 3. Measured percentage of dietary components (means) and those estimated (means) for each group. [1]Corrected with treatment mean faecal recoveries; [2]without recovery correction. Values in the same line for each group with different letters are significantly different ($P < 0.05$); SEM: standard error of mean; Sig: Significance; *: $P < 0.05$; **: $P < 0.01$; ***: $P < 0.001$; ns: non-significant. Diet comp.: diet components; T.r: *Trifolium repens*; L.f/s: *Lolium* leaf/stem; L.s.: *Lolium* spike; U.g.: *Ulex gallii*; E.: *Erica* sp.

Groups	G1					G2					G3				
Diet comp.	T.r	L.fs	L.s.	U.g.	E.	T.r	L.f/s	L.s.	U.g	E.	T.r	L.f/s	L.s.	U.g.	E.
Measured	0a	28b	12b	30a	30b	30b	39a	0a	0a	30b	30b	70a	0a	0a	0a
Estimated[1]	1.8a	25b	12b	32a	30b	31b	36a	0.4a	2a	30b	29b	69a	0.5a	0.7a	0.1a
Estimated[2]	0a	12a	4a	57b	28a	6.4a	54b	1.2a	12b	26a	9a	83b	0a	7b	0.2a
SEM	0.6	2.1	0.6	1.8	0.5	2	3	0.4	1.5	1.1	3.1	3.4	0.3	0.7	0.1
Sig	ns	**	***	***	*	***	**	ns	***	*	**	*	ns	***	ns

Conclusions

This study confirms that alkanes are useful markers to estimate the proportion of different heathland vegetation components offered to sheep and goats in mixed diets and hence can be applied to silvopastoral systems composed of gorse-heathland with areas of improved pastures. However, animal species and diet composition influence the faecal recovery of alkanes; therefore the application of these markers for estimating the diet selection of sheep and goats under grazing conditions should be preceded by a calculation of the actual alkane faecal recoveries for each condition.

References

Ferreira, L.M.M., Oliván, M., Rodrigues, M.A.M., Dias-da-Silva, A. and Osoro, K. (2003) The use of alkanes as markers for estimating diet composition in sheep and goats. Wild and domestic herbivore diet characterization. In: Ku Vera, J.C. and Rymer, C. (eds) *Satellite Meeting of the VI International Symposium on the Nutrition of Herbivores*. Universidad Autónoma de Yucatán, Mérida, Mexico, pp. 47-49.

Mayes, R.W., Lamb, C.S. and Colgrove, P.M. (1986) The use of dosed and herbage n-alkanes as markers for the determination of herbage intake. *Journal of Agriculture Science* 107, 161-170.

Oliván, M. and Osoro, K. (1999) Effect of temperature on alkane extraction from faeces and herbage. *Journal of Agriculture Science* 132, 305-312.

Effects of grazing on *Quercus faginea* Lam. forests in Navarra (Spain)

V. Ferrer [1], C. Ferrer [2], A. Broca [2] and M. Maestro [3]

[1] Consultoría de Estudios y Proyectos de Pastizales, C/Batondoa, nº 3. 31006 Burlada, Navarra. Spain. [2] Dept. Agricultura y Economía Agraria, Universidad de Zaragoza, Miguel Server 177, E-50013 Zaragoza, Spain. [3] Instituto Pirenaico de Ecología, CSIC, Apartado 202, 50080 Zaragoza. Spain. saltus@teleline.es.

Abstract

We report the effect of grazing in four dense forests of *Quercus faginea* Lam. in central Navarra. Tree density was 2550 trees/ha and canopy cover was 93%. Grazing significantly increased soil denudation. The cover of *Helictrotrichon cantabricum* decreased significantly although other grasses such as *Bromus erectus* and *Festuca rubra* increased. Overall, species richness and diversity of the herbaceous canopy increased. Eventually, livestock activity increased the grazing value of the area.

Key words: silvopasture, floristic change, biodiversity, grazing value

Introduction

Until the 1980s, the Navarra government developed the "Programme for amelioration of grasslands in communal areas". One of the aims of this programme was the development of silvopastures in abandoned crop fields and forest areas. For that purpose, several areas with different types of vegetation (grasslands, forests, shrublands, etc) were fenced and facilities for livestock management were established. These areas were managed with livestock for 20 years with the double purpose of production and conservation, with especial regard to maintenance of diversity.

Despite the interest in this project, data on the results are scarce. This chapter reports on the data from a 2-year record on four forest areas. These forests had a sub-Mediterranean climate, and differed in both the stocking rates and the starting dates of the silvopastoral practice. Results in this chapter include analysis of the dynamics of the herbaceous and shrubland vegetation under grazing compared to controls established in neighbouring, non-grazed areas.

Materials and Methods

The study sites were four forests of *Quercus faginea* (*Spiraeo obovatae-Quercetum fagineae*). Selected areas had a density of 2550 trees/ha, with a mean height of 6 m, a mean diameter of 12 cm (at 1.3 m height) and a tree canopy cover of 93% (Ferrer, 1997). These forests originated by vegetative means 30-40 years ago. The substrate was calcareous, altitudes ranged from 575 to 880 m and mean annual rainfall was 600 mm. In the control areas, mean tree canopy cover was 9% and herbaceous cover reached 86%, with the remainder (5%) being bare soil. Two of the areas were grazed by cattle all year round for 7 and 11 years, respectively (Aibar and Lerga). The other two areas were grazed by horses from November till May for 8 and 12 years, respectively (San Martín and Leache).

In each of the four pastured forests, sample plots of 20 m × 20 m were set up. For each area a control plot, without grazing, was established in order to compare and simulate the initial conditions of the grazed forests. Shrubs and herbaceous vegetation were sampled for 2 years in the spring and summer months following the "point quadrat" method (Daget and Poissonet, 1971) along 20 m transects. Along each transect, 100 points were recorded for the specific frequency of each species (Fs) and the percentage of bare soil (SD). The percentage specific contribution of each species was calculated (Cs = Fs × (100 - SD)/ΣFs), species richness and the diversity index of Shannon (H′ = - $\Sigma pi \times \log_2 pi$, pi being the number of individuals of a species in relation to the total number of individuals). Three samplings were made in each control area and an overall of 25 samplings in the four grazed areas. The grazing value (VP) was determined according to the methodology of Daget and Poissonet (1972) (VP = 0.2 × ΣCsIs, where Is is the "specific index of quality", which varies according to literature from 0 to 5 in function of the productivity, nutritious value, digestibility, etc.).

To estimate grazing pressure, the variables recorded were: number of years during which grazing took place and grazing intensity (I) determined by the "differences method" (De Simiane and Damiani, 1981; Lambert and Senn, 1984) by means of the equation I = (N - (C + P)/S, where the extraction of the herbaceous biomass is expressed as FU/ha per year (FU = forage units) collected by livestock during grazing, N nutritive needs of livestock, C and P the complementary feeding and grazing meadows, respectively, all them expressed in FU/year, and S the surface area of grazed forest areas. Other variables recorded were the degree of grazing influence, which included the two former ones (Figure 1) dividing the field of coordinates (x = years, y = intensity), by means of perpendicular bands to the straight line $x = y$ (Ferrer, 1997).

Figure 1. Grazing "degree of influence", in relation to years during which livestock took place at the four sites and "intensity" of the same (following Ferrer, 1997). Statistical analyses of significant differences (Student test) among the mean values of all data obtained in the controls and the grazed plots were carried out.

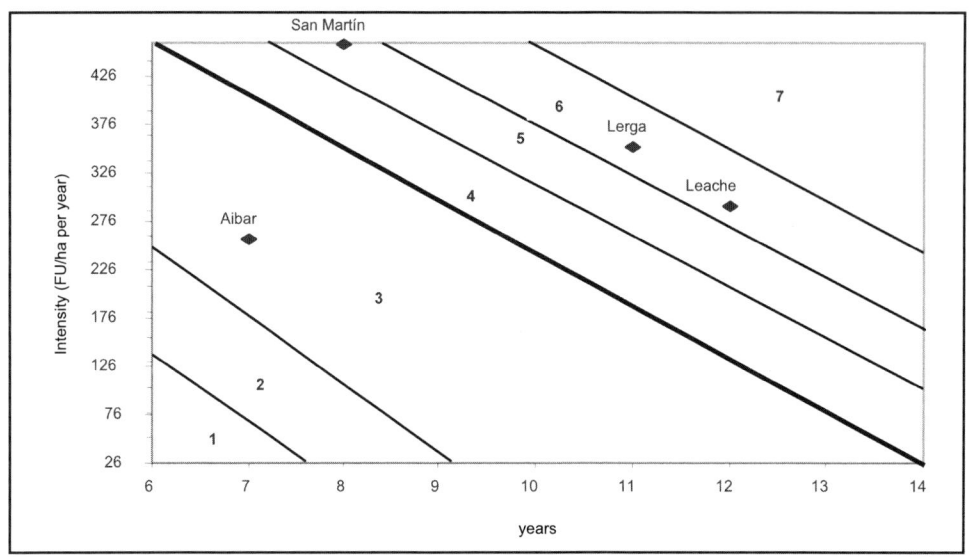

Results and Discussion

The grazing intensity ranged between 257 and 458 FU/ha per year (Table 1). One of the effects of grazing was a decrease in the cover of the herbaceous vegetation and a consequent, significant increase in the percentage of bare soil (Table 2). These effects increased in line with the degree of grazing influence. Within herbaceous species, dominant perennial grasses decreased significantly when the grazing intensity increased. This phenomenon has already been observed by Naveh and Whittaker (1980), Allen et al. (1995) and Rosenstock (1996). A significant decrease in cover is reported for *Helictrotichon cantabricum*. Similar results are also reported by Bobbink et al. (1987) in calcareous grasslands dominated by *Brachypodium rupestre*. The decrease in cover of *H. cantabricum* is compensated by significant increases in the cover of *Bromus erectus*, *Festuca rubra* and "other herbaceous species" (*Avenula bromoides*, *Bupleurum rigidum*, *Centaurea montana*, *Inula salicina*, *Melittis melissophyllum*, *Thalictrum tuberosum*, etc.). According to Grime (1979) in disturbed (grazed) areas, short-lived species with high growth rates and possessing conspicuous and small seeds have better chances for establishment. Differences reported between *Brachypodium rupestre* and *Carex flacca* in control and grazed areas are of little interest.

Table 1. Calculation of "intensity" (I) of grazing (FU/ha per year) in wooded areas. N = needs of livestock; C = complementary feeding; P = grazing meadows; S = surface area of grazed wooded areas (Ferrer, 1997).

	N (FU/year)	C (FU/year)	P (FU/year)	C + P (FU/year)	N - (C + P) (FU/year)	S (ha)	I = (N - (C + P))/S (FU/ha/year)
Aibar	119.429	5.799	91.338	97.137	22.292	87	257
Leache	46.308	0	0	0	46.308	159	292
Lerga	120.510	64.010	47.334	111.344	9.166	26	353
S. Martín	42.172	0	1.686	1.686	40.486	88	458

Both species richness and diversity had a tendency to increase in the four grazed areas, although the tendency was significant only in the two most intensely grazed plots. It is well-known that moderate levels of disturbance (like grazing) increase diversity and that, at high disturbance levels, diversity decreases (Naveh and Whittaker, 1980; Lepart and Escarré, 1983; Collins, 1987). Since the most intensely grazed areas displayed the highest diversity values, we conclude that the grazing threshold has not been reached in our plots from a diversity point of view. The grazing value (VP) increased significantly in all locations and correlated with grazing intensity.

Table 2. Mean values of control plots (T) and grazed plots (P) and statistical significances (S): ** ($P < 0.05$), *($P < 0.2$). NS: Number of samplings; Bs: Bare soil; Sh: Shrubs; H: Herbaceous species; G: Grasses; L: Legumes; Co: Compositae; Oth: Other families; Bra: *Brachypodium rupestre*; Br: *Bromus erectus*; Ca: *Carex flacca*; F: *Festuca rubra* s.l.; He: *Helictotrichon cantabricum*; OH: Other herbaceous species; N: Number of species; SI: Shannon diversity index; V: Grazing value.

		NS	Bs%	Sh	H	G	L	Co	Oth	Bra	Br	Ca	F	He	OH	N	SI	V
Aibar	T	3	4	6	90	68	1	0	27	14	6	12	23	19	16	17	3	7
	P	6	13	3	83	59	1	3	23	21	7	10	12	16	17	21	3	19
	S		*		*	*		*					*					**
Leache	T	3	4	6	90	68	1	0	27	14	6	12	23	19	16	17	3	7
	P	10	18	7	75	54	2	1	25	15	10	10	26	10	15	20	3	20
	S		**		**	**		*					*					**
S. Martin	T	3	5	13	82	74	4	1	16	6	1	1	2	66	7	13	2	2
	P	6	18	11	71	45	3	2	33	4	18	10	7	10	22	22	3	20
	S		**		**	**			**	**	*	*	**	**	**	**	**	**
Lerga	T	3	5	13	82	74	4	1	16	6	1	1	2	66	7	13	2	1
	P	3	30	6	64	49	1	2	18	3	15	4	22	9	12	19	2	19
	S		**	*	*	**	*		*		**	**	**	**	*	**	**	**

Conclusions

We conclude that actual livestock grazing and management enhance in many ways the forest of *Quercus faginea*: grazing opens the vegetation, lowers the accumulation of necromass, promotes the growth of many species, increases species richness and diversity, and enhances the grazing value of the community, which in turn allows a progressive increase in stocking rates.

References

Allen, R.B., Wilson, J.B. and Manson, C.R. (1995) Vegetation change following exclusion of grazing animals in depleted grassland, Central Otago, New Zealand. *Journal of Vegetation Science* 6, 616-626.

Bobbink, R., During, H.J., Shreurs, J., Willems, J. and Zielman, R. (1987) Effects of selective clipping and mowing time on species diversity in chalk grassland. *Folia Geobotnanic Phytotaxon* 22, 363-376.

Collins, S.L. (1987) Interaction of disturbances in a tallgrass prairie: a field experiment. *Ecology* 68, 1243-1250.

Daget, Ph. and Poissonet, J. (1971) Phytological analysis method for meadows (Une méthode d'analyse phytologique des prairies). *Annales d'Agronomie* 22, 5-41. (In French.)

Daget, Ph. and Poissonet, J. (1972) Assessment of pastoral value of grazing lands (Un procédé d'estimation de la valeur pastorale des pâturages). *Fourrages* 49, 31-39. (In French.)

De Simiane, M. and Damiani, C. (1981) Productivity of a feed system using grazing lands with pasture (Système d'alimentation avec utilisation de pâturages à faible productivité). In: Morfando-Fehv, P., Bourbouze, A. and De Simiane, M. (eds) *Symposium Nutrition et Système d'Alimentation de la Chèvre International.* ITOVIC-INRA, Tours, France, pp. 448-495. (In French.)

Ferrer, V. (1997) Effects of grazing in tree and shrub ecosystems on vegetation, flora and soil in Central Navarra (Efectos del pastoreo en ecosistemas arbolados y arbustivos de la Navarra Media, sobre la vegetación, la flora y el suelo). MSc thesis. The University of Zaragoza, Zaragoza, Spain. (In Spanish.)

Grime, J.P. (1979) *Plant Strategies and Vegetation Processes.* John Wiley, London, United Kingdom.

Lambert, B. and Senn, O. (1984) *Report on the Use of Pastures by Ovine Livestock in the Dry Pre-Alps* (Rapport sur l'utilisation des parcours par les ovins dans les Préalpes sèches). ADEO. Gap. (In French.)

Lepart, J. and Escarré, J. (1983) Vegetation succession, mechanisms and modelling: literature review (La succession végétale, mécanismes and modèles: analyse bibliographique). *Bulletin D'Ecologie* 14, 133-178. (In French.)

Naveh, Z. and Whittaker, R.H. (1980) Structural and floristic diversity of shrublands and woodlands in northern Israel and another Mediterranean areas. *Vegetatio* 41, 171-190.

Rosenstock, L. (1996) NIOSH Testimony to the USA Department of Labor. *Air Quality, Applied Occupational Environmental Hygiene* 11 (12), 1368.

Effect of breed and stocking rate on live weight changes of goats grazing heath-gorse vegetation community

U. García, R. Celaya, B. M. Jáuregui, A. Martínez and K. Osoro

Servicio Regional de Investigación y desarrollo Agroalimentario (SERIDA). Apdo. 13, 33300 Villaviciosa, Asturias, Spain. rcelaya@serida.org

Abstract

Large areas of marginal lands in temperate areas are covered by heath-gorse vegetation. These communities have very poor nutritive value and high combustibility and goats, due to their browser behaviour, should be the most appropiate domestic species for their sustainable management. Two breeds of goats were managed: Cashmere goats (34 kg LW) under two stocking rates, high (12 animals/ha) and medium (6 animals/ha), and local Celtiberian goats (45 kg LW) only under high stocking rate. Replicates in three blocks were set up. Live weight changes were studied during two consecutive (2002 and 2003) grazing seasons. Significant differences between breeds in live weight changes were observed. These differences varied along the grazing season, increasing in the last period (September-November) in both years. Results indicate a better adaptation to these severe conditions of Cashmere goats, which have lower body weight than local goats.

Key words: heathland, Cashmere goats, animal performance

Introduction

Heathlands cover a large part of the less favoured areas in the Cantabrian region (N Spain) due to deforestation and abandonment of livestock management. Their woody vegetation has a very poor nutritive value and high combustibility. This limits animal production and the sustainability of livestock farming systems in these areas and increases the fire hazard. Thus, alternative management strategies need to be developed to match the sustainability of pastoral and silvopastoral systems together with the maintenance or enhancement of biodiversity. Goats browse, utilizing these plant communities more efficiently than cattle or sheep, and therefore should be the most appropiate species for the management of these heath-gorse communities (Osoro *et al.,* 2000). Cashmere goats may enhance the profitability of silvopastoral systems in these marginal lands as they diversify production and their fibre has high quality value to supplement meat production.

In 2002 an integrated experimental programme was set up with other European sites (EU project FORBIOBEN, QLRT-1999-30130) to study the effects of stocking rate and goat breed (one local *vs* one commercial foraging) on the biodiversity of different natural or semi-natural grasslands and shrublands, and on the socio-economic outcome of these livestock systems (Rook *et al.,* 2002). The objective of this work is to study the performance (live weight and body condition score changes) of local Celtiberian goats (45 kg LW) at high stocking rate, and cashmere goats (34 kg LW) at high or low stocking rate, grazing these heathland communities.

Materials and Methods

The experimental farm is located in Illano, Asturias (NW Spain), at 900 m asl. The vegetation is a shrubland dominated by heaths (*Erica* sp, *Calluna, Daboecia*) and gorse (*Ulex gallii*). Nine plots of 0.6 ha with three grazing treatments (3-replicated) were established: low stocking rate with four Cashmere goats (LC), high stocking rate with eight Cashmere goats (HC) and high stocking rate with seven local Celtiberian goats (HL). Plots were grazed from July to November in 2002 and from May to November in 2003. There was one more goat in HC than HL because of the differences in the adult live weight between breeds. The females were not suckling kids.

Animals were weighed and their body condition score was assessed (Russel *et al.,* 1969) at the beginning and at the end of each period within the experimental grazing seasons. Two periods were considered in year one (2002): July-August and September-November; and three periods in year two (2003): May-June, July-August and September-November. Treatment and replicate effects on the live weight and body condition changes were analysed by analysis of variance (SPSS, 1989).

Results and Discussion

Cashmere goats had better ($P < 0.001$) live weight changes than local goats, either in absolute values or in percentages of their live weight, except for the period 2 (July) of the first year (Table 1). Stocking rate and replicate had no significant effects. There were better ($P < 0.001$) live weight changes in year 2 (-4 g/day) than in year 1 (-16 g/day). The interactions between the main factors, treatment, replicate and year were not significant.

Table 1. Live weight (LW) changes of Cashmere goats at low stocking (LC) or high stocking rate (HC) and of local Celtiberian goats at high stocking rate (HL) grazing heathland vegetation in two consecutive years. a, b, c, Row means with common superscripts in each year do not differ ($P > 0.05$).

Year	2002			2003		
Treatment	LC	HC	HL	LC	HC	HL
Initial LW (kg)	35.6a	35.1a	45.1b	33.9a	34.8a	41.9b
LW changes (g/day)						
Period 1 (20/5-3/7)	–	–	–	100a	118a	43b
Period 2 (3/7-29/8)	31a	26a	33a	-21a	-29a	-40a
Period 3 (29/8-6/11)	-32a	-38a	-73b	-29a	-44a	-71b
Total grazing season	-2a	-8a	-23b	12a	9a	-28b
LW change (%)						
Total grazing season	-1.1a	-2.8a	-6.8b	7.3a	4.9a	-10.6b

Changes in body condition score followed similar trends to live weight changes, with greater decreases in body condition in local than in Cashmere goats (Figure 1). Osoro *et al.* (1999, 2002) found better performances of the smaller sheep breed when the availability of nutrient resources was low compared to the bigger breed. In the present work, a similar effect of body size of goats was observed in conditions where the available vegetation had poor nutritive value. Differences in diet selection have been reported between local and cashmere goats in the same experimental farm (CIATA, 1996), with higher percentages of shrubs in the diet of the former breed. This different behaviour is causing different vegetation dynamics between treatments, resulting in higher decreases of shrubs (heaths and gorse) and higher increases of available grasses under local than under cashmere goat grazing (see Celaya *et al.*, 2005, this volume), which could affect the subsequent performance of the goats.

Figure 1. Changes in the body condition score (0-5) of goats grazing heathland vegetation in two consecutive years. LC: low stocking rate, Cashmere breed; HC: high stocking rate, Cashmere breed; HL: high stocking rate, local breed.

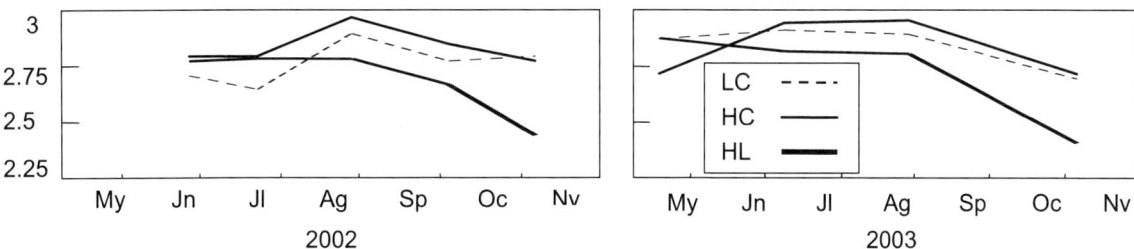

Conclusions

Cashmere goats, due to their smaller body size and lower maintenance requirements, are better adapted to the gorse-heath shrubland communities than the heavier local goats and lose less weight during the grazing season. This exotic breed could diversify production and add to the value of high-quality fibre and meat.

References

CIATA (1996) *Food and Agriculture Research.* Memoria del Centro de Investigación Aplicada y Tecnología Agroalimentaria, Villaviciosa, Spain. (In Spanish.)

Osoro, K., Oliván, M., Celaya, R. and Martínez, A. (1999) Effects of genotype on the performance and intake characteristics of sheep grazing contrasting hill vegetation communities. *Animal Science* 69, 419-426.

Osoro, K., Celaya, R., Martínez, A. and Zorita, E. (2000) Grazing mountainous vegetation communities by domestic ruminants: animal production and vegetation dynamics (Pastoreo de las comunidades vegetales de montaña por rumiantes domésticos: producción animal y dinámica vegetal). *Pastos* XXX (1), 3-50. (In Spanish.)

Osoro, K., Martínez, A. and Celaya, R. (2002) Effect of breed and sward height on sheep performance and production per hectare during the spring and autumn in Northern Spain. *Grass and Forage Science* 57, 137-146.

Rook, A.J., Petit, M., Isselstein, J., Osoro, K., Wallis de Vries, M.F., Parente, G. and Gaskell, P. (2002) Integrating foraging attributes of domestic livestock breeds into sustainable systems for grassland biodiversity. *Grassland Science in Europe* 7, 1068-1069.

Russel, A.J.F., Doney, J.M. and Gunn, R.G. (1969) Subjective assessment of body fat in live sheep. *Journal of Agricultural Science* 72, 451-454.

SPSS (1989) *SPSS for Windows*, Release 5.0.1. SPSS, Chicago, Illinois, USA.

Response of sheep fed with tropical tree legume foliages and Taiwan grass (*Pennisetum purpureum*)

D. Grande [1], F. Reyes [2], H. Losada [1], J. Nahed [3], N. Romero [1], G. Valdivieso [1] and F. Pérez-Gil [4]

[1]Área de desarrollo Agropecuario Sustentable, División de Ciencias Biológicas y de la Salud, Universidad Autónoma Metropolitana Iztapalapa, Mexico, D.F., Mexico. [2]Centro Regional Universitario del Sureste, Universidad Autónoma Chapingo, Teapa, Tabasco, Mexico. [3]División de Sistemas de Producción Alternativos, El Colegio de la Frontera Sur, San Cristóbal de Las Casas, Chiapas, Mexico. [4]Dept. de Nutrición Animal, Instituto Nacional de Ciencias Médicas y Nutrición "Salvador Zubirán", Mexico, D.F., Mexico. ifig@xanum.uam.mx, freyes@taurus1.chapingo.mx, jnahed@sclc.ecosur.mx, fernandoperezgil@hotmail.com

Abstract

Sixteen growing intact male Pelibuey sheep, 16.2 kg initial live weight, were randomly allocated to pens. Four treatments with four replications were applied in order to evaluate the response of sheep to tree legume foliages. The treatments (on dry matter basis) were: T1 = 100% Taiwan (*Pennisetum purpureum*) grass (TG); T2 = 50% *Erythrina* sp leaves (EL) + 50% TG; T3 = 50% *Gliricidia sepium* leaves (GSL) + 50% TG; and T4 =50% *Erythrina poeppigiana* leaves (EPL) + 50% TG. The highest ($P < 0.05$) total dry matter intake (DMI) were for T1 and T2 (453.2 and 517.4 g/sheep per day, respectively), and both were higher ($P < 0.05$) than T3 and T4 (419.9 and 416.6 g/sheep per day, respectively). Animals in all the treatments evaluated lost weight (40, 23, 28 and 28 g/sheep per day for T1, T2, T3 and T4, respectively), with no differences among treatments. The results showed a poor growth response of sheep fed with diets containing tree legume foliages.

Key words: *Erythrina, Gliricidia sepium, Pennisetum*, sheep feeding

Introduction

Fodder shrub and tree species are used in animal feeding by livestock producers worldwide. In various regions of Mexico, the producers recognize the fodder potential of woody species (Nahed *et al.*, 1997) and previous reports (Camacho *et al.*, 1999; Ku *et al.*, 2001) have registered the potential of diverse tree species for animal feeding. Nevertheless, due to the high levels of deforestation, the intensification of land use and the scarcity of fodder during drought periods, it is necessary to carry out research into the recognition, evaluation and selection of potential shrub and tree species in different areas. This is to study and plan their incorporation into animal production, as a part of sustainable development within an integrated strategy for resource management. The objective of this study was to evaluate the use of tropical tree legume foliages and Taiwan grass (*Pennisetum purpureum*) in sheep feeding. In this study, the tree legumes evaluated were *Erythrina* sp, *Gliricidia sepium* and *Erythrina poeppigiana*. These were selected on the basis of their abundance and common use as multipurpose species, as has been reported for the region (Llera and Meléndez, 1989; Reyes and Jiménez, 1999; Maldonado, 2001; Meléndez, 2001) or worldwide (Kass, 1994; Simons and Stewart, 1994).

Materials and Methods

This experiment was carried out at the Centro Regional Universitario del Sureste facilities in Teapa, Tabasco, Mexico. Sixteen growing intact male Pelibuey sheep, 16.2 kg initial live weight, were randomly allocated to pens. The experiment had a randomized design with four treatments and four replications per treatment. The treatments evaluated (on dry matter basis) were: T1 = 100% chopped Taiwan grass (*Pennisetum purpureum*) (TG); T2 = 50% *Erythrina* sp leaves (EL) + 50% TG; T3 = 50% *Gliricidia sepium* leaves (GSL) + 50% TG; and T4 = 50% *Erythrina poeppigiana* leaves (EPL) + 50% TG. Taiwan grass was chopped and ground, whereas the legume tree foliages were provided whole. The experimental period was 45 days. Chemical analysis included determination of crude protein (CP) and ash (AOAC, 1990). Neutral detergent fibre (NDF), acid detergent fibre (ADF), cellulose and lignin concentrations were evaluated based on Goering and Van Soest (1970) and Van Soest *et al.* (1991).

Results and Discussion

Table 1 shows the chemical composition of fodders utilized in this study. The crude protein (CP) contents of the fodders used (dry matter basis) were 10.4, 16.3, 23.3 and 26.8% for TG, EL, GSL and EPL, respectively.

Table 1. Chemical composition of fodder utilized (CP: crude protein; NDF: neutral detergent fibre; ADF: acid detergent fibre).

	CP%	Ash%	NDF%	ADF%	Hemicellulose%	Lignin%
Taiwan grass	10.4	13.1	66.4	33.1	33.3	8.2
Erythrina sp leaves	16.3	7.3	54.3	33.6	20.6	14.7
Gliricidia sepium leaves	23.3	6.1	45.0	28.9	16.0	18.5
E. poeppigiana leaves	26.8	8.8	52.0	26.8	25.2	11.5

The highest total dry matter intakes (DMI) were for T1 and T2 (453.2 and 517.4 g/sheep per day, respectively), and both were significantly higher ($P < 0.05$) than T3 and T4 (419.9 and 416.6 g/sheep per day, respectively). Grass and legume DMI were not statistically different ($P > 0.05$) between any of the treatments (T2, T3 and T4) that included tree legume foliages. Animals in all the treatments evaluated lost weight (40, 23, 28 and 28 g/sheep per day for T1, T2, T3 and T4, respectively), with no differences among treatments (Table 2). Protein intake improved in diets that included tree legume leaves; nevertheless, the results showed a poor growth response of sheep fed with diets containing tree legume foliages. This could be explained by the high levels of intake and inclusion (50% of DM) of legume tree foliages in the diets, coupled with other factors such as high content of toxic compounds, which have been reported for these species (Lowry, 1990; Norton, 1994; Kumar and D'Mello, 1995; Shelton, 2001). On the other hand, the poor response of sheep shows the need to improve the quality of the diets, through the utilization of other local and low-cost fodder resources or by-products (for example, cassava, molasses, citrus pulp, rice polishings, rejected bananas), as has been recommended by Benavides (2001).

Table 2. Voluntary dry matter intake of sheep fed with Taiwan grass and tree legume foliages in different DM proportions (TG: Taiwan grass; E: *Erythrina* sp; GS: *Gliricidia sepium*; EP: *Erythrina poeppigiana*). Different letters (a, b, c) in the same row are statistically different ($P < 0.05$).

	Treatments			
	TG	TG + E	TG + GS	TG + EP
Live weight gain				
Initial live weight, kg	16.0	16.2	16.5	16.2
Final live weight, kg	14.5	15.5	15.5	15.2
Daily live weight gain, g/day	-42	-20	-28	-28
Dry matter intake, g/head day				
Taiwan grass	453.2 a	219.4 b	209.5 b	193.4 b
Tree legume leaves	–	298.00	210.40	223.20
Total intake	453.2 a	517.4 a	419.9 b	416.6 b
Protein, g/day	47.1 a	71.3 b	70.7 b	79.9 b

Conclusions

The results of this experiment showed a poor growth response in sheep fed with diets containing 50% of dry matter of tropical tree legume foliages. This performance could be explained by the high levels of inclusion and intake of legume tree foliages in the diets, coupled with other factors such as the high content of toxic compounds.

References

AOAC, (1990) *Official Methods of Analysis of the Association of Official Analytical Chemists*, 15th edn. AOAC, Washington, DC, USA.

Benavides, J. (2001) Forage trees and shrubs: an agroforestry alternative for husbandry (Árboles y arbustos forrajeros: una alternativa agroforestal para la ganadería). In: Rosales, M., Murgueitio, E., Osorio, H., Sánchez, M. and Speedy, A. (eds) *Memorias de la Primera Conferencia Electrónica FAO-CIPAV sobre Agroforestería para la Producción Animal en Latinoamérica*. FAO, Rome, Italy, pp. 367-394. (In Spanish.)

Camacho, D., Nahed, J., Ochoa, S., Jiménez, G., Soto, L., Grande, D., Pérez-Gil, F., Carmona, J. and Aguilar, C. (1999) Traditional knowledge and fodder potential of the genus *Buddleia* in the Highlands of Chiapas, Mexico. *Animal Feed Science and Technology* 80, 123-134.

Goering, H.K. and Van Soest, P.J. (1970) *Forage Fibre Analysis*. Agriculture Research Handbook No. 379, ARS-USDA, Washington, DC, USA.

Kass, D.L. (1994) *Erythrina* species – pantropical multipurpose tree legumes. In: Gutteridge, R.C. and Shelton, H.M. (eds) *Forage Tree Legumes in Tropical Agriculture*. CAB International, Wallingford, United Kingdom, pp. 84-96.

Ku, J.C., Ramírez, L., Jiménez, G., Alayón, J.A. and Ramírez, L. (2001) Trees and shrubs for animal production in the Mexican tropics (Árboles y arbustos para la producción animal en el trópico Mexicano). In: Rosales, M., Murgueitio, E., Osorio, H., Sánchez, M.D. and Speedy, A. (eds) *Memorias de la Primera. Conferencia Electrónica FAO-CIPAV sobre Agroforestería para la producción animal en Latinoamérica*. FAO, Rome, Italy, pp. 231-258. (In Spanish.)

Kumar, R. and D'Mello, J.P.F. (1995) Anti-nutritional factors in forage legumes. In: D'Mello, J.P.F. and Devendra, C. (eds) *Tropical Legumes in Animal Nutrition*. CAB International, Wallingford, United Kingdom, pp. 95-97.

Llera, Z.M. and Meléndez, N.F. (1989) Survival of forest species as living fences under different drainage conditions in Tabasco (Supervivencia de especies forestales para cercos vivos, bajo diferentes condiciones de drenaje en Tabasco). In: *Resúmenes de Ponencias del Simposio Agroforestal en Mexico*. Universidad Autónoma de Nuevo León, Linares, Mexico, p. 13. (In Spanish.)

Lowry, J.B. (1990) Toxic factors and problems: methods of alleviating them in animals. In: Devendra, C. (ed.) *Shrubs and Tree Fodders for Farm Animals: Proceedings of a Workshop*. IDRC, Ottawa, Canada, pp. 76-88.

Maldonado, M. (2001) Assessment of tropical forage woody species for ruminants feed in Tabasco (Evaluación de leñosas forrajeras tropicales para la alimentación de rumiantes en el estado de Tabasco). MSc thesis. University Nacional Autónoma de Mexico, Mexico, D.F. (In Spanish.)

Meléndez, N.F. (2001) Fodder potential of some tropical trees in Tabasco (Potencial forrajero de algunos árboles tropicales en Tabasco). In: *Memorias de la II Reunión Nacional sobre Sistemas Agro y Silvopastoriles*. Universidad Autónoma Chapingo, Villahermosa, Mexico. (In Spanish.)

Nahed, J., Villafuerte, L., Grande, D., Pérez-Gil, F., Alemán, T. and Carmona, J. (1997) Fodder shrub and tree species in the Highlands of Southern Mexico. *Animal Feed Science and Technology* 68, 213-223.

Norton, B.W. (1994) Anti-nutritive and toxic factors in forage tree legumes. In: Gutteridge, R.C. and Shelton, H.M. (eds) *Forage Tree Legumes in Tropical Agriculture*. CAB International, Wallingford, United Kingdom, pp. 202-215.

Reyes, M.F. and Jiménez, F.G. (1999) Trees and shrubs with forage potential in the Sierra region, Tabasco, Mexico (Arboles y arbustos con potencial forrajero en la región de la Sierra, Tabasco, Mexico). In: *First National Meeting on Agrosilvopastoral Systems* (I Reunión Nacional sobre Sistemas Agro y Silvopastoriles). Universidad Autónoma Chapingo, Huatusco, Mexico. (In Spanish.)

Shelton, M. (2001) Advances in forage legumes: shrub legumes. In: Gomidde, J.A., Mattos, W.R.S. and Carneiro da Silva, S. (eds) *19th International Grassland Congress*. Fundacao de Estudos Agrarios Luz Queiroz, Piracicaba, Brazil.

Simons, A.J. and Stewart, J.L. (1994) *Gliricidia sepium* – a multipurpose forage tree legume. In: Gutteridge, R.C. and Shelton, H.M. (eds) *Forage Tree Legumes in Tropical Agriculture*. CAB International, Wallingford, United Kingdom, pp. 30-48.

Van Soest, P.J., Robertson, J.B. and Lewis, B.A. (1991) Methods for dietary fibre, neutral detergent fibre, and nonstarch polysaccharides in relation to animal nutrition. *Journal of Dairy Science* 74 (10), 3583-3597.

Nutritional characterization of some Mediterranean forestry resources

A. Hajer [1], S. Lopez [2] and A. Chermiti [3]

[1]Ecole Nationale de Médecine Vétérinaire de Sidi Thabet, Dpt. Production Animale, Tunisia. [2]Dept. de Producción Animal, Universidad de Leon, Spain. [3]Institut National de Recherche Agronomique de Tunis, Av Hedí Karray, Ariana, Tunisia. ammarhajer@yahoo.com

Abstract

Some forestry resources were important dietary components for small ruminants in the silvopastoral areas and may play a significant role in animal nutrition during a prolonged dry season. This chapter deals with the nutritional characteristics of some forestry resources and their practical uses in small ruminant nutrition. Chemical composition and *in vitro* digestibility are considered a useful tool in the determination of the nutritive value of *Arbutus unedo*, *Calycotome villosa*, *Erica arborea*, *Myrtus communis*, *Phyllerea angustifolia*, *Pistacia lentiscus* and *Quercus suber*. The ranking order of shrub species was: *C. villosa* > *Ph. Angustifolia* > *M. communis* > *A. unedo* > *Q. suber* > *E. arborea* > *P. lentiscus*. However, based on their palatability the ranking order of these species was *C. villosa* > *Q. suber* > *Ph. Angustifolia* > *E. arborea* > *M. communis* > *P. lentiscus* > *A. unedo*. Feeding experiments are the best way to evaluate the true nutritive potential of shrubs.

Key words: shrub, palatability, goats, sheep

Introduction

Feeding tree and shrub leaves to small ruminants is important for nutrition and could represent an important strategy in developing countries. In Tunisia many browse species were introduced in ruminant feeding systems and approved to have a high potential feeding value and palatability, particularly during dry periods. Despite their high crude protein content, some species thought to be palatable are rejected by sheep and goats. In this present study the potential nutritive value (in terms of chemical composition, secondary compounds, digestibility) and palatability of different shrubs commonly used in the silvopastoral systems in northern Tunisia are reviewed.

Materials and Methods

Leaves and current season's twigs from seven shrub species - *Arbutus unedo*, *Calycotome villosa*, *Erica arborea*, *Myrtus communis*, *Phyllerea angustifolia*, *Pistacia lentiscus* and *Quercus suber*, - were collected from the Taref uplands in the district of Nefza (northwestern Tunisia) in spring of 1998. Following milling, crude protein (CP) was determined (AOAC, 1995) and, whereas neutral detergent fibre (NDF), acid detergent fibre (ADF) and acid detergent lignin (ADL) were determined (Van Soest *et al.*, 1991), *in vitro* dry matter digestibility (DMD) (Goering and Van Soest, 1970) and *in vitro* organic matter digestibility (OMD) were assessed (Theodorou *et al.*, 1994). *In vitro* incubations were performed in one triplicated run (three repetitions for each sample). Statistical differences between species for DMD and OMD were evaluated by one-way analysis of variance (Steel and Torrie, 1980).

Results and Discussion

Although many Mediterranean shrubs gain increasing significance as the nutritional value of grass drops, usually browse does not play a prominent role in diet for a number of reasons such as the low crude protein (CP) content of browse from many shrubby species. In the present work, CP content was particularly low (< 100 g/kg DM) in many of the browses (Table 1). The low CP content in *A. unedo*, *E. arborea*, *P. lentiscus* and *Q. suber* could be due to either high proportions of mature leaves or twigs in the sample. These low levels of N would be unlikely to be rectified even by selective grazing of the younger leaves with a higher N content. This may be more of a problem for sheep and cattle than for goats. The CP content was highest in the foliage of the legume *C. villosa*; probably it can fix atmospheric nitrogen. *C. villosa* will normally provide sufficient CP to meet requirements of sheep, goats and cattle. Caution is needed in the interpretation of these data since CP content cannot be considered a uniform fraction and this may bias the results upwards.

It is well known that shrubs contain a wide range of phenolics, from low molecular weight to polyphenols. Tannins are the most common polyphenols, but their implications in animal feeding are not clear. A major characteristic is their propensity to form chemical complexes with proteins and many other compounds. This may provoke an astringent reaction in the mouth of the animal, impairing salivation. Therefore, an inverse relationship between tannin concentrations in browse sources and voluntary feed intake by herbivores is established but the exact mechanism is poorly understood. Hence, when tannin-rich leaves are offered as a sole feed to sheep and goats they may not provide the maintenance requirement despite their relatively high protein and low fibre contents (Silanikove *et al.*, 1996). These authors reported that tannin levels of approximately 20% of DM drastically reduced leaf intake of *P. lentiscus* and nitrogen balance for goats fed such a diet was negative. With the exception of *C. villosa* and *Ph. angustifolia*, total condensed tannin contents were high and varied from 10% (*M. communis*) to 38% (*P. lentiscus*) (unpublished data). Thus inhibitory properties of tannins might be a possible reason for lower estimates of digestibility, particularly of organic matter (Table 1).

Table 1. Chemical composition (DM g/kg), *in vitro* dry matter (DM), dry matter digestibility (DMD), organic matter (OM), crude protein (CP), organic matter digestibility (OMD), neutral detergent fibre (NDF), acid detergent fibre (ADF), acid detergent lignin (ADL), of several Mediterranean shrub species. a, b, c, d, e within the same column, means without a common superscript letter are significantly different ($P < 0.05$); s.e.d. standard error of the difference.

Species	DM	OM	CP	NDF	ADF	ADL	OMD	DMD
A. unedo	440	947	55	410	250	135	0.459 c	0.647 a
C. villosa	261	948	221	572	204	112	0.607 a	0.653 a
E. arborea	528	965	80	524	398	297	0.337 d	0.528 b
M. communis	200	957	130	383	200	110	0.523 b	0.674 a
Ph. angustifolia	533	953	110	445	353	260	0.561ab	0.681 a
P. lentiscus	478	946	60	433	252	240	0.286 d	0.523 b
Q. suber	490	959	70	551	395	360	0.420 c	0.543 b
s.e.d.							0.05700	0.01630

Based on digestibility and crude protein data, shrubs are ranked as follow: *C. villosa* > *Ph. angustifolia* > *M. communis* > *A. unedo* > *Q. suber* > *E. arborea* > *P. lentiscus*. In addition to common criteria, knowledge of plant species commonly selected by grazing and browsing herbivores at different time of the year is fundamental to render rangeland rehabilitation and grazing management more effective. The relationship between shrub palatability and nutritive value has been reported (Papachristou *et al.*, 1999) and contrasting effects of the nutritive characteristics of shrub species on their palatability to sheep and goats are reported in the literature (Nolan and Nastis, 1997). The shrubs could be classified into three different groups according to their palatability to grazing goats (Table 2). The first class is *C. villosa* and *Q. suber*. The high intake of *C. villosa* was expected since its low tannin content would generally be considered unlikely to significantly affect digestion of nutrients in ruminants. However, *Ph. angustifolia*, *M. communis*, *P. lentiscus* and *E. arborea* make an intermediate group with less palatability. *A. unedo* was the least palatable shrub: it was almost avoided by goats and was browsed for only 0.03 h/day.

Table 2. Goat preferences for some shrubs on Tunisian rangelands at two stocking rates (C1: 1 goat/ha and C2: 9 goats/ha).

	Number of bites		Duration of browsing (h)	
Species	C1	C2	C1	C2
C. villosa	69	71	1.15	1.15
Ph. angustifolia	28	37	0.38	0.51
M. communis	12	23	0.23	0.30
P. lentiscus	12	18	0.19	0.23
E. arborea	33	28	0.43	0.41
A. unedo	2	4	0.01	0.03
Q. suber	65	67	1.43	0.98

Conclusions

Although all of the methods mentioned and used to understand the nutritional potential of a new plant species have some merit, none replaces, or is likely to replace, practical feeding experiments with the target ruminant species. Further studies are therefore recommended to determine whether the observed superiority in the nutritional value of some shrubs like *C. villosa* could be translated into greater animal output, which is a better determinant of forage quality. For the remaining shrubs with high tannin content, research on the improvement of their nutritive value is needed to make them palatable for different ruminant species.

References

AOAC Association of Official Analytical Chemist (1995) *Official Methods of Analysis*, 16th edn. AOAC International, Arlington, USA.

Goering, H.K. and Van Soest, P.J. (1970) *Forage Fiber Analysis (Apparatus, Reagents, Procedures and Some Applications)*. USDA Handbook No. 379, Washington, DC.

Nolan, T. and Nastis, A. (1997) Some aspects of the use of vegetation by grazing sheep and goats. *Options Méditerranéennes Série A* 34, 11-25.

Papachristou, T.G., Platis, P.D., Papanastis, V.P. and Tsiouvaras, C.N. (1999) Use of deciduous woody species as a diet supplement for goats grazing Mediterranean shrublands during the dry season. *Animal Feed Science Technology* 80, 267-279.

Silanikove, N., Gilboa, N., Perevolotsky, A. and Nitsan, Z. (1996) Goats fed tannin-containing leaves do not exhibit toxic syndromes. *Small Ruminant Research* 21, 195-201.

Steel, R.G.D. and Torrie, J.H. (1980). *Principles and Procedures of Statistics*. McGraw-Hill, New York, USA.

Theodorou, M.K., Williams, B.A., Dhanoa, M.S., McAllan, A.B. and France, J. (1994) A simple gas production method using a pressure transducer to determine the fermentation kinetics of ruminant feeds. *Animal Feed Science Technology* 48, 185-197.

Van Soest, P.J., Roberston, J.B. And Lewis, B.A. (1991) Methods for dietary fibre, and nonstarch polysaccharides in relation to animal nutrition. *Journal of Dairy Science* 74, 3583-3597.

Behaviour at grazing of calves with Galician Blonde suckler cows

A. Iglesias [1], J. A. Carballo [2], A. López [1] and L. Monserrat [2]
[1]Dept. Anatomía y Producción Animal, Facultad de Veterinaria, 27002 Lugo, Spain. [2]Centro de Investigaciones Agrarias (CIAM), Apto 10, 15080 A Coruña, Spain. anigbe@lugo.usc.es

Abstract
The knowledge of aspects related to the behaviour of bovine grazing animals and the interaction of plants and animals in production systems based on meadowland is very important for rational use of the resources available for production. The aim of this chapter was to determine the behaviour of suckler cows with calves of the Galician Blonde breed grazing in meadows. The experiment was carried out in Galicia (NW Spain), on an experimental farm (CIAM) located in the province of Coruña. The study took place during the mid-spring and summer of the years 2000 and 2001. Eighteen calves of the "Rubia Galega" breed with an average weight of 170 kg in spring and 236 kg in summer were monitored. The stocking rate was 4 LU/ha in spring and 2 LU/ha in summer. Average weight of the cows was 650 kg. Observations took place in a 0.94 ha field with some scattered trees, whose main function is to give shelter and shadow/shade to the cattle and to maintain biodiversity. The amount of time used by the animals in suckling, grazing, standing and resting was also measured. The percentage of time used by the animals in these activities is different for each of the periods under study.

Key words: cattle breed, silvopasture

Introduction
Grazing suckler cows and calves have different nutritional needs and a very different position in the hierarchy of the herd. The calves must be provided with long grass if they are to grow adequately and compete for food with cows. The aim of this chapter is to determine the behaviour of calves of the Galician Blonde breed under grazing, considering the time of year and the number of days spent on the same plot of land.

Materials and Methods
The test was carried out in the CIAM of Galicia (Agrarian Research Centre in Mabegondo) (NW. Spain) during the second and third rotation (mid-spring and early summer, respectively) on a group of animals from the experimental herd. Eighteen calves of the Galician Blonde breed were monitored along with their mothers in a mixed ryegrass-white clover pasture sown (< 10 trees/ha) with *Quercus robur* trees in a low density before 1995 in a 0.94 ha meadow.

Average initial weight of calves varied from 170 kg for those that started grazing in spring to 236 kg for those starting in early summer. Average weight of the cows was 650 kg. The times used by the animals to suckle, graze, eat from the feeder and rest were measured on farm every 10 min until 10 p.m. during days one and three of each of the periods under study. Samples of the grass offered to the animals were taken from the selected fields and analysed to measured their dry matter content (OM), crude fibre (CF), crude protein (CP) (AOAC, 1990). Calculations of digestibility were also done using equations of prediction (Tilley and Terry, 1963; Alexander, 1969). The neutral detergent fibre (NDF) and acid detergent fibre (ADF) (Goering and Van Soest, 1970) were also calculated. All statistical studies were carried out using the programme JMP (SAS, 2003).

Results and Discussion
The chemical composition of the grass is shown (Table 1). The height of growth, indicative of the degree of defoliation, was 10.56 cm after the third day of grazing on the plot occupied by cows and calves in the spring control, representing 37.1% of the initial height. In the summer control the height of the grass rejected by the cattle was 7.66 cm. Hence, the percentage of the height before grazing was 26.4%.

The average digestibility of the grass offered to the cows and calves was greater in spring (82.1%) than in summer (75.6%) (Table 1).

The average digestibility of the grass consumed by the animals was greater at the start of grazing, when the animals are able to choose and consume the tops of the green leaves, than at the end.

Table 1. Average quality of the meadows in spring and summer.

	Period	
	Spring	Summer
Chemical composition of grass		
Dry matter DM (%)	17.5	44
Organic matter (% DM)	92.1	93.1
Crude protein (% DM)	13.5	11.9
Crude fibre (% DM)	23.4	30.6
Neutral detergent fibre (% DM)	49.8	61.6
Acid detergent fibre (% DM)	27.3	33.5
FU/kg DM	1.06	0.97
Organic matter digestibility (%)	82.1	75.6

As regards the distribution of the main activities of the calves during the day and taking into account the number of days already spent on the plot (Table 2), the time spent grazing was longer in the morning (7 a.m. to 12 a.m.) and towards nightfall (5 p.m. to 10 p.m.) than in the middle of the day both in spring and in summer, except during the third day in summer.

Table 2. Distribution of activities of calves during the observation period taking into account the time of year and the ordinal number of the day spent on the plot of land.

		Spring		Summer	
	Time	Day 1 Mean ± s.e.	Day 3 Mean ± s.e.	Day Mean ± s.e.	Day 3 Mean ± s.e.
Grazing	7-12 h	136.0 ± 21.64	94.0 ± 32.83	108.7 ± 25.96	40.66 ± 16.24
	12-17 h	102.7 ± 39.36	48.0 ± 22.42	88.0 ± 26.24	57.99 ± 22.74
	17-22 h	141.3 ± 16.84	107.3 ± 40.74	100.0 ± 33.81	89.99 ± 33.59
Resting	7-12 h	154.00 ± 30.44	102.00 ± 34.06	135.33 ± 20.66	174.00 ± 14.74
	12-17 h	170.00 ± 40.44	143.33 ± 30.86	72.00 ± 20.07	91.33 ± 24.16
	17-22 h	118.00 ± 42.20	102.00 ± 43,95	123.66 ± 38.26	134.66 ± 23.86
Standing	7-12 h	20.66 ± 24.63	103.33 ± 28.95	46.66 ± 19.15	81.33 ± 17.67
	12-17 h	16.00 ± 14.54	99.33 ± 25.48	90.66 ± 25.49	134.7 ± 33.81
	17-22 h	26.66 ± 23.80	70.66 ± 24.34	38.66 ± 17.26	53.33 ± 24.69
Suckling	7-12 h	7.33 ± 09.61	8.66 ± 09.90	7.33 ± 05.94	7.33 ± 08.84
	12-17 h	6.00 ± 07.37	9.33 ± 04.57	8.66 ± 10.60	6.66 ± 11.10
	17-22 h	6.00 ± 08.28	10.00 ± 10.00	13.33 ± 09.76	10.66 ± 10.90

In the same way the distribution of the time spent resting/ruminating or walking/standing follows a similar pattern during the first and third days to that of the rest of the observation period, i.e. a longer rest period at noon than during the rest of the day in spring and a longer walking/standing period than resting period in summer. The difference between the time spent grazing between the studied periods was due to the fact that in summer the calves spent a large part standing in the shade or wandering around looking for shade. The quality of the grass decreased rapidly during both seasons (spring-summer) for the periods of grazing between day 1 (when animals entered the plot) and day 3 (when defoliation of the grass had taken place). The time of year and the number of days already spent on the same plot influenced the grazing behaviour of the calves. The percentage of time used by the animals in spring was: grazing 35%, resting 44%, walking 19% and suckling 2%. In summer the distribution of the time utilized in the different activities was: grazing 28%, resting 42%, walking 27% and suckling 3%.

Conclusions

The study has increased knowledge of animal behaviour. This is important in the sustainable use of mountain pasture, where an indicator system should be established to study evolution and control.

References

Alexander, R.M. (1969) *The Establishment of a Laboratory Procedure for "in-vitro" Determination of Digestibility*. The West of Scotland College, Research Bull. no. 42, Ayrshire, United Kingdom.

Association of Official Analytical Chemist (AOAC) (1990) *Official Methods of Analysis*. S. Kennet Helrich (ed.). AOAC, Arlington, USA.

Goering, H.K. and Van Soest, P.J. (1970) *Forage Fibre Analyses. Apparatus, Reagents, Procedures, and some Applications*. Agriculture Handbook no. 379 USDA, Washington, USA.

SAS (2003) *Statistics and Graphics Guide version 5.1. Statistical Discovery Software*. SAS Institute, Cary, USA.

Tilley, J.M.A. and Terry, R.A. (1963) A two-stage technique for "*in vitro*" digestion of forage crops. *Journal of the British Grassland Society* 18, 104-111.

Vegetation dynamics of burned heath-gorse communities grazed by sheep or goats

B. M. Jáuregui, R. Celaya, U. García and K. Osoro
Servicio Regional de Investigación y desarrollo Agroalimentario (SERIDA), Apdo. 13-33300, Villaviciosa, Asturias, Spain. bmartinez@serida.org

Abstract

Vegetation dynamics in previously burned shrublands dominated by gorse (*Ulex gallii*) under sheep (Gallega breed) or goat (Cashmere or local breed) grazing were studied. Four plots, two each, were grazed either with goats or sheep. After two grazing seasons, the plots were halved and the treatments were reversed in half of the subplots (those previously grazed by sheep were replaced by goats and vice versa). The results show that goats are an effective tool to control the regrowth of gorse, since in the plots grazed by sheep the gorse cover and biomass were significantly higher ($P < 0.001$) than in plots grazed by goats. After the change of the treatments, the management of the previous years buffered the differences between sheep and goats even though some differences in the herbaceous cover were observed.

Key words: burning, goat, sheep

Introduction

The west of Asturias is a good example of the less favoured areas in the Cantabrian region where fires have become common, mainly in the areas covered by shrublands. Similarly, the accumulation of combustible biomass in the forest undergrowth increases fire risk, preventing the sustainability of silvopastoral systems. In recent years several studies focused on the increase in profitability of the grazing systems in these areas have shown that goats are an effective tool for controlling accumulation of woody biomass in partially improved land (Osoro *et al.*, 2000, 2003). In the present work the main objective was to study changes in a previously burned heath-gorse community under sheep or goat grazing.

Materials and Methods

The experimental field is located in Illano (west Asturias), at 900 m altitude. The vegetation is dominated by heaths (*Erica* sp, *Calluna vulgaris*) and gorse (*Ulex gallii*). In spring of 2001, the experimental plots were superficially burned without severely affecting soil structure, resulting in a gorse-dominated stand (Jáuregui *et al.*, 2003). In September 2001, grazing treatments were imposed on four 1.2 ha plots, with sheep (Gallega breed, average weight 43 kg) in two of the plots and goats (Cashmere, 31 kg, and local breed, 49 kg) in the other two. In both cases the stocking rate was 10 adult females/ha (12 per plot). The grazing period extended from September 2001 to January 2002, and from May to November 2002 (Period 1). In the spring of 2003, at the start of Period 2, the four plots were halved, resulting in eight subplots of 0.6 ha. In four of them the livestock species were the same as in previous years, whereas in the other four subplots the livestock species were changed, those previously grazed by sheep being replaced by goats and vice versa. The stocking rate in Period 2 was reduced to 6.7 head/ha (four per subplot) because of the low animal performance achieved in Period 1. The animals grazed from 19 May to 5 November 2003. Plant cover was estimated, in six transects of 13 m length per plot (three per subplot), recording 100 vertical hits per transect with the "point-quadrat" technique (Grant, 1981), in October 2001, June 2002 and May, August and October of 2003. Five samples of 0.2 m × 1 m of vegetation cut to ground level were randomly collected per subplot in May, August and November 2003. They were manually sorted into heaths, gorse and herbaceous vegetation, dried and weighed. Significance of treatment effects (animal species) on vegetation was tested by analysis of variance. In Period 2 a factorial design of two current grazing species × two previous species was considered.

Results and Discussion

At the end of Period 1 there were significant differences in the cover and biomass percentages of gorse and herbaceous pecies between sheep- and goat-grazed plots. The percentage cover of live gorse increased from 18 to 27% under sheep grazing whereas it decreased from 20 to 14% under goat grazing (Table 1). The herbaceous species, mainly grasses such as *Pseudarrhenatherum longifolium* and *Agrostis curtisii*, increased more in the plots grazed by goats (from 22 to 42%) than in those grazed by sheep (from 22 to 27%). Similar results have been observed in New Zealand for *Ulex europaeus* dominated pastures (Clark *et al.*, 1982; Radcliffe, 1986).

Table 1. Live gorse and herbaceous cover percentages for each treatment in Period 1. ***: significant at $P < 0.001$; ns: not significant.

Date	Gorse (%)			Herbaceous (%)		
	Sheep	Goat	Sig.	Sheep	Goat	Sig.
October 2001	17.8	19.6	ns	22.1	21.5	ns
June 2002	20.9	17.3	ns	24.2	28.8	ns
May 2003	27.0	14.2	***	26.6	42.1	***

During Period 2, significant differences between previous grazing species of Period 1 remained for these plots but differences between species grazing in Period 2 were not significant (Table 2), even though the herbaceous percentage in October 2003 was higher ($P < 0.05$) under goat than under sheep grazing in those plots previously grazed by goats in Period 1.

Table 2. Cover percentages of live gorse and herbaceous vegetation for the different grazing treatments in Period 2, taking into account previous grazing in Period 1. $+ P < 0.1$; $* P < 0.05$; $** P < 0.01$; $*** P < 0.001$; abc: different letters in each row mean significant ($P < 0.05$) differences between treatments (LSD test). Period 1: Sept 2001-Jan, 2002, May-Nov 2002; Period 2: May-Nov 2003.

Period 1		Sheep		Goat		Effects		
Period 2		Sheep	Goat	Sheep	Goat	Period 1	Period 2	P1 x P2
Gorse (%)	May 03	26.2 a	27.8 a	14.7 b	13.7 b	***	ns	ns
	Aug 03	42.2 a	39.5 a	28.7 b	25.7 b	***	ns	ns
	Oct 03	43.3 a	39.8 a	29.0 b	22.8 b	***	ns	ns
Herbaceous (%)	May 03	29.5 bc	23.7 c	37.9 ab	46.3 a	***	ns	*
	Aug 03	24.8 b	16.2 c	20.5 ab	34.5 a	*	ns	***
	Oct 03	17.8 b	13.3 b	19.0 b	32.5 a	***	+	**
Dead matter (%)	May 03	34.20	33.20	34.80	34.00	ns	ns	ns
	Aug 03	26.7 b	32.5 ab	36.3 a	34.3 ab	ns	ns	ns
	Oct 03	32.70	32.80	35.50	39.20	ns	ns	ns
Bare ground (%)	May 03	9.50	12.20	11.00	5.00	ns	ns	ns
	Aug 03	5.7 b	9.5 ab	12.8 a	5.2 b	ns	*	*
	Oct 03	5.2 bc	11.2 ab	14.8 a	4.8 c	*	*	***

Percentage cover of dead matter (burned woody bushes and litter) ranged from 27 to 39% but no significant differences were observed between treatments. Bare ground accounted for 5-15% cover, being higher in those subplots where animal species were changed, so the interaction between the species that grazed in Period 1 and in Period 2 was significant in August ($P < 0.05$) and in October 2003 ($P < 0.001$).

Table 3. Accumulation of biomass and its composition in Period 2 under sheep or goat grazing, taking account of previous grazing in Period 1. $+ P < 0.1$; $* P < 0.05$; $** P < 0.01$; $*** P < 0.001$; abc: different letters in each row mean significant ($P < 0.05$) differences between treatments (LSD test). Period 1: Sept 2001-Jan 2002, May-Nov 2002; Period 2: May-Nov 2003.

Period 1	Sheep		Goat		Effects		
Period 2	Sheep	Goat	Sheep	Goat	Period 1	Period 2	P1xP2
Total biomass (kg DM/ha)							
May 03	8,482 b	14,522 a	8,524 b	8,286 b	**	**	**
Aug 03	12,219 b	16,916 a	8,678 c	10,536 bc	***	**	ns
Nov 03	10,176 ab	11,682 a	10,396 ab	7,850 b	+	ns	+
% Gorse							
May 03	68.1 a	72.2 a	49.1 b	43.0 b	***	ns	ns
Aug 03	66.8 a	67.1 a	50.3 b	49.2 b	***	ns	ns
Nov 03	76.2 a	75.0 a	55.4 b	45.7 b	***	ns	ns
% Heaths							
May-03	11.0 b	15.0 b	19.8 ab	26.0 a	**	ns	ns
Aug-03	14.1 ab	21.1 ab	23.3 a	12.0 b	ns	ns	*
Nov-03	10.2 b	11.9 b	26.8 a	15.9 b	***	+	*
% Herbaceous							
May 03	20.8 ab	12.6 b	31.0 a	30.8 a	***	ns	ns
Aug 03	19.1 bc	11.7 c	26.2 b	38.8 a	***	ns	**
Nov 03	13.6 b	13.1 b	17.7 b	38.4 a	***	***	***

At the end of Period 1 (May 2003), significantly ($P < 0.01$) higher accumulation of biomass was observed in sheep-grazed plots (11.5 t DM/ha) than in goat-grazed plots (8.4 t DM/ha). Gorse percentage was also higher ($P < 0.001$) in the sheep than in the goat treatments (70% vs 46%) (Table 3). In those sheep-grazed plots, higher biomass accumulation resulted in plots that would be grazed by goats in Period 2, thus resulting in a significant interaction ($P < 0.01$). However, that biomass was similar between the treatments at the end of 2003, because of the greater reduction due to goat grazing than sheep grazing from August to November (Table 3).

As with cover data, the percentage of herbaceous biomass was higher in those plots grazed by goats in Period 1 and in Period 2. Percentage of heath was higher in plots that had been grazed by goats in Period 1 but were grazed by sheep in Period 2.

Conclusions

Goats are more effective than sheep in controlling post-fire regrowth of gorse, thus leading to a higher presence of herbaceous plants in the canopy and preventing subsequent fire risk, a major problem in forest systems, which can be avoided by silvopastoral systems in this region. Nevertheless, after two grazing years, gorse remains the dominant species in all treatments, which could limit the animal performance in these burned areas.

References

Clark, D.A., Lambert, M.G., Rolston, M.P. and Dymock, N. (1982) Diet selection by goats and sheep on hill country. *Proceedings of the New Zealand Society of Animal Production* 42, 155-157.

Grant, S.A. (1981) Sward components. In: Hodgson, J., Baker, R.D., Davies, A., Laidlaw, A.S. and Leaver, J.D. (eds) *Sward Measurement Handbook*. B.G.S. British Grassland Society, Reading, Berkshire, United Kingdom, pp. 71-92.

Jáuregui, B.M., Celaya, R., García, U. and Osoro, K. (2003) Sprouting of heath-gorse lands after burning and evolution under grazing by ovine and caprine livestock (Rebrote del brezal-tojal tras una quema y su evolución posterior con pastoreo de ovino o caprino). In: Robles, A.B., Ramos, M.E., Morales, M.C., Simón, E., González, J.L. and Boza, J. (eds) *Pastos, Desarrollo y Conservación*. SEEP Junta de Andalucía, Granada, Spain, pp. 495-500. (In Spanish.)

Osoro, K., Celaya, R., Martínez, A. and Zorita, E. (2000) Grazing of the mountainous vegetation by domestic ruminants: animal production and vegetation dynamics (Pastoreo de las comunidades vegetales de montaña por rumiantes domésticos: producción animal y dinámica vegetal). *Pastos* 30 (1), 3-50. (In Spanish.)

Osoro, K., Oliván, M., Martínez, A., García, U. and Celaya, R. (2003) Diet selection and live weight changes in domestic ruminants grazing on heathland vegetation with areas of improved pastures. In: *VI International Symposium on the Nutrition of Herbivores*. Universidad Autónoma de Yucatán, Mérida, Mexico, pp. 491-494.

Radcliffe, J.E. (1986) Gorse - a resource for goats? *New Zealand Journal of Experimental Agriculture* 14, 399-410.

Adaptation of herbaceous plant species in the understorey of *Pinus brutia*

Z. Koukoura and A. Kyriazopoulos
Aristotle University of Thessaloniki, Faculty of Forestry and Natural Environment, Laboratory of Range Science (235), 54124 Thessaloniki, Greece. zoikouk@for.auth.gr, apkyr@for.auth.gr

Abstract
The purpose of this study was to examine the adaptation of herbaceous plant species under *Pinus brutia* canopy, in monocultures. The experiment was conducted in northern Greece. Monocultures of four plant groups (annual and perennial grasses, annual and perennial legumes) were sown in experimental areas of 50% and 70% light intensity. The demographic changes indicated that annual and perennial grasses as well as perennial legumes were affected by the shade cast by the *Pinus brutia* canopy. The annual legumes maintained their population density almost unchanged under different light intensities of the measured species: the annual grasses *Bromus mollis* and *Lolium rigidum*; the perennial grasses *Dactylis glomerata* var. *palestina*, *Festuca arundinacea* and *Agropyron cristatum*; the perennial legume *Medicago sativa* var. *romana*; the annual legumes *Medicago lupulina* and all the tested varieties of *Trifolium subterraneum* adapted well to the 50% light intensity habitat.

Key words: plant densities, light intensity, mortality

Introduction
In the natural warm coniferous forests of *Pinus halepensis*, *Pinus brutia*, the relatively high tree density results in increased competition for water and depression of the understorey vegetation. Opening up the tree canopy by thinning is a very effective way to increase understorey plant fitness. The competitive ability affects the efficiency with which plants capture and utilize available light resources (Chazdon *et al.*, 1996). However, little attention has been paid to the demographic dynamic of the understorey vegetation even though it is strongly affected by the absorbed radiation (Harper, 1977; Larcher, 1980). The range value of pine silvopastoral systems may be improved if shade-tolerant herbaceous species are introduced. The objective of this research was to describe the demographic temporal changes in grasses and legumes plant monocultures in the understorey of *Pinus brutia* stands.

Materials and Methods
The research was conducted in northern Greece in a 60-year-old *Pinus brutia* forest at about 600 m asl. Mean annual precipitation in the area is 630 mm and mean annual temperature is 13°C. The initial density of 400 trees/ha was reduced to 250 trees/ha by thinning. In the understorey of the unthinned and of the thinned stand, two sites of 50% and 70% light intensity were selected. In addition, another site in an adjacent open area was selected. Light intensity was measured by a Licor quantum sensor (Li 190 SB). In October 1998, equal seed quantities (20 g) from each plant species were sown in different quadrats (50 cm × 50 cm), one quadrat per species. Each site was replicated three times. Plant species density (plants per m^2) was measured monthly after the sowing date during the first growing season with quadrats (10 cm ×10 cm) applied around permanent sticks that were put in the middle of each species cover area. Plant mortality rate (PMR) was calculated using the formula PMR = dD/dt where dD is the difference of densities for the period dt.

Table 1. Plant species from each plant group that were sown.

Perennial grasses	Perennial legumes
Agropyrum cristatum	Medicago sativa var. romana
Agropyrum trichophorum	Medicago sativa var. iliki
Dactylis glomerata var. palestina	Onobrychis sativa
Dactylis glomerata var. chrisopigi	Trifolium repens
Festuca arundinacea	**Annual legumes**
Lolium perenne	Medicago lupulina
Annual grasses	Melilotus officinalis
Bromus mollis	Trifolium alexandrinum
Lolium rigidum	Trifolium subterraneum var. trikala
	Trifolium subterraneum var. baker
	Trifolium subterraneum var. woogenelup

The experimental design was split-split plot with various light intensities as plots and the groups of plant species (perennial grasses, annual grasses, perennial legumes and annual legumes) as subplots and sampling dates (5/11,

5/12, 5/01, 5/02, 5/03, 05/04, 05/05, 05/06) as sub-subplots. The data were subjected to analysis of variance using the MSTAT program while the LSD test was used for means comparison (Steel and Torrie, 1980).

Results and Discussion

Plant densities differed among the treatments, dates and the plant groups. More plants were found in sites with 50% and 70% than in 100% light intensity (Figure 1).

Figure 1. Changes of density (plants/m^2) in four groups of plants (perennial grasses, annual grasses, perennial legumes and annual legumes) at eight dates (5/11, 5/12, 5/01, 5/02, 5/03, 05/04, 05/05, 05/06) over the growing period at 50%, 70% and 100% light intensities.

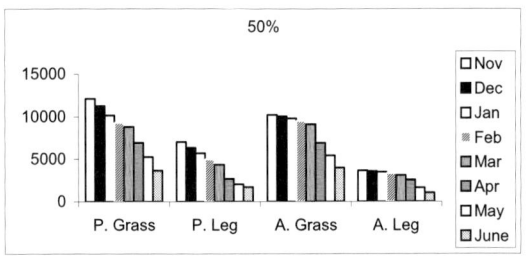
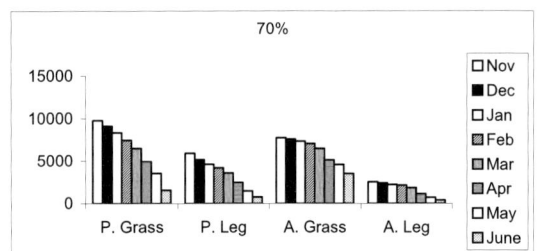
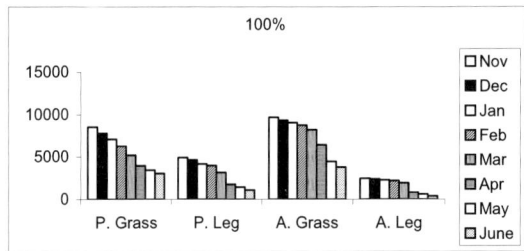

The shade and microclimate under the trees seem to favour plant density in all groups except for the annual legumes. The density of annual legumes was unaffected by light intensity. It is known that under Mediterranean conditions tree cover canopy is considered as a determinant climatic factor (Ovalle and Avendano, 1988). This may be because the demographic changes in the grasses and perennial legumes at three different levels of light intensity seem to be similar to the leaf area changes during leaf morphogenesis under different light intensities and qualities (Lamaire and Chapman, 1996; Lamaire and Agnusdei, 2000). It is obvious that plant density of all groups tends to decrease with the coming of the drought period. According to Lamaire and Chapman's (1996) model, climatic variables of tree canopies which affect tiller density could become limiting factors towards the end of the growing period. Tiller number was reduced from five and three to one and one for annual and perennial grasses, respectively. The decrease of density implies a relative increase in the PMR to the 50% and 70% light intensity treatments compared to the 100% one. The PMR was significantly higher at the 50% and 70% than at the 100% light intensity treatment (Figure 2) in all groups of plants with the exception of the annual legumes. This indicates that these legumes may be more sensitive to high light intensities. The higher values of PMR were observed during April in all treatments of plant groups. The best adaptive species in the understorey treatments (Table 2) were those with the higher densities at the end of the growing season. The use of these species can be suggested for the improvement of silvopastoral systems.

Figure 2. Changes of plant mortality rate (PMR) in four groups of plants (perennial grasses, annual grasses, perennial legumes and annual legumes) at seven dates (5/12, 5/01, 5/02, 5/03, 05/04, 05/05, 05/06) over the growing period at 50%, 70% and 100% light intensities.

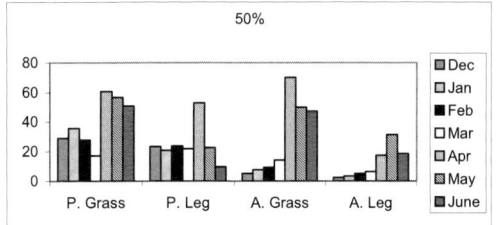
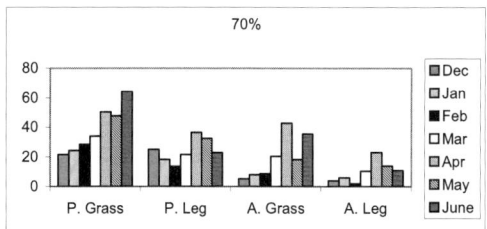
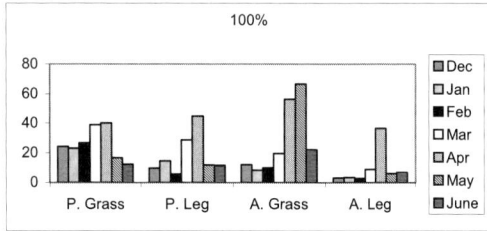

Table 2. Survival (%) of the best adaptive species to the shade treatments at the end of the growing season.

Plant species	Light intensities	
	50%	70%
Perennial grasses		
Agropyrum cristatum	28.6	19.6
Dactylis glomerata var. palestina	43.0	37.9
Festuca arundinacea	19.2	16.0
Perennial legumes		
Medicago sativa var. romana	25.1	21.3
Annual grasses		
Bromus mollis	72.3	48.7
Lolium rigidum	67.8	59.4
Annual legumes		
Trifolium subterraneum var. trikala	19.5	19.1
Trifolium subterraneum var. baker	24.0	22.4
Trifolium subterraneum var. woogenelup	19.8	17.2
Medicago lupulina	19.2	16.6

Conclusions

The demographic changes indicated that annual and perennial grasses as well as perennial legumes were affected by shade cast by the *Pinus brutia* canopy. The annual legumes maintained their population density almost unchanged under different light intensities.

References

Chazdon, R.L., Pearcy, R.W., Lee, D.W. and Fetser, N. (1996) Photosynthetic responses of tropical forests plant to contrasting light environments. In: Mulkey, S., Chazdon, R.L. and A.P. (eds) *Tropical Plant Ecophysiology*. Chapman and Hall, New York, USA, pp. 5-55.
Harper, J.L. (1977) *Population Biology of Plants*. Academic Press, London, United Kingdom.
Lamaire, G. and Agnusdei, M. (2000) Leaf tissue turnover and efficiency of herbage utilization. In: Lamaire, G., Hodgson, J., de Moraes, A. de F., Carvalho, P.C. and Nabinger, C. (eds) *Grassland Ecophysiology and Grazing Ecology*. CAB International, Wallingford, UK, pp. 265-288.
Lamaire, G. and Chapman, D.E. (1996) Tissue flows in grazed communities. In: Hodgson, J. and Illius, A.W. (eds) *The Ecology and Management of Grazing Systems*. CAB International, Wallingford, United Kingdom, pp. 3-37.
Larcher, W. (1980) *Physiological Plant Ecology*. Springer-Verlag-Berlin, New York, USA.
Ovalle, C. and Avendano, J. (1988) Interaction strate of woody and herbaceous strata in the formations of *Acacia carven* in Chili: I. Free influence on the "Houevs" composition, production and phenology of strata herbaceus (Interaction de la strate ligneuse avec la strate herbacee dans les formations d'*Acacia carven* au Chili: I. Influence de l'arbre sur la composition floristique, la production et la phénologie de la strate herbacée.) *Ecologia Plantarum* 22 (4), 385-404. (In French.)
Steel, R.G.D. and Torrie, J.H. (1980) *Principles and Procedures of Statistics*, 2nd edn. McGraw-Hill Book Coorporation, New York, USA.

Transition of an abandoned Dutch agrosilvopastoral landscape to 'new wilderness' by extensive grazing with free-ranging cattle and horses

A. T. Kuiters
Centre for Ecosystem Studies - Alterra Green World Research, P.O. Box 47, NL-6700 AA, Wageningen, The Netherlands. loek.kuiters@wur.nl

Abstract
The Veluwezoom National Park, a former agrosilvopastoral landscape covering 4900 ha (central part of The Netherlands), was designated in 1930 as a national park after most agricultural activities had been stopped due to changed agroeconomic conditions. The area comprised a mosaic of dry grass-heaths, pastures, abandoned arable fields, scrub and coniferous and deciduous woodland. From the 1980s, transition to 'new wilderness' became the main management aim, applying large-scale extensive grazing. Free-ranging cattle and horses were introduced. It was questioned if the high habitat diversity of this semi-open landscape was maintained by extensive grazing, without additional management measures. Research has been carried out on habitat use and interaction of domestic and wild herbivore grazers, and the impact of grazing on woody regeneration dynamics. Possibilities and constraints of extensive, year-round grazing with free-ranging cattle and horses as a management option in the transition of an agrosilvopastoral landscape to 'new wilderness' are discussed.

Key words: wild and domestic grazers, woody regeneration, agrosilvopastoralism

Introduction
The Veluwe, a forest-heathland covering *circa* 90,000 ha, is the largest former pastoral landscape in The Netherlands. From the 13th till the 19th century sheep-grazing on vast heaths with herded flocks was one of the main land-use forms. Sheep dung, collected in the sheepfold and mixed with 'plaggen' from heath, was applied to fertilize arable fields, where rye, buckwheat and potatoes were cropped. Cattle and pigs were grazed in the woodlands. Periodically, heaths were burnt to maintain *Calluna* dwarf-shrub vegetation. Due to changes in agroeconomic conditions, traditional cropping and pastoral sheep-herding activities were stopped in the late 19th century. Domestic grazers were removed, oak-coppicing was stopped and most heaths were planted with coniferous trees. The area is now composed of a mosaic of heaths, pastures and coniferous and deciduous woodlands, and is of major importance for the conservation of natural, culture-historical and recreational values. The maintenance of man-made nature by burning, mowing, sod-cutting of heath and opening up the tree canopy by periodically cutting is a management strategy experienced in certain parts of the Veluwe. Veluwezoom National Park, however, situated at the south-eastern part of the Veluwe, has been managed as 'new wilderness' area since the early 1980s, allowing natural processes to occur with minimal human interference (Riecken *et al.*, 2002). Reintroduction of extensive grazing with free-ranging cattle and horses is part of this alternative management strategy to promote naturalness (Vera, 2000). Experiences of 20-year grazing with free-ranging herds of cattle and horses in the Veluwezoom National Park are summarized and prospects are given for maintaining habitat diversity and related nature conservation values.

Materials and Methods
The study was conducted in the Veluwezoom National Park, a site covering 4900 ha and designated in 1930 as a National Park after most agricultural activities had been stopped. Mean annual precipitation was 750-780 mm and mean annual temperature is 9.5°C. Main vegetation types are summarized in Table 1. Scottish Highland cattle and Iceland ponies were introduced in 1982 and 1986, respectively, in separate compartments with densities varying from 5 to 12 animals per 100 ha (Table 1). Year-round visual observations were made of the habitat use of cattle, ponies, red deer (*Cervus elaphus*) and roe deer (*Capreolus capreolus*). Red deer and roe deer observations were made weekly by car from 2 h before to 2 h after sunset from two fixed transects. The length of each transect was *circa* 10 km. They were situated at different locations in the area, crossing all main habitat types. Observations on habitat use by cattle and ponies were made biweekly by following a randomly chosen observation animal for 4 hours during each observation session. Location and activity of the observed animals were registered. Impact on woody regeneration was established using exclosures (40 m × 40 m) and unfenced control plots, and vegetation surveys in plots (10 m × 10 m) along transects (n = 8), varying in length between 100 and 300 m. A total of 90 plots were surveyed.

Results and Discussion
Habitat use of ungulate grazers
Cattle, ponies and red deer highly preferred foraging in pastures and abandoned fields. Roe deer preferentially used deciduous woodland for foraging. Cattle grazing facilitated winter feeding by red deer and wild boar in pastures and former arable land. The wild ungulates preferred swards previously grazed by cattle.

Table 1. Main vegetation types in Veluwezoom NP and their use by the ungulate grazers (average number per 100 ha). Data are based on year-round visual observations. There was a lack of data for wild boar.

Vegetation type	% Area	Cattle	Ponies	Red deer	Roe deer
Pasture/abandoned field	6	16	83	6	3
Grass-heath	31	9	6	3	3
Deciduous woodland	17	2	7	3	10
Coniferous woodland	46	6	5	4	4
Total	100	6.5	11.9	3.7	5.0

Vegetation dynamics and woody regeneration
From the vegetation surveys and an exclosure study it was concluded that scrub and tree regeneration in pastures and former arable fields are hindered for a long time, because grazers concentrate here for foraging. Succession of grass-heath towards scrub and woodland, with Scots pine (*Pinus sylvestris*), silver birch (*Betula pendula*) and pedunculate oak (*Quercus robur*) as most common species, will proceed. Heath as a favourite foraging site will decrease in cover. Therefore, management should anticipate lowering the number of grazers in the near future. This may favour the establishment of woody vegetation in pastures and abandoned fields, starting with the spread of less palatable species such as bramble (*Rubus* sp) and elder (*Sambucus nigra*) (Kuiters and Slim, 2003). In woodland, less-preferred species such as Scots pine and beech (*Fagus sylvatica*) are favoured by grazing (Kuiters and Slim, 2002). Woody regeneration in storm-gaps may be hindered by ungulate grazers, eventually resulting in open vegetation if grazing pressure is sufficiently high. Horses might contribute to the break-up of the forest canopy by debarking trees, beech in particular.

Table 2. Total density (number per ha) and species composition (most dominant species) of natural regeneration in different habitats in Veluwezoom NP.

	Pasture/former arable field	Grass-heath	Scots pine woodland	Oak woodland
Number/ha	<10	1300	750	1800
Dominant woody species	*Sambucus nigra* *Fagus sylvatica*	*Betula pendula* *Pinus sylvestris* *Frangula alnus* *Quercus robur*	*Sorbus aucuparia* *Pinus sylvestris* *Quercus robur* *Fagus sylvatica*	*Frangula alnus* *Sorbus aucaparia* *Fagus sylvatica* *Betula pendula*

Conclusions
Perspectives for biodiversity preservation
In many regions all over Europe, traditional land-use forms such as agro and silvopastoralism are under pressure due to changes in socio-economic conditions. Grazing with free-ranging herds is a good management strategy to preserve biodiversity by restoring naturalness in abandoned agro- and silvopastoral landscapes. A temporary loss of biodiversity might occur, especially of plant and animal species connected with man-made nature. It is hypothesized that the success of the 'new wilderness' strategy largely depends on the ratio between short vegetation and woodland cover. With a relatively small cover of open pastures or heaths, ungulate densities will be too low to open up woodlands in the long run. Creating gaps by scrub- and wood-cutting might then be temporally necessary as an additional management measure.

References

Kuiters, A.T. and Slim, P.A. (2002) Regeneration of mixed deciduous forest in a Dutch forest-heathland, following a reduction of ungulate densities. *Biological Conservation* 105, 65-74.

Kuiters, A.T. and Slim, P.A. (2003) Tree colonisation of abandoned arable land after 27 years of horse-grazing: the role of bramble-shrub as a facilitator of oak wood regeneration. *Forest Ecology and Management* 181, 239-251.

Riecken, U.P., Finck, P. and Schröder, E. (2002) Significance of pasture landscapes for nature conservation and extensive agriculture. In: Redecker, B., Finck, P., Härdtle, W., Riecken, U. and Schröder, E. (eds) *Pasture Landscapes and Nature Conservation.* Springer, Berlin, pp. 423-435.

Vera, F.W.M. (2000) *Grazing Ecology and Forest History.* CAB International, Wallingford, United Kingdom.

Live fences as a means of sustainable integration of livestock, trees and crops in Ségou, Mali

V. Levasseur [1], A. Niang [2] and A. Olivier [3]
[1]AVRDC – ADRAO, Bamako, Mali. [2]ICRAF, Bamako, Mali. [3]Dept. de phytologie, Université Laval, Québec, Canada.
v.levasseur@cgiar.org, a.niang@icrisatml.org, alain.olivier@plg.ulaval.ca

Abstract
For hundreds of years, agriculture and livestock production developed alongside each other in the Sahel. During the rainy season, the land was devoted to crops while, in the dry season, it became available for livestock. During the last decades, however, farmers began to produce crops during the dry season. In order to protect their crops from roaming livestock, fences composed of living trees were promoted. A study conducted with 180 farmers in 11 villages of the Ségou region (Mali) showed that live fences are not always desired, since they are often perceived as a sign of conflict for land use. Nevertheless, as population density increases, live fences are seen by farmers as a valuable way of marking their fields' boundaries, intensifying crop production, as well as allowing livestock to roam around freely. The use of live fences, therefore, represents an effort to better integrate crops, trees and livestock production.

Key words: adoption, land use, market crops, silvopastoralism

Introduction
The last decades have seen tremendous changes in the patterns of land use in the Sahel. In order to respond to the increased food demand resulting from population growth and urbanization, farmers have started to grow different market crops such as cassava and vegetables during the dry season (Yamba et al., 1997, Chaléard, 1998). Small dry-season gardens are not new in Mali. They used to be located close to the houses and protected by dead fences, e.g. fences made of branches from thorny trees. As their size increased, however, farmers started to move them into the fields that were normally devoted to livestock during the dry season. Thus, dead fences are needed more than ever. Unfortunately, building a dead fence is time consuming since deforestation has made the quest for thorny branches difficult. Moreover, dead fences are fragile and can easily be destroyed by roaming livestock (Ayuk, 1997).

In 1996, in order to lower the pressure on forest resources, the International Centre for Research in Agroforestry (ICRAF) has begun promoting the use of live fences, that is, living trees planted close to one another on the boundary of the cultivated plot, as a replacement for dead fences (Bonkoungou et al., 1998). The species chosen include *Ziziphus mauritiana*, *Acacia nilotica*, *Acacia senegal*, *Lawsonia inermis* and *Bauhinia rufescens*. ICRAF suggests that trees be planted in two rows, spaced 0.5 m apart, and be pruned to a height of 1.5 m. Carefully managed, the live fences protect the crops efficiently and provide useful products from trees for farmers (Levasseur et al., 2004). Therefore, they could be seen as a means of sustainable integration of livestock, trees and crops, which could help develop silvopastoral systems without affecting the dry-season crop production. The present study was undertaken in order to understand the adoption process of these live fences by farmers in the Ségou region (Mali).

Materials and Methods
The study was conducted between November 2000 and November 2001, in 11 villages of the Ségou region, in Mali: Brambiela, Bougounina, Dakala, Djigo, Dougoukouna, N'Tobougou, Pendia Were, Sama, Sikila, Tesseribougou and Zogofina. A qualitative approach, including group interviews, village transect, mental mapping and semi-structured interviews with 180 farmers, was used. Data collected included village history, land and tree tenure, evolution in the use of natural resources, pasture and livestock populations, agricultural production during wet and dry season, land allocation, significance of tree planting and the use of the different types of fences to protect the crops during the dry season. Data interpretation was carried out by means of content analysis.

Results and Discussion
The villages of the sample had various population densities, ranging from 4.7 to 80.9 people per km^2. Our results show that, in the low population density villages (35.6 and less), live fences were not readily accepted. In fact, most farmers perceived them as a sign of conflict over cultivated land. Indeed, in these villages, planting trees on the field boundaries usually means that there is a conflict in the use of the land between the neighbours of the adjacent cultivated fields. Live fences were also seen as a technology that would bring village land fragmentation.

In the villages exposed to denser population pressure, however, planting trees on the boundaries of cultivated fields was not an uncommon phenomenon. In these villages, land suitable for cultivation is scarce. Many of the farmers owning primary rights to the land - the only people allowed to plant trees under the customary type of land tenure - do not hesitate to plant trees on the boundaries of their fields to clearly indicate their property rights.

In those villages, live fences were not seen as a sign of conflict but were used to prevent them. Moreover, many farmers told us that live fences enable them to intensify their crop production within the protected area, since livestock are not able to break down the trees and consequently destroy the crops inside. It is worth knowing that, in those villages, livestock population was very prevalent. From the farmers' point of view, it is to their advantage to keep their livestock roaming around in the dry season so they can find their food themselves. Most farmers could not afford, in terms of labour, land and capital, to grow animal feed in addition to crops for human consumption, which is what they would have to do if they kept their livestock in cattle pens.

Conclusions

Although live fences are widespread in other areas of Africa where populations are dense (Gauthier, 1992; Pélissier, 1995), this is not the case for the Sahel, where space, low population density and availability of woody products have allowed for extensive agriculture for many years. Our study demonstrates, however, that, because of the increase in the human and livestock population, many farmers of the Ségou region are seeking tools to secure their land and protect their cultivated fields. Live fences thus appear as an effective and sustainable means to address those issues. They could also help to enhance silvopastoral systems without affecting the dry-season crop production. Indeed, since their use reduces the pressure on tree resources, live fences help to preserve an important source of forage for roaming livestock. Moreover, some products from such fences can also be used to feed livestock. Thus live fences, which seem to be part of a movement of land use intensification, represent an effort to better integrate crops, trees and livestock in the changing landscape of the Sahel.

References

Ayuk, E.T. (1997) Adoption of agroforestry technology: the case of live hedges in the Central Plateau of Burkina Faso. *Agricultural Systems* 54 (2), 189-206.

Bonkoungou, E., Djimdé, M., Ayuk, E.T., Zoungrana, I. and Tchoundjeu, Z. (1998) *Taking Stock of Agroforestry in the Sahel - Harvesting Results for the Future: End of the Phase Report 1989-1996*. ICRAF, Nairobi, Kenya.

Chaléard, J.L. (1998) Urban growth and production (Croissance urbaine et production vivrière). *Afrique Contemporaine* 185, 3-17. (In french).

Gauthier, D. (1992) Bamileke hedges and systems of production: the example of the cheffery Bafou (west-Cameroon) (Haies bamilékés et systèmes de production: l'exemple de la cheftainship Bafou (Ouest-Cameroun)). *Les Cahiers de la Recherche Développement* 31 (1), 65-78. (In French.)

Levasseur, V., Djimdé, M. and Olivier, A. (2004) Live fences in Segou, Mali: an evaluation by their early users. *Agroforestry Systems* 60, 131-136.

Pélissier, P. (1995) *African Countryside in Evolution* (Campagne Africaine en devenir). Arguments, Paris, France. (In French.)

Yamba, B., Bouzou, I.M. and Amadou, B. (1997) The dynamics of the agrarian systems in South-West Niger: the case of the out-of.season growing in the region of Boboye (La dynamique des systèmes agraires dans le Sud-Ouest Nigérien: le cas des cultures de contre-saison dans la région du Boboye). In: Singaravélou (ed.) *Pratiques de Gestion de l'Environnement dans les Pays Tropicaux*. DYMSET and CRET, Talence (Bordeaux), pp. 295-309. (In French.)

Comparison of terrestrial and aerial oversowing of ski lanes grazed by sheep in the northern Apennines (central Italy)

F. Longhi, A. Pardini, L. Ghiselli and R. Tallarico
DiSAT, University of Florence, Italy. andrea.pardini@unifi.it

Abstract

Ski lane oversowing is used to reduce soil erosion and to maintain soft snow cover during the skiing season. Sown swards are colonized by native species and, in some cases, persistency of introduced mixtures is limited to a few years; however, sowing has a very positive influence on soil erosion, especially when the lane has just been opened. The use of terrestrial means is frequently preferred in lanes with relatively easy slope and aerial means (helicopters) are used in enhanced slopes with rough land. Animal grazing is often considered useful to maintain biodiversity; nonetheless, some local managers are against this practice because animals dig plants out from shallow soils where root systems are superficial. The persistency of mixtures hand sown by workers or sparsely by helicopters in an area of Tuscan-Emilian Apennine (Abetone and Cimone mountains) was compared. Both lanes were grazed by sheep and there were no un-grazed controls. Hand sowing is preferable to helicopter sowing because germination and survival of seedlings is higher in the first case and, consequently, the sward persists longer. Animal grazing did not affect survival of sown mixtures, even if several plants were seen dug out of soil. This could have a negative influence on soil erosion, and also on animal stocking rate.

Key words: sheep grazing, native species, helicopter sowing, sward persistency

Introduction

Pastures are found in some forest districts in the Italian northern Apennines and they are grazed by small herds of sheep or cattle. The pasture area is nowadays larger than needed by animals and herds are kept also for the benefit of tourists and land conservation. However, the low animal stocking rate allows shrub ingression in pastures and in ski lanes. For these reasons it has been suggested that grazing be concentrated in ski lanes. Unfortunately only some managers agree with animal grazing due to variability of effects on the grass cover and increased growth and higher erosion have been reported. However, forest clearing and landscaping in ski lanes cause serious soil erosion. Sowing sward mixtures can accelerate soil cover before native species colonize the ground (Talamucci, 1984; Pardini *et al.*, 1997; Argenti *et al.*, 2000). Hand sowing is normally cheaper than with helicopters, although this method is preferred on steep slopes and rough topography (Tallarico *et al.*, 2002). In this work the persistence of hand sown or helicopter sown mixtures was compared with or without grazing.

Materials and Methods

Two ski lanes in northern Tuscany, with almost the same altitude, slope and soil, were over-sown, one by hand and one by helicopter. The forests in the area are mainly composed of *Abies alba*, *Pieca abies* and *Fagus sylvatica*. They are dense stands with herbs adapted to shade conditions not found in cleared lanes. This results in slow colonization of the slope and makes sowing desirable. Both lanes were grazed by sheep before sowing to reduce sward height and grazed again after sowing to help seed burying by animal trampling. Grazing by sheep was every year in July-August. Seed quantity was 25 kg/ha; the proportion of species in the sown mixture was the same for the two lanes (Table 1).

Table 1. Species, cultivar, origin and percentage of weight in sowing mixture.

Species	Cultivar	Origin	% in seeds
Festuca rubra	Franklin	DK	35
Lolium perenne	Repell	F	35
Poa pratense	Geronimo	DK	20
Agrostis tenuis	Highland	USA	5
Lotus corniculatus	-	Native	5

The species were chosen on the basis of their adaptation to climate and soil parameters of the area as found in earlier trials which had compared different species and mixtures in the same area. The results from the earlier trials proved that soil erosion in unsown lanes is double that sown lanes, even 2 years after forest clearing and ski lane landscaping, so this trial did not repeat this work. Three years after the sowing (summer 2003): (i) soil cover by sown and native species (linear analysis); (ii) botanical composition (as for soil cover); and (iii) sward height (average of height measured on analysis lines) were measured. Grazing was minimal (85 sheep for 3 days on July 15th and again August 15th) and normal management for the area due to the small number of animals reared.

Results and Discussion

Hand sowing resulted in significantly better plant cover than sowing by helicopter (Table 2), probably due to air flotation of part of the seed that might have been blown out of the lane. The presence of sown species (either sown or native) was higher in the hand sown lane than where sowing was by helicopter. Differences due to grazing were not significant; nonetheless, grazing seems to favour sown species in hand sown areas and to favour native species where the sowing was done by helicopter. This is due to lower adaptation of selected cultivars in comparison to native ecotypes. Hand sowing provides better distribution of seeds which can germinate better and consequently compete with native ecotypes; these are favoured if the sowing is done with minimal care, as with helicopters. As would be expected sward height was reduced by grazing. Differences due to sowing technique were not significant.

Table 2. Soil cover (%) total and by sown species (ecotypes + cultivar), sward height (cm) (values having different letters in columns are significantly different at $P = 0.05$).

	Total soil cover %	Soil cover % by sown species	Sward height (cm)
Hand sown - grazed	87 a	68 a	21 b
Hand sown - not grazed	88 a	63 a	36 a
Heli-sown - grazed	83 b	37 b	27 b
Heli-sown - not grazed	81 b	47 b	37 a

The number of species was not different in the swards (Table 3). The most frequent species was always found in the sown mixture; however, this was more frequent in the hand sown lane. Sown species survived better in hand sown lanes. Grazing caused changes only in frequency of minor species. Some of the native species that colonized can be used in the new sown mixture (above all *Phleum pratense*, *Plantago lanceolata*, *Trifolium repens*).

Table 3. Number of species, specific contribution (SC) of the most frequent species (% - may not be 100 because of minor species). Species in bold were in the sowing mixture. N.sp: Number of species.

	N.sp	Species	SC (%)		N.sp	Species	SC (%)
Hand sown - grazed	32 a	**Festuca rubra**	21	Heli-sown - grazed	29a	**Festuca rubra**	16
		Lolium perenne	19			**Poa pratensis**	13
		Poa pratensis	17			Phleum pratensis	12
Hand sown - not grazed	30 a	**Festuca rubra**	24	Heli-sown - not grazed	30a	**Festuca rubra**	17
		Lolium perenne	14			Phleum pratensis	13
		Poa pratensis	21			**Poa pratensis**	11

Conclusions

The presence of productive and palatable pasture can help prevent shrub ingress from forest edges into ski lanes by encouraging more grazing. Moreover, soil cover was rapid from the beginning in both ski lanes; this confirms the value of seed in limiting initial soil erosion. However, soil erosion was not controlled in this trial and further investigation is needed for this. Hand sowing has favoured sward establishment more than helicopter sowing. However, native species have increased their contribution in comparison to sown species in both lanes, so hand sowing is preferable to aerial means in all cases where topography is not very steep or rough. The sown species persisted well; this suggests that the sowing mixture was well chosen. *Phleum pratensis* can be added because it is one of the native species that is becoming more common. It was not possible to distinguish certainly sown cultivars from ecotypes in the sward; however, morphological differences were observed and this suggests that native genotypes are spreading and gaining advantage over the sowing mixture. Grazing was of no particular benefit in the initial years; however, over a longer period it might influence the rate of survival of cultivars/ecotypes and also the colonization by native species that were not included in the sowing mixture.

References

Argenti, G., Merati, M., Stagliano, N. and Talamucci, P. (2000) Herb sowing and evolution in ski lanes in alpine areas (Insediamento ed evoluzione di inerbimenti tecnici di piste da sci in ambiente alpino). *Rivista Agronomica* 34, 1. (In Italian.)

Pardini, A., Pazzi, G., Piemontese, S. and Talamucci, P. (1997) Establishment, root development speed and anti-erosion capacity of several species when sowing ski lanes (Velocità di insediamento, sviluppo radicale e azione antierosiva di alcune specie da impiegare nell'inerbimento di piste da sci). *Rivista Agronomica* 31 (1) suppl. 246-249. (In Italian.)

Talamucci, P. (1984) Herb cover and soil conservation (Cotiche erbose e conservazione del suolo). *Rivista Agronomica* 18, 182-198. (In Italian.)

Tallarico, R., Ghiselli, L., Pardini, A. and Argenti, G. (2002) Cover in lanes (Coperture in pista). *Acer* 1, 69-73. (In Italian.)

Tree growth and pasture production under sewage sludge fertilization

M. L. López-Díaz [1], M. R. Mosquera-Losada [2] and A. Rigueiro-Rodríguez [2]

[1]Dept. Biología y Producción de los Vegetales, Centro Universitario de Plasencia, C.P. 10600, Plasencia (Cáceres), Spain. [2]Dept. Producción Vegetal, Escola Politécnica Superior de Lugo, C.P. 27002, Lugo, Spain. lurdesld@unex.es, romos@correo.lugo.usc.es

Abstract
The objective of these experiments was to compare the effect on pasture production and tree growth in a silvopastoral system of no fertilization, three doses of sewage sludge, with or without liming, and the fertilization usually used in the region. The experiment was conducted over 3 years in the north-west of Spain. The soil was acid (pH = 4.5) and had very low nutrient levels. It was sown with a grass and legume mixture (25 kg/ha of *Lolium perenne*, 10 kg/ha of *Dactylis glomerata* and 4 kg/ha of *Trifolium repens*) in autumn 1997 under a 5-year-old plantation of *Pinus radiata* at a density of 1667 trees/ha. Medium and high sewage sludge doses increased pasture production during the first 2 years. The lower pasture production was obtained with inorganic fertilization, which produced greater height and diameter increments.

Key words: sewage sludge, *Pinus radiata*, pasture production

Introduction
It is estimated that, by 2005, the production of sewage sludge in Europe will be about 8,200 million tons (EEA, 2000). Disposal of sewage sludge is an important problem for the European Union. Sewage sludge can be used as a fertilizer in silvopastoral systems. This is an ecological way of using it, on account of its nutrient content (N, P and K). The main problem of this residue is its heavy metal content, which should be studied, as these elements could reach the human food chain. In Galicia, there is a large area of forest (1.5 million ha) (Ministerio de Medio Ambiente, 2000). The introduction of animals and the establishment of agroforestry systems will reduce fire risk as the shrubs will be controlled by the animals and will also yield an economic return as meat (Mosquera-Losada *et al.*, 2001). When trees are young, fertilization affects the competition between pasture and trees (Campbell *et al.*, 1994).

Materials and Methods
The experiment was conducted in Lugo, in the north-west of Spain on highland, at 510 m asl, on an acid soil (pH = 4.5) with very low nutrient levels. Mean annual precipitation is 1350 mm. The experimental area was sown with a grass and legume mixture (25 kg/ha of *Lolium perenne*, 10 kg/ha of *Dactylis glomerata* and 4 kg/ha of *Trifolium repens*) in autumn of 1997 under a 5-year-old plantation of *Pinus radiata* at a density of 1667 trees/ha. Treatments consisted of three sludge doses, equivalent to total nitrogen applications of 160 kg N/ha (L1), 320 kg N/ha (L2) and 480 kg N/ha (L3), and the same doses with 2.5 t CO_3Ca (L1C, L2C and L3C). In all cases, heavy metal soil values were lower than the limits of Spanish legislation with respect to sewage sludge application as fertilizer. The percentage of the mineralized nitrogen applied with sewage sludge was estimated as 25% (Serna and Pomares, 1992). Two control treatments were established: no fertilization (NF) and the fertilization usually used in the region (500 kg 8:24:16/ha) (MIN). These treatments were applied each spring for 3 years, in plots of 96 m^2, with a completely randomized block design with three replicates. Eight harvests were taken during 1998 (in July and December), 1999 (May, July and November) and 2000 (May, July and November). In each cut, four randomized samples of grass (0.09 m^2) per plot were taken. Samples were transported to the laboratory and they were dried in the oven (60°C) to determine annual production. Once a year, basal diameter and total height were measured. All data relating to the different plots were analysed by ANOVA and means separated by LSD.

Results and Discussion
For the first 2 years, medium and high sewage sludge doses increased pasture production with respect to the other fertilization treatments (Figure 1). The response of pasture production to mineral fertilization was lower than from sewage sludge, maybe because inorganic treatment decreased soil pH, which was already very low. In 2000, there were no significant difference between treatments. This may be because trees limited pasture growth more than nutrients in soil could supply (Balocchi and Phillips, 1997).

Figure 1. Annual pasture production (t/ha) in 1998, 1999 and 2000. NF: no fertilization; L1: low sewage sludge doses (160 kg N/ha); L2: medium sewage sludge doses (320 kg N/ha); L3: high sewage sludge doses (480 kg N/ha); MIN: mineral fertilization (500 kg 8:24:16/ha). Different letters indicate significantly different means (at $P ≤ 0.5$).

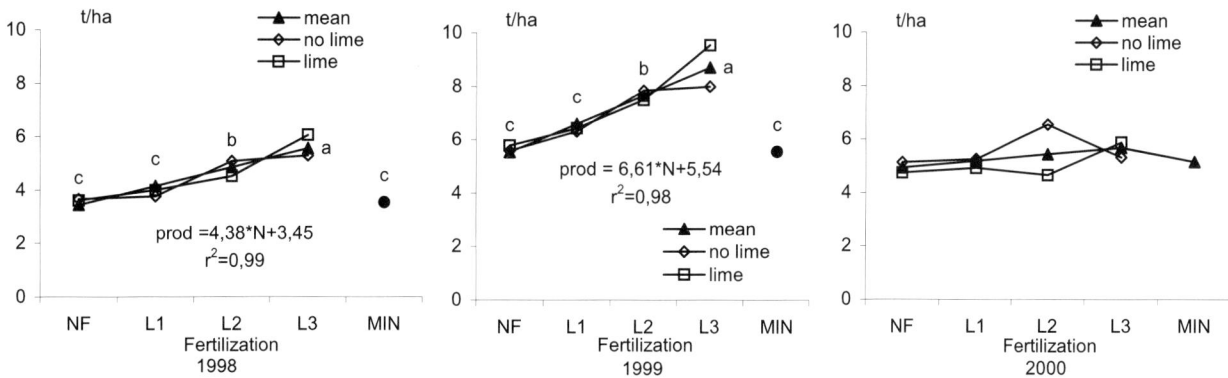

With respect to the trees, mineral fertilization (MIN) produced greater height and diameter increments than sewage sludge (Figure 2) in all years, because the N and P from this treatment were easily mineralizable. The increased tree growth could explain the low pasture production obtained with this treatment. Sánchez-Rodríguez *et al.* (2002) reported content in low-P soils in Galician forest soils (Calvo-de-Anta *et al.*, 1992), with respect to site index. During the third year of the trial, diameter increments were similar for organic and inorganic fertilization.

Figure 2. Height and diameter increment in 2000. NF: no fertilization; L1: low sewage sludge doses (160 kg N/ha); L2: medium sewage sludge doses (320 kg N/ha); L3: high sewage sludge doses (480 kg N/ha); MIN: mineral fertilization (500 kg 8:24:16/ha). Different letters indicate significantly different means (at $P ≤ 0.05$).

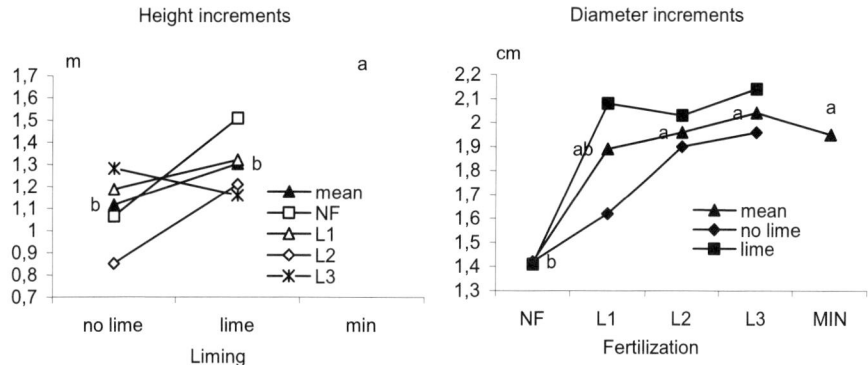

Conclusions

Medium and high sewage sludge doses increased pasture production the first 2 years. In the last year, there was no significant difference between treatments because the trees limited pasture growth. Lower pasture production and most tree height and diameter increments were found with inorganic fertilization.

Acknowledgements

We are grateful to CICYT, XUNTA de Galicia, GESTAGUA S.A and AGROAMBPRODALT for financial assistance, to Escuela Politécnica Superior for facilities, to Divina Vázquez Varela, José Javier Santiago-Freijanes, Teresa Piñeiro López, Antonio Rodríguez Rigueiro for helping in processing, laboratory and field.

References

Balocchi, O.A. and Phillips, C.J.C. (1997) Grazing and fertilizer management for establishment of *Lotus uliginosus* and *Trifolium subterraneum* under *Pinus radiata* in southern Chile. *Agroforestry Systems* 37, 1-14.

Calvo-de-Anta, R., Macías, F. and Riveiro Cruz, A. (1992) *Agronomic Viability in the Province of La Coruña (Crops, Pine, Oak, Eucalyptus and Chestnut Tree)*. Departamento de Edafología y Química Agraria, University of Santiago de Compostela, Santiago de Compostela, Spain. (In Spanish.)

Campbell, C.D., Atkinson, D., Jarvis, P.G. and Newbould, P. (1994) Effects of nitrogen fertilizer on tree/pasture competition during the establishment phase of a silvopastoral system. *Annals of Applied Biology* 124, 83-96.

European Environment Agency (EEA) (2000) *Report of the Commission for the Council and the European Parliament on the Application of the Community Law on Waste Management During the Period 1995-1997*. EEA, Brussels, Belgium.

Ministerio de Medio Ambiente (2000) *III National Forest Inventory - Galicia*. MMA, Madrid, Spain. (In Spanish.)

Mosquera-Losada, M.R., Rigueiro-Rodríguez, A. and Villarino-Urtiaga, J.J. (2001) *Establishment of Silvopastoral Systems*. Consellería de Agricultura, Ganadería y Política Agroalimentaria - Xunta de Galicia, Santiago de Compostela, Spain. (In Galician).

Sánchez-Rodríguez, F., Rodríguez-Soalleiro, R., Español, E., López, C.A. and Merino, A. (2002) Influence of edaphic factor and tree nutritive status on the productivity of *P. radiata* D. plantations in NW Spain. *Forest Ecology and Management* 171, 181-189.

Serna, M.D. and Pomares, F. (1992) Indexes of assessing N availability in sewage sludge. *Plant and Soil* 139, 15-21.

Utilization of gorse and heather communities by cattle in mountain grazing

N. Mandaluniz [1], L. M. Oregui [1] and A. Aldezabal [2]
[1]Agrosystems and Animal Production Dept, NEIKER A.B., P.O. Box 46, E-01080 Vitoria-Gasteiz, Spain.
[2]Landare-Biologia eta Ekologia Saila, UPV-EHU, P.K. 644, E-48080, Bilbo, Spain. loregi@neiker.net

Abstract

Mountain pastures are made up of a mosaic of different vegetation communities, including shrubby ones. In this study the utilization of two management units of Gorbeia Natural Park (Basque Country) by cows was studied. Five different vegetation communities were considered in the study: open pasture; heather, with less than 50% of covering; heather-gorse-fern, with more than 50% cover; woody; and rocky pastures. Habitat and diet selection were assessed by animal observation and microscopic faeces analysis in two herds throughout two grazing seasons. Open pasture was selected in spring and summer, but at the end of summer and autumn the shrubby communities were positively selected, resulting in a modification of the diet selected. Although the herbaceous component was predominant in both periods, in the second one grass of lower value was the major component in the diet, and the shrubs made a significant contribution. Hence shrubby communities seem to play a significant role as a reservoir nutritive resource for later grazing periods. This could be related to the protection of shrubs over herbage species in periods of climatic stress.

Key words: mountain pastures, grazing behaviour, shrubs

Introduction

Mountain pastures are made up of a mosaic of different vegetation communities, open pastures, woody areas and shrubby areas. Their nature and composition are affected by the soil, climate, management imposed conditions. The contribution of shrub communities to animal feeding is usually undervalued in arid zones (Morales, 1993). Studies of these communities are principally related to their control more than their grazing value, which is only considered for small ruminants studies. In this work the utilization by cows of different communities and vegetation components and the role of shrubs in these systems were studied in mountain pastures of Gorbeia Natural Park.

Materials and Methods

The work was carried out in two management units (Steward and Eno, 1998) of the north side of Gorbeia Natural Park: Aldamiñape, at 900-1100 m asl and 180 ha area, and Egiriñao at 1000-1300 m asl and 125 ha. The vegetation communities in both management units (MU) were defined from the Forest Inventory 1996 (Gobierno Vasco, 1998) and their boundaries assessed with GPS. Communities differentiated (with their principal components - Aizpuru *et al.*, 1999) were: open pasture (Op), corresponding to an *Agrostis capillaris* (L.) Linné and *Festuca rubra* (L) pasture with *Trifolium repens* (L.) and other dicot species (Mendarte, 1998); heather community (H), a mosaic of grass, *Erica vagans* (L.), *Erica cinerea* (L.) and gorse (*Ulex europeaus* (L)), with shrub cover being less than 50%; heather-gorse-fern (HGF) community, a mixture of grass, heather, gorse (*Ulex eurepeaus* (L.)) and fern (*Pteridium aquilinun* (L.)) with more than 50% of the surface covered by shrubs; rocky pastures (R), areas with lime outcrops with species-rich pasture; wooded areas (W) of *Fagus silvatica* (L.) or *Pinus radiata* (D.) Don, with undergrowth formed by pasture, brambles and dead leaves. The floristic composition of communities and sward heights were evaluated at the same time as grazing occurred.

For each of the MU a herd with 14 to 16 cows and three to five heifers was selected and controlled throughout two consecutive grazing seasons on seven different occasions in four periods: spring (May-June); summer (July-August); end of summer (September); and autumn (October-November). Cow activity (grazing, resting or walking) was monitored by scan-sampling (Altmann, 1974) every 15 min during daylight. Biting rate (bites/min) was assessed in the same controls, for periods longer than 1.5 min, and distinguishing the vegetation eaten (grass in Op, grass in shrubs communities or shrubs). The flock localization through the control was plotted on a 1:10,000 map every hour. All this information was digitalized and processed using the Arc-Info software. Jacobs Index (J) was calculated as $J = (Ui - Ai)/\{(Ui + Ai) - [2 \times (Ui \times Ai)/100]\}$, where Ui is the percentage of i component in the faeces and Ai the availability of i component in the management unit. The comparison between availability and utilization of different communities was done by a chi-square test. Individual faeces were collected 2 days after the activity records, from 50% of the adult cows, and pooled for microscopic analysis (Sparks and Malechek, 1968).

The effect of the year, MU and period on the studied variables was analysed by a Generalized Lineal Model (SAS, 2001), taking into account principal variables and all double interactions.

Results and Discussion

Cattle grazing activity showed a bimodal pattern with a resting period at midday, which was reduced or even disappeared as daylight length decreased. Consequently resting time was abruptly reduced ($P < 0.001$), with a drop in the percentage of daylight time dedicated to this activity (37% in spring and 24% in autumn). Total grazing time was significantly affected ($P < 0.001$) by control, resulting in a decrease throughout the grazing season (Table 1), in parallel to a reduction in daylight length. This reduction in grazing time was observed even though the percentage of daylight time devoted to grazing increased from 54% in spring to 64% in autumn.

Table 1. Time devoted to different activities (mean ± s.e.) throughout the grazing season and percentage of grazing time on different vegetation components (mean ± s.e.). Values with different superscripts denote significant differences ($P < 0.05$) within row.

Season/Activity	Spring	Summer	End of summer	Autumn
Grazing (min)	509ab ± 45	529a ± 34	450b ± 98	433b ± 86
Herbaceous (%)	98.0a ± 1.0	96.3a ± 1.0	91.7b ± 2.0	88.8b ± 2.0
Shrubs (%)	1.9a ± 1.0	3.7a ± 1.0	8.3b ± 2.0	11.1b ± 2.0
Resting (min)	344a ± 64	294a ± 39	194b ± 53	157b ± 64
Walking (min)	87 ± 10	69 ± 9	68 ± 12	76 ± 16

A change in the communities grazed throughout the grazing season was also observed. Op was positively selected ($J > 0.5$), especially in Egiriñao, where this type is relatively more abundant. The selection of shrub communities tended to increase at the end of summer and autumn, reaching values of Jacob's index (J) greater than 0.5. This community selection could be related to the differences in pasture height as swards were taller ($P < 0.001$) in shrub communities than in open pastures. The increase of time devoted to grazing on shrubs (Table 1) was also observed. These modifications were confirmed by faeces analysis, with a reduction of dicots at the end of the grazing season ($P < 0.001$), especially in the MU with a smaller area of the H community. Also an increase of woody species in the diet in spring-summer in comparison with the end of summer-autumn (9.8 ± 0.1-13.0 ± 0.1 vs 17.5 ± 0.1-19.3 ± 0.1; $P < 0.05$, respectively), was observed. This woody component of the diet is based on heather, and a reduction of tannin concentration in autumn has been shown in this species (Hervás *et al.*, 2003). Grasses can be subdivided into more palatable (i.e. *Agrostis capillaris* (L.) or *Festuca rubra* (L.)) or less palatable (i.e. *Agrostis curtisii* Kerguélen or *Nardus stricta* (L.)), and there was a reduction in the former as season progressed ($P < 0.01$). Palatable species dropped throughout the grazing season from more than 80% of total graminoids in spring and summer to only 21% in autumn ($P<0.05$). At the end of the summer they had an intermediate contribution (57%), but different ($P < 0.05$) from other periods. That observed graminoid utilization goes with communities use, as more palatable grasses are usually found in open pastures whereas the less palatable ones were found principally in shrubby communities.

Conclusions

Even though grass and herbs were the main components of the cattle diets, the presence of shrubs was significant, especially later in the grazing season. Shrub communities, with taller pasture than Op, play an important role at the end of the grazing season, when the available herbage is reduced. Their contribution is confirmed by the differences in the diet composition between areas with different composition of shrubby communities.

Acknowledgements

The authors express their gratitude to the INIA (Project RTA02-086-C2-1) for the financial support provided for this study.

References

Aizpuru, I., Aseginolaza, C., Uribe-Echevarria, P.M., Urrutia, P. and Zorrakin, I. (1999) *Claves Ilustradas de la Flora del País Vasco y Territorios Limítrofes*. Eusko Jaurlaritzaren Argitalpen-Zerbitzua, Gasteiz, Spain.

Altman, J. (1974) Observational study of behaviour: sampling methods. *Behaviour* 49, 227-267.

Gobierno Vasco-Eusko Jaurlaritza (1998) *Inventario Forestal del País Vasco 1986*. Departamento de Agricultura - Nekazaritza eta Arrantza Saila. Vitoria-Gasteiz, Spain.

Hervás, G., Mandaluniz, N., Oregui, L.M., Mantecón, A.R. and Frutos, P. (2003) Evolución anual del contenido de taninos del brezo (*Erica vagans*) y relación con otros parámetros indicativos de su valor nutritivo. *ITEA* 99A, 69-84.

Mendarte, S. (1998) Gorbeiako larreen landaredi-egirura eta faktore edafikoak. MSc thesis, Euskal Herriko Univertsitatea (EHU-UPV), Leioa, Spain. (In Basque.)

Morales, M.C. (1993) La flora bética: Su interés como fuente de recursos ganaderos. In: Aguilera, J.F. (ed.) *Nutrición de Rumiantes en Zonas Áridas y de Montaña y su Relación con la Conservación del Medio Natural*. Servicio Publicaciones y Divulgación-Junta de Andalucia, Sevilla, Spain, pp. 49-56.

SAS (2001) *S.A.S./Stat User's Guide*. Cary, NC, USA.

Sparks, D.R. and Malechek, J.C. (1968) Estimating percentage of dry weight in diets using a microscopic technique. *Journal of Range Management* 21, 264-265.

Steward, F.E. and Eno, S.G. (1998) *Grazing Management Planning for Upland Natura 2000 Sites: a Practical Manual.* The National Trust for Scotland, Edinburgh, United Kingdom.

Importance of holm-oak (*Quercus ilex*) as a woody food resource in Spanish ibex (*Capra pyrenaica*) diet in Mediterranean forest

T. Martínez

Instituto Madrileño de Investigación Agraria (IMIA), El Encín, Apdo 127, Alcalá de Henares, Madrid, Spain.
teodora.martinez@imia.madrid.org

Abstract

The role of holm-oak (*Quercus ilex*) in Spanish ibex (*Capra pyrenaica*) diet and as part of its consumed woody resources was analysed in south-eastern Spain (Cazorla and Segura Ranges). The content of holm-oak in the annual, seasonal, altitude zonal, and sex and age classes diet was analysed. Holm-oak was a relevant part of the annual diet and comprised more than 22% of the woody vegetation consumed. Consumption was greatest in autumn and winter and in spring holm-oak was also important in relation to the rest of the woody plants consumed. Lowest consumption was recorded in summer. Holm-oak was consumed a little more at low altitude than at high altitude and was important for both males and females. The holm-oak selection index in the annual diet was close to zero, suggesting a close relationship between consumption and availability. Selection was positive in the upper altitude zone and negative in the low zone. Males selected the species positively, while females showed a negative selection index that was close to zero. Optimum control of stocking loads by the various ruminant species in the study area is recommended in order to safeguard *Quercus ilex* from heavy impact by browsing and to benefit silvopastoral systems.

Key words: consumption, selection index, silvopastoral systems

Introduction

Mediterranean forest covers large areas of the Iberian península. These areas are expected to spread under the effect of new policies on land-use change, farm abandonment and reforestation programmes. In many of these environments, the implementation of silvopastoral systems involving sustainable management would lead to landscape diversification, the prevention of biodiversity loss and the improvement of income in a range of grazing systems. Wild ungulates play an important ecological role in Mediterranean woodlands, and are also highly rated as game resources. They are well adapted to a habitat where holm-oak woods are important and play a role as a food resource for large herbivores, particularly deer and goats, thus enhancing silvopastoral systems. Consequently, the aim of this study was to assess the importance of holm-oak in the Spanish ibex diet, especially as part of woody food resources. The Spanish ibex (*Capra pyrenaica*) is a wild herbivore of considerable importance in these environments, particularly in the study area.

Materials and Methods

The study area was in southeastern Spain (Cazorla and Segura Ranges), where two zones with different altitudes were defined: low (800-1500 m) and high (1500-2000 m). Botanical analysis of the rumen was carried out (Martínez, 1992, 2001, 2002): 105 rumen samples were collected in April, June, July, August, November and December. A sample of 1 l was taken from each rumen, washed and separated into fragments corresponding to each species; then the volume and dry weight were obtained. Final data were expressed as percentage dry weights of plant species in the total sample. The number of samples collected from each zone, season and sex and age classes (males, females and young (males and females younger than 2 two years) is showed in Table 1. Holm-oak availability in the Cazorla and Segura Ranges was 23.7% of woody resources in the overall study area, and 31% and 11% in the low and high zones, respectively (Martínez, 1992). Selection index was estimated by the Ivlev Selection Index: ISI = Consumption - Availability/Consumption + Availability.

Results and Discussion

In the annual diet of the overall study area, holm-oak formed 22% of woody vegetation consumed (Table 1a). This is a considerable amount bearing undoubtedly the woody species richness of the diet. The selection index was negative, though close to zero, suggesting a close relationship between consumption and availability. In the low and high zones holm-oak had a similar consumption to the overall study area (Table 1a). However, the selection index was positive in the high zone and negative in the low zone, influenced in the latter by the abundance of holm-oak. Amongst the different sex and age classes, males were the group with the highest holm-oak consumption and this was also the plant species with the heaviest consumption and positive selection index. Females also consumed a large amount, with a negative selection index that was close to zero. Holm-oak was considerably less important in the diet of the young animals than for the other two groups. In all seasons, holm-oak was practically the most heavily consumed resource

(Martínez, 1992). Its consumption was only slightly surpassed in spring and winter by two species of woody plants, *Phillyrea latifolia* (9.6%) and *Juniperus oxycedrus* (16.8%), respectively. Autumn had the heaviest holm-oak consumption, with positive selection. In the rest of the seasons, selection was negative, with the lowest index in summer (Table 1a). In spring, holm-oak formed practically the same percentage of the diet woody component as in winter, when woody vegetation was heavily consumed (77.6% of diet) in contrast to 41.6% in spring, suggesting the importance of the species in this period, possibly due to consumption of new shoots with higher quality than in other periods.

Table 1. Holm-oak as a percentage of biomass in the diet in terms of the total woody vegetation consumed (TWV), and selection index of holm-oak by Spanish ibex (SI) in Cazorla and Segura Ranges. Z = Zone, Y = Young
(a) Annual (overall area study, zones, sex and age classes) and seasonal diet.

	Area study	Low Z	High Z	♂	♀	Y	Spring	Summer	Autumn	Winter
Samples	105	57	38	52	34	19	22	28	20	35
Diet (%)	13.4	14.1	11.2	16.6	11.6	4.1	73	7.6	24.7	14
TWV (%)	22	22	21.9	26.4	19.9	7.1	17.5	12.9	37.1	18
SI	-0.03	-0.17	0.33	0.05	-0.08	-0.54	-0.15	-0.29	0.22	-0.13

(b) Seasonal diet by zones (high and low).

	Spring		Summer		Autumn		Winter	
	Low Z	High Z	Low Z	High Z	Low Z	High Z	Low Z	High Z
Samples	9	12	15	11	13	4	20	11
Diet (%)	12.4	3.6	7.6	7.6	16.2	43.4	15.2	13.2
TWV (%)	23.1	10.4	13	13.4	25.1	52.6	19.2	18.3
SI	-0.14	-0.02	-0.28	0.1	-0.09	0.65	-0.23	0.25

(c) Seasonal diet of males, females and young.

	Spring			Summer			Autumn			Winter		
	♂	♀	Y	♂	♀	Y	♂	♀	Y	♂	♀	Y
Samples	10	8	4	14	6	8	11	7	2	17	13	5
Diet (%)	11.6	3.9	3.5	9.3	7.6	4.0	21.5	33.9	9	17.5	10.5	6.7
TWV (%)	21.7	10.7	9.9	17.5	11.1	6.7	32.4	46.2	15	21.9	14.5	7.7
SI	-0.04	-0.37	-0.40	-0.15	-0.36	-0.6	0.15	0.32	-0.22	-0.04	-0.24	-0.51

In all seasons holm-oak was a secondary food source, after *Phillyrea latifolia*, in the low zones. In spring and autumn holm-oak was more relevant in terms of consumed woody material, and less important in summer (Table 1b). Selection was negative in all seasons, assisted in this respect by the abundance of holm-oak in this zone. At high altitude, holm-oak was the most heavily consumed species in all seasons, and was particularly important in autumn. The selection index was positive in all periods except in spring. Holm-oak was the most heavily consumed species by males and females in all seasons except spring (Table 1c). Autumn was the season when it was most consumed by both groups and selection was positive. In spring and winter the selection index was close to zero in males. Young ibex made the least use of holm-oak in all seasons; this low importance may be related to its lower accessibility, its low digestibility and its high lignin content, 33.8% and 38%, respectively (Martínez, 1992).

Conclusions

Holm-oak plays a crucial role for Spanish ibex (*Capra pyrenaica*) in the Cazorla and Segura Ranges, and also in other Mediterranean habitats (Martínez, 1994). It is also known to be attractive for deer and domestic goat (Cuartas, 1992; Martínez, 2002) and, albeit to a lower degree, for other wild and domestic herbivores. Optimum control of stocking loads by the various ruminant species in the study area is recommended in order to safeguard holm-oak from heavy impact by browsing. This would benefit silvopastoral systems with appropriate management of herbivore populations while also encouraging regrowth and avoiding depletion and stripping of the plants that are accessible to the large herbivores. This chapter shows how the relationship between forest trees and wild goats influences sustainable development of natural areas in the Iberian woodlands and the management of goats and deer to the benefit of silvopastoral systems.

References

Cuartas, P. (1992) Herbivorism of large mammals in a Mediterranean mountain ecosystem (Herbivorismo de grandes mamíferos en un ecosistema de montaña mediterránea). MSc thesis. U. de Oviedo, Oviedo, Spain. (In Spanish.)

Martínez, T. (1992) Feed strategy of Spanish ibex (*Capra pyrenaica*) and its relation with wild and domestic ungulates in Sierra Nevada, Sierra de Gredos and Sierra de Cazorla (Estrategia alimentaria de la cabra montés (*Capra pyrenaica*) y sus relaciones tróficas con los ungulados silvestres y domésticos en Sierra Nevada, Sierra de Gredos y Sierra de Cazorla). MSc thesis. UCM, Madrid, Spain. (In Spanish.)

Martínez, T. (1994) Seasonal diet of the Spanish ibex (*Capra pyrenaica*) in Tortosa and Beceite highlands (Mediterranean area in North-eastern Spain) (Dieta estacional de la cabra montés (*Capra pyrenaica*) en los Puertos de Tortosa y Beceite (Área Mediterránea del Noreste de Spain)). *Ecología* 8, 373-380. (In Spanish.)

Martinez, T. (2001) The feeding strategy of Spanish ibex (*Capra pyrenaica*) in the northern Sierra de Gredos (Central Spain). *Folia Zoologica* 50 (4), 257-270.

Martínez, T. (2002) Comparison and overlap of sympatric wild ungulate diet in Cazorla, Segura and Las Villas Natural Park. *Pirineos* 157, 103-115.

Woodland grazing in Northern Ireland: effects on botanical diversity and tree regeneration

P. McEvoy [1] and J. H. McAdam [2]

[1]Dept of Applied Plant Science, Queen's University, Newforge Lane, Belfast, BT9 5PX, Northern Ireland. [2]Dept of Applied Plant Science, Department of Agriculture and Rural Development, Newforge Lane, Belfast, BT9 5PX, Northern Ireland. p.m.mcevoy@qub.ac.uk, jim.mcadam@dardni.gov.uk

Abstract

A survey of over 100 areas of semi-natural woodland was carried out over two field seasons (2002, 2003) in lowland and upland areas of Northern Ireland. General grazing status was recorded and a detailed survey of ground flora and regeneration recorded.

Ecological analyses found that grazed woods tended to have more botanical diversity than ungrazed woods. This was suggested as being related to the reduction in dominant vegetation height via grazing and trampling.

Seedling and sapling species were found to differ in their response to browsing and trampling. A model of sapling age and height for four key species was developed to predict the height at which the terminal leader would be out of browse height and grazing could be considered.

Key words: silvopasture, browsing, regeneration, botanical diversity, sapling damage

Introduction

Northern Ireland is the least-wooded area of the EC (Cooper and McCann, 2002). Semi-natural woodland exists as small fragments adjacent to areas of pasture. These remaining fragments of semi-natural woodland usually survive adjacent to grazing land and include many woods which have been grazed by deer and domestic stock for many hundreds of years. Concerns over lack of regeneration in these woodlands have prompted agri-environment policy to exclude all grazing livestock from woodlands in agri-environment agreements (DARD, 2000).

Exclusion of large herbivores can produce dramatic changes in the structure and composition of the woodland, which can lead to a reduction in diversity and abundance of both plant and animal communities (Kirby *et al.*, 1994). These changes can also arrest tree seedling recruitment, therefore nullifying the main reason for exclosure. Grazing by large herbivores can help to achieve and maintain diversity in the structure and composition of vegetative communities, which in turn promotes diversity in woodland faunas (Margules and Usher, 1981).

Results of a 2-year survey of semi-natural woodlands in Northern Ireland will be presented with respect to the effects of livestock grazing and exclosure from grazing on the ground flora and regeneration.

Materials and Methods

Woodland surveys were carried out throughout Northern Ireland in conjunction with the agri-environment monitoring programme at Queen's University, Belfast, in spring of 2002 and 2003 (Cameron *et al.*, 2001).

The grazing status, i.e. grazer species and approximate graze intensity, was recorded. Botanical data were gathered using a nested quadrat method, with an internal 2 m × 2 m quadrat in which percentage cover of all plant species was recorded. In the outer 14 m × 14 m quadrat additional species were recorded as present. Average heights of vegetation layers, soil type and other physical characteristics of the woodland were recorded. A photo of the quadrat was also taken for future remonitoring.

Seedlings and saplings were counted and recorded by species in both the inner and outer quadrats.

Saplings were collected from a range of woods and height correlated with age, which was derived from counting annual growth rings.

Results and Discussion

Grazing was observed to have an effect on the abundance of many key features of the woodland ground flora. For a summary of results see Table 1.

Table 1. The effects of grazing on key components of the woodland ground flora.

Increase	No effect	Decrease
Grasses	Mosses	Leaf litter
Ruderal/weed species	Deadwood	Bluebell
Bare ground	Many typical	Competitor sp, e.g. bramble
Patch heterogeneity	Woodland species	Many woody spp.
		Vegetation height

Grazing was found to have significant effects on the abundance of a number of tree seedling species (Table 2). Grazing had no significant effects on tree sapling numbers.

Table 2. Effects of grazing on tree seedling abundance.

	Grazed	Ungrazed	Significance
All species	69.50	52.30	ns
Acer pseudoplatanus	2.24	20.30	*
Betula sp	18.02	0.00	*
Fraxinus excelsior	32.20	29.20	ns
Quercus sp	0.08	0.15	ns

Average sapling growth rates are detailed in Table 3, with suggested time-periods that would be required for the terminal leader of a sapling to grow beyond browse height.

Table 3. Average growth rates of some common sapling species in Northern Ireland.

		Age until critical browse height	
Sapling species	Growth rate (cm/year)	Sheep (1.5 m)	Cattle (2 m)
Acer pseudoplatanus	26	7.0	9
Betula sp	18	8.5	11
Fraxinus excelsior	17	9.0	12
Quercus sp	11	12.5	17

Conclusion

Woodland grazing can have significant effects on the woodland vegetation. Knowledge of sapling growth rates can assist in determining the period of exclosure required to ensure saplings can grow beyond browse height and become established if stock are to be reintroduced to the wood. Knowledge of these effects can assist in formulating woodland management policy; however, other factors still require research, e.g. seasonality of stocking and stocking density.

References

Cameron, A., Flexen, M. and Johnston, R.J. (2001) Remonitoring of the Mournes and Slieve Croob, the Antril coast, glens and Rathlin, the Sperrins and Slieve Gullion: biological evaluation of the ESA scheme between 1984 and 2000. Unpublished report to DARD, Queen's University, Belfast, United Kingdom.

Cooper, A. and McCann, T. (2002) *Habitat Change in the Northern Ireland Countryside.* Technical Report of the Northern Ireland Countryside Survey 2000, University of Ulster, Coleraine, United Kingdom.

DARD (2000) *Countryside Management Scheme Explanatory Booklet.* Department of Agriculture and Rural Development, Northern Ireland, Belfast, United Kingdom.

Kirby, K.J., Mitchell, F.J. and Hester, A.J. (1994) A role for large herbivores in nature conservation management in British seminatural woods. *Arboricultural Journal* 18, 381-399.

Margules, C. and Usher, M.B. (1981) Criteria used in assessing wildlife conservation potential: a review. *Biological Conservation* 21, 79-109.

Application of limed sewage sludge to a young *Pinus radiata* plantation on acid soil

B. Omil-Ignacio [1], M. R. Mosquera-Losada [2], A. Rigueiro-Rodriguez [2] and A. Merino-García [2]

[1]Soil Science Dept., Santiago de Compostela University, E-27002 Lugo, Spain. [2]Crop Production Dept., Santiago de Compostela University, E-27002 Lugo, Spain. bomilig@lugo.usc.es, romos@lugo.usc.es, amerino@lugo.usc.es

Abstract

In this study the effectiveness of addition of sewage sludge (alkaline, rich in N, P and Ca, and with low levels of heavy metals) originating from a dairy industry to fertilize a silvopastoral system, established on a 10-year-old *Pinus radiata* plantation, was assessed. The experiment consisted of applying two doses of the dairy sludge - 10.8 t DM/ha and 21.6 t DM/ha and subsequent sowing with *Dactylis glomerata* and *Trifolium repens*. The low rates of nitrification along with the high demand of plants prevented the release of NO_3^- Application of the sludge resulted in increases in soil pH and available P, K, Ca and Mg. In the forest vegetation, it also led to higher foliar P and N concentrations and higher growth. During the first 2 years, the treatments favoured the growth of *Agrostis* sp, *Dactylis glomerata* and *Pseudarrhenatherum longifolium* whereas it reduced that of shrubs *Erica* sp and *Ulex* sp.

Key words: silvopastoral system, forest nutrition, biosolids

Introduction

During the last decade, land application has become an interesting option to manage municipal and industrial residual wastes. Sewage sludge contains nutrients beneficial to plant growth and organic matter. When properly managed, in silvopastoral systems, its application can increase the growth of trees and sown pasture. However, the application of these alkaline and rich N residues to forest soils may have ecological and environmental risk associated with an increase in mineralization of organic matter and nitrate leaching. While a number of studies have been conducted on municipal biosolids, there is little information on the effects of dewatered alkaline stabilized biosolids on silvopastoral systems. This research has been undertaken to assess the responses of forest ecosystems to applied alkaline biosolids on (a) soil biological and chemical properties, and (b) plant response (growth and nutrition of trees and understorey vegetation).

Materials and Methods

Treatments were established in November 1998, in a 10-year-old *Pinus radiata* plantation. The soil, developed on granite rock, was sandy (sand = 56%), rich in organic matter (9.4%), strongly acid (pH KCl = 4.0) and had a large capacity for P retention. The sewage sludge was obtained from the wastewater treatment plant of a milk factory located near the experimental field. The sludge was dewatered and stabilized by adding CaO. It was alkaline (pH 10), rich in N (64 mg/kg), P (13 mg/kg), Ca (23 mg/kg) and the heavy metal concentrations were below the current limits established for EU. Two treatments were applied in a randomized block design with four replicates (surface area 100 m^2). The sludge was added to the plots at rates of 10.8 and 21.6 t/ha on a dry matter basis (200 and 400 kg N/ha). The sludge was spread and disked to a depth of approximately 20 cm. Pasture sown in the autumn of 1998 (25 kg/ha of *Dactylis glomerata* and 4 kg/ha of *Trifolium repens*) after clearing. Throughout the study period (November 1998-October 2003) samples of soils and *Pinus radiata* needles were analysed for total macro- and micronutrients with an spectrophotometer IPC (Inductive Plasma Coupling) after microwave digestion with nitric acid (1:50). Soils were analysed for pH, C, N and S in soil (LECO-2000) and Mellich 3-extractable cations determined with an IPC (Perkin Elmer, Wellesley, MA, USA).The diameter and height were measured in all trees within the subplots in the autumns of 1998 onwards. Diameter at breast height (dbh) was determined and the ratio of the volume of timber to 7.5 cm thin-end stem diameter was calculated. Botanical composition was monitored through the 5 years of the study.

Results and Discussion

Soil properties (Figure 1). Slight initial increases in pH were observed after the application of sludge, although the effect was limited to the few first months. The sludge application led to significant increases in Ca and P, whereas those of K were limited to the first months after application. The concentrations of Mellich 3-extractable Fe, Mn, Cr, Pb, Cu, Zn, Ni were not increased after the sludge application. On the contrary, the extractable levels of Mn and Zn were slightly decreased during the first sample dates. Both KCl-extractable NH_4-N and NO_3^--N concentrations increased for a few months after sludge treatment. The highest effect was observed for NH_4-N, for which maximum concentrations of 80 mg/kg were recorded at the highest addition of sludge. NO_3^--N concentrations in the soil were always less than 2 mg/kg of dry soil with small variations throughout the study period. Compared with untreated soil, the sludge addition led to lower mineralization rates (92-143 kg N/ha). This behaviour was probably due to the easily decomposable organic C compounds in the sludge (Quemada and Cabrera, 1995). In all soils, the yearly net nitrification was 6-9 mg/kg (9-14 kg N/ha), which represented 3-15% of the net mineralization. Reduced nitrification rates, such as those found in this soil, are normal in undisturbed acid forest soils, and are attributed to low soil pH, low initial populations of nitrifying bacteria or low soil NH^+_4 availability (Vitousek and Matson, 1985).

Figure 1. Selected properties of the upper mineral soils in the *Pinus radiata* plots subjected to biosolids treatments (10.80 and 21.6 t DM/ha). Significantly different treatment means (Tukey-Kramer test, $P < 0.05$) are indicated by *.

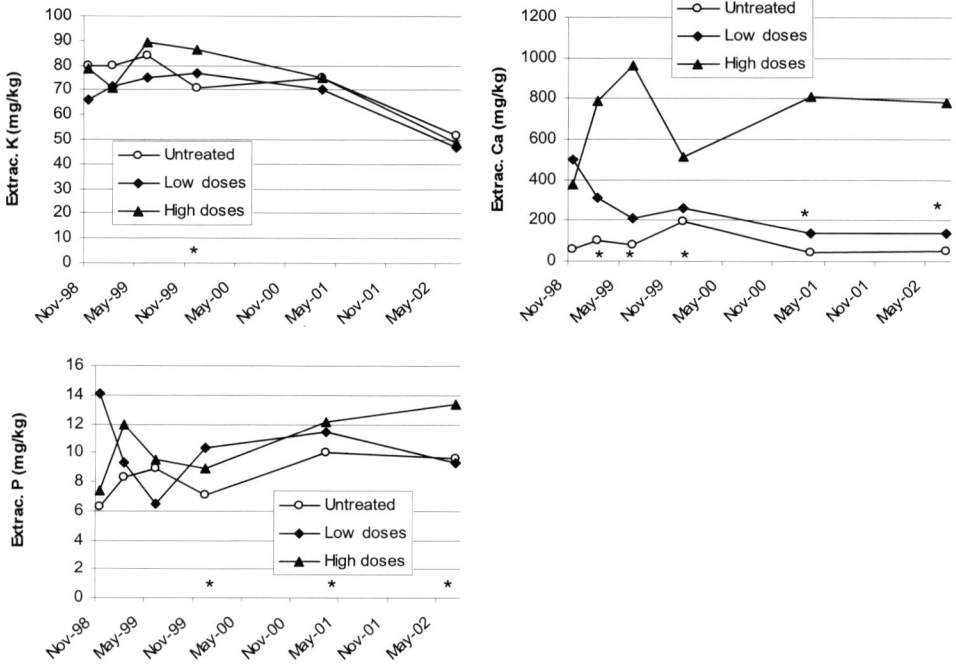

Response of tree vegetation. Fresh pine needles from untreated plots were low in P, low to medium in Ca and K and medium in Mg. Sludge addition increased the needle concentrations of Ca and decreased those of Al, Mn, Ni and Cu. The treatment also had an effect on the N/P ratio, although it was more marked in some months than others (Figure 2). The application of the alkaline sludge to this young plantation has a positive effect on tree growth (Table 1). After 5 years the difference in tree height between untreated plots and treated plots was 2.4 m. A combination of differences in total height, dbh and survival resulted in an increase in timber volume of over 45%.

Figure 2. Change in N/P and Al concentration in *Pinus radiata* needles sampled after treatments with biosolids (10.8 and 21.6 t/ha). Significantly different treatment means (Tukey-Kramer test, $P < 0.05$) are indicated by *.

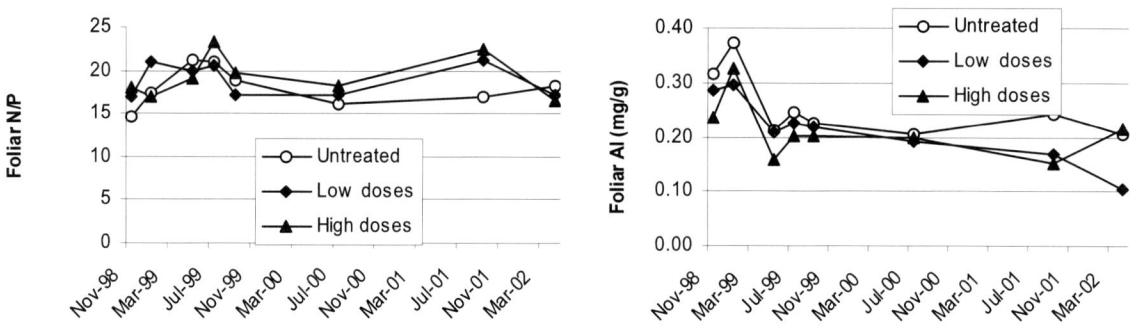

Table 1. Growth (height and diameter) and productivity (volume) of *Pinus radiata* in response to the application of sewage sludge. Significantly different treatment means are indicated by different lower-case letters (Duncan test, $P < 0.05$).

		1998	1999	2000	2001	2002	2003
Ht (m)	Untreated	3.75	4.54	5.1	6.03	6.69a	7.17a
	Low doses	3.83	4.79	5.55	6.84	7.72ab	7.82ab
	High doses	3.76	5.18	5.83	6.94	7.89b	9.58b
Diam (cm)	Untreated	7.53	7.54	8.84	9.14a	9.94a	10.30a
	Low doses	7.97	8.7	9.81	11.93b	11.92b	13.08b
	High doses	7.27	8.82	9.98	12b	11.65ab	11.82ab
V (m^3/ha)	Untreated	33.54	41.54	52.49	69.36a	96.47	115.05
	Low doses	36.62	58.17	80.49	135.05b	131.23	167.78
	High doses	27.78	50.23	65.29	116.44b	132.56	185.38

Development of understorey vegetation. Pasture production was not significantly affected by treatments, because the not fertilized treatment had higher shrub content than fertilized treatments; however, an important increment of pasture quality was found, as was previously described by Omil *et al.* (2000). During the two first years, the addition of sludge increased the percentage of *Dactylis glomerata* and reduced that of shrubs (*Erica* sp and *Ulex* sp). Afterwards, there was a progressive gain in proportion of shrub species (Table 2).

Table 2. Botanical composition of different treaments in the 3 years experiment.

		Jun-99	Dec-99	Jul-02	Nov-03
Dactylis (%)	Untreated	10.7	3.7a	<0.1	<0.1
	Low doses	18.4	32.5b	<0.1	<0.1
	High doses	21.5	45.4b	<0.1	<0.1
Shrubs (%)	Untreated	14.2	1.8	26.1	31.7
	Low doses	16.4	2.2	3.0	12.8
	High doses	28.4	1.1	7.1	5.9
Unsown herbage (%)	Untreated	49.6	57.6	40.8	24.9
	Low doses	66.3	61.6	63.9	25.9
	High doses	62.6	70.2	64.8	42.1

Acknowledgements

We are grateful to CICYT, XUNTA de Galicia, GESTAGUA S.A and AGROAMBPRODALT for financial assistance, to Escuela Politécnica Superior for facilities, to Divina Vázquez Varela, José Javier Santiago-Freijanes, Teresa Piñeiro López and Antonio Rodríguez Rigueiro for helping in processing, laboratory and field.

References

Omil, B., Mosquera, M.R., Rigueiro, A. and Merino, A. (2000) Chemical and biological properties of agroforestry soil treated with dairy-plant waste. In. *International Symposium Managing Forest Soils for Sustainable Productivity.* International Union of Soil Science, Vilareal, Portugal, p. 231.

Quemada, M. and Cabrera, M.L. (1995) Carbon and nitrogen mineralization from leaves and stems of four cover crops. *Proceedings of American Soil Science Society* 59, 471-477.

Vitousek, P.M. and Matson, P.A. (1985) Disturbance, nitrogen availability, and nitrogen losses in an intensively managed Loblolly Pine plantation. *Ecology* 66, 1360-1376.

Effect of season on the ingestive behaviour of cattle grazing *Leucaena leucocephala*

R. L. Ortega, H. J. Castillo and P. F. Rivas
C.E Mocochá, CIR-Sureste INIFAP, km 25.5 Carretera Mérida-Motul, Mocochá, Yuc, CP 97454, Mocochá, Yucatán, Mexico. ortega.luis@inifap.gob.mx

Abstract
The ingestive behaviour of zebu cattle was investigated at two different plant heights (90 cm = L90 and 120 cm = L120) and three seasons of the year (dry, wet and *nortes*). There was a significant effect ($P < 0.05$) for the main factors. Cattle took more bites/min when grazing L90 (37) than L120 (35). They also had a bigger bite size (0.48 g/bite) in L90 compared to L120 (0.44 g/bite). This resulted in a higher intake rate for cattle grazing L90. A higher biting rate was observed during the wet season than in the dry and *nortes* season. The amount of forage cattle ingested per bite was higher during *nortes* and dry seasons and lower in the wet season. The combined results of bite size and biting rate resulted in similar forage intakes among seasons. Managing *Leucaena* at 90 cm favoured higher shoot production compared to L120. As a result, there was a significant ($P < 0.05$) height by season interaction for all the variables studied. Time of year and plant structure, as a result of grazing management, affected intake of *Leucaena* by cattle. A higher grazing efficiency was observed in L90 than L120.

Key words: shrub, tropical pasture, silvopasture

Introduction
Leucaena (*Leucaena leucocephala*) is a tropical shrub legume that produces excellent yields of high-quality forage, and is widely used for grazing animals all year round. However, because forage production and plant structure change during the year, selection and diet intake depend, not only on the available plant resources, but also on the animal's capacity to efficiently harvest the forage (Forbes, 1988; Laca and Demment, 1996). Different studies have shown that feeding behaviour is controlled by animal and plant interactions. Animal factors include species, physiological status, previous experience with vegetation and habitat exploration, while plant factors include availability and the physical and qualitative characteristics of plant biomass (Dicko and Sikena, 1992). A better understanding of feeding behaviour of cattle grazing *Leucaena* will allow the development of management strategies aimed at maximizing the use of silvopastoral systems to increase animal production. The aim of this study is to evaluate how managing *Leucaena* at two different heights influences the ingestive behaviour by zebu cattle, during the dry, wet and *nortes* seasons. The *nortes* season (December, January, February and April) is a distinct time of the year in which cold north winds and tropical moisture come together.

Materials and Methods
At San Jose Kuche, a silvopastoral farm, 2 ha of *Leucaena* were divided in two equal parts and cut at 90 cm (L90) and 120 cm (L120) height during the study. Ten zebu cows were used to record the ingestive behaviour during the key times of the year when climatic variations affected vegetation (dry, wet and *nortes* seasons). The dry season usually occurs between the months of March and May whereas the wet season occurs between June and September. Variables recorded were: bites/minute (b/min), using a hand counter, grazing time per feeding station (GTFS) and movement time between feeding stations (MTFS). An animal's feeding station is established when it stops walking, lowers its head and bites a plant (Stuth, 1991). GTFS and MTFS were recorded by measuring the time in seconds using a chronometer. Variables were recorded four times per animal and day, during five consecutive days in each season. Bite size (g/bite) was indirectly estimated by weighing extrusa samples collected by means of three fistulated cattle from the study area and divided by the number of bites during the sampling period. Intake rate (g/min) was the product of weight of extrusa divided by sampling min. Data were analysed by a factorial design (2 heights × 3 seasons) using the GLM procedure of the SAS (1985) and an LSD test was used to differentiate the means.

Results and Discussion
Cattle had significantly ($P < 0.05$) more bites/min when grazing L90 (37) than L120 (35), but higher GTFS in L120 (46 s) than L90 (43 s) (Table 1). The same pattern was observed for MTFS, 2.3 s for L120 and 2.0 s for L90. When comparing seasons, a higher ($P < 0.05$) biting rate was observed during the wet season than in other seasons. The combined results of bite size and biting rate resulted in similar forage intakes among seasons. The results of biting rate found in this study were higher than those reported for cattle grazing different shrub types (Hernández *et al.*, 2001). When forage was scarce (dry season), cattle spent less time foraging in one place (30 s) than during the more abundant seasons, *nortes* (49 s) and wet (54 s) (Table 1).

Table 1. The effects of variables on the ingestive behaviour of zebu cattle grazing *Leucaena* plantations. Within plant height or season, mean values in the same row with different letters are statistically different ($P < 0.05$) (b: bite, GTFS: grazing time per feeding season, MTFS: movement time between feeding stations).

Variables	Plant height		Season		
	90 cm	120 cm	Dry	*Nortes*	Wet
b/min	37a	35b	36a	33b	38c
g/b	0.48a	0.44b	0.48a	0.49a	0.42b
g/min	17a	15b	17a	16a	16a
g/h	1051a	931b	1048a	974a	952a
GTFS (s)	43a	46a	30a	49b	54b
MTFS (s)	2.0a	2.3b	2.1a	2.1a	2.2a

There was a significant ($P < 0.05$) height season interaction for all variables. For both heights, a higher number of bites was observed during the wet season than in other seasons, when at least 85% of the plant material consumed by cattle was current season leaves and twigs. During the dry season a higher biting rate was observed in L90 compared with L120. This was because the amount of available forage for consumption in L90 was twice that in L120. It is well known that bite mass is largely constrained by plant characteristics (Orr *et al.*, 2001). The distribution and plant structure of L90 allowed cattle to obtain a higher bite size in the dry and *nortes* seasons than in wet seasons and the values were different ($P < 0.05$) from those obtained in L120 (Table 2). Cattle had a higher GTFS ($P < 0.05$), compared to the other treatments, when grazing L120 during the wet season (58 s). MTFS varied from 1.7 s for L90 (dry) to 2.5 s for L120 (dry).

Table 2. Ingestive behaviour of cattle in *Leucaena* plantations. Mean values in the same row with different letters are statistically different ($P < 0.05$).

Variables	90 cm			120 cm		
	Dry	*Nortes*	Wet	Dry	*Nortes*	Wet
b/min	38ab	33c	39a	33c	34c	37b
g/bite	0.52a	0.53a	0.40c	0.44bc	0.45b	0.44bc
g/min	20a	17b	15b	15b	15b	16b
g/h	1197a	1027b	930b	900b	920b	974b
GTFE (s)	29c	50b	49b	31c	49b	58a
MTFS (s)	1.7b	2.1ab	2.2a	2.5a	2.1ab	2.2a

Conclusions

Time of the year and plant structure, as a result of height management, affected feeding behaviour of cattle in *Leucaena* plantations. Managing *Leucaena* at 90 cm favoured higher shoot production and hence better grazing efficiency. The results of this study will allow the development of management strategies for cattle and *Leucaena* in silvopastoral systems.

References

Dicko, M.S. and Sikena, L.K. (1992) Feeding behaviour, quantitative and qualitative intake of browse by domestic ruminants. In: Speedy, A. and Pugliese, P.L. (eds) *Legume Trees and Other Fodder Trees as Protein Sources for Livestock.* FAO Animal Production and Health Paper 102, FAO, Rome, Italy.

Forbes, T.D.A. (1988) Researching the plant-animal interface: the investigation of ingestive behaviour in grazing animals. Stuth, J. W. (1991) Foraging Behaviour. *Journal of Animal Science* 66, 2369-2379.

Hernández, A., Pinto, R.R., Ramírez, A.L., Ortega R.L., Macias, W., Muñoz, J. and Trejo, J. (2001) Grazing habits, ingestive behaviour and voluntary feeding of bulls in silvopastoral systems in the central valley of Chiapas (Hábitos de pastoreo, conducta ingestiva y consumo voluntario de toretes en silvopastoreo en el valle central de Chiapas). In: *II Reunión Nacional sobre Sistemas Agro y Silvopastoriles.* Centro Regional Universitario del Sureste, Universidad Autónoma Chapingo,Villahermosa, Mexico. (In Spanish.)

Laca, E.M. and Demment, M.W. (1996) Foraging strategies of grazing animals. In: Hodgson, J. and Illius, A.W. (eds) *The Ecology and Management of Grazing Systems.* CABI International, Wallingford, United Kingdom, pp. 137-158.

Orr, R.J., Rutter, S.M., Penning, P.D. and Rook, A.J. (2001) Matching grass supply to grazing patterns for dairy cows. *Grass and Forage Science* 56, 352-361.

SAS (1985) *SAS User's Guide: Statistics,* 5th edn. SAS Institute, Cary, NC, USA.

Stuth, J.W. (1991) Foraging behaviour. In: Heitschmidt, R.K. and Stuth, J. W. (eds) *Grazing Management: an Ecological Perspective.* Portland, USA, pp. 65-84.

Horse grazing on a mixture of *Trifolium brachycalycinum* and *Cynodon dactylon* in firebreaks of Tuscan Maremma (central Italy)

A. Pardini and F. Natali
DiSAT, University of Florence, Italy. andrea.pardini@unifi.it

Abstract
Firebreaks are necessary to prevent and fight forest fires. Unfortunately bare soil increases soil erosion and the eventual presence of a native sward can even increase the risk of fire. A sown mixture of *Trifolium subterraneum brachycalycinum* + *Cynodon dactylon* has been shown to have very useful parameters for both fire prevention (short grass that remains green in summer, easy passage of service tracks) and forage production (the clover can give high yields with good quality). However, this pasture needs to be grazed short to prevent build-up of dead biomass in summer. Unfortunately, cattle and sheep rearing in Italy is being steadily reduced and market trends do not appear to indicate a revival. However, the development of rural tourism allows for a certain amount of grazing horses. This chapter refers to results of a new trial carried out in central Italy to evaluate the use of horses to maintain the sown sward. Soil cover, sward height, total and residual biomass, dead fuel biomass in summer were assessed. The mixture was intensively grazed and very little dead herbage accumulated over summer; consequently horse grazing is suggested for these mixtures on farms which have firebreaks and offer horseback riding to rural tourists.

Key words: fuel biomass, farm tourism, territory multiple use, forage

Introduction
Many Italian pastures are under-grazed, which can lead to forest fires (Kyriakakis *et al.*, 1999), and this problem can also affect firebreaks. Firebreaks management is expensive and animal grazing is therefore preferred to keep the grass short, especially now that some Tuscan farmers use horses for tourist trekking and the same animals to keep the grass short in pastures (Pardini *et al.*, 2002b). Firebreaks must be sown with low-growing species which produce little fuel biomass. Research has already established the ideal mixture - *Cynodon dactylon* + *Trifolium subterraneum* (Pardini, 2002a). As the warm season grass remains green in summer and is very well adapted to animal trampling and intensive grazing, subclover is used for its good productivity and palatability. The adaptation of the sown mixture to horse grazing was investigated and the results of this research can be diffused to many Italian farms with pastures in the forest area, where it can contribute to better pasture management and, at the same time, reduce forest fires, hence improving the silvopastoral system.

Materials and Methods
The trial was carried out in Mediterranean central Italy. The mixture *Cynodon dactylon* + *Trifolium brachycalycinum* cv Clare was sown in a 20 m wide firebreak at 7.5 kg/ha *C. dactylon* + 12.5 kg/ha subterranean clover. The grass was sown in April 2000 following disk-harrowing; the clover was oversown in early October 2000. Data presented are the average over the period January 2001-January 2003. Grass production in grazed and ungrazed stands was compared. The pasture in grazed plots was grazed by horses at mid-spring and at the beginning of the summer, each time with six animals for 5 days. The experimental design was a split-plot with plots 20 m × 30 m and subplots 10 m × 15 m, each treatment repeated three times. The following data were measured halfway through each season: (1) soil cover (linear analysis, each 20 m long with measuring points at each 20 cm, three lines in each transect per plot), (2) sward height (average of the transect heights measured at each recording point on analysis lines), (3) total dry biomass (sward biomass cut from 1 m^2 quadrat adjacent to each transect analysis line and then oven dried) and (4) fuel biomass (separated dead component of herbage). Data were analysed using ANOVA.

Results and Discussion
Soil cover. The sown mixture covered 89% of soil surface over all seasons, the grass contributed most in the summer, the clover from the autumn to the spring. Differences between grazed and ungrazed plots were not significant.
Sward height (Table 1). The pasture always remained short enough to avoid flame transmission to trees. Grazed swards were shorter than in spring and summer, animal grazing in these two seasons also influenced height in autumn.

Table 1. Mean pasture height (cm) per season over years 2001-2003. Numbers having different letters in columns are significantly different at $P = 0.05$.

	Winter	Spring	Summer	Autumn
Grazed plots (residuals)	12 (not grazed)	8 b (grazed)	4 b (grazed)	7 b (not grazed)
Ungrazed plots	14	19 a	16 a	10 a

Biomass (Table 2). Biomass was the same in all plots in winter due to an absence of grazing. Biomass was reduced in plots grazed by horses during spring, with an even greater reduction in summer. In autumn there was no grazing; however, the plants in the grazed plots were probably stressed and produced less than in the ungrazed plots.

Table 2. Mean biomass (t/ha) over years 2001-2003. Numbers having different letters in columns are significantly different at $P = 0.05$.

	Winter	Spring	Summer	Autumn
Grazed plots (residuals)	1.06 (not grazed)	0.98 b (grazed)	0.42 b (grazed)	1.58 b (not grazed)
Ungrazed plots	1.22	3.63 a	1.15 a	1.87 a

Fuel biomass (Table 3). Dry herbage biomass in grazed plots was lower than in the ungrazed all the time except during winter. In spring and summer there was less combustible herbage in grazed plots than in the ungrazed ones.

Table 3. Mean fuel biomass (t/ha) per season over years 2001-2003. Numbers having different letters in columns are significantly different at $P = 0.05$.

	Winter	Spring	Summer	Autumn
Grazed plots (residuals)	0.87 a (not grazed)	0.23 b (grazed)	0.20 b (grazed)	1.30 b (not grazed)
Ungrazed plots	1.11 a	1.98 a	0.56 a	1.50 a

Conclusions

The sown mixture provided good soil cover whether grazed or ungrazed. However, it is possible that horses will have an influence over a longer number of years. Grazed and ungrazed swards were short due to the prostrate habit of both grass and legumes. Forage production was low; however, this is not important in farms where the animal stocking rate is much lower than the pasture carrying capacity. Seasonal availability of forage was well balanced. Fuel biomass in spring and summer was significantly reduced by horses. Horses proved to be good grazers of the sown mixture and to keep swards low, thus reducing fire risks and maintaining a silvopastoral system. Further research is necessary to ensure persistency of the sown sward over a long period and to find a good technique for the regeneration of an exhausted pasture of this kind once the contribution of the sown species has decreased.

References

Kyriakakis, S., Kazakis, G., Abid, M., Doulis, A. and Papanastasis, V.P. (1999) Effects of grazing and burning on woody vegetation of Mediterranean rangelands of Crete. *Grasslands Science in Europe* 4, 79-84.

Pardini, A., Tallarico, R. and Ghiselli, L. (2002a) *Trifolium brachycalycinum* and *Cynodon dactylon* sowing mixture for grazed firebreaks in Central Italy. *Herba* 13, 57-63.

Pardini, A., Mosquera, M.R. and Rigueiro, A. (2002b) Land management to develop naturalistic tourism. In: *Proceedings V International IFSA*. International Farming Systems Association, Florence, Italy, pp. 399-407.

Dry matter production and nutritive value of cocksfoot (*Dactylis glomerata* L.) grown under different light regimes

P. L. Peri [1], R. J. Lucas [1] and D. J. Moot [2]

[1] Área Forestal, Universidad Nacional de la Patagonia Austral – INTA, CP 9400, Río Gallegos, Santa Cruz, Argentina.
[2] Soil, Plant and Ecological Division, Lincoln University, PO Box 84, New Zealand.

Abstract

Dry matter (DM) production, crude protein (CP) and organic matter digestibility (OMD) were measured in a grazed cocksfoot (*Dactylis glomerata* L.) pasture at the Lincoln University silvopastoral experimental area and an adjacent site without trees in Canterbury, New Zealand. The forest consisted of 200 10-year-old *Pinus radiata* stems per hectare. Four levels of light intensity were compared: full sunlight (100% photosynthetic photon flux density - PPFD), open + wooden slats (45% PPFD), trees (60% PPFD) and trees + slats (25% PPFD). The mean total DM production was 8200 kg DM/ha per year in the open, 7300 kg DM/ha per year in open pasture under slatted shade, 600 kg DM/ha per year under trees shade and 3800 kg DM/ha per year in the trees + slats treatment. The interaction between treatments and time (rotations) was expressed by seasonal fluctuations in pasture DM growth rates. CP increased as PPFD declined. In contrast, the intensity of fluctuating shade had little effect on OMD.

Key words: crude protein, digestibility, fluctuating light regime, *Pinus radiata*

Introduction

Plant production and nutritive value under shading can differ between temperate grasses such as cocksfoot (Devkota *et al.*, 1997). Foliar nitrogen concentration has been shown to increase in shaded cocksfoot (Sheehy and Cooper, 1973) but results are variable for herbage *in vitro* digestibility of grasses (Wong, 1991). Most of these results have been generated by using continuous shading by cloth (Devkota *et al.*, 1997). However, in silvopastoral systems understorey plants experience fluctuations in irradiance from full sun to shade (Peri *et al.*, 2002) in silvopastoral systems with low tree density but the effects of this have not been reported. Thus, the aims of this study were to quantify the response of cocksfoot dry matter (DM) production and nutritive value under different intensities of fluctuating light regimes compared with an adjacent open site.

Materials and Methods

Dry matter (DM) production, crude protein (CP) and organic matter digestibility (OMD) were measured in a grazed cocksfoot (*Dactylis glomerata* L.) pasture (28 - day rotation with 21 ± 1 days regrowth) from September 1999 to May 2001 at the Lincoln University silvopastoral experimental area (5.2 ha) and an adjacent site without trees (1 ha) in Canterbury, New Zealand (43° 38'S and 172° 28'E). The climate is temperate and sub-humid with average rainfall of 660 mm but annual evapotranspiration about double this. Mean annual temperature is 11.4°C. Pastures were sown in September 1990 and herbage was cut and carried off the site for silage during the first 3 years but all plots were grazed by sheep from spring 1993. Stocking rate during grazing periods under trees and in the open averaged 21 lambs/ha and was adjusted when necessary after each live weight measurement (37 ± 5 day intervals). Neither fertilizer, lime nor irrigation has been applied to the experimental area since its establishment. The forest consisted of 200 10-year-old *Pinus radiata* stems per hectare high pruned to 6 m with 7 m between tree rows. The daily photosynthetic photon flux density (PPFD) integral under trees, measured in the middle of rows, was 60% of the open PPFD over a sunny day in summer, with alternating periods of full sunlight (1900 μmol^{-2} m/s PPFD) to deep shade (130 μmol^{-2} m/s PPFD). Within each of the three 0.2 ha main cocksfoot plots, a study plot of 14.0 m × 5.0 m was located in the middle of the inter-row under trees and also in the adjacent open pasture plots (0.05 ha). Within these areas, slatted shade structures measuring 3.0 m × 2.1 m covered with pine wood slats (150 mm wide) with 150 mm gaps were used to reduce the total incidence of light by approximately 50% and create a bimodal light regime. This structure is representative of the light regime of a pine tree silvopastoral system and was used to represent increased shade from more developed pine stands. These gave four levels of light intensity using a randomized block design with three replicates: full sunlight in the open (100% PPFD), open + wooden slats (45% PPFD), trees (60% PPFD) and trees + slats (25% PPFD). The use of the artificial structures widened the duration of severe shade. Statistical analyses were carried out using the Genstat statistical package. Standard error of means (sem) was used to evaluate least significant differences (lsd) at the 0.05 probability level for means separation of the pasture variables. Significant differences for the experiment with four light regimes were determined for each rotation by analysis of variance (ANOVA).

Results and Discussion

The mean total DM production was 8200 kg DM/ha per year in open, 700 kg DM/ha per year in open + wooden slats, 6300 kg DM/ha per year under trees shade and 3800 kg DM/ha per year in the trees + slats treatment. Differences in DM production were driven by DM growth rates (Figure 1) being lower under trees and trees + slats compared with the full sunlight treatment in all seasons. The interaction ($P < 0.05$) between treatments and time (rotations) showed DM production decreased with shade intensity but the differences were lowest when other factors restricted growth such as temperature < 8°C (winter) and soil volumetric water content - VWC < 15%. Similarly, Korte et al. (1987) reported that low levels of solar radiation do not appear to limit unshaded pasture production in winter.

Figure 1. Cocksfoot dry matter growth rates (21 ± 1 days regrowth) over time for four shade treatments: open (○) (100% PPFD), open + slats (▽) (45% PPFD), under trees (●) (60% PPFD) and trees + slats (▼) (25% PPFD). Bars indicate standard error of the mean (sem).

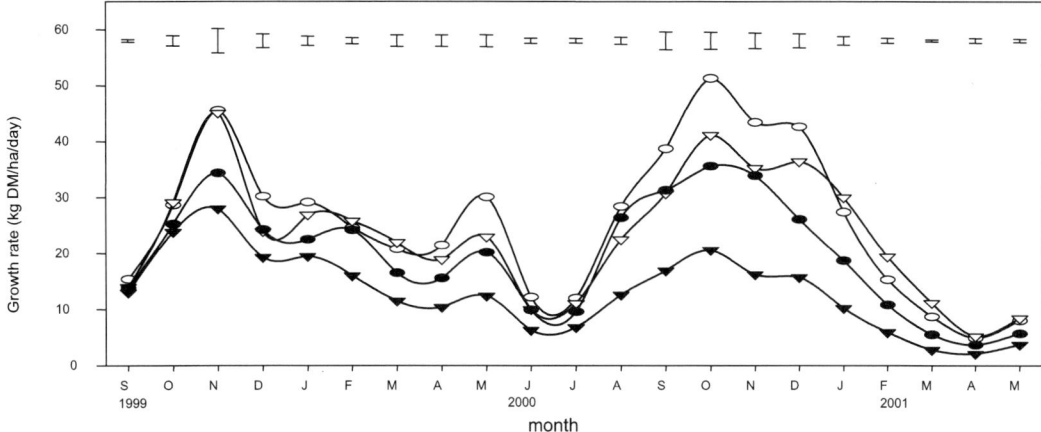

CP increased as PPFD declined (Figure 2) with mean values of 18.6% in open, 21.2% in open + wooden slats, 19.5% under trees shade and 22.5% in the trees + slats treatment. This may be attributed to either a decrease in photosynthates, with a consequent rise in the nitrogen concentration, or to an increase in soil organic matter mineralization under trees that provided greater nitrogen for grass uptake. The interaction ($P < 0.05$) between treatments and time showed that the increase in CP with shade under trees showed less difference during severe drought in summer (Figure 2). In contrast, the intensity of fluctuating shade had little effect on OMD with a mean value of 79 ± 3.2%.

Figure 2. Cocksfoot crude protein content over time for four shade treatments: open (○), open + slats (▽), under trees (●) and trees+slats (▼). Bars indicate standard error of the mean (sem).

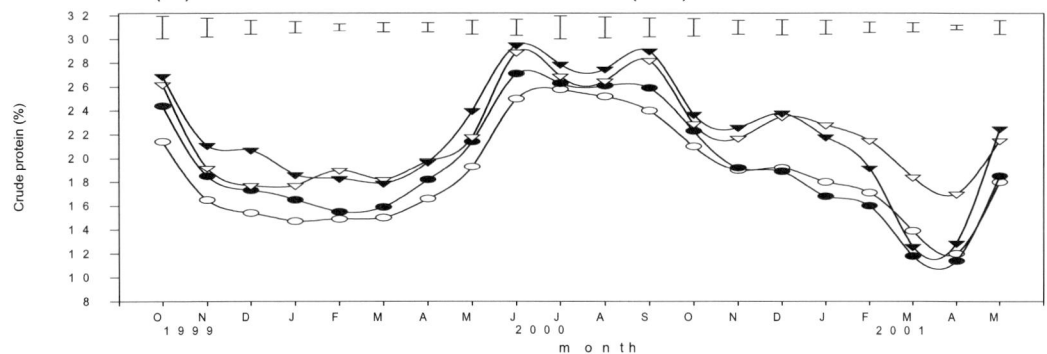

References

Devkota, N.R., Kemp, P.D. and Hodgson, J. (1997) Screening pasture species for shade tolerance. *Proceedings of the Agronomy Society of New Zealand* 27, 119-128.

Korte, C.J., Chu, A.C.P. and Field, T.R.O. (1987) Pasture production. In: Nicol, A.M. (ed.) *Feeding Livestock on Pasture*. New Zealand Society of Animal Production, Hamilton, New Zealand, pp.7-20.

Peri, P.L., McNeil, D.L., Moot, D.J., Varella, A.C. and Lucas, R.J. (2002) Net photosynthetic rate of cocksfoot leaves under continuous and fluctuating shade conditions in the field. *Grass and Forage Science* 57, 157-170.

Sheehy, J.E. and Cooper, J.P. (1973) Light interception, photosynthetic activity, and crop growth rate in canopies of six temperate forage grasses. *Journal of Applied Ecology* 10, 239-250.

Wong, C.C. (1991) Shade tolerance of tropical forage: a review. In: Shelton, H.M. and Stur, W.W. (eds) Forages for Plantation Crops. ACIAR Proceedings (32), Canberra, Australia, pp.64-69.

Evaluation of the production of acorns of the oak (*Quercus ilex* Lam. ssp *ballota*) from south-west of Extremadura (Spain)

M. J. Poblaciones, R. López-Bellido, L. Olea and C. Benito
Dept. de Biología y Producción de los Vegetales, Universidad de Extremadura, ctra Cáceres s/n, C.P. 06071, Badajoz, Spain. majops@unex.es

Abstract
Over 2 years (1997-98 and 1998-99) the total production and seasonal distribution of the holm-oak acorn in the montanera (October-January) was measured in five dehesas in the south-west of Extremadura (Spain). The average total production of the five dehesas varied between 600 and 830 kg of acorns/ha, with large variations between the different dehesas. The availability of acorns was much larger in the first half of the montanera in most of the dehesas. The month of highest production changed with the year and dehesa. The average proportion of nut/shell was 71%, showing variation between farms and years.

Key words: montanera, extensive pig-farming

Introduction
The main species of tree of the dehesa ecosystem, of which 1.2 million ha exist in Extremadura, are the holm-oak (*Quercus ilex* Lam. ssp *ballota*) and/or the cork-oak (*Quercus suber*). The fruit of holm-oak and cork-oak is used in extensive feeding in the montanera (October-January) principally for pig-farming in the Iberian peninsula (Olea and Paredes, 1997). Originally acorns would be knocked from the tree artificially with a stick 'vareo' nowadays this is unthinkable, because it is very costly and difficult, and the acorn must fall naturally to be useful. The acorn production has great variations between different dehesas and between years, which in the holm-oak can reach 70% (Martín et al., 1998; Zulueta and Cañellas, 1989). There is agreement in the literature that climatological factors (wind, rain, etc.) and fruit pathological factors (plagues) are the most important conditioners of the distribution and timing of the acorn falling. The objective of this work is to estimate the annual production of the acorn and its distribution on falling to the ground to be used for extensive pig-farming in montanera in the dehesas in the SW of Extremadura.

Materials and Methods
The study was carried out in five representative dehesas in the SW of the province of Badajoz (Extremadura), over the montaneras of 1997-98 and 1998-99. The dehesa ecosystem is a unique silvopastoral system that must be considered as an ecosystem resulting from extensive stocking. The edapho-climatic characteristics of these dehesas are typical of this area with loam to sandy-loam texture, acid pH (6 to 6.9 in water), limited organic material (0.7 to 1.9%) and semi-arid Mediterranean climate (L'Houerou, 1975) with a density of 20 to 45 trees/ha.

For the sampling of the fallen acorn the distribution and extrapolation method of Zulueta and Cañellas (1989) has been used, based on the placing of containers distributed, following this methodology, below the tree tops, taking up a surface of 0.283 m^2/tree, with four replications. We have recorded: climatic data, pathological attacks and production, proportion of shell and distribution of the acorn-fall production during the montanera (every 15 days from October 1 until January 15 of each year).

Results and Discussion
Climatology: the two periods of study (1997/98 and 1998/99) were climatologically different: while the second period was very similar to the average year expected from every dehesa, except for the autumn rain measurement, which was around 65-75% less, the first year was rainier in general (30-40% more than the yearly average), especially in autumn with twice the average rainfall The annual acorn-fall production is shown in Table 1 which presents the 2 years of study. There is considerable uniformity between the different dehesas, even between the 2 years, the average production varying between approximately 600 and 830 kg DM/ha of acorns.

Table.1 Distribution of the fall of the acorn in the years 1997/98, 1998/99 and their average.

Plantation	Tapada	Mampolin	Crespa	Quintana	Bujardo
Annual production (kg/ha)					
1997/98	673	522	536	708	888
1998/99	776	660	678	529	773
Average	724.5	591	607	618.5	830.5

The distribution of the fall of the acorns from the trees is presented in Figure 1, which shows the distribution for the montaneras of 1997/98 and 1998/99, and their average in the five dehesas. The second year presents the ideal distribution (a progressive fall of acorns), forming a Gaussian curve and quite uniform throughout the year; meanwhile, in the first year we can see important imbalances with different intensity in the five dehesas due to the very rainy autumn, which accentuated the appearance of a plague, the *Balaninus elephas*, which provoked the sudden acorn fall in October ($\approx 38\%$) and the limited fall in November ($\approx 23\%$), the contrary of a normal year.

Figure 1. Distribution of the fall of the acorn in the years 1997/98, 1998/99 and their average.

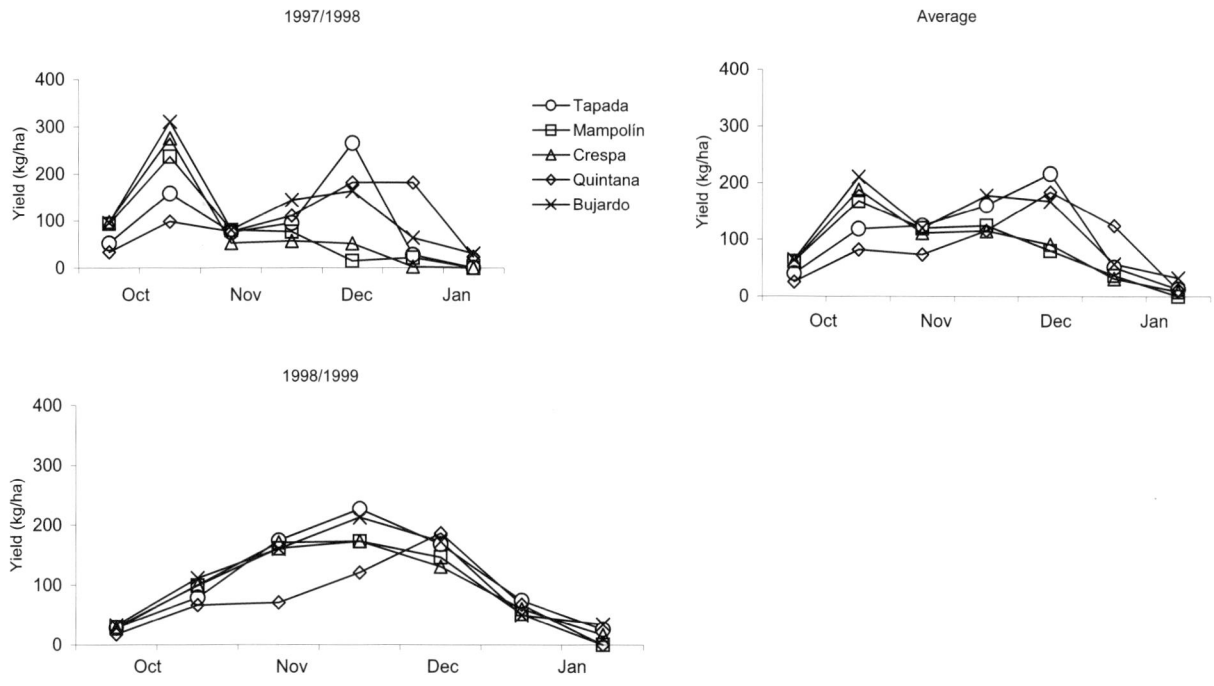

The attacks of the Balaninus elephas (Table 2) in the first year (24-49% of the fallen acorns in October) were far greater than those of the second year (7-17%).

The fruit is attacked and the inner presence of Balaninus elephas causes a loss of alimentary substance and of resistance against other pathological attacks in time. The existence of such an interaction (heavy rain in October and Balaninus elephas attack) is quite frequent in the SW of Extremadura as demonstrated by data evaluated from rain measurement reports and plague controls in the dehesa.

Table 2. Evolution of the *Balaninus elephas* attack on the fallen acorn in October.

Plantation	Tapada	Mampolin	Crespa	Quintana	Bujardo
Affected acorn (%)					
1997/98	30	39	49	24	41
1998/99	7	17	11	8	16

Conclusions

Acorn production in the dehesas in the SW of Extremadura (Spain) varies between 600 and 830 kg/ha per year. Autumn rain provokes early acorn falling. Maximum levels of acorn-fall production are reached between November 15 and December 15. The *Balaninus elephas* incidence is of great importance in these dehesas, provoking, when the autumn rain is early and abundant (October), abundant falls in October ($\approx 37\%$ of the years total) and limited falls in November ($\approx 23\%$ of the years total), which provokes a loss of production and an imbalance in its availability for cattle.

References

L'Houerou, H.N. (1975) *Bioclimatology of the Mediterranean Region* (Bioclimatología de la región mediterránea). Seminario CRIDA 08, Badajoz, Spain. (In Spanish.)

Martín, A., Infante, J.M., García Gordo, J., Merino, J.A. and Fernández, R. (1998) Acorn production in forests and dehesas of southwestern Spain (Producción de bellotas en montes y dehesas del suroeste español). *Pastos* XXVIII (2), 237-248. (In Spanish.)

Olea, L. and Paredes, J. (1997) Influence of the available area and the flock size on the improved pastures and the production of the dehesa in south-western Spain (Influencia de la superficie disponible y del tamaño del rebaño en los pastos mejorados y con la producción de la dehesa en el S.O. de Spain). *Pastos XXVII* (2), 219-247. (In Spanish.)

Zulueta, J. and Cañellas, I. (1989) Method to estimate the real acorn production in cork-oak stands (Método para estimar la producción real de bellota en un alcornocal). *Scientia Germodensis* I, 115-119. (In Spanish.)

Effect of different tree and shrub densities on the form, agronomic performance and quality of *Panicum maximum* in the Chaqueño Mountains, Bolivia

P. P. del Pozo [1], R. Rizzo [2] and E. Fernández [1]

[1] Faculty of Veterinary Medicine, Agrarian University of Havana, Cuba. [2] Faculty of Agricultural and Forest Sciences, University of Tarija, Bolivia. delpozo@isch.edu.cu

Abstract

The effect of tree and shrub densities on the form, agronomic performance and quality of *Panicum maximum* cv Gatton Panic during its establishment was studied in an experiment with a random block design and four replicates. The treatments consisted of 40%, 60% and 100% of natural communities, with a density of 700 shrub and 180 tree species per hectare. Over one year the height, length, leaf width, leaf area, total number of shoots per plant, the distance between nodes, the yield of the total dry matter (TDMY), the yield of its components (leaves (LDMY) and stems, (SDMY)) and the quality of biomass were measured. The population density modified ($P < 0.01$) growth and morphological development of the plant. The percentage of leaves increased ($P < 0.001$) and stems decreased when density reduced values between 29.5 and 41.55 and 70.0 and 54.5% for 40% and 100% of the population, respectively. Light availability in the natural mountain area (100%) reduced ($P < 0.01$) TDMY and its components and were related ($P < 0.01$) through the TDMY = -270.52 + 0.161 × LUX, LDMY = -2673.41 + 342.27 × Ln (LUX) and SDMY= -297.53 + 0.119 × LUX with $r^2 > 0.98$. The content of CP, IVOMD and CF averaged 8.76, 55.5 and 37.3%, respectively. The reduction of shrub and tree cover from 100% up to 40% increased the productivity of the agroecosystem by 3.4 times and the carrying capacity up to 0.71 animals/ha.

Key words: profitability, herbaceous strata quality, woodland improvement, silvopasture

Introduction

The Bolivia Chaqueña region (160,000 km^2) is classified as a fragile ecosystem, and livestock mismanagement (i.e. continuous grazing with no stocking rate control) has caused a change in vegetation and reduced forage production. The establishment and management of grasses in silvopastoral system with associated trees and shrubs in the natural mountain are the most important alternative for the livestock region, because this increases the biodiversity, productivity of the agroecosystem and improves husbandry indicators of livestock (Joaquín, 1991; Pezo and Ibrahim, 1999; Schimpf *et al.*, 1999). *P. maximum* cv Gatton Panic performed well in this region, in both deforested and reforested areas. The establishment of this species in the region for its adaptability to shade (Saravia *et al.*, 1995; Castillo, 2001) is recommended although ecophysiological studies are still required to explain various aspects related with the space design of tree and shrub components in the system which would help achieve a sustained and integrated utilization of all resources. The objective of the present work was to evaluate the effect of different tree and shrub densities on the form, agronomic performance and quality of *Panicum maximum* cv Gatton Panic in the Chaqueño Mountains of Bolivia.

Materials and Methods

The experimental work was carried out in Villamontes (Bolivian Chaco), department of Tarija, on a Haplic Calcisol soil, with a content of potassium (0.77 cmol (+)/100 g), organic matter (2.77%) and phosphorus (11.25 cmol (+)/100 g) (IBTA, 2000). The climate of this region is classified as semi-arid (ZONISIG, 2000), with an annual average temperature of 24.5°C (min 17.7°C - max 30.4°C). Annual average rainfall is 751.5 mm with a rainy season from November to April with 91% of the total rain and May to October with only 9% of the total. The native vegetation of this area is characterized by the presence of three strata of vegetation: tree, shrub and herbaceous. The tree stratum has a varied canopy cover that could be scarce or dense in sectors with emergent trees of between 12 and 15 m (*Amburana cearensis*, *Caesalpinia paraguarensis*, *Calycophyllum multiflorum*, *Anadenanthera colubrina*, *Phyllostylon rhamnoides*). The shrub stratum is dense with average heights from 4 to 6 m and the predominant species is *Ruprechtia triflora*. The herbaceous stratum (*Sida* sp, *Bouteloa* sp, *Petivenia alliaceae*, *Smilax officinalis*, *Manettia cordifolia*) is generally low, with a variable covering from semi-dense to disperse.

According to the number of perennial species in the area (arrangement and population) trees and shrubs were cut, maintaining 60% and 40% of the natural cover on the mountain, a density of 700 and 180 shrubs and trees per hectare, respectively. The experimental plots were sown with cv Gatton Panic on 5/01/2000 at 3.0 kg/ha and cutting began at 120 days after sowing. At the first two harvests (4/05/2000 and 1/12/2000), 4 m^2 of grass was harvested per plot at 5 cm high. Samples were dried at 60°C to determine dry matter content, dry matter yield (TDMY) and its morphological components (leaves (LDMY) and stems (SDMY)). The content of crude protein (CP) and crude fibre (CF) was determined (AOAC, 1995) as well as *in vitro* organic matter digestibility (IVOMD) (Tilley and Terry, 1963). Plant morphological measurements were carried out on 12 randomly chosen shoots per treatment. In each *Panicum* plant, height (cm), leaf length (from the upper side of the ligule to the apex), leaf width (on the medium longitudinal point), number of leaves, total shoots per plant, and the internode distance in each stem (cm) were measured. The level of light in each treatment (lux/day) was systematically sampled (36 per treatment) at 7:00, 12:00, 17:00 each day. A randomized block design with four replications was used, where experimental treatments were shrub and tree densities. The relationship between TDMY, LDMY, SDMY and lux level were established by regression analysis. Analysis of Variance and Student-Newman-Keuls's Multiple Range test were used to compare the performance of treatments. The carrying capacity was determined based on the daily intake of an adult cow weighing 500 kg live weight (LW) and grass yield per annum and treatment.

Results and Discussion

Morphological development and growth of *Panicum* (Table 1), expressed by plant height and productivity, were modified by tree and shrub density ($P < 0.001$). The highest values of all indicators studied were found in the 40% density treatment, i.e. at higher levels of light reaching the herbaceous layer: range 13,048 - 13,340 lux, with maximum values of up to 26,937 at noon. Leaf percentage increased ($P < 0.001$) when tree and shrub density decreased and it showed an inverse relationship with stems. Leaf and stem percentages ranged between 29.5 and 41.55% for 40% treatment and between 70.0 and 54.5% for 100% population treatment.

Table 1. Effect of different tree and shrub densities on the performance of some morphological indicators. a,b,c Means in the same row followed by different letters are significantly different at $P < 0.001$; *** $P < 0.001$. Note: The average light intensity values per treatment of each tree and shrub density were 4812, 11,477 and 13,048 for 100%, 60% and 40%, respectively.

Variables	Population (Treatment)			SE (±)
	100%	60%	40%	
Height (cm)	41.0a	127.5b	132.5c	0.33***
Leaf length (cm)	30.0a	48.0b	51.0c	0.22***
Leaf width (cm)	1.8a	2.0b	2.1c	0.02***
Leaf area (cm^2)	38.3a	68.2b	73.1c	1.11***
Number of leaves/plant	18.0a	112.0b	130.0c	1.09***
Number of shoots/plant	6.0a	21.0b	26.0c	0.21***
Internode distance (cm)	8.0a	14.0b	14.5c	0.11***

Under natural light in the mountains leaf, stem and total dry matter yields of *Panicum* were reduced ($P < 0.01$) (Figure 1). Results indicated the need to reduce the density of tree and shrub populations in the ecosystem if more herbaceous production is desired, so that *Panicum* may show its potential growth and maximum productivity. These results are confirmed in Table 2, where total dry matter yields and stocking rate capacity reached values of 3685 kg/ha per year and 0.47 animals/ha, respectively.

Figure 1. Relationship between accumulated aerial dry matter (leaves (LDMY= -2673.41 + 342.273 × Ln (LUX); $r^2 = 0.99$***, SE ± 25.87), stem (SDMY= -297.535 + 0.119 × LUX; $r^2 = 0.98$***, SE ± 26.47) and total dry matter (TDMY= -270.519 + 0.161315 × LUX ; $r^2 = 0.99$***, SE± 12.19) yields) and levels of illumination (lux) reached in herbaceous stratum.

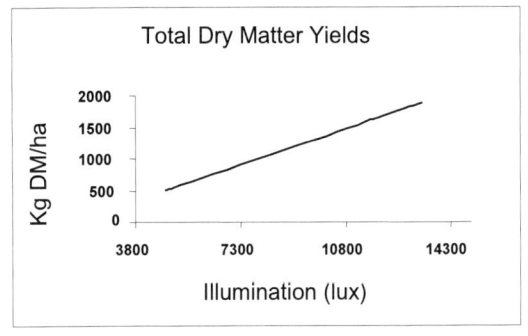
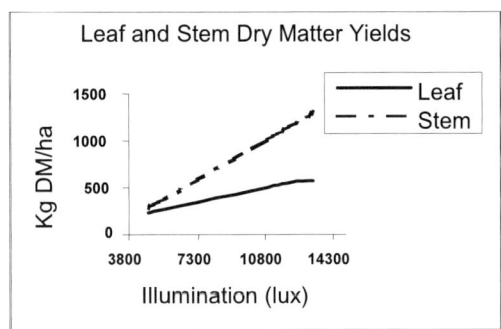

Table 2. Influence of different tree and shrub densities on total dry matter yield and stocking rate capacity of system. Different letters in the same column indicate significant differences betweent treatments at $P < 0.001$; ** $P < 0.01$.

Treatments (% cover)	Total dry matter yield (kg/ha)	Stocking rate (animals/ha)
100	1097.5a	0.14a
60	3250b	0.40b
40	3685c	0.47c
SE (±)	8.71 **	0.22 **

Average IVOMD, CP and CF contents of *Panicum* are given in Table 3. The best change in biochemical indicators evaluated was obtained in the 100% cover treatment; however, no significant differences were observed among treatments.

Table 3. Changes in IVOMD, CP and CF contents of *Panicum* with density of trees and shrubs.

Treatments	IVOMD (%)	CP (%)	CF (%)
100%	56.54	9.19	36.0
60%	53.56	8.32	38.6
40%	55.26	8.76	37.4
SE (±)	0.40	0.03	0.28

Further studies on energy balance and recycling of materials are necessary under controlled conditions, to obtain basic information on efficient management of the upland silvopastoral ecosystem.

Conclusions

Panicum grew best (3.4 times more) and carried more stock (0.71 animals/ha) under 40% reduction in canopy cover than in full light. However, there were no significant changes in the quality of the grass studied when the density of the trees and shrubs was modified in the natural mountains of the Chaco region.

References

AOAC (1995) *Official Methods of Analysis,* 16th edn. Association of Official Analytical Chemists, Arlington, VA.

Castillo, F.I. (2001) Effect of the sowing proportion on the establishment and initial phase of use of the association *Desmanthus virgatus - Panicum maximum* cv *Gatton panic* in the Bolivian Chaco (Efecto de la proporción de siembra en el establecimiento y fase inicial de explotación de la asociación *Desmanthus virgatus - Panicum maximum* cv *Gatton panic* en condiciones del Chaco Boliviano). MSc thesis. Universidad Juan Misael Saracho, Tarija, Bolivia. (In Spanish.)

IBTA (2000). *Soil Classification and Analysis* (Clasificación y Análisis de suelos). Instituto Boliviano de Tecnología Agropecuaria, Tarija, Bolivia. (In Spanish.)

Joaquín, J.N. (1991) *Management Systems of the Chaqueño Mountain Forest for Livestock Fodder* (Sistemas de manejo de Bosque Chaqueño para forraje ganadero). Estación Experimental "El Salvador", Chuquisaca, Bolivia. (In Spanish.)

Pezo, D. and Ibrahim, M. (1999) *Dispersed Trees and Shrubs in Pastures* (Árboles y arbustos dispersos en potreros). Sistemas silvopastoriles, Modulo de enseñanza agroforestal No 2, CATIE-GTZ, Turrialba, Costa Rica. (In Spanish.)

Saravia, T.C.J.,Virieux, M., Segovia, L.G. and Salas, G.E. (1995) *Guidelines for Husbandry in the Bolivian Chaco* (Manual de Ganadería del Chaco Boliviano). Fegasacruz, Santa Cruz, Bolivia. (In Spanish.)

Schimpf, J.H., García, F.E., Kopsir, A.N. and Abarzar, S.V. (1999) Fodder assessment in natural and selective forests in temperate Jujuy valleys in Argentina (Evaluación forrajera de monte natural y selectivo en valles templados de Jujuy, Argentina). In: *Proceedings of III Jornadas Regionales de Información Científico Técnicas de las Facultades de Ciencias Agrarias en Potosí Bolivia.* Universidad Autónoma Tomás Frías-Proyecto AUTAPO-DFID. Potosí, Bolivia, pp.13-18. (In Spanish.)

Tilley, J.M.A. and Terry, R.A. (1963) A two stage technique for the in-vitro digestion of forages crops. *Journal of the British Grassland Society* 18, 104-111.

ZONISIG (2000) *Forest Inventory and Classification in the Province Gran Chaco and O'Connor in the Department of Tarija-Bolivia* (Inventario y clasificación tipológica de bosques en la Provincia Gran Chaco y O'Connor del Departamento de Tarija-Bolivia). Cooperación del Gobierno de los Países Bajos. 265 pp. (In Spanish.)

Effects of management on acorn production and viability in holm-oak dehesas

F. J. Pulido, E. García, J. J. Obrador and M. J. Montero
Dept. of Plant Biology and Plant Production, School of Forestry, University of Extremadura, Avenida Virgen del Puerto 2, E-10600 Plasencia, Cáceres, Spain. nando@unex.es

Abstract
In this work we analyse the effect of different management treatments of dehesas, namely rotational cereal cultivation (C), continuous grazing (G) and reduced grazing with invasion by *matorral* (M), on acorn crop and percentage of seeds lost to abortion or infestation by weevil (*Curculio*) and moth (*Cydia*) larvae. Management had a significant effect on acorn production. Viable fruit production was higher in C plots and lower in M. Acorn viability was similarly affected by management, the percentage of non-viable acorns being higher in M and lower in C. Early abortion of fruits was the main cause of fruit loss and the relative incidence of different causes did not depend on management. Overall, these results confirm that dehesa cultivation favours acorn production and viability and that reduced grazing causes the opposite effect.

Key words: acorn production, acorn viability, dehesa, management

Introduction
Dehesa is a kind of agrosilvopastoral system that covers 3.1 million ha in western Spain and Portugal. Acorn production is the primary winter source of food for livestock and wildlife, yet little is known about factors explaining spatial and temporal variation in mast production (reviews in Vázquez, 1998; Pulido and Díaz, 2002). For a given year and locality, management of the understorey and the mature oak trees is supposed to influence the number, size and viability of acorns produced, but no study has addressed these effects explicitly (see Martín-Vicente *et al.*, 2000, for a partial study). In this work we analyse the effect of different management treatments on the acorn crop and percentage of seeds lost to abortion or infestation by weevil (*Curculio*) and moth (*Cydia*) larvae.

Materials and Methods
The study area is located in Cuatro Lugares county (Cáceres, central Spain). Climate is of a Mediterranean type with annual temperature averaging 16°C and precipitation around 600 mm. In this area dehesas are private farms ranging in size from 300 to 600 ha and with a mosaic composed of cereal cultivation (C), permanently grazed pastures (G) and areas with reduced grazing that are colonized by shrubs (*matorral*, M). We selected three representative farms and in each of them we selected one cultivated (ungrazed), one grazed and one shrubby (with *matorral*, M) plot (three plots for each management treatment). In each plot we randomly selected seven holm-oak trees. Acorn production was estimated by means of four seed traps (0.12 m^2 in area) per tree, installed in May 2002 and monitored until January 2003. We inspected seed traps biweekly and acorns found were classified as non-fertilized, aborted, drippy, *Cydia*-infested, *Curculio*-infested and full-sized viable.

Results and Discussion
Phenology of acorn fall is characterized by two peaks, the first one in July reflecting early abortion of fruits and the second one in November reflecting acorn maturation and dispersal (Figure 1). Another peak is found in the case of M plots, in which acorn fall occurs earlier as a consequence of high pre-dispersal losses. Management had no significant effect on total production of propagules per square metre of canopy area (flowers plus fruits: $F = 4.58$, $P = 0.011$; two-way ANOVA for the effects of land use and plot). Considering only full-grown viable acorns, there was a significant effect of management ($F = 5.016$, $P = 0.007$), with acorn production per square metre being highest in C (49.39), intermediate in P (32.33) and lowest in M (13.39). Acorn viability was similarly affected by management, the percentage of viable acorns being higher in P and lower in M ($F = 5.067$, $P = 0.0096$; Figure 2). Early abortion of fruits was the main cause of fruit loss, followed by fruit infestation by weevils. Management had no significant effect on the incidence of different factors with the exception of drippy fruits, which were much more frequent in M plots ($F = 14.280$, $P < 0.001$).

Figure 1. Seasonal changes in the intensity of seed rain (mean value of total number of flowers/fruits for three plots) in each management type. Labels in the x axis refer to the first and second 15-day period in each month.

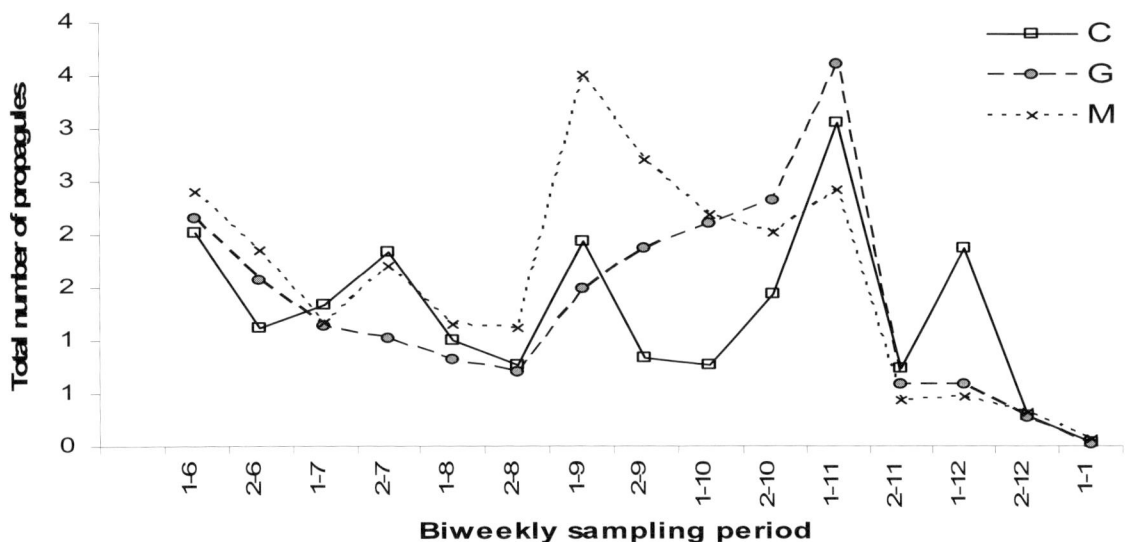

Overall, these results show that dehesa cultivation favours acorn production and that reduced grazing causes the opposite effect. These effects have been assumed by many authors but our data are the first supporting evidence. The effect of cultivation may be explained by increased water infiltration since water stress has been shown to favour acorn abortion in holm-oaks (Siscart *et al.*, 1999). This explanation would be also valid for the negative effect of shrub competition in M plots. The detrimental effect of shrub encroachment was also found for acorn viability as a consequence of increased abortion and infestation rates (Pulido, 1999). Finally, the effect of cultivation on acorn viability remains unclear, as the grazed treatment showed significantly higher values of viability.

Figure 2. Viability of acorn production according to management types. Bars show the mean percentage of acorns in each viability class for cultivated (C), grazed (G) and shrubby (M) dehesa plots.

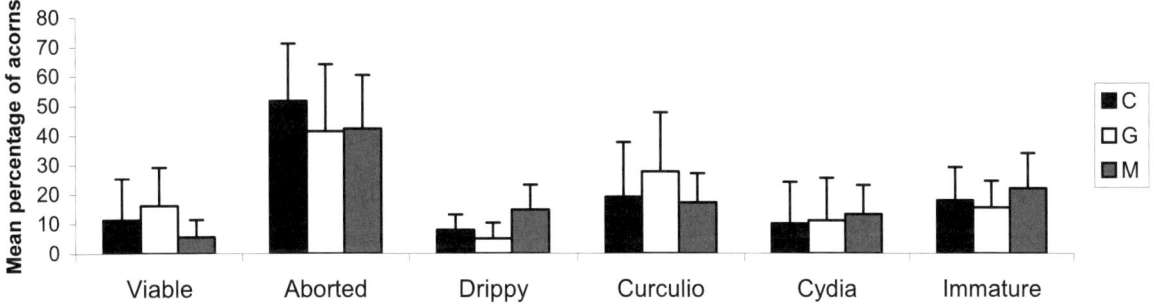

References

Pulido, F.J. (1999) Herbivores and regeneration of holm-oaks in forests and dehesas (Herbivorismo y regeneración de la encina en bosques y dehesas). MSc thesis. Universidad de Extremadura, Badajoz, Spain. (In Spanish.)

Pulido, F.J. and Díaz, M. (2002) Natural dynamic of regeneration of holm and cork-oaks (Dinámica natural de la regeneración del arbolado de encina y alcornoque). In: Pulido, F.J., Campos, P. and Montero, G. (eds) *La Gestión Forestal de las Dehesas. Historia, Ecología, Selvicultura y Economía.* Instituto del Corcho, la Madera y el Carbón, Junta de Extremadura, Mérida, Spain, pp. 39-62. (In Spanish.)

Martín-Vicente, A., Infante, J.M., García-Gordo, J., Merino, J. and Fernández-Alés, R. (2000) Acorn production in forests and dehesas in south-western Spain (Producción de bellotas en montes y dehesas del suroeste español). *Pastos* 28, 237-248. (In Spanish.)

Siscart, D., Diego, V. and Lloret, F. (1999) Acorn ecology. In: Rodá, F., Gracia, C., Retana, J. and Bellot, J. (eds) *The Ecology of Mediterranean Evergreen Oak Forests.* Springer-Verlag. Berlin, Germany, pp. 75-87.

Vázquez, F.M. (1998) *Quercus Seeds. Biology, Ecology and Management* (Semillas de *Quercus*. Biología, ecología y manejo). Consejería de Agricultura y Comercio, Badajoz, Spain. (In Spanish.)

Effect of fertilization, tree species, plantation density and sowing mixture on pasture production and soil characteristics in silvopastoral systems

A. Rigueiro-Rodríguez, E. Fernández-Núñez and M. R. Mosquera-Losada
Dept. of Plant Production, EPS, Universidad de Santiago de Compostela, 27002, Lugo, Spain. romos@lugo.usc.es; anriro@lugo.usc.es.

Abstract
The aim of the experiment was to test the effect of three fertilizer treatments (no fertilizer, organic fertilizer and inorganic fertilizer) on cocksfoot pasture established under two tree species *(Betula alba* and *Pinus radiata*) at two densities (2500 and 833 trees/ha). After 7 years the effects of fertilizer treatments on tree growth and pasture production are more important in plots with forest species with a high growth rate and planted at a high density; in these cases soil pH and fertility are altered. Organic fertilizer enhanced tree growth in sandy soils with a low organic matter content.

Introduction
Galicia is a region with a temperate climate and is influenced by the Atlantic Ocean; therefore it has warmer temperatures and mean rainfall figures of between 800 and 2500 mm in the coastal provinces. In Spain, forestry and dairy products are produced in distributed woods and pastures. It is also one of the regions in Europe with the greatest forest fire risk, and silvopastoral systems can reduce understorey fuel and therefore fire risk (see Rigueiro-Rodríguez *et al.*, this volume). Most of the productive forestland in Galicia is made up of species which grow quickly - *Pinus pinaster*, *Pinus radiata* and *Eucalyptus globulus*. However, some regional forestry strategies promote the use of broadleaved species such as *Quercus*, *Castanaea* and *Betula* in new plantations to improve the quality of forest products and because they enhance biodiversity and landscapes, enhancing the natural vegetation of the area. The aim of this study was to evaluate the effects of two different species, pine (*Pinus radiata*) and birch (*Betula alba*), at two densities on pasture and tree production and soil pH.

Materials and Methods
The experiment was established in the Spanish Atlantic area, in Lugo (NW Spain), in spring 1995. Treatments were established following a randomized block design with three replicates. Pine and birch trees were planted at a density of 833 (3 m × 4 m) and 2500 (2 m × 2 m) trees/ha on abandoned farmland with an initial soil pH of 6.8. Each plot consisted of 25 trees planted in a square of 5 × 5 trees. Pasture was established after soil preparation (clearing and ploughing) with 25 kg/ha *Dactylis glomerata* var. Saborto + 4 kg/ha *Trifolium repens* var. Ladino, 1 kg/ha *Trifolium pratense* var. Marino. Fertilizer treatments consisted of no fertilizer (NF), mineral fertilizer (M) (application of 500 kg/ha of compound fertilizer 8:24:16 (March) and 40 kg N/ha (May)) and organic fertilizer (L), with 154 m^3/ha of dairy sewage sludge only at establishment (giving a total input of 160 kg N, 85.9 kg P_2O_5 and 23.4 kg K_2O per ha). No fertilizer was applied in 1996 and 1997 in the organic treatment, but the same doses used with M treatment were applied from 1998 on. Each year, soil pH was measured (analysed in water at 1:2.5), pasture production was estimated by harvesting an area between the four inner trees in the experimental plots, three times in spring and once in autumn, with the exception of the first year, when two harvests, one per season, were made. The height and diameter of the nine inner trees of each experimental plot were determined annually by using a calliper and meter. This chapter presents the results 7 years after planting. ANOVA was used for statistical analyses and the Duncan test for mean separation.

Results and Discussion
Density affected tree height and diameter in a different way in both species (Figure 1). Pines were higher when planted at 2 m × 2 m, compared to low density, due perhaps to the more intense aerial competition for light. However, *Pinus* diameter was higher at low densities than high densities because tree competition for light, water and nutrients was lower. However, *Betula* height and diameter were higher at low densities than at high densities, possibly due to the preference for less intensive planting shown by this species and the lack of aerial competition at high density. This can also probably be explained by the fact that birch is well adapted to grow at high density, as happens in natural stands. Height was affected by treatments in the same way as diameter, with the exception of birch diameter at both densities, which was not modified by treatment. Pine height and diameter were larger than in birch due to the different growth rates of these two species. Both species responded differently to fertilizer treatment. Height of pines was reduced with inorganic fertilizer at both densities, but this treatment increased birch height at a high density, compared to no fertilizer treatment, probably due to different growth rates.

Figure 1. Pasture production (bars), tree height (continuous line) and diameter (discontinuous line) at high and low density for the different treatments evaluated. L: organic fertilization M: mineral fertilization and NF: no fertilization). Different letters indicated significant differences between fertilizer treatments for pasture production for both tree species and in each tree species for tree height and diameter.

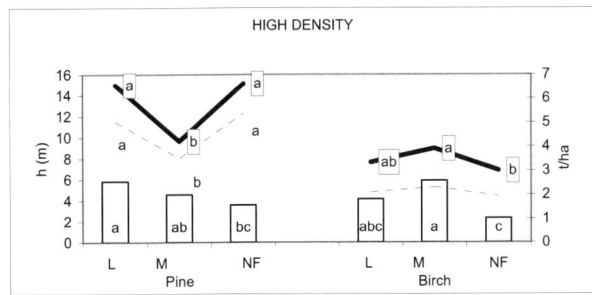
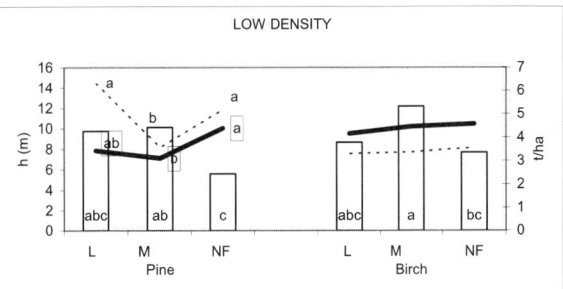

Organic fertilizer enhanced pine height and diameter at both densities and the same occurred with birch at high density. Pasture production was higher at low densities under both tree species. It was 59, 45 and 64% higher for L, M and NF treatments, respectively, under pine and 47, 48 and 30% under birch, respectively. This indicates the different effect of competition from birch and pine production. Tree canopy cover was higher under high densities for pine (123.0, 62.3 and 97.8 at high density, and 17.7, 25.3 and 27.7 at low density for L, M and NF treatments, respectively) and birch (70.5, 106.3 and 68.7 at low density, and 44.9, 60.7 and 51.0 at high density for L, M and NF treatments, respectively). This indicates the lower level of interception of light under the birch canopy. Pasture production was increased by fertilizer, the response being greater under low densities, due to the higher light input. An interaction was found between pasture and tree production and fertilizer treatment. Organic fertilizer affected pasture production and tree growth positively, as did mineral fertilizer under birch. However, inorganic fertilizer reduced pine growth. This is because of a lower level of competition between the tree and grass in birch than in pine, due to the lower growth rate of the former. However, when milk sewage sludge was used it increased pine growth, probably due to the slow nutrient release from this kind of fertilizer, and the increase in soil organic matter content, which enhances water retention in sandy soils and allows tree growth during the summer (Rigueiro-Rodríguez et al., 2000). Soil pH reduction (Figure 2) since the start of the experiment is the result of cation extraction of crops. Overall productivity (tree and pasture) was lower under mineral treatments, which made pH reduction slower. Birch growth was not affected by fertilizer treatments at low densities and for this reason had no effect on soil pH.

Figure 2. Organic matter (bars) and water soil pH (continuous line) at high and low density for the different treatments evaluated. L: organic fertilization M: mineral fertilization and NF: no fertilization). Different letters indicated significant differences between fertilizer treatments for both tree species.

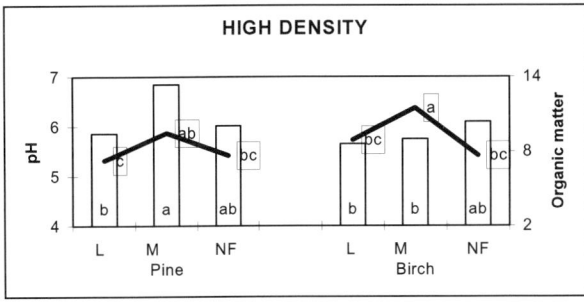
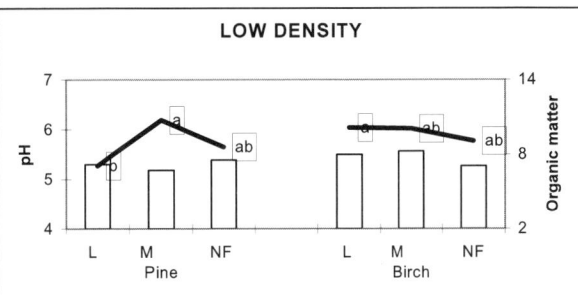

Conclusions

The effects of fertilizer treatments on tree growth in silvopastoral systems are more important in species which grow fast at high densities. Such species have a greater effect on soil pH and fertility. Organic fertilizer will enhance tree growth in sandy soils with a low organic matter content and water retention capacity.

Acknowledgements

We are grateful to CICYT, XUNTA de Galicia, GESTAGUA S.A and AGROAMBPRODALT for financial assistance, to Escuela Politécnica Superior for facilities, to Divina Vázquez Varela, José Javier Santiago-Freijanes, Teresa Piñeiro López, Antonio Rodríguez Rigueiro for helping in processing, laboratory and field.

Reference

Rigueiro-Rodríguez, A., Mosquera-Losada, M.R. and Gatica-Trabanini, E. (2000) Pasture production and tree growth in a young pine plantation fertilized with inorganic fertilizers and milk sewage sludge in north-western Spain. *Agroforestry Systems* 48, 245-256.

Structure characterization of *Quercus pyrenaica* Willd open woodlands in the Spanish Central Mountain: implications for silvopastoral management

S. Roig, M. Río, I. Cañellas, A. Bravo and G. Montero
Centre for Forestry Research (CIFOR-INIA), Ctra. A Coruña km 7,5, 28040 Madrid, Spain. sroig@inia.es

Abstract
This chapter presents a preliminary characterization of *Quercus pyrenaica* Willd. open woodlands in Madrid province. From data from the second and third National Forest Inventory we compare main stand variables (N, Dg, Hm, Ho, G, V and Hart-Becking index) at the two inventories for open woodlands with low density (N <100 trees/ha). A trend in recruitment of new trees from natural regeneration or growth of coppices under the main stand has been found, as well as a reduction in density of old dehesas of large trees.

Key words: dasometric characterization, grazing, recruitment

Introduction
Quercus pyrenaica Willd. (rebollo oak) open woodlands are extensive silvopastoral systems in many acid soil areas of Spain, of which the Spanish Central Mountain stands are a good example. The traditional management of these stands was aimed at producing diverse products: wood, pasture, timber…, mainly wood and charcoal, which resulted in coppice as the most used silvicultural treatment. Nowadays interest in these products has changed and the aim of rebollo oak open woodlands is mainly for maintaining landscape, recreation areas, ecological functions and some production functions, mainly livestock grazing. Changes in use and the difficulty of defining silvicultural managements to reach transformation of coppices have caused diverse ecological, silvicultural and socio-economic problems as regeneration is lacking, growth ceases, stumps decay, forest fires are a high risk, livestock causes damage, etc. (Serrada et al., 1991; González del Tánago et al., 2000). In order to build silvicultural rules for this species, the objective of this work is to characterize the structure of open woodlands of rebollo oak in Madrid estate and analyse the change over 10 years.

Materials and Methods
Data from the second and third National Forest Inventory (ICONA, 1990) from the Madrid estate have been characterized dendrometrically. To select second NFI plots and manage data of the forest inventory, BASIFOR software (Río et al., 2002) was used. We selected the plots with at least one tree with dbh bigger than 7.5 cm (191 plots in Madrid) of *Quercus pyrenaica* and N < 100 trees/ha for dehesa stands (41 plots) (Serrada et al., 1991). Plots were classified as pure (basal area (G) of *Q. pyrenaica* > 90%, 29 plots), mixed dominated by rebollo oak (90% > G *Q. pyrenaica* > 50%, four plots) and mixed not dominated by rebollo oak (G *Q. pyrenaica* < 50%, six plots). A dendrometric characterization was done for the main variables for each stand type: number of stems (N); quadratic diameter (Dg); mean height (Hm); dominant height (Ho), calculated as the mean height of the 30 largest trees per hectare; basal area (G); volume (V) and Hart-Becking index (I_H). The same stand variables were calculated for the third NFI for the same selected plots that we studied at the second NFI and all dehesa-like plots (N < 100 trees/ha). To calculate volumes, equations from the NFI were used.

Results and Discussion
Rebollo oak woodlands on the Spanish Central Mountain are mainly pure stands of low height. Stands can be very diverse (Table 1); the most common structure is coppice or coppice with standards (Dg = 25.5 cm) with small diameters and variable density, but we can also find old and large trees in open woodlands (D = 100 cm, G = 15 m^2/ha, V = 63 m^3/ha). Sometimes, rebollo oak forms mixed stands with *Pinus sylvestris*, *Pinus pinea*, *Quercus ilex*, *Fraxinus* sp or *Populus nigra*, showing the wide ecological range of the species. Canopy cover is about 50% (51% for pure, 47% for mixed dominated by rebollo stands, and 53% in mixed but not dominated by rebollo stands).

Table 1. Mean, minimum and maximum values of mean dasometric variables for pure and mixed woodlands of *Quercus pyrenaica* in Central Mountain (second NFI) in Madrid estate. [a] Mean height of the largest 30 trees/ha. (1) Dominated by *Q. pyrenaica* stands. (2) Stands not dominated by *Q. pyrenaica*. In parentheses, number of plots. Variables defined in the text.

	N (trees/ha)	Dg (cm)	Hm (m)	Ho[a] (m)	G (m^2/ha)	V (m^3/ha)	I_H (%)
Pure stands (29 p)	49.6	25.5	9	9	2.3	7.9	201.7
	5.1-95.5	12.8-65.9	5.5-16	5.5-16	0.4-15.6	1.1-63.9	86.4-509
Mixed stands (1) (4 p)	81.3	21	8.1	9.5	3	11.3	163.4
	36.9-99	14.3-29.1	5.8-12.3	6-13.3	1.3-6.6	2.9-21.6	94.7-260
Mixed stands (2) (6 p)	59.9	40.7	10.2	11	8.4	21.2	157.8
	29.4-79.2	28.7-52.1	5-19.9	5-20	2.2-16.6	7.9-36	58.3-238

We found 64 plots in the third NFI of rebollo oak open woodlands (Table 2). 62.5% of them are plots that did not appear in the selection of the second NFI; half of these (23) are plots with small diameters, with no trees (D > 7.5 cm) at second NFI. The rest of these new open woodland plots are the result of density reduction in old stands (large diameters) due to natural mortality or silvocultural treatments (thinning, plant health treatments, etc.) that have been classified as dehesa-like stands at the third NFI but not at the second. From the plots considered as open woodlands at the second NFI, 38% have strong recruitment of trees with diameter under 7.5 cm, resulting in them actually having higher densities (N > 100 trees/ha).

The 69.2% of the open woodlands plots of 2nd NFI that remained classified as open woodlands at the 3rd NFI have clearly had a positive increasement in G, Dg, Ho and N, but on the other hand the 22.2% of open woodlands plots a N slightly decrease has been observed. In the open woodland plots of 2nd NFI that were not considered as dehesa-like stands at 3rd NFI we have observed an increment in G and N values and Dg clearly due to the small diameter tree recruitment.

Recruitment of small trees can be the result of processes of regeneration of the stands or growth of the coppices underneath the main stratum of larger trees due to changes in use. Changes in grazing management practices can cause severe damage to woody vegetation in overgrazed zones, and loss of herbaceous pastures and recovery of woody vegetation in lightly grazed zones (González del Tánago et al., 2000).

Table 2. Mean, minimum and maximum values of mean dasometric variables for pure and mixed woodlands of *Quercus pyrenaica* in Central Mountain (third NFI) in Madrid estate. A: low density stands at inventories 2NFI and 3NFI; B: low density stands only at second NFI; C: low density stands only at third NFI. [a] Mean height of the largest 30 trees/ha. (1) Dominated by *Q. pyrenaica* stands. (2) Stands not dominated by *Q. pyrenaica*. In parentheses, number of plots. * Damaged or very old trees do not have volume equation at NFI. Variables defined in the text.

		N (trees/ha)	Dg (cm)	Hm (m)	Ho[a] (m)	G (m^2/ha)	V (m^3/ha)	I_H (%)
Pure stands	A (16 p)	57	27.5	9	9.1	2.8	8.2	187.2
		8-95.5	14.5-64	6-11.7	6-12.5	0.5-6.4	1.3-19.9	109-381
	B (13 p)	492.5	17.0	9.4	10.8	8.5	28.2	70.6
		102-1539	9.4-28.4	6.3-15.2	6.8-18.7	3-30.5	9.6-107.1	33.7-141
	C (40 p)	47.9	29.0	10.1	10.2	2.7	8.8	196.6
		8-95.5	9.9-72.1	5-17.6	5-17.6	0.5-12.3	0*-70.7	64-448
Mixed stands (1)	A (3 p)	62.8	33.5	9.9	10	3.7	9.3	207.4
		8-95.5	17.2-50.5	7.7-11.9	7.7-12.3	1.6-7.2	2.4-18.6	98-381
	B (1 p)	456.4	13.4	10.4	13.8	6.4	26.7	48.6
		-	-	-	-	-	-	-
Mixed stands (2)	A (5 p)	18.7	33.3	11.2	11.2	1.2	4.4	268.7
		8-31.8	19.4-48	6-18	6-18	0.9-1.4	1.9-10.2	185-381
	B (1 p)	127.3	16.4	8.3	8.4	2.7	9.6	114.4
		-	-	-	-	-	-	-

Conclusions

In spite of the lack of management of coppices or woodlands of rebollo oak in recent decades, there is a clear growth in density of these stands and we do not have evidence of stand stagnation. In some cases of old and large tree woodlands (between the 25 and 30% of dehesa-like plots at the third NFI) there has been a reduction in density (N, G) due to different reasons into which it would be interesting to research. It is necessary to develop suitable management techniques for both types of stands to guarantee the persistence and stability of those forestry formations.

Acknowledgements

This work was carried out within the framework of the project 07M-0035-2002.

References

González del Tánago, M., Roig, S., Antón, N., Fernández, M. and Pérez-Carral, C. (2000) Grazing effects on soil infiltration and forest recovery in Canencia Mountains (NE Madrid, Spain). In: Rubio, J.L.., Morgan, R.P.C., Asins, S. and Andreu, V. (eds) *Proceedings III International Congress of the European Society for Soil Conservation*. Geoforma, Logroño, Spain.

ICONA (1990) *Second National Forest Inventory. Explanations and Methods, 1986-1995 (Segundo Inventario Forestal Nacional. Explicaciones y métodos, 1986-1995)*. ICONA, Madrid, Spain. (In Spanish.)

Río, M., Rivas, J., Condés, S., Martínez Millán, J., Montero, G., Cañellas, I., Ordóñez, C., Pando, V., San Martín, R. and Bravo, F. (2002) BASIFOR: computer application for managing the databases of the Second NFI (BASIFOR: aplicación informática para el manejo de bases de datos del Segundo IFN). In: Bravo, F., Río, M. and del Peso, C. (eds) *El Inventario Forestal Nacional. Elemento Clave para la Gestión Forestal Sostenible*. Ministerio de Medio Ambiente y Universidad de Valladolid, Madrid, Spain. (In Spanish.)

Serrada, R., González Doncel, I., López Peña, C., Marchal, B., San Miguel, A. and Tolosana, E. (1991) *Dasometric Characterization of "Rebollares" (Quercus pyrenaica Willd.) in the Region of Madrid. Silvopastoral Alternatives. Design of an Experimental Plan* (Tipificación dasométrica de los rebollares (*Quercus pyrenaica* Willd.) de la Comunidad de Madrid. Alternativas silvopastorales. Diseño de un plan experimental). EUIT Forestal and ETSI Montes, UPM, Madrid, Spain. (In Spanish.)

Pasture production under different tree species and spacing in an Atlantic silvopastoral system

M. J. Rozados [1], M. P. González-Hernández [2] and F. J. Silva-Pando [1,2]
[1] Forestry and Environmental Research Centre Lourizán, CDS-Medio Ambiente-Xunta de Galicia, Apdo. 127, CP 36080 Pontevedra, Spain. [2]Dept of Crop Production, Santiago de Compostela University, CP 27002, Lugo, Spain. mjrozados.cifal@siam-cma.org, pilargh@lugo.usc.es

Abstract
The effect on pasture production of six different tree densities and six species (*Pseudotsuga menziesii*, *Pinus pinaster*, *Pinus radiata*, *Betula celtiberica*, *Quercus rubra* and *Castanea sativa*) was investigated during 3 years. Under *P. menziesii* and *P. pinaster*, pasture production decreases starting from 1048 trees/ha, while under *P. radiata* decreases appeared at 424 trees/ha. All the deciduous species studied showed a density effect from 2000 trees/ha. Pasture production was studied under different tree canopies to determine the relationship between leaf area index (LAI) and herbage production. At low tree spacing, 2 m × 2 m, LAI was negatively correlated ($r^2 = 72.4\%$) with annual pasture production.

Key words: canopy, tree spacing, herbage production, leaf area index, temperate area

Introduction
There is uncertainty concerning the extent to which herbage yield and quality are affected by tree spacing (Burner and Brauer, 2003). Recently, important contributions to tree-pasture relationships in a temperate climate were published, mainly under conifers (Fernández *et al.*, 2002; Silva-Pando *et al.*, 2002; Burner and Brauer, 2003), and under hardwood species (González-Hernández and Silva-Pando, 1996; González-Hernández *et al.*, 1998; Silva-Pando *et al.*, 1998; Teklehaimanot *et al.*, 2002). This study was conducted to provide information about pasture production under different tree species and its relationship with different densities and tree canopy. Enhancement of silvopastoral systems could be achieved combining tree species and adequate spacing, looking for an equilibrium between pasture production and tree growth.

Materials and Methods
The site is located in Pontecaldelas, Pontevedra, NW Spain (42°22'N, 8°29'W, 386 m asl), and has a temperate and humid climate. The substrate is granitic and the soil is a sandy brown acidic earth. Overall chemical and physical properties of the top 0.20 m before planting were: 0.602% total N; 17.1% OM; 11.8 mg/kg available P (Bray II); 27, 22 and 74 mg/kg exchangeable Ca, Mg and K, respectively (ammonium acetate extraction and flame atomic spectrophotometry), pH 4.6; 81.5% sand; 10.25% silt; 8.25% clay. Five tree species were planted in 1996, *Pseudotsuga menziesii*, *Pinus pinaster*, *Pinus radiata*, *Quercus rubra* and *Castanea sativa*, while *Betula celtiberica* was planted in 1997, at different densities in a parallel row design with systematic sampling, in an area of 0.8 ha. Densities were 222, 256, 424, 518, 1048, 2000 and 2500 stems/ha, corresponding to 7 m × 8 m, 6 m × 7 m, 5 m × 6 m, 4 m × 5 m, 3 m × 4 m, 2 m × 3 m and 2 m × 2 m. A mixture of *Dactylis glomerata* var. 'Artabro' and *Trifolium repens* var. 'Huia' was sown in early spring. Yearly, a first dose of a granular fertilizer was applied at the rate of 5 kg N/ha, 40 kg Ca/ha, 1 kg Mg/ha and 16 kg P/ha in March, and a second dose was applied in May at the rate of 5 kg N/ha, 2 kg Ca/ha and 1 kg Mg/ha. In 2003, granular fertilizer was replaced by three-slow release tablets per tree (Medramás® 11-18-11-4 MgO). Leaf area index was measured with LAI2000 (Licor, USA), during a cloudy homogeneous day in August 2002, 30 cm above ground, between tree rows 2 m × 2 m spaced.

Table 1. Root collar diameter (d) and mean height (h) in year 2000, excluding border trees.

Stems/ha	P. menziesii		C. sativa		P. radiata		B. celtiberica		Quercus rubra		P. pinaster	
	d (cm)	h (m)	d (cm)	h (m)	d (cm)	h (m)	d (cm)	h (m)	d (cm)	h (m)	d (cm)	h (m)
222	5.3	2.1	2.9	1.7	15.8	5.3	5.5	2.9	1.6	1.1	13.0	3.9
256	3.4	1.2	4.0	1.8	15.9	5.7	5.7	3.3	1.6	1.3	6.9	2.4
424	5.4	1.6	3.5	1.7	15.9	5.7	5.3	2.8	2.3	1.5	10.6	3.3
518	4.1	1.1	3.7	2.1	17.1	6.2	5.1	2.5	1.9	1.6	11.0	3.7
1048	5.5	1.6	3.6	2.1	15.3	6.0	4.0	2.4	2.2	1.8	7.8	2.9
2000	4.6	1.5	3.8	2.0	12.8	6.0	3.3	1.8	1.9	1.5	10.2	3.4
2500	4.0	1.4	4.9	2.4	12.4	6.3	3.1	1.9	1.8	1.3	8.8	3.4

Herbage yield was measured by harvesting four quadrats 30 cm × 30 cm between two consecutive tree rows, then drying in an air-forced oven at 80°C to determine dry matter. Quadrats were harvested in June and December 2000,

2001 and 2002. Inmediately after June sampling, all the area was cut, to prevent the shadow effect of tall grasses on the quadrat production. Root collar diameter and height were measured in 2000. Pasture data were analysed using PROC GLM (SAS Institute, 1990) as a repeated measured analysis of variance, with sampling date as a within factor and species and densities as between factors. Regression was made to correlate LAI values with pasture production. The possibility of grouping evergreen (*P. radiata*, *P. pinaster* and *P. menziesii*) and broadleaved (*Q. rubra*, *B. celtiberica* and *C. sativa*) species for statistical treatment to avoid significant species × density interactions was tested.

Results and Discussion

Herbage annual production ranged from 5.18 to 10.67 t/ha (Figure 1). Lower values, considering overall densities, were observed under *P. radiata*. In 2001, some *P. pinaster* trees died and gaps appeared in the canopy, yielding the highest values shown.

Figure 1. Mean annual pasture production over all densities in each species stand.

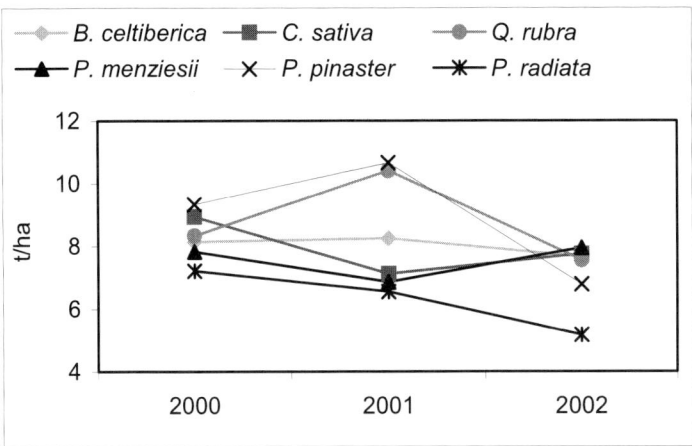

By density, average pasture production over all 3 years, showed similar patterns. Under *P. menziesii* and *P. pinaster*, pasture production decreases started from 1048 trees/ha, while under *P. radiata* decreases appeared at 424 trees/ha.

Figure 2. Mean pasture production over all years in each species stand.

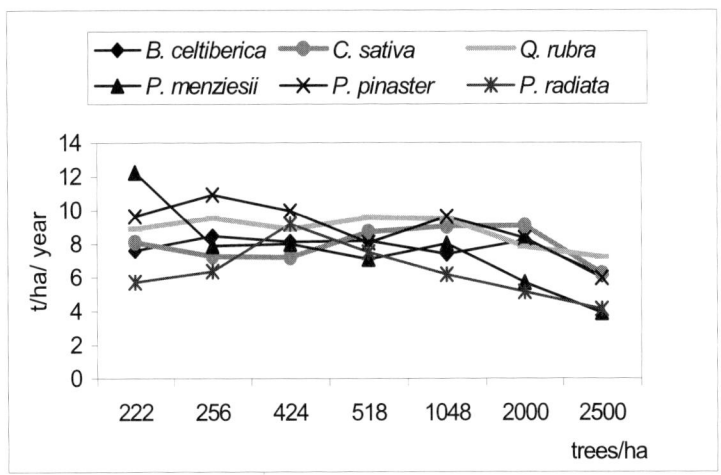

All the deciduous species studied showed density effect at 2000 trees/ha (Figure 2). Significant differences among species were observed at the highest and lowest densities, while no differences were found for intermediate values (Table 2).

Table 2. Species effect separated by spacing on each year. * $P < 0.05$, ** $P < 0.01$, *** $P < 0.001$, ns $P > 0.05$.

Tree spacing m	2 x 2	2 x 3	3 x 4	4 x 5	5 x 6	6 x 7	7 x 8
Density trees/ha	2500	2000	1048	518	424	256	222
2000	ns	*	ns	ns	ns	ns	*
2001	**	*	***	ns	ns	***	*
2002	ns	ns	*	ns	ns	ns	***

Differences among densities were detected in pasture production under evergreen species. Production under deciduous species seemed to be slightly affected by the stand density, probably due to slower growth and then smaller stems, compared with conifer species (Table 3). Burner and Brauer (2003) found that loblolly pine spacing, at the fifth to sixth growing season, affected herbage productivity at densities lower than 840 trees/ha, while Pearson *et al.* (1995) have not found significant decreases in herbage yield with loblolly pine densities 4451 trees/ha 10 years after planting.

Table 3. Species effect separated by species on each year. * $P < 0.05$, ** $P < 0.01$, *** $P < 0.001$, ns $P > 0.05$.

Species	Betula celtiberica	Castanea sativa	Quercus rubra	Pseudotsuga menziesii	Pinus pinaster	Pinus radiata
2000	ns	ns	ns	**	ns	*
2001	ns	ns	*	*	***	*
2002	*	ns	ns	***	*	ns

Figure 3. Leaf area index measured in August 2002 against annual production in the same year, under trees planted at a spacing of 2 m × 2 m with fitted regression line.

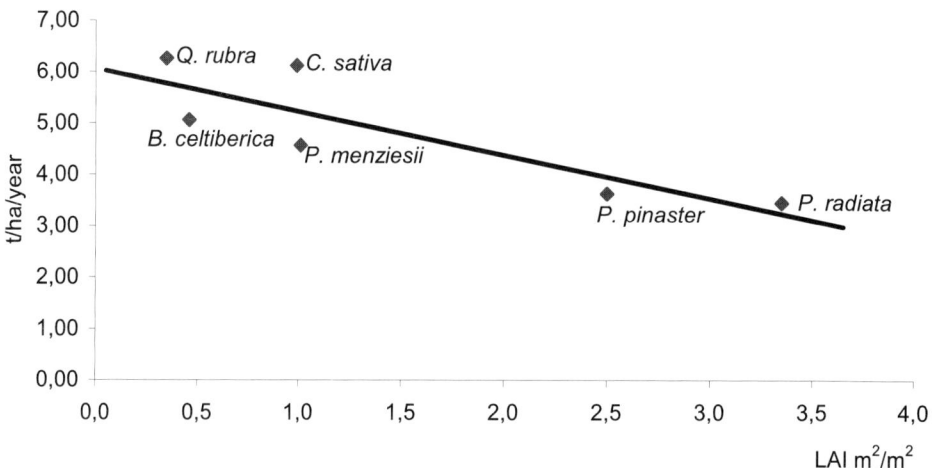

The linear regression of pasture production on LAI explains most of the variation in our data (Figure 3). The fitted line is: pasture production = 6.065 – 0.843 LAI ($r^2 = 72.4\%$, $P < 0.0316$).

Conclusions

In a temperate Atlantic area, pasture production under different tree species is inversely correlated with leaf area index, at the sixth growing season. Under conifers, results suggested that spacing of 4 m × 5 m (*Pinus radiata*) and 5 m × 6 m (*Pinus pinaster* and *Pseudotsuga menziesii*) maintained the most consistent pasture production over time. Generally, over all broadleaved species studied, none of the differences in tree spacing significantly reduced pasture production.

Acknowledgements

This research was funded through INIA SC96-032 and INIA PD99-002 projects. The authors thank M. Alonso and F. Ignacio for soil analysis and Jose Ríos, Aurea Pazos and Kike Diz for excellent assistance in the field.

References

Burner, D.M. and Brauer, D.K. (2003) Herbage response to spacing of loblolly pine trees in a minimal management silvopasture in southeastern USA. *Agroforestry Systems* 57, 69-77.

Fernández, M.E., Gyenge, J.E., Dalla Salda, G. and Schlichter, T.M. (2002) Silvopastoral systems in northwestern Patagonia I: growth and photosynthesis of *Stipa speciosa* under different levels of *Pinus ponderosa* cover. *Agroforestry Systems* 55, 27-35.

González-Hernández, M.P. and Silva-Pando, F.J. (1996) Grazing effects of ungulates in a Galician oak forest (northwest Spain). *Forest Ecology and Management* 88, 65-70.

González-Hernández, M.P., Silva-Pando, F.J. and Casal Jiménez, M. (1998) Production patterns of understorey layers in several Galician (NW Spain) woodlands. *Forest Ecology and Management* 109, 251-259.

Pearson, H.A., Wolters, G.L., Thill, R.E., Martin, A.J. and Baldwin, V.C. (1995) Plant response to soils, site preparation, and initial pine planting density. *Journal of Range Management* 48, 511-516.

SAS Institute Incorporated (1990) *SAS/STAT User's Guide, Version* 6, 4th edn. Cary, NC, USA.

Silva-Pando, F.J., González-Hernández, M.P. and Rigueiro, A. (1998) Grazing by livestock in pinewood and eucalyptus forests: multiple use in northwest Spain. *Agroforestry Forum* 9 (1), 36-43.

Silva-Pando, F.J., González-Hernández, M.P. and Rozados, M.J. (2002) Pasture production in a silvopastoral system in relation with microclimate variables in the Atlantic coast of Spain. *Agroforestry Systems* 56, 203-211.

Teklehaimanot, Z., Jones, M. and Sinclair, F.L. (2002) Tree and livestock productivity in relation to tree planting configuration in a silvopastoral system in North Wales, United Kingdom. *Agroforestry Systems* 56, 47-55.

Effects of dairy sludge application on pasture production and arbuscular mycorrhizae in a sown meadow established in a hill soil

M. J. Sainz [1], X. A. Alonso [1], A. Vilariño [2] and M. E. López-Mosquera [1]
[1]Plant Production Dept. Santiago de Compostela University, CP 27002, Lugo, Spain. [2]Instituto de Investigaciones Agrobiológicas de Galicia, CSIC, POB 122, CP 17580, Santiago de Compostela, Spain. mjsainz@lugo.usc.es, ellomo@lugo.usc.es, AnVilar@iiag.cesga.es

Abstract
A field experiment was carried out during 2 years to study the effects of applying a pelletized biofertilizer containing arbuscular mycorrhizal (AM) fungal propagules on forage production and arbuscular mycorrhiza formation in a sown meadow in a *Quercus/Betula* forest under the following treatments: control, NPK mineral fertilization and dairy sludge application. In the spring when silage cuts were made, the highest forage yield was achieved under the dairy sludge treatment. The biofertilizer did not affect forage production in any treatment in the first year, but increased yields in the dairy sludge plots in the second. Colonization of roots by AM fungi was high in all treatments and was not affected by AM pellets.

Key words: botanical composition, forage production, silvopasture

Introduction
Most plants form arbuscular mycorrhizae, a symbiosis between their roots and certain fungi that enhances plant growth, health and mineral nutrition (mainly P), particularly in soils with low or moderate fertility (Marschner and Dell, 1994). In Galicia (NW Spain), establishment of sown meadows in hill soils usually requires intensive tillage and fertilization, which can have detrimental effects on survival of indigenous AM fungi (Thompson, 1994). The inoculation of effective AM fungal isolates might be a useful tool to re-establish soil mycorrhizal potential in silvopastoral systems. The aim of this work was to study the effects of applying a pelletized biofertilizer containing AM fungal propagules on forage production and arbuscular mycorrhiza formation in a sown meadow under mineral and dairy sludge fertilization.

Materials and Methods
A field experiment was carried out in 1998 at Goiriz-Lugo, NW Spain (43° 19' 13" N, 7° 36' 19" W; 530.6 m asl). Mean annual maximum temperature was 17.2°C, mean annual minimum temperature 6.3°C, and mean annual precipitation 1300 mm. The study was established on a forest of *Quercus robur* L. and *Betula* sp, with shrubs (*Ulex* sp, *Cytisus* sp and *Erica* sp) and a herbaceous cover. One hectare was cleared and cultivated. The soil was a humic Umbrisol with the following characteristics: pH 4.99, OM 25.5%, 15.4 mg Olsen-P/kg, 0.46 cmol K/kg, 0.96 cmol(+) Ca/kg, 2.8 cmol(+) Mg/kg, and 3.0 cmol(+) Al/kg (cations were extracted by 1M NH_4Cl, following Peech *et al.,* 1947).

Plots (4 m^2) were established away from the trees in the cleared area, limed and three fertilization treatments (eight plots per treatment) applied: an unfertilized control, mineral NPK fertilization, and dairy sludge application. The mineral fertilization consisted in 500 kg/ha of a 8:24:16 NPK fertilizer applied before autumn sowing, and fractionated doses of N applied 2 months before the spring silage cut (60 kg N/ha both in 1999 and 2000) and immediately after this cut (30 kg N/ha in 1999). The dairy sludge treatment was done by adding 80 m^3/ha of a dairy sludge (with total contents of 200 kg N/ha, 160 kg P_2O_5/ha and 26 kg K_2O/ha, of which 60% were supposed to be available for plant growth in the short term), 78 kg P_2O_5/ha and 60 kg K_2O/ha before sowing, and another 80 m^3/ha of dairy sludge (with total contents of 100 kg N/ha, 125 kg P_2O_5/ha and 20 kg K_2O/ha) in March 2000. For every fertilization treatment, half of the plots were randomly selected and inoculated with 120 sand-vermiculite based pellets (88 kg/ha) containing propagules of the AM fungus *Glomus macrocarpum*. Plots were seeded with a mixture of 20 kg/ha of *Lolium perenne cv* Nui, 10 kg/ha of *Lolium hybridum cv* Balto and 3 kg/ha of *Trifolium repens cv* Huia. In May 1999 and 2000, at the beginning of grass ear emergence, forage was mown in each plot to a height of 5 cm above ground level. Forage samples of 500-1000 g were taken for dry matter determination and to assess botanical composition of forage. Also root samples were collected to determine the extent of root colonization by AM fungi (Giovannetti and Mosse, 1980). Data were subjected to one-way analysis of variance.

Results and Discussion
In the spring silage cut, year 1, the highest forage yield was achieved under the dairy sludge treatment (Table 1), probably due to a good availability of nitrogen from this product. The biofertilizer did not affect forage production in any treatment.

Table 1. Forage production (kg DM/ha) in the spring silage cuts made in 1999 and 2000. For each year, values followed by a different letter are significantly different for $P \leq 0.05$.

		Fertilization treatment					
		Control	Control + AM pellets	Dairy sludge	Dairy sludge + AM pellets	Mineral	Mineral + AM pellets
Dry matter yield (kg/ha)	May 99	108.2a	89.4a	8882.0c	8661.8c	4776.1b	4022.4b
	May 00	178.1a	279.0a	2663.8c	3573.8d	906.4b	1059.1b

In the second year, dry matter yields were again highest in the dairy sludge treatment (Table 1); forage production was significantly increased by the application of AM pellets only under the dairy sludge treatment.

Table 2. Botanical composition of forage (%) in 1999 and 2000 spring silage cuts. For each column, values followed by a different letter are significantly different for $P \leq 0.05$. Colonization of roots by AM fungi was high in most treatments and was not affected by AM pellets (Figure 1).

	May 1999			May 2000		
Fertilization treatment	Grass	Clover	Weeds	Grass	Clover	Weeds
Control	100	0.0	0.0	58.0	4.7	37.3 b
Control + AM pellets	99.9	0.1	0.0	65.4	9.2	25.4 ab
Dairy sludge	97.1	2.4	0.5	76.2	11.4	12.4 a
Dairy sludge + AM pellets	97.3	2.3	0.4	74.7	11.4	13.9 a
Mineral	95.6	4.1	0.3	70.3	10.3	16.2 a
Mineral + AM pellets	97.3	0.9	1.8	73.0	8.2	*19.8 a*

In both years, botanical composition of forage was dominated by sown grasses in all cuts, white clover showing low percentages in all treatments, particularly in 1999; in the second year, a significant re-establishment of natural vegetation in control plots was observed (Table 2).

Figure 1. Percentage of AM root colonization in 1999 and 2000 spring silage cuts. For each year, bars followed by a different letter are significantly different for $P \leq 0.05$.

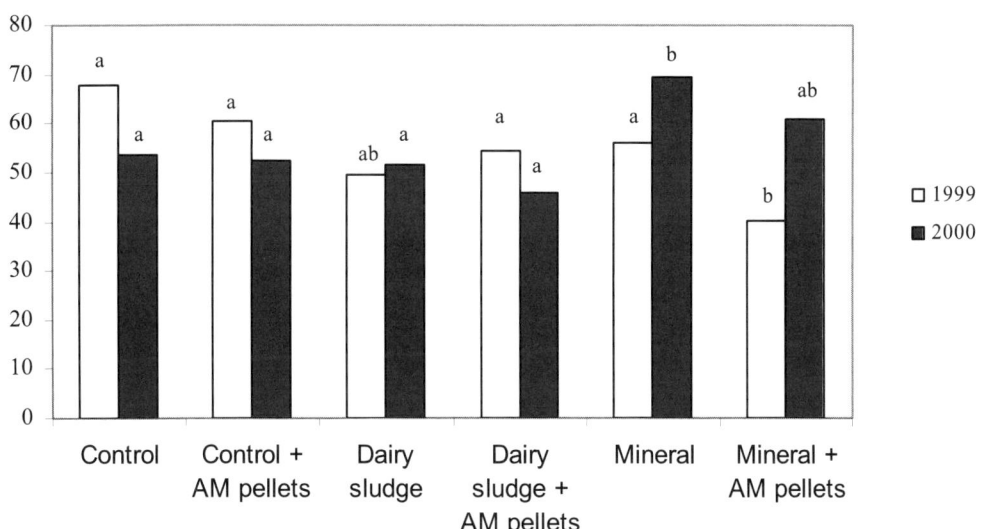

Conclusion

Dairy sludge, supplemented with P and K, increased pasture production compared to the mineral NPK treatment in silage cuts made in spring, probably because forage benefited from most of the nitrogen applied with this product. The inoculation of AM fungi can be effective to increase forage yield when using organic fertilizers within silvopastoral systems.

Acknowledgements

We are grateful to Moisés Carballeira for kindly allowing us to carry out our research at his farm. We thank Isabel Rivas, Cristina Vázquez and Susana Dopico for their skilful technical assistance. Financial support of this work by the Spanish Ministry of Science and Technology (projects AGF99-0418-C02-01 and AGF99-0418-C02-02), Xunta de Galicia (project PGIDT99PXI40001B) and FEDER project 1FD-0334 is acknowledged and appreciated.

References

Giovannetti, M. and Mosse, B. (1980) An evaluation of techniques for measuring vesicular-arbuscular mycorrhizal infection in roots. *New Phytologist* 84, 489-500.

Marschner, H. and Dell, B. (1994) Nutrient uptake in mycorrhizal symbiosis. In: Robson, A.D., Abbott, L.K. and Malacjzuk, N. (eds) *Management of Mycorrhizas in Agriculture, Horticulture and Forestry*. Kluwer Academic Publishers, Dordrecht, The Netherlands, pp. 89-102.

Peech, M., Alexander, L.T., Dean, L.A. and Reed, J.F. (1947) *Methods of Soil Analysis for Soil-Fertility Investigations*. USDA Circular no. 757, Washington, DC, USA.

Thompson, J.P. (1994) What is the potential for management of mycorrhizas in agriculture? In: Robson, A.D., Abbott, L.K. and Malacjzuk, N. (eds) *Management of Mycorrhizas in Agriculture, Horticulture and Forestry*. Kluwer Academic Publishers, Dordrecht, The Netherlands, pp. 191-200.

Forage yield and botanical composition of a hill sown meadow fertilized with dried pelletized broiler litter

M. J. Sainz, M. J. Bande, F. Cabaleiro and M. E. López-Mosquera
Plant Production Dept., Santiago de Compostela University, CP 27002, Lugo, Spain. mjsainz@lugo.usc.es, ellomo@lugo.usc.es

Abstract

A field experiment was established to study the effects of applying a dried pelletized broiler litter (BIOF) as fertilizer on forage production and botanical composition in a sown meadow located in a mountain area with stands of *Quercus robur*. Three fertilization managements were compared: (i) NPK fertilizer + cattle slurry application, (ii) 1000 kg/ha BIOF, and (iii) 2000 kg/ha BIOF. At the first cut, no significant differences were found either in dry matter forage production or botanical composition among treatments. However, when pasture was grazed in July, forage production was twice as high in plots fertilized with the highest dose of BIOF with respect to the other two treatments. A significant increase of white clover in forage was observed in the July grazing as a result of BIOF application, particularly when 2000 kg/ha were applied.

Keywords: grazing, pasture, silage cut, white clover

Introduction

Broiler litter (a mixture of manure, wasted feed, feathers and wood shavings or other crop residues) is widely applied as a fertilizer for forage used by cattle grazing (Pederson *et al.*, 2002). Contents of N, P and K in broiler litter is twice or three times higher than that in cattle and pig manure (Lampkin, 1994). Attention must be paid to the broiler litter dose and timing of application. When applied in excess of plant requirements for N, significant leaching of soil nitrate-N and accumulation of excessive levels of soil P will occur (Wood *et al.*, 1996). Conditioning of broiler litter by drying and pelletizing would facilitate handling and field application, minimizing problems found with fresh broiler litter management (John *et al.*, 1996). Litter obtained from broiler faeces and rice hull as bedding material has been dried and pelletized in a plant located in NW Spain to produce an organic fertilizer called BIOF. In this work, the effects of BIOF on forage production and botanical composition in a sown meadow established in a hill soil have been studied. Although variable in composition, pelletized broiler litter could be a good alternative to mineral or other organic fertilizers in sustainable silvopastoral systems.

Materials and Methods

A field experiment was established in Reigosa, located in a mountain area of NW Spain (42° 44' N, 7° 15' W; 831 m asl), mostly covered by stands of *Quercus robur*. The study was established in a meadow sown in 1996 with a mixture of *Lolium perenne* (10 kg/ha), *Dactylis glomerata* (10 kg/ha) and *Trifolium repens* (3 kg/ha), which had been managed by a silage cut at the end of spring and rotational beef grazing in summer and autumn, and fertilized with cattle slurry for the last 3 years (last application of 24,500 l/ha in October 2001).

In February 2002, the main characteristics of the soil were: pH 5.5, OM 5.2%, 10.3 mg Olsen-P/kg, 0.26 cmol(+) K/kg, 2.3 cmol(+) Ca/kg, 0.4 cmol(+) Mg/kg, and 0.2 cmol(+) Al/kg (cations were extracted by 1M NH_4Cl, following Peech *et al.* (1947)). In March, three treatments were applied in 250 m^2 plots: (i) 400 kg/ha of a 8:24:16 NPK fertilizer, (ii) 1000 kg/ha of the dried pelletized broiler litter BIOF, and (iii) 2000 kg/ha BIOF. Main properties of the BIOF fertilizer were: humidity 14.5%, OM 68.8%, N 3.5%, P 1.6% and K 2.6%. To obtain forage for silage, all plots were harvested at a 5 cm stubble height using a rotary mower on 20th June. A strip of 3 m^2 was cut at random from each plot, and used for dry weight and botanical composition asessment. At the end of July, forage regrowth was grazed down by Rubia beef cattle to a meadow surface height of 5 cm. In September plots of treatment i received 10,500 l/ha of cattle slurry. All plots were grazed again at the end of October. Just before grazing, four 0.1 m^2 quadrats were randomly harvested from each plot. Forage in each quadrat was cut with electric shears to a height of 5 cm for dry matter and botanical analysis. Data were subjected to one-way analysis of variance.

Results and Discussion

In the silage cut, no significant differences were found either in dry matter production or botanical composition among treatments (Figures 1 and 2). However, in July, forage production was two times higher in plots fertilized with the highest dose of BIOF than in the other two treatments. Despite cattle slurry being applied to the NPK-fertilized plots in September, no differences in yield were found with respect to the BIOF treatments in the forage cut in October.

Figure 1. Forage production (kg DM/ha) in 2002. For each cut, bars followed by a different letter are significantly different for $P \leq 0.05$.

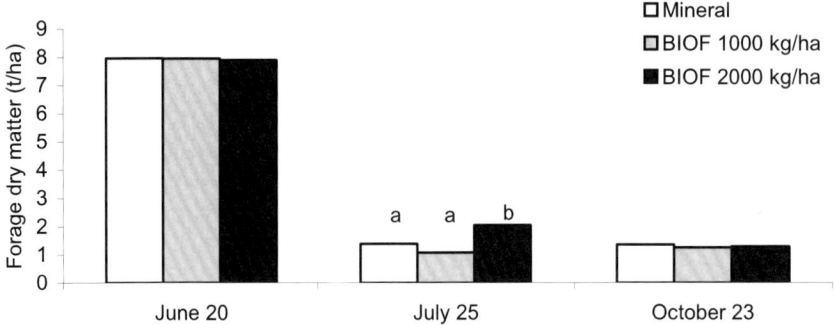

The low production in October can be largely attributed to weather conditions in the preceding months. For July, August and September, precipitation values were only of 22, 32 and 55 mm, respectively, whereas mean maximum temperature was 23.4, 21.9 and 23.3°C, and mean minimum temperature was 9.6, 8.1 and 10.1°C. Forage was mainly grasses in all cuts (Figure 2). A significant increase in white clover in forage was observed in the July grazing as a result of BIOF application, particularly when 2000 kg/ha was applied.

Figure 2. Botanical composition of forage in 2002. Within each cutting date and for each plant material, bars with a different letter are significantly different for $P \leq 0.05$.

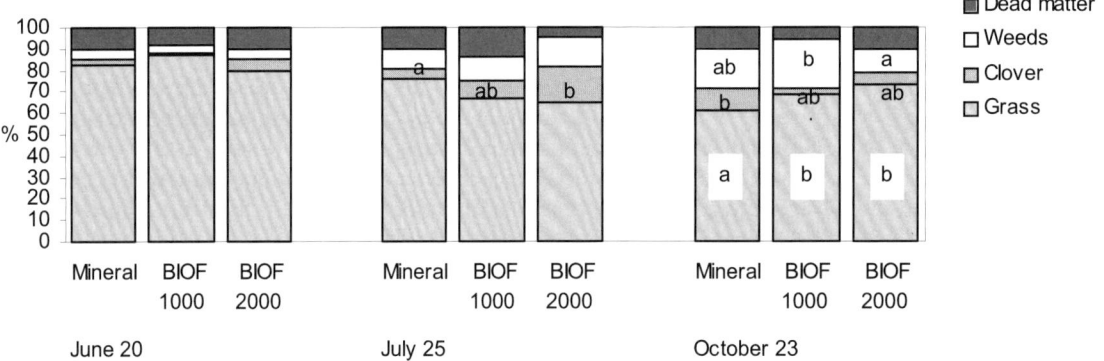

Conclusions
Pelletized broiler litter used as pasture fertilizer can result in similar forage yield to that achieved by mineral NPK + cattle slurry fertilization. Results suggest that nutrients needed for pasture growth are continuously released from the pelleted form of broiler litter, being available over the yearly forage production period.

Acknowledgements
We are grateful to Casa Arza for kindly allowing us to carry out our research at their farm. We also thank Susana Dopico and Cristina Vázquez for capable and skilful technical assistance. Financial support of this work by Xunta de Galicia (project PGIDT01AGR02E) is acknowledged and appreciated.

References

John, N.M., Adeoye, G.O. and Sridhar, M.K.C. (1996) Compost pelletization eases and use in Nigeria. *Biocycle* 37, 55-56.
Lampkin, N. (1994) *Organic Farming*. Farming Press, Tonbridge, United Kingdom.
Pederson, G.A., Brink, G.E. and Fairbrother, T.E. (2002) Nutrient uptake in plant parts of sixteen forages fertilized with poultry litter: nitrogen, phosphorus, potassium, copper and zinc. *Agronomy Journal* 94, 895-904.
Peech, M., Alexander, L.T., Dean, L.A. and Reed, J.F. (1947) *Methods of Soil Analysis for Soil-Fertility Investigations*. USDA Circular no. 757, Washington, DC, USA.
Wood, B.H., Wood, C.W., Yoo, K.H., Yoon, K.S. and Delaney, D.P. (1996) Nutrient accumulation and nitrate leaching under broiler litter amended corn fields. *Communications in Soil Science and Plant Analysis* 27, 2875-2894.

Establishment of a bovine/*Quercus* silvopastoral experiment in lowland Ireland

I. Short [1], J. McAdam [2], N. Culleton [1] and G. Douglas [3]

[1]Teagasc, Johnstown Castle Research Centre, Wexford, Co. Wexford, Ireland, [2]Dept. of Applied Biology, Queen's University, Belfast, N. Ireland, [3]Teagasc, Kinsealy Research Centre, Malahide Road, Dublin 17, Ireland.
IShort@johnstown.teagasc.ie; NCulleton@johnstown.teagasc.ie; jim.McAdam@dardni.gov.uk; GDouglas@kinsealy.teagasc.ie

Abstract

A silvopastoral experiment was established at Teagasc in County Wexford, Ireland, in 2002 with oak (*Quercus robur* L.) in an alley design and bovines. The experiment includes some treatments with trees produced with an enhanced root system (RPM). The treatments are: (1) control pasture plots; (2) RPM agroforestry (400 stems/ha); (3) conventional agroforestry (400 stems/ha); (4) RPM forestry (6600 stems/ha); and (5) conventional forestry (6600 stems/ha).

The trees were successfully established and cattle were successfully managed in combination with the trees. In the first year, height growth of bare-root oaks was significantly greater in the forestry treatment compared to the agroforestry treatment and, overall, RPM oaks were taller than bare-root plants. Among the RPM trees, the agroforestry system resulted in a greater stem diameter than those in the forestry plots. Height increment was greater for RPM trees than for bare-root trees.

Key words: bovine, system design, *Quercus*

Introduction

The planting targets in the *Strategic Plan for the Development of the Forestry Sector in Ireland* aim to increase the forest estate to 1.2 million hectares (17% of the land area) by 2030 (Anon, 1996). The level of the estate is currently 10%. Current afforestation rates are approximately 12,000 ha per year, most of which takes place on private farmers' land. Farmers in Ireland have not adopted agroforestry, possibly as a result of the lack of grant aid for agroforestry and the lack of knowledge and available information, though there are several recent reports which indicate the feasibility and advantages of silvopasture with sheep (e.g. Crowe and McAdam, 1999). Since over 70% of farms in Ireland have bovines as a main enterprise (Connolly *et al.*, 2002), we initiated a bovine silvopastoral experiment.

The objectives of the experiment are:

- To establish a viable silvopastoral experiment consisting of oak trees with cattle that may be used for production research.
- To compare the potential of agroforestry with conventional land use (pastoral and forestry).
- To collect baseline biological and soil data that can be used to describe and manage the systems.
- To compare tree growth at different planting densities.
- To compare the growth potential of Root Production Method™ (RPM) oak with bare-root oak.

The RPM oak trees are the same species and provenance as the bare-root trees but were produced in the nursery by a different process that resulted in a greater root:shoot ratio than those produced by conventional nursery practices (Lovelace, 1998). First season height and stem diameter results will be reported here.

Materials and Methods

The 3.2 ha experimental site was established in County Wexford (54.27 lat., 6.50 long.), in the south-east of Ireland in March 2002. The land is at an altitude of 80 m and receives a mean annual precipitation of 1000 mm. The experiment has five treatments with three replications in a randomized complete block design (Figure 1). Plot dimensions are given in Table 1. The treatments are: conventional agroforestry, i.e. bare-root oak trees planted at agroforestry spacings (CAF), conventional forestry (CF), RPM agroforestry (RAF), RPM forestry (RF) and pasture control (P).

The RPM plants were 1 year old and the bare-root plants were 2 years old.

Trees in the forestry treatments are planted at the spacing of 0.75 m × 2 m (6600 stems/ha). The trees in the agroforestry treatments are in an alley arrangement at 2.5 m × 10 m intra- and inter-row spacing, respectively (400 stems/ha). Each tree row is protected by a double-strand electric fence enclosing a width of 2.5 m to allow for tree protection in later years (Figure 2).

Table 1. Treatment dimensions and available pasture area.

System	Plot dimensions (m)	Plot area (ha)	Pasture area (ha)	Pasture area (%)
Pasture	21 x 60	0.126	0.126	100
Agroforestry	52.5 x 60	0.315	0.240	76
Forestry	21 x 30	0.063	-	-

Results and Discussion

The trees were successfully established with > 90% survival rate and the cattle were successfully managed in combination with the trees except for one incursion into a tree row. Tree height and stem diameter of the RPM trees were significantly greater ($P < 0.05$) than for the bare-root trees after the first growing season. Height growth increment of the RPM trees for the first growth season was significantly greater ($P < 0.05$) than for the bare-root trees (Figure 3). The negative increment for the bare-root trees is as a result of dieback. The RPM trees were not as affected by dieback, possibly due to the enhanced root system or the planting media associated with them. Trees grown in the forestry treatments were significantly taller ($P < 0.05$) than those grown in the agroforestry treatments after the first season, perhaps benefiting from intra-species sheltering or competition for light. Stem diameter of the RPM trees grown in the agroforestry treatment was significantly greater ($P < 0.05$) than those in the forestry system after the first season (Figure 4). This may be due to less shelter relative to the forestry treatment resulting in increased stem flexing and the formation of more reaction wood. There was no similar significant difference between the forestry and agroforestry bare-root treatments.

Figure 1. Site design.

Figure 2. Tree row protection.

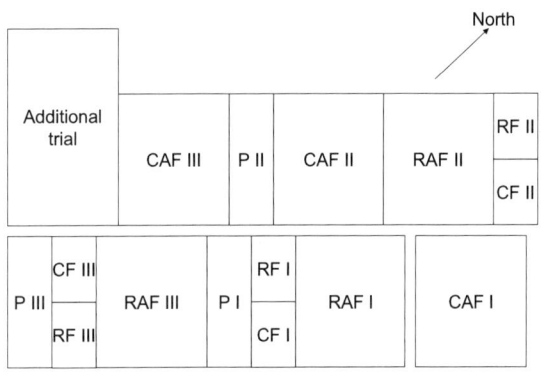

Figure 3. Height increment.

Figure 4. Stem diameter.

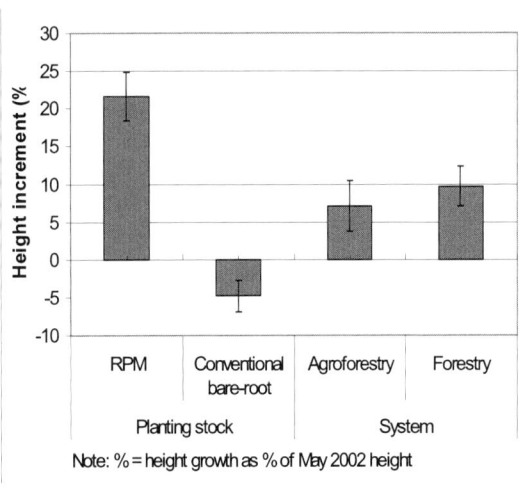

Note: % = height growth as % of May 2002 height

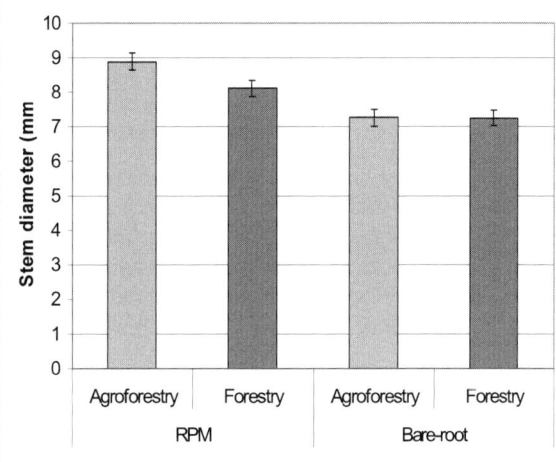

Conclusion

Quercus robur plants produced by RPM technology can be successfully established in forestry and agroforestry spacings in Ireland. They show greater height increment and stem diameter than bare-root plants and should be further investigated. A bovine silvopastoral alley system has been successfully established and should continue to be investigated.

References

Anon (1996) *Growing for the Future: A Strategic Plan for the Development of the Forestry Sector in Ireland.* Department of Agriculture, Food and Forestry, Stationery Office, Dublin, Ireland.

Connolly, L., Finnerty, E., Kinsella, A. and Quinlan, G. (2002) *National Farm Survey 2001*. Teagasc, Dublin, Ireland.

Crowe, S.R. and McAdam, J.H. (1999) Silvopastoral practice on farm agroforestry in Northern Ireland. *Scottish Forestry* 53 (1), 33-36.

Lovelace, W. (1998) The Root Production Method (RPM) system for producing container trees. *International Plant Propagation Society* 46, 556-557.

Integration of horses to orange tree plantations

L. Simón [1], M. D. Sánchez [2], M. Hernández [1], S. Sánchez [1] and C. Mendoza [1]

[1]Estación Experimental de Pastos y Forrajes "Indio Hatuey" Central Spain Republicana, C.P. 44280, Matanzas, Cuba. LSimón@indio.atenas.inf.cu [2]FAO, Dirección de Producción y Sanidad Vegetal, Rome, Italy

Abstract

The trial was carried out with the objective of quantifying the effect of horse rearing in orange tree plantations on fruit production and the soil, as well as determining the productive performance of the horses integrated to this system regarding their meat production potential. For this purpose 58 newly weaned fillies were used, which were randomly distributed in two treatments: (A) continuous grazing with irrigation and (B) continuous grazing without irrigation, in addition to (C) traditional method of weed control (control without animals). The stocking rate used was one animal/ha and the weight gains differed significantly ($P < 0.001$), favouring the treatment with irrigation (0.569 vs 0.431 kg), which was favoured by a higher availability of DM and nutrients in the pasture. Fruit yield and quality were slightly higher in the control, while the best soil indicators corresponded to irrigation. The possibility of integrating horse production to orange tree plantations is demonstrated.

Key words: grazing, citrus fruits

Introduction

Weed control in citrus fruit plantations by means of mechanical cutting is a very costly activity, due to the high prices of fuel and machinery maintenance and use. That is why an initial research was carried out with the aim of determining the usefulness of horse grazing in citrus fruit plantations (Simón *et al.*, 1994), for which mares and their offspring were used. With these antecedents, an experiment was performed in order to study the effects of horse grazing in orange tree plantations on fruit production and quality, soil characteristics and the growth of fillies.

Materials and Methods

A total of 58 newly weaned fillies, with an average weight of 220 kg, were randomly distributed in two treatments: (A) continuous grazing with irrigation and (B) continuous grazing without irrigation, which were compared to (C) control (without animals). Irrigation was applied in treatments A and C four times. Each treatment covered 30 ha located on a Ferralitic Red soil (Hernández, 1999). The animals were managed in continuous grazing with a stocking rate of one filly/ha. Dry matter availability per hectare, chemical composition, botanical composition of the pasture, live weight gain of the animals and chemical and biological composition of the soil were determined.

Results and Discussion

The performance of the fillies regarding live weight and daily gain is shown in Table 1. The accumulated and daily gains in the animals differed significantly among themselves ($P < 0.001$), favouring the treatment with irrigation, which could have been associated to the higher DM availability and pasture quality during the whole development of the experimental work (1812-1642 kg DM/ha) for treatments A and B, respectively. Simón *et al.* (1994) reported much higher availabilities (up to 122.6 kg DM/animal) than the ones in this work, because the area had never been grazed and *P. maximum* was the prevailing pasture on the site instead of *P. notatum*, while, in this study, regarding botanical composition, highly significant differences ($P < 0.001$) were observed in favour of *D. annulatum* and *P. maximum* in treatment C and the highest percentages of depopulation and weeds were found in B. On the other hand, the legumes did not differ among themselves and had the lowest percentages in all the treatments.

Table 1. Live weight and daily gain of the horses (kg). a,b Values with different letters differ at $P < 0.05$, *** $P < 0.001$.

Treatments	Initial live weight	Final live weight	Accumulated gain	Daily gain
A	217.80	314.20	96.36a	0.569a
B	227.90	301.90	74.00b	0.431b
SE ±	10.51	12.52	3.06***	0.023***

The results as a whole showed once more the effectiveness of horse grazing in the control and elimination of *P. maximum*, previously reported by Simón *et al.* (1994). The difference between the use or not of horses for weed control in the citrus fruit plantation meant the decrease of orange production by only 670 kg/ha (13,780 kg DM/ha vs. 14,450 kg DM/ha). However, the use of horses at a rate of one filly per hectare provided a daily gain of about 0.5 kg of meat, in addition to the economy of machinery and fuel in the crop weeding, which represented a remarkable decrease in production costs. Fruit quality of the last harvest, according to the quality control laboratory, is shown in Table 2. No remarkable differences were observed either in the various indicators or in fruit size in the treatment with irrigation and horses (A) and the control (C), but there seem to be some differences between them and the treatment with no irrigation and horses (B) in favour of the former ones.

Table 2. Some indicators of fruit quality.

Area covered	Treatment		
	A	B	C
Soluble solids	12.80	12.70	12.80
Acidity (%)	0.49	0.71	0.45
Maturity index	27.60	18.10	28.50
Juice content (%)	46.60	46.00	48.40
Size 82-108 mm (%)	9.40	5.00	12.50
Size 60-82 mm (%)	81.70	85.00	87.50
Size < 60 mm (%)	8.90	10.00	0.00
Weight size 82-108 mm (g)	404.00	372.40	420.00
Weight size 60-82 mm (g)	211.50	198.60	255.00
Weight size < 60 mm (g)	138.00	120.60	140.50

In the soil the K_2O, cations, as well as N and OM contents were observed to increase in treatment (A) with irrigation. This could have occurred because of the effect of water on the mobilization of nutrients and their interaction with the animals, which contribute to recycling through the dung. The beneficial effect could also be observed in treatment B, in which OM and N contents were slightly higher than in the control. In both cases the edaphic macrofauna was benefited, which is related to the role of biota in the processes of decomposition and nutrient recycling (Lavelle *et al.*, 1994).

Conclusions

The results support the possibility of horse production in areas dedicated to orange tree plantations, in a more sustainable economic and ecological production system, in which live weight gains could increase if males were used instead of females as the former have higher corpulence.

References

Hernández, A. (1999) *Genetic Classification of the Soils in Cuba* (Clasificación genética de los suelos de Cuba). Instituto de Suelos, Ministerio de la Agricultura, AGRINFOR, Ciudad de La Habana, Cuba. (In Spanish.)

Lavelle, P., Angerfield, M., Ragoso, C., Eschenbrenner, V., López-Hernández, D., Phashanasi, B. and Brussaard, L. (1994) The relationship between soil macrofauna and tropical soil fertility. In: Woomer, P.L. and Swift, M.J. (eds) *The Biological Management of Tropical Soil Fertility*. TSBF, A Wiley-Sayce Publication, New York, USA.

Simón, L., Iglesias, R., Cáceres, O. and Duquesne, P. (1994) Equidae in citrus stands. Utility of equidae grazing in citrus stands (Equinos en los cítricos. Utilidad del pastoreo de equinos en los cítricos). *ACPA 1 (94)*, 41. (In Spanish.)

Facilitation of tree regeneration in pasture woodlands

C. Smit [1], D. Béguin [2], A. Buttler [2] and H. Müller-Schärer [1]
[1]Dept. of Biology, Ecology and Evolution, University of Fribourg, Chemin du Musée 10, CH-1700 Fribourg, Switzerland, christian.smit@unifr.ch [2]Swiss Federal Research Institute WSL-Antenne Romande, CP 96, CH-1015 Lausanne, Switzerland.

Abstract
Tree rejuvenation was examined in pasture woodlands in the Swiss Jura Mountains and the distribution of tree saplings hypothesized to be spatially associated with nursing structures or plant species. A correlative study was conducted in two sites on two spatial scales to test this hypothesis. In each site (1 ha) 150 plots of 4 m^2 and 1 m^2 were sampled and number of *Picea abies* saplings (< 40 cm), and cover of: (i) rocks, (ii) shrubs, (iii) unpalatable plant species, (iv) tree stumps and (v) overhanging tree branches recorded. The effects of these five variables on number of *Picea abies* saplings per "full" plot (with saplings) differed with scale but presence of *Picea* saplings was positively related to rock and unpalatable cover in both 4 m^2 and 1 m^2 plots. Experiments are under way to further explore the role of unpalatable plants as facilitators of *Picea abies* in these habitats.

Key words: nurse plants, tree establishment, extensive grazing

Introduction
Spatial associations of tree saplings with other plant species, often shrubs, have been found in a wide range of ecosystems, such as savannahs and grasslands (Weltzin and McPherson, 1999), woodlands (Chambers, 2001), old fields (Li and Wilson, 1998), montane steppes (Callaway *et al.*, 1996), chaparral (Dunne and Parker, 1999) and laurel forests (Arevalo and Fernandez-Palacios, 2003). These associations often suggest nurse effects of the neighbouring plants, as they directly or indirectly contribute positively to the performance of the tree saplings. The few recent studies that focused on tree establishment in grazed ecosystems showed the importance of unpalatable species (plants defended by spines, toxic elements, etc.) for protecting spatially associated tree saplings from cattle grazing (Rousset and Lepart, 1999, 2000). This process of spatial association with unpalatable plants (associational resistance or defence guilds) is considered as a key process in the dynamics of grazed temperate woodlands, leading to cyclic shifts in vegetation mosaics from open grasslands towards woodland (Olff *et al.*, 1999). These cattle-driven shifts are highly important for the maintenance of spatial and temporal variation in vegetation structure, leading to high biodiversity (Olff and Ritchie, 1998). In this chapter spatial associations of naturally established tree saplings in pastured woodlands of the Swiss Jura Mountains were focused on. These silvopastoral systems have declined drastically in Europe during the last few decades and are presently threatened (Gillet and Gallandat, 1996). Understanding tree regeneration is a prerequisite for sensitive management interventions. Tree saplings were hypothesized to be spatially associated with facilitative structures or vascular plants and this was tested in a correlative study.

Materials and Methods
The study was conducted in pastured woodlands in the central part of the Swiss Jura Mountains (900-1400 m asl). Mean yearly temperature varies between 3 and 5°C, with an annual precipitation between 1400-2000 mm. Soils are considered relatively dry because of the high permeability of the underlying calcareous layers. This combined with low winter temperature leads to a relatively short growing season of about 4 months. Norway spruce (*Picea abies* L.) is the dominating tree species in this area. Two study sites were selected, one with free grazing cattle (50 livestock units (LU) day/ha) and one with rotational grazing (82 LU day/ha). In each site (1 ha) 150 plots of 4 m^2 and six blocks with 50 1 m^2 plots were sampled. Of these, the number of tree saplings (< 40 cm) were counted and the percentage coverage of: rocks, shrubs, overhanging tree branches (< 150 cm), tree stumps and unpalatable plants (from a priori list) were sampled. Logistic regressions using Wald statistics and ANOVA (using only plots containing saplings) to explain presence-absence and number of *Picea abies* saplings by the measured variables were used.

Results and Discussion
Logistic regressions on the presence-absence data of *Picea* saplings in both the 4 m^2 and 1 m^2 plots revealed significant positive effects of rock and unpalatable cover (Table 1). ANOVA showed in the 4 m^2 plots a significant effect of rock, tree stump and unpalatable cover, and of tree cover on the number of *Picea* saplings in the 1 m^2 plots (Table 2). No interactions were significant in the models. The 1 m^2 resolution might well be too fine grained to detect spatial associations, especially when relating number of tree saplings with the studied structures, explaining the spatial differences.

Table 1. Results of logistic regressions on presence-absence of *Picea abies* saplings per 4 m² and 1 m² plots. Independent variables included site, block (only at 1m² plots) and rock, shrub, tree stump, tree branch and unpalatable cover (covers log+1 transformed). (*) $P < 0.1$, * $P < 0.05$, ** $P < 0.01$, ***$P < 0.001$, n.p. = not present.

	4 m² plots (n = 294)		1 m² plots (n = 300)	
Parameter	Estimate + se	Wald stat	Estimate + se	Wald stat
Site	1.736 ± 0.293	34.613***	1.235 ± 0.291	17.962***
Block	n.p.	n.p.	-	17.770**
Rock	0.516 ± 0.127	16.610***	0.837 ± 0.288	8.433**
Shrub	0.004 ± 0.154	0.001	0.158 ± 0.554	0.081
Stump	2.412 ± 1.941	1.544	n.p.	n.p.
Tree	0.098 ± 0.098	0.992	-0.065 ± 0.224	0.083
Unpalatable	0.485 ± 0.133	13.382***	0.469 ± 0.282	2.764(*)
Constant	-2.394 ± 0.374	41.000***	-1.865 ± 0.426	19.177***

Table 2. ANOVA table for the effects of site, block (only at 1 m² plots) and rock, shrub, tree stump, tree branch and unpalatable cover (all fixed factors) on the number of *Picea abies* saplings (log+1 transformed). (*) $P < 0.1$, * $P < 0.05$, ** $P < 0.01$, ***$P < 0.001$, n.p. = not present.

	4 m² plots (n = 139)			1 m² plots (n = 114)		
Source	SS type III	Df	F	SS type III	Df	F
Corrected model	8.776	9	3.815***	1.918	13	2.953**
Intercept	45.041	1	176.205***	5.891	1	117.877***
Site	3.555	1	13.909***	0.452	1	9.045**
Block	n.p.	-	n.p.	0.525	5	2.100(*)
Rock	1.489	2	2.912(*)	0.040	2	0.402
Shrub	0.0356	1	0.140	0.119	1	2.387
Stump	1.016	1	3.975*	n.p.	-	n.p.
Tree	0.661	2	1.293	0.214	1	4.285*
Unpalatable	1.705	2	3.335*	0.182	3	1.213
Error	32.975	129		4.998	100	
Total	257.625	139		36.507	114	
Corrected total	41.750	138		6.916	113	

Conclusions

In this study spatial associations of *Picea abies* saplings with mainly rock outcrops and unpalatable plants were observed. This was found in both sites, which varied in grazing system, grazing intensity and altitude (1150 m and 1340 m asl), and on two spatial scales (4 m² and 1 m²). These results clearly suggest a facilitative effect of unpalatable plants and rocks protecting tree saplings against the grazing and browsing of cattle, which is influential on the dynamics of the system. The 4 m² scale is more appropriate for detecting spatial associations than the 1 m² scale, especially when relating number of saplings with the occupancy of the relatively large studied structures.

References

Arevalo, J.R. and Fernandez-Palacios, J.M. (2003) Spatial patterns of trees and juveniles in a laurel forest of Tenerife, Canary Islands. *Plant Ecology* 165, 1-10.
Callaway, R.M., de Lucia, E.H., Moore, D., Nowak, R. and Schlesinger, W.H. (1996) Competition and facilitation: contrasting effects of *Artemisia tridentata* on desert *vs* montane pines. *Ecology* 77, 2130-2141.
Chambers, J.C. (2001) *Pinus monophylla* establishment in an expanding *Pinus-Juniperus* woodland: environmental conditions, facilitation and interacting factors. *Journal of Vegetation Science* 12, 27-40.
Dunne, J.A. and Parker, V.T. (1999) Species-mediated soil moisture availability and patchy establishment of *Pseudotsuga menziesii* in chaparral. *Oecologia* 119, 36-45.
Gillet, F. and Gallandat, J.D. (1996) Wooded pastures of the Jura Mountains. In: Etienne, M. (ed.) *Western European Silvopastoral Systems*. INRA, Paris, France, pp. 37-53.
Li, X.D. and Wilson, S.D. (1998) Facilitation among woody plants establishing in an old field. *Ecology* 79, 2694-2705.
Olff, H. and Ritchie, M.E. (1998) Effects of herbivores on grassland plant diversity. *Trends in Ecology and Evolution* 13, 261-265.
Olff, H., Vera, F.W.M., Bokdam, J., Bakker, E.S., Gleichman, J.M., Maeyer, K.D. and Smit, R. (1999) Shifting mosaics in grazed woodlands driven by the alternation of plant facilitation and competition. *Plant Biology* 1, 127-137.
Rousset, O. and Lepart, J. (1999) Shrub facilitation of *Quercus humilis* regeneration in succession on calcareous grasslands. *Journal of Vegetation Science* 10, 493-502.
Rousset, O. and Lepart, J. (2000) Positive and negative interactions at different life stages of a colonizing species (*Quercus humilis*). *Journal of Ecology* 88, 401-412.
Weltzin, F. and McPherson, G.R. (1999) Facilitation of conspecific seedling recruitment and shifts in temperate savanna ecotones. *Ecological Monographs* 69, 513-534.

Effect of sheep grazing on coffee quality

J. A. Torres
Centro Regional Universitario Oriente, Universidad Autónoma Chapingo, Apdo. 49, CP 94100, Huatusco, Ver., Mexico. tora_sheep@hotmail.com

Abstract
The effect of sheep grazing on the quality of fruit, bean and liquor of coffee was evaluated in order to provide scientific information as to whether or not grazing could be recommended to sustain diversification in coffee plantations. Samples of mature fruit were taken from plantations which had been grazed for more than 5 years and were compared with fruit from conventionally managed plantations. All coffee plantations were located in Veracruz, Mexico. The samples were processed following Mexican Official Norm PROY-NOM-149-SCFI-2000. A panel of professional coffee tasters carried out sensory evaluation of the liquor prepared with roasted and milled beans. The caffeine content was determined in green beans, by high performance liquid chromatography. Sheep grazing does not adversely affect the parameters of quality of coffee in its different forms, and in some aspects the coffee is even better than that produced from conventional plantations.

Key words: bean, caffeine, liquor, silvopastoral system, tasting

Introduction
It has been demonstrated that sheep grazing fruit plantations present ecological and socio-economic benefits compared to conventional systems of production without sheep, mainly when by lack of money the producers do not do a good control of weeds, they do not fertilize and do not execute other necessary works to maintain the good functioning of the system (Torres, 2002). Nevertheless, there are doubts about the quality of products obtained from silvopastoral systems. This is the case with coffee that is produced and traded for its qualities as a beverage that satisfies the tastes which consumers demand. Coffee quality is a parameter influenced by a wide range of factors, such as the environment where it is produced, variety, husbandry, harvest and processing (Njoroge, 1998). There is information on the effect of the application of cattle and sheep manure on coffee yield, and sheep grazing in citrus orchards, but no references could be found on the effect of sheep grazing on the coffee quality. A study in Kenya, referring to the continuous use of cattle manure, found that there was an increase in the proportion of brown coffee beans and an adverse effect on liquor quality (Kamau, 1976, cited by Njoroge, 1998). Another study carried out in an organic coffee farm in India, using compost prepared with sheep manure, indicated highest percentage of "A" grade beans; also the quality of coffee in the cup was reported to be good with fair to good body, fair acidity sometimes associated with a special flavour (Kamala *et al.*, 2002). In Veracruz, Mexico, sheep grazing during 4 years reduced the size and weight of oranges but increased the proportion of juice with low acid content (Torres, 1996). It is important to know if grazing produces adverse effects on the quality of commercial products, because, if it does, producers may not want to adopt this type of silvopastoral system. The objective of the present study was to investigate the effect of sheep grazing in coffee plantations on the quality of cherry coffee (ripe fruits with red skin), parchment bean (coffee pulped and dried), green bean (without husk parchment and silver skin), roasting bean and liquor or brewed flavour, in order to provide scientific information as to whether or not grazing could be recommended to sustain diversification in coffee plantations.

Materials and Methods
The study was conducted in January 2003 in the central zone of the state of Veracruz, in the west of Mexico. The plantations used were between 19°08' and 19°12' N, 96°50' and 96°56' W, and at an altitude of 800 to 1240 m, with a semi-warm humid climate, volcanic soils and hilly relief. There were 12 coffee plantations on private property, half of which had been grazed with sheep for more than 5 years, the rest having been conventionally managed without grazing. Each plantation was managed either as a commercial mixed-cultivation system or as a system with specialized shade trees (mainly *Citrus*, *Erythrina*, *Inga*) or solely coffee plants system, under rotational grazing and stocking rate from 1.5 to 9.4 sheep/ha depending on the prevalence of forage weeds. In conventional plantations the control of weeds was done with machetes or herbicides. In each plantation only mature fruits of *Coffea arabica* L. var. Tipica were harvested in grazed and excluded areas. The samples were processed following the Mexican Official Norm PROY-NOM-149-SCFI-2000 (DGN, 2001). A sensory panel of two professional coffee tasters evaluated the liquor prepared with roasted and milled beans, taking into account the parameters: aroma, acidity, body and bad flavours. The caffeine content was determined by high-performance liquid chromatography. Data were statistically analysed using Statistical Analysis System (SAS, 1993), with GLM procedure for analysis of variance, and multiple comparison of means by the Duncan test ($\alpha = 0.05$).

Results and Discussion

The cherries from plantations with grazing were less heavy compared to those from plantations without grazing ($P \leq 0.05$). The lighter weight probably was due to a reduced content of total soluble solids or water in the skin and mucilage. There were no significant differences in the shape of the fruit, in the percentage of stained and floated fruits, or in the yield of cherry coffee to parchment bean and the cherry coffee to green bean. The content of total soluble solids (degree Brix) and pH in mucilage were reduced, which indicates that sheep faeces contributed to correcting potassium deficiencies in the coffee plants (Carvajal, 1984). With sheep grazing, a greater percentage of large beans was obtained. It is probable that the higher quality is due to weed control by animals, reducing competition for nutrients and water in the soil. The smaller size of the beans from the plantations without grazing may be due to loss in the fertility of the soil (Blore, 1965), as a consequence of reduction in the application of fertilizers as producers have less capital to spend but sheep somehow contribute to a solution. There were no differences in the shape of the beans; nevertheless it is important to emphasize that grazing reduced the percentage of defective beans, particularly brocades, shells and triangles. There were no differences in the colour of beans, in all cases being 5835 U. Grazing did not have an effect on the caffeine content. There were no differences between treatments ($P > 0.05$) for any of the sensory evaluation variables. Nevertheless, in 83% of the plantations greater aroma, acidity and body were obtained on average when the sheep grazing was implemented. There were no bad flavours due to sheep grazing.

Conclusions

The study suggests that sheep grazing in coffee plantations can be recommended to improve or at least maintain the functionality of the system, because the parameters of quality of the coffee, in its different forms, are better in sheep-grazed plantations or at least equal to the conventionally managed plantation. More specific studies are necessary with other varieties of coffee and in different agroecological zones.

References

Blore, T.W.D. (1965) Some agronomic practices affecting the quality of Kenya coffee. *Turrialba* 15 (3), 111-118.
Carvajal, J.F. (1984) *Coffee Cultivation and Fertilization* (Cafeto-cultivo y fertilización). Instituto Internacional de la Potasa. Worblaufen-Bern, Switzerland. (In Spanish.)
DGN (2001) *Project of Official Mexican Rules, Coffee Veracruz – Specifications and Testing Methods* (Proyecto de norma oficial mexicana PROY-NOM-149.SCFI-2000, café Veracruz- Especificaciones y métodos de prueba). Dirección General de Normas, Secretaría de Economía, Mexico D.F., Mexico. (In Spanish.)
Kamala, S., Hariyappa, N., Mani, S.D., Seetharam, H.G., Shivaram, G.T. and Raghuramulu, Y. (2002) Organic coffee farming: a case study. *Indian Coffee* 66(5), 21-22.
Njoroge, J.M. (1998) Agronomic and processing factors affecting coffee quality. *Outlook on Agriculture* 27 (3), 163-166.
SAS (1993) *SAS/User's Guide*, Version 6, 4th edn. SAS Institute, NC, USA, p.1686.
Torres, J.A. (1996) Diagnosis of the agroecosystem citrus-ovine livestock in Tlapacoyan (Diagnóstico del agroecosistema cítricos-ovinos en Tlapacoyan). MSc thesis. Colegio de Postgraduados Campus Veracruz, Tepetates, Mexico. (In Spanish.)Torres, J.A. (2002) Diagnosis of sustainability of the agroecosystem coffee-ovine livestock versus the conventional coffee system in Veracruz, Mexico (Diagnóstico de sostenibilidad del agroecosistema café-ovinos respecto a los sistemas cafetaleros convencionales en Veracruz, Mexico). *Café Cacao* 3 (1), 14-17. (In Spanish.)

The influence of spacing and coppice height on herbage mass and other growth characteristics of *Robinia pseudoacacia* in a south-eastern USA silvopastoral system

L. J. Unruh Snyder, J. P. Mueller [1], J.M. Luginbuhl and C. Brownie
[1]North Carolina State University, Campus Box 7620, Raleigh, NC 27695-7620 USA. Paul_Mueller@ncsu.edu

Abstract
Data were collected for 2 years (1999 and 2000) to evaluate growth characteristics of a 5-year-old stand of *Robinia pseudoacacia* for goat (*Capra hircus*) browse in a randomized complete block design (intra-row spacing 0.5 or 1.0 m and coppice height 0.25 or 0.5 m) replicated six times. Main branch size was not affected by spacing when trees were coppiced at 0.25 m; however, when coppiced at 0.5 m, trees spaced at 1.0 m had larger branches ($P < 0.05$) than trees spaced at 0.5 m. Herbage biomass was not influenced by spacing when trees were coppiced at 0.5 m whereas trees spaced at 1.0 m produced less woody and herbaceous biomass than trees spaced at 0.5 m when coppiced at 0.25 m ($P < 0.05$). These results suggest that *Robinia pseudoacacia* would be an excellent candidate as a silvopastoral component in the south-eastern USA based on growth characteristics and herbage biomass production.

Key words: herbage biomass, goats, plant population

Introduction
Interest in goat (*Capra hircus*) meat production in the eastern USA is expanding as a result of increased demand from several ethnic communities. Including browse (woody brush and tree foliage) as a component of a grazing system makes use of the natural tendency of goats to select as much as 50% of their diet from browse sources (Luginbuhl *et al.*, 1996). The incorporation of a browse species, such as black locust (BL; *Robinia pseudoacacia* L.), could fill a feed gap that occurs when cool-season grasses and legumes are semi-dormant in the hot summer months. The decline in quality and productivity of cool-season forages in the summer usually occurs when demand for quality forage by growing and lactating animals is high. In North Carolina and much of the south-east region, BL is a native tree that possesses the potential to contribute such a complement. Furthermore, BL appears to be a desirable candidate for a silvopastoral system, due to its ability to fix atmospheric N, 75 to 150 kg N/ha year (Boring *et al.*, 1981), and supply forage high in crude protein (CP), ranging from 20 to 24% (Addlestone *et al.*, 1999). Two BL ecotypes have been identified based on leaf morphology: oligophyllous possessing few large leaflets and small thorns; and polyphyllous with many small leaflets and large thorns (Dini-Papanastasi and Panetsos, 2000). Although most BL exhibit formidable paired, thorny stipules in the juvenile growth, these spines pose no particular problem for browsing goats.

The incorporation of BL as a system component may be beneficial to farmers seeking alternative feed resources for their livestock. The objectives of this experiment were to determine the effects of coppice height and intra-row spacing on main branch size and herbage mass (HM) production.

Materials and Methods
The study was conducted at North Carolina State University in Raleigh, NC, at approximately 35.75° N latitude and 78.75° W longitude. The climate is temperate with 1191 mm long-term mean annual precipitation and average annual maximum and minimum temperatures of 21.1 and 10.5°C, respectively. Soils of the study area were Cecil Series, (Clayey, Kaolinitic, Thermic, Typic Hapludult) on slopes ranging from 6 to 10% (USDA, 1970). In November 1999, one soil sample was taken (15 cm depth) for each plot (72 total), pooled by replication, and sent to the North Carolina Department of Agriculture Soil Testing Lab to derive soil chemistry data. Soil fertility at the experimental site was characterized by high base saturation (82%) and a pH of 6.1. Due to adequate P and K levels, soil base saturation and the ability of BL to fix N, no fertilizer or lime was applied during 1997 to 2000.

The field study was established by planting bare-root seedlings from a nursery stock in March 1995. Plots were 90.3 m length with 3 m between rows and 3 m alleys between blocks. Trees were planted at a spacing of 0.5 m (6296 trees/ha) or 1.0 m (3333 trees/ha) within rows and coppiced at heights of 0.25 or 0.5 m above ground level each February. The primary main branches with a diameter greater than 10 mm were counted and measured with a caliper at the coppice height. Destructive sampling of the BL trees was performed to determine biomass. Two trees were randomly sampled per plot for the 0.5 m row spacing and one tree for the 1.0 m row spacing. The herbage mass, a composite sample of leaflets, petioles and herbaceous stem tips (soft tissue petioles), was stripped by hand. All BL measurements were taken on 19 July 1999 and 2000. Composite samples of dried herbage were chemically analysed for Kjeldahl N according to AOAC (1999). Kjeldahl N was then multiplied by 6.25 to estimate crude protein (CP). The samples were also analysed for neutral detergent fibre (NDF), acid detergent fibre (ADF) and 72% sulphuric acid lignin

(ADL) sequentially according to Van Soest *et al.* (1991) as modified by Komarek *et al.* (1994). Acid detergent lignin was corrected for mineral matter by ashing the ADL residue in a muffle furnace at 500°C. *In vitro* true DM disappearance (IVTDMD) was determined using ruminal contents collected from a ruminally cannulated Hereford steer maintained on high-quality alfalfa (*Medicago sativa* L) hay (Goering and Van Soest, 1970). After 48 h of incubation with ruminal inoculum in a batch processor (Ankom Technology Corp., Fairport, NY), samples were extracted with neutral detergent solution for IVTDMD estimation.

Results and Discussion

Based on analysis of foliar samples collected in June 1999 and 2000, herbage quality estimates of composite samples of leaflets, petioles and herbaceous stem tips were high. Foliar CP of BL averaged 25%. Black locust leaf CP levels have been reported to be similar to lucerne, ranging from 20 to 25% (Addlestone, 1996; Papachristou *et al.*, 1999). Assuming adequate intake (> 3% body weight (BW)) and no anti-quality factors, this level of CP exceeds the nutritional requirements (12 to 14% CP) of a 20 kg goat kid gaining 150 g/day by approximately 13 percentage units (NRC, 1981).

The average foliar NDF and ADF values were low (38 and 21%, respectively), indicating high digestibility. Lignin concentration averaged 11%. Horton and Christensen (1981) reported higher values for ADF (28.5%) and NDF (53.6%), and higher ADL values (13.5%) (for BL), indicating more advanced maturity herbage was analysed compared with our study. Also, for this study IVTDMD concentration averaged 68%, whereas Papachristou *et al.* (1999) reported *in vitro* organic matter disappearance (IVOMD) as 52%, which is roughly equivalent to 64% IVTDMD (Holden, 1999). Spacing distance and coppice height showed similar trends in both years; therefore, data were combined over years. There was an interaction between spacing distance and coppice height ($P = 0.01$) for the sum of main branch diameters (SMBD). Spacing had no influence on the SMBD of trees coppiced at 0.25 m, whereas, at a coppice height of 0.5 m, trees spaced at 1.0 m had a greater SMBD than trees spaced at 0.5 m (Figure 1). Addlestone *et al.* (1999) reported a similar trend for trees with these spacing and coppice height. Mean HM was 2142 kg DM/ha (1999: 2286 kg DM/ha; 2000: 1998 kg DM/ha). HM was very similar to that (2390 kg DM/ha) reported by Addlestone *et al.* (1999). Papanastasis *et al.* (1997) reported a mean total biomass of 2.28 t DM/ha and 1.026 t DM/ha of grazeable leaves and twigs, respectively. An interaction was observed (Figure 2) between spacing distance and coppice height ($P < 0.01$) for HM. At the 1.0 m spacing there was a large difference in HM due to coppice height whereas coppice height had no influence on HM at the 0.5 m spacing. The wide spacing treatment (1.0 m) and lowest coppice height (0.25 m) produced the least ($P = 0.05$) HM (1.311 t DM/ha). Overall, a coppice height of 0.5 m appeared to be less stressful to the trees.

Figure 1. Sum of main branch diameters greater than 10 mm per plant of black locust averaged over years (1999 and 2000). $LSD_{0.05} = 21.1$.

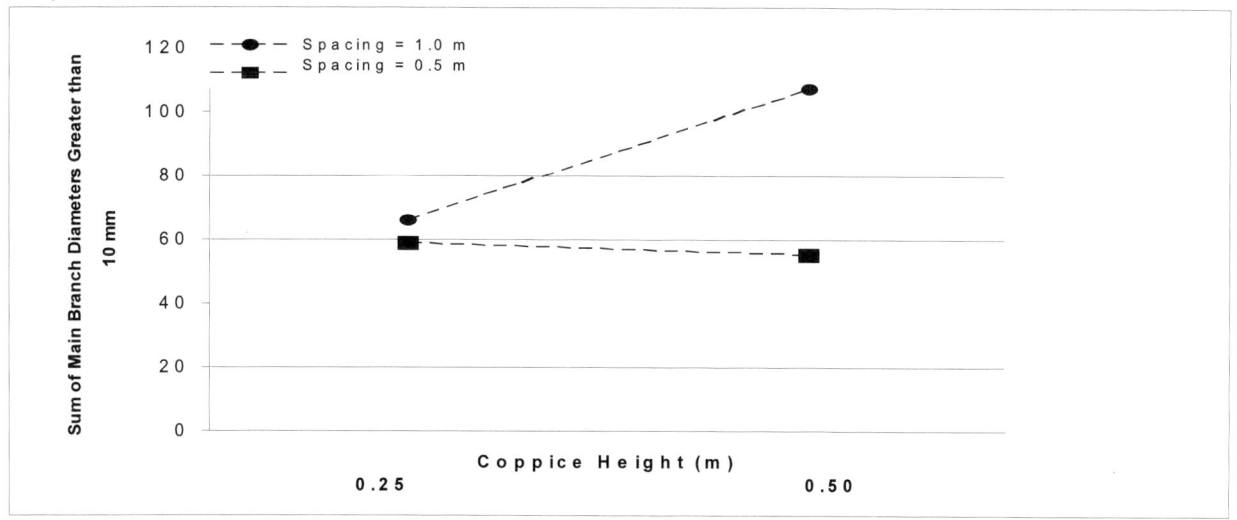

Figure 2. Mean herbage mass of black locust per unit area averaged over years 1999 and 2000. $LSD_{0.05} = 616.1$.

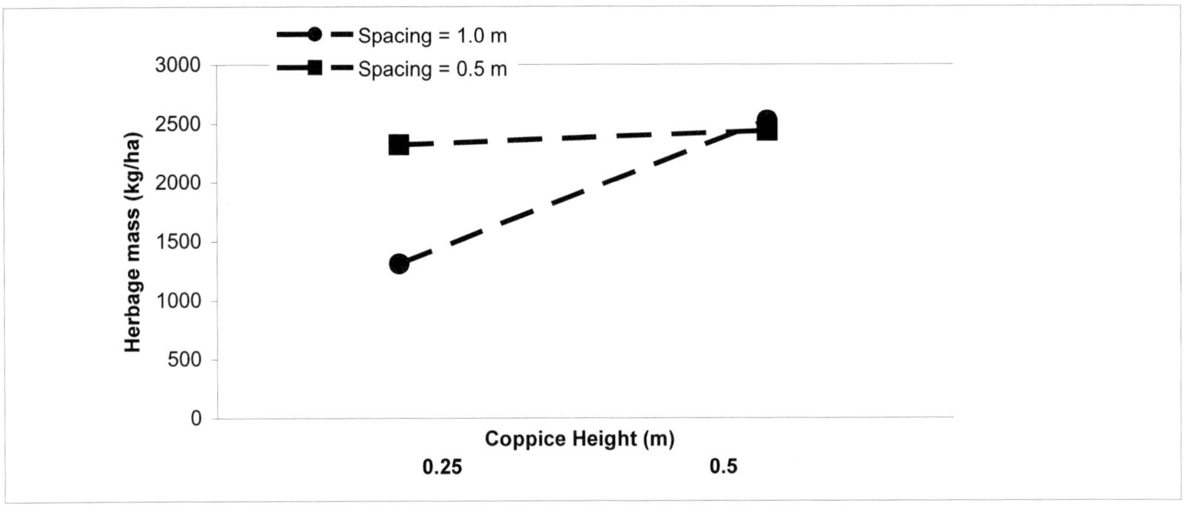

Conclusions

Biomass production was greater at the 0.5 m tree spacing due to increased number of trees per unit area compared with the 1.0 m spacing. In both years, SMBD and HM were measured from a single harvest. It is possible that appreciably higher seasonal yields (perhaps double) could be obtained from a multiple harvest system where trees were cut or browsed two or three times per year. Overall, this research suggests that BL would be an excellent candidate as a silvopastoral component in the south-eastern USA, based on estimated BL herbage production and estimated quality.

References

Addlestone, B.J. (1996) density and cutting height effects on the herbage mass of three tree legumes grown for meat goat production. MSc thesis. North Carolina State Univ., Raleigh, USA.

Addlestone, B.J., Mueller, J.P. and Luginbuhl, J.M. (1999) The establishment and early growth of three leguminous tree species for use in silvopastoral systems of the southeastern USA. *Agroforestry Systems* 44, 253-265.

AOAC (1999) *International Official Methods of Analysis* 16th edn. Association of Official Analytical Chemists, Arlington, VA, USA.

Boring, L.R., Monk, C.D. and Swank, W.T. (1981) Early regeneration of a clear-coppiced southern Appalachian forest. *Journal of Ecology* 62, 1244-1253.

Dini-Papanastasi, O. and Panetsos, C.P. (2000) Relation between growth and morphological traits and genetic parameters of *Robinia pseudoacacia* var. *monophylla* in northern Greece. *Silvae-Genetica* 49, 37-44.

Goering, H.K. and Van Soest, P.J. (1970) *Forage Fiber Analyses (Apparatus Reagents, Procedures and some Applications).* Agriculture Handbook 379, USA Government Printing Office, Washington, DC, USA.

Holden, L.A. (1999) Comparison of methods of *in vitro* dry matter digestibility for ten feeds. *Journal of Dairy Science* 82, 1791-1794.

Horton, G.M.J. and Christensen, D.A. (1981) Nutritional value of black locust tree leaf meal (*Robinia pseudeoacacia*) and alfalfa meal. *Canadian Journal of Animal Science* 61 (2), 503-506.

Komarek, A.R., Robertson, J.B. and Van Soest, J.B. (1994) Comparison of the filter bag technique to conventional filtration in the Van Soest analysis of 21 feeds. In: Fahey, G., Collins, M., Merrens, D. and Moser, L.E. (eds) *Proceedings of the National Conference on Forage Quality, Evaluation and Utilization.* Lincoln, NE, USA, p. 78.

Luginbuhl, J.M., Green, J.T., Mueller, J.P. and Poore, M.H. (1996) Meat goats in land and forage management. In: *Proceedings of the south-east Regional Meat Goat Production Symposium. "Meat Goat Production in the south-east Today and Tomorrow".* Florida University, Tallahassee, FL, USA.

National Research Council (NRC) (1981) *Nutrient Requirements for Goats.* National Academy Press, Washington, DC.

Papachristou, T.G., Platis, P.D., Papanastasis, V.P. and Tsiouvaris, C.N. (1999) Use of deciduous woody species as a diet supplement for goat grazing Mediterranean shrublands during the dry season. *Animal and Feeding Science Technology* 80, 267-279.

Papanastasis, V.P., Platis, P.D. and Dini-Papanastasi, O. (1997) Productivity of deciduous woody and fodder species in relation to air temperature and precipitation in a Mediterranean environment. *Agroforestry Systems* 37, 187-198.

USDA (1970) *Soil Survey of Wake County North Carolina.* USA Government Printing Office, Washington, DC, USA.

Van Soest, P.J., Robertson, J.B. and Lewis, B.A. (1991) Methods for dietary fiber, neutral detergent fiber, and nonstarch polysaccharides in relation to animal nutrition. *Journal of Dairy Science* 74, 3583-3597.

**Session 3.
Ecological implications of the silvopastoral systems:
biodiversity and sustainable management**

How much carbon can be stored in Canadian agroecosystems using a silvopastoral approach?

A. M. Gordon, R. P. F. Naresh and V. Thevathasan
Dept. of Environmental Biology, University of Guelph, Guelph, Ontario, Canada, N1G 2W1

Abstract

In Canada, greenhouse gas (GHG) emissions from the agricultural sector amount to approximately 64.4 million tonnes of CO_2-e (carbon dioxide equivalent) per year, representing about 10% of all GHG emissions. To become carbon neutral within the agricultural sector would mean that these emissions would have to be sequestered on agricultural lands, which include approximately 51 million ha of pasture and range-land across the country. In this study, carbon (C) dynamics were examined in both silvopastoral and monoculture pasture systems. In a silvopastoral system using hybrid poplar (*Populus* sp) at a density of 111 trees/ha, the net annual C sequestration potential could be as high as 2.7 t/ha per year, whereas, in a monoculture pasture system, the net annual C sequestration potential might be less than 1.0 t/ha per year. Silvopastoral systems with fast-growing tree species therefore have the potential to sequester approximately 2.7 to 3 times more C than that of monoculture pasture systems. A net annual C sequestration rate of 2.7 t/ha per year is equivalent to an immobilization rate of 9.9 t of atmospheric CO_2/ha per year. Based on the results of this study, it is estimated that the total GHG emissions of the Canadian agricultural sector could potentially be sequestered in as little as 6.4 million ha of pastureland, as long as these lands were under silvopastoral systems with fast-growing tree species. Using the Ontario sheep industry as an example, it is also estimated that, if each sheep farmer in the province put less than 1 ha of pastureland into a silvopastoral system, the entire Ontario sheep industry (225,000 animals) would become carbon neutral. Although the mitigation effect will vary depending upon climatic region, soils, tree species and age and other factors, there remains vast potential for this type of land-use system in Canada, largely in southern regions of the country. In addition to climate change issues, silvopastoral systems can also address issues related to soil erosion, nutrient management, animal welfare and biodiversity.

Key words: agroforestry, carbon sequestration, greenhouse gases, soil carbon, sheep, poplar

Introduction

Silvopasture is a form of agroforestry whereby trees are incorporated into managed pasture. The benefits of silvopastoral systems are numerous and include diversified farm income, soil/nutrient conservation, provision of shade for animals, enhanced field diversity and an increased potential for carbon sequestration (Gordon and Newman, 1997; Garrett *et al.*, 2000). With respect to the latter, trees are perennial photosynthetic plants and, as such, act as a long-term sink for atmospheric CO_2. The introduction of trees into pasture systems is thus one possible mechanism by which carbon capture can be enhanced. In this chapter, hypotheses are presented about the carbon balance (including sequestration) in both a silvopastoral system and a monocropped pasture system. An existing hybrid poplar (*Populus* sp) and Norway spruce (*Picea abies*) intercropping system was utilised as a surrogate for a silvopastoral system, since very few of the latter exist in Canada. It should be noted that the tree density utilized (111 trees/ha) would be similar to that found in many operational silvopastoral systems. The reason for including both a fast-growing tree species (hybrid poplar) and a slower-growing tree species (Norway spruce) was to determine the high and the lower end of C sequestration potentials in silvopastoral systems. It was also decided to develop the system using small ruminants because methane emissions by these animals are comparatively lower than emissions from large ruminants.

Silvopastoral land-use practices are not widely adopted in Canada although there is great potential for their implementation. Estimates of marginal or degraded land technically suitable for establishment of silvopastoral in Canada range to 51 million ha (McKeague, 1975). Successful establishment of silvopastoral systems on these lands may contribute to a significant long-term reduction in atmospheric CO_2 levels in Canada in particular and improved animal welfare in general.

In this chapter quantitative measurement of C sequestration in a hypothetical silvopastoral system is described in order to identify all possible sinks of C and to compare these parameters with those of conventional pasture systems, by creating carbon flow models for both systems. This will be helpful in the future design and development of silvopastoral systems, from a Kyoto perspective, on marginal lands in Canada and elsewhere.

Materials and Methods

Experimental site
A long-term tree-based intercropping research site (Figure 1A), which was initiated in 1987 on a 30 ha field at the University of Guelph Agroforestry Research Station, Ontario, Canada (43° 32' 28" N latitude, 80° 12' 32" W longitude), was used as a surrogate for a silvopastoral system since very few of the latter exist in Canada. Ten tree species, namely,

silver maple (*Acer saccharinum*), hazelnut (*Corylus avellana*), white ash (*Fraxinus americana*), black walnut (*Juglans nigra*), Norway spruce, hybrid poplar, red oak (*Quercus rubra*), black locust (*Robinia pseudoacacia*), willow (*Salix discolor*) and white cedar (*Thuja occidentalis*), were planted and annually intercropped with maize (corn) (*Zea mays*), soybean (*Glycine max*) and winter wheat (*Triticum aestivum*) or barley (*Hordeum vulgare*). Tree rows were spaced at 12.5 or 15 m apart with a within-row tree spacing of 3 m or 6 m. For the study presented here, 15 m wide poplar tree rows were utilized with a within-row spacing of 6 m. Grass was grown in the tree rows, and data on monoculture pasture (ryegrass (*Lolium perenne*)) was collected where trees were not present. The soil type is sandy loam (Typic Hapludalf), the length of the growing season is 105 days with a mean temperature of 17.4°C, and the annual total precipitation is 833 mm with approximately 334 mm falling as rain during the growing season.

Carbon quantification in trees
Both above- and below-ground C quantification was undertaken by destructive sampling. In order to obtain above-ground oven-dry biomass, sub-samples were collected from each tree component and the moisture content determined. Moisture content derived from these respective tree components was then used to convert the fresh weights of tree components into oven-dry biomass, which was then converted to carbon content.

In order to quantify tree below-ground biomass data, heavy machinery was used to dig a pit to a depth of 2 m around a 3 m × 3 m block of earth containing the tree stump (Figure 1B). This block of earth (soil) was then washed off with high-pressure water from a fire hose. For each tree, 45,000 to 55,000 litres of fresh water were used to expose the below-ground root system (Figure 1C,D). The same technique was also used to quantify C in a slower-growing Norway spruce tree.

Figure 1. (A) The University of Guelph Agroforestry Research Site, used as a surrogate for a silvopastoral system. The tree density of approximately 111 trees/ha is typical of low-density silvopastoral systems. (B) 3 m wide and 2 m deep pit being excavated by a backhoe, leaving a 3 m by 3 m by 2 m block of soil containing the lower portion of the tree trunk and the root system. (C) The soil around the roots is being removed via pressure-washing in order to expose the rooting system. (D) Exposed tree roots for biomass and C quantification.

Above-ground biomass and carbon of pasture grass
Above-ground annual biomass production of grass was quantified by manually harvesting (three harvests per year) above-ground biomass from three randomly selected 0.25 m² quadrats permanently established in both 'silvopastoral' and monoculture systems. Dry biomass was determined by drying the harvested fresh grass in a conventional drying oven at 65°C for 5 days.

Below-ground biomass and carbon estimation of pasture roots
Ryegrass was manually excavated so that most of the roots remained intact. Samples were then washed in slow running water to remove soil from roots, and were dried in a conventional drying oven at 65°C for 5 days. The root/shoot ratio was then determined from sample dry weights. Using this ratio in combination with the above-ground biomass of ryegrass, below-ground root biomass was estimated. The C concentration in both above- and below-ground components of dry pasture was determined by analysing 0.1-0.2 g of ground pasture root sample in a LECO CR-12 analyser using standard methods. C content was then calculated multiplying the C concentration by the estimated or measured biomass.

Litter decomposition
In order to determine the annual addition of C to soil via poplar leaf litter (only), litter decomposition was quantified. To measure litter decomposition, litterbags, 30 cm × 30 cm, were filled with 100 g (wet weight) of poplar leaf material, closed with staples and secured to the ground at random locations within the 'silvopastoral' system. The litterbag mesh size (0.2 cm × 0.2 cm) corresponded to that utilized by Tian *et al.* (1992). A total of 27 bags were used, with three bags retrieved on a random basis every month after placement. The material remaining in each collected bag was cleaned, dried and weighed as described by Anderson and Ingram (1993). Wet/dry ratios were calculated for the initial material by drying material at 65°C for 48 h.

Soil organic carbon (SOC)
Soil samples were collected from a depth of 0-5 cm from both systems to determine SOC content. In the 'silvopastoral' system, five random samples were collected from under tree canopies and five were taken between trees. In the monoculture pasture system, five random samples were taken. Soil samples were prepared for SOC analysis in the lab using standard techniques as outlined by Carter (1993). Prepared soil samples were analysed for SOC using a LECO CR12 Carbon Analyser.

Soil leachate carbon
In order to determine the carbon concentration of soil solution leaching below the pasture root zone (> 30 cm depth) from both systems, tension lysimeters were installed to a 30 cm depth and soil solution samples were taken each month from June to October 2002 and then again in May 2003.

Soil respiration
Soil respiration from both systems was quantified using the soda-lime (lime ($Ca(OH)_2$) with 5-20% sodium hydroxide (NaOH)) absorption method of Edwards (1982). In this method, CO_2 released from the soil is captured in darkened chambers (to prevent plant photosynthesis) inverted upon the ground. The CO_2 is subsequently absorbed by open pans of soda-lime under the chambers over a measured period of time. The CO_2 evolved is calculated as weight gain of the soda-lime.

Results and Discussion

The combination of trees and pasture (a silvopastoral system), as a form of agroforestry, creates a dynamic agroecosystem that, when properly designed, can increase and diversify farm income, enhance animal welfare and wildlife habitat, abate soil erosion and nutrient loading to waterways, protect watersheds from erosion and enhance sinks for GHG in agroecosystems. Silvopastoral land-use systems have not been widely adopted in Canada mainly because of a lack of understanding about interactions between system components (trees, pasture and animals). Research results from the investigations presented here will increase our knowledge of C sequestration potentials in silvopastoral systems.

Thirteen-year-old poplar and Norway spruce were sampled in the current study. The total mean C sequestered in the permanent woody components of the fast-growing hybrid poplar was 14 t/ha (111 trees per hectare; 6 m within-row by 15 m between-row spacing, Table 1). In addition, the C contribution to soil from leaf litter and fine-root turnover for the last 13 years totals approximately 25 t/ha. The total contribution in terms of carbon sequestration over the last 13 years at this experimental site is therefore approximately 39 t C/ha. Theoretically, this also implies that this system has immobilized 143 t of CO_2/h. However, 67.5% of the C added via leaf litter and fine roots was released back into the atmosphere through microbial decomposition (September 2002-September 2003), and therefore the net annual sequestration potential from the trees alone is 1.70 t C/ha per year or approximately 6 t of CO_2/ha per year.

Table 1. Carbon content of individual tree components for hybrid poplar growing in the 'silvopastoral' system. The total C sequestered per tree in permanent tree components, excluding leaves, is 125.9 ± 16.3 (kg). d: diameter.

Tree components	Biomass dry weight (kg)	Carbon concentration (%)	Carbon content (kg)
Leaves	26.8 ± 7.5	43	11.5 ± 3.5
Twigs	14.0 ± 4.1	44	6.2 ± 2.3
Small branches (< 7 cm d)	33.9 ± 11.3	45	15.3 ± 6.2
Large branches (> 7 cm d)	62.0 ± 31.2	45	27.9 ± 17.3
Trunk	135.9 ± 68.8	40	54.4 ± 33.5
Total above-ground biomass	272.6 ± 43.6	-	115.3 ± 17.2
Roots	51.5 ± 10.9	43	22.1 ± 4.7
TOTAL	324.1		137.4

It is also interesting to note that if fast-growing trees (poplars) were replaced with slower-growing trees (e.g. Norway spruce), at the same tree density per hectare (111 trees/ha), the C sequestration in the permanent woody components of Norway spruce was 39.6% of that seen in the poplar (5.55 t/ha compared to 14 t/ha – Table 2).

Table 2. Carbon content of individual tree components for Norway spruce growing in the 'silvopastoral' system. The total C sequestered per tree in permanent tree components, excluding needles, is 49.8 ± 8.4 kg.

Tree components	Biomass dry weight (kg)	Carbon concentration (%)	Carbon content (kg)
Needles	14.7 ± 6.9	51.2	7.5 ± 3.5
Branches	55.7 ± 16.8	51.3	28.6 ± 8.6
Stem	21.7 ± 10.7	49.2	10.7 ± 5.3
Total above-ground biomass	92.1 ± 33.9		46.8 ± 17.3
Roots	21.0 ± 9.2	50.8	10.5 ± 4.3
TOTAL	113.1		57.3

However, in silvopastoral systems, conifer tree densities could be increased to 150 to 170 trees per hectare with little or no adverse effects on pasture production (Gordon and Newman, 1997). In this context, a density of 167 Norway spruce trees per hectare (4 m within-row by 15 m between-row spacing) may prove suitable for Canadian pasture landscapes. At 167 trees/hectare, C sequestration can potentially be enhanced to ~60% of that seen in faster-growing trees (8.35 t/ha compared to 14 t/ha). Pasture (ryegrass) above-ground annual dry biomass yields are given in Table 3 for both 'silvopastoral' (10.82 ± 0.30 t/ha) and monoculture pasture (10.89 ± 0.36 t/ha) systems.

Table 3. Above-ground dry biomass and carbon content of ryegrass in the 'silvopastoral' and monoculture systems, 2002.

	Above-ground biomass (g m^2)		Carbon content (g m^2)	
Sampling time	Silvopastoral	Monoculture	Silvopastoral	Monoculture
June 11	583.1 ± 77.5	537.3 ± 81.5	291.6 ± 38.8	268.7 ± 40.8
July 15	203.3 ± 15.3	243.2 ± 14.5	101.7 ± 7.7	121.6 ± 7.3
Sept. 20	295.3 ± 32.3	308.9 ± 35.5	147.7 ± 16.2	154.5 ± 17.8
Total	1081.7 ± 161.7	1089.4 ± 126.0	540.9 ± 80.9	544.7 ± 63.0

There were no significant yield differences observed between systems. The root/shoot ratio was also calculated and the below-ground annual root dry biomass for both systems was 46.1% of the above-ground dry biomass, on sample plants. Pasture root biomass was therefore 4.99 and 5.03 t/ha for silvopastoral and monoculture pasture systems, respectively. The carbon content of pasture roots was 49.67% and, as livestock will consume most of the above-ground biomass, only root C will be added to the soil carbon pool. Based on a 40% addition to the recalcitrant soil carbon pool (Falk, 1976), the net annual C sequestration by pasture alone in either the silvopastoral or monoculture pasture systems will be 0.99 t/ha per year. As indicated above, the net annual C sequestration with poplars alone was 1.7 t/ha per year. Therefore, in a silvopastoral system with poplars, the net annual C sequestration potential will be 2.69 t/ha per year, whereas, in a monoculture pasture system, the data from this study suggest that the net annual C sequestration potential will only be 0.99 t/ha per year. The data also suggest that silvopastoral systems with fast-growing tree species have the potential to sequester 2.71 times more C than monoculture pasture systems. An annual C sequestration rate of 2.69 t/ha per year in a silvopastoral system would also mean an immobilization rate of 9.87 t of atmospheric CO_2/ ha per year.

There was no significant difference in soil organic carbon (SOC) between the two systems. Soils in the silvopastoral system had a SOC concentration of 2.34% ± 0.17 and in the monoculture pasture system 2.19% ± 0.14 (Table 4).

Table 4. Soil organic carbon (%) in the 'silvopastoral' and monoculture pasture systems.

	Soil organic carbon (%)		
	Silvopastoral		Monoculture pasture
	Under tree canopy	Away from trees	2.19 ± 0.14
	2.46 ± 0.22	2.22 ± 0.12	
Mean	2.34 ± 0.17		2.19 ± 0.14

Although the difference (0.15%) is negligible on a percentage basis, on a per hectare soil weight basis an increase of 0.15% SOC, if significant, will result in 3 t/ha of additional SOC in the first 15 cm of the plough layer (assumption: 15 cm plough layer of soil on 1 ha of land weighs 2000 t). At this time of system development, although the difference in SOC is not statistically significant, this potential addition of SOC in the silvopastoral system is likely due to annual leaf biomass and fine root turnover, which will continue to increase as the system ages.

Table 5. Monthly mean soil respiration rates (CO_2 (g/month m^2) (SE)) in 'silvopastoral' and pasture monoculture systems (soil respiration rates during the winter months were estimated from summer and autumn empirical data).

Month	Silvopastoral	Pasture monoculture
January 2002	34.30	18.99
February 2002	30.98	17.16
March 2002	32.10	17.77
April 2002	33.21	18.39
May 2002	70.66 (0.8)	39.13 (0.4)
June 2002	95.51 (0.8)	52.89 (0.9)
July 2002	113.37 (6.7)	62.79 (3.7)
August 2002	111.16 (1.5)	61.56 (5.6)
September 2002	78.40 (0.4)	43.42 (0.2)
October 2002	56.22 (0.2)	31.14 (0.1)
November 2002	33.20	18.39
December 2002	34.30	19.20
Total yearly emissions (g/m^2)	723.40 (31.0)	400.65 (17.2)

Total yearly emission from the 'silvopastoral' system was significantly higher than that of monoculture pasture (Table 5). The presence of tree roots (as we could not separate tree root respiration from soil respiration) and decomposing litter (leaves and fine root turnover) likely contributed towards the higher respiration rates in the 'silvopastoral' system.

Table 6. Leached soil C concentrations in both systems in order to estimate the total annual C loss[1] via soil leachate (SE). [1]The annual leaching loss was taken to be 200 mm (Gisi, 1997).

	Silvopastoral	Monoculture pasture
Month	C (%)	C (%)
June 2002	0.06 (0.01)	0.04 (0.02)
July 2002	0.07 (0.01)	0.06 (0.00)
August 2002	0.07 (0.02)	0.04 (0.03)
September 2002	0.08 (0.02)	0.07 (0.01)
October 2002	0.08 (0.01)	0.04 (0.01)
May 2003	0.06 (0.01)	0.04 (0.02)
Mean leaching over 6 months	0.07 (0.01)	0.05 (0.02)

There was no statistical difference in leached C concentration between the 'silvopastoral' and monoculture pasture systems (Table 6). However, the C concentration in leachate from the 'silvopastoral' system was slightly higher than that from the monoculture pasture system. A higher amount of organic carbon inputs via leaf litter and tree fine root turnover in the 'silvopastoral' system could have caused this slight difference. These mean values were used to estimate the annual leached C losses from both systems in conjunction with total annual leaching losses (please see model construction section).

In order to determine the net carbon input from poplar leaf litter, a poplar leaf decomposition study was undertaken (Table 7).

Table 7. Poplar leaf litter decomposition (% loss) (2002/2003) in the silvopastoral system.

Time	Remaining mean weights in the litter bag (g) ± SE	Mass loss (%)
September 2002	100 (initial weight)	0
October 2002	72.82 (4.5)	27.2
November 2002	66.77 (3.4)	33.2
March 2003	61.02 (3.0)	39.0
April 2003	56.74 (2.0)	43.3
May 2003	51.35 (5.2)	48.7
June 2003	50.02 (5.0)	50.0
July 2003	43.33 (4.2)	56.7
August 2003	36.50 (2.5)	63.5
September 2003	32.48 (4.6)	67.5

A constant weight (no further weight loss) of decomposing poplar leaves was only achieved after one full year, when 67.5% of the leaf mass was lost. In conjunction with leaf and fine root biomass, this was used to calculate the annual C input from these sources in a 'silvopastoral' system.

Figure 2. Hypothesized carbon budget for poplar-based silvopastoral system. All fluxes are expressed as t C/ha/year and all pools (within parentheses) as t C/ha.

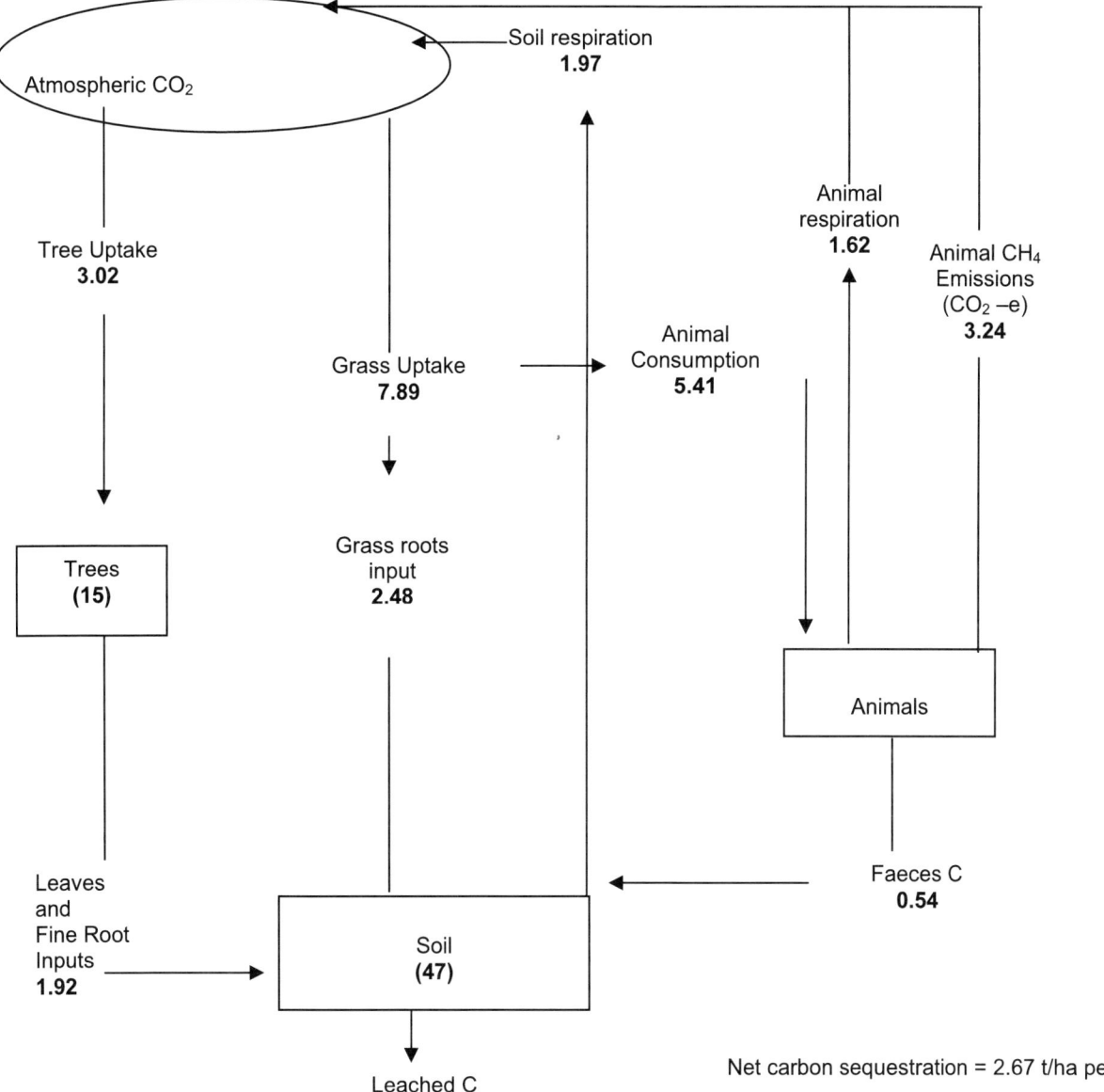

Based on the results derived, C models for a poplar-based 'silvopastoral' system (Figure 2) and for a monoculture pasture system (Figure 3), were constructed using the following calculations, estimations and assumptions. Net above- and below-ground C for poplar was calculated using the data from Table 1 (per tree C, excluding leaves =

125.9 kg × 111 = 13,975/13 = 1075 kg/ha per year or 1.1 t/ha per year). C input from annual leaves and fine roots was calculated from leaf biomass input per year per tree (Table 1). The trees were harvested at year 13. Leaf biomass input for the first 6 years was estimated to be one-half of the current input and the input over the latter 7 years was taken as current-year input. Fine root addition was estimated to be the same as leaf biomass addition (Gray, 2000). Based on the aforementioned calculations, approximately 25 t C/ha has been added via leaves and fine roots in the last 13 years. Therefore, the annual addition was calculated as 1.92 t/ha per year. Above- and below-ground C additions from pasture grass were calculated using data from Table 3 and from pasture root biomass, which was 4.99 and 5.03 t/ha for 'silvopastoral' and monoculture pasture systems, respectively. (Note: C content of oven-dry grass shoots and roots was 50.0% and 49.67%, respectively). It was also assumed that all above-ground grass C will be consumed by sheep and 60% of that will be lost as methane, 30% through animal respiration and 10% will be added to the soil via faeces (Takahashi and Young, 2002). C loss via soil respiration was calculated using data from Table 5 and C loss via soil leachate was calculated using data from Table 6. Note: The annual leaching loss was taken to be 200 mm (Gisi, 1997).

Figure 3. Hypothesized carbon budget for monoculture pasture system. All fluxes are expressed as t C/ha/year and soil pool (within parentheses) as t C/ha.

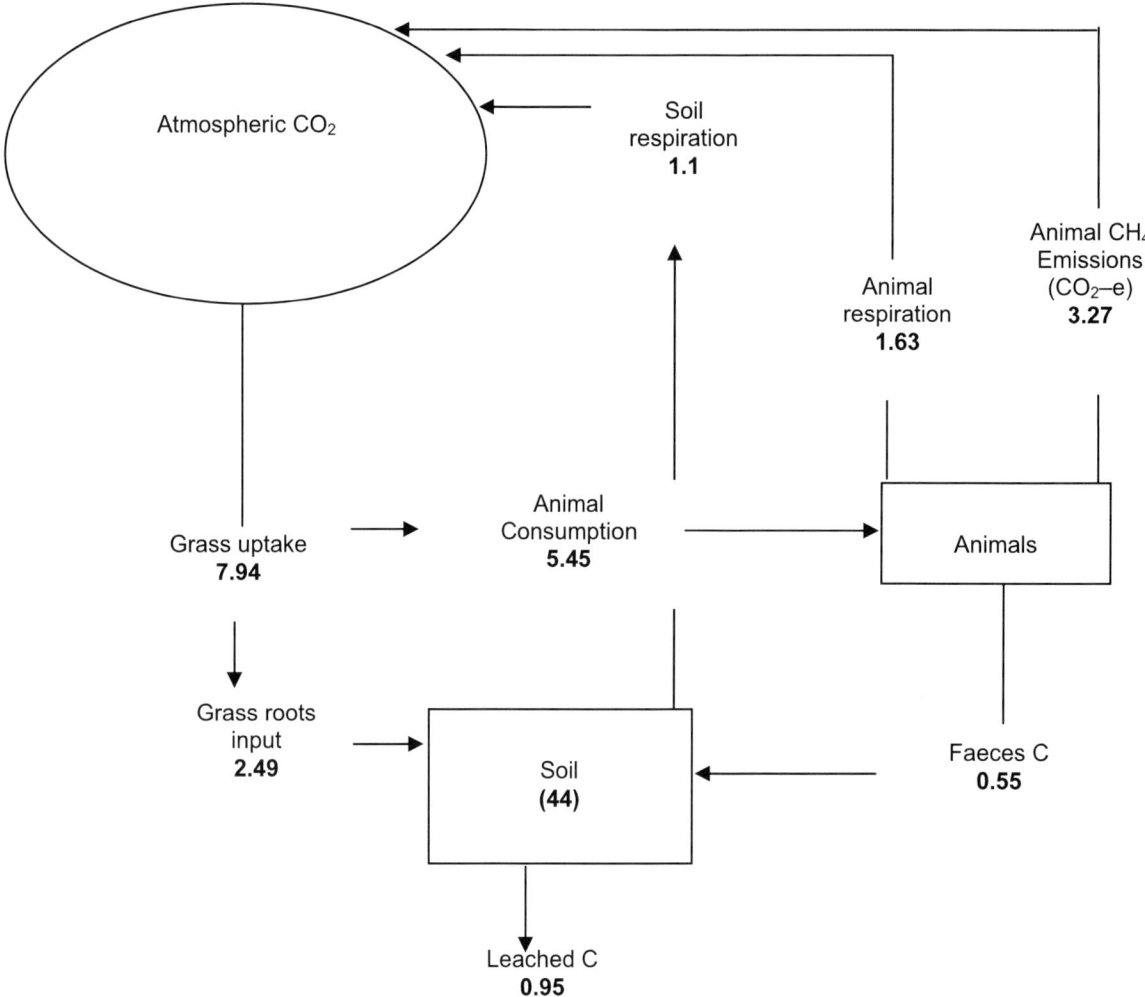

Net carbon sequestration = 0.99 t/ha per year

In order to examine C sequestration values in relation to global warming potentials (GWP), methane emission values for sheep were obtained from the literature. Hegarty (2002) suggests that, depending upon percentage digestibility and dry matter intake, methane emissions from a single sheep could vary between 20 and 25 g/day. Based on a 25 g/day emission and a GWP value of 21 for methane (GWP 21 for methane indicates that this greenhouse gas is 21 times more deleterious than CO_2, in terms of its global warming potential), results suggest that a silvopastoral system with fast-growing tree species should be 'carbon neutral' with 52 sheep per hectare (on an annual basis, excluding GHG emissions from other farm operations). However, the sheep density drops to a minimum of 19 sheep per hectare if there are no trees incorporated into the system. It should be noted that these herd densities are derived based on results from this study, when trees were 13 years old and at peak sequestration potential. During the early tree establishment phase

(0-5 years) and during the late maturity stage the carrying capacity of sheep per hectare will be lower than these suggested herd densities. In addition, conifer-based silvopastoral systems will support lower carrying capacity numbers. Other dairy or beef cattle operations will also have lowered carrying capacities due to higher methane emissions. Further research is also needed on carrying capacity for sheep and other animals in silvopastoral systems; forage production will obviously decline at high tree densities (e.g. Hawke and Knowles, 1997).

In order to set the results obtained from this study into a greenhouse gas mitigation context, sheep operations in the province of Ontario were selected to demonstrate how livestock industries in Canada could potentially become carbon neutral. In 1999 there were 885,000 sheep in Canada, and 225,000 in Ontario. Sheep numbers have been increasing steadily since 1995, even as the numbers of farms decline. There are 4500 registered sheep producers in Ontario and the average herd size on Ontario farms is approximately 66 animals (Ontario Farm Animal Council, 2001). To maintain the average flock size of 66 animals in a carbon-neutral situation, only 1.3 ha of silvopastoral land with a fast-growing fibre tree species is required. Developing silvopastoral systems over a large land base should not be a threat or a problem in Ontario since more than 5 million hectares are currently available for afforestation (Afforestation Studies, 1999). Of this, almost 2.5 million hectares are either pastured or abandoned land suitable for the development of silvopastoral systems. Considering the sheep population of Ontario (225,000) and the net peak sequestration potential (2.7 t C/ha per year), as shown in this study, only 4351 ha of silvopastoral land are required in Ontario to develop a C neutral sheep industry. Note: 225,000 × 25g (methane emission per sheep per day) × 365 (days in a year) × 21 (GWP index)/1000 g (to convert g to kg)/3.67 (to convert CO_2 to C)/2700 (kg C/ha) = 4351.16 ha.

Conclusions

At the national level, Canadian greenhouse gas emissions have been estimated to be approximately 675 million tonnes of CO_2 equivalent (CO_2-e) per year at present. The contribution from the agricultural sector is around 64.4 million tonnes of CO_2-e per year (~10% of the total emissions) (Hastie, 1999). It has also been estimated that the total extent of pasture- or rangeland in Canada is approximately 51 million ha (McKeague, 1975). 64.4 million tonnes of CO_2-e per year could technically be sequestered in 6.4 million hectare of pastureland in Canada, as long as these lands are under silvopastoral systems with fast-growing tree species. Although these numbers will vary depending upon climatic regions, soils, tree species and age, etc., there remains vast potential for this type of land use in Canada, which will also address, in addition to climate change mitigation endeavours, aspects of the animal welfare issue.

Little debate remains on the benefits of providing livestock with an environment that limits physical stress and society in general has expressed a desire for more humane treatment of livestock. However, there are few established systems that can address these concerns and at the same time provide suitable economic returns to the producer. Shade from trees in silvopastoral systems can significantly reduce physiological stress on animals while providing additional economic return from the accrual of equity in wood products. It can be concluded that the incorporation of trees into pasture, especially fast-growing tree species such as hybrid poplars, could significantly contribute towards maintaining carbon-neutral livestock operations in Ontario and nationally. Silvopastoral systems might also provide opportunities to diversify farm economies while contributing to improved air quality, from the perspective of greenhouse gases, and the general health and welfare of livestock.

Acknowledgements

The authors wish to thank the Agriculture and Agri-Food Canada's Livestock Environmental Initiative (LEI) programme and the Ontario Ministry of Agriculture and Food (OMAF) for the financial assistance received. The opinions expressed in this chapter are those of the authors and not necessarily the opinions of the funding agents. We also would like to thank the organizing committee of the Silvopastoralism and Sustainable Management Congress (Lugo, Spain, April 2004) for the opportunity to present this work to the Congress.

References

Afforestation Studies No. 2 and No. 3 (1999) *Forest Sector Table and Sinks Table.* Canadian National Climate Change Process, Ontario, Canada.
Anderson, J.M. and Ingram, J.S. (1993) *Tropical Soil Biology and Fertility. A Handbook of Methods.* CAB International, Wallingford, United Kingdom.
Carter, M.R. (1993) *Soil Sampling and Methods of Analysis.* CRC Press, LLC, Boca Ratón, Florida, USA.
Edwards, N.T. (1982) The use of soda-lime for measuring respiration rates in terrestrial systems. *Pedobiologia* 23, 321-330.
Falk, J. H. (1976) Energetics of a suburban lawn ecosystem. *Ecology* 57, 141-150.
Garrett, H.E., Rietveld, W.J., Fisher, R.F., Kral, D.M. and Viney, M.K. (2000) *North American Agroforestry: An Integrated Science and Practices.* American Society of Agronomy, Madison, WI, USA.
Gisi, U. (1997) *Bodenökologie. 2. Auflage.* Georg Thieme Verlag, Stuttgart, Germany.
Gordon, A.M. and Newman, S.M. (1997) *Temperate Agroforestry Systems.* CAB International, Wallingford, United Kingdom.
Gray, G.R.A. (2000) Root distribution of hybrid polar in a temperate agroforestry intercropping system. MSc. thesis. Department of Environmental Biology, University of Guelph, Guelph, Canada, USA.

Hastie, J. (1999) *Contact'99 – Science and Climate Change.* Soil and Water Conservation Society (SWCS), Autumn Meeting, November 23 1999, Guelph, Canada, USA.

Hawke, M.F. and Knowles, R.L. (1997) Temperate agroforestry systems in New Zealand. In: Gordon, A.M. and Newman, S.M. *Temperate Agroforestry Sistens,* CAB International, Wallingford, United Kingdom, pp. 85-118.

Hegarty, R.S. (2002) Strategies for mitigating methane emissions from livestock – Australian options and opportunities. In: Takahashi, J. and Young, B.A. (eds) *Greenhouse Gases and Animal Agriculture.* Elsevier Science, Amsterdam, The Netherlands, pp. 61-65.

McKeague, J.A. (1975) Canadian Inventory: How much land do we have? *Agrologist* Autumn, 10-12 Ontario Farm Animal Council (2001) http://www.ofac.org/factsheets/fact15.html.

Takahashi, J. and Young, B.A. (2002) *Greenhouse Gases and Animal Agriculture.* Elsevier Science, Amsterdam, The Netherlands.

Tian, G., Kang, B.T. and Brussard, L. (1992) Biological effects of plant residues with contrasting chemical compositions under humid tropical conditions – decomposition and nutrient release. *Soil Biology and Biochemistry* 2, 1051-1060.

Silvopastoral systems in Latin America and their contribution to sustainable development and biodiversity

I. Hernández [1] and M. D. Sánchez [2]
[1]Researcher and International Consultant, Estación Experimental de Pastos y Forrajes "Indio Hatuey", Matanzas, Cuba.
[2]Secretary of Agricultural Development and Hydraulic Resources of the State of San Luis Potosí, Mexico.

Abstract

Since the 1950s, more than half the natural forests of Latin America and the Caribbean have been replaced by pastureland, which can only be maintained by a system which implies the use of important quantities of agricultural consumables in order to main productivity. The majority of this pastureland became impoverished as a result of overgrazing and soil flattening, uncontrolled burning, inefficient use of fertility and the use of inadequate pasture, as well as other activities inappropriate for the development of productive farming. This change in the use of the land had serious environmental consequences, including the loss of soil and biodiversity and availability of water.

Negative environmental consequences, the result of the reconversion of land, at times in which significant technological changes were occurring to modernize the farming industry would seem to leave us with two options: one is to develop intensive farming production systems with modern technology (dependent on a high level of agricultural consumables, fossil fuels, imported grain and a large number of veterinary products) or take advantage of this opportunity to promote an alternative method of production, based on techniques which are beneficial to the environment. These techniques could lead eventually to conservationist or organic farming production systems, whose market is growing internationally.

In this context, it would seem logical that the integration of perennial trees and shrubs on farms is a good means of reducing negative environmental consequences characteristic of traditional production systems, as well as obtaining high levels of production per area, that is, making them more sustainable. Silvopastoral systems incorporate perennial trees and shrubs into farms, those which have a significant interaction with the pasture and/or livestock and can be managed, therefore, as agro-ecosystems.

In countries such as Colombia, Costa Rica, Cuba and Mexico, among others, this is one of farming's productive alternatives. In some countries, significant areas using sustainable farming production models already exist.

Introduction

Since the 1950s, over half of the natural forests in Latin America and the Caribbean have been substituted by pastures which can only be maintained through the utilization of significant amounts of agricultural inputs to sustain their productivity. Most of these pastures were degraded due to overgrazing and soil compaction, controlled burning, nonefficient management of soil fertility and the use of poorly adapted grasses, amongst other inappropriate activities for the development of a productive livestock. This substitution in land use has had serious environmental consequences, including loss of soils, biodiversity and water availability.

The negative environmental impacts of this land transformation, at a time when significant technological changes have "modernized" the livestock industry, seem to leave only two options: to develop intensive animal production systems with modern technology (dependent on high agricultural inputs, fossil fuel, imported grains and a wide range of veterinary products); or, conversely, to take this opportunity to promote an alternative mode of production, based on environmentally friendly technologies. The latter could eventually lead to the development of conservationist or organic production systems, which have found an increasing market niche internationally.

Within this context, it seems logical to integrate woody perennials (trees and shrubs) within cattle farms as a sound way to minimize negative environmental impacts characteristic of traditional production systems, and to increase productivity per unit area, in turn making the system more sustainable. Silvopastoral systems (SPS) integrate woody perennials within farms, in such a way that they interact significantly with the animals and/or grasses, and are therefore managed as an agroecosystem. Some countries already have significant areas under a model of sustainable livestock production. These include Colombia, Costa Rica, Cuba and Mexico, amongst others.

Sustainability of current SPS

From an agroecological perspective, sustainable development implies keeping agroecosystems healthy, i.e. allowing these systems to maintain a functional integrity so they can, over time, provide humans and all living organisms within their boundaries with the basic needs of food, shelter and the other resources they need.

However, development has three fundamental components: economic, social and environmental. None can be ignored when considering the sustainability of a system. Addressing the environment first does not imply a predominance of this element; on the contrary, only when the economic and social objectives are met can the environmental objectives be fulfilled (Steinfeld, 2000).

The integration of woody perennials within animal production systems meets many of these premises. Trees can be established within farms in different spatial arrangements: as a compact block, planted in rows, dispersed within the grassland, used as living fences or as windbreaks. These trees in turn can have multiple uses, and provide a range of environmental, social and economic benefits for the producers.

Soil Productivity. In some cases, trees (roots and organic matter) can improve the physico-chemical properties of soils, reducing loss of soil fertility and erosion, with the obvious long-term benefits for soil productivity.

Climatic stress. Tree shade can reduce the temperature underneath by 2-3°C, improving the comfort zone for the animals, thus influencing their behaviour, productivity, reproduction and survival rate. It also reduces incidence of skin cancer, and other problems related to photosensitivity.

Biodiversity conservation. An evident environmental benefit is the increase and conservation of biodiversity. Trees offer food and shelter to a wide range of animals including mammals, birds, arthropods, etc.

Carbon sequestration. Trees have the potential to capture and fix C from the atmosphere. With a 50-year rotation, an SPS in the humid tropics can fix between 150 and 198 Pg of C/ha in that period of time (Dixon, 1995).

Animal feed. Feeding animals with tree foliage is a tradition in tropical and subtropical America in the case of smaller animals like deer, goats, rabbits, etc. Since the 1980s systematic studies have been undertaken regarding tree nutrient quality, agronomical management and potential to intensify animal production systems.

Cultural continuity. SPS are part of an ancient tropical agricultural tradition: the association of trees and crops. Some of the most extended practices include shaded coffee and cocoa.

Production risks. An economic benefit is that of economic risk reduction through the system's diversification. Besides animal products, a farm can produce in the same area fruits, timber, firewood, folk medicine, fibres, etc.

Biodiversity in SPS

Biological diversity, or biodiversity, refers to the variability of all living organisms and the ecological complexes in which they occur (ecosytems). In SPS this variability can be planned, where the producer makes a decision on the species and animal breeds or grasses, trees and other plants he/she wishes to introduce into the system; or it can be spontaneous, i.e. where the previous vegetation is invaded and replaced by other species of plants (e.g. weeds) and other animal species, like birds, mammals and insects, are also introduced. Although the spontaneous introduction of other species does not always have a beneficial effect, many species and communities can provide a range of direct economic benefits or perform other services beneficial to human life (Harvey, 2001).

Numerous publications quantify biodiversity of different living organisms within SPS. These systems may integrate woody perennials with different purposes and diverse spatial arrangements. Usually protein banks and trees dispersed in grasslands for browsing or other purposes are considered part of SPS. In some cases, woodlots, windbreaks and other arrangements associated with grasses and/or animals are also considered part of an SPS.

A study on young windbreaks in Monteverde, Costa Rica, showed the presence of over 90 species of tree seedlings associated within the tree rows, which represent a quarter of all tree species found in the region. In addition, they were visited by over 80 species of birds, including frugivorous, insectivorous and nectarivorous, birds as well as numerous species of insects, many of which are beneficial like pollinators and pest parasites (Harvey, 2001).

Similar studies were carried out on trees interspersed with pastures in Veracruz, Mexico, and Monteverde, Costa Rica. In both cases a large number of arboreal species was found (98 and 190, respectively), as well as epiphytes, lianas, mosses and other flora. In addition, studies in Belize with the same design reflected tree capacity to attract birds and other animals (Harvey, 2001).

SPS provide edaphic and climatic conditions which favour the development of a rich soil fauna. When comparing ecological diversity indexes of soil organisms, Sánchez *et al.* (1998) observed higher numbers of species, relative abundance, diversity and dominance in SPS than in areas without trees (Table 1).

Table 1. Soil macrofauna in systems with and without trees. Source: adapted from Sánchez et al., 1998.

Class	Order	SPS	System without trees
Insecta	Coleoptera	180	88
	Orthoptera	10	-
	Dermaptera	8	2
Miryapoda	Dyplopoda	10	4
Crustacea	Isopoda	26	8
Oligochaeta	-	66	68
TOTAL		300	170

Soil biodiversity, in turn, can act positively in other aspects of animal production *per se*. For example, there is a considerable reduction in the number of eggs of gastrointestinal nematodes in bovine manure within SPS when compared with those in grass monocrops (Table 2). This is probably due to the fact that contaminated manure is eliminated by coprophagous organisms before the parasites reach their infectious stage (Soca and Arece, 2002).

Table 2. Parasitic infestation (nematodes) in young animals Holstein x Cebu in SPS and grasslands without trees under Cuban conditions. Source: adapted from Soca and Arece (2002). * Values transformed by log 10.

Systems	Egg numbers (per g)*		
	Annual	Dry season	Rainy season
Silvopastoral	2.48a	2.36a	2.64a
Grassland	3.013b	2.97b	3.14b
SD±	0.06	0.11	0.07

Advances in SPS and successful development models

SPS in Latin America and the Caribbean are no longer a concept exclusively relating timber trees with pastures and animals, and their importance has extended to include the value of foliage, fruits and even tree bark as a source of animal feed. This area has advanced both in research and in extension to production farms.

Cuba currently has livestock farms covering over 16,000 ha of woody perennials used as fodder both in direct grazing and cut-and-carry systems. The most commonly used trees are *Leucaena leucocephala* and mulberry (*Morus alba*), as well as sugar cane banks. Cuban data suggest that *Leucaena*-fed cattle can increase their daily weight when compared with grass-fed animals (550-800 g/day in the rainy season and 350-670 g/day in the dry season). In other experiments in Colombia, milk production values reached 17,026 kg/ha per year (Table 3), which resemble those of temperate climates (Molina et al., 2001). An additional evident benefit of trees is that they maintain their foliage during the dry season when the grass dries out. In Veracruz and other Mexican states, SPS based on *Leucaena* associated with pastures and mulberry protein banks are being increasingly accepted by dairy farmers in the lowland areas. In Costa Rica and Mexico these systems are being promoted in highland dairy farms (over 800 m asl) and combine cut-and-carry SPS with fodder from mulberry and sugar cane banks, with confined or semi-confined animals (Table 3).

Promising work is being developed on farms in Costa Rica, Cuba and Panama using agroforestry modules with mulberry banks and other sources of fodder for confined dairy goats, which can produce an average of 2 kg of milk/animal per day; in turn, goat manure is used to fertilize the fodder banks (Table 3).

Table 3. Successful models of SPS developed in Latin America and the Caribbean.

Country	Type and purpose of the production	System	Production results	Other products
Colombia, Cuba, Mexico	Large and medium-size farms. Meat and cow milk in grazing/browsing systems	*Leucaena* associated with grass, fodder banks of sugar cane and mulberry	7-11 kg of milk and weight gains of 400-800 g/animal per day	Seeds, organic fertilizer, agro-ecotourism
Costa Rica, Mexico	Large and medium size highland farms (over 800 m asl). Cow milk in confined and semi-confined systems	Integrated systems with banks of sugar cane and mulberry, and grazing	12-15 kg of milk/animal per day	Biogas, organic fertilizer, agro-ecotourism
Costa Rica, Cuba, Panama	Small and medium size producers. Goat milk in confined systems.	Agroforestry systems with goat milk with mulberry banks and other food sources	1-3 kg of milk/animal per day	organic fertilizer, cheese, yogurt, agro-ecotourism

An increasing phenomenon with SPS is the interest expressed by the private sector in their research and adoption. For example, in Colombia, Costa Rica, Mexico and Panama there is manifest interest in saving external inputs, on the basis of a good cost/benefit ratio, with output diversification which include agroecotourism, energy production, value added of dairy products (cheese and yogurt) and a greater interest in environmental conservation.

Although highly promising, the methodology of farms which include cattle and trees still needs to be improved. More information is needed on how to manage trees to maximize biomass production, minimize competition for resources (e.g. light, water, and nutrients) with the grasses, how and how frequently to prune them and how to deal with potential toxicity problems. At both farm and regional level it would be important to quantify other economic and environmental benefits that trees could provide. It is still necessary to conduct a cost/benefit analysis for the majority of tree species (Hernández et al., 2000).

Conclusions

SPS are an alternative for animal production in Latin America and the Caribbean which can bolster environmental policies and form part of the working tools of diverse institutions and commercial enterprises linked to the region's public and private sectors. This is due, among other aspects, to their ability to function efficiently in different agroecosystems because they provide coherent technical proposals for the farmer's needs and for the current interest in conservation and organic production. All this magnifies the current paradigm of resource management and environmental protection in Latin America.

References

Dixon, R.K. (1995) Agroforestry systems and greenhouse gases (Sistemas agroforestales y gases de invernadero). *Agroforestería en las Américas* 2 (7), 22. (In Spanish.)

Harvey, C.A. (2001) Agroforestry and biodiversity (Agroforestería y biodiversidad). In: Jiménez, F., Muschler, R. and Köpsell, E. (eds) *Funciones y Aplicaciones de los Sistemas Agroforestales*. CATIE, Costa Rica, p. 187. (In Spanish.)

Hernández, I., Benavides, J.E. and Martín, G. (2000) Fodder trees as alternative for an environmental and intensive husbandry (El corte y acarreo de los árboles forrajeros como una alternativa en una ganadería ambiental e intensiva). In: *Memorias. IV Taller Internacional Silvopastoril "Los Arboles y Arbustos en la Ganadería Tropical"*. EEPF "Indio Hatuey", Matanzas, Cuba. (In Spanish.)

Molina, C.H., Molina, C.H., Molina, E.J., Molina, J.P. and Navas, A. (2001) Advances in the implementation of high tree density silvopastoral systems. In: Ibrahim, M. (ed.) *International Symposium of Silvopastoral Systems and Second Congress on Agroforestry and Livestock Production in Latin America*. CATIE, San José, Costa Rica, p. 299.

Sánchez, S., Hernández, M. and Simón, L. (1998) Diversity of soil organisms under a silvopastoral system (Diversidad de los organismos del suelo bajo un sistema silvopastoril). In: *Memorias III Taller Internacional Silvopastoril*. EEPF "Indio Hatuey", Matanzas, Cuba, pp. 295-297. (In Spanish.)

Soca, M. and Arece, J. (2002) Parasitic diseases in bovine and ovine livestock in silvopastoral systems (Comportamiento de las enfermedades parasitarias en bovinos y ovinos bajo condiciones de sistemas silvopastoriles). In: *II Conferencia Electrónica Nacional "Los sistemas silvopastoriles una alternativa para la producción animal sostenible"*, EEPF "Indio Hatuey". Matanzas, Cuba. (In Spanish.)

Steinfeld, H. (2000) Animal production and environment in Central America (Producción animal y medio ambiente en Centroamérica). In: Pomareda, C. and Steinfeld, H. (eds) *Intensificación de la Ganadería en Centroamérica*. CATIE, FAO, SIDE. Nuestra Tierra Editorial, San José, Costa Rica, p. 334. (In Spanish.)

Compaction and erosion: effects on soil ecology and soil quality

A. Paz González and E. Vidal Vázquez

Área de Edafología y Q. Agrícola, Facultad de Ciencias, Universidade da Coruña, A Zapateira, 15.071, Coruña, Spain.

Abstract

Soil degradation is a severe global problem. It has notable adverse economic and ecological impacts on local, regional, national and global scales. Important soil degradative (degradation) processes with severe adverse impacts are soil compaction and soil erosion among others. Decline in inherent soil quality can occur because of compaction and erosion, through adverse effects on many physical chemical and biological properties of the soil. This chapter focuses on physical and biological or ecological effects of soil compaction and erosion. As agriculture becomes more and more mechanized concerns for compaction of both the surface soil layer and the subsoil continue to increase. Soil compaction alters basic soil properties such as pore volume and pore size distribution, macropore continuity and soil strength. These properties have a large influence on elongation of plant roots and on movement of water, air and heat in soils. Soil erosion is considered to be the major cause of soil degradation. One-sixth of the world's soils have already been degraded by water and wind erosion. In Europe, soil degradation due to erosion is probably the most important environmental problem caused by conventional agriculture. Soil erosion reduces land productivity, degrades soil functions and challenges agricultural sustainability. Soil erosion on agricultural land is affected by different management practices and varies with landscape position and land use. Erosion can play an important role in the sequestration or release of carbon, nitrogen and phosphorus. The nutrient enrichment or eutrophication in surface waters is most commonly related to erosion from agricultural soils; in some regions this may be the most important reason to protect soil resources. Erosion leads to depletion of plant-available water and a reduction in water use efficiency. Climate change together with changes in land use, farming practice and population pressure on the land is thought to constitute a threat of increased erosion. Maintaining and improving soil quality require the protection of soil resources from degradation; this issue is discussed with particular emphasis on soil erosion, the most widespread and deleterious among the degradation processes. The most important possibilities to reduce compaction by heavy machinery are outlined briefly. Integrated soil conservation measures at the farm and catchment level have also been developed successfully and they have proved to be the most reliable options to limit soil erosion.

Introduction

Soil degradation is a severe global problem. It has notable adverse economic and ecological impacts on local, regional, national and global scales (Zoebish and Dexter, 2002; Lal *et al.*, 2003). Sustainable management of soil depends on a thorough understanding of its attributes, the processes moderating its ecosystem services or functions of terrestrial importance, and transformations that occur through its interaction with the environment. Important attributes of soil include the following:

- It is non-renewable over the human timescale of decades to centuries.
- It is unequally distributed over the landscape and among biomes/ecoregions.
- It is susceptible to misuse and mismanagement.

Degradation means adverse changes in attributes, leading to a reduced capacity to function. Thus, soil degradation means adverse changes in soil properties and processes over time. These adverse changes can be set in motion by disturbance of the dynamic equilibrium of soil with its environment by either natural or anthropogenic perturbations.

Soil degradation is defined as diminution of soil's potential or actual utility, and reduction in its ability to perform ecosystem functions (Poch, 2002). In other words, it is the decline in soil quality leading to a reduction in biomass productivity and in its ability to moderate its environment (water and air). Two types of soil degradation processes - physical, and biological - are shown in Table 1. Note that soil compaction and soil erosion by water are two of the major threats to the soil resource. Both processes reduce the main functions of a soil. Also chemical degradation processes, i.e. acidification, nutrient depletion, contamination, have to be taken into account.

Table 1. Types of soil physical and biological degradation (Lal *et al.*, 2003).

Type	Degradation process
Physical	Breakdown of soil structure. Crusting and surface sealing
	Compaction, surface and subsoil
	Reduction in water infiltration capacity. Increase in runoff rate and amount.
	Inundation, waterlogging and anaerobiosis
	Accelerated erosion
	Desertification
Biological	Depletion of the soil organic carbon pool. Decline in soil biodiversity. Increase in soil-borne pathogens

Soil quality refers to the capacity of the soil to perform specific functions (Doran *et al.*, 1996; Carter, 2002; Gregorich, 2002). According to Karlen *et al.* (1997), the main functions of a soil include: providing a physical, chemical and biological setting for living organisms, regulating and partitioning water flows, storing and cycling nutrients and other elements, supporting biological activity and diversity for plant and animal productivity, filtering, buffering, degrading, immobilizing and detoxifying organic and inorganic materials, and providing mechanical support for living organisms and their structures.

Soil quality cannot be expressed by a single property. It has been recommended that the evaluation of soil quality should be based on soil functions. A general sequence of how to evaluate soil quality is to (1) define the soil functions of concern, (2) identify specific soil processes associated with those functions, (3) identify soil properties and indicators that are sensitive enough to detect changes in the functions or soil processes of concern, (4) evaluate changes in soil functions based on indicators, and (5) combine evaluations of defined soil functions to assess soil quality.

Soil quality indicators are physical, chemical and biological properties, processes and characteristics that can be measured to monitor changes in the soil. Soil physical, chemical and biological degradative processes interact between them and exacerbate the adverse impacts on biomass production and the environment.

Figure 1. Interaction between soil degradation processes and decline of soil quality.

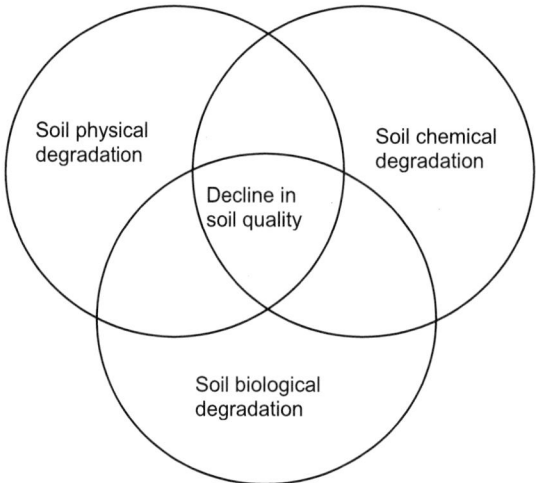

Accelerated erosion driven by anthropogenic forces is the most widespread soil degradation process. One-sixth of the world's soils have already been degraded by water and wind erosion. In Europe, soil degradation due to erosion is probably the most important environmental problem caused by conventional agriculture, seriously affecting nearly 157 million ha (16% of Europe). Most EU countries are affected by this problem, with around 25 million ha threatened by erosion in Western and Central Europe and over 50% of the agricultural land classified as having a medium to high risk of erosion (Ludwig *et al.*, 1995; EEA, 2000; Valcárcel *et al.*, 2003). In the Mediterranean area, soil erosion and degradation can result in losses of 20 to 40 tons of soil per hectare during extreme events.

It is often reported that more than 30 million ha of farmland are classified as irreversibly degraded by soil compaction even only in Europe, while worldwide more than 80 million ha are deformed as a result of such non-site- and time-adjusted agricultural and/or perhaps also forest management strategies (Horn *et al.*, 2000).

In agriculture, soil compaction and soil erosion by wind and water are classified as the most harmful processes. They do not end just in a reduction in productivity of a site, but are also responsible for groundwater pollution, gas emission and a higher requirement of energy input in order to obtain a target crop yield. This chapter focuses on physical and biological or ecological effects of soil compaction and erosion.

Soil compaction

Soil compaction as considered from the viewpoint of the plant ecologist and agronomist is an undesirable consequence of mechanization that reduces the soil's biological productivity and - in extreme cases - makes it unfit for plant growth (Soane and van Ouwverkerk, 1994; Alakkuku, 1996). In this chapter, the term *compaction* is used for a process in a three-phase soil system induced by a mechanical stress, often caused by machinery traffic and characterized by a decrease in volume (an increase in density), mainly under extrusion of air. The term *compactness* is used for the state of the soil, being the net result of various loosening, compaction and natural processes.

Compaction caused by farm equipment or grazing animals has increased dramatically during recent decades, the most important source being wheel traffic by off-road vehicles. However, not only has the mass of the agricultural and forestry machinery been enlarged three- or fourfold, but also the frequency of wheeling has increased by the same proportion. In mechanized agriculture, subsoil compaction by machinery with high axle load is thought to be one of the major long-term threats to soil productivity. Several aspects of machinery-induced soil compaction have been extensively reviewed (Soane and van Ouwverkerk, 1994).

Soil compaction alters basic soil properties such as pore volume, pore size distribution, macropore continuity and soil strength. Increased bulk density and reduced pore space have a large influence on elongation of plant roots, activity of soil fauna and microorganisms and on storage and movement of water, air and heat in soil. Moreover, hard and compacted layers require more energy for tillage (Håkansson et al., 1998; Hartge and Horn, 1999).

Critical limits have been established for specific plant growth factors affected by compaction. For example, penetration resistance indicates the relative difficulty of root elongation through soil and is greatly influenced by compaction. It is also relatively fast and easy to measure. A value of 2-2.5 MPa is often cited as a critical penetration resistance (cone index) beyond which plant root elongation is severely restricted (Taylor, 1971). The aeration of the soil is also greatly affected by compaction, and an air-filled porosity of 10% is frequently cited as the minimum value that can allow sufficient oxygen supply to sustain plant growth (Stepniewski et al., 2002). However, critical limits depend on several other factors. For example, plants growing under high evaporative demand will be more sensitive to limiting root development and/or low unsaturated hydraulic conductivity than plants growing under low evaporative demand.

While the assessment of compaction effects on individual soil properties can be relatively easy, the assessment of compaction effects on biological systems (such as plant growth) is extremely difficult. This is due to (1) the high degree of interaction between several soil factors, (2) a very dynamic environment in which future responses are mediated by past responses, and (3) variable and unpredictable climatic conditions. All these factors interact to make it impossible to determine one critical limit for all aspects of soil compaction. Rather, different limits are required for various aspects and environments. Soil compaction generally results in a suboptimal use of crop production inputs such as fertilizers, herbicides or fuel. It increases the demand for tillage and the energy required for each operation. It reduces the uptake of plant nutrients by the roots, thus leaving larger quantities of nutrients unused and prone to leaching. It may increase denitrification under wet conditions. It will decrease water infiltration and increase runoff, erosion and transport of chemicals to the aquatic systems. However, a soil that is too loose may be more vulnerable to water erosion (Lal et al., 1997).

Although ploughing loosens a compacted soil, it may not alleviate all effects of compaction on soil structure. In spite of annual ploughing, negative effects on crop growth may persist for a 5-year period, and they increase with the traffic intensity, ground contact pressure, soil water content and clay content (Håkansson et al., 1998). The reasons are poorly understood, because of complex relationships between different factors. Compaction depth varies with equipment, soil type, moisture content, organic matter and texture. In the upper part of the soil profile, the incidence of compaction is determined mainly by the ground contact pressure of the running gear; however, at greater depths the axle load is a more important factor (Horn and Rostek, 2000). Soils that are low in organic matter with a high content of silt or fine sand are most susceptible to compaction. Traffic on wet arable soils typically causes significant compaction to depths > 30 cm.

Subsoil compaction is characterized as one of the most harmful and persisting degradation phenomena. Traffic by vehicles with high axle loads often causes compaction in deep subsoil layers and very persistent crop yield reductions (Håkansson and Reeder, 1994; Håkansson et al., 1998). Moreover, traffic-induced subsoil compaction cannot be completely alleviated by subsoiling. The consequences of subsoil compaction phenomena also result in a more pronounced mass transport by water in the traffic lanes. In addition the ploughpan impermeability also causes water interflow, with an induced consecutive submerged soil suspension mass transfer.

In a system with reduced tillage, compaction effects tend to accumulate and be more persistent than in a system with ploughing. Such a system implies that either intensity, depth or frequency of mechanical loosening is decreased or, in the case of direct drilling, the soil is not loosened at all. In clay soils, conditions may still be adequate due to improved continuity of the macropore system, but compaction may prevent continuous use of reduced tillage or direct drilling, especially on sandy soils.

Soil organisms as a part of the edaphon suffer in the compacted soil from the diminished pore space, the reduced pore continuity, the change in water and air conductivity and, if they burrow through the soil, from the increased energy demand (Larink and Schrader, 2000).

Soil erosion

Erosion is a process that removes and redistributes soil. Both water and wind erosion remove topsoil, which is the soil layer best fitted to support life. Loss of all or part of this surface layer impairs the ability of the soil to produce a crop by reducing its fertility and its ability to accept and store water and air. Although some erosion takes place gradually, most results from extreme weather events, such as a windstorm or heavy rainfall.

The materials removed may be redeposited in some nearby leeward or low-lying area with little or no apparent consequence to the environment at that site. In some areas, however, the materials are transported to streams, rivers and lakes - even to oceans -, which may considerably affect the water quality. Each further loss of topsoil compounds the effects of erosion, so the soil increasingly loses its ability to produce crops and to regulate and partition the flow of water in the environment. As soil fertility declines, lost nutrients are often replaced by applying fertilizer, and the chance of nutrient loading in the runoff during subsequent erosion events increases. Water erosion selectively removes finer, lighter particles from the soil surface, leaving coarser particles behind. As organic matter and clay particles are removed from the soil surface through erosion, attached nutrients are relocated in the landscape. Loss of these fine particles also impairs the ability to store nutrients, reducing soil fertility. As erosion proceeds, the food source for organisms is reduced, leading to decline in soil populations (Lal et al., 1997).

Soil erosion is a natural process and all soils have an inherent erodibility, based on soil features, topography and climate. However, human activity such as tillage, livestock grazing and urban development can greatly accelerate natural rates of erosion. Most changes in land use are a direct and immediate cause of increased risk of erosion. Disturbances, including forest affected by fires, construction and mine sites and roads, also accelerate erosion rates.

Soil erosion usually degrades soil quality and a soil of poorer quality is less able to withstand further erosion, thus creating a downward spiral of soil degradation (Ditzler, 2002). Examples of negative disturbances leading to degradation or decline in soil productivity and environmental moderation capacity are: (1) Water erosion rates accelerate with increasing erosion. As a consequence, erosion reduces water infiltration, which causes more runoff and hence more erosion. (2) Erosion reduces plant cover and plant residue cover, promoting higher erosion rates. (3) Human pressures on the land convert crop rotations to monocultures, creating more soil erosion, creating more pressures on the land, creating more monoculture. (4) Sheet erosion leads to rill erosion (faster erosion), leading to gully erosion (even faster rates). (5) Overgrazing producing soil erosion creating less grass, creating a greater degree of over-grazing. (6) Population pressures forcing croplands onto steeper slopes, causing increased runoff and erosion debris to damage croplands down on the valley floor. (7) Human pressures on the land force more overgrazing on adjacent mountainsides causing increased runoff and erosion debris to damage croplands down on the valley floor.

Many studies have investigated effects of erosion on physical and chemical properties. Erosion degrades the soil of nitrogen, phosphorus, potassium (Kilmer, 1963; Sharpley and Rekolainen, 1997) as eroded sediment usually contains higher amounts of plant nutrients than do bulk soils (Stoltenberg and White, 1953). Erosion reduces organic carbon (Mannering et al., 1985; Jones et al., 1989; Magdoff, 1992) and leads to depletion of plant-available water (Ebeid et al., 1995; Aguilar et al., 1998;) and a reduction in water use efficiency (Tenge et al., 1998). Thus, erosion can also play a very important role in the sequestration or release of carbon, nitrogen and phosphorus. Carbon and nitrogen are strongly associated with eroded organic soil, while much phosphorus is lost through adsorption on clay soil fractions.

The accelerated nutrient enrichment or eutrophication of surface waters has become a significant environment problem in many developed countries and agriculture has been identified as a significant phosphorus source. Phosphorus enrichment of fresh waters is considered to be the most serious problem facing the ecology of fresh waters in many countries and diffuse sources in runoff and erosion make a major, and increasingly the dominant, contribution to phosphorus inputs to water (Ditzler, 2002; Lal et al., 2003).

Reductions in organic matter are closely linked to reductions in water-stable aggregates. A high significant correlation was obtained for the relationship between aggregate stability and organic matter and some properties associated with it, suggesting that organic matter is the main factor responsible for the stabilization of aggregates in these soils (Tisdall and Oades, 1982; Chaney and Swift, 1984). Le Bissonnais (1996) referred to the soil organic matter content as the most important property in influencing aggregate stability. Water-stable aggregation is closely linked to soil erodibility (Le Bissonnais and Arrouays, 1997; Barthes et al., 1999). Water-stable aggregates are also important in protecting soil carbon compounds from microbial attack; when aggregates collapse, the organic compounds held in close association with clays are exposed to attack by organisms such as soil bacteria (Angers et al., 1997).

Organic matter loss may significantly affect also detachment, infiltration rate and soil losses of crusted surfaces. In general, the lower the aggregate stability at the soil surface, the higher the susceptibility of the soil to detachment and to seal formation (Le Bissonnais, 1996). Consequently, under seal formation conditions, the infiltration rate and the inherent susceptibility of the soil to inter-rill erosion are affected by the upper soil layer properties and they should be correlative. Ben-Hur and Agassi (1997) found that inter-rill erodibility factor decreased exponentially as the final infiltration rate of the soil increased, for a wide range of soils with differing soil properties. Using soils with different organic matter contents, a significant decrease in soil detachment or an increase in aggregate stability was evidenced as the organic matter content increased (Ekwe, 1991; Fullen, 1991; Guerra, 1994; Whalen et al., 2003). Organic matter content of 3% was found to be a threshold value below which the aggregates were unstable and the soil erodibility was high. The effect of organic matter content on soil sealing and soil infiltration rate was analysed by comparison of soil samples collected from adjacent cultivated and grassland field with an organic carbon content of 2.3% and 3.5%, respectively (Lado *et al.*, 2004). The soil in both fields was sandy loam. Samples with different aggregate sizes from each of the two soils were subjected to 80 mm of distilled water via a rainfall simulator. Figure 2 shows surface crust obtained at the end of the experience for aggregate sizes < 2 mm. The final infiltration rates ranged from 4.2 to 5.2 mm/h in the low-organic matter soil and from 5.8 to 10.8 mm/h in the high-organic matter soil. A higher organic matter content also reduced aggregate breakdown and prevented soil dispersion, which, in turn, reduced seal formation and soil loss and increased the infiltration rate.

Figure 2. Water erosion on cropland, December 2003, Mabegondo, Corunna, Spain.

In characterizing soil quality, biological properties have received less emphasis than chemical and physical properties, because their effects are difficult to measure, predict or quantify. As these biological/ecological indexes of soil are dynamic, they will require effective monitoring and assessment. Information about effects of farming systems and erosion risk/rate on soil biological parameters is therefore very limited. Soil biological properties include, for example, soil microbial biomass and enzymes and their activity. Soil microbial biomass is an important component of the soil organic matter that regulates transformation and storage of nutrients (Dalal, 1998). The effects of tillage, crop rotations and soil type on organic C and nutrient turnover can be assessed by following nutrient pools and activity associated with the soil microbial biomass. The latter also has been shown to be a sensitive indicator of differences in sustaining cropping systems (Anderson and Domsch, 1989). The toxicity of pollutants and the degradation of organic compounds, like pesticides, can be monitored by following changes in the soil microbial biomass. Microbial activity (Chan and Heenan, 1999) is also associated with aggregate stability through specific structure-stabilizing compounds (Haynes, 1999; Martens, 2000). Any reductions in soil fauna associated with erosion will also have a detrimental influence on aggregate stability (Katterings *et al.*, 1997; Marinissen and Hillenaar, 1997). There is evidence that enzyme activity in sediments is higher than in corresponding topsoil, and that with increasing slope steepness microbial activity decreases (Liu *et al.*, 1995).

Soil microflora is a potential indicator of soil quality and is influenced by different farming systems. Especially twine shaped and hypha growing microorganisms (bacteria and fungi) are able to stabilize soil aggregates and reduce the susceptibility of soils to erosion. In forest soils hypha length up to 3 km/g soil occur, while intensively used arable soils reach only 5 to 10 m hypha length/g soil. Hypha length measured by a microscopic method (Hartman *et al.*, 1997) seems to be a capable and workable indicator of biological aggregate stability.

In summary, the interaction of accelerated soil erosion and soil quality is complex. Soil erosion usually reduces soil quality and a soil of poorer quality is prone to further erosion and further degradation. Anthropogenic induced soil erosion on agricultural land is a result of farming system affected by different management practices and varies with landscape position. It influences most of the soil attributes that determine soil quality, such as cation exchange capacity, soil organic matter and structure stability (Ditzler, 2002). The main consequences of erosion are not only on-site: soil degradation, declining soil fertility, limiting infiltration capacity and water storage. Off-site impacts include eutrophication of watercourses and lakes, destruction of wildlife habitats, silting of reservoirs and rivers and flood damage to property. Runoff prevention and soil protection have beneficial effects in reducing flood risk, especially against a background of climate change.

Soil conservation for improving soil quality

Although soil degradation has taken place since cultivation first began, in the past the high value placed on maximum crop yields came frequently at the expense of soil health. Nowadays the problem of soil degradation receives more attention and is better understood. However, soil quality will not be maintained or improved unless corrective action is taken, which must involve both stopping practices that degrade soil and starting practices that enhance it.

One of the reasons to protect soil resources from degradation is the reduced ability of society to produce sufficient food due to loss of quality and depth of soils and the effects of off-site pollution associated with erosion. These include silting of dams, pollution of watercourses by agricultural chemicals and damage to property by soil-laden runoff. Soil productivity is not the only, and in some regions even not the most important, reason to protect soil resources. Soil and water quality are intrinsically linked. Preventing water pollution by nutrients, pesticides, sediments and other pollutants is difficult and significantly more expensive if soil degradation is not controlled (Sharpley and Rekolainen, 1997).

Practices that enhance soil quality build up and protect soil organic matter and soil structure. These practices include: conservation tillage, residue management, contour cultivation on hilly land, application of organic amendments, reducing fallow by extending crop rotations, water management and erosion control. Most agricultural soil degradation is the result of inappropriate farming techniques. The management practices needed to improve soil quality not only exist, but are already being used. Trends in soil quality will depend on how quickly these practices are adopted by producers who are still using conventional farming methods. Conservation tillage generally improves soil health by reducing the mechanical disturbance of the soil, protecting the soil surface with residue cover and adding more organic matter to the soil.

Conventional tillage breaks down soil structure and increases the risk of water erosion. Tillage machinery causes tillage erosion and compaction of the soil. Some traditional cultivation practices, as for example monoculture, fallowing or up- and down-slope cultivation, also result in deteriorating soil structure because they promote erosion and the loss of soil organic matter.

In the past decade promotion of conservation farming has focused on individual management practices, such as conservation tillage. However, because each area of land has its own characteristics, not all conservation practices are suitable for all areas. Conventional tillage may be superior to conservation tillage in some areas where the soil benefits from being broken up occasionally. Furthermore, conservation practices are usually most beneficial when used in combination, such as conservation practices plus continuous cropping with an extended rotation.

Aggravation of the problem of erosion is not universally due to the intensification of agricultural activities. In many regions, socio-economic problems and erosion are intertwined: agricultural systems in marginal, i.e. sloping or mountainous areas, have a hard time competing with those from lowlands, resulting in bad agricultural practices, migration and abandonment. As an example, the scale of grazing-land degradation has become so extreme that the word "desertification" has been devised mainly to describe it. Grazing animal populations tend to expand in parallel with human populations, doubling every five decades. Carrying capacity of grazing lands decrease in parallel with topsoil loss. Topsoil loss increases with increased bare soil and water runoff. Runoff increases with soil compaction by animal hooves. Grazing-land degradation rates are at least proportional to the difference between grazing-animal population and carrying capacity. Hence degradation rates must continually accelerate until action is taken (Lal *et al.,* 2003).

Management practices should be combined in farm systems that are aimed at keeping soils healthy and productive. These systems require the usual decisions about the variety, pattern and sequence of crops, tillage methods and levels of inputs, including agrochemicals and organic amendments. But these decisions are made with the understanding that they affect soil health and the environment, along with productivity (Carter, 2002; Ditzler, 2002). The adoption of conservation farming practices, policies and programme that encourage beneficial farming systems requires carefully study of the factors that cause producers to select certain farming practices over others. Producers usually implement conservation farming practices only when they are convinced of their economic benefits. Conservation farming is practised more easily when producers realize that continued use of conventional methods will inevitably result in further soil degradation and declining profitability.

Agricultural policy has typically focused on high production. The new policies for sustainable agriculture and resource conservation should have as their main objective to maintain and to improve agricultural soil quality as an essential step in maintaining and improving environmental health and ensuring long-term profitability. Sustainable land management should promote practices that conserve resources, including recycling organic wastes and conserving water quality (EEA, 2000). Many possibilities exist to reduce compaction by heavy machines. The most important is to avoid traffic on wet soil, but this may require a major change of the whole crop production system. Good drainage will improve soil drying. It is also helpful to use machines with large and soft low-pressure tyres or dual tyres, and to adjust the tyre inflation pressure to the lowest possible for the work being done. Four-wheel drive or front-wheel assisted rather than two-wheel drive tractors will provide more traction at a certain total weight. To limit subsoil compaction, vehicles should have a low axle load. For transport and spreading operations, good planning of the traffic, suitable positioning of entrances to the fields, a large working width, matching the load to the field length, and special transport lanes in large fields all help to reduce the distance that heavy vehicles move in the fields. Whenever feasible, different vehicles should be used for road and field transports. Controlled traffic, with all machines using the same tracks, may sometimes be useful, but, if the aim is to reduce subsoil compaction, these tracks must always be in the same position. Organic matter improves soil structure, and can help the soil to resist compaction stresses (Hartge and Horn, 1999; Horn *et al.,* 2000).

Soil erosion and degradation processes have extensively been studied during the last decades and soil conservation measures have successfully been developed. Now that runoff and erosion have been identified as major causes of water quality deterioration and loss of habitat diversity, there is a need for action at the farm level, where many of the decisions have to be taken. Improving the soil physical condition is a key issue, as previously discussed. Different types of filter strips can be used to reduce runoff velocity and provide filtering. Engineering erosion control measures include contouring, terraces and waterways (Lal *et al.,* 2003).

On the one hand, soil protection and conservation measures on agricultural lands have to fit within the farm organization, which involves simultaneous consideration of various social, economic and environmental concerns. On the other hand, soil conservation measures are increasingly required at catchment and regional scales, and for planning time horizons of a few years to several decades. Soil degradation may not always adversely affect all uses at the same time, as long as some land use is possible and some soil functions are achievable. As a consequence, management of agricultural and agroforestry catchments should be a very important tool for mitigating the effects of compaction and erosion and to protect and restore soil quality.

Management at the scale of the river basin provides a natural unit for the flow of water, sediment, nutrients and pollutants. At the scale of meso-catchments (c. 100 km^2), it is both possible and rational to accumulate the effects of farm-scale conservation management, retaining a clear understanding of the underlying physical and chemical processes and bringing them together within an area which is large enough to be relevant for regional planning and within which it is essential to be aware of reservoir sedimentation, river water quality and associated issues. At this scale it is also appropriate to set agricultural impacts on water quality.

References

Aguilar, R., Kelly, E.F. and Heil, R.D. (1998) Effects of cultivation on soils in Northern Great Plains rangeland. *Soil Science Society of American Journal* 52, 1081-1085.

Alakkuku, L. (1996) Persistence of soil compaction due to high axle load. I: Short term effects. II: Long term effects. *Soil and Tillage Research* 37, 211-238.

Anderson, J.P.E. and Domsch, K.H. (1989) Quantification of bacterial and fungal contribution to soil respiration. *Archive Mikrobiol* 93, 113-127.

Angers, D.A., Recous, S. and Aita, C. (1997) Fate of carbon and nitrogen in water-stable aggregates during decomposition of 13C15N tracing and soil particle size fractionation. *European Journal of Soil Science* 48, 295-300.

Barthes, B., Albrecht, A., Asseline, J., de Noni, G. and Roose, E. (1999) Relationship between soil erodibility and topsoil aggregate stability or carbon content in a cultivated Mediterranean highland (Aveyron, France). *Communication Soil Science Plant Anal* 30, 1929-1938.

Ben-Hur, M. and Agassi, M. (1997) Predicting interrill erodibility factor from measured infiltration rate. *Water Resources Research* 33, 2409-2415.

Carter, M.R. (2002) Quality, critical limits and standardization. In: Lal, R. (ed.) *Encyclopedia of Soil Science*. Marcel Dekker, New York, USA, pp. 1062-1065.

Chan, K.Y. and Heenan, D.P. (1999) Microbial-induced soil aggregate stability under different crop rotations. *Biology and Fertility of Soils* 30, 29-32.

Chaney, K. and Swift, R.S. (1984) The influence of organic matter on aggregate stability in some British soils. *Journal of Soil Science* 35, 223-230.

Dalal, R.C. (1998) Soil microbial biomass. What do the numbers really mean? *Australian Journal of Experimental Agriculture* 38, 649-665.

Ditzler, C. (2002) Quality and erosion. In: Lal, R. (ed.). *Encyclopedia of Soil Science*. Marcel Dekker, New York, USA, pp. 1066-1068.

Doran, J.W., Sarrantonio, M. and Liebig, M.A. (1996) Soil health and sustainability. *Advances in Agronomy* 56, 1-54.

Ebeid, M.M., Lal, R., Hall, G.F. and Miller, E. (1995) Erosion effects on soil properties and soybean yield of a Miamian soil in Western Ohio in a season with below normal rainfall. *Soil Technology* 8 (2), 97-108.

Ekwe, E.I. (1991) The effects of soil organic matter content, rainfall duration and aggregates size on soil detachment. *Soil Technologies* 4, 197-207.

European Environment Agency (EEA) (2000) Down to the earth: soil degradation and sustainable development in Europe. *Environmental Issue Series* 16, 31.

Fullen, M.A. (1991) Soil organic matter and erosion processes on arable loamy sand soils in the West Midlands of England. *Soil Technologies* 4, 19-31.

Gregorich, E.G. (2002) Quality. In: Lal, R. (ed.) *Encyclopedia of Soil Science*. Marcel Dekker, New York, USA, pp. 1058-1061.

Guerra, A. (1994) The effect of organic matter content on soil erosion in simulated rainfall experiments in W. Sussex, United Kingdom. *Soil Use and Management* 10, 60-64.

Håkansson, I. and Reeder, R.C. (1994) Subsoil compaction by vehicles with high axle load-extent, persistence and crop response. *Soil Tillage Research* 29, 277-304.

Håkansson, I., Voorhes, W.B. and Riley, H. (1998) Vehicle and wheel factors influencing soil compaction and crop response in different traffic regimes. *Soil and Tillage Research* 11, 239-282.

Hartge, K.H. and Horn, R. (1999) *Introduction to Soil Physics* (Einführung in die Bodenphysik) Enke Verlag, Stuttgart, 3rd edn Germany. (In German.)

Hartman, A., Assmus, B., Kirchhof, G. and Scholter, M. (1997) Direct approaches for studying soil microbes. In: van Elsas, J.D., Trevors, J.T. and Wellington, E.M.H. (eds) *Modern Soil Microbiology,* Marcel Dekker, New York, USA, pp. 279-309.

Haynes, R.J. (1999) Size and activity of the soil microbial biomass under grass and arable management. *Biology and Fertility of Soils* 30, 210-216.

Horn, R. and Rostek, J. (2000) Subsoil compaction processes. In: Horn, R., van den Akker, J.J.H. and Arvidsson, J. (eds) *Subsoil Compaction: Distribution, Processes and Consequences. Advances in Geoecology* 32, 44-54.

Horn, R., van den Akker, J.J.H. and Arvidsson, J. (2000) Subsoil compaction: distribution, processes and consequences. *Advances in Geoecology* 32, 462.

Jones, A.J., Mielke, L.N., Barles, C.A. and Miller, C.A. (1989) Relationship of landscape prosition to crop production and profitability. *Journal of Soil and Water Conservation* 44, 328-332.

Karlen, D.L., Mausbach, M.J., Doran, J.W., Cline, R.G., Harris, R.F. and Schuman, G.E. (1997) Soil quality: a concept, definition, and framework for evaluation (a guest editorial). *Soil Science Society of America Journal* 64, 4-10.

Katterings, Q.M., Blair, J.M. and Marinissen, J.C.Y. (1997) Effects of earthworms on soil aggregate stability and carbon and nitrogen storage in a legume cover crop agroecosystem. *Soil Biology and Biochemistry* 29, 401-408.

Kilmer, L. (1963) Plant nutrient losses from soils by water erosion. *Advances in Agronomy* 15, 303-316.

Lado, M., Paz, A. and Ben-Hur, M. (2004) Organic matter and aggregate-size interactions in saturated hydraulic conductivity. *Soil Science Society of America Journal* 68, 234-242.

Lal, R., Blum, W.H., Valentine, C. and Stewart, B.A. (1997) *Methods for Assessment of Soil Degradation*. Advances in Soil Science, CRC Press, Boca Raton, USA, p. 558.

Lal, R., Sobecki, T.M., Iivari, T. and Kimble, J.M. (2003) *Soil Degradation in the United States. Extent, Severity and Trends*. Lewis Publishers, Boca Raton, USA, p. 204.

Larink, O. and Schrader, St. (2000) Rehabilitation of degraded compacted soil by earthworms. In: Horn, R., Van den Akker, J.J.H. and Arvidsson, J. (eds) *Subsoil Compaction: Distribution, Processes and Consequences. Advances in Geoecology* 32, 284-294.

Le Bissonnais, Y. (1996) Aggregate stability and assessment of soil crusting and erodibility: I. Theory and methodology. *European Journal Soil Science* 47, 425-437.

Le Bissonnais, Y. and Arrouyas, D. (1997) Aggregate stability and assessment of soil crustability and erodibility. *European Journal Soil Science* 48, 39-48.

Liu, B., Li, G., Wu, F. and Zhao, X. (1995) Rule of soil nutrient loss on the Southern Loess Plateau. *Journal of Soil and Water Conservation* 9 (2), 77-86.

Ludwig, B., Boiffin, J., Chadoeuf, J. and Auzet, V. (1995) Hydrological structure and erosion damage caused by concentrated flow in cultivated catchments. *Catena* 25, 227-252.

Magdoff, F. (1992) *Building Soils for Better Crops*. University of Nebraska Press, Lincoln, USA, p. 176.

Mannering, J.V., Franzmeier, D.P., Schertz, D.L., Moldenhauer, W.C. and Norton, L.D. (1985) Regional effects of soil erosion on crop productivity in the Midwest. In: Follet, R.F. and Stewart, BAA (eds) *Soil Erosion and Cropland Productivity*. American Society of Agronomy, Madison, WI, USA. pp. 189-211.

Marinissen, J.C.Y. and Hillenaar, S.I. (1997) Earthworm-induced distribution of organic matter in macro-aggregates from differently managed arable fields. *Soil Biology and Biochemistry* 29, 391-395.

Martens, D.A. (2000) Management and crop residue influence soil aggregate stability. *Journal of Environmental Quality* 29, 723-727.

Poch, R.M. (2002) Degradation. In: Lal, R. (ed.) *Encyclopedia of Soil Science*. Marcel Dekker, New York, pp. 260-263.

Sharpley, A.N. and Rekolainen, S. (1997) Phosphorus in agriculture and its environmental implications. In: Tunney, H., Carton, O. T., Brookes, P.C. and Johnston, A.E. (eds) *Phosphorus Losses from Soil to Water*. CAB International, Wallingford, United Kingdom, pp. 1-54.

Soane, B.D. and van Ouwerkerk, C. (1994) *Soil Compaction in Crop Production*. Elsevier, Amsterdam, The Netherlands, p.622.

Stepniewski, W., Horn, R. and Martyniuk, S. (2002) Managing soil biophysical properties for environmental protection. *Agriculture of Ecosysten Environmental* 88, 175-181.

Stoltenberg, N.L. and White, J.L. (1953) Selective loss of plant nutrients by soil erosion. *Soil Science Society of America Proceeding* 17, 406-410.

Taylor, H.M. (1971) Effects of soil strength on seedling emergence, root growth and crop yield. In: *Compaction of Agricultural Soils*. American Society of Agricultural Energy Monograph. St. Joseph, MI, USA, pp. 292-305.

Tenge, A.J., Kaihura, F.B.S., Lal, R. and Singh, B.R. (1998) Erosion effects on soil moisture and corn yield on two soils at Mlingano, Tanzania. *American Journal of Alternative Agriculture* 13 (2), 83-89.

Tisdall, J.M., and Oades, J.M. (1982) Organic matter and water-stable aggregates in soils. *Journal Soil Science* 33, 141-163.

Valcárcel, M., Taboada, M.T., Paz, A. and Dafonte, J. (2003) Ephemeral gully erosion in northwestern Spain. *Catena* 50, 199-216.

Whalen, J.K., Hu, Q. and Liu, A. (2003) Compost applications increase water-stable aggregates in conventional and no-tillage systems. *Soil Science Society of America Journal* 67, 1842-1847.

Zoebish, M.A. and Dexter, A.R. (2002) Degradation, physical. In: Lal, R. (ed.) *Encyclopedia of Soil Science*. Marcel Dekker, New York, USA, pp. 311-316.

Indigenous breeds and silvopastoral systems

L. Sánchez
Dept. of Animal Anatomy and Production, Faculty of Veterinary Medicine, 27002 Lugo, Spain. lusaga@lugo.usc.es

Abstract
We have analysed a particular situation within the agricultural sector in Galicia involving the increasing substitution of agriculture by cattle breeding to understand how the resources can be used more efficiently and the sustainable use of the natural environment guaranteed, thus meeting new social functions and demands. Beef production using local traditional breeds is a good opportunity to do this in extensive systems, based on making better economic use of shrub and herbaceous vegetation by managing livestock in pastoral ecosystems. The Rubia Gallega breed provided the best results under the following conditions: autumn calving, the winter silage diet included grass and hay, grazing with rotation before the calves entered the plot when the height of the grass was at least 15 cm and when they left it was 6 cm, the calves were supplemented when grazing up until 2 months before they were killed and during the last 2 months they were fattened using a concentrated feed with hay as well as their mothers' milk. This type of production gave an average net margin per suckler cow of 439.08 €. Grazing with dark breeds was more difficult (qualified by a soil and botanical study) and results were less clear. This study also assessed vegetation cover in an area with a wide range of tree species and dense undergrowth after undergrowth clearance. The possible damage caused to the different tree species was assessed.

Key words: extensive systems

Introduction
Traditional pasture exploitation systems are as old as civilization itself, and information is needed on their sustainable management. Extensive animal production has very clear economic aims. In production models of beef, use is made of unique plant resources in a long-term sustainable system.

An economic solution based on a fundamentally ecological approach to extensive production has been the most successful for farming organizations and traditional animal production systems in Spain. The economic and ecological positive aspects of extensive beef production systems remain today, but with other more concrete and real connotations.

The choice of rustic breeds with a high grazing capacity is important. The adaptation of the animals to a certain environment is essential if they have to survive and maintain adequate production levels. Many environmental factors come into play in the adaptation process and functional adaptations to each specific environment that are transmitted genetically must exist. Extreme or adverse environmental conditions lead to a reduction in growth, production and the capacity to reproduce. It is clear that the animals must adapt to the natural vegetation available and tolerate adverse climatic conditions. There is evidence of the influence of genetic factors on the characteristics of rusticity: resistance to wide ranges of differences in temperature, solar radiation, low sustenance needs and periodic drought, as well as resistance to insects and diseases. It is clear that in an adverse environment these breeds are capable of producing more than exotic breeds. This is the case of the Rubia Gallega, a breed which produces high-quality beef and has important maternal qualities. At present, the Rubia Gallega is used on family-run farms using traditional management systems (Sánchez, 1985; Sánchez, 2000).

Agricultural production in Galicia has changed substantially. There has been a decline in agriculture overall but an increase in bovine stockbreeding in general, particularly for beef production. Over the last 25 years beef production has increased by 85%, a situation that not only is being maintained but continues its upward trend. This may be as a result of the decision of the CAP to give incentives to farmers for the land used, as well as the commitment to a more extensive production system compatible with the environment. This helps maintain rural populations so that farming activity can be compatible with agri-environmental activity. Perhaps in this way the necessary balance between economic development, ecological concerns and an adequate population of farming families can be found (Monserrat *et al.*, 2000).

In view of this it is important to address how rustic breeds can best exploit the shrub and herbaceous resources available.

The rational exploitation of indigenous cattle breeds in Spain takes place in extensive, year-round grazing systems. In these circumstances, the presence of trees will improve the productivity and well-being of the animals, through protection against the sun and extreme temperatures, reducing their need for food to combat the cold and removing the need for energy expenditure to look for shade in summer.

The presence of trees is not a problem for stockbreeding in the south and east of Spain, where the dehesa ecosystem is predominant. However, it is a problem in the north-west, where there is a lack or scarcity of trees in fields or pastureland, although a lot of wooded mountainous land that exists in the area could constitute an excellent refuge for the animals, which would also clear the scrubland and undergrowth (García Salmerón, 1991; Madrigal *et al.*, 1999). For this reason, Rigueiro *et al.* (1997) consider that livestock on silvopastoral systems help prevent forest fires by greatly reducing the amount of combustible undergrowth. However, the tree-covered areas are usually out of bounds for the livestock as it is thought that they do too much damage to the trees, eating the bark or scratching themselves on them (Sánchez *et al.*, 2002).

Bearing these circumstances in mind, the Rubia Gallega breed was tested in an area with a wide range of tree species and dense scrubland in order to study land occupation, the clearing of the undergrowth and the possible damage to the different tree species. This study helps to derive the economic benefits realizable to farming families involved in stockbreeding using indigenous breeds under extensive systems for beef production under silvopastoral systems. We will also show the results relating to the effect of the livestock on the tree species and undergrowth in pastoral systems.

Materials and Methods
(a) Experiments with family-run farmland using extensive beef-producing systems
The overall aim was to test, *in situ*, the possibility of producing the traditional high-quality bovine calf in a grazing system in developed mountainous inland areas of Galicia.

The plots were made up of 27 and 22 ha of pastureland, respectively: one in Fonsagrada (Lugo), at 800 m asl and in one of the most mountainous areas in Galicia, and the other in Palas del Rey (Lugo), at 500 m asl, both in inland Galicia. They were chosen as they belonged to farmers convinced of how useful their management systems are (where they apply many of their practices) and are included in management discussion groups where they can debate the project, and they are respected farmers in their area. Also the farms are representative and located in a meat-producing area in the mountains of Lugo.

The Fonsagrada herd was made up of one stud, four heifers, 34 cows and 24 calves. The Palas de Rey herd was made up of five heifers, 40 cows and 30 calves. The minimum installations or facilities were: wire fences to fence the plot, drinking troughs, holding pens and the old stables, to be used for the fattening phase and, if necessary, to isolate stock with health problems. Machinery and equipment necessary were a tractor with all its tools and scales to weigh the animals. The owner and family provided the labour. The cows were kept in the field where they grazed rotationally in the spring and autumn, eating the silage and hay which was given to them in the plot in the winter and summer. The calves were kept with their mothers until 5 or 6 months old, when they were taken to the stables for fattening with concentrated feed and their mothers' milk, when the mothers had unrestricted access and were accustomed to going to feed them. All the farmer had to do was to open and close the stable door when it was time for them to feed. For mating, a bull was put in with the cows at a certain time so that calving would take place in the autumn and winter.

Assessments made were:
(a) Technical: Calving date, identifying the cow and the calf, data on the difficulty or otherwise of the birth, perinatal mortality and any other observation that might be considered abnormal, the weight of the calves when born, when entering the fattening phase and when sold. From this herd fertility percentages were calculated (interval between calving and percentage of cows which calved again the following year), growth rate for both calf sexes both in the field and during the fattening phase, difficulties in calving and perinatal mortality.

(b) Economic: Annual inventory of the farm. Sale of animals, purchase of fertilizers, herbicides, fodder for the livestock and products for preventive and curative treatments for the animals. Veterinary costs, repair of buildings and maintenance of machinery as well as any other farm expenses. This enabled the net profit margins, economic gross and net indexes per hectare and per work unit and livestock unit, and the technical farm structure indexes, stocking rate, production in terms of kilograms of meat per animal, consumption of fertilizer per hectare and feed concentrate per calf to be calculated.

(b) Behaviour of rustic breeds on mountain pastureland
In order to study the behaviour of weaned calves grazing on mountain pastureland the Dark Galician breeds (Cachena, Caldelana, Frieiresa and Vianesa) were used. They were studied in two phases. First, the growth of the calves from weaning (6 months) until puberty (14 months) was monitored. They were weighed every 30 days. As this is the stage when growth rate is more inportant, they were given a supplement of 2 kg of feed/animal per day. Second, the study consisted of controlling their growth between the age of 14 and 18 months, when they were checked at the same time intervals, under difficult continuous mountain grazing conditions, when they had to take advantage of pasture resources, which at present have a reduced stocking rate (1 LU/29 ha). Soil and botanical studies were carried out using the samples obtained from the grazing areas placed on the mountains of eastern Ourense. The chemical composition of the forage and the productivity were also studied. Soil pH, organic matter, nitrogen, phosphorus, potassium and calcium carbonate were determined. For the botanical study, the floral composition and incidence of herbaceous species were determined. The analysis of the chemical composition of the forage was carried out in accordance with AOAC methods (1975).

(c) Effect of the herd on the trees and undergrowth in the pasture systems

A 1.66 ha woodland with a tree density of 830 trees/ha per m^2 containing birch (*Betula pubescens*), alder (*Alnus glutinosa*), chestnut (*Castanea sativa*), eucalyptus (*Eucalyptus globulus*), laurel (*Laurus nobilis*), pine (*Pinus pinaster*), oak (*Quercus robur*), willow (*Salix caprea*) and elder (*Sambucus nigra*) was studied. The trees were covered with ivy and the dense impenetrable undergrowth was made up of common gorse (*Ulex europaeus* L.), bramble (*Rubus* sp) and common fern (*Pteridium aquilinum*).

Thirteen cows and one bull 2 and 7 years old, respectively, from the Mabegondo Agricultural Research Centre (Xunta de Galicia) were used. The woodland was fenced off and 1900 m^2 (11.4% of the total surface area) was cleared at one end to open up a path and provide a place to put the feeding trough, and to provide the animals with a place where they could be given supplementary feed during the 62 days they were on the trial between the end of the autumn and beginning of winter. Animal grazing behaviours and introduction with trees were monitored five times per week. The study period ended when it was considered that if the herd stayed any longer the trees could be in serious danger of being damaged. The tests carried out were as follows: clearing of the undergrowth (the percentage of bare land being estimated), inventory of the wood (identification of species, measurement of the diameter and height of the treetops and damage to the bark - measurement of damaged surface area of trunk, branches and roots).

Results and Discussion
(a) Experiences with extensive beef-producing family-run farms

All of the objectives set out have been achieved. The test farms showed that it is possible to obtain a high-quality calf in a semi-extensive grazing management system, with net income per cow similar to that obtained using a traditional system, but with less work. In the test farms, the calves remained on the land with their mothers from birth until 164 ± 36 days old, and were sold when they were 247 ± 25 days old, having spent 101 days in the fattening house where they gained on average 1617 ± 265 g/day. The weight gained per sex and the weight of the animals at sale from the test farms were compared with traditional farms (Table 1). It can be seen that there are few differences in weight gain from birth to weaning.

Table 1. Weight gain of calves from birth to sale and weight and age of calves from the test farms and all of the farms monitored.

Farms:	Males		Females	
	Total	Test	Total	Test
No. animals	175	90	75	47
Weight at sale (kg)	377 ± 35	350 ± 18	304 ± 32	289 ± 19
Age at sale (days)	253 ± 16	250 ± 21	247 ± 15	243 ± 15
Weight gain Birth-sale (g)	1268 ± 147	1221 ± 35	1068 ± 86	1023 ± 43

The average net margin per nurse cow in the two test farms was 439.08 €, a rate higher than the average within the management group they belonged to (Table 2). The net margin per hectare in the test farms was higher than that for the other 32 management group farms, as their stocking rate (1.38 cows/ha) was higher than that for the other farms (1.21 cows/ha). The net margin per work unit among the semi-extensive or traditional farms cannot be compared properly as the number of cows per herd is too low to note any advantages in a family-run farm, with 1.5 UTH, as is the case, although we ought to point out that the work volume is lower in the test farms, where there was more free time for other activities.

Table 2. Classification per net margin index per cow from the meat-producing farms using suckler cows in inland mountainous areas of Lugo.

Farms:	High level	Low level	Average
Items	€/cow	€/cow	€/cow
Number of farms	8	18	8
Sale of animals	446.05	428.65	439.99
Inventory difference	456.79	57.04	199.58
Sale of milk	171.61	0	58.69
Total gross income	1074.46	485.68	701.27
Purchase of feed	139.75	157.19	111.73
Fertilizers and land	75.45	72.54	55.57
Health and reproduction	24.43	42.09	24.18
Total variable costs	239.63	272.25	191.48
Technical amortizations:			
Buildings and installations	62.87	52.80	44.17
Machinery and equipment	37.75	31.90	38.04
Maintenance of buildings	5.28	0	0.01
Equipment maintenance	36.97	31.97	34.76
Labour	2.35	2.83	5.57
Financial costs	0	0	0
General costs	45.93	110.80	61.24
Total fixed costs	191.15	230.31	183.63
Net margin	643.68	-16.46	375.55

(b) Behaviour of the rustic breeds on mountain pastureland

The results from the first phase show clearly different growth rates in favour of the breeds that are larger in size and have greater nutritional needs. The average weight gains (kg/day) were 0.457, 0.609, 0.604, 0.714 and 0.609, respectively, in the order set out above. In the second phase, the results were the opposite; the Cachena breed showed a greater rusticity and capacity to exploit the pasture under adverse conditions, regardless of compensatory growth. However, the Frieiresa breed showed very low average gains during this phase (Sánchez, 1985). The results obtained from the soil analysis indicate that it is highly acidic and has a high organic matter content and a silt-sand texture, i.e. a net deficiency of assimilable elements. The botanical study shows that flora composition is quite uniform because of the low stocking rate. The predominant flora is Gramineae, compound Leguminosae and Cistaceae. The analysis of the chemical composition shows the consequences of the low summer rainfall, giving low levels of humidity, protein differences and high levels of gross fibre (Iglesias *et al.*, 1991).

(c) Effect of the livestock on the trees and undergrowth in pasture systems

At the end of the experiment, it was found that 77% of the area had been left bare and only one area with a steep hill maintained half of its surface area covered by undergrowth. There were differential effects of stock on trees: elder (100%), alder (69%) and willow (65.4%) were most affected and the least affected were eucalyptus (16.0%), pine (18.2%) and oak (18.8%). The area which suffered the most was the area in and around the feeding trough: 55.7% of trees were affected, possibly because the animals rested there in a group after eating. The area least affected was the highest area, which was the most difficult for the animals to access: only 5.14% of the trees there were damaged.

The most damaged part was the bark (91.2% of the trees) and the least damaged the branches (1.67%). The scale of the damage or surface area of the damage to the bark, in relation to the total surface area of the tree up to a height of 1.6 m, the maximum height it was calculated the animals could reach with their head, was 11.54% (for all the species of trees).

Conclusions

Woodland can be integrated into pasture systems, which could be beneficial to the animals' well-being and considerably reduce the fire hazard of the undergrowth, as long as the animals do not remain on a particular plot long enough to cause damage to the trees.

The economic studies carried out in Galicia on extensive beef-producing family-run farms showed an average net margin per suckler cow of 439.08 €. The results for dark breeds on pastureland were less clear.

The species which suffered the most damage due to the action of the animals were elder (*Sambucus nigra*) and alder (*Alnus glutinosa*) and those which suffer the least were laurel (*Laurus nobilis*) and eucalyptus (*Eucalyptus globulus*). The tree species which suffered the most damage were those which had a diameter of between 0.12 and 0.8 m.

References

AOAC (1975) Official methods of analysis of the association of official agricultural chemists. Tenth edition. Washington

García Salmerón, J. (1991) *Forestry afforestation Handbook* (Manual de Repoblaciones Forestales). Escuela Técnica Superior de Ingenieros de Montes, Fundación Conde del Valle de Salazar, Madrid, Spain.

Iglesias, A., Sánchez, L. and Vallejo, M. (1991) Evaluation of rangeland resources in stressed environments utilized by local cattle breed in North West Spain. In: Gaston, A., Kernick, M. and Le Houerou, H. (eds) *IV International Rangeland Congress*. INRA, Montpellier, France.

Madrigal Collazo, A., Fernández-Cavada Labat, J.L., Ortuño Pérez, S.F. and Notario Gómez, A. (1999) *El Sector Forestal Español*. Fundación Conde del Valle de Salazar, Madrid, Spain.

Monserrat, L. and Sánchez, L. (2000) Sistemas de producción de carne en pastoreo con Rubia Gallega. *Bovis* 92, 23-34.

Rigueiro, A., Silva, F.J., Rodríguez, R., Castillón, P.A., Álvarez, P., Mosquera, R., Romero, R. and Gonzalez, M.P. (1997) *Silvopastoral Systems Handbook*, (Manual de Sistemas Silvopastorales). Universidad de Santiago de Compostela, Lugo, Spain.

Sánchez, L. (1985) Bovine autochthonous breeds in NW and subutilized resources use (Las razas bovinas autóctonas del Noroeste en el aprovechamiento de recursos infrautilizados). *Buiatría Española* 1 (1), 39-54.

Sánchez, L. (2000) Modelo de la empresa familiar en la producción de carne de vacuno en extensivo. In: *Congreso Internacional de Producción y Sanidad Animal*. Barcelona, Spain (in press.)

Sánchez, L., Vallejo, M., Iglesias, A., Álvarez, F., Fernández, M. and Salgado, L.M. (1992) *Bovine Autochtonous Breeds of Galicia* (Razas bovinas autóctonas de Galicia). l. Razas Morenas Gallegas (Cachena, Caldelana, Frieiresa, Limiana y Vianesa). Recursos genéticos a conservar. Xunta de Galicia, Santiago de Compostela, Spain.

Sánchez, L., Monserrat, L. and Moreno, T. (2002) Farm model for beef extensive production (Economía, ganadería e medio ambiente). *Revista Galega de Economía* 2, 307-322.

Silvopastoral systems to prevent soil losses in sustainable livestock systems

Z. G. Acosta[1], G. Reyes[1] and J. L. Montejo[2]
[1] Centro de Investigaciones de Medio Ambiente de Camagüey, Cisneros No. 105 (altos) e/Ángel y Pobre, C.P. 70 100 Camagüey, Cuba. [2] Dirección Provincial de Suelos Camagüey, Cuba. cimac@cimac.cmw.inf.cu

Abstract
The universal soil loss equation (USLE) was incorporated into a geographical information system (GIS) and used to determine soil losses in a specific livestock region of 7646 ha in Jimaguayú, Camagüey, Cuba. One hundred and twelve management units were defined according to their use and soils characteristics. In 3198.5 ha there were soil losses between 10 and 147 t/ha per year distributed in five zones. In order to reduce the losses to less than 10 t/ha per year in these zones, alternative management is proposed - taking into account the results of regionalization programmes of trees, grasses and legumes. The use of silvopastoral systems (SPS) was considered as the main alternative in zones where soil limiting factors allowed it and in other places the establishment of a more productive grass species covering was proposed. In all cases, the most convenient species were selected according to the soil and locally limiting factors.

Introduction
To maintain the natural landscape for livestock production in Cuba, sustainable practices must be adopted to allow soils to recover and to improve and guarantee a supply of quality fodder for animals (Paretas, 2002). Silvopastoral systems meet these requirements (Murgueitio, 1999; Sánchez et al., 2001) because research to develop this technology in Cuba has given encouraging results (Renda et al., 1997; Hernández et al., 1998; ICA, 1998; Oquendo, 2002; Paretas, 2002). The objective of this chapter is to determine soil losses in a livestock-interesting region in Camagüey, Cuba, and to plan the establishment of SPS in order to reduce such losses as well as providing a livestock sustainable development.

Materials and Methods
The project was carried out in an area of 7646.7 ha in the Najasa River watershed, located between 21° 11′ 45″-21° 5′ 44″ North and 77° 56′ 7″-77° 46′ 23″ West in the region of Jimaguayú, Camagüey, Cuba (Figure 1).

Figure 1. The location of the area studied in central Cuba.

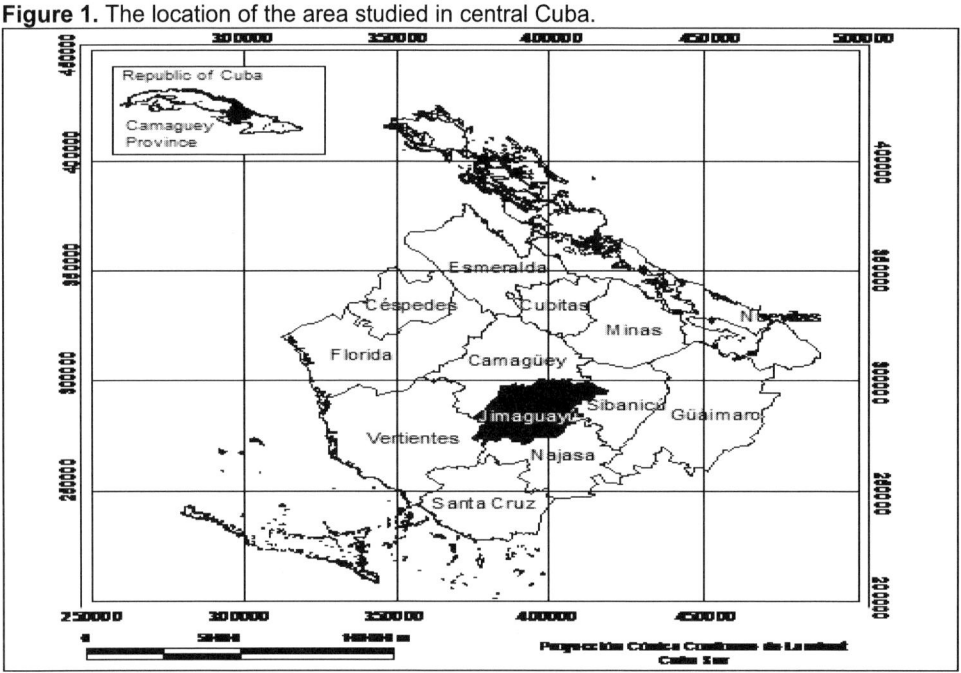

USLE (I) was used in order to determinate soil erosion (Planas, 1986). This equation was incorporated into a GIS with graphics and attribute databases of the area.

(I) A = R K L S C P

Where A = soil losses (t/ha), R = erosion from the rain (t/ha), K = erosion caused by soil characteristics (t/ha), L = length of the slope (m), S = degree of soil slope (%), C = vegetation cover and P = mechanical cutting to reduce or delay erosion.

Soil database types and uses were handled establishing units of management to achieve the highest precision in the evaluation of soil losses for better planning of SPS. From the results of USLE, critical zones where soil losses were beyond 10 t/ha years were chosen and convenient alternatives to mitigate the erosion were described. Types and depth of the soils were taken into account as well as affinity, growth rate and possible use of suitable tree species, according to the results of regionalization of tree species (Paretas et al., 2002).

Results and Discussion

From the GIS, 112 units of management were defined, of which three, (198.51 ha) had losses between 10-147 t/ha and per year, which represents 41.8% of the total area distributed in five zones, a, b, c, d and e. The main characteristics of the soil in each zone are shown in Table 1. The highest loss records correspond to a zone dominated by grasses (zone d), which give a low cover for soil protection, due to the slope of the soil surface (5.58°C).

Table 1. Characteristics of the soil in critical zones.

Zones	Area (ha)	Depth (cm)	Slope (degrees)	Loss of soil (t/ha per year)
A	5.63	51-100	1.42	39.84
B	0.81	51-100	1.55	29.22
C	90.44	51-100	0.52	26.29
D	2.19	20-50	5.58	148.93
E	3099.44	20-50	0.71	10.54
Total	3198.51			

Recommendations for use of sustainable cattle systems for each zone are as follows.

Zone A
Change soil use. Plant in an E-W direction with fruit trees *Spondias purpurea* L., *Citrus limon* and *Psidium guajava* at 5 m × 5 m, the timber trees *Samanea saman*, *Cordia gerascanthus* and *Eucalyptus saligna* or fruit trees *Persea americana* var. *americana*, *Melicocca bijugatus* and *Manguifera indica* at 12 m × 12 m spacing. Grass cover should be removed to introduce more productive, local species such as *Cynodon dactylon* (L.) Pers. cv '67', '68', 'cruzada 1'and 'cruzada 2', *C. nlemfuensis* Vanderyst cv 'jamaicano', 'panameño' and 'tocumen', *Cenchrus ciliaris* L. cv 'biloela and 'formidable', *Panicum maximum* Jacq. cv 'común' and 'likoni', *Digitaria decumbens*, cv 'común' and 'p-32' *Brachiaria mutica*, *Centrosema pubescens* cv 'CIAT 438', *Neonatonia wightii*, *Pueraria phaseoloides* Benth. cv 'CIAT-9900', *Leucaena leucocephala*, *Macroptilium atropurpureum*, *Stylosanthes guianensis* cv '136' and '184' and *Teranmus labiales*. This system would be for concurrent use of trees and grasses in a silvopastoral system.

Zone B
Mitigate soil losses as for Zone A, although, due to its small size, the variants of *S. purpurea*, *C. limon* and *P. guajava* seem to be the most convenient.

Zone C
To reduce soil losses in this zone, plant a mixed forage bank with *Saccharum officinarum*, *Pennisetum purpureum* and *Gliricidia sepium*. The high quality of this forage allows its utilization as an animal supplement with management either as a "cut and carry" system or under grazing. The rest of the species (*Pennisetum purpureum and Gliricidia sepium*) can be introduced using 0.5 m between plants and 1 m between rows in an E-W direction, in such way that the *S. officinarum* fills the spaces between the transects.

Zone D
In this area, where grasses occur, the establishment of *S. purpurea*, *C. limon* and *P. guajava* is proposed to mitigate soil loss because they have shallow roots and grow fast according to the soil characteristics. A planting configuration of 5 m × 5 m is recommended to allow the entrance of sunlight for good development of pasture and the management of both layers simultaneously within an SPS. For the grass layer, it is recommended to follow the suggestions made for Zones A and B.

Zone E
This is the largest zone with least soil loss, where alternatives relating to soil depth and the options available to the owners should be developed. It is recommended to use the same species and procedures indicated for Zone D.

Conclusions

The USLE incorporated into a GIS, containing graphic and attribute bases with information about the area studied, allowed the calculation and definition of five zones with soil losses higher than 10 t/ha year in Jimaguayú, Camagüey. This allowed the planning of a strategy to mitigate soil losses and to develop a sustainable livestock in the region.

References

Hernández, I., Milera, M., Simón, L., Hernández, D., Iglesias, J., Lamela, L., Toral, O., Matías, C. and Francisco, G. (1998) Advances of silvopastoral systems research in Cuba. (Avances de las investigaciones en sistemas silvopastoriles en Cuba) In: *I Conferencia Electrónica sobre Agroforestería para la Producción Animal en América Latina.* AGROFOR1, CIPAV, Cali, Colombia, pp. 47-59. http://www.cipav.org.co/cipav/indexsp.htm. (In Spanish.)

Instituto de Ciencia Animal. (ICA) Grupo Multidisciplinario de Leguminosas (1998) Enfoque acerca del trabajo sobre árboles y arbustos desarrollados por el instituto de ciencia animal de Cuba. In: *I Conferencia Electrónica sobre Agroforestería para la Producción Animal en América Latina.* AGROFOR1, CIPAV, Cali, Colombia, pp. 309-323. http://www.cipav.org.co/cipav/confr/indexsp.html. (In Spanish.)

Murgueitio, E. (1999) Reconversión ambiental y social de la ganadería bovina en Colombia. *World Animal Review* 93, 2-15. (In Spanish.)

Oquendo, G. (2002) *Technology to Promote the Use of Pastures and Forages* (Tecnologías para el fomento y explotación de pastos y forrajes), Agro Acción Alemana, Ministerio Federal para la Cooperación Económica y el Desarrollo y Asociación Cubana de Producción Animal, Habana, Cuba.

Paretas, J.J. (2002) Sostenibilidad ganadería-medio ambiente basada en PAF. *ACPA* 4, 22-23. (In Spanish.)

Paretas, J.J., López, M., Acosta, R., Serrano, R. and Gallardo, L. (2002) Regionalización de árboles multipropósitos. In: *Memorias del V Taller Internacional Silvopastoril y I Reunión Regional de Morera*, Matanzas. Cuba. CD rom ISBN 959-16-0172-7. (In Spanish.)

Planas, G. (1986) *Hydric Erosion Losses in Cuban Soils* (Pérdidas por Erosión Hídrica de los Suelos de Cuba). Instituto de Hidroeconomía, Departamento de Investigaciones Aplicadas, Habana, Cuba. (In Spanish.)

Renda, A., Calzadilla, E., Jiménez, M. and Sánchez, J. (1977) *Agroforestry in Cuba* (La Agroforestería en Cuba), Red Latinoamericana de Cooperación Técnica en Sistemas Agroforestales. Dirección de Recursos Forestales, FAO, Rome, Italy. (In Spanish.)

Sánchez, M.D., Rosales, M. and Murgueitio, E. (2001) Agroforestería pecuaria en América Latina. In: *II Conferencia Electrónica sobre Agroforestería para la Producción Animal en América Latina.* AGROFOR2, CIPAV. http://www.cipav.org.co/cipav/confr/indexsp.html. (In Spanish.)

Spatial dependence and seasonal patterns of cattle activity

A. Buttler [1,2,3], F. Kohler [1,4], H. Wagner [5] and F. Gillet [1,3,4]

[1] Swiss Federal Research Institute WSL, Antenne romande, 1015 Lausanne, Switzerland. [2] Dept. of Chrono-ecology, University of Franche-Comté, 25030 Besançon, France. [3] Laboratory of Ecological Systems, EPFL, 1015 Lausanne, Switzerland. [4] Dept. of Plant Ecology, University of Neuchâtel, 2007 Neuchâtel, Switzerland. [5] Swiss Federal Research Institute WSL, 8903 Birmensdorf, Switzerland. alexandre.buttler@epfl.ch, florian.kohler@unine.ch, francois.gillet@epfl.ch, helene.wagner@wsl.ch

Abstract

Cattle activity or grazing *sensu lato* can be subdivided into three components: dung deposition, herbage removal (foraging or grazing *sensu stricto*) and trampling. All these actions modify vegetation. Our purpose was to observe the medium-scale distributions of dung-pat density, trampling effect and herbage removal in a mountain wooded pasture and to relate these distributions to 'natural structures' of the landscape and 'management-induced structures'. The scale-dependent response to landscape structure was also investigated. Results showed that the three variables describing cattle activity exhibited significantly different spatio-temporal patterns. The relative influence of environmental factors was different for each activity and the scale of cattle response to landscape structure depended on the activity.

Key words: herbage removal, trampling, dunging, vegetation dynamics, mountain pasture

Introduction

Cattle activity or grazing *sensu lato* can be subdivided into three components: dung deposition, herbage removal (foraging or grazing *sensu stricto*) and trampling. All these actions have various effects on vegetation dynamics at a fine spatial scale (Kohler *et al.*, 2004). At the medium or large scale, the pattern of cattle activity is generally described only from the foraging behaviour, by considering implicitly grazing as an emergent behaviour of the three primary activities. The purpose of this work was to assess the validity of this assumption by observing the medium-scale distributions of dung-pat density, trampling effect and herbage removal in a mountain wooded pasture and try to attribute these distributions to various environmental factors. Secondly, the scale-dependent response to landscape structure was tested. This may have important consequences at an ecosystem level. The understanding of patterns of cattle activities is essential to predict vegetation dynamics in this type of ecosystem where shrubs and trees regenerate naturally.

Materials and Methods

First, to determine the medium-scale distributions of cattle activities, a paddock was subdivided into a regular grid of 393 square cells of 25 m × 25 m. In each cell, response data about cattle activities after every grazing rotation were collected, three times in one year: (1) dung pat density by counting the number of dung pats, (2) herbage removal, also called grazing hereafter, by applying an index of foraging intensity with three levels, and (3) trampling effect by estimating the percentage of bare soil and flattened vegetation due to trampling. Furthermore, to relate these distributions to various environmental factors in each cell, the 'natural structures', such as slope, vegetation openness, cover of trees, shrubs and rock outcrops, fodder potential, and the 'management-induced structures', such as distance to fence or to the nearest watering place, were recorded. For each response data set, the aggregation level of explanatory data sets ($X_{25m \times 25m}$, $X_{75m \times 75m}$, ..., $X_{425m \times 425m}$) with the largest square root of the proportion of explained variance in the constrained redundancy analysis was interpreted as the main scale of response, and tests for residual autocorrelation and multi-scale response (Wagner, H.H. and Wiens *et al.*, submitted) were applied at that scale. The Spearman correlations between the response variables were calculated for both the original response variables and their residuals at the main scale of response, and tested using the Dutilleul (1993) method of correcting for autocorrelation. The study was conducted in the Jura Mountains of north-western Switzerland.

Results and Discussion

At the medium scale, results showed that the three variables describing cattle activity exhibited significantly different spatio-temporal patterns (Figure 1). Many high correlations were observed between activities and 'natural structures' or 'management-induced structures' (Kohler *et al.*, 2004). Dunging was concentrated in flat, smooth areas without rock outcrops, with high fodder potential and generally in the centre and the north of the paddock. In the first rotational period, herbage removal was concentrated in areas near the fence and in the southern part of the paddock. Later in the season, cattle foraged the paddock more evenly, only avoiding the most wooded areas. Trampling effects, on the other hand, were strongest in the wooded areas, where cattle tend to seek shelter.

Figure 1. Aerial photography of the field and maps of the cattle activity assessed by its three components at the second rotation. The paddock is subdivided in 393 cells. Darker shading corresponds to higher value of the variable for each activity, the map border corresponds to the fence, and the points represent watering places (adapted from Kohler et al., 2004).

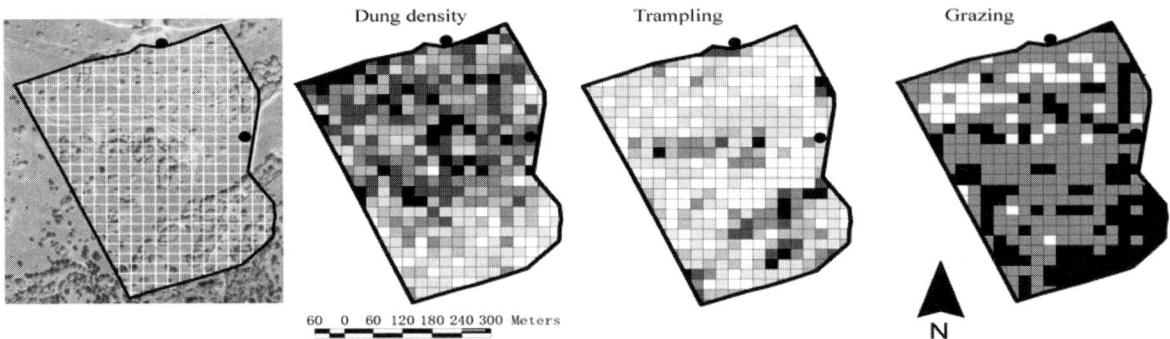

The combination of bivariate scaling analysis and multi-scale ordination showed that the scale of cattle response to landscape structure depended on the activity, with a local response at 25 m for trampling patterns, an intermediate-scale response at 125 m for herbage removal, and a large-scale response at 275 m for dung deposition (Wagner, H. Gillet, F., Kohler F. and Buttler, A., 2004, unpublished results; Wagner, H.H. and Wiens, 2004, unpublished results). All three cattle activities were autocorrelated within 25-50 m (Table 1). The gregarious behaviour of cattle, which are rarely observed in groups of less than ten individuals, could explain this pattern. During the second rotation, after accounting for autocorrelation (Dutilleul, 1993), there was a negative correlation between dunging and grazing. This spatial segregation could lead in the long run to a transfer of nutrients like phosphorus or nitrogen from the feeding places to areas with high dung density. This negative correlation disappeared after accounting for activity-dependent response to landscape structure. This means that the spatial segregation of the two activities can be explained by the landscape structure and may even be counteracted by an optimal placement of fences and watering places.

Table 1. Range of autocorrelation, defined as an uninterrupted sequence of distance classes with significant positive autocorrelation, starting at the smallest distance class of 25 m (500 permutations, Bonferroni correction with significance level $P = 0.05/14$).

	Dunging	Trampling	Grazing
Range of original autocorrelation	225 m	50 m	175 m
Range of residual autocorrelation constrained by X_{25x25}	100 m	50 m	50 m
Range of residual autocorrelation constrained by X at the main scale of response	25 m	50 m	25 m
Main scale of response	$X_{275x275}$	X_{25x25}	$X_{125x125}$

Conclusions

In heterogeneous wooded pastures, the relative independence between grazing, dunging and trampling activities may lead to various dynamic combinations of biotic constraints and perturbations affecting vegetation dynamics, with implication for tree regeneration. By exhibiting different spatial and seasonal patterns of habitat use, cattle maintain complex interactions with vegetation, particularly with shrubs and trees, and thus contribute to its variability and its heterogeneity at a landscape scale. It might therefore be necessary to consider dunging, trampling and grazing separately in spatially explicit models of vegetation dynamics in wooded pastures. This type of model is an essential management tool for landscape conservation.

References

Dutilleul, P. (1993) Modifying the t test for assessing the correlation between two spatial processes. *Biometrics* 49, 305-314.
Kohler, F., Gillet, F., Gobat, J.-M. and Buttler, A. (2004) Seasonal vegetation changes in mountain pastures due to simulated effects of cattle grazing. *Journal of Vegetation Science* 15, 143-150.

Impacts of cutting and fertilization on pasture systems in the Cantabrian mountain range in León province (NW Spain)

L. Calvo [1], A. Fernández [1], E. Marcos [2], L. Valbuena [2], R. Tárrega [1] and E. Luis [1]
[1] Área de Ecología, Facultad de Ciencias Biológicas y Ambientales, Universidad de León, 24071 León, Spain. [2] Área de Ecología, ESTIA, Campus Ponferrada, Universidad de León, Ponferrada (León), Spain. deglcg@unileon.es

Abstract

The aim of this study was to investigate the structural changes in a community of heathland when subjected to experimental cutting and nitrogen fertilization. Random patches with significant proportions of *Calluna vulgaris* and *Erica tetralix* were selected in three areas (mountain passes) in the Cantabrian mountains, Spain. In April 1998 experimental treatments (cutting and fertilizing with 5.6 g/m^2 ammonium nitrate) were imposed in each of the three areas. Fertilization does not have a significant influence on species number. Cutting increased the diversity of woody and perennial forb species. Cutting plus fertilization significantly increased the number of woody, perennial forbs and perennial graminoid species.

Key words: *Calluna vulgaris*, mountain pastures, biodiversity

Introduction

Heathlands dominated by *Calluna vulgaris* are important communities in the Cantabrian mountains. Historically, a great part of the pastures and associated vegetation in this area was used in transhumance pastoral systems. Heathlands were regularly cut and burnt to provide pasture (Calvo *et al.*, 2002). However, during the last few decades traditional management has almost disappeared due to changes in agricultural practices and for socio-economic reasons (Webb, 1998). Nowadays, patches dominated by *Calluna vulgaris* and *Erica tetralix* L. are rare and the remaining areas of heathland are in a mature to degenerate state (Watt, 1955), which may affect the regenerative potential. Also in the Cantabrian mountain range, there has been an increase in the deposition of atmospheric nitrogen (Rivero Fernández *et al.*, 1996). Increases in the long-term rate of nitrogen supply are likely to lead to changes in the structure and composition of the vegetation (Pitcairn *et al.*, 1995; Alonso *et al.*, 2001). Thus, this study aims to determine the changes in the vegetation diversity in heathlands after experimental cutting and nitrogen fertilization. Cutting has been proposed as an alternative to grazing in areas where the latter is not economically or socially viable.

Materials and Methods

Three mountain passes in the province of León (NW Spain) were selected: San Isidro (1600 m asl, 43º03'N, 1º40' W), Tarna (1625 m asl, 43º04'40''N, 1º33'W) and Vegarada (1585 m asl, 43º02'20''N, 1º48'20''W). The average annual temperature is 5.5ºC and the average annual rainfall is 1319.5 mm in all areas. In April 1998, a random area of approximately 1-2 ha containing a significant proportion of both *Calluna* and *Erica* was selected in each pass. Twenty experimental plots (1 m^2) were selected and marked in each pass. Experimental treatments were: five control plots (0-uF), five annual fertilization plots (0-F), five removal by cutting *Calluna* and *Erica* and non-fertilization (CE-uF) and five removal by cutting *Calluna* and *Erica* plus annual fertilization (CE-F). Fertilizer (granules of NH_4NO_3) was spread by hand in May of each year from 1998 until 2002. The fertilizer level (5.6 g N/m^2 per year as weight of fertilizer) is equivalent to twice the estimated current background pollution levels in this area. The percentage cover of each vascular plant species present was estimated visually in each plot before treatment and annually after treatments. The data were used to determine diversity (H') (Shannon Index) (Shannon and Weaver, 1949) and its two components, richness (S = species number) and evenness (J' = H'/H'max). Data from structural parameters were analysed by means of factorial analysis of variance. Scheffe F-tests (Scheffe, 1959) were performed to determine the significance of the differences.

Results and Discussion

Figure 1 shows the variation in richness and evenness in Vegarada (other passes are similar and not shown). Fertilizer increased species richness ($P < 0.005$), particularly in the Vegarada and Tarna passes. Woody species are the most frequent in the control plots (Table 1). However, fertilization favoured the presence of perennial forbs. Annual forbs and graminoids were not significantly affected by fertilization.

Cutting treatments (CE-uF) significantly increased ($P < 0.05$) richness values in all the study areas, through the elimination of dominant species, allowing an opportunity for other woody species and perennial forbs (Table 1). This increase in richness was shown throughout the study period as the regeneration speed of the dominant species, *Calluna vulgaris*, is very low. This is partly because of the abiotic conditioning, which determines a short vegetative period. Likewise, *Calluna vulgaris* is at a mature stage of development (over 20 years old) and this has a negative influence on the speed of vegetative regeneration (Mohamed and Gimingham, 1970). Such plants use germination as the main recovery mechanism. Most richness increase is promoted by the cutting plus fertilization treatment (CE-F) because this favours woody species, perennial forbs and graminoids (Table 1). The uniformity values (Figure 1) did not differ significantly between either passes or treatments.

Figure 1. Variations in the richness and evenness in Vegarada in the original situation and after experimental treatments. 0-uF: control non-fertilization; 0-F: Control with annual fertilization; CE-Uf = cut *Calluna vulgaris* and *Erica tetralix* non-fertilization, CE-F = cut *Calluna vulgaris* and *Erica tetralix* with annual fertilization.

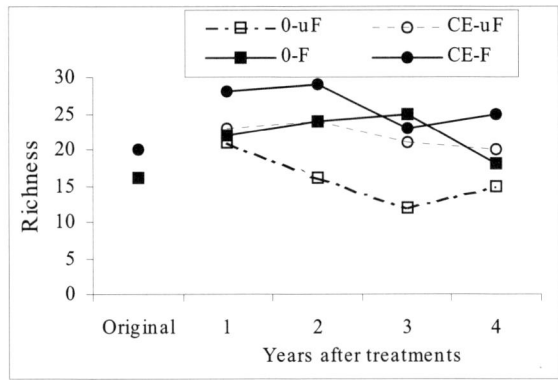
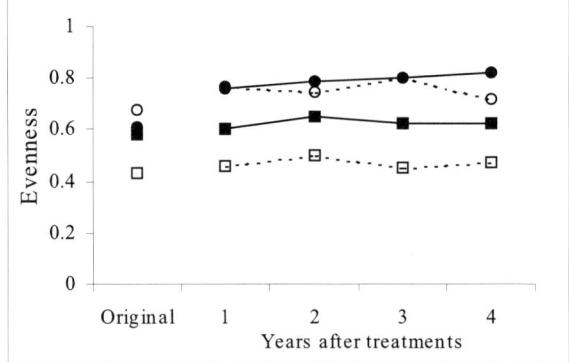

Cutting treatments, with and without fertilization, had a significant ($P < 0.05$) effect on diversity (Figure 2).

The interactive effects of disturbances such as cutting and increased atmospheric deposition have been shown to favour the entry of new species in the community without any clear dominant. Both aspects led to increased diversity values. There is experimental evidence of the impact of nitrogen deposits on the palatability and stress sensitivity of *Calluna* for vertebrate and invertebrate herbivores, including sheep (Power *et al.*, 1998). This study has found that the quantity of fertilizer deposited in the Cantabrian mountain range does not produce any short-term drastic changes in the community. The main effect recorded is the increase in diversity, primarily based on the presence of graminoid herbs, some of which are highly palatable to herbivores. The positive effects of fertilization on the herbaceous vegetation are boosted by the negative effect that it has on the germination of the *Calluna vulgaris* seeds.

Figure 2. Variations in the diversity values in Vegarada in the original situation and after experimental treatments. 0-uF: control non-fertilization; 0-F: control with annual fertilization; CE-uF = cut *Calluna vulgaris* and *Erica tetralix* non-fertilization, CE-F = cut *Calluna vulgaris* and *Erica tetralix* with annual fertilization.

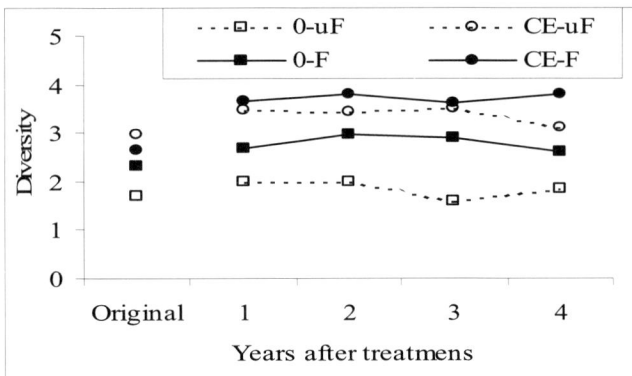

Table 1. Mean species number for life forms (woody, annual graminoids, annual forbs, perennial graminoids, perennial forbs) in Vegarada area. 0-uF: control non-fertilization; 0-F: control with annual fertilization; CE-uF = cut *Calluna vulgaris* and *Erica tetralix* non-fertilization, CE-F = cut *Calluna vulgaris* and *Erica tetralix* with annual fertilization.

	NON-FERTILIZED CONTROL					FERTILIZED CONTROL				
	Years after treatments					Years after treatments				
	Original	1	2	3	4	Original	1	2	3	4
Woody	2.2 (0.4)	4 (0.2)	3 (0.2)	2 (0.2)	2.6 (0.2)	2.6 (0.4)	4 (0.3)	4 (0.4)	3 (0.2)	2.4 (0.2)
Annual graminoids	0	0	0	0	0	0	0	0	0	0
Annual forbs	0	1 (0.2)	1 (0.2)	1 (0.2)	0.8 (0.2)	0.8 (0.2)	0 (0.4)	1 (0.3)	1 (0.5)	0.2 (0.2)
Perennial graminoids	3 (0.8)	4 (0.9)	2 (0.6)	1 (0.2)	1.4 (0.7)	2 (0.4)	3 (0.3)	3 (0.2)	4 (0.6)	2.8 (0.6)
Perennial forbs	1.6 (0.6)	4 (0.4)	2 (0.5)	2 (0.3)	2.8 (0.4)	2.6 (0.4)	6 (1.3)	4 (0.7)	4 (0.8)	3.6 (0.6)

	CE UNFERTILIZED					CE FERTILIZED				
Woody	3.2 (0.2)	3 (0.6)	4 (0.3)	3 (0.2)	3 (0)	3.4 (0.2)	3 (0.2)	4 (0.2)	3 (0.2)	3
Annual graminoids	0	0	1 (0.3)	1 (0.2)	0.6 (0.2)	0	0	1 (0.2)	0	0.2 (0.2)
Annual forbs	0.2 (0.2)	1 (0.4)	1 (0.2)	0	0.4 (0.2)	0.2 (0.2)	1 (0.2)	0	1 (0.2)	0.8 (0.2)
Perennial graminoids	2.6 (0.7)	4 (0.8)	4 (0.6)	4 (0.4)	3.4 (0.8)	2.2 (0.7)	5 (0.5)	6 (0.2)	6 (0.6)	6.2 (0.3)
Perennial forbs	4.8 (1.1)	4 (1.0)	4 (0.9)	4 (0.7)	3.8 (0.4)	4 (0.8)	5 (0.4)	5 (0.8)	6 (0.5)	5 (0.5)

Conclusions

Cutting has been proposed as an alternative mechanism to the traditional management of these heathlands to obtain more plant diversity. Cutting plus fertilization favours an increase in diversity and pasture quality of these heathlands.

References

Alonso, I., Hartley, S.E. and Thurlow, M. (2001) Competition between heather and grasses on Scottish moorlands: interacting effects of nutrient and grazing regime. *Journal of Vegetation Science* 12, 249-260.
Calvo, L., Tárrega, R. and Luis, E. (2002) Regeneration patterns in a *Calluna vulgaris* heathland in the Cantabrian mountains (NW Spain): effects of burning, cutting and ploughing. *Acta Oecologica* 23, 81-90.
Mohamed, B.F. and Gimingham, C.H. (1970) The morphology of vegetative regeneration in *Calluna vulgaris*. *New Phytologist*, 69 743-750.
Pitcairn, C.E.R., Fowler, D. and Grace, J. (1995) Deposition of fixed atmospheric nitrogen and foliar nitrogen content of bryophytes and *Calluna vulgaris* (L.) Hull. *Environmental Pollution* 88, 193-205.
Power, S.A., Ashmore, M.R., Cousins, D.A. and Sheppard, L.J. (1998) Effects of nitrogen addition on the stress sensitivity of *Calluna vulgaris*. *New Phytologist* 13, 663-673.
Rivero Fernández, C., Rabago, I., Sousa Carrera, M., Lorente Ibáñez, M. and Schmid, T. (1996) *Calculation and Mapping Critical Loads for Spain* (Cálculo y cartografía de cargas críticas para Spain). Aplicación del modelo SMB, Centro de Investigaciones Energéticas Medioambientales y Tecnológicas, Madrid, Spain. (In Spanish.)
Scheffe, H. (1959) *The Analysis of Variance*. John Wiley and Sons, New York, USA.
Shannon, C.E. and Weaver, W. (1949) *The Mathematical Theory of Communication*. University of Illinois Press, Urbana, USA.
Watt, A.S. (1955) Bracken versus heather, a study in plant sociology. *Journal of Ecology* 43, 490-506.
Webb, N. (1998) The traditional management of European heathlands. *Journal of Applied Ecology* 35, 987-990.

Floristic stability of pastures in the Sierra Mágina nature reserve, Andalusia, Spain

A. Cano-Ortiz, A. García-Fuentes, J. A. Torres, R. Montilla, L. Ruiz, C. Salazar and E. Cano
Dept. Biología Animal, Biología Vegetal y Ecología, Área de Botánica, Universidad de Jaén, Las Lagunillas s/n 23071 Jaén, Spain. ecano@ujaen.es

Abstract

In this research project, a new faster and more efficient method of assessing the biodiversity and the conservation value of pastures was derived. The methodology proposed is based on the phytosociological method. Taking the Shannon-Weaver index as a starting point, a new phytocoenotic stability index (I_{ef}) is suggested for each plant association. I_{ef} is a good index for the management of pastureland because it shows the conservation value of the pasture phytocoenosis and how they tend, in time, to change into different pasture associations of variable quality if the external disruptive factors continue to exist.

Key words: biodiversity, stability, pasture dynamics

Introduction

Pasture communities are plant formations with a particular floristic combination, where the characteristic, dominant species of the association are surrounded by a number of other companions. These phytocoenoses are determined by clearly defined ecological limits because they and the environmental and anthropozoogenic factors affecting them can change with time. These pasture formations deserve more attention: Spain has about 12,872.6 million ha of productive pastureland (San Miguel Ayanz, 2001), the livestock being mostly sheep and cattle.

This chapter aims to present an integrated methodology using the phytosociological method to describe and identify pastures and their dynamics. In addition, the research makes use of biodiversity indices for pastureland, such as the Shannon-Weaver index (Shannon and Weaver, 1981), and describes a new index called the phytocoenotical stability index (I_{ef}), which aims at showing whether the floristic composition of the pasture community under study is in an optimal state or in a transitional state towards a different coenosis. This method of assessing plant communities allows the establishment of objective criteria to optimize use and management of pastureland.

Materials and Methods

The area under study is the nature reserve of Sierra Mágina and the surrounding area (Jaén, S Spain). This area of 28,000 ha has a rugged topography, with considerable escarpments and steep slopes. The dominant soil materials are limestones, marls, dolomites and gypsum substrates. Using the phytosociological method (Braun-Blanquet, 1979), ten pasture communities were chosen for identification and sampling purposes. In order to obtain statistically relevant results a minimum of ten phytosociological inventories were carried out for each association. Before the sampling were made, the minimal sampling area for each community was estimated by means of the species area curve. This area was between 0.5 and 3 m² for the ten associations (Figure 1).

Figure 1. Fitting curve of the number of species as regards the sampling area.

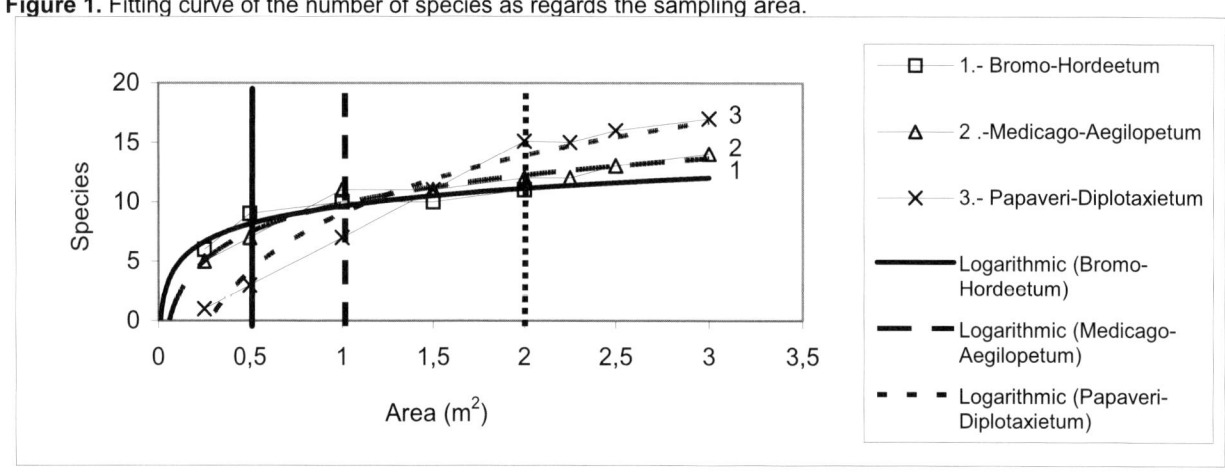

The Shannon-Weaver index is calculated according to the number of species and their relative abundance in each phytosociological inventory.

Results and Discussion

The characteristic species (spc) of each association are compared with the non-characteristic species or companions (spco) (Table 1). The Shannon-Weaver index was calculated for all the species (I_{sht}) and the characteristic species (I_{shc}). The phytocoenotical stability index I_{ef} is the ratio of the Shannon index for the total of species (I_{sht}) and the Shannon index for the characteristic species of the community (I_{shc}).

Table 1. Analysed parameters in each phytosociological association. Am: minimal area. No. spt: total number of species. No. spc: number of characteristic species. No. spco: number of companion species. I_{sht}: Shannon index for all the species of the phytocoenosis. I_{shc}: Shannon index for the characteristic species. I_{ef}: phytocoenotical stability index.

Pastureland communities	Am	No. spt	No. spc	No. spco	I_{sht}	I_{shc}	I_{ef}
Bromo-Hordeetum leporini	0.5	68	37	31	2.3	2.0	0.9
Anacyclo-Hordeetum leporini	0.5	59	43	16	2.2	2.0	0.9
Papaveri-Diplotaxietum virgatae	2	38	33	5	2.3	2.2	1.0
Torilido-Scandicetum australis	1	30	22	8	2.1	1.8	0.9
Aegilopo-Stipetum capensis	1	64	36	28	2.3	1.9	0.8
Medicagini-Aegilopetum geniculatae	1	91	52	39	2.3	2.0	0.8
Velezio-Asteriscetum aquatici	1	58	22	36	2.2	1.5	0.7
Saxifrago-Hornungietum petraeae	0.5	31	13	18	2.0	1.5	0.7
Poo-Astragaletum sesamei	1	44	4	40	2.2	0.4	0.2
Coronillo-Astragaletum nummularioidis	1	32	4	28	2.1	1.1	0.5

When the Shannon-Weaver index is applied to the ten inventories of each association, either to all the species or only to those characteristic of the community, it shows different values for I_{sht} and I_{shc}. The smaller the difference (I_{sht} - I_{shc}), the greater the number of characteristic species in the association. These are better adapted to the specific enviromental conditions than the general species. Therefore the pasture community is more stable in space and time. There is a direct correlation between the number of characteristic species in a plant community and the ecological factors. Consequently, if the ratio $I_{ef} = I_{shc}/I_{sht}$ is close or equal to 1, the pasture is at its ecological optimum. This would condition the maximum stability of the phytocoenosis, since, as long as the value of the proposed new index I_{ef} is equal to 1, the stability of the plant community is at its highest and can undergo no dynamic changes. By contrast, when I_{ef} < 1 and is close to zero, this reveals a decrease in the number of characteristic species and an increase in the number of companions. This occurs when the environmental factors have been modified, resulting in a negative tendency for the association to turn into a different phytocoenosis.

Conclusions

A new index, namely, the phytocoenotical stability index, I_{ef}, is presented. This index can be applied to the pasture formations of a given territory and can efficiently estimate whether the community under study has an optimal floristic composition (which could be stable under similar environmental factors) or, by contrast, tends to change both in its structure and composition and finally give way to a different phytocoenosis. The phytocoenotical stability index allows the validation of information about the conservation value of pastureland and permits the establishment of criteria for its management, especially when these plant communities are deteriorating in quality.

References

Braun-Blanquet, J. (1979) *Phytosociology. Basis for the Study of Vegetation Communities* (Fitosociología. Bases para el estudio de las comunidades vegetales). Blume, Madrid, Spain. (In Spanish.)

San Miguel Ayanz, A. (2001) *Spanish Natural Pastures. Characterization, Management and Improvement Possibilities* (Pastos naturales españoles. Caracterización, aprovechamiento y posibilidades de mejora). Fundación Conde del Valle Salazar y Mundi-Prensa, Madrid, Spain. (In Spanish.)

Shannon, C.E. and Weaver, W. (1981) *Mathematical Theory of Communication* (Teoría matemática de la comunicación (primera edición en ingles: 1949)). Forja, Madrid, Spain. (In Spanish.)

Effects of breed and stocking rate on vegetation dynamics and biodiversity in heath-gorse communities grazed by goats

R. Celaya, B. M. Jáuregui, U. García and K. Osoro
Servicio Regional de Investigación y desarrollo Agroalimentario (SERIDA), Apdo. 13, 33300 Villaviciosa, Asturias, Spain. rcelaya@serida.org

Abstract

The objective of this work is to study the changes in plant and animal biodiversity of heathlands under goats' grazing in a Cantabrian range. Three treatments (in three replicates) were established: low (LC) and high stocking rate with Cashmere goats (HC), and high stocking rate with local goats (HL). Botanical composition and height of the canopy were assessed during 2002 and 2003. Since 2003 animal biodiversity has also being controlled, counting birds and sampling butterflies, grasshoppers and ground-dwelling invertebrates.

Preliminary results indicate a greater decrease in both frequency and mean height of the shrubs (mainly heaths and gorse) in HL than in HC, leading to a greater increase in herbaceous species in the former. Among the cashmere treatments, the woody plants were better controlled at the higher stocking rate. The number of birds and butterflies varied according to the grazing period, but no clear effects of the established treatments were observed.

Key words: biodiversity, breed, goat, heath-gorse, stocking rate

Introduction

Heathlands are widespread in the Cantabrian region (N Spain) as a consequence of deforestation and abandonment of livestock managements, mainly in the less favoured areas with poor and acid soils. Their poor nutritive value limits animal production and the accumulation of woody biomass involves serious problems of uncontrolled fires. Goats utilize these plant communities more efficiently than cattle or sheep, controlling biomass accumulation and improving the nutritive value of the available vegetation due to an increase in herbaceous species in the canopy (Osoro *et al.*, 2000; Celaya and Osoro, 2002).

In recent years there has been increasing interest in the conservation and enhancement of biodiversity of pastoral resources, coupled moreover with production objectives for the sustainability of the pastoral and silvopastoral systems. This work has been carried out since 2002 in an integrated experimental programme with other European sites (EU project FORBIOBEN, QLRT-1999-30130) to study the effects of stocking rate and goat breed (one local *vs* one commercial foreign breed) on plant and animal biodiversity of different natural or semi-natural grasslands and shrublands, as well as on the socio-economic outcome of these livestock systems (Rook *et al.*, 2002).

Materials and Methods

The experimental farm is located in Illano, Asturias (NW Spain), at 900 m asl. The vegetation is shrubland dominated by heaths (*Erica sp*, *Calluna*, *Daboecia*) and gorse (*Ulex gallii*). Nine plots of 0.6 ha with three grazing treatments (in three replicates) were established: low stocking rate with Cashmere goats (LC), high stocking rate with Cashmere goats (HC) and high stocking rate with local Celtiberian goats (HL). The animals, four goats per plot in LC, eight in HC and seven in HL, grazed the experimental plots from July to October in 2002 and from May to November in 2003. Botanical composition was assesed using a sward stick in July and October 2002 and in May, August and October 2003 by two sampling procedures: recording the two higher hits (species, growth stage and height) at 10 cm intervals along one 50 m long fixed transect per plot (1000 contacts per transect), and recording 100 hits per plot at random. Birds, butterflies and grasshoppers were counted on five occasions in the 2003 grazing season. Also ground-dwelling invertebrates were sampled in May, August and October 2003 in 12 pifall traps per plot, though these have still to be identified. Treatment effects on the plant frequencies were analysed by a chi-squared test.

Results and Discussion

The frequency of live shrubs decreased more in the HC than in the LC treatment, but it decreased even more in the HL than in the HC treatment (Figures 1 and 2). This difference was restricted mainly to heaths while gorse percentage was not significantly affected by breed. The greater control of shrubs by local goats was associated with higher increases in herbaceous species in the canopy in these treatments, mainly in the new gaps formed among the shrubs by browsing and trampling, reaching 30-33% of the sward stick hits in May 2003. The effect of stocking rate on the frequency percentage of herbaceous species was lower than breed effect, with the percentage being greater in HC than in LC in the transects but this difference already appeared in the first control. With random sampling, the herbaceous percentage was significantly higher in HC than in LC in May 2003 only. The percentages of bare ground remained relatively constant along the five controls in all treatments while the percentage of litter, as well as the dead shoots of shrubs, increased more under local goat grazing than under Cashmere goat grazing, and obviously under high stocking than under low stocking.

The mean height of the shrubs also decreased more in the plots grazed by local goats than in those grazed by Cashmere goats, mainly in the case of gorse. Among the Cashmere treatments, the mean height of shrubs in general was more reduced under high stocking than under low stocking. However, with random sampling the differences in gorse height were very small.

Figure 1. Frequency percentages of live shrubs and herbaceous species along 50 m transects in plots grazed by goats (n = 3000). LC: low stocking rate, Cashmere breed; HC: high stocking rate, Cashmere breed; HL: high stocking rate, local breed. a,b,c: different letters in each sampling date and plant component mean significant differences ($P < 0.05$) between treatments.

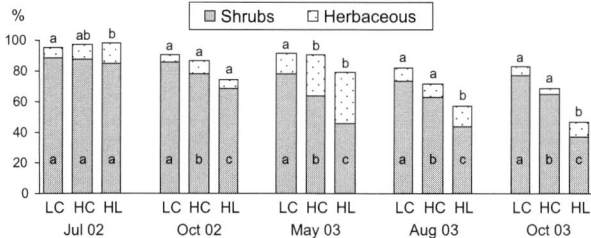

The changes in the number of species and diversity (H'Shannon) were more conditioned by the time of sampling than by the grazing treatments. The higher indices were recorded in July 2002 and May 2003 while the number of plant species was reduced in August and October according to the greater drought and senescence rates in these months.

Figure 2. Random frequency percentages of live shrubs and herbaceous in plots grazed by goats (n = 300). Abbreviations and symbols as in Figure 1.

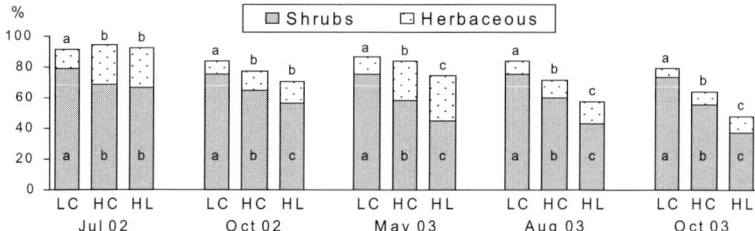

The number of birds and butterflies varied according to the grazing period, but no clear effects of the established treatments were observed. The abundance of grasshoppers was lower in LC than in HC, being intermediate in HL. As vegetation structure and composition are diverging between treatments it is expected that greater differences in the diversity of the fauna will emerge associated with the different grazing managements.

Conclusions

The woody plants are more intensively utilized by local compared to Cashmere goats, leading to a higher presence of herbaceous species in the former. The percentage of live shrubs is maintained near to their initial values at the lower stocking rate with Cashmere goats.

References

Celaya, R. and Osoro, K. (2002) Effect of the ovine and caprine livestock flock share on the vegetation dynamics of improved heath-gorse lands (Efecto de la proporción de ovinos y caprinos en el rebaño sobre la dinámica vegetal de brezales-tojales parcialmente mejorados). In: Chocarro, C., Santiveri, F., Fanlo, R., Bovet, I. and Lloveras, J. (eds) *Producción de Pastos, Forrajes y Céspedes. Actas de la XLII Reunión Científica de la Sociedad Española para el Estudio de los Pastos*. Edicions de la Universitat de Lleida, Lleida, Spain, pp. 537-542. (In Spanish.)

Osoro, K., Celaya, R., Martínez, A. and Zorita, E. (2000) Grazing of mountainous vegetation communities by domestic ruminants: animal production and vegetation dynamics (Pastoreo de las comunidades vegetales de montaña por rumiantes domésticos: producción animal y dinámica vegetal). *Pastos* XXX (1), 3-50. (In Spanish.)

Rook, A.J., Petit, M., Isselstein, J., Osoro, K., Wallis de Vries, M.F., Parente, G. and Gaskell, P. (2002) Integrating foraging attributes of domestic livestock breeds into sustainable systems for grassland biodiversity. *Grassland Science in Europe* 7, 1068-1069.

Biodiversity and dynamics of traditional silvopastoral systems in Galicia (north-west Spain): *Cytisus* scrubs

T. Cornide[1], E. Díaz-Vizcaíno[2] and M. Casal[3]

[1,2]Dept. de Botánica, Escuela Politécnica Superior, Universidad de Santiago, Campus de Lugo, 27002 Lugo, Spain.
[3]Dept. de Biología Celular y Ecología, Facultad de Biología, Universidad de Santiago, 15782 Santiago de Compostela (A Coruña), Spain. bvtcpaz@lugo.usc.es, bvlueadv@lugo.usc.es, bfmcasal@usc.es

Abstract

In inland Galicia, north-west Spain broom scrubs ("*xesteiras*") dominated by *Cytisus striatus* or *C. multiflorus* often form part of traditional silvopastoral systems, with non-woody components providing grazing and forage, and woody components providing firewood and bedding. In some cases the communities may form part of crop/pasture rotations. This chapter reports a quantitative ecological study of *C. striatus* and *C. multiflorus* formations in the foothills of the Sierra de Oribio (Lugo Province, Spain). The study analysed variations in floristic composition, floristic abundance and physionomic structure at different stages in community development, with the aim of understanding community dynamics and relating them to traditional uses. All samples contained one of the two species of *Cytisus* as dominant legume, ensuring the fertility of the soil and favouring the development of a herb layer whose composition changes as the community develops. Useful pasture species frequently present in the herb layer included *Holcus mollis* L., *Agrostis capillaris* L., *Vulpia bromoides* (L.) S.F. Gray, *Anthoxanthum aristatum* Boiss, *Dactylis glomerata* L. and *Luzula campestris* (L.) D.C. Although these species are present in *Cytisus* scrubs at lower frequencies than in mowed meadows in humid regions of Spain, they clearly represent an important pasture resource.

Key words: pasture value, silvopasture

Introduction

Among the most representative scrub communities in inland areas of the region of Galicia in north-west Spain are broom scrubs ("*xesteiras*") dominated by *Cytisus striatus* (Hill) Rothm. or *C. multiflorus* (L'Hér.) Sweet. *Cytisus striatus* is endemic to Portugal and west/central Spain, and *C. multiflorus* to north/central Portugal and north-west/central Spain.

These communities are an important element of the landscape of eastern Galicia, and are in some cases indicative of oceanic influence, in others of the Atlantic-Mediterranean transition (Izco, 1987). They are serial communities, very frequently as part of degraded forestland and planted forest, as well as serial shrublands distributed in mosaics with forest communities (*Quercus pyrenaica* L. or *Quercus robur* Wild.) which are grazed together. Here we report a quantitative ecological study of *C. striatus* and *C. multiflorus* formations in the foothills of the Sierra de Oribio in Lugo Province in north-west Spain.

Materials and Methods

In a quantitative ecological study of *C. striatus* and *C. multiflorus* communities in the Sierra de Oribio, 75 samples representing well-differentiated stages in the development of these communities were surveyed (Cornide *et al.*, 1991; Reyes *et al.*, 2000). The analysis aimed to identify variations in composition and floristic abundance, with the aim of characterizing community dynamics, and relating dynamics to traditional habitat uses. Hence, in each sample frequency data for all species present, as well as height and density of *Cytisus sp*, were recorded.

Frequency was estimated by quadrant sampling, height by measurement of ten individuals randomly selected within each sample, and density as the mean of ten measurements of the number of main plants rooted in a 1-square-metre quadrant within each sample (Greig-Smith, 1983; Goldsmith *et al.*, 1986).

Results and Discussion

Table 1 shows mean species richnesses in the two communities at different developmental stages, with separate estimates for major species groups, together with data on canopy height and *Cytisus* density. In the *C. striatus* communities, juvenile-phase samples (mean height 1.12 m, cover about 40% and density 4.44 plants m^2) contained on average 24 species (five woody species, 19 herb species, nine perennial herbs, ten annual herbs). Mature-phase samples (mean height 2.78 m, cover about 70%, density 1.98 plants m^2) contained on average 21 species (five woody species, 16 herb species, nine perennial herbs, seven annual herbs). Senescent-phase samples (mean height 4.31 m, cover about 60%, density 1.31 plants m^2) contained on average only 19 species (five woody species, 14 herb species, eight perennial herbs, six annual herbs). The number of species with a mean frequency of more than 10% was 15 in juvenile samples (two woody species, seven perennial herbs, six annual herbs), eight in mature samples (one woody species, five perennial herbs, two annual herbs), and nine in senescent samples (one woody species, seven perennial herbs, one annual herb).

Table 1. Mean species richnesses and structural descriptors of samples from juvenile J, mature M and senescent S phase *Cytisus striatus* and *Cytisus multiflorus* communities.

Community	Cytisus striatus			Cytisus multiflorus		
	J	M	S	J	M	S
Species richness						
All species	23.66 ± 1.50	20.67± 1.54	19.08 ± 1.24	25.17± 1.28	3.00 ± 1.99	17.89 ± 2.34
Woody species	4.58 ± 0.80	4.80 ± 0.73	4.54 ± 0.58	2.50 ± 0.66	4.28 ± 0.65	5.89 ± 1.11
Herb species	19.08 ± 1.19	15.87± 5.34	14.54 ± 1.24	22.67 ± 1.29	18.71 ± 2.10	12.00 ±1.68
Annual herbs	9.92 ± 1.07	7.27 ± 0.94	6.23 ± 0.89	12.50 ± 1.46	10.35 ± 1.42	4.89 ± 0.74
Annual herbs: grasses	1.75 ± 1.13	1.26 ± 1.09	1.69 ± 1.10	2.91 ± 1.24	2.07 ± 1.11	1.60 ± 0.96
Annual herbs: legumes	1.33 ± 1.43	1.13 ± 1.59	0.46 ± 1.12	1.66 ± 1.07	1.15 ± 1.28	0.30 ± 0.48
Other annual herbs	6.83 ± 2.62	4.86 ± 2.32	4.30 ± 2.52	7.91 ± 3.94	6.38 ± 3.04	4.50 ± 4.74
Perennial herbs	9.16 ± 0.72	8.60 ± 0.78	8.31 ± 0.70	10.17 ± 0.89	8.36 ± 0.99	7.11 ± 1.14
Perennial herbs: gras.	3.83 ± 1.02	3.33 ± 0.72	3.84 ± 0.80	3.83 ± 1.11	3.84 ± 1.46	3.30 ± 1.15
Perennial herbs: legumes	0.41 ± 0.66	0.13 ± 0.35	0.00 ± 0.00	0.25 ± 0.45	0.00 ± 0.00	0.10 ± 0.31
Other perennial herbs	4.91 ± 2.06	5.86 ± 2.89	4.46 ± 2.29	6.08 ± 2.35	4.15 ± 2.88	4.30 ± 2.94
Structural parameters						
Height (cm)	112.1 ± 39.45	278.1 ± 78.9	431.0 ± 5.76	69.48 ± 28.3	30.00 ± 26.6	180.5 ± 24.4
Cytisus density	4.44 ± 3.98	1.98 ± 1.04	1.31 ± 0.65	3.11 ± 0.93	3.01 ±1.33	1.62 ± 0.09

Figure 1. Range-abundance diagrams for mature communities of *Cytisus striatus*, considering all species with frequency ≥ 1%. Species sequence by abundance order: 1: *Cytisus striatus*; 2: *Holcus mollis*; 3: *Anthoxanthum aristatum*; 4: *Agrostis capillaris*.; 5: *Hypochoeris radicata*; 6: *Rumex acetosella*; 7: *Agrostis x foulladei*; 8: *Dactylis glomerata*; 9: *Quercus robur*; 10: *Senecio vulgaris*; 11: *Galium saxatile*; 12: *Holcus lanatus*; 13: *Geranium lucidum*; 14: *Poa pratensis*; 15: *Trifolium dubium*; 16: *Arenaria montana*; 17: *Crepis capillaris*; 18: *Digitalis purpurea*; 19:*Ornithopus compressus*; 20: *Vulpia bromoides*; 21: *Jasione montana*; 22: *Plantago lanceolata*; 23: *Hypericum humifusum*; 24: *Rubus* sp; 25: *Galium aparine*; 26: *Stellaria holostea*; 27: *Cardamine hirsuta*; 28: *Lonicera periclymenum*; 29: *Ulex europaeus*; 30: *Vicia sativa*; 31: *Adenocarpus complicatus*; 32: *Agrostis delicatula*; 33: *Aira praecox*; 34: *Ornithopus sativus*; 35: *Stellaria media*; 36: *Andryala integrifolia*; 37: *Bromus hordeaceus*; 38: *Corydalis claviculata*; 39: *Cytisus multiflorus*; 40: *Frangula alnus*; 41: *Luzula campestris*; 42: *Orobanche rapumgenistae*; 43: *Raphanus raphanistrum*; 44: *Sedum arenarium*; 45: *Castanea sativa*; 46: *Lotus uliginosus*; 47: *Pteridium aquilinum*; 48: *Taraxacum officinale*; 49: *Ulex gallii*; 50: *Veronica beccabunga*.

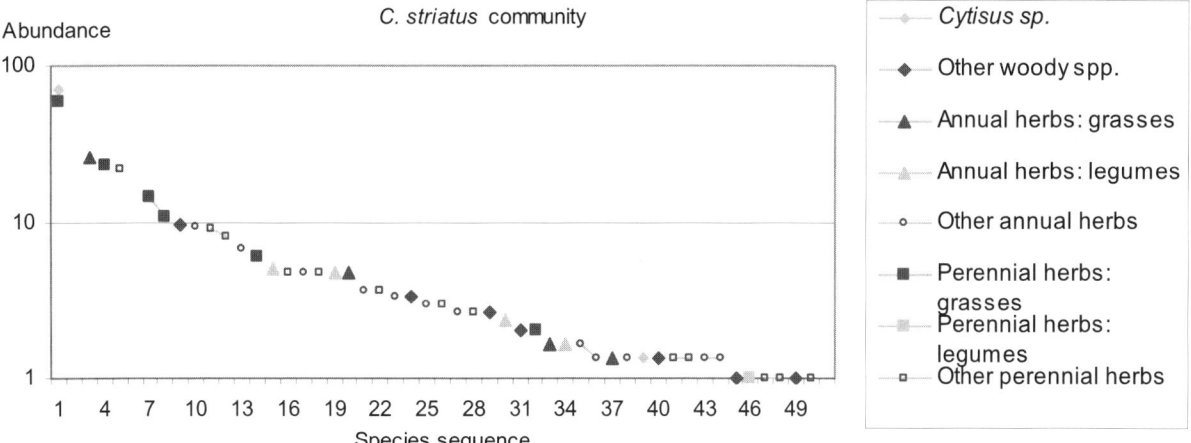

Species richnesses in both communities was higher than described by Basanta *et al.* (1989) for broom scrub in other regions of Galicia, though physiognomic structure is similar (Basanta *et al.*, 1988). Both communities contain a number of species not commonly seen in other Galician scrub.

Figures 1 and 2 show range-abundance diagrams for mature-phase samples of both formations, showing only species with mean frequency greater than 1%. *Cytisus striatus* and *C. multiflorus* were the most frequent legumes in their respective communities. These shrubs are noteworthy for their high protein content and lack of tannins to reduce protein digestibility (González-Hernández *et al.*, 2000), making them a useful nutritional resource for browsing ruminants (Mosquera-Losada *et al.*, 2000). Perennial and annual leguminous herbs are infrequent in these communities. However, a number of perennial and annual herbs of agronomic importance (Grime *et al.*, 1988) were present with frequencies of more than 10% in both communities, namely *Holcus mollis*, *Anthoxanthum aristatum*, *Agrostis capillaris*, *Agrostis × foulladei*, *Dactylis glomerata* and *Vulpia bromoides*. These herb species are palatable to livestock, and have low to moderate concentrations of important nutritional elements like P, Ca, Mg and K (Gónzalez-Hernández et al., 2000).

Figure 2. Range-abundance diagrams for mature communities of *Cytisus multiflorus*, considering all species with frequency ≥ 1%. Species sequence by abundance order: 1: *Cytisus multiflorus*; 2: *Holcus mollis*; 3: *Vulpia bromoides*; 4: *Agrostis capillaris*; 5: *Anthoxanthum aristatum*; 6: *Agrostis x foulladei*; 7: *Hypochoeris radicata*; 8: *Rumex acetosella*; 9: *Daboecia cantabrica*; 10: *Erica cinerea*; 11: *Jasione montana*; 12: *Halimium alyssoides*; 13: *Plantago lanceolata*; 14: *Sedum arenarium*; 15: *Agrostis delicatula*; 16: *Dactylis glomerata*; 17: *Avenula marginata*; 18: *Hypericum linarifolium*; 19: *Rubus* sp; 20: *Quercus robur*; 21: *Andryala integrifolia*; 22: *Bromus hordeaceus*; 23: *Calluna vulgaris*; 24: *Senecio vulgaris*; 25: *Digitalis purpurea*; 26: *Hieracium pilosella*; 27: *Pteridium aquilinum*; 28: *Ornithopus compressus*; 29: *Teesdalia nudicaulis*; 30: *Tuberaria guttata*; 31: *Chamaemelum mixtum*; 32: *Holcus lanatus*; 33: *Orobanche rapumgenistae*; 34: *Cardamine hirsuta*; 35: *Crepis capillaris*; 36: *Luzula campestris*; 37: *Narduroides salzmannii*; 38: *Potentilla erecta*; 39: *Anarrhinum bellidifolium*; 40: *Arenaria montana*; 41: *Ulex europaeus*; 42: *Aphanes arvensis*; 43: *Geranium molle*; 44: *Helianthemum nummularium*; 45: *Vicia sativa*; 46: *Cytisus striatus*; 47: *Geranium lucidum*; 48: *Ranunculus repens*; 49: *Trifolium dubium*.

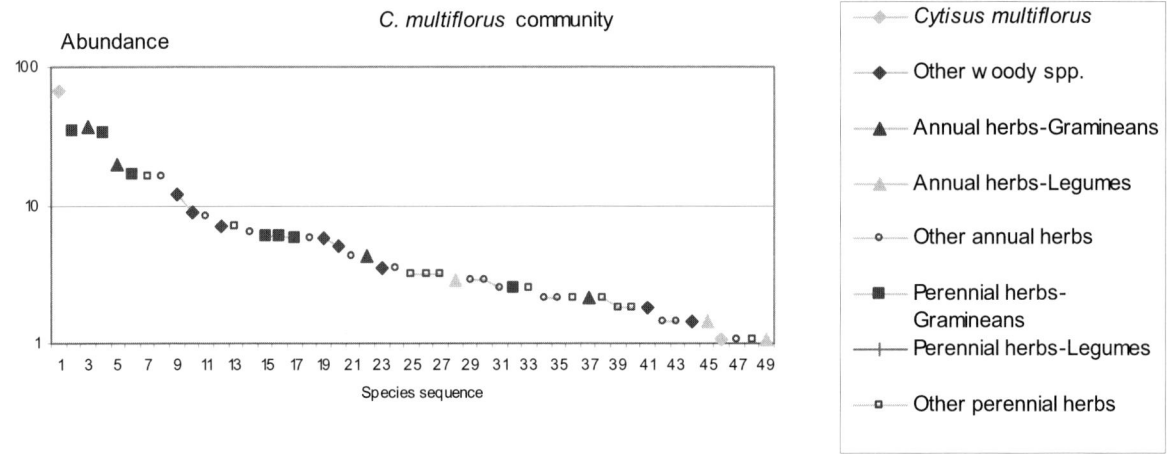

Conclusions

The present findings reveal the high species richnesses of *Cytisus striatus* and *C. multiflorus* broom scrub in north-west Spain. The two *Cytisus* species are abundant in their respective communities, at all developmental stages. A small number of grasses which are palatable and nutritional to livestock are also present.

References

Basanta, M., Díaz Vizcaíno, E. and Casal, M. (1988) Structure of shrubland communities in Galicia (NW Spain). In: Werger, H.J.A. and Willens, J.H. (eds) *Diversity and Pattern in Plant Communities*. SPB Academic Publishing, The Hague, The Netherlands, pp. 25-36.

Basanta, M., Díaz Vizcaíno, E., Casal, M. and Morey, M. (1989) Diversity measurements in shrubland communities of Galicia (NW Spain). *Vegetatio* 82, 105-112.

Cornide, T., Díaz Vizcaíno, E. and Casal, M. (1991) Relation between structure and life cycle in "retamares" in Galicia (Relación entre estructura y ciclo de vida en retamares de Galicia). *Studia Oecologica* VIII, 107-117. (In Spanish.)

González-Hernández, M.P., Starkey, E.E. and Karchesy, J. (2000) Presence of tannins in plants in Galician forests (NW Spain) and their importance in pasture lands management (Presencia de taninos en las plantas del monte gallego (NW Spain) y su importancia en la gestión de ecosistemas pascícolas). In: *III Reunión Ibérica de Pastos y Forrajes*. Xunta de Galicia, A Coruña, Spain, pp. 683-688. (In Spanish.)

Greig-Smith, P. (1983) *Quantitative Plant Ecology*. Butterworths, London, United Kingdom.

Grime, J.P., Hodgson, J.G. and Hunt, R. (1988) *Comparative Plant Ecology*. Unwin Hyman, London, United Kingdom.

Izco, J. (1987) Galicia. In: Peinado Lorca, M. and Rivas-Martínez, R. (eds) *La Vegetación de España*. Universidad de Alcalá de Henares. Madrid, Spain, pp. 387-418. (In Spanish)

Goldsmith, F.B., Harrison, C.M. and Morton, A.J. (1986) Description and analysis of vegetation. In: Moore, P.D. and Chapman, S.B. (eds) *Methods in Plant Ecology*. Blackwell Scientific Publications, Oxford, United Kingdom, pp. 437-524.

Mosquera-Losada, M.R., Rigueiro-Rodríguez, A. and Jardon-Bouzas, B. (2000) Content of proteins, phosphorus, calcium, potassium and magnesium in several shrub species (Contenidos en proteína, fósforo, calcio, potasio y magnesio de distintas especies arbustivas). In: *III Reunión Ibérica de Pastos y Forrajes*. Xunta de Galicia, Santiago de Compostela, Spain, pp. 665-670. (In Spanish.)

Reyes, O., Basanta, M., Casal, M. and Díaz Vizcaíno, E. (2000) Functioning and dynamics of woody plant ecosystems in Galicia (NW Spain). In: Trabaud, L. (ed.) *Life and Environment in the Mediterranean*. WIT Press, Southampton, United Kingdom, pp. 1-42.

Root density and soil water relationships of a silvopastoral system of the tropical region of Yucatan, Mexico

H. G. Delgado [1], L. A. Ramírez [2], J. V. Ku [2], P. M. Velásquez [3] and J. B. Escamilla [4]

[1]Universidad del Zulia, Facultad de Ciencias Veterinarias, Núcleo Agropecuario, Apartado 15252, Maracaibo 4500-A, Estado Zulia, Venezuela. [2]Universidad Autónoma de Yucatán, Facultad de Medicina Veterinaria y Zootecnia, km 15.5 carretera Mérida-Xmatkuil, Apartado 4-116 Itzimná, Mérida 97100 Yucatán, Mexico. [3]Instituto Tecnológico Agropecuario No 2, Conkal Yucatán, Mexico. [4]Centro de Investigación Científica de Yucatán, calle 43 N° 140 col, Chuburná, Mérida 97200 Yucatán, Mexico. gomezdel@latinmail.com, raviles@tunku.uady.mx, kuvera@tunku.uady.mx, avelaz@itaconkal.edu.mx, jae@cicy.mx

Abstract

In the Yucatán Peninsula of Mexico, cattle production is limited by forage availability during the dry season since water and soil fertility are the main factors limiting production. The objective of this study was to assess the effect of *Leucaena leucocephala* grown at 5000 trees/ha on forage production (DM), crude protein content (CP), fine root density (RD) and soil gravimetric moisture content (SWC) of *Panicum maximum cv* Colonial. The experiment was conducted during the rainy seasons of 2001 and 2002 in Mexico. Forage biomass yield and CP were higher in silvopasture compared with *P. maximum* alone (35,000 kg DM/ha of grass + 4000 kg DM/ha of *L. leucocephala vs* 28,000 kg DM/ha per year) ($P < 0.001$); CP increased ($P < 0.01$) in the grass close to stems of *L. leucocephala* and SWC (soil water content) increased with soil depth ($P < 0.01$) in all sampling periods. RD diminished ($P < 0.01$) with distance from the line of trees and soil depth increased during the dry season. This might be due to the effect of the shadow of the tree canopy, which could enhanced the nitrogen availability as well as improving the microhabitat of the grass. In the silvopasture *L. leucocephala* increased the growth of *P. maximum* in terms of both biomass production and rooting density.

Key words: forage yield, root dynamics, soil moisture

Introduction

Silvopastoral systems are an alternative to clearing land for pasture in cattle production. The principle of silvopastoral systems is that trees must acquire resources that pasture would not get. Also trees may contribute to improving soil productivity by having a more closed nutrient cycle than pasture systems and enrich the soil with nutrients and organic matter (Young, 1995). In the Yucatán Peninsula, cattle production is limited by forage availability during the dry season since water and soil fertility are the main factors limiting production. Also shallow, poor soils are common, therefore forage production declines over time, resulting in the abandonment of degraded land and the clearing of new land for pasture. *L. leucocephala* is a native tree of the Yucatán Peninsula and has good nutritive value (24-30% CP). It can withstand drought and grazing, therefore its introduction in pasture production systems is recommended as an alternative to forage quality and production. However, in associations both species could compete for water and nutrients. There is now a more pressing need to understand how the introduction of trees into grass production systems influences the potential for pasture growth under a variety of conditions. The study of roots allows evaluation of compatibility among species (Schroth and Zech, 1995).

Materials and Methods

The experiment was located in southern Yucatan, Mexico, at the Center for Technical Development, Tantakin (20° 01' 46" N, 89° 02' 49" W, 55 m asl), with mean annual rainfall below 1200 mm/year and Luvisol soil. Treatments were: a silvopastoral system for cattle production, combining the grass *P. maximum cv* Colonial and the legume *L. leucocephala*; and open pasture of *P. maximum*. The trees were established in hedgerows and the planting density of *L. leucocephala* was equivalent to 5000 trees/ha. The experiment was carried out between 21 September 2001 and 1 November 2002. Treatments were established in a fully randomized plot design, with four replicate plots of each treatment. Forage production from the silvopastoral system was calculated to include tree and pasture biomass on a surface area basis. Biomass harvesting was done every 42 days during the rainy season and every 62 days during the dry season. The RD was estimated by the soil core-break method (Escamilla *et al.*, 1991), and SWC soil samples were taken every week at 15, 30 and 45 cm depth.

Results and Discussion

Soil gravimetric moisture content (SWC) increased with soil depth ($P < 0.01$) across all sampling periods (Figure 1). The increase was larger when SWC was below soil permanent wilting point (PWP) (27.8%). Forage biomass yield (Figure 2) in silvopasture increased 20% this was twice as much as in the monoculture of *P. maximum* (35,600 kg

DM/ha of grass plus 4400 kg DM/ha of *L. leucocephala vs* 28,600 kg DM/ha per year) ($P < 0.001$). During the rainy season DM was twice as much as in the dry season. DM and CP of pasture were affected by distance from *Leucaena* trees, being greater at 1 m from the stems than at 3 m, and greater than in the monoculture of grass (Figures 3 and 4). Fine root density of *L. leucocephala* decreased in different seasons and the presence of pasture roots in proximity to tree roots can have detrimental effects on the growth and distribution of the latter (Odhiambo *et al.*, 2001). *L. leucocephala* roots did not compete for water and nutrients with grass roots in Luvisol soil. Fine root density of pasture decreased with increasing distance from *L. leucocephala* and it was greater in the monoculture than at 3 m distance from the row of trees and lower than at 1 m (Figure 5). Increased growth of pastures adjacent to *L. leucocephala* can be largely attributed to enhanced nitrogen availability. Introduction of legume trees in pasture leads to a faster mineralization of organic reserves in soil than in monoculture (Wilson and Wild, 1991). In addition, greater growth could also be due to improved microclimate owing to increased organic matter content (Belsky *et al.*, 1989; Wilson, 1998); SWC changes between successive biomass samplings were correlated to changes in *P. maximum* DM. From the yield equation dry matter yield (DM yield = $805.0 + 1.73 SWC - 2.4 SWC^2$, $r^2 = 0.143$; $P < 0.01$) shows an increased yield until 36.01% SWC (a figure very close to field capacity, 36.3%); after that point, DM yield decreases.

Figure 1. Soil water content (SWC) (%) at three depths (cm) measured gravimetrically (mean of eight samples) and the monthly rainfall at the study site. Permanent wilting point and field capacity are plotted as reference.

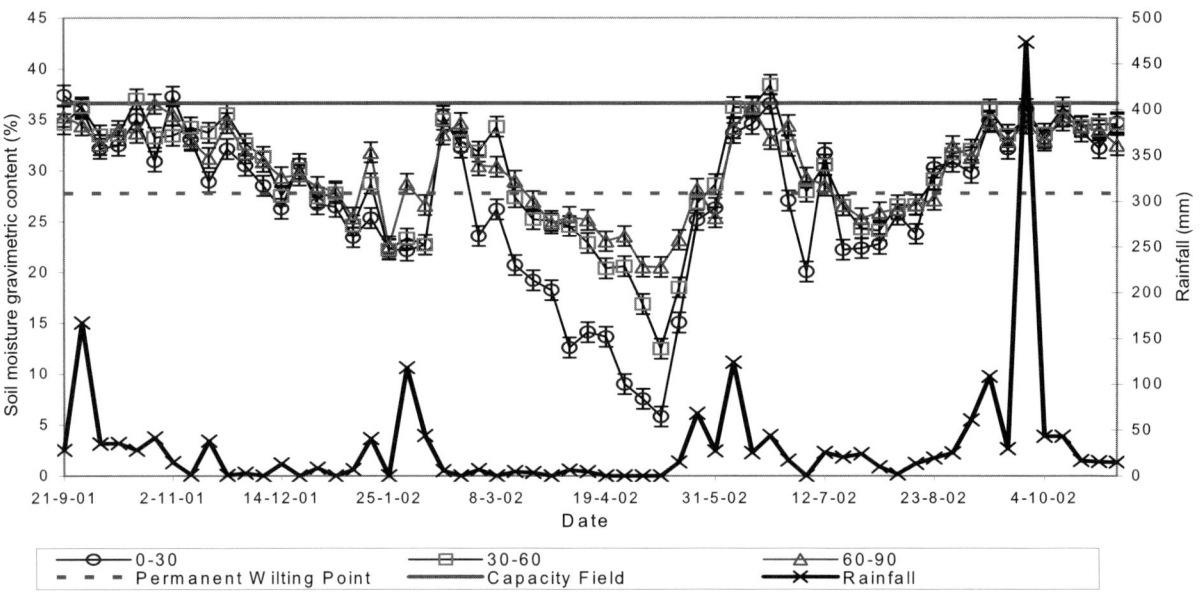

Figure 2. Mean biomass yield ($n = 4$) of *L. leucocephala*, *P. maximum* in monopasture and the association *P. maximum* - *L. leucocephala*.

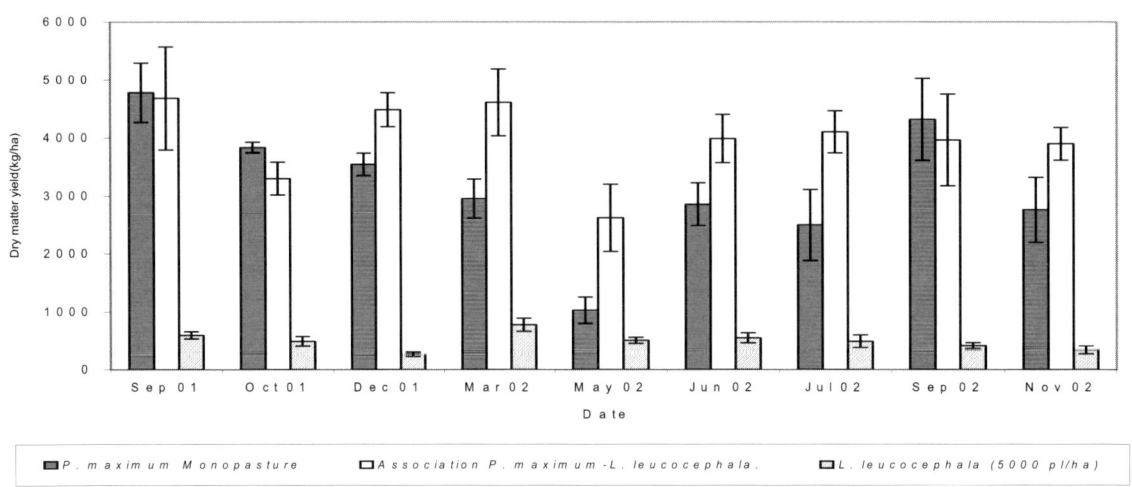

Figure 3. Mean biomass yield (*n* = 4) of *L. leucocephala*, *P. maximum* in monopasture and 1 m and 3 m distance from *L. leucocephala* hedgerow in a silvopastoral system.

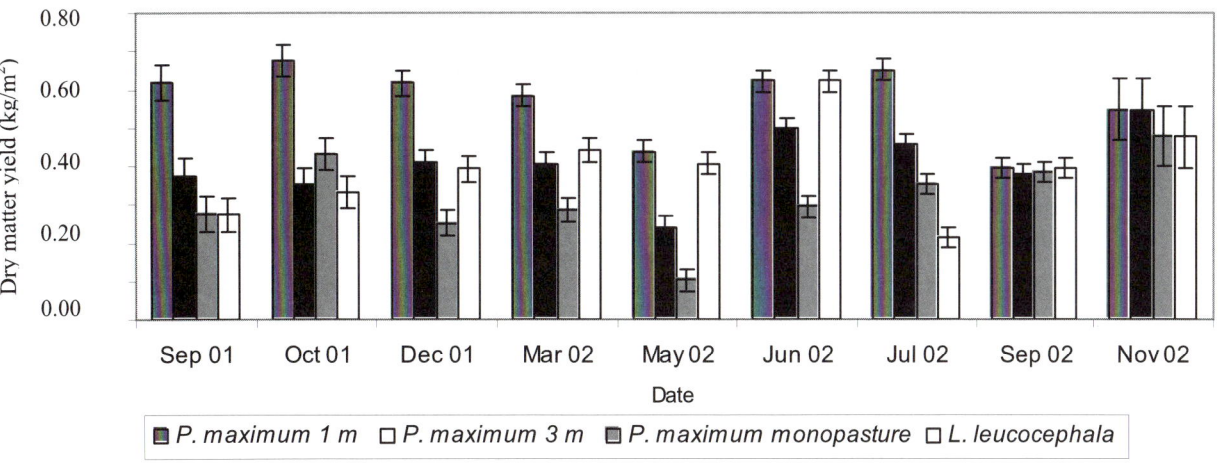

Figure 4. Crude protein content mean (n = 4) of *L. leucocephala*, *P. maximum* in monoculture, 1 m and 3 m distance from the *L. leucocephala* hedgerow in a silvopastoral system.

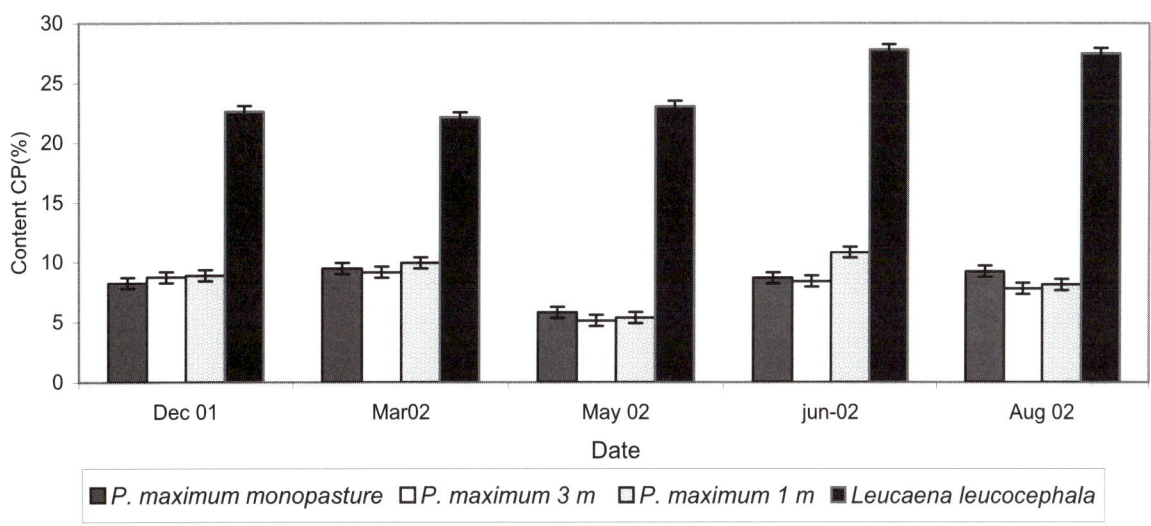

Figure 5. Root density mean distribution (*n* = 4) of *P. maximum* in monoculture, and at 1 m and 3 m distance from *L. leucocephala* hedgerow in a silvopastoral system.

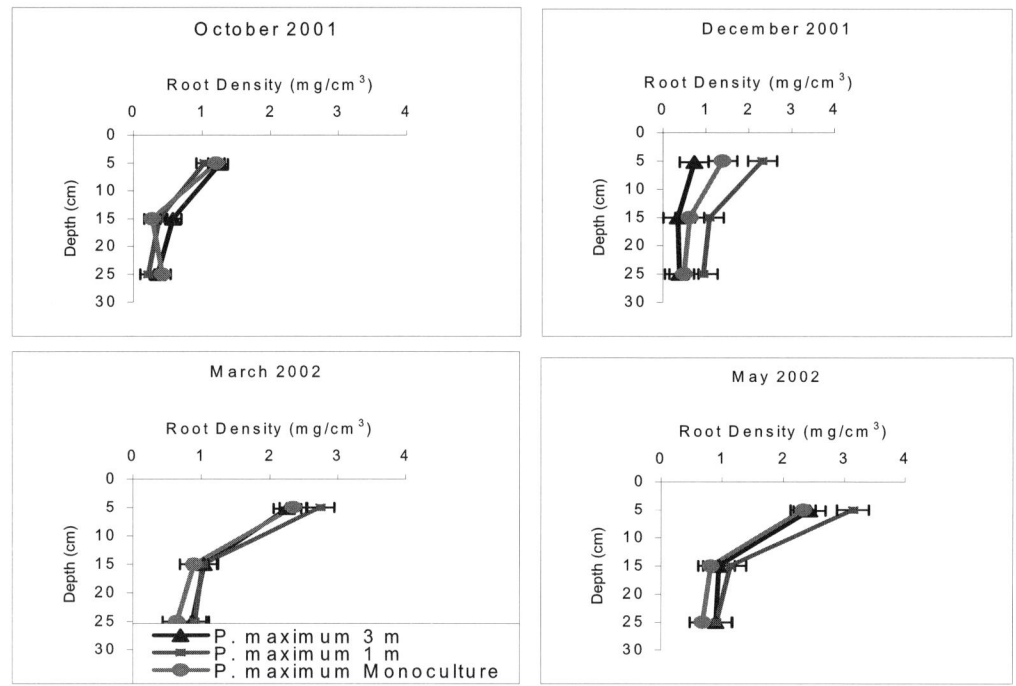

Conclusions

P. maximum cv Colonial in Luvisol soil yielded a higher forage biomass (35,000 *vs* 28,000 kg DM/ha per year) and crude protein content with the inclusion of *L. leucocephala*, which increased soil fertility through a higher nitrogen availability by the shadow action of tree canopies. Growing both species together also resulted in an increase in root density in pasture next to *L. leucocephala* sytems. Root density was higher in the dry than in the rainy season. These data suggest that the silvopastoral system studied has a more closed nutrient cycle compared to the *P. maximum* monoculture. The roots of the association of *P. maximum* with *L. leucocephala* explored a larger volume of soil, using growth resources in lower soil layers more efficiently than the monoculture system. *L. leucocephala* increased the biomass production and rooting density of *P. maximum*.

Acknowledgements

Research was supported by project CONACYT – 31661-B of the National Council of Science and Technology. We thank Ing. Juan M. Sol Zapotitla from TANTAKIN for technical assistance and financial support by FIRA. This work is part of the doctorate project of the first author.

References

Belsky, A.J., Amundson, R.G., Duxbury, J.M., Riha, S.J., Alí, A.R. and Mwonga, S.M. (1989) The effect of trees on their physical, chemical and biological environments in a semi arid savanna in Kenya. *Journal of Applied Ecology* 26, 1005-1024.

Escamilla, J.A., Comerford, N.B. and Neary, D.G. (1991) Soil-core break method to estimate pine root distribution. *Soil Science Society of America Journal* 55, 1722-1726.

Odhiambo, H.O., Ong, C.K., Deans, J.D., Wilson, A., Khan, A.H. and Sprent, J.I. (2001) Roots, soil water and crop yield: tree crop interactions in a semi-arid agroforestry systems in Kenya. *Plant and Soil* 235, 221-233.

Schroth, G. and Zech, W. (1995) Above and below ground biomass amics in a sole cropping and alley cropping system with *Gliricidia sepium* in the semi-deciduos rainforest zone of West Africa. *Agroforestry Systems* 31, 181-198.

Wilson, J.R. (1998) Influence of planting four tree species on the yield and soil water status of green panic pasture in sub-humid south-east Queensland. *Tropical Grassland* 32, 209-220.

Wilson, J.R. and Wild, D.W.M. (1991) Improvement of nitrogen nutrition and grass growth under shading. In: Shelton, H.M. and Stur, W.W. (eds) *Forages for Plantation Crops*. ACIAR Proceedings No. 32, pp. 77-82.

Young, A. (1995) Ten hypotheses for soil agroforestry research. In: *IV curso Internacional de Entrenamiento de la Agroforestería para el Ecodesarrollo*. Universidad de Chapingo, Chapingo, Mexico, pp. 123-126.

Cover crops effects on plant and insect biodiversity in Western Australian vineyards

A. Dinatale, A. Pardini and G. Argenti
DiSAT, University of Florence, Italy. andrea.pardini@unifi.it

Abstract
The clearing of large *Eucalyptus* forests in Mediterranean Australia caused negative effects on hydrological balance. The reintroduction of trees and shrubs and the use of perennial herbaceous species in association with woody plants are considered necessary to limit further damages to natural and agricultural ecosystems. The introduction of cover crops in vineyards can be useful to maintain the productivity of the land. Cover crops affect vegetation cover and insect biodiversity but this has not been sufficiently investigated. This chapter reports results from Western Australian vineyards: two vineyard managements (with cover crop and mown once a year at the end of spring; without cover crop and soil disc-harrowed once a year at the end of spring) were compared at two sites characterized by large differences in land use and vegetation diversity (areas near natural forests or in the specialized vineyard area). Data collection concerned botanical composition of cover crops and density of ground insects captured in the vineyard. Preliminary results show an increase in the number of plant species and an increase in the number of insects (individuals and species) in the vineyards in the area with higher plant diversity. It is possible that in well-conserved environments (from an ecological and botanical point of view) the number of insects which can colonize the cover crops is higher than in specialized vineyard areas, where their density is highly reduced and they are mainly represented by parasites.

Key words: botanical composition, land use, vine parasites, agricultural ecosystems

Introduction
In many areas of Australia *Eucalyptus* forest clearing has caused negative effects on hydrological balances. Annual plants absorb less water than the native forest, the water table is higher and upwelling water has carried salt deposits to the surface. As a result, large areas are now unproductive due to salinization. It is considered necessary to introduce new trees and shrubs associated with perennial herbaceous plants to limit further damage to the natural environment (Dinatale, 2003). Cover cropping in vineyards can be part of this strategy and will also help to develop an environmentally friendly agriculture (Faiello *et al.*, 2003). The use of low-input techniques is preferable to using chemicals, to preserve soil fertility, organic matter content, and to reduce the spread of specialized pests (Parlevliet and McCoy, 2001). Cover cropping also affects the level of plant and animal biodiversity, including the population of insects; however, this aspect has not been sufficiently investigated and data available in the literature are fragmented.

This chapter reports effects of botanical composition and insect populations experimental trials in two areas with different land uses in the Mediterranean part of Western Australia. The aim is to support higher insect diversity under diversified land use such as silvopasture than in specialized crops.

Materials and Methods
Trials were carried out in vineyards located in two areas of Western Australia: the surroundings of Perth (Perth Hills district) and Swan Valley. The former site is an area near forests with differentiated land use; the latter site is a more specialized vineyard area with little differentiation in land uses. Two vineyards were chosen in each area, with and without cover crops. Cover crops sown in Perth Hills were three *Trifolium* species (*T. subterraneum*, *T. campestre* and *T. dubium*), and in the Swan Valley a grass/legumes mixture (*Trifolium incarnatum*, *Lupinus cosentinii* and *Avena sativa*) was sown. In each vineyard botanical composition of sown and native species and presence of insects caught in pitfall traps placed at ground level and filled with the liquid solution of water, ethanol and glycerol were recorded. The results reported are the average of measurements repeated at four dates during spring. Average values were compared by Student's *t* test.

Results and Discussion
The botanical analyses are reported only for the cover-cropped vineyards (Table 1), because very few plants were present in un-cover-cropped vineyards due to the soil cultivation. Remarkable differences were found in percentage presence of sown species (14 *vs* 40) and in both areas one of the sown plants (first *T. subterraneum* and secondly *T. incarnatum*) made almost the total specific contribution to the sown species. Substantial differences in total plant biodiversity were found within sites, due mainly to native species colonizing the vineyard (25 *vs* 16).

Table 1. Specific contribution (%) and number of plant species in cover-cropped vineyards. Values in the same columns with the same letters are not significantly different for *t* test at $P < 0.05$.

Site	Specific contribution (%)		Number of species		
	Sown species	Native sp	Sown species	Native sp	Total
Forest area (Perth Hills)	14 b	86 a	3	25 a	28 a
Vineyard area (Swan Valley)	40 a	60 b	2	16 b	18 b

The presence of cover crop had no effect on the number of insect individuals in Perth Hills, while there was an effect in Swan Valley (Figure 1). On average, higher numbers of insect individuals and species were recorded in the area with less specialized land use.

Figure 1. Number of insects (individuals and species) in sampled vineyards.

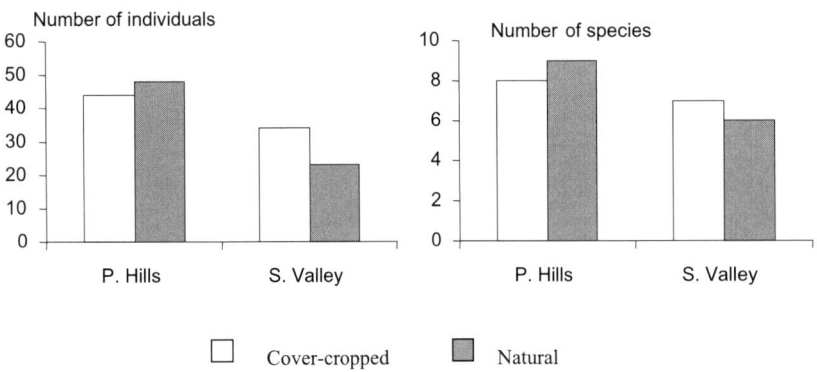

The proportions of insects found at both locations (irrespective of presence of cover crop) and grouped on the basis of their impact on the vineyard or other crops (Table 2), showed that more (+50%) useful insects (especially *Coccinella trasversalis*, *Plautia affinis* and spider species which eat vine parasites) were found in the forest area (Perth Hills) than in the specialized area (Swan Valley), with a corresponding reduction of insects dangerous to crops (especially *Carpophilus hemipterus* and *Platypus* sp).

Table 2. Distribution of insect types on each trial site. Values in the same columns with the same letters are not significantly different for *t* test at $P < 0.05$.

Site	Insect types		
	Useful	Neutral	Dangerous
Perth Hills	30 a	37 ns	33 b
Swan Valley	20 b	30 ns	49 a

Conclusions

Our preliminary results showed an increase in the number and diversity of plants and insects (both individuals and species) in the vineyards located in forest areas with high diversity of land use. Moreover, in these areas insect species could be described as beneficial. The effect of cover crop on insect diversity was notable only in the specialized vineyard area. Hence, these results indicate that in ecologically well-preserved environments (in our case the forest area) the number of insects which can colonize the cover crops is higher than in specialized vineyards areas, where they are mainly represented by crop parasites.

References

Dinatale, A. (2003) Effect of the establishment of vineyards on biodiversity in Western Australia (Effetti dell'inerbimento sulla biodiversità in vigneti del Western Australia). MSc thesis, University of Florence, Florence, Italy. (In Italian.)

Faiello, C., Pardini, A. and Litjens, L.C. (2003) Current socio-economic trends of organic plant and animal produce in Western Australia. In: *Socio-economic Aspects of Animal Health and Food Safety in Organic Farming Systems*. Havi, M., Martini, A. and Padel, S. (eds) *Proceedings of the 1st SAFO, Workshop*. (in press.)

Parlevliet, G. and McCoy, S. (2001) *Organic Grapes and Wine, a Production Guide*. Bulletin 4516,. Western Australian Agriculture Department, Perth, Australia.

Live weight changes of sheep and goats grazing a burned heath-gorse vegetation community

U. García, B. M. Jáuregui, R. Celaya, A. Martínez and K. Osoro

Servicio Regional de Investigación y desarrollo Agroalimentario (SERIDA), Apdo. 13-33300 Villaviciosa, Asturias, Spain. rcelaya@serida.org

Abstract

Sheep and goat performance was studied on previously burned heathlands. Initially sheep and goats were allocated individually to two replicated plots, and managed during two consecutive grazing seasons (autumn 2001 and spring-autumn 2002) at 10 animals/ha. For the third grazing season (2003) each plot was subdivided and sheep and goats were allocated to each half in a factorial design at 6.7 animals/ha. The results show better performances of goats than sheep at most times on this gorse-dominated vegetation. During the 2002 grazing season, local goats had greater weight losses than Cashmere goats and similar to those of sheep. During 2003 sheep gained more weight in plots previously grazed by sheep than in those previously grazed by goats, although more grasses were available in the latter.

Key words: animal performance, heathland, silvopastoral system

Introduction

Controlled burning has often been practised to improve vegetation in many marginal lands in the Cantabrian region. However, this is now out of control and many fires burn large areas of heath-gorse shrublands near forests, causing serious ecological and economical problems with deleterious effects on soil conservation and biodiversity and limiting the sustainability of pastoral and silvopastoral systems. In this case, in acid, poor soils, gorse (*Ulex* sp) increases its dominance and the heath vegetation regrowth is very slow. Gorse is well grazed by goats, but, as a consequence of the changes from gorse to grass due to goat grazing, sheep could graze these areas and could benefit the most from this grass. However, the optimum balance of gorse and grass to suit either goat or sheep performance is not known. This information would be useful for the establishment of more suitable grazing managements for the enhancement of pastoral and silvopastoral systems on the less favoured areas. The objective of this work is to study the performance of sheep and goats on previously burned heathlands following either sheep or goat grazing.

Materials and Methods

The experimental field, located in Illano (west of Asturias) at 900 m asl was superficially burned in the 2001 spring, resulting in a gorse (*Ulex gallii*) dominated stand (Jáuregui *et al.*, 2003). Grazing treatments started in September 2001 in four 1.2 ha plots, with Gallega sheep in two of the plots and Cashmere goats (together with local Celtiberian goats in 2002) in the other two. In both cases the stocking rate was 10 adult females/ha (12 per plot). Grazing was from September 2001 to January 2002, and from May to November 2002 (Phase 1). In the spring of 2003 the four plots were halved. In four of them the livestock species introduced were the same as in previous years, whereas in the other four subplots the species were changed and the stocking rate of 6.7 animals/ha (4 per subplot) introduced. The animals grazed from 19 May to 5 November 2003 (Phase 2). Animals were weighed and body condition (BC) score assessed (0-5 scale; Russel *et al.*, 1969) at the beginning and at the end of each grazing season, with intermediate controls at approximately monthly intervals. In the case of sheep, live weights (LW) were corrected for losses due to the wool shearing. LW and BC changes were analysed by ANOVA for treatment effects (species and replicate, with goat breed in 2002, and with previous grazing species in 2003).

Results

During Phase 1, the LW changes showed that goats are better adapted to this woody vegetation dominated by gorse than sheep, though in the first period of 2001 autumn grazing (24/9-28/11), when more grasses were available in the post-fire regrowth, ewes gained more weight than goats (59 *vs* 29 g/d; $P < 0.001$). In the following months (28/11-30/1) this difference was compensated and sheep lost significantly ($P < 0.001$) more weight (-91 g/day) than goats (-48 g/day). There were no significant differences in BC score changes between sheep and goats.

During the 2002 grazing season there were no significant differences in the LW or BC changes between sheep and goats in the total period (9/5-12/11). Nevertheless, in most periods till October, goats had better LW changes whereas sheep lost less weight than goats from 28/9 to 12/11 (Table 1). Cashmere goats, with a lower weight, performed significantly better than the bigger local goats. This effect of body size was also observed in unburned heathlands (see García *et al.*, this volume). Replicates (plot) had minor significant effects on animal LW changes, depending on the available vegetation (gorse/grass) in each plot.

Table 1. Live weight and body condition score changes of sheep and goats (Cashmere or local breed) on burned heathland during 2002 grazing season (Phase 1).

	Animal species			Goat breed		
	Sheep	Goat	Sig.	Cashmere	Local	Sig.
Initial LW (kg) 9 May	44.40	39.80	ns	31.20	49.20	***
Initial BC score 9 May	3.11	2.57	***	2.45	2.68	$P = 0.05$
LW change (g/day)						
9 May-6 June	30	35	ns	76	-3	***
6 June-24 July	3	37	**	44	31	ns
24 July-21 Aug	-7	15	ns	20	9	ns
21 Aug-28 Sep	-60	-66	ns	-44	-87	**
28 Sep-12 Nov	-55	-96	**	-78	-112	*
9 May-12 Nov	-22	-20	ns	-2	-36	***
BC change (units)						
9 May-12 Nov	-0.56	-0.53	ns	-0.30	-0.75	*

During the 2003 grazing season (Phase 2) there was a significant ($P < 0.05$) interaction between the actual and previous grazing species, mainly in the first period from May to July (Table 2). In the plots previously grazed by sheep, ewes grew more ($P < 0.05$) than goats whereas in the plots previously grazed by goats the reverse occurred (ns). Both sheep and goats lost more weight and body condition when they were allocated to plots previously grazed by the other species. Thus, the former hypothesis of a possible beneficial effect of goat grazing on subsequent sheep performance was not fullfilled in the present work in this grazing season, even though significant differences in the available vegetation could be observed (see Jáuregui et al., this volume). Contrary to the expected, ewes gained more LW as the available gorse to grass ratio in the plot increased.

Table 2. Live weight and body condition score changes of sheep and Cashmere goats on burned heathland during 2003 grazing season after sheep or goat previous grazing (Phase 2).

Previous species in Phase 1:	Sheep		Goat		Sig. of treatment effects		
Grazing species in Phase 2:	Sheep	Goat	Sheep	Goat	Phase 1	Phase 2	P1 x P2
Initial LW (kg) 20 May	40.70	36.20	41.00	37.20	ns	p=0.10	ns
Initial BC score 20 May	2.96	2.71	2.94	2.65	ns	**	ns
LW change (g/day)							
20 May-8 July	74	-20	-10	25	ns	ns	***
8 July-29 Aug	26	6	22	11	ns	ns	ns
29 Aug-4 Nov	-35	-31	-51	-40	ns	ns	ns
20 May-4 Nov	16	-16	-16	-5	ns	ns	*
BC change (units)							
20 May-4 Nov	-0.12	-0.25	-0.28	0.06	ns	ns	$P = 0.06$

Conclusions

Cashmere goats grew better than sheep or local goats in gorse-dominated shrublands. Nevertheless, the performance of all animals was notably restricted, rendering the production systems unsuitable on this poor vegetation.

References

Jáuregui, B.M., Celaya, R., García, U. and Osoro, K. (2003) Sprouting of heath-gorse communities after burning and their evolution under grazing by sheep and goats (Rebrote del brezal-tojal tras una quema y su evolución posterior con pastoreo de ovino o caprino). In: Robles, A.B., Ramos, M.E., Morales, M.C., Simón, E., González, J.L. and Boza, J. (eds) *Pastos, Desarrollo y Conservación*. SEEP Junta de Andalucía, Granada, Spain, pp. 495-500. (In Spanish.)

Russsel, A.J.F., Doney, J.M. and Gunn, R.G. (1969) Subjective assessment of body fat in live sheep. *Journal of Agriculture Science* 72, 451-454.

Management and ecological implications of silvopastoral systems in the Alps

A. C. Mayer
WSL Swiss Federal Institute for Snow and Avalanche Research SLF, Flüelastrasse 11, CH-7260 Davos Dorf, Switzerland. mayer@slf.ch

Abstract
In the Alps, 15% of the mountain forests are grazed, mainly by cattle browsing young trees, thus reducing the protection function of the forests against natural hazards. In this study, the condition of young trees on subalpine wood-pastures was assessed, the selection of herbaceous species was recorded and the digestibility of the herbage was calculated. Additionally, the long-term influence on the forest structure was studied using dendroecological techniques. The results suggest that grazing cattle may not severely damage young trees if stocking density is low and the ranges are sufficiently large. Grazed forests tend to have a more heterogeneous age structure than ungrazed forests. Cattle are able to select herbage of medium digestibility on subalpine wood-pastures and prefer grass species. These results demonstrate that forest grazing can be compliant with other very important functions of subalpine forests, such as avalanche protection.

Key words: browsing, digestibility, forest grazing, protection forest, sustainable land use

Introduction
Mountain forests fulfil several benefits at the same time, e.g. timber and fuelwood production, protection against erosion and natural hazards. Fifteen per cent of Swiss mountain forests are grazed during summer by domestic animals, mainly by cattle (Brassel and Brändli, 1999).

Swiss foresters fear that forest grazing hinders tree regeneration and reduces the protective function of mountain forests (Delucchi, 1993). Others claim that forest grazing provides the opportunity for extensive agricultural and forest production and that wood-pastures have a high structural diversity (Ten Klooster, 2000). In this study, the effects of grazing on the dynamic of mountain forests and the benefits of forest grazing for cattle were studied, using comparative field studies and experiments.

Materials and Methods
In a field study, seven traditional subalpine wood-pastures were grazed at stocking rates of 0.4 to 2.8 livestock units per hectare and during grazing periods of 12 to 114 days. In an experiment, the pasture site was separated into four fields of 0.51 ha each and fenced. One of these fields was left ungrazed, while the other three fields were stocked with three, six and nine 1-year-old heifers, respectively. Quantity and quality of the herbage available and consumed by cattle was studied by collecting representative samples of the biomass cut approximately 3 cm above ground on 20 cm by 20 cm squares and by analysing them as previously described (Mayer et al., 2003). Vegetation sampling was performed before and after grazing to determine the frequency of functional botanical groups in the herbage available and in the herbage consumed by the cattle. Herbage intake, selection and digestibility were determined using the double alkane technique (Mayer et al., 2003).

Damage to young trees by grazing cattle was studied by assessing the condition of young trees (Norway spruce: *Picea abies* (L.) Karst., European larch: *Larix decidua* Miller and rowan: *Sorbus aucuparia* L.). Tree condition was assessed by various shoot length measurements and by recording tree lesions (e.g. browsing of apical and lateral shoots or trampling). As the test areas had been grazed before, all existing tree lesions were recorded immediately before grazing commenced. After the cattle grazing period, tree condition was assessed again, and changes in tree condition were interpreted as damage mainly caused by cattle. In the experiment, additionally position and activity of the heifers were recorded.

To analyse the long-term influence of summer grazing (during several decades) on forest structure, 30 225 m^2 plots were located on a south-facing slope (1600 to 1950 m asl). Fifteen plots were selected randomly in forests that had been grazed each summer at least since 1930, and another 15 plots were randomly selected in forests that had not been grazed since 1930. In each plot, tree species, height and diameter of all trees were recorded and tree cores were taken at a height of 0 to 30 cm or bud-scale scars and the yearly shoot whirls were counted. To compare stand structure of grazed and ungrazed forests in terms of number of trees, species composition and tree height, Wilcoxon rank-sum tests were performed in S-Plus (MathSoft Incorporated, 1999). To compare the age structure of grazed and ungrazed plots, analyses of variance (ANOVA) and multiple comparisons between the plots were performed in S-Plus (MathSoft Incorporated, 1999).

Results and Discussion

In the comparative field study, grass species were grazed on nearly half of the plots on which they occurred, whereas legumes, forbs and shrubs were grazed on approximately 20% of the plots. Although the herbage quantity provided by the pastures was quite small in some cases, cattle selected herbage of relatively constant digestibility (Mayer *et al.*, 2003). The cattle used the forest disproportionately for resting and grazing. The percentage of young trees damaged on the traditional ranges was quite low on average (9%) and variation could almost completely be explained by the stocking rate (Mayer *et al.*, 2002). Additionally, low range size, low herbage biomass supply and high fibre contents of the herbage provided by the pastures enhanced the frequency of tree browsing by cattle. In the experiment, the browsing intensity of rowan and European larch was higher than that of Norway spruce and strongly correlated with stocking density and grazing period. At stocking densities > 3 LU/ha (livestock units per hectare) also Norway spruce was frequently browsed. Individual animals that have developed a specific preference for spruce needles may play a decisive role in the extent of browsing on spruce (Mayer, A.C. *et al.* unpublished results). The comparative dendroecological study showed, that the grazed forests were less dense than the ungrazed forests and had a higher percentage of European larch. The gap size (areas without trees $\geq$ 3 m) within the plots of grazed forest (mean size = 39 m^2) and ungrazed forest (mean size = 24 m^2) differed significantly. However, the density of the forests investigated was high enough to impede the release of avalanches. The young trees grew faster in the grazed forest. There were fewer dominant trees (> 25 m) in the grazed forests, and these trees were much older than in the ungrazed forests. The most frequently found structure type in the grazed forest was multi-layered open, whereas in the ungrazed forests the most frequently found structure type was uniform dense. Usually, subalpine forests tend to form stands of uniform height, although the age of the trees may differ greatly (Ott *et al.*, 1997). By encouraging forest owners to conduct selective logging operations, the utilization of subalpine forests as wood-pastures counteracts the natural tendency of the subalpine spruce-dominated forests to form uniform stands of the same height, thus favouring the development of a multi-layered structure, and consequently facilitating the regeneration of the forest.

Conclusions

Forest grazing can encourage the development of heterogeneously structured forest stands, which can regenerate more easily while fulfilling an avalanche protective function. Thus, the management of mountain forests as wood-pastures can be compliant with both the nutrition of livestock and the sustainable protection against natural hazards. However, a sustainable use of wood-pastures requires restrictions in stocking density and length of grazing period, in order to minimize the extent of damage to young trees by grazing cattle. The maximum grazing intensity should be assessed by measuring the actual damage on the young trees. The regeneration of the forest stands influenced by grazing animals should be compared with a target value for the regeneration necessary in the respective forest type. Based on this, adequate measures can be taken, such as reduction of stocking density or grazing period. If these preconditions are fulfilled, forest grazing is a land use highly adapted to the conditions in the subalpine stage. Concerning avalanche protection it is reasonable to combine forest and pasture on the same unit of land, rather than to create large open pastures on the one hand and to let the surrounding forests grow more and more densely on the other hand.

References

Brassel, P. and Brändli, U.B. (1999) *Swiss National Forest Inventory. Results of the Second Inventory 1993-1995* (Schweizerisches Landesforstinventar. Ergebnisse der Zweitaufnahme 1993-1995). Haupt, Berne, Stuttgart, Vienna, Austria. (In German.)
Delucchi, M., (1993) Wood-pastures from a forestry viewpoint (Waldweide aus forstlicher Sicht). *Bündner Wald* 46, 12-15. (In German.)
MathSoft Incorporated (1999) *S-Plus 2000 User's Guide.* MathSoft, Seattle, USA.
Mayer, A.C. (2003) Range management on wood-pastures of an Alpine valley. *Australian Journal of Forest Science* 120 (1), 19-28.
Mayer, A.C., Stöckli, V., Konold, W., Estermann, B.L. and Kreuzer, M. (2002). Effects of grazing cattle on subalpine forests. In: Bottarin, R. and Tappeiner, U. (eds) *Interdisciplinary Mountain Research*. Blackwell, Berlin, Germany, pp. 208-218.
Mayer, A.C., Stöckli, V., Huovinen, C., Estermann, B.L., Konold, W. and Kreuzer, M. (2003) Herbage selection by cattle on sub-alpine wood-pastures. *Forest Ecology and Management* 181, 39-50.
Ott, E., Frehner, M., Frey, H.U. and Lüscher, P. (1997) Gebirgsnadelwälder. Ein praxisorientierter Leitfaden für eine standortgerechte Waldbehandlung. Verlag Paul Haupt, Bern, Germany.
Ten Klooster, L. (2000) Waldweide als Teil der alpinen Kulturlandschaft. Diploma thesis, Universität für Bodenkultur Wien, Wien/Wageningen, The Netherlands.

Light availability for understorey pasture in holm-oak dehesas

M. J. Montero [1] and G. Moreno [2]
[1]Centro Universitario, Universidad de Extremadura, Plasencia 10600, Cáceres, Spain. [2]Centro Universitario, Universidad de Extremadura, Plasencia 10600, Cáceres, Spain. cmontero@unex.es, gmoreno@unex.es

Abstract
The percentage of radiation transmitted through the tree layer to the understorey pasture in dehesas of west-central Spain has been modelled. To achieve that, different allometric relationships between tree ages and stem diameter (DBH), and between stem diameter and canopy size (tree height, canopy height and canopy width) have been established. The amount of radiation available for understorey grasses was estimated by fish-eye photography (FEP). There was a rapid increase in light availability with distance from the tree, except in the proximity of the tree stem, where the increment was not so sharp; over 20 m from the tree, radiation levels were constant. Applying a multivariable regression, distance, DBH and canopy width explained 94.9% of the light variability. From the equations obtained a different radiation map in dehesas differing in tree density and tree age (size) has been generated.

Key words: allometric relationships, light transmission, *Quercus ilex*

Introduction
Dehesa (oak-savannah) is the most important silvopastoral system in the south-western part of the Iberian peninsula. Dehesas are a consequence of human activity, especially during the second half of the 19th century and the beginning of the 20th century (Pulido *et al.*, 2001). Grazing is considered as an important element of productivity and stability control in these ecosystems; thus a good knowledge of the performance of the herbaceous component, as well as of the tree-crop interactions could be useful in assessing the optimum productivity and sustainable management of dehesas. In general, where soil nutrients, water and temperature are not limiting, crop growth and yields are dependent on the total solar radiation intercepted during the growing season (Monteith, 1978; quoted in Bellow and Nair, 2003). Although studies on radiation availability for understorey species have been reported from other agroforestry systems (McMurtrie and Wolf, 1983; Sinoquet and Bonhomme, 1992; Sinoquet and Caldwell, 1995; Bellow and Nair, 2003), no such reports are available for holm-oak (*Quercus ilex*) dehesas. Thus, the main objective of the present study was to model the percentage of radiation transmitted through the holm-oak canopy to the pasture beneath dehesas of west-central Spain. The model will be part of a more general model of agroforestry system functioning (HySAFE), which is intended to be a tool to optimize agroforestry system design, management and profitability.

Materials and Methods
The study was conducted in an experimental farm, "El Baldío" situated in "Cuatro Lugares" county, west-central Spain (39° 41′ N-6° 13′ W; elevation 380 m), with a climate classified as subtropical Mediterranean. The relationship between tree age (A) (number of rings) and tree stem diameter (DBH) was established using data from 78 trees of DBH > 10 cm (Plieninger *et al.*, 2002) in the same area, and 200 trees with DBH < 10 cm measured for the present project. A second allometric relationship was determined by measuring stem diameter (DBH), tree height (Th), canopy height (tree height minus stem height; Ch), and canopy width (mean diameter of canopy; Cw) in 360 trees (covering all the size classes). By combination of both relationships, tree growth with age was modelled. The percentage of transmitted radiation was studied by means of fish-eye photographs (FEP) (Sinoquet and Caldwell, 1995). These were taken for 28 trees covering all the diameter classes found in the farm (2 to 73.2 cm of DBH), four orientations (N, S, E, W) and six distances from the tree trunk (0.5 to 30 m). Photos were taken in spring 2003 early in the morning (before sunrise) and in the evening (after sunset), at a height of 70 cm from the ground, giving a total of 577 photos, covering approximately 4 trees/day. In every FEP the percentage of total transmitted radiation was analysed using the software "Gap Light Analyser" (Frazer and Canham, 1999) from 1 November to 31 May as the integrating period (the growing season for grasses). The relationship between intercepted radiation (1%) and DBH (cm), canopy width (Cw, m) and distance from the trunk (D, m) was then estimated by multiple regression.

Results and Discussion
Both allometric equations had high r^2 values (> 90%), indicating high goodness of fit.

$$DBH = 98.73 \times (1 - EXP(-0.00546 \times AGE)); \quad (r^2 = 91.47\%) \quad (1)$$
$$Cw = 0.598 \times DBH^{0.736}; \quad (r^2 = 90.3\%) \quad (2)$$

Transmitted radiation was heavily related to canopy width, with slightly lower r^2 values for DBH and tree height. The sigmoid trend levelled off with the distance from the tree trunk; the percentage of transmitted radiation was close to

100% at distances more than 20 m, irrespective of tree size. The increment of transmitted radiation with distance followed an exponential trend, regardless of orientation and tree size, indicating a rapid increase in light availability with distance. Applying a multivariable non-linear regression including the four variables (DBH, Cw, Th, Ch) defining the tree size and the distance to tree trunk, only distance (D), DBH, and canopy width (Cw) explained the light interception (I):

$$I = (19.32/(1+64.4 \times e^{-0.62 \times DBH})) \times (17.96/(1+5.82 \times e^{-0.05 \times Cw})) \times 1.26^{-D}; \quad (r^2 = 94.4\%) \quad (3)$$

Applying equations 1, 2 and 3, the intercepted radiation by trees can be calculated and mapped using an interpolation software (SURFER, Golden Software, 1999), considering different scenarios based on tree density and age. An example of a simulation (75-year-old trees, 25 trees/ha) is shown (Figure 1). The average percentages of radiation intercepted in 1 ha for every simulated scenario are summarized (Figure 2).

From these results the equation relating the average transmitted radiation (I) with tree age (A) and tree density (dens) has been estimated by means of a multiple regression:

$$I = 0.297 \times dens^{0.641} \times A^{0.521}; \quad (r^2 = 92.89\%; n = 30)$$

This equation is useful in determining the optimum tree density at different ages of a tree plantation to maintain a specific level of light availability for pasture.

Figure 1. Map of intercepted radiation in a plantation (1 ha) of 75, year-old holm-oak, 25 trees/ha.

Figure 2. Percentage of intercepted radiation related to tree age for different densities of plantation (25-400 trees/ha).

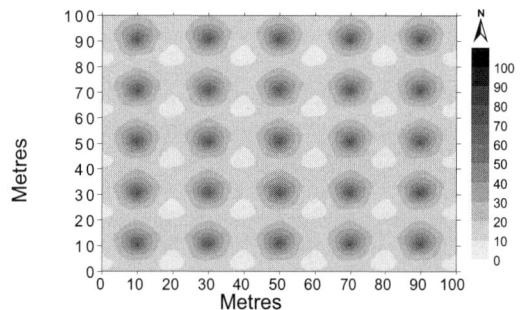

Conclusions

The model presented is a simple tool to determine the light availability for the understorey in dehesas of holm-oak from a very easy measurement (DBH). The amount of light available for the understorey is highly dependent on the tree age and density because light reduction is only significant close to the trees.

Acknowledgements

This study was supported by the EU (SAFE project, QLX-2001-0560), Spanish government (MICASA project, AGL-2001-0850) and Consejería de Educación de Extremadura (CASA project, contract 2PR02C012).

References

Bellow, J.G. and Nair, P.K.R. (2003) Comparing common methods for assessing understorey light availability in shaded-perennial agroforestry systems. *Agricultural and Forest Meteorology* 114, 197-211.

Frazer, G.W. and Canham, C.D. (1999) *GLA version 2: Gap Light Analyser*. Simon Frazer University, Burnaby, British Columbia, and the Institute of Ecosystem Studies, Millbrook, New York, USA. (http:// www.ecostudies.org)

Golden Software, Inc. (1999) *SURFER version 8: 2D and 3D Surface Modelling Package*. Golden Software, Inc., Colorado, USA. (htpp:// www.goldensoftware.com)

McMurtrie, R. and Wolf, L. (1983) A model of competition between trees and grass for radiation, water and nutrients. *Annals of Botany* 52, 449-458.

Plieninger, T., Pulido, J.F. and Konold, W. (2002) Effects of land-use history on size structure of holm-oak stands in Spanish dehesas: implications for conservation and restoration. *Environmental Conservation* 30, 61-70.

Pulido, F.J., Díaz, M. and Hidalgo de Trucios, S.J. (2001) Size structure and regeneration of Spanish holm-oak *Quercus ilex* forests and dehesas: effects of agroforestry use on their long-term sustainability. *Forest Ecology and Management* 146, 1-13.

Sinoquet, H. and Bonhomme, R. (1992) Modelling radiative transfer in mixed and row intercropping systems. *Agricultural and Forest Meteorology* 62, 219-240.

Sinoquet, H. and Caldwell, R.M. (1995) Estimation of light capture and partitioning in intercropping systems. In: Sinoquet, H. and Cruz, P. (eds) *Ecophysiology of Tropical Intercropping*. INRA, Paris, France, pp. 79-97.

Consequences of dehesa management on tree-understorey interactions

G. Moreno, J. Obrador, E. García, E. Cubera, M. J. Montero and F. Pulido
Centro Universitario, Universidad de Extremadura, Plasencia 10600, Cáceres, Spain. gmoreno@unex.es

Abstract
The distribution of resources (light, microclimate, soil moisture and soil nutrients) and fine roots (herbaceous plants and trees) around trees were studied in four dehesas with three types of management: cropped (forage), grazed (native grasses) and encroached (*matorral*). Additionally, the forage yield and the physiological status, growth and acorn production of holm-oaks were measured. There was a low overlap of the root systems of the herbaceous plants and the trees. This could help avoid competition. Soil water consumption by trees did not reduce forage productivity. Light interception by trees was limited to the close vicinity of the trees, with very little effect outside the canopy. On the other hand, both microclimate and soil fertility improved significantly in the vicinity of the trees, although they only contributed to increased forage yield in less fertile soils. Dehesa management significantly affected the physiological status, growth and productivity of trees.

Key words: above-ground resources, below-ground interactions, root system, silvopasture

Introduction
Dehesa generally consists of a mosaic of scattered trees combined with crops, pasture and shrubs, as a result of the different management practices, namely agriculture, livestock husbandry, forestry and hunting. Some early work on dehesa functioning showed the positive effects of the trees on soil nutrient content (Escudero, 1985), soil water storage capacity (Joffre and Rambal, 1988) and pasture production (Puerto *et al.,* 1987). On the other hand, an improved physiological status and productivity of the trees in dehesas with respect to the forest has been also shown by Infante *et al.* (1999) and Pulido and Díaz (2003), respectively.

In this chapter, the main results found in a comprehensive study of the distribution of the main above- and below-ground resources in dehesas with three main types of management are presented and analysed in order to understand the functioning of the whole system. This information will be use, as a next step, to build a model (HySAFE) based on the intensity of the tree-understorey interactions.

Materials and Methods
The study has been carried out in four dehesas located in "Curator Lugar's" county, in CW Spain (39° 41' N - 6° 13' W; altitude: 380 m asl). In the four farms (T, CL, ST and BA in decreasing order of soil fertility), three main type of management were defined and studied: trees (holm-oak) combined (i) with intercropped cereal forage (C), (ii) with grazed native grasses (G), and (iii) with abundant shrubs (*matorral*) (M). The distribution of several above- and below-ground parameters (Table 1) were studied around trees (from 1 to 30 m from the tree trunk) in three types of plots (dehesa managements): transmitted radiation by fish-eye photographs (28 trees, four orientations, from 0.5 to 30 m distance), air temperature and humidity by microdataloggers (two trees, from 1 to 30 m distance), soil temperature (two trees, beneath and outside the tree canopy), soil moisture by TDR technique (33 trees, from 0 to 200 cm depth, from 2.5 to 20 m distance), soil fertility (pH, organic carbon, ECC, total N, P-Olsen, base cations (exchangeable Ca^{2+}, Mg^{2+}, K^+, Na^+) in samples taken from the first 20 cm of soil depth (47 trees, from 2 to 20 m distance), herbaceous plant and tree rooting system by soil cores (13 trees, from 0 to 20 m distance and to 200 cm depth) and studying 51 profiles located in seven different road cuttings. Additionally, tree nutritional (N and P) and physiological status (leaf water potential and CO_2 assimilation rates) were measured in 269 trees and 28 trees, respectively. Finally, tree growth was measured through current year shoots in 188 trees and acorn production was measured by hung traps in 84 trees.

Figure 1. Mean values of forage yield at different distances to the tree stem in fertile (e.g. T-2002) and infertile (e.g. SO 2002) intercropped dehesas.

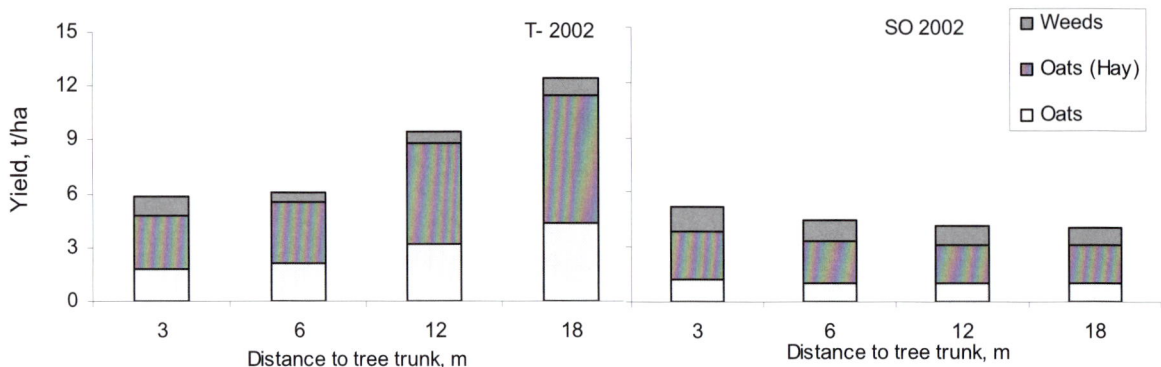

Table 1. Spatial variation of the several resources around holm-oaks in dehesa of CW Spain.

Parameter		Effect of distance
Light availability		Sigmoid increase of light availability with distance Light,% = 100/(1+4.18×EXP(-0.43×DIST))
Microclimate	Soil temperature (maximum)	Maximum measured temperatures of 29.6 versus 46°C under and out of the tree, respectively
	DmxT: Daily maximum air temp. DmnT: Daily minimum air temp. Mdv: Mean daily air temp. variation	DmxT: 1 m (27.8) < 10 m (28.8) ≈ 20 m (29.0) ≈ 30 m (29.5) DmnT: 1 m (7.4) > 10 m (6.6) ≈ 20 m (6.3) ≈ 30 m (6.6) Mdv: 1 m (14.2) < 10 m (16.1) ≈ 20 m (16.5) ≈ 30 m (16.6)
	DmxAR: Daily max. air humidity DmnAR: Daily min. air humidity Mdv: Mean daily air humidity variation	DmxAR: 1 m (91.2) < 10 m (95.1) ≈ 20 m (94.7) ≈ 30 m (95.6) DmnAR: No significant differences Mdv: 1 m (42.9) < 10 m (46.9) ≈ 20 m (47.3) ≈ 30 m (46.8)
Soil water content		Overall: No significant differences between distances Cropped: SWC slightly higher out of the tree Grazed: SWC slightly higher beneath the tree Matorral: SWC significantly higher beneath the tree
Soil fertility	CEC, Exch Ca^{2+} and K^+	Negative exponential decrease
	Organic C and total N	Logarithmitc decrease with distance
	pH, P and exch. Mg^{2+} and Na^+	No clear tendency with distance
Root system	Herbaceous - root length density	Exponential decrease with depth. d_{50} = 10.7, d_{95} = 46.2 No significant differences between distances
	Tree - root length density	Linear and slow decrease with depth and distance d_{50} = 96.4, d_{95} = 416.6, maximum distance around 25 m

Figure 2. Effect of the dehesa management on tree growth and productivity.

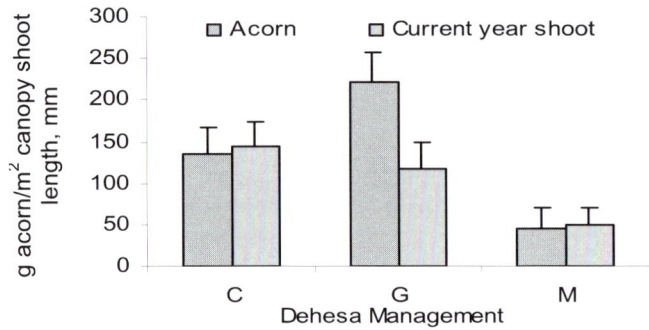

Results and Discussion

Leaf water status, photosynthetic rate (Table 2), growth and acorn production (Figure 2) were significantly higher in C and G trees than in M and forest ones. This could be explained by the large amount of soil explored by tree roots (more than 20 m of distance and more than 4 m of depth). This extensive root system involves an unusually low density in the first 80 cm of the soil, which indicates a low intensity of herbaceous plants for resource competition. As a result, forage production was independent of distance, except in more fertile soils, where yield increased with distance, indicating that light interception by trees is only relevant in terms of competition in very fertile soils (Figure 1).

Table 2. Effect of dehesa management on resource availability and tree physiological status. [1] 5/9/2003.

Parameter			Effect of management
Soil water content (% v/v)			C ≈ P >> M
Soil fertility	SOM (%)		M (2.52) > C (2.17) ≈ G (2.12)
	Total N (g/kg)		M (1.07) > C (0.97) > G (0.81)
	Available P (ppm)		C (8.56) > M (7.84) ≈ P (7.45)
	Exchangeable cations		M (6.1) > C (4.0) > G (3.0 cmol/kg)
	ECC		No significant differences (14.0 cmol/kg)
	pH		No significant differences (5.45)
Tree nutritional status	N (%)		C (1.13) > G (1.07) > M (0.96)
	P (%)		M (0.065) > G (0.054) > C (0.048)
Tree physiological status	Leaf water potential[1]	Predawn	C (-0.43) ≈ P (-0.45) > M (-0.79) > F (-1.34) MPa
		Midday	C (-1.7) ≈ P (-2.05) > M (-2.57) > F (-3.57) MPa
	Photosynthesis[1]		G (10.0) ≈ C (9.2) ≈ M (8.6) >> F (2.41) µmol $CO_2/m^2/s^1$

Conclusions

Dehesa management significantly affected the physiological status, growth and productivity of trees. Moreover, trees affected more positively than negatively forage yield in most of the cases studied.

Acknowledgements

This study was supported by the EU (SAFE project, QLX-2001-0560), Spanish government (MICASA project, AGL-2001-0850) and Junta de Extremadura (CASA project, contract 2PR02C012).

References

Escudero, A. (1985) Effect of dispersed trees on chemical soil properties (Efectos de árboles aislados sobre las propiedades químicas del suelo). *Revue d'Ecologie et de Biologie du Sol* 22 (2), 149-159. (In Spanish.)

Infante, J.M., Damesin, C., Rambal, S. and Fernández-Alés, R. (1999) Modelling leaf gas exchange in holm-oak trees in southern Spain. *Agricultural and Forest Meteorology* 95, 203-223.

Joffre, R. and Rambal, S. (1988) Soil water improvement by trees in the rangelands of southern Spain. *Oecologia Plantarum* 9, 405-422.

Puerto, A., García, J.A. and García, A. (1987) Hill systems as clarifying element of certain effects of trees on pastures (El sistema de ladera como elemento esclarecedor de algunos efectos del arbolado sobre el pasto). *Anuario del CEBA de Salamanca* 12, 297-312. (In Spanish.)

Pulido, F.J. and Díaz, M. (2003) Dynamics of tree natural regeneration in holm and cork-oaks (Dinámica de la regeneración natural del arbolado de encina y alcornoque). In: Pulido, F.J., Campos, P. and Montero, G. (eds) *La Gestión Forestal de las Dehesa.*, IPROCOR, Merida, Junta de Extremadura, Instituto del Corcho, la Madera y el Carbón Vegetal, Spain, pp. 39-62. (In Spanish.)

Pasture establishment for extensive systems

M. R. Mosquera-Losada, S. Rodríguez-Barreira and A. Rigueiro-Rodríguez
Dept. Producción Vegetal, Escuela Politécnica Superior, Universidad de Santiago de Compostela, 27002, Lugo, Spain.
romos@lugo.usc.es, anriro@lugo.usc.es

Abstract
It is important to determine the effect of tree cover (shade) on pasture establishment and productivity in silvopastoral systems.

An experiment was carried out in Galicia to evaluate the effect of spring and autumn sowing and shading (0 and 50% reduction of incident radiation) on pasture production and the botanical composition of swards established with a monoculture of *Agrostis tenuis* cv Highland. Pasture production, persistence and productivity under open conditions were worse with spring sowing than autumn sowing due to the encroachment of *Capsella bursa-pastoris*. However, pasture performance was lower under shade than in the open when sowing took place in the autumn due to encroachment by *Stellaria media*.

Key words: sowing period, *Agrostis*, shading, silvopastoral systems

Introduction
Small farms in Galicia are tending to be abandoned, while the larger farms are increasing the number of animals they carry (López-Garrido *et al.*, 1998). Small farm owners usually plant trees on their land as a result of EU agri-environmental measures to increase timber production from the land. Unfortunately, this land use sustains a lower rural population, as the level of employment is low and return from the land is long-term. Silvopastoral systems, as practised in New Zealand and Australia with *Pinus radiata*, and established by planting trees into swards, can be an alternative land use which makes the production of wood and farm products compatible.

It is important to introduce grass species adapted to shading conditions in the sown mixture as tree growth will reduce the input of light to the system and limit the persistence of more desirable species, such as ryegrass, in temperate pastures. The evaluation of different sowing dates and shading conditions will improve sward establishment and management. The objective of this experiment was to evaluate the effect of two different sowing periods (spring and autumn) and shading on pasture production and the changes in composition of a sward established with a monoculture of *Agrostis tenuis*.

Materials and Methods
The experiment was carried out in Lugo (NW Spain) during 1996 and 1997. The climate is Atlantic and there is a drought during the summer period with around 910 mm/year of precipitation and a mean temperature of around 14.5°C. Soil is sandy with a pH of 6.5 (water), 2.6% organic matter, and cations of interchange complex (ammonium acetate) of 0.05 cmol (+)/kg of Na, 0.22 cmol (+)/kg K, 5.85 cmol (+)/kg Ca and 0.45 cmol (+)/kg Mg. The experimental design was a randomized block with three replicates. Treatments consisted of two sowing dates of a monoculture of 8.5 kg/ha *Agrostis tenuis* cv Highland under shading and open sward conditions. Sowing took place in the spring of 1996 and the autumn of 1996 in 1 m^2 plots. Shading was simulated with a plastic mesh that intercepted 50% of incident radiation and was located 0.5 m above the soil. Soil was disc-ploughed, and fertilized at 500 kg/ha of 8:24:16, and rotavated. Seed was broadcast by hand and the seed bed was rolled at the end. The seed germination test gave a result close to 100%.

Harvesting took place in May, July, September and November 1996 and April, June, July, October and November 1997. Samples from plots were taken by cutting two areas of 0.3 m × 0.3 m in each experimental plot using electric hand clippers. Samples were transported to the laboratory to determine the dry matter concentration and pasture production by drying for 48 h at 60°C.

Data were analysed using ANOVA with the statistics program for Windows SAS 6.11 (1985), and the averages were separated using the LSD test.

Results and Discussion
Pasture production in each harvest of 1996 and 1997 and treatment can be seen in Figure 1. Annual pasture production in Galicia is usually higher during the first year after establishment when sowing takes place in the autumn rather than in the spring due to the autumn and winter temperatures; this explains the low productivity of pasture during the spring of 1996 after sowing compared to the spring of 1997, when pasture was sown in the autumn of 1996. It was also found that spring production in 1997 was higher when sowing took place in the autumn of 1996 instead of spring 1996. Shading treatment reduced pasture production more for some harvests and for those plots previously sown in the spring rather than in the autumn.

Figure 1. Total pasture production per harvest in plots sown in spring (a and b) and autumn (c). The shaded plots are represented by black bars and the unshaded plots with white bars. The total pasture production includes sown and weed species. Different letters indicate significant differences (P < 0.05).

(a) 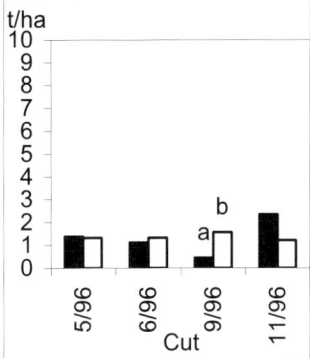 (b) 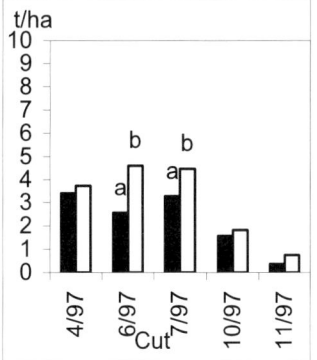 (c)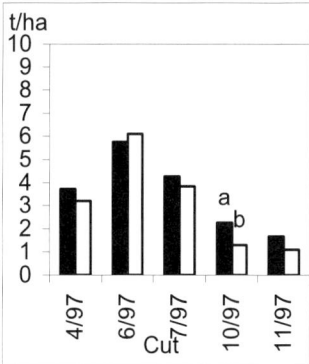

Pasture production after the summer was higher in those plots established in the autumn as compared to the spring sowing treatment. *Agrostis tenuis* establishment was better under shaded conditions (Figure 2) when sowing was made during the spring, as it represented around 80% of the total dry weight. However, when sowing took place in the autumn the establishment and persistence of *Agrostis* was better in the unshaded treatments.

Figure 2. *Agrostis tenuis* percentage in each harvest in those plots sown in spring (a and b) and autumn (c). Shaded plots are represented by black bars and unshaded plots with white bars. Different letters indicate significant differences (P < 0.05).

(a) (b) 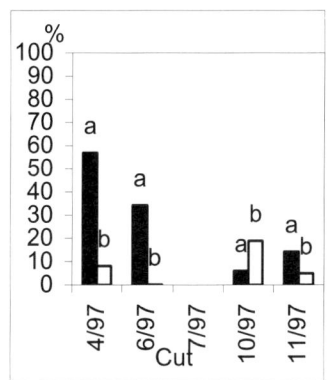 (c)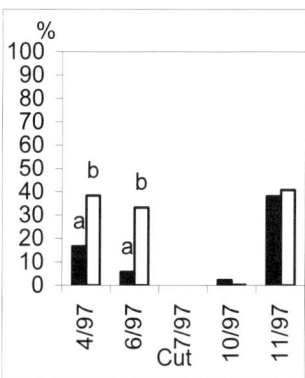

In spite of the better establishment of *Agrostis* in shaded conditions when it was sown in spring, the effect of the shading was not apparent during the first year, due to the restricted light input. Moreover, in unshaded plots the development of unsown species limited the development and production of the sown species.

The percentage of *Capsella bursa-pastoris* in the treatments is shown in Figure 3. This species was an important contributor (more than 60% in some treatments) to the open (unshaded) swards during the first year when sowing took place in the spring, disappearing in the second year of the study. This can be explained by the annual nature of this species, which allows it to establish itself quickly in recently ploughed soils, when the sowing species has not yet had time to develop (Remón-Eraso, 1991). This explains also the low productivity and worse establishment of *Agrostis* under open conditions than in shade when establishment took place in spring, due to the prostrate growth habit of *Capsella bursa-pastoris* (Mosquera-Losada et al., 2000), which limits the development of *Agrostis*.

Figure 3. Percentage of *Capsella bursa-pastoris* in each harvest in those plots sown in spring (a and b) and autumn (c). Shaded plots are represented by black bars and unshaded plots with white bars. Different letters indicate significant differences ($P < 0.05$).

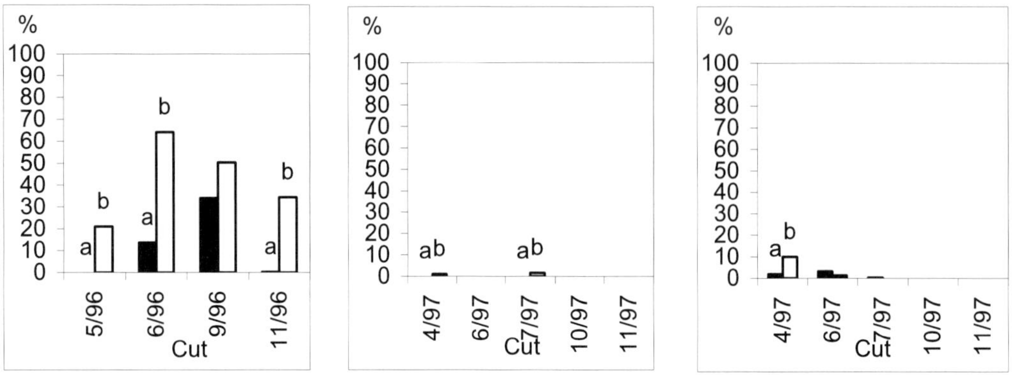

However, when sowing took place during the spring the percentage of *Capsella bursa-pastoris* was very low (under 10%) during the first year after establishment.

Stellaria media percentage (Figure 4) was higher in those plots established in the autumn under shaded conditions. However, this limited the development of *Agrostis* under shaded conditions and reduced pasture production.

Figure 4. *Stellaria media* percentage in each harvest in those plots sown in spring (a and b) and autumn (c). Shaded plots are represented by black bars and unshaded plots with white bars. Different letters indicate significant differences ($P < 0.05$).

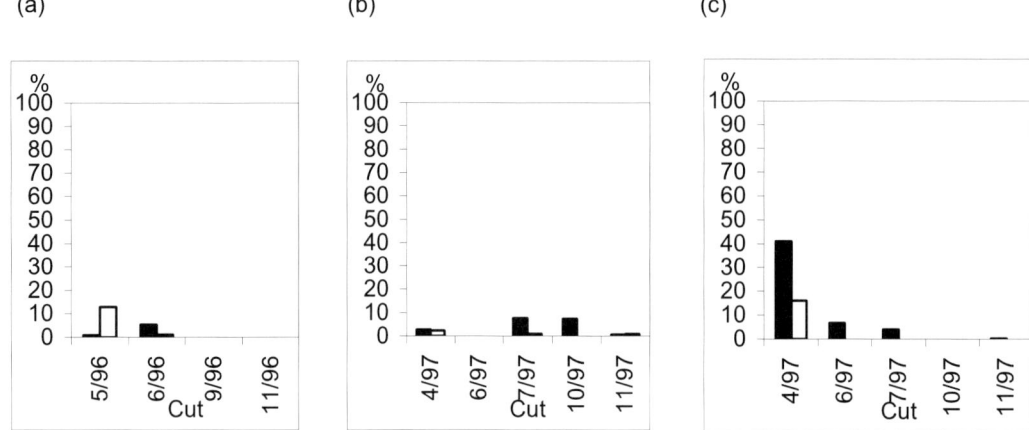

Conclusions

Agrostis tenuis development, which is a species with a slow establishment speed and associated with shaded conditions, depends on the timing of sowing due to the competitive relationship that is established with annual species.

Acknowledgements

We are grateful to CICYT and XUNTA de Galicia for financial assitance, to Escuela Politécnica Superior for facilities, to Divina Vázquez Varela, José Javier Santiago-Freijanes, Teresa Piñeiro López and Antonio Rodríguez Rigueiro for helping in processing, laboratory and field.

References

López-Garrido, C., Flores-Calvete, G., Estévez-Feijoo, E., Amor-Fernández, J.M., Vázquez-Yañez, O., López-Iglesia, E. and Lois-Mosquera, M. (1998) Análisis del sector gallego y su posible evolución tras la asignación definitiva de cuotas e identificación de los obstáculos para su reorientación. In: *Memoria del Centro de Investigaciones Agrarias de Mabegondo*. Xunta de Galicia, Galicia, Spain, pp. 463-491.

Mosquera-Losada, M.R., Romero-Franco, R., Rigueiro-Rodríguez, A., Piñeiro-Andión, J., González-Hernández, M.P. and López-Díaz, M.L. (2000) *Plantas de los Prados del Noroeste de la Península Ibérica*. Universidad de Santiago de Compostela, Spain.

Remón-Eraso, J. (1991) *Las Plantas de Nuestros Prados*. Mundi-Prensa, Madrid, Spain.

Biological diversity in communities of *Erica ciliaris* and *Erica tetralix*: analysis of their spatial and temporal variation

A. Muñoz [1], X. M. Pesqueira [1], R. Álvarez [1], O. Reyes [1,2] and M. Casal [1]

[1]Área de Ecología, Dept. de Biología Celular y Ecología, Facultad de Biología, Campus Sur, 15782, Santiago de Compostela, Spain. txokyto@hotmail.com [2]Área de Ecología, Dept. de Biología Celular y Ecología, E.P.S. de Lugo, 27002, Lugo, Spain. bfreyes@usc.es

Abstract

The shrubland communities of *Erica ciliaris* and *Erica tetralix*, as in other types of West European shrublands, are so linked to human activity that the abandonment of their traditional uses in recent times has altered their areas. These communities form part of the typical mountain mosaic landscape (open natural and artificial forest, dense natural and artificial forest, shrublands, meadows...) in some areas of Galicia where animals can seasonally graze. The wet heathland communities are still subject to human exploitation, such as clearing and burning, to create cattle pastures. In two wet heathland sites located in different areas of Galicia, Masgalán (Pontevedra) and Baio (A Coruña), four replicate plots were fixed in each community - two burned to simulate a cultural fire and two used as controls. Sampling was carried out every spring and autumn from spring 2002 to evaluate the changes produced in the biological diversity of the vegetation after burning. A quantified temporal and spatial comparison of biological diversity using different variables is made in this study.

Key words: diversity, burning

Introduction

Heathland communities occupy a large part of Western Europe, from south of the Scandinavian peninsula to the northwest of the Iberian peninsula (Gimingham *et al.*, 1981). In Galicia, seven different shrubland communities have been structurally described, one of which is *E. tetralix* and *E. ciliaris* (Basanta *et al.*, 1988), which is included in the group of wet heaths and moors. The importance of the *E. tetralix* and *E. ciliaris* communities is recognized by the European Community and by Spain, through the inclusion of this type of habitat within Annex I of the Habitat Directive (92/43/EEC) and through its consideration as a priority interest habitat (PIH). Fire is the perturbation that affects the most shrublands worldwide. Traditional exploitation of this community is extensive animal husbandry, with intensive grazing on the shrubland and herbaceous formations (Izco and Ramil, 2001). Galicia is the autonomous community which has the greatest numbers of fires (Pereiras, 2001), due, principally, to anthropic causes. Most of the surface areas burned are shrubland communities; hence the response of this type of ecosystem to fire is of great interest. These communities form part of a typical mountain mosaic landscape (open natural and artificial forest, dense natural and artifical forests, shrublands, meadows…) found in some areas of Galicia where animals can seasonally graze. In this work, we study the effect of fire on the diversity of communities of *E. tetralix* and *E. ciliaris*.

Materials and Methods

In two ecosystems of *E. tetralix* and *E. ciliaris* situated in Baio (A Coruña) and Masgalán (Pontevedra) four experimental plots were established: two controls and two burnt. The burning of the plots was done in autumn 2001 (Baio) and in winter 2002 (Masgalán). Before burning control plots were sampled and the burnt plots were sampled in spring (2002) and autumn, until spring 2003. In these, 30 samples of presence-absence were collected and expressed as frequency values. The following parameters were calculated for total species, woody species and herbaceous species to fully describe diversity (Basanta, 1984; Basanta *et al.*, 1989): (a) specific richness (S), S being the total number of species; (b) the Shannon diversity index: $H' = \sum p_i \times \log_2 p_i$, where H' is the diversity and p_i = frequency of species i/frequency of all species (Basanta *et al.*, 1989; Smith and Smith, 2001; Terradas, 2001); the Pielou diversity index: $H_2 = -\log_2 \sum p_i^2$ (c) the Simpson dominance index: $\lambda = \sum p_i^2$ (d) the Pielou equitability index: $J' = H'/H'max$, where H' is the Shannon diversity index and $H'max$ corresponds to the formula $H'max = \log_2 S$. Once the values of the different indices were obtained, a spatial and temporal comparison was carried out using the average values of the two plots.

Results and Discussion

In both communities woody species contributed to most of the diversity, with H' and H_2 values remaining practically constant over time (Figure 1). The herbaceous species, despite a lower degree of diversity, varied much more in the first stages of secondary succession. These species tend to increase their H' and H_2 values. J' exceeds λ in both communities and this difference becomes greater over time. In the control plots the values of both indices of herbaceous species are equal and the diversity of the herbaceous species is low. This may be because, although the number of herbaceous species increases at first, the specific richness (S) decreases over time to the point when a few species become dominant and therefore λ increases and J′ decreases. Both communities had very similar values of S before they were burnt (Baio = 9.5; Masgalán = 8.5). However, important differences post-fire were observed, such as a much higher value of S in Masgalán, due to the increase in herbaceous species in the first sampling. This important difference could be due to different abiotic causes (management, groundwater regime) and other differences (percentage of regenerating species, number of predisturbance species) at the level of the specific composition of the community.

Figure 1. Temporal changes in the biological diversity in two sites of *E. ciliaris* and *E. tetralix*, total number of species, number of woody species and number of herbaceous species.

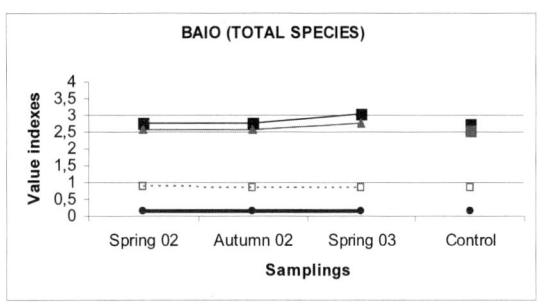
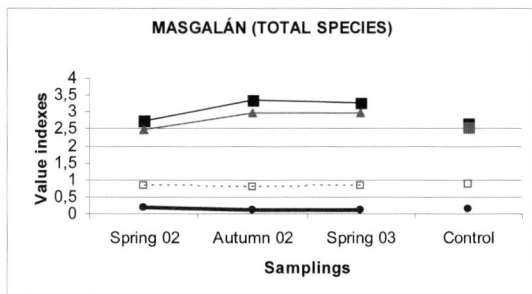
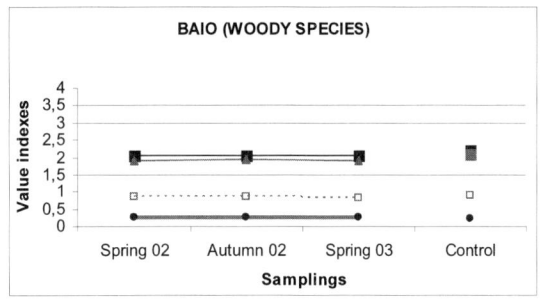
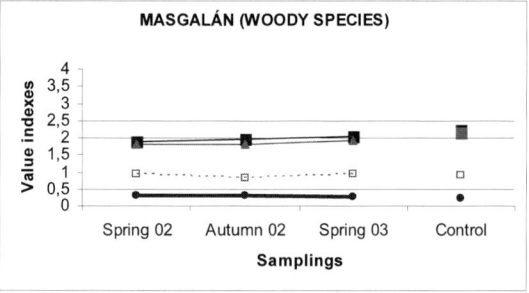
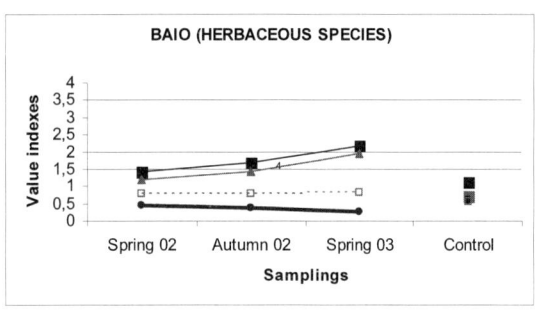
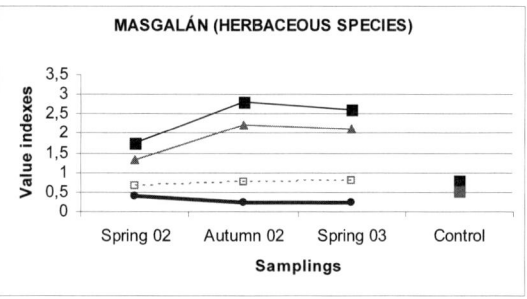

- ■ - Shannon Diversity Index
- ▲ - Pielou Diversity Index
- □ - Pielou Equability Index
- ● - Simpson Index of Dominance

Conclusions

In the first stages of post-fire succession the diversity of the herbaceous species, increases considerably. Diversity of woody species is higher and remains practically unchanged over time in both the sites studied.

References

Basanta, M. (1984) Structure of shrublands in Tambre river basin (La Coruña) (Estructura de los sistemas de matorral de la cuenca del Tambre (La Coruña)). MSc thesis. Universidad de Santiago de Compostela, Santiago de Compostela, Spain. (In Spanish.)

Basanta, M., Díaz Vizcaíno, E. and Casal, M. (1988) Structure of shrubland communities in Galicia (NW Spain). In: During, H.J., Werger, M.J.A. and Willems, J.H. (eds) *Diversity and Pattern in Plant Communities*. Academic Publishing, The Hague, The Netherlands.

Basanta, M., Díaz Vizcaíno, E., Casal, M. and Morey, M. (1989) Diversity measurements in shrubland communities of Galicia (NW Spain). *Vegetatio* 82, 105-118.

Gimingham, C.H., Chapman, S.B. and Webb, N.R. (1981) European heathlands. In: Specht, R.L. (ed.) *Heathlands and Related Shrublands*. Elsevier Scientific Publishing Company, Amsterdam, The Netherlands, pp. 365-409.

Izco, J. and Ramil, P. (2001) *Analysis and Valuation of the Range O Xistral: An Example of Applying Habitats Directive in Galicia* (Análisis y valoración de la Sierra de O Xistral: Un modelo de aplicación de la Directiva Hábitat en Galicia). Xunta de Galicia, Santiago de Compostela, Spain. (In Spanish.)

Pereiras, X. (2001) *Forest Fires in Galicia* (Os incendios forestais en Galicia). Ediciones Bahía, A Coruña, Spain. (In Galician.)

Smith, R.L. and Smith, T.M. (2001) *Ecology* (Ecología), 4th edn. Pearson Educación, Madrid, Spain. (In Spanish.)

Terradas, J. (2001) *Vegetation Ecology. From Plant Ecophysiology to Communities and Landscape Dynamics* (Ecología de la vegetación. de la ecofisiología de las plantas a la dinámica de comunidades y paisajes). Ediciones Omega, Barcelona, Spain. (In Spanish.)

Silvopasture as an approach to reducing nutrient loading of surface water from farms

V. D. Nair [1] and R. S. Kalmbacher [2]

[1]Soil and Water Science Dept. University of Florida, 106 Newell Hall, PO Box 110510, Gainesville, FL 32611-0510, USA. [2]Range Management and Forage Crops, Range Cattle REC – Ona, 3401 Experiment Station, University of Florida, Ona, FL 33865-9706, USA. vdna@ifas.ufl.edu, rskalmbacher@ifas.ufl.edu

Abstract
Nitrogen and P concentrations within soil profiles are being monitored in three pastures at Ona, Florida (27° 23'N, 81° 57'W). Two of the pastures – a treeless pasture and a pasture with slash pine (*Pinus elliotti*) (silvopasture) - consist of bahiagrass (*Paspalum notatum*) with legumes *Desmodium heterocarpon* and *Vigna parkeri*. The third pasture consists of native vegetation (native silvopasture) with the same tree configuration as the bahiagrass silvopasture. Measurements of soil nutrient concentrations in ten soil profiles (0-5, 5-15, 15-30, 30-50, 50-75 and 75-100 cm depths) within each pasture showed that P concentrations were in the order: pasture > silvopasture > native silvopasture. Ammonium-N and NO_3-N concentrations were higher in the surface horizon of the treeless pasture. The differential rooting zones of the pine and forage crops may absorb the nutrients more completely in the silvopastures than in the treeless system, suggesting that silvopasture association is better than the treeless pasture in reducing nutrient loss from soil to the surface water.

Key words: nitrogen, phosphorus, pine, soil profile

Introduction
The problem of P loss from soils could be particularly serious in coarse-textured, poorly drained soils with either surface or tile drainage in which soil drainage water ultimately enters surface waters. Soils of this nature are prevalent in several parts of the world including the south-eastern and mid-Atlantic states of the United States, The Netherlands and the Po region of Italy. Silvopasture is one of the most prevalent forms of agroforestry found in the United States and Canada, with the largest blocks of grazed forests occurring in the southern and south-eastern United States (Clason and Sharrow, 2000). Such systems, which include both tree and pasture components, are expected to minimize nutrient losses from the soil because of enhanced nutrient uptake by tree and forage crop roots from varying soil depths, compared to more localized and shallow rooting depths found in pasture without trees (Nair and Graetz, 2004).

Materials and Methods
Three pastures were selected for this study at Ona, Florida (27° 23'N, 81° 57'W). Two of the pastures consist of bahiagrass (*Paspalum notatum*) with legumes *Desmodium heterocarpon* and *Vigna parkeri*. One of the two pastures has slash pine (*Pinus elliotti*) initially planted at 1100 trees/ha in double rows, 2.4 m apart with a 12.2 m alley (silvopasture). The third pasture consists of native vegetation (native silvopasture) with the same configuration as the bahiagrass silvopasture, but never having received lime or fertilizer. All pastures are on Spodosols, Ona fine sand (sandy, siliceous, hyperthermic Aeric Alaquod). Spodosols typically have a sandy A horizon, followed by an eluted E horizon, below which is a highly P-retentive spodic (Bh) horizon.

The treeless pasture is 20 years older than the silvopasture. The treeless pasture has had a single inorganic P application (6 kg P/ha) in 2003, which was the only P application during the past 20 years. The silvopasture has been fertilized annually with ~6 kg P/ha since 1998. The native silvopasture was never fertilized, but has been subjected to grazing activities for the past 5 years. Historically, cattle have grazed the native silvopasture periodically for > 60 years. Pines were planted in December 1988-January 1989 and grazing was deferred until ~1997. At all three pastures, ten soil profiles each were sampled by depth, 0-5, 5-15, 15-30, 30-50, 50-75 and 75-100 cm. For the silvopasture and the native silvopasture, the sampling was in the alley of the silvopastoral systems. All soils were analysed for soluble reactive P (SRP), Mehlich 1-P (soil test P), NO_3-N and NH_4-N.

Results and Discussion
Both SRP (Figure 1) and Mehlich 1-P (Figure 2) concentrations were higher throughout the soil profile of the treeless pasture compared to either the silvopasture or the native silvopasture.

Figure 1. Soluble reactive P as a function of depth for a silvopasture, a native silvopasture and a treeless pasture. Error bars designate standard errors.

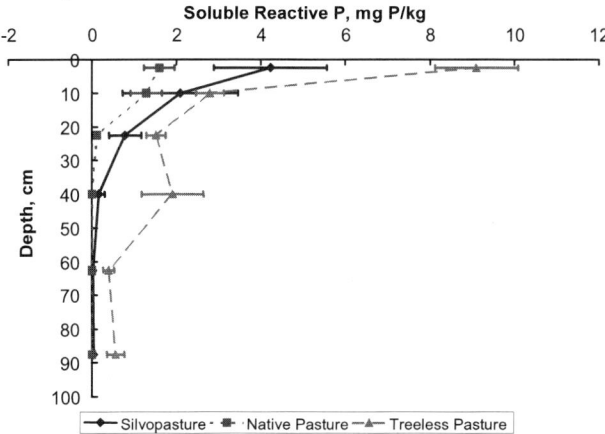

The retentive Bh horizon for the soil profile was at an average depth of 30 cm for all three sites and would explain the generally reduced P concentrations beyond this depth, except when the spodic horizon was not continuous. Because of the high and fluctuating water table in these soils, there is the likelihood of lateral P loss before reaching the spodic horizon at these sites.

Figure 2. Mehlich 1-P as a function of depth for a silvopasture, a native silvopasture and a treeless pasture. Error bars designate standard errors.

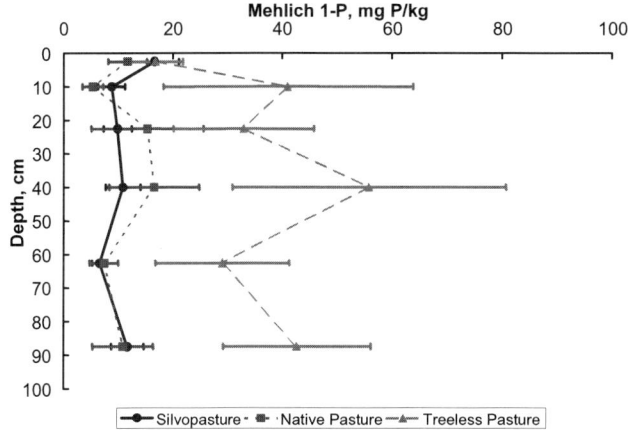

Table 1. Nitrate-N and ammonium-N concentrations with depth at a native silvopasture, a silvopasture and an adjacent treeless pasture at Ona, Florida, USA. Note: Ammonium-N and NO_3-N concentrations were higher in the surface horizon of the treeless pasture although there appeared to be little difference in concentrations at the lower depths for all sites. Numbers in parentheses are standard errors.

Depth, cm	Native silvopasture	Silvopasture	Treeless pasture
		NO_3-N, mg/kg	
0-5	0.83 (0.26)	0.87 (0.37)	2.43 (0.64)
5-5	0.55 (0.18)	0.74 (0.26)	0.50 (0.09)
15-30	0.23 (0.06)	0.49 (0.09)	0.38 (0.04)
30-50	0.23 (0.07)	0.35 (0.06)	0.28 (0.07)
50-75	0.20 (0.06)	0.40 (0.10)	0.23 (0.08)
75-100	0.17 (0.06)	0.34 (0.06)	0.23 (0.09)
		NH_4-N, mg/kg	
0-5	6.36 (0.80)	6.87 (0.47)	9.77 (1.53)
5-15	5.41 (0.49)	5.15 (0.54)	4.07 (0.30)
15-30	3.32 (0.34)	3.86 (0.34)	2.98 (0.28)
30-50	2.28 (0.17)	3.39 (0.50)	2.24 (0.15)
50-75	1.96 (0.22)	2.98 (0.33)	2.14 (0.25)
75-100	1.93 (0.13)	2.71 (0.32)	2.10 (0.27)

Conclusion

The differential rooting zones of the pine and forage crops may absorb the P more completely in the silvopastures than in the treeless system, suggesting that the silvopasture association is better than the treeless pasture in reducing P loss from soil to the surface water. A similar trend in NH_4-N and NO_3-N decreases with depth was not noted for the silvopastures (Table 1).

References

Clason, T.R. and Sharrow, S.H. (2000) Silvopastoral practices. In: Garrett, H.E., Rietveld, W.J. and Fisher, R.F. (eds) *North American Agroforestry: An Integrated Science and Practice*. American Society of Agronomy, Madison, Wisconsin, USA, pp. 119-147.

Nair, V.D. and Graetz, D.A. (2004) Agroforestry as an approach to minimizing nutrient loss from heavily fertilized soils: the Florida experience. In: Nair, P.K.R, Rao, M.R. and Buck, L.E (eds) *New Vistas in Agroforestry: A Compendium for the 1st World Congress of Agroforestry*. Kluwer, Dordrecht, The Netherlands (in press.)

Some ecological impacts of *Quercus rotundifolia* trees on the understorey environment in the "Montado" agrosilvopastoral system, southern Portugal

J. Nunes [1], M. Madeira [2] and L. Gazarini [1]

[1]Dept. de Biologia, Universidade de Évora, Apartado 94, 7002-554 Évora, Portugal. [2]Dept. de Ciências do Ambiente, Instituto Superior de Agronomia, Tapada da Ajuda, 1349-017 Lisbon, Portugal. jdnunes@uevora.pt

Abstract

The influence of *Quercus rotundifolia* trees on precipitation chemistry, soil physical and chemical characteristics, light, nutrient and water availability and biomass production of the understorey was studied in a "montado" at Évora (southern Portugal). Ca and K were added to the soil by throughfall (21.05 and 38.49 kg/ha, respectively) and gross rainfall (3.49 and 3.98 kg/ha). Thickness of the soil organic layer and nutrient availability decreased with the distance from the tree trunk and were higher in ungrazed areas. Soil temperature was higher in areas without canopy, especially in winter. Soil moisture tended to decrease and refill faster in areas without canopy influence. PAR (photosynthetic active radiation) was intercepted around 75 to 90% by the canopy depending on the time of the year, and showed significant diurnal differences between inside and outside the canopy. Biomass production of the herbaceous layer responded more to water and fertilizer in the areas outside the tree canopy and without artificial shading than under the canopy. However, when the rate between the distinct biomass and the control was calculated, the response to water was more efficient under the canopy for the three treatments in the second year.

Key words: herbage production, tree-grass interaction, microclimate, fertilizer, soil moisture

Introduction

The main agroforestry system of southern Portugal is an open *Quercus* woodland, called "montado", which corresponds to a savanna-like formation (Joffre and Rambal, 1993). The interactions between the trees and the underlayer vegetation may have an important role in the management of these systems for diverse uses. There is insufficient knowledge about the effect of the *Quercus rotundifolia* Lam. on herbaceous productivity and on the understorey microclimatic characteristics. In this context the effects of the canopy on nutrient and water availability, radiation decrease and herbaceous productivity and microclimate were studied.

Materials and Methods

The study was carried out in the Évora region, southern Portugal (Centro de Estudos e Experimentação da Mitra, da Universidade de Évora - 38°32'N, 8°01'W, 240 m asl). The area has a typical Mediterranean-type climate, mean annual temperature 15.6°C and mean annual rainfall 655.6 mm. Soils were mainly Eutric Leptosol derived from gneiss. Vegetation is sparse woodland of *Q. suber* and *Q. rotundifolia*, with an understorey of shrubs, dominated by *Cistus salvifolius*. This "montado" has a tree density of 35-45 trees/ha (0.49 m of DBH) in a 5-16% slope area and is grazed mainly by goats and cattle. Subplots were isolated from grazing to carrying on the different studies.

The amounts of gross rainfall, throughfall (at 1.5, 5.2, 7.0, 10.0 and 15.0 m from the tree trunk) and stem flow were measured continuously in one large isolated tree, following the N, S, E and W orientation. Samples for chemical analysis were collected weekly from November 1996 to December 2000. Organic layer and mineral soil samples were collected at 33, 66, 100, 133 and 200% of the crown ratio, at the four cardinal points, in ten isolated trees for mass quantification and chemical characterization. Microclimatic parameters under and outside the tree canopies (soil moisture at 5 and 20 cm deep, soil temperature at 2.5, 5, 10 and 20 cm deep and PAR at 1.0 m above the soil surface) were logged and measured hourly.

The herbaceous biomass was measured in 36 microplots (1 m × 1 m) established in the study area, the sampling design being as follows: 12 microplots were located under the tree canopies (UC), 12 under artificial shadow (AS - a green plastic net with 80% interception of G) and the remaining 12 in areas without shadow influence (IO). Each of these three microplot groups was submitted to the following treatments: irrigation (I), NP fertilization (F), irrigation with NP fertilization (IF) and without irrigation and fertilization (C). Biomass was collected in April (maximum herbaceous production in this area) of 2000 and 2001, separated into grasses, forbs and legumes and dried. Subsamples of all the collected plant material were ground and analysed for N, P, K, Ca, Mg and Mn content. Differences in the herbaceous production were tested by the H test from Kruskal-Wallis.

Results and Discussion

The litter layer mass decreased from 3200 kg/ha near the trunk to 1600 kg/ha at the canopy projection limit. The amount of nutrients returned to the soil by throughfall was much higher than that returned through bulk precipitation (Table 1). The quantity of nutrients returned to the soil by stem flow was much lower, probably because it influenced a more restricted area.

Table 1. Average amount (kg/ha per year) of nutrients returned to the soil by bulk precipitation (BP, mm), throughfall (TF, mm) and stem flow (ST, mm).

	PR	Ca	Mg	Na	K	$N-NH_4^+$	$N-NO_3^-$	P	Cl	$S-SO_4^{2-}$
BP	654	3.49	1.84	8.66	3.98	2.82	2.07	0.77	17.94	3.88
TF	505	21.05	6.54	22.35	38.42	4.20	5.37	1.35	54.18	7.05
ST	3	0.86	0.13	0.16	0.63	0.01	0.01	<0.01	0.35	0.06

Soil total C and N contents were higher under the tree canopies than in the areas outside their influence. The same tendency was observed for the exchangeable base cations (Ca, Mg and K) and for extractable K and P. All of these values decreased from the areas near the trunk to those located at the canopy projection limit. Both pH and exchangeable Mg values were not influenced by the tree canopies. Photosynthetically active radiation (PAR) in the open was higher than under the canopy: PAR interception by the tree and shrub layers reached 95%. The pattern determined for the radiation decrease is in accordance with that found by other authors (Belsky *et al.*, 1993) in savannah formations.

The temperatures obtained from 2.5 cm to 20 cm depth were always higher in the open than under the canopy. However, the differences found between the soil in the open and under the canopy were greater during the summer months. A similar pattern was found by Rorison (1991), who compared north and south slopes, and by Belsky *et al.* (1993) in their studies with *Acacia tortilis* and *Adansonia digitata*.

The pattern of soil moisture decrease was the same for all situations, although the highest values were obtained for the open areas. The same tendency was found during the rainy period after summer drought, having been similar for both areas in the end. Tree canopy had no significant effect on soil moisture.

Biomass production of the herbaceous layer responded more to water and fertilizer in the open than under the canopy (Table 2). However, the response to I was greater for the first year, and for F, I and IF in the second year under the tree canopy, in relation to C. A similar response to water was found by Ludwing *et al.* (2001) in studies with *Acacia tortilis*.

Table 2. Average herbaceous biomass (g/m) production in the open (IO), under tree canopy (UC), with artificial shading (AS); with irrigation (I), NP fertilization (F), irrigation with NP fertilization (IF) and without irrigation and fertilization (C). Different letters mean significant differences between treatments ($P < 0.05$).

	IO		UC		As	
	2001	2002	2001	2002	2001	2002
C	302.40a	383.79a	126.88a	91.87a	112.64a	136.53a
F	514.24b	520.53a	190.08a	421.02b	222.61a	217.97a
I	363.47c	386.83a	224.59a	179.04ac	161.23a	129.82a
IF	700.11b	606.74a	290.51a	292.05c	410.03a	191.47a

In general the biomass N, P, K, Ca, Mg and Mn content was higher in the microplots where radiation was decreased, either under the tree canopies or in artificial shade, than in the open (Table 3).

Table 3. Herbaceous biomass N, P, K, Ca, Mg and Mn concentration (mg/g) by group of plants (grasses, forbs and legumes) in the open (IO), under tree canopy (UC), with artificial shadowing (AS); with irrigation (I), NP fertilization (F), irrigation with NP fertilization (IF) and without irrigation and fertilization (C).

		IO						UC						AS					
	Grasses		Forbs		Legm.		Grasses		Forbs		Legm.		Grasses		Forbs		Legm.	
	2001	2002	2001	2002	2001	2002	2001	2002	2001	2002	2001	2002	2001	2002	2001	2002	2001	2002
									mg g^{-1}									
N																		
C	7.34	6.79	8.23	9.61	22.59	17.67	9.44	9.49	10.87	10.05	23.53	16.59	10.14	7.80	10.46	9.39	23.65	18.63
I	6.83	6.76	8.59	8.85	22.61	19.35	11.80	9.50	8.81	10.70	21.08	19.02	9.93	10.33	10.23	12.00	0.00	0.00
F	6.34	6.59	10.34	10.36	19.19	18.87	8.97	6.10	9.31	11.92	25.88	19.03	10.98	10.09	11.65	10.75	24.82	20.76
IF	5.65	8.02	13.75	9.38	20.65	21.13	6.62	8.44	8.93	8.95	27.14	19.98	8.77	10.02	11.98	13.48	26.93	0.00
P																		
C	1.03	1.44	1.52	1.74	1.20	1.27	1.33	1.19	1.72	2.03	1.30	1.36	1.31	1.59	1.55	1.50	1.41	1.29
I	1.24	1.99	1.69	2.02	1.79	1.85	2.21	2.90	2.49	2.98	1.86	1.98	1.25	2.69	1.26	2.77	0.00	0.00
F	1.05	1.04	1.68	1.64	1.35	1.44	1.10	0.88	1.40	1.18	1.19	2.37	1.04	1.52	1.54	1.53	1.75	1.32
IF	1.46	2.36	2.08	3.32	1.73	2.40	1.63	2.29	2.43	2.78	1.91	1.10	2.04	2.79	1.88	3.13	2.05	0.00
K																		
C	10.21	11.87	14.06	15.55	9.73	14.33	15.07	19.07	20.69	25.78	18.75	19.71	11.96	16.37	22.27	19.31	10.79	14.48
I	9.68	11.80	11.68	15.59	6.78	13.08	23.34	25.10	24.97	28.98	18.72	25.81	12.52	11.77	12.34	13.00	0.00	0.00
F	7.12	10.47	13.49	16.45	10.11	14.94	14.36	21.72	22.12	23.92	18.49	16.53	14.89	16.42	18.46	18.23	13.00	9.45
IF	7.33	10.59	12.29	17.39	6.59	13.44	12.74	16.38	20.71	20.65	19.80	21.37	13.15	14.32	16.46	17.81	10.17	0.00
Ca																		
C	1.21	1.19	7.40	7.11	9.83	6.61	1.79	1.17	5.85	6.17	8.84	6.30	2.02	1.39	8.06	4.95	9.32	5.17
I	1.29	1.37	8.11	6.00	12.28	7.60	2.02	1.60	5.69	7.16	8.59	6.92	1.80	2.00	10.11	7.99	0.00	0.00
F	1.20	0.93	7.75	7.46	7.06	5.65	1.88	0.91	6.35	7.55	9.36	8.11	1.95	1.51	7.77	6.43	10.96	7.26
IF	1.44	1.42	6.92	7.18	7.92	5.82	1.85	1.49	6.11	6.11	11.70	5.74	1.88	1.60	8.66	7.06	16.92	0.00
Mg																		
C	0.66	0.56	1.83	1.83	1.90	1.88	1.02	0.97	2.71	2.54	2.29	2.07	0.75	0.73	2.18	1.87	2.01	1.67
I	0.68	0.61	2.06	1.79	2.00	1.99	1.10	1.00	2.58	2.38	2.71	2.88	0.65	0.69	2.63	1.49	0.00	0.00
F	0.65	0.51	1.90	2.01	1.75	1.48	1.02	0.81	2.52	2.50	2.08	2.33	0.99	0.84	2.28	2.17	2.24	1.98
IF	0.66	0.59	1.81	2.07	1.70	1.56	0.83	0.77	2.04	1.92	2.86	1.53	0.89	0.81	2.58	2.06	2.96	0.00
Mn																		
C	0.09	0.11	0.05	0.09	0.02	0.09	0.16	0.14	0.13	0.12	0.05	0.02	0.08	0.07	0.04	0.10	0.00	0.03
I	0.07	0.07	0.07	0.06	0.05	0.02	0.12	0.20	0.10	0.09	0.01	0.02	0.06	0.06	0.05	0.06	0.00	0.00
F	0.10	0.12	0.13	0.14	0.03	0.06	0.10	0.12	0.11	0.07	0.01	0.02	0.08	0.14	0.07	0.06	0.04	0.00
IF	0.08	0.08	0.10	0.12	0.03	0.00	0.17	0.17	0.17	0.14	0.04	0.00	0.09	0.08	0.06	0.09	0.04	0.00

Conclusions

The response to water and fertilizer application was more efficient under tree canopy than in the open. The herbaceous vegetation biomass production tended to be greater in the irrigated and NP fertilized plots. The biomass N, P, K, Ca, Mg and Mn contents reflected the effects of all treatments, particularly the effect of the radiation decrease combined with that of the increase of nutrient and water availability. This information suggests that water and fertilizer resources should be managed differently in agrosilvopastoral systems with sparse trees like "montado".

Acknowledgement

The first author has a PhD grant from PRAXIS XXI/BD/18451/98.

References

Belsky, A.J., Mwonga, S.M., Amundson, R.G., Duxbury, J.M. and Ali, A.R. (1993) Comparative effects of isolated trees on their undercanopy environments in high and low-rainfall savannas. *Journal of Applied Ecology* 30, 143-155.
Joffre, R. and Rambal, S. (1993) How tree cover influences the water balance of Mediterranean rangelands. *Ecology* 74, 570-582.
Ludwing, F., Kroon, H., de Berendese, F. and Pins, H.H.T. (2001) Savanna tree influences on nutrient, water and light availability and how this affects productivity and composition of the understorey. *Tropical Resource Management Papers* 39, 15-34.
Rorison, I.H. (1991) Ecophysiological aspects of nutrition. In: Porter, J.R. and Lawlor, D.W. (eds) *Plant Growth Interactions with Nutrition and Environment.* Cambridge University Press, Cambridge, United Kingdom, pp. 157-176.

Soil nutrient status and forage yield at varying distances from trees in four dehesas in Extremadura, Spain

J. J. Obrador [1] and G. Moreno [2]

[1]Colegio de Postgraduados, Campus Tabasco, Ap. Post 24. H. Cárdenas, Tabasco, Mexico. [2] Centro Universitario, Universidad de Extremadura, Plasencia 10600, Cáceres, Spain. obradoro@colpos.colpos.mx, gmoreno@unex.es

Abstract

The aim of this study was to understand the effect of holm-oak (*Quercus ilex*) on the soil nutrient concentration and its consequence for the yield of understorey forage (*Avena sativa*) in four dehesas of CW Spain. The soils of the dehesas varied in soil fertility (chromic Luvisols and Achrisols, and eutric Leptosols). Forage dry matter yields were determined from 1 m^2 sample plots at distances ranging from 2 to 20 m from the tree (nine trees/farm per year). Soil samples (0-30 cm depth) were also collected from the same sampling locations, and were analysed for pH, electrical conductivity, organic C, CEC, total N, available N and base cations. Soil analysis results showed that organic C, total N, CEC and exchangeable Ca^{2+} and K^+ increased in the vicinity of the tree. Differences in forage yield were mainly explained by fertilization dosage, light availability and soil CEC. In more fertile soils, forage production was negatively affected by the presence of the trees, as a consequence of light reduction (competence), while, in more oligotrophic soils, forage production was positively affected by trees (facilitation).

Key words: soil nutrient heterogeneity, competence, facilitation, silvopasture

Introduction

Dehesas are silvopastoral systems of extensive utilization, where autochthonous pastures and periodical crops are combined with scattered trees. Periodical crops (usually cereal) aim to control shrub encroachment and to provide an additional fodder complement for cattle.

Information on the effect of trees on understorey forages in the dehesas is required to model agroforestry functioning based on tree-crop interactions (HySAFE; see Dupraz, this volume), which is expected to be useful in improving the management of these agroforestry systems. This study was therefore undertaken to gather data on the effect of *Quercus ilex* on the soil nutrient status and its consequence for the forage yield in four dehesas of Extremadura, Spain.

Materials and Methods

The study was carried out in four holm-oak dehesas of the Cuatro Lugares County, Cáceres (west-central Spain: 34°4'N, 6°13'W). The climate is Mediterranean, with dry and hot summers and cool, rainy winters, mean annual rainfall of 579 mm, and mean annual temperature 16°C. Soils are mainly chromic Luvisols in CL and chromic Acrisols in SO and BA, developed over tertiary sediments (Miocene feldspathic sands) with abundant quartzite. Both types of soils show a low chemical fertility (Obrador Olán *et al.*, 2004). Eutric Leptosols (in T, more fertile soils) have developed on slates from where the sediments have been eroded. Common management practices in dehesa are cattle raising with native pasture and cereal intercrops. Tree density varies from 15 to 50 tree/ha, depending on their main use (lower densities in intercropped areas and higher densities in areas reserved for hunting). Dehesas were dominated by mature trees, ranging from 80 to 120 years old and 7 to 12 m canopy width.

Forage dry matter yields were estimated from 1 m^2 samples taken at the physiological maturity stage of forage in four intercropped dehesa farms, with different soil fertility (in decreasing order: T, CL, SO and BA; see CEC in Figure 1) and fertilization dosage (0 to 250 kg/ha of NPK 7-12-7; noted as susbcript after farm name: BA_{250}, BA_0, T_{200}, CL_{150}, SO_{50}, SO_0). Samples were taken in 2002 and 2003, around nine trees per farm, in two orientations and four distances (from beneath the tree (2 m) to 20 m). Soil samples (composed of five subsamples; 0-30 cm depth) were taken from the same locations in early March. Soil samples were analysed for pH (1:2.5 water), electrical conductivity, organic C, cation exchange capacity, total N, available P and exchangeable base cations (Ca^{2+}, Mg^{2+}, K^+ and Na^+), following Bigham and Bartels (1996). The use of broadcast spreaders to fertilize with scattered trees can produce differences in the dose of fertilizer received at different distances from the trees; thus the amount of fertilizer applied at each distance and site was measured by means of a collector placed on the soil surface when fertilizer was applied.

Results and Discussion

From soil analysis, most parameters (CEC, exchangeable Ca^+ and K^+, electrical conductivity, organic C and total N) decreased with distance away from the tree (see CEC in Figure 1 as an example). Similar results have been found in dehesas by Escudero (1985), Joffre (1987), Puerto *et al.* (1987) and Obrador *et al.* (2004). No definite pattern was observed for changes in pH, base saturation, available P and exchangeable Mg^{2+} and Na^+ with distance from trees.

Forage productivity showed a very irregular trend (Figure 1) as can be seen in some of the evaluated dehesas, increasing significantly with distance in T_{200}-2002, BA_{250}-2003 and SO_0-2003. Forage yield only decreased significantly with distance in one case (SO_{50}-2002). Other cases (CL_{150}-2002, CL_{150}-2003, T_{200}-2003, SO_{50}-2003 and BA_0-2003) did not show any significant tendency. Puerto *et al.* (1987) also showed that the pattern of pasture yield around oaks varied with soil quality.

Figure 1. Mean values of forage yield (including both oats (grain and straw) and weeds) at different distances from the tree stem in four intercropped dehesas, which vary in soil fertility (dashed lines show the mean values of CEC) and in doses of fertilization (50, 150, 200 and 250 kg/ha of N+P+K 7-12-7 in SO, CL, T and BA, respectively; solid lines show the variation of N+P+K doses with distance, measured at the moment of the fertilizer application).

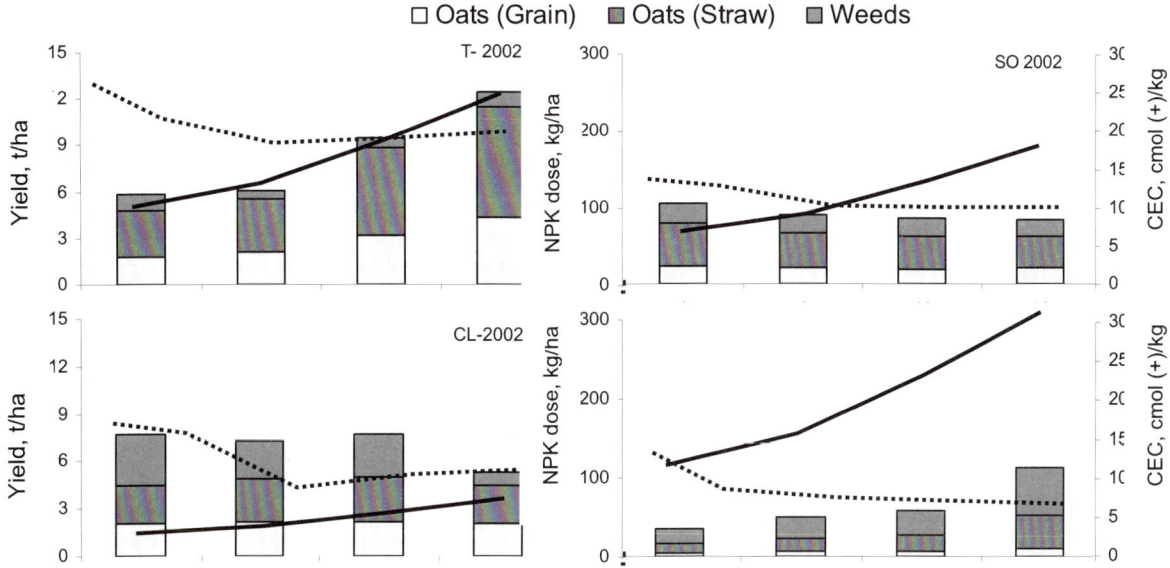

Multiple regressions have been applied in order to discriminate the role of the different parameters in forage production (Table 1). The highest part of the variablity was explained by the fertilization (N+P+K) dosage, which increased with the distance to the tree stem (Figure 1). Light played a minor role (r^2 around 7%). Among the soil parameters, CEC explained around 16% of the oats yield variability (both straw and grain). Surprisingly, SOM and total N did not contribute significantly to explain the forage variability.

Table 1. Percentage of forage yield variability explained by different parameters. Values have been estimated by multiple regression (r^2 values). In both cases, only significant variables are included. #See Montero and Moreno, this volume.

Parameter		Total biomass		Oats biomass		Grain biomass	
		r^2	(r)	r^2	(r)	r^2	(r)
Other	Distance to tree		(+)		(+)		(+)
	Light #	0.066	(+)	0.079	(+)	0.072	(+)
	Fertilization dose	0.358	(+)	0.346	(+)	0.231	(+)
Soil properties	pH-water		(-)	0.049	(-)	0.053	(-)
	Electrical conductivity	0.030	(+)		(+)		(+)
	Soil organic matter (SOM)						
	Total N				(+)		
	Available P	0.032	(+)		(+)		(+)
	CEC		(+)	0.167	(+)	0.160	(+)
	Exchangeable Ca^{2+}				(+)		(+)
	Exchangeable Mg^{2+}		(+)	0.037	(+)	0.032	(+)
	Exchangeable K^+	0.030	(+)		(+)	0.042	(+)
	Exchangeable Na^+	0.167	(-)				(-)
	Base saturation						
Total explained variability		0.74		0.77		0.59	

Conclusions

Soil fertility was significantly higher in the vicinity of the trees than further away, in spite of the amount of fertilizer applied increased with the distance from the trees. This result indicates a positive effect of the trees on the soil fertility. However, it seems that the increased fertility had no effect on forage productivity, except in unfertilized oligrotrophic soils, where forage production was very low, being slightly higher beneath trees than outside the trees.

Acknowledgements

This study was supported by the EU (SAFE project, QLX-2001-0560) and Spanish government (MICASA project, AGL-2001-0850). Jesus Obrador was awarded grants by ANUIES and Colegio de Posgraduados (Mexico).

References

Bigham, J.M. and Bartels, J.M. (1996) *Methods of Soils Analysis, SSSA Book Series 5 (part 3): Chemical Methods.* American Society of Agronomy, Madison, Wisconsin, USA.

Escudero, A. (1985) Effect of dispersed trees on the chemical soil properties (Efectos de árboles aislados sobre las propiedades químicas del suelo). *Revue d'Ecologie et de Biologie du Sol* 22, 149-159. (In Spanish.)

Joffre, R. (1987) Constraints of the environment and response of the herbaceous vegetation in the dehesas of the Sierra Norte (Andalusia, Spain) (Contraintes du milieu et réponses de la végétation herbacée dans les dehesas de la Sierra Norte (Andalousie, Espagne). MSc thesis, CNRS-CEPE, Montpellier, France. (In French.)

Obrador Olán, J.J., García López, E. and Moreno, G. (2004) Consequences of dehesa management practices on nutritional status of vegetation in Cuatro Lugares, Extremadura, Spain. In: Schnabel, S. and Gonçalves, A. (eds) *Sustainability of Agrosilvopastoral Systems*. Advances in Geoecology, 37, Catena Verlag, Reiskirchen, Germany. (in press.)

Puerto, A., García, J.A. and García, A. (1987) Hill systems as clarifying element of certain effects of trees on pastures (El sistema de ladera como elemento esclarecedor de algunos efectos del arbolado sobre el pasto). *Anuario del CEBA de Salamanca* 12, 297-312. (In Spanish.)

Rangeland health assessment in silvopastoral systems of northern Greece

Ch. I. Pantazopoulos, M. S. Vrahnakis, D. Chouvardas, M. Papadimitriou and V. P. Papanastasis
Laboratory of Rangeland Ecology, Faculty of Forestry and Natural Environment, Aristotle University of Thessaloniki (286), GR - 54124, Greece. (Part of the EU Project GeoRange EVK2-CT-2000-0091.) chpantaz@auth.gr

Abstract
Silvopastoral areas in Greece dominated by deciduous oaks are mainly utilized by domestic animals but irrational grazing results in their degradation. The aim of this study was to identify which are the most important indicators for assessing rangeland health. In Lagadas county, northern Greece, 17 representative plots of 0.1 ha each were selected, where 19 indicators related with soil stability, hydrologic function and biotic integrity were measured. Data were analysed statistically using the principal component analysis (PCA). It was found that 16 out of the 19 indicators measured accounted for 89.6% of the total variance. It is suggested that these indicators are the most important to be used in the assessment of rangeland health.

Key words: biotic indicators, PCA analysis, Pearson's correlation, physical indicators

Introduction
Rangeland health is an important concept in the assessment and monitoring of rangeland ecosystems. It indicates the degree of integrity of the soil and ecological processes that are most important in sustaining the capacity of rangelands to satisfy values and produce commodities (National Research Council, 1994). Its assessment is based on a wide range of biotic and abiotic indicators related to soil stability, hydrologic function and integrity of rangelands (Pellant *et al.*, 2000).

In the Mediterranean region, silvopastoral systems are widespread grazing resources, invaluable in providing feed for animals, especially in the long, dry summer period. In addition, they are complex systems with many more products and services than forage production, while their management requires an integrated approach (Papanastasis, 1996). In this chapter, an attempt is made to identify the most suitable indicators for assessing rangeland health in Mediterranean silvopastoral system in Greece.

Materials and Methods
The study was conducted in Lagadas county (40°47′ N, 23°12′ E) in central Macedonia, northern Greece, during the spring of 2003. The dominant vegetation was deciduous oaks (e.g. *Quercus pubescens, Q. petrea*). In each of 17 silvopastoral sites (0.1 ha each) 16 quadrats (50 cm × 50 cm each) were placed along two diagonals. In each quadrat, 19 indicators were assessed (Table 1). Cover indicators were ocularly estimated by two independent observers, plant height was measured and the quantitive indicators were rated on a scale from 1 to 6. The relative palatability factor (RPF) of the dominant species for sheep (s) and goats (g) was estimated. In each quadrat, the cover of the three dominant species was determined. For each of these species, a value for their palatability for sheep and goats was assigned. After multiplying this value with the cover of each dominant species, their RPF was calculated. To select the most important indicators, the PCA in a varimax rotated space was applied and the 19 indicators were classified into factors. PCA analysis was based on the statistical package SPSS/PC (version 11.0).

Table 1. List of indicators used in assessment (E: estimated, M: measured, C: calculated).

Physical	Mode	Units	Biotic	Mode	Units
Cover of litter	E	(%)	Height of trees	M	cm
Cover of bare ground	E	(%)	Height of shrubs	M	cm
Cover of rock, gravel	E	(%)	Height of herbs	M	cm
Cover of cryptogams	E	(%)	Plant vigour	E	1-6
Cover of annuals	E	(%)	Number of seed heads	E	1-6
Cover of perennials	E	(%)	Age-class distribution	E	1-6
Cover of shrubs	E	(%)	Presence of legumes	E	1-6
Cover of trees	E	(%)	RPF (s)	C	(%)
Erosion	E	1-6	RPF (g)	C	(%)
Animal trail	E	1-6			

Results and Discussion

From the PCA analysis seven factors had eigenvalues higher than 1. These factors accounted for 89.6% of the total variance and included 16 out of the 19 indicators used in the analysis. The positive or negative correlation of each indicator with each of the seven factors is presented (Table 2). Indicator loadings less than 0.6 (i.e. loadings related to low strength correlations) were excluded. Such loadings were given by the cover of rock and gravel, the cover of cryptogams and the height of shrubs. These three indicators seem not to be important in the assessment of rangeland health in the silvopastoral systems studied.

By using Pearson's correlation (data not shown) it was found that the cover of perennial grasses was negatively correlated ($P < 0.05$) with the animal trail and erosion indicators. Animal trails, on the other hand, were positively correlated with the cover of shrubs and erosion. There was also a strong positive correlation among plant vigour, age-class distribution and number of seed heads. It is important to note that, as the cover of bare ground increased, the height of herbs decreased. The RPF (s) was positively correlated with the height of trees, while the RPF (g) was positively correlated with the cover of perennial herbs. On the contrary, the presence of legumes did not correlate with any other indicator.

Table 2. Rotated component matrix.

	Component						
	1	2	3	4	5	6	7
Cover of annuals	0.843						
RPF (g)	-0.784						
Cover of perennials	-0.653						
Cover of litter		0.879					
Cover of trees		0.872					
Cover of shrubs			0.869				
Animal trail			0.855				
Erosion			0.605				
Height of herbs				0.858			
Bare ground cover				-0.837			
Plant vigour					0.867		
Age-class distribution					0.818		
Number of seed heads					0.705		
Height of trees						0.899	
RPF (s)						0.639	
Presence of legumes							0.863

Conclusions

Indicators related with cover of various vegetation components as well as the erosional status of the soil seem to be important in assessing rangeland health in Mediterranean silvopastoral systems dominated by deciduous oaks. This is also supported by the fact that indicators which are connected with erosional status have a strong correlation with ground cover.

References

National Research Council (1994) *Rangeland Health: New Methods to Classify, Inventory, and Monitor Rangelands.* Committee on Rangeland Classification, Board of Agriculture, National Academy Press, Washington, DC, USA.

Papanastasis, V.P. (1996) Silvopastoral systems and range management in the Mediterranean region. In: Etienne, M. (ed) *Western European Silvopastoral Systems.* INRA, Paris, France, pp. 143-156.

Pellant, M., Shaver, P., Pyke, D.A. and Herrick, J.E. (2000) *Interpreting Indicators of Rangeland Health, version 3. Technical Reference 1734-6, USDI, BLM.* National Scientific and Technical Center, Denver, USA.

Post-fire ecological and dynamic characterization of shrubland communities in Galicia using structural variables

X. M. Pesqueira [1], A. Muñoz [1], R. Álvarez [1], O. Reyes [1,2] and M. Casal [1]

[1] Área de Ecología, Dept. de Biología Celular y Ecología, Facultad de Biología, Campus Sur. 15782, Santiago de Compostela, Spain. [2] Área de Ecología, Dept. de Biología Celular y Ecología, E.P.S. de Lugo, 27002, Lugo, Spain. xmpesque@usc.es, bfreyes@usc.es.

Abstract

Two Galician moorland communities, one dominated by *Pterospartum tridentatum* and *Erica umbellata* and the other by *Cytisus striatus*, were ecologically characterized, and their recovery after a traditional use of fire to obtain pasture for grazing was evaluated. The study was carried out in A Serra do Candán (Pontevedra), a site recognized by the European Community and included in Natura 2000. Two control plots and two that were burned in the autumn of 1999 were selected. Plant dynamics were studied by taking two annual samples, one in springtime and one in autumn, from the spring of 2001 to 2003. Measurements of the principal structural variables were obtained from these samples. In the first 2 years after fire, herbaceous species peaked in the *C. striatus* community, followed by a clearly marked descent in the third year, probably due to the competition of shrub species. Moreover, there was a notable increase in the dominance of *Rubus fruticosus*. On the other hand, in the community of *P. tridentatum*, the herbaceous species did not reach high cover at any time and become scarce after 3 years.

Key words: succession, *Cytisus striatus* community, *Pterospartum tridentatum* and *Erica umbellata* community, wild-land fire, cover, vertical structure

Introduction

Controlled burning of scrub, Galician forest and/or shrubland used as a technique for forest fire prevention encourages the regrowth of more palatable herbaceous plants. However, this technique is controversial because of ecological effects and fire risk in adjacent areas. The objective is to maintain the vegetation in a stage of arrested succession which is adequate for grazing. This structurally modifies and conditions the development of the plant communities, as they are subjected to periodic burning. Shrublands suffer exceptionally because of this, as they are forced to adapt to a high frequency of burning. Both the edible herbaceous species phytomass and grazing value change after fire (Trabaud and Casal, 1989). This study addresses the ecological characterization of two shrubland communities subjected to these perturbations, as well as an understanding of their post-fire development.

Materials and Methods

The plant communities studied are found in A Serra do Candán (Pontevedra), a site recognized by the European Community and included in Natura 2000. One is a community of *C. striatus* and the other of *P. tridentatum* and *E. umbellata*. Both suffered a fire in autumn 1999.

Ecological characterization was carried out from 2001 to 2003 by sampling, in spring and in autumn. Four fixed plots, two in unburned shrubland (control plots) and two in burned shrubland, were set up in each zone, each with an area of 25 m^2. Data for different structural variables were obtained from each of the plots, such as:

Vertical structure of the vegetation, as % cover per horizon, in accordance with Basanta *et al.* (1988), with the following stratification scale:

Horizon	Limits	Horizon	Limits
1	0-5 cm	4	50-100 cm
2	5-25 cm	5	1-2 m
3	25-50 cm	6	>2 m

Linear cover, in five transects of 5 m each: the cover of the shrub species was registered, with the herbaceous species recorded as a variable only when they did not grow underneath shrub species.

Results and discussion

The control plots of *C. striatus* had a typical vertical structure, with 100% vegetation cover in strata 1 and 2, decreasing in strata 3 and 4 until reaching strata 5 and 6, where they increased due to the development of a form of *C. striatus*. After burning, a rapid recuperation of cover in the first strata was observed, due to the establishment of herbaceous species such as *Agrostis* sp and rapid growth of woody species such as *R. fruticosus*. Strata 5 and 6 did not recover within the period of this study, although a tendency towards the state of cover shown by the control plots was noted.

A high degree of cover was observed in strata 1 and 3 in the control plots of *E. umbellata* and *P. tridentatum*, which disappeared in the strata above 1 metre. After the fire, a progressive recuperation of the structure was observed, until it reached the values of the control plots in the final samples.

Legumes normally associated with *C. striatus* had high cover values (Figure 1a) in that community. It should be noted that this group made a rapid recovery, reaching values similar to those of the controls in the first years of succession. This is not only due to the development of *C. striatus*, but also to a large increase in *Ulex gallii*, which has an important capacity for resprouting after fire (Reyes *et al.*, 2000). The appearance of different woody species, such as *Halimium alyssoides*, *Lithodora difusa* and *R. fruticosus*, is notable. These do not appear in the control plots. Edible herbaceous species cover was high over the 2 years after fire.

In the control plots of the *E. umbellata* and *P. tridentatum* community (Figure 1b), the high value of the group of *Ericaceae* is notable. This is due to *E. umbellata*. After burning, *E. umbellata* disappeared; it does not resprout after fire and its regeneration from seed is very slow (Fernandes Paulo and Rego, 1996). A notable increase in the *Leguminosae* is observed, mainly due to *P. tridentatum* and *U. gallii*, which are species that resprout very well after fire (Reyes *et al.*, 2000). The development of other woody species, such as *H. alyssoides*, is also notable. In this community herbaceous species developed, but in no case was it comparable to that of the community of *C. striatus*.

Figure 1. a,b: Cover of the different groups of shrub, herbaceous species and base ground in the two shrubland communities of O Candán, during the 3 years of this study.

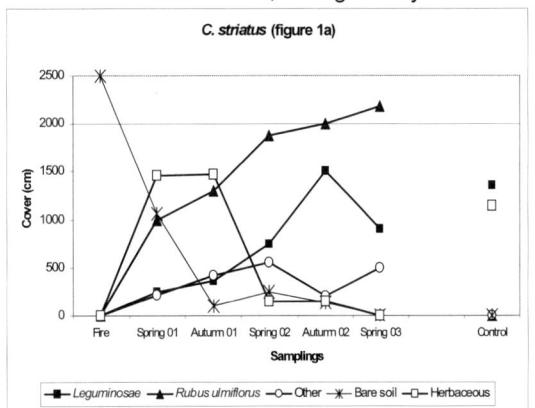
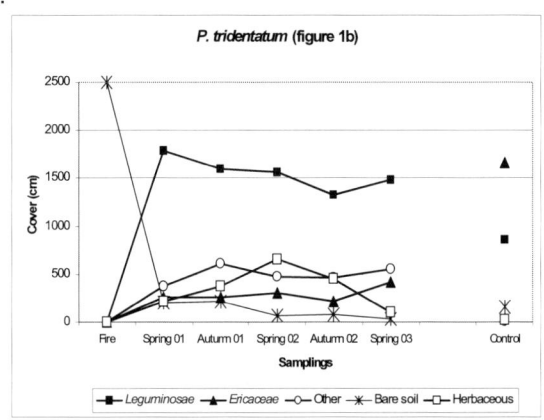

Conclusions

Both communities show a slow recovery of woody and herbaceous components after fire. Herbaceous vegetation develops better in the *C. striatus* community than in the *P. tridentatum* one. As cover is a legume, this probably improves soil nitrogen content and gives a higher grazing value.

References

Basanta, M., Díaz Vizcaíno, E. and Casal, M. (1988) Structure of shrubland communities in Galicia (NW Spain). In: Durin, H.J., Werger, M.J.A. and Willems, J.H. (eds) *Diversity and Pattern in Plant Communities*. SPB Academic Publ. The Hague, The Netherlands, pp. 25-36.

Fernandes Paulo, M. and Rego, F. (1996) Changes in fuel structure and fire behaviour with heathland aging in northern Portugal. In: Weber, R. (ed.) *Proceeding of 13th Fire and Forest Meteorology Conference.*, R. IAWF (International Association of Wildland Fire). Washington, USA.

Reyes, O., Basanta, M., Casal M. and Díaz-Vizcaino, E. (2000) Functioning and dynamics of woody plant ecosystems in Galicia (NW Spain). In: Trabaud, L. (ed) *Life and Environment in the Mediterranean*. Wit Press, Southampton, United Kingdon, pp. 1-42.

Trabaud, L. and Casal, M. (1989) Fire and pastoralism in southern Europe. In: *Landscape Ecology: Study of Mediterranean Ecosystems*. MAB (Unesco Man and the Biosphere programme), Niza, Italy, pp. 120-132.

Extensive livestock systems as tools for environmental management: impact of grazing on the vegetation of a protected mountain area

J. L. Riedel, I. Casasús, A. Sanz, M. Blanco, R. Revilla and A. Bernués
Centro de Investigación y Tecnología Agroalimentaria de Aragón, Gobierno de Aragón, Apdo. 727, 50080 Zaragoza, Spain. jlriedel@aragob.es

Abstract
The study was conducted in the 'Sierra de Guara' Natural Park (80,739 ha, Huesca, Spain), a protected area dominated by shrub and forest pastures. The effect of livestock grazing on herbaceous and shrub vegetation was determined in six locations representative of pasture types in the park. Vegetation characteristics were measured in adjacent grazed and non-grazed areas before and after the grazing season in three consecutive years. Sward height, biomass, green-dead ratio and chemical composition were analysed, and shrub vegetation was characterized in terms of species composition, volume and biomass. The evolution patterns in non-grazed and grazed areas were significantly different, with higher herbage biomass accumulation in the former, both in absolute terms (2609 vs 1070 kg DM/ha in the third year, $P < 0.001$) and when only dead material was considered (1898 vs 499 kg DM/ha, $P < 0.001$), and with lower forage quality. Results concerning shrub vegetation revealed a similar trend. These results underline the role that livestock can play in biomass control and reduction of landscape degradation, which has particular relevance in protected natural areas.

Key words: rangeland management, herbage biomass, shrub pasture, "Sierra de Guara" Natural Park

Introduction
The abandonment of farming activities in a number of European agrosilvopastoral areas has originated a process of evolution of existing vegetation towards climactic types. In medium-altitude mountain pastures the main consequence of this process is shrub invasion, as an early stage of succession towards the climactic forest. This phenomenon often increases environmental risks, such as fire hazards, and landscape degradation. The objective of this study was to determine the effect of livestock grazing on the evolution of herbage and shrub vegetation in the 'Sierra de Guara' Natural Park rangeland areas.

Materials and Methods
The study was conducted in the 'Sierra de Guara' Natural Park (80,739 ha), which constitutes a representative example of middle-altitude Mediterranean mountain areas. The park is grazed at very low stocking rates (0.11-0.64 LU/ha, depending on pasture type), mainly by sheep (75.8% of total LU) but also by cattle (18.6%), mares (4.0%) and goats (1.6%) (Bernués et al., 2002). Six zones representative of the park pasture types were selected. In these zones, shrub cover varied between 15 and 48% (main species were *Genista scorpius*, *Echinospartum horridum* and *Thymus vulgaris*) and tree cover varied between 3 and 10% (main species were *Pinus sylvestris* and *Quercus faginea*). Two plots of 10 m × 10 m were fenced in spring 2001 in each zone to prevent animal grazing thereafter. Measurements of herbaceous and shrub vegetation were taken in the following three grazing seasons, before and after the grazing period (spring and autumn), inside and outside the fenced areas ($n = 24$). Total sward biomass was estimated from average sward height (measured with the HFRO sward stick on 60 points at random), using a prediction equation obtained for these conditions (Casasús et al., 2002). Sward samples were obtained, separated into green and dead herbage fractions and oven-dried at 60°C. Neutral detergent fibre (NDF), acid detergent fibre (ADF) and lignin (Goering and Van Soest, 1970) and crude protein (AOAC, 1995) were determined. For the study of woody vegetation, two fixed transects of 10 m × 1 m were placed inside and outside the fenced areas. All shrub individuals within the transects were identified to calculate species contribution. The height and two whole plant diameters (major and minor) of five tagged individuals per species were measured each spring. Shrub biomass was then estimated from these measurements using existing species-specific equations (Torrano, 2001). Data were analysed using a General Linear Model (SAS, 1990).

Results and Discussion
Total herbage biomass increased significantly ($P < 0.001$) in non-grazed areas, while it remained constant or decreased slightly in grazed areas (ns). Therefore, when comparing grazed vs non-grazed areas significant differences were found, especially in the last samplings (Figure 1).

Figure 1. Total herbage biomass in grazed and non-grazed areas in spring and autumn 2001-2003. Notation: (ns) non significant; * ($P < 0.05$); ** ($P < 0.01$); *** ($P < 0.001$). deviation bars: standard error.

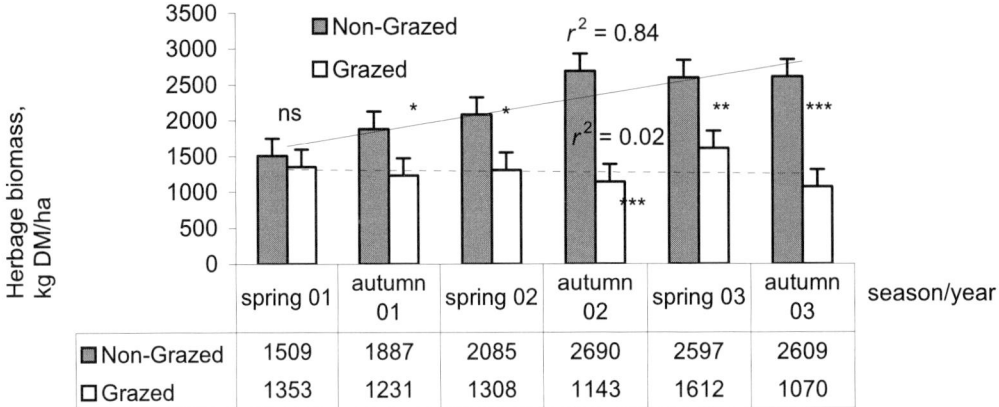

Concerning the green and dead herbage fractions, the total amount of green herbage remained constant in grazed and non-grazed areas, but dead biomass increased significantly in non-grazed areas, while it decreased in grazed areas (Figure 2).

Figure 2. Evolution of green and dead herbage biomass through the period of study. Notation: (ns) non-significant; * ($P < 0.05$); ** ($P < 0.01$); *** ($P < 0.001$). s = spring; a = autumn.

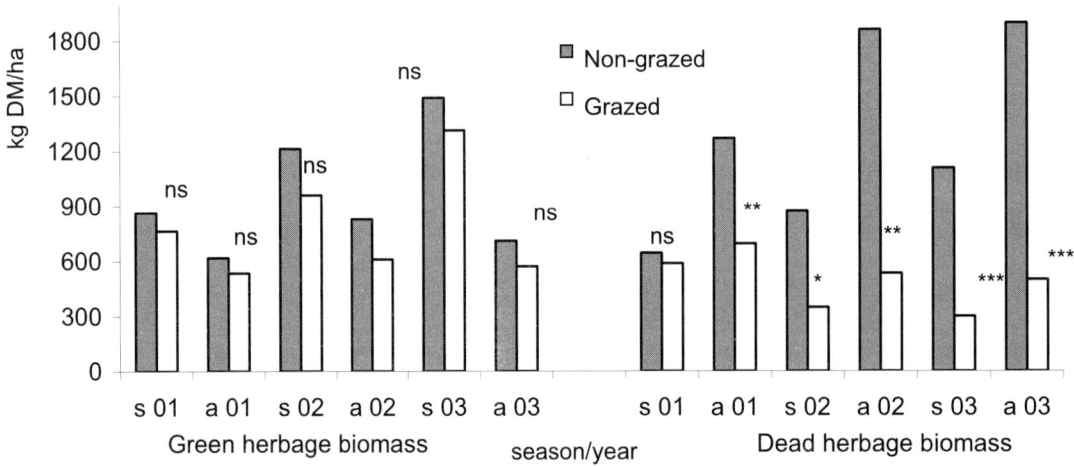

Along with the increase of dead material in non-grazed areas, herbage nutritional quality was significantly lower at the end of the study. Fibre content was always higher in non-grazed areas, both in terms of ADF (36.38 *vs* 31.52%, $P < 0.001$) and NDF (68.83 *vs* 65.05%, ns). Lignin content was also higher in non-grazed areas (5.60 *vs* 4.56%, $P < 0.01$). Crude protein content was higher in grazed areas (10.74 *vs* 7.13%, $P < 0.001$).

Total shrub biomass increased both in grazed and non-grazed areas, with significant differences between initial and final values ($P < 0.01$), although the increments were higher in absence of grazing (+80% *vs* +42% of increase relative to initial volume, $P < 0.01$). The magnitude of this effect varied in different shrub species and in areas with different stocking rates.

Conclusions

The results show how livestock grazing can control both herbage and shrub vegetation in Mediterranean agrosilvopastoral mountain areas. In the absence of grazing, both herbage and shrub biomass increased significantly, and forage quality was reduced. As a consequence, potential use by livestock in the future would be compromised and landscape degradation could occur. Hence, livestock grazing systems can be considered as tools to preserve landscape and environmental values, while contributing to the sustainability of economic activities in rural areas.

Acknowledgements

Research work funded by INIA (RTA 02-086-C2-2) and Gobierno de Aragón (CONSI+D P085/2000 and CTPI 03/2001). The first and fourth authors want to acknowledge the support of Gobierno de Aragón - Fondo Social Europeo (CONSI+D: B108/2003) and INIA, respectively.

References

AOAC (1995) *Official Methods of Analysis*, 16th edn. Association of Official Analytical Chemists, Arlington, USA.

Bernués, A., Olaizola, A., Casasús, I., Ammar, A., Flores, N. and Manrique, E. (2002) Livestock farming systems and conservation of Spanish Mediterranean mountain areas: the case of the 'Sierra de Guara Natural Park'. 1. Characterization of farming systems. In: *11th Meeting of the FAO-CIHEAM Sub-Network on Mediterranean Pastures and Fodder Crops. "Rangeland and Pastures Rehabilitation in Mediterranean Areas"*. FAO-CIHEAM, Djerba, Tunisia. (In press.)

Casasús, I., Bernués, A., Flores, N., Sanz, A., Valderrábano, J. and Revilla, R. (2002) Livestock farming systems and conservation of Spanish Mediterranean mountain areas: the case of the 'Sierra de Guara Natural Park'. 2. Effects of grazing on vegetation. In: *11th Meeting of the FAO-CIHEAM Sub-Network on Mediterranean Pastures and Fodder Crops. "Rangeland and Pastures Rehabilitation in Mediterranean Areas"*. FAO-CIHEAM, Djerba, Tunisia.

Goering, H.K. and Van Soest, P.J. (1970) *Forage Fibre Analysis*. Agricultural Handbook No. 379, USDA, Washington, DC, USA.

SAS (1990). *SAS/STAT User's Guide. Version 6*, 4th edn. SAS Institute Incorporated, North Carolina, Cary, USA.

Torrano, L. (2001) Use of encroached forest areas by goats and impact on the understorey vegetation (Utilización por el ganado caprino de espacios forestales invadidos por el matorral y su impacto sobre la vegatación del sotobosque). MSc thesis. University of Zaragoza, Spain. (In Spanish.)

Historical effects of grazing on tree establishment in the Cantabrian lowlands, northern Spain: a dendroecological analysis in two old-growth forests

V. Rozas

Dept. de Biología de Organismos y Sistemas, Universidad de Oviedo, Catedrático Rodrigo Uría, 33071 Oviedo, Spain. vrozas@correo.uniovi.es

Abstract

Tree regeneration patterns were reconstructed in two old-growth deciduous forests in the Cantabrian lowlands, northern Spain: the Caviedes forest and the Tragamón woodlands. Caviedes is dominated by beech (*Fagus sylvatica* L.) and pedunculate oak (*Quercus robur* L.), while Tragamón is constituted by pedunculate oak only. For several centuries, Caviedes was administered by the Spanish Royal Navy in order to produce oak wood for warship building. From the early 19th century to the early 20th, grazing intensity by domestic animals was intense. Beech exhibited continuous recruitment in periods of forest protection against grazing, while pedunculate oak recruitment was dependent on both low grazing intensities and disturbances that resulted in the expansion of canopy gaps. Tragamón was subjected to human management for centuries by pollarding trees and the understorey was grazed. Oak age distributions in Tragamón displayed three separate cohorts: < 50, 150–200 and 250–500 years. The uneven spatial distribution of these cohorts could be a consequence of the contrasting alternation of heavy and lighter grazing intensities in different woodland sectors. These results suggest that browsing-sensitive species, such as beech and pedunculate oak, will successfully regenerate in intensely grazed woodlands only if temporal and spatial variation in browsing pressure is allowed to occur.

Key words: tree rings, age distribution, *Fagus sylvatica*, forest-use history, *Quercus robur*

Introduction

Forest structure and composition are mainly a consequence of past tree recruitment patterns, which are controlled by several factors such as specific traits in the life cycle, the physical and biological characteristics of the environment, and chance events that change over space and time (Veblen, 1992). Information on the importance of these different factors on tree regeneration is critical for the restoration of natural forests in preserved areas, or for sustainable timber production. Age structures of trees provide information on the causes of successful regeneration. Regeneration patterns of trees in dense temperate forests are interpreted as a consequence of the frequency and magnitude of disturbances that killed dominant canopy trees.

The disturbance history of a forest can be reconstructed using dendroecological techniques (Nowacki and Abrams, 1997), and the coupling with land-use data provides an improved understanding of the processes that affect tree recruitment (Ruffner and Abrams, 1998). Some rare fragments of native forest persist in the Cantabrian lowlands in northern Spain, which are usually former silvopastoral systems. The objective of this study was to reconstruct tree age distributions and disturbance regimes in two old-growth woodlands, and compare them with forest-use data, especially grazing pressure, in order to infer those factors that affected tree recruitment.

Materials and Methods

Table 1. Environmental and structural characteristics of Caviedes and Tragamón woodlands.

	Caviedes area	Tragamón area
Mean precipitation (mm)	1210	977
Mean temperature (°C)	14.0	14.8
Mean altitude (m)	140	27
Total area sampled (ha)	1.35	4.03
Mean density (trees/ha)	1057	118
Mean basal area (m²/ha)	33.1	42.0

Two old-growth woodlands were studied: the Caviedes forest area located in western Cantabria, and the Tragamón woodland area found near Gijón, Asturias. Caviedes is dominated by beech (*Fagus sylvatica* L.) and pedunculate oak (*Quercus robur* L.), while Tragamón comprises only pedunculate oak. The environmental and structural characteristics of both study areas are shown in Table 1. In Caviedes, all living trees were cored for age estimation and tree-ring analysis. All those living oaks without external evidence of bole rottenness were cored in Tragamón. Tree age estimates were achieved according to methods previously described (Rozas, 2003). Canopy history was reconstructed through tree-ring width analyses. Growth releases were identified using the percentage growth change filter (Nowacki and Abrams, 1997), and the percentage of trees with suppressed growth was calculated. An indicator of grazing intensity is the volume of milk destined for industrial use (Domínguez and de la Puente, 1996). This indicator showed the historical evolution of Holsteins, which replaced the traditional races that extensively grazed in woodlands and forests.

Results and Discussion

Figure 1. Tree age distributions, percentage of trees with suppressed growth and growth releases in Caviedes (left), in Tragamón (right) and miles of litres (Ml) of industrial milk produced in Cantabria.

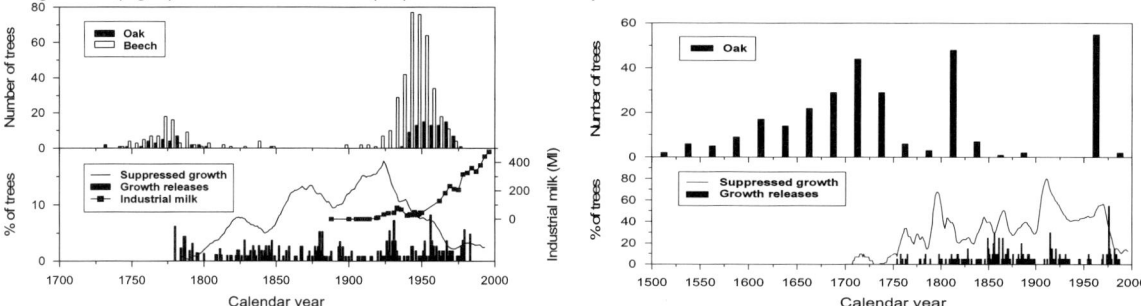

In Caviedes, beech exhibited continuous recruitment in periods of forest protection against grazing (1740–1830) or when a large amount of industrial milk was produced (1920–1970) (Figure 1). Oak recruitment was dependent on both low grazing intensities and disturbances that resulted in the expansion of canopy gaps (1943, 1956 and 1967). Tragamón was subjected to human management for centuries by pollarding trees and grazing of the understorey. Oak age distributions in Tragamón displayed three separate cohorts (< 50 years, 150–200 years and 250–500 years) regardless which established independent of forest history. The uneven spatial distribution of these cohorts could be a consequence of the contrasting alternation of heavy and lighter grazing intensities in different sectors of Tragamón.

Conclusions

The results obtained suggested that browsing-sensitive species, such as beech and pedunculate oak, will successfully regenerate in intensely grazed woodlands only if temporal and spatial variation in browsing pressure is allowed to occur. This conclusion agreed with those previously reported based on tree establishment data following a reduction of grazing intensity (Kuiters and Slim, 2002).

References

Domínguez, R. and de la Puente, L. (1996) History of a leadership: technical change and technology evolution in husbandry in Cantabria 1850-1950 (Historia de un liderazgo: cambio técnico y trayectorias de la tecnología en la ganadería de Cantabria, 1850-1950) In: Domínguez, R. (ed.) *La Vocación Ganadera del Norte de Spain: del Modelo Tradicional a los Desafíos del Mercado Mundial*. MAPA, Ministerio de Agricultura, Pesca y Alimentación, Madrid, Spain, pp. 89-146. (In Spanish.)
Kuiters, A.T. and Slim, P.A. (2002) Regeneration of mixed deciduous forest in a Dutch forest-heathland, following a reduction of ungulate densities. *Biological Conservation* 105, 65-74.
Nowacki, G.J. and Abrams, M.D. (1997) Radial-growth averaging criteria for reconstructing disturbance histories from presettlement-origin oaks. *Ecological Monographs* 67, 225-249.
Rozas, V. (2003) Tree age estimates in *Fagus sylvatica* and *Quercus robur*: testing previous and improved methods. *Plant Ecology* 167, 193-212.
Ruffner, C.M. and Abrams, M.D. (1998) Relating land-use history and climate to the dendroecology of a 326-year-old *Quercus prinus* talus slope forest. *Canadian Journal of Forest Research* 28, 347-358.
Veblen, T.T. (1992) Regeneration dynamics. In: Glenn-Lewin, D.C., Peet, R.K. and Veblen, T.T. (eds) *Plant Succession: Theory and Prediction*. Chapman and Hall, London, United Kingdom, pp. 152-187.

Changes in biodiversity after abandonment in dehesa systems in the province of León

R. Tárrega, L. Calvo, C. Diez, E. Luis, L. Valbuena and E. Marcos
Area de Ecología, Universidad de León, 24071 León, Spain. deglcg@unileon.es

Abstract
The aim of this work was to determine the changes of diversity of plant community in relation to silvopastoral use. For this purpose seven zones with a different livestock pressure were compared, the cover of all understorey species was estimated and the diversity per quadrat (alpha diversity), the global diversity of each zone (gamma diversity) and the spatial heterogeneity (beta diversity) were compared. The pasture zones show great diversity on a small scale, but they are far more homogeneous (less beta diversity). The zones dominated by woody species show less alpha and gamma diversity, but greater heterogeneity. Finally greater gamma diversity was observed in the zones of intermediate grazing pressure.

Key words: *Quercus pyrenaica*, shrubland, dehesas, changes in use

Introduction
The effect of herbivores on plant community diversity depends not only on grazing pressure but also on the type of community (Zamora *et al.*, 2001). Livestock in Mediterranean pastures and dehesas increases herbaceous diversity (Montalvo *et al.*, 1993; Sternberg *et al.*, 2000), whilst the effect seems to be the opposite in forest ecosystems (Meson and Montoya, 1993; González-Hernández and Silva-Pando, 1996). In the last few decades many of the traditional pasture systems have been gradually abandoned in Spain. This has caused the spread of shrub communities, which are considered marginal lands of low economic or ecological value and with little biodiversity. However, several authors have pointed out that the concept of biodiversity implies much more than total number of species (Magurran, 1989; Gordillo, 2002). Diversity also varies in space and this spatial dimension, although known, is not always taken into account in the studies on effects of change in silvopastoral systems use. Therefore the aim of this study is to determine the effect of grazing on the diversity of plants on different spatial scales, comparing ecosystems with a different animal stocking rate.

Materials and Methods
Seven zones with different livestock stocking rates in the province of León, close to each other and with similar environmental parameters, were investigated. They are at 950-1100 m asl, in a sub-humid Mediterranean climate, with cambisol soils.

- Three *Quercus pyrenaica* dehesas, used as communal pasturage for sheep and cattle. Two (Almanza dehesa = AD and Rueda dehesa = RD) were kept in traditional pasture use. In the third (Corcos dehesa = CD) livestock number was reduced and shrubs began to spread. Oak density was 25-50 trees/ha.
- Two oak (*Quercus pyrenaica*) forests. One (Vegaquemada oak forest = VF) was an extensive sheep pasture. The other (Corcos oak forest = CF) showed no use. Oak density was 1500-2500 trees/ha.
- Two shrublands (Rueda shrubland = RS and Corcos shrubland = CS), dominated by *Erica australis*, in old pasture zones abandoned at least 20 years before.

Therefore, there were two zones with high animal stocking rate (AD, RD), two with intermediate (CD, VF), two abandoned (RS, CS) and the oak forest (CF). The understorey vegetation was studied in all these zones. The sampling unit was a quadrat of 1 m^2 where in the understorey shrub species were abundant (CF, RS, CS) and a quadrat of 0.25 m^2 where herbaceous species were dominant (AD, RD, CD, VF). The cover percentage of the species present in each quadrat was estimated; the total cover values above 100% are caused by overlapped plant layers. Richness (number of species) was used as diversity index, since other indices which also consider evenness, like Shannon index, can be confusing if used to compare samples with only herbaceous species and those which show both herbaceous and woody species. The alpha diversity ($S\alpha$), or diversity on a small scale, was calculated as the mean of the number of species found per quadrat in each zone; gamma diversity ($S\gamma$) as the total species found in each zone; and beta diversity ($S\beta$), or spatial heterogeneity, by comparing the two previous ones using the Whittaker formula (Magurran, 1989): $S\beta = (S\gamma/ \text{mean } S\alpha) - 1$. Data from alpha diversity were compared by analysis of variance.

Results and Discussion
The diversity on a small scale (alpha diversity) had highest values in the dehesa zones, including that less used by livestock (CD). The Vegaquemada oak forest also had a similar value. However, low mean values were found in the two shrublands with the minimum in the Corcos oak forest (Figure 1). Statistically significant differences were found

between the three zones CF, RS, CS and the first four (AD, RD, CD, VF). Gamma diversity (total number of species recorded in each zone) had different trends. The highest diversity was in the zones with low animal stocking rate (CD, VF) and the lowest in the oak forest not used as pasture (CF). The two dehesas used by livestock (AD, RD) and one of the shrublands (CS) had the same value. Spatial heterogeneity (beta diversity) is maximum in CF (that of lowest alpha and gamma diversity) and minimum in the dehesas, above all in those with a greater livestock load. All the study zones (except VF) were continuous forest previously, under active management. When analysing diversity at landscape scale, a total of 91 species were recorded in the sampling carried out in these six zones, 38 appearing only in one. It is remarkable that in the dehesa systems only nine appear (six in RD, three in CD and none in AD), seven in Corcos oak forest and 22 in the shrublands (eight in RS and 14 in CS). Another 16 additional new species appeared in the Vegaquemada oak forest.

No woody species appeared in the understorey in grazed dehesas (AD, RD). Herbaceous species also dominated the two zones with less grazing pressure (CD, VF). The woody species clearly dominated in non-grazed zones, Corcos oak forest and the two shrublands (Figure 2). There were patches of bare soil on RS; the opposite occurred in CS, where woody cover was almost 150%, due to the presence of *Arctostaphylos uva-ursi* (a prostrate species) and taller species like *Erica australis* and *Calluna vulgaris*.

Figure 1. Alpha diversity (empty circles, mean values of species in each area, standard error), gamma (solid circles, total species number) and beta diversity (rhombus, space heterogeneity) in the study areas. (RD = Rueda dehesa, AD = Almanza dehesa, CD = Corcos dehesa, VF = Vegaquemada forest, CF = Corcos forest, RS = Rueda shrub, CS = Corcos shrub).

Figure 2. Cover by herbaceous and woody species in the study areas. (RD = Rueda dehesa, AD = Almanza dehesa, CD = Corcos dehesa, VF = Vegaquemada forest, CF = Corcos forest, RS = Rueda shrub, CS = Corcos shrub).

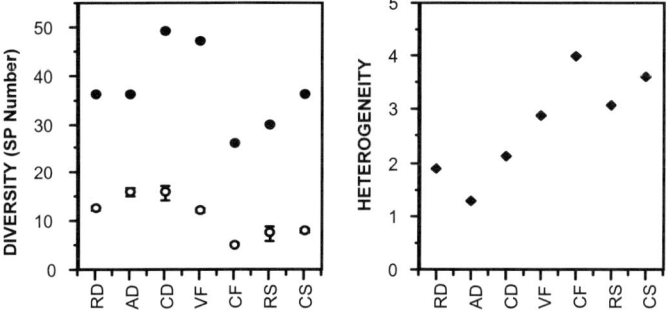
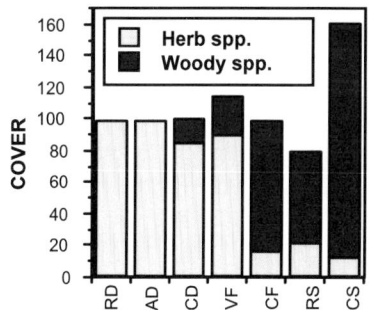

Table 1. Correlation analysis between values of the measured variables in the study areas. * ($P < 0.05$, $n = 7$); ** ($P < 0.01$, $n = 7$).

	Sα	Sγ	Sβ	Herb	Wood
Alpha diversity (Sα)	----				
Gamma diversity (Sγ)	0.73	----			
Beta diversity (Sβ)	0.91	0.42	----		
Herbaceous sp cover (Herb)	0.90	0.64	0.87	----	
Woody sp cover (Wood)	0.76*	0.40	0.82**	0.89**	----

There was a positive correlation between alpha diversity and herbaceous cover, and negative between beta diversity and woody cover (Table 1). Although the trend is the same for gamma diversity, the values were not significant. Beta diversity was positively correlated with woody cover and negatively with herb cover. The results showed that the grazed zones had great diversity on a small scale, but were much more homogeneous in their spatial structure. In contrast the greatest spatial heterogeneity was observed in the zones with the lowest alpha and gamma diversity, the two shrublands and Corcos oak forest, the last being the most mature and least disturbed community. Finally, the two zones with an intermediate grazing pressure were the most diverse overall. Both had the greatest gamma diversity, conditioned by the presence of small patches with pasture, which favours the herbs and maintains relatively high alpha diversity, and others with shrub dominance, which determines different microhabitats and therefore different species. It must also be considered, however, that the three communities with lower alpha and gamma diversity indices contribute more species (which only appear in one of the seven zones studied) to the landscape than the three dehesas, in spite of the greater diversity of the latter on a small scale.

Conclusions

The highest alpha diversity and lowest beta diversity appeared in the areas with the greatest grazing pressure, while the highest gamma diversity was in zones with an intermediate grazing pressure, and the highest spatial heterogeneity appeared when grazing was abandoned. It is concluded that greater diversity is attained on a regional scale by maintaining a mosaic with all the types of ecosystems considered, with different pressure on the pasture from stock, since neither the species appearing in them nor their structural diversity, dynamics or functioning is exactly the same.

References

González-Hernández, M.P. and Silva-Pando, F.J. (1996) Grazing effects of ungulates of a Galician oak forest (northwest Spain). *Forest Ecology and Management* 88, 65-70.

Gordillo, E. (2002) Methodology for biodiversity assessment within the National Forest Inventory (Metodología para la caracterización de la biodiversidad en el Inventario Forestal Nacional). In: Bravo, F., Del Río, M. and Del Peso, C. (eds) *El Inventario Forestal Nacional. Elemento Clave para la Gestión Forestal Sostenible*. Ministerio de Medio Ambiente (DGCN), Madrid, Spain, pp .37-55. (In Spanish.)

Magurran, A.E. (1989) *Ecological Diversity and its Measurement* (Diversidad ecológica y su medición). Vedrà, Barcelona, Spain. (In Spanish.)

Meson, M. and Montoya, M. (1993) *Mediterranean Silviculture* (Selvicultura Mediterranea). Mundi-Prensa, Madrid, Spain. (In Spanish.)

Montalvo, J., Casado, M.A., Levasor, C. and Pineda, F.D. (1993) Species diversity in Mediterranean grasslands. *Journal of Vegetation Science* 4, 213-222.

Sternberg, M., Gutman, M., Perevolotsky, A., Ungar, E.G. and Kigel, J. (2000) Vegetation response to grazing management in a Mediterranean herbaceous community: a functional group approach. *Journal of Applied Ecology* 37, 224-237.

Zamora, R., Gómez, J.M. and Hódar, J.A. (2001) Interaction between plants and animals in the Mediterranean area: relevance of the ecological context and the organization level (Las interacciones entre plantas y animales en el Mediterráneo: importancia del contexto ecológico y el nivel de organización). In: Zamora, R. and Pugnaire, F.I. (eds) *Ecosistemas Mediterráneos. Análisis Funcional*. CSIC and AEET, Madrid, Spain, pp. 237-268. (In Spanish.)

Session 4.
Economic, social and cultural benefits of the silvopastoral systems

Economic considerations of silvopastoralism in California oak woodlands

R. B. Standiford [1], L. Huntsinger [1], P. Campos-Palacín [2] and A. Caparrós [2]

[1]College of Natural Resources, University of California, 145 Mulford Hall, MC 3114, Berkeley, CA, 94720-3114, USA. standifo@nature.berkeley.edu; buckaroo@nature.berkeley.edu [2]Institute of Economics and Geography (IEG), Spanish Council for Scientific Research (CSIC), Pinar 25, 28006, Madrid, Spain. pcampos@ieg.csic.es, acaparros@ieg.csic.es

Abstract

Oak woodlands in California cover 4 million hectares and are over 80% privately owned. Silvopastoralism is the dominant land use, providing a working landscape with the highest biodiversity levels in the state. However, high opportunity costs from alternative land uses threaten to fragment these areas. Despite these economic pressures, landowners make management decisions reflecting their utility for environmental services. Positive mathematical programming, an optimization technique that constrains solutions with actual producer behaviour, is used to determine these values. These derived environmental service values are incorporated into a dynamic optimal control model illustrating the interaction of livestock grazing, hunt clubs and firewood harvest for different risk and land productivity scenarios. The value of silvopastoral management of oak woodlands is further demonstrated with contingent valuation and hedonic pricing. Individual parcels with oak cover have higher value than bare land. Oak woodland open space adds value to individual parcel owners and the entire community. These results are discussed in the context of conservation policies for oak woodlands.

Key words: optimal control, working landscapes, resource economics

Introduction

California has approximately 4 million hectares of oak woodlands, which are the most biologically diverse broad habitat in the state (Standiford and Tinnin, 1996). Most of the state's water flows through these lands and they supply both aesthetics and recreational values. These public values are mainly supplied by private landowners, who own over 80% of the state's oak woodlands (Standiford and Tinnin, 1996). Over two-thirds of all oak woodlands are grazed by domestic livestock and managed as silvopastoral enterprises (Huntsinger et al., 1997).

The continued supply of public values from these private lands depends on the value of silvopastoral enterprises and the opportunity costs of competing land uses, such as urban developments, intensive agricultural enterprises and rural subdivisions. Economic institutions such as conservation easements and property tax policies provide incentives to private owners to supply public values. Broadened markets for oak woodlands products, including fee hunting, recreational leasing and mitigation banking, increase returns and help maintain extensively managed silvopastoral working landscapes. Economic quantification of the ecological services from oak woodlands demonstrates the importance of conservation policies.

Landowner investment in environmental value

Models of likely silvopastoral management decisions must incorporate landowners' utility for environmental services produced on their lands. Poorly specified production models understate a manager's self-consumption of amenity and environmental services, and lead to erroneous conclusions about likely management strategies and appropriate public polices.

Figure 1. Net firewood return per cubic metre as function of amount of wood harvested (Standiford and Howitt, 1992).

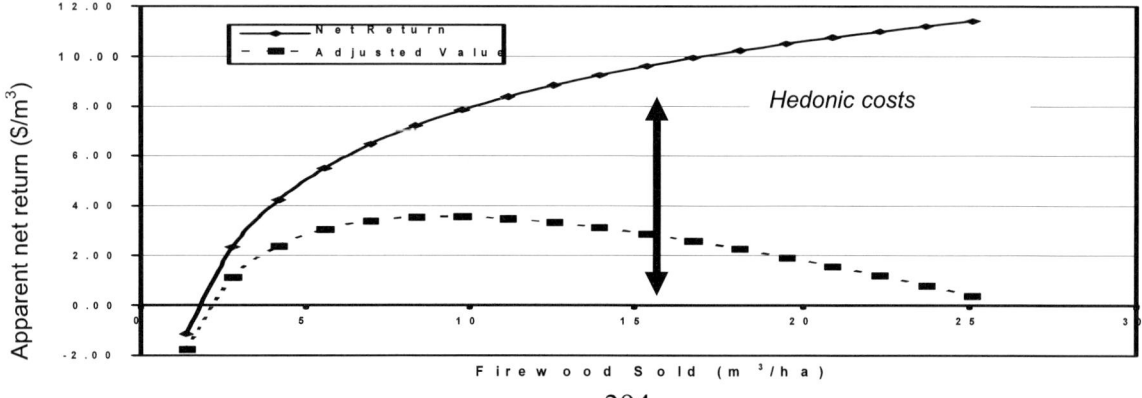

A normative dynamic oak woodland optimization model including cattle, firewood and hunting concluded that markets at that time would lead to oak clearing to increase forage yield for livestock production (Standiford and Howitt, 1992). Although common in the 1940s to 1970s, this behaviour was rare in recent years (Standiford *et al.*, 1996). The model shortcomings were due to failure to accurately account for a landowner's utility from retaining oaks for their amenity value. A positive mathematical programming (PMP) approach (Howitt, 1995) was used to derive missing elements of the true costs and returns of oak harvest omitted from the normative model. The dynamic optimization model was constrained by actual landowner behaviour to derive these missing values. The shadow prices from the behaviour constraint represent the marginal benefit of retaining trees from what might otherwise be predicted from an engineering approach to harvest values.

Figure 1 shows the firewood revenue model developed from market information, and the hedonic pricing model calibrated from the actual behaviour of oak woodland owners. The difference between the two curves represents the environmental self-consumption value of retaining trees. Figure 2 shows how this specification, which incorporates actual landowner behaviour, gives a more realistic assessment of actual landowner behaviour than a model which omits the value a landowner places on tree retention (Standiford and Howitt, 1992).

Figure 2. Oak volume levels in California oak woodlands under normative and positive modelling approaches (Standiford and Howitt, 1993).

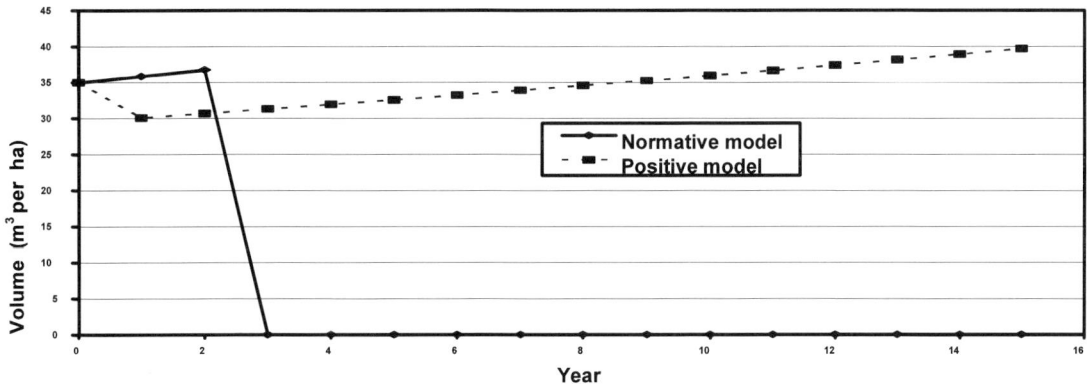

Commercial values of silvopastoral systems

This optimization model, incorporating landowner utility, is used to evaluate oak cover, firewood harvest and cattle grazing under different risk and land productivity conditions (Standiford and Howitt, 1992). Figure 3 shows the contribution of the three major commercial enterprises to total net present value of California oak woodlands with an initial oak volume of 50 m^3/ha (Standiford and Howitt, 1993). Cow-calf enterprises on average have a positive economic value. Fee hunting can be an important enterprise, contributing from 40% (on good range sites) to 70% (on poor range sites) of the total silvopastoral value. The economic contribution of wood harvest is low.

Figure 3. Net present value of California oak woodlands from various commercial enterprises (Standiford and Howitt, 1993).

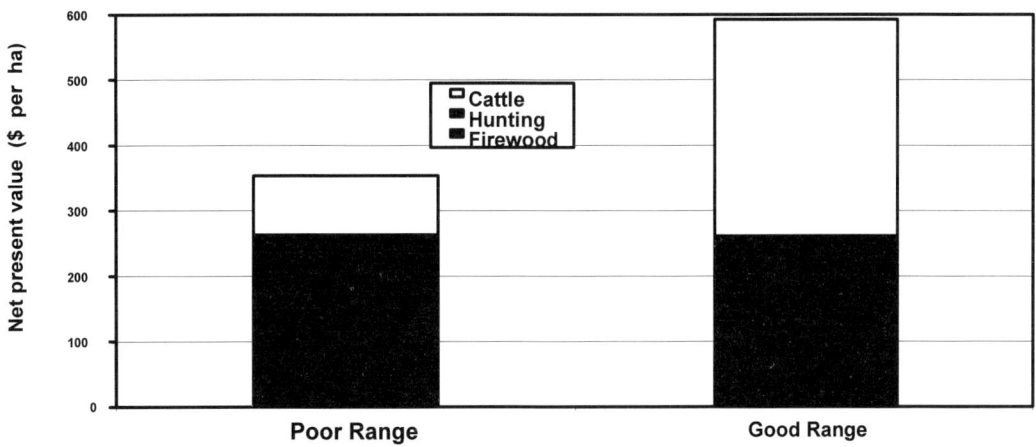

The model showed that diversification of silvopastoral enterprises reduced tree harvesting and cattle grazing. The marginal value of retaining oaks for wildlife habitat for hunt clubs exceeded the marginal value of the extra forage or firewood harvest (Standiford and Howitt, 1992). Wood harvest is used in years with poor forage production or low livestock prices. The capital value of the trees is a hedge against years with low livestock profitability. Inclusion of a risk term shows that firewood harvest and livestock grazing intensity both increase. Policies reducing landowners' risk, such as a subsidized loan programme during poor forage production or low livestock price years, might reduce the need to liquidate oak tree capital assets.

Opportunity costs of oak woodlands

In many areas of California, the commercial values from silvopastoral management represent only a small fraction of the actual land value. Alternative land uses such as intensively managed agricultural products or subdivision for residential housing usually have much higher market values than extensively managed oak woodlands. Many of these higher value land uses, unless carefully planned, convert and fragment oak woodland habitats, diminishing their capacity to supply public amenity values (Merenlender *et al.*, 1998).

These alternative land uses create an opportunity cost for owners. For example, in the central coast of California, grazing land value may be worth less than 10% of the value of the land for intensive agricultural use for wine grapes (Figure 4), or less than 1% of its value for residential uses (CALASFMRA, 2001), creating tremendous pressure to move to land uses that may cause higher environmental costs.

The California Land Conservation Act (CLCA), also known as the Williamson Act, is one attempt to reduce the conversion pressure by basing annual property tax on current land use, rather than its "highest and best use" (Carman, 1977). This policy requires landowners to maintain their extensive agricultural use for 10 years.

Estate taxes of oak woodland parcels are determined by their "highest and best use" derived through the land market. High estate taxes, driven by these opportunity costs, have been identified as one of the largest constraints to inter-generational transfers of large, extensively managed oak woodland parcels (Johnson, 1997). USA estate tax reform is being considered to reduce conversion pressures on agricultural lands, including oak woodlands.

Figure 4. Typical value of central coast California oak woodlands for different land uses (CALASFMRA, 2001).

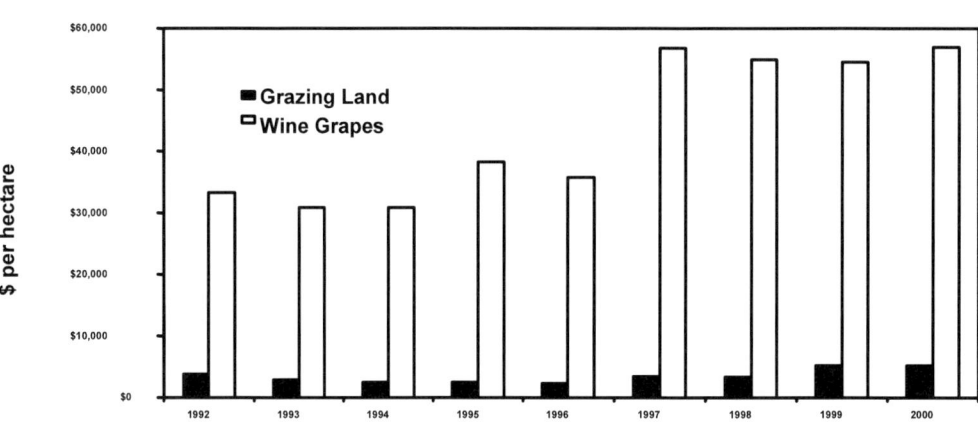

Environmental service values from oak woodlands

One of the reasons for the migration of Californians from urban areas to oak woodlands is because of their amenity values. Land markets for oak woodlands include amenity values. The oaks on the property, the presence of oaks in a surrounding neighbourhood and the presence of oak woodland in open space adjacent to a property all affect property values.

Contingent valuation was done on different spatial arrangements of oak stands to evaluate how oak cover affects property value (Diamond *et al.*, 1987). On 2 hectare lots, rangeland with at least 100 oaks per hectare was worth 27% more than open land. There was a similar value for open to heavy tree stocking (100 to 1140 trees/ha) on these 2 hectare lots. Similar trends were also observed on 0.8 hectare lots, with 100 trees/ha being worth 22% more than bare land. Denser areas (over 100 trees/ha) were not worth as much as the more open stands, but still had higher value than bare land.

The effect of a 3400 hectare oak woodland open space in southern California on overall community land and home value was evaluated using hedonic pricing. A decrease of 10% in the distance to the nearest oak stands and to the edge of the permanent open space land resulted in an increase of $4 million in the total home value and an increase of $16 million in total land value in the community (Standiford and Scott, 2001). Evaluation of over 3000 individual home and land parcel values showed the effect of adjacent oak stands and open space land. The average home immediately adjacent to a native oak stand was 12% more valuable than a home located 0.25 km from an oak stand. Land prices for lots immediately adjacent to the oak woodland open space is valued 17% higher than the same land characteristics set

0.25 km from the edge of the open space area. Private owners receive a premium by being located adjacent to land that will remain as dedicated open space.

Conservation of oak woodland open space increases overall land and home value of an entire community. The overall assessed property value of a community is higher because of the value added by these environmental assets. The resulting increases in annual property tax accruing to local government can be used to justify public financing of local oak restoration efforts, or the purchase of development rights for permanent open space or extensively managed working landscapes.

Conservation easements and land trusts

With the large private ownership of California's oak woodlands, and high opportunity costs of maintaining silvopastoral working landscapes, new approaches are needed to conserve these lands. Currently, one of the largest sources of funding for oak woodland conservation comes through a diverse set of institutions known as "the Land Trust Movement". Land trusts are organizations that act directly to conserve land. These vary in scale from localized groups, operating with volunteer staffs and little to no direct budget, to regional groups with staffs and some funding, to large international groups, such as the Nature Conservancy. In California, there are 132 land trusts, conserving over 400,000 ha of land (LTA, 2002).

Land trusts purchase directly or accept donations of conservation easements. Conservation easements are contracts that divide the bundle of rights involved in land ownership (development rights, grazing rights, mineral rights, water rights, etc.) between the landowner and the holder of the easement, in this case a land trust. The conservation easement creates a permanently deeded restriction on the limits and kinds of development for a property. For example, the urban development rights for an oak woodland property may be sold or donated to a land trust. The development rights are held in perpetuity by the land trust. The landowner receives benefits from the capital value of the rights donated or sold, and society benefits by the maintenance of the ecological value of the land. The area can continue to be used for silvopastoralism.

Funding for conservation easement purchases comes from private sources, such as foundations, as well as from public sources. Considerable oak woodland area has been placed into conservation easements in Sonoma County in California, funded by a local sales tax surcharge for the county (Mackenzie and Merenlender, 2000). In the Northern Sierra, the Nature Conservancy, working with a state organization called the Rangeland Trust, has acquired conservation easements on blue oak woodlands using private foundation funding sources (Reiner et al., 2002). An oak woodland conservation easement to preserve unique habitat for several threatened and endangered species was acquired in the San Francisco Bay Area, funded by fees provided by a private developer as mitigation for habitat being lost elsewhere as part of a development project. Landowners were compensated for having their oak woodland serve as a mitigation bank for unique habitats.

Another type of conservation easement transaction involves donations of the easement to a land trust. The market value of the portion of the property rights donated represents a reduction in the land's basis. This can be considered a charitable donation, reducing the landowner's taxable income. Lowering the land basis also reduces the inheritance tax as the land passes from generation to generation. This reduces the need to liquidate some of the opportunity costs of the land in order to pay inheritance taxes.

Conclusions

Silvopastoral management of California's oak woodlands provides a diverse array of products. These economic enterprises serve as working, privately managed landscapes, supplying high environmental service values to Californians. Much of the production of these environmental values comes from landowners' utility for these values. These amenity and environmental values of oak woodlands create incentives for landowners to maintain the health and vigour of trees.

Owners have motivation to maintain oak stands in areas that may be developed, because of the higher value for these lots. Also, since forested neighbourhoods have higher value, it may be wise for homeowner associations to utilize covenants, codes and restrictions (CCRs) to maintain overall oak stands in a neighbourhood. The effect of extensively managed open space on enhancing adjacent property values points to the role of compensation of large ownerships through land trusts, because of the economic as well as the conservation value of these types of lands. Silvopastoral management strategies can offer cost-effective means to provide environmental services to the adjacent community.

Many oak woodland ecological values can be correlated with economic value as these resources become increasingly scarce. Outright compensation of landowners through the purchase of development rights or tax and estate planning benefits through donation of the land value differences provide additional economic incentives for landowners to maintain the conservation value of their lands. However, the opportunity cost of conversion to higher value land use is driving a decrease in large expanses of oak woodland open space.

Silvopastoral management strategies for oak woodlands in California provide important conservation values to the public. However, these values have been undervalued by traditional agricultural production models. New approaches to evaluating the self-consumption of environmental services and the quantification of the real utility of amenity values observed in the market offer promising approaches to better represent their value to landowners and society. These tools can be used to evaluate new conservation policies for California's oak woodland resources.

References

CALASFMRA (American Society of Farm Managers and Rural Appraisers: California Chapter) (2001) Trends in agricultural land and lease values-California land prices: Central Coast. www.calasfmra.com/landvalues/2002 (accessed April 2004).

Carman, H.F. (1977) California landowner's adoption of a use-value assessment program. *Land Economics* 53, 275-287.

Diamond, N.K., Standiford, R.B., Passof, P.C. and LeBlanc, J. (1987) Oak trees have varied effect on land values. *California Agriculture* 41 (9-10), 4-6.

Howitt, R. (1995) Positive mathematical programming. *American Journal of Agricultural Economics* 77, 329-342.

Huntsinger, L., Buttoloph, L. and Hopkinson, P. (1997) Ownership and management changes on California hardwood rangelands: 1985 to 1992. *Journal of Range Management* 50 (4), 423-430.

Johnson, S.G. (1997) Factors contributing to land-use change in the hardwood rangelands of two central Sierra Nevadan counties. In: Pillsbury, N.H., Vecner, J. and Tietje, W.D. (eds) *Proceedings of a Symposium on Oak Woodlands: Ecology, Management, and Urban Interface Issues*. USDA Forest Service General Technical Report PSW-GTR-160, San Luis Obispo, USA, pp. 593-602.

LTA (Land Trust Alliance) (2002) *The Land Trust Census*. LTA, Washington, DC, USA.

Mackenzie, A. and Merenlender, A. (2000) Sonoma County Acquisition Plan 2000: a tool for conserving oak woodlands. *UC-IHRMP Oaks 'n Folks* 16, 2.

Merenlender, A.M., Heise, K.L. and Brooks, C. (1998) Effects of sub-dividing private property on biodiversity in California's north coast oak woodlands. *Transactions of the Wildlife Society* 34, 9-20.

Reiner, R., Underwood, E. and Niles, J.O. (2002) Monitoring conservation success in a large oak woodland landscape. In: Standiford, R.B., McCreary, D. and Purcell, K.L. (eds) *Proceedings of the Fifth Symposium on Oak Woodland: Oaks in California's Changing Landscape*. USDA Forest Service General Technical Report PSW-GTR-184, San Diego, USA, pp. 639-650.

Standiford, R.B. and Howitt, R.E. (1992) Solving empirical bioeconomic models: a rangeland management application. *American Journal of Agricultural Economics* 74, 421-433.

Standiford, R.B. and Howitt, R.E. (1993) Multiple use management of California's hardwood rangelands. *Journal of Range Management* 46, 176-181.

Standiford, R.B. and Scott, T.A. (2001) Value of oak woodlands and open space on private property values in Southern California. *Investigación Agraria: Sistemas y Recursos Forestales* 1, 137-152.

Standiford, R.B. and Tinnin, P. (1996) *Guidelines for Managing California's Hardwood Rangelands*. Leaflet no. 3368, University of California, Berkeley, USA.

Standiford, R.B., McCreary, D., Gaertner, S. and Forero, L. (1996) Impact of firewood harvesting on hardwood rangelands varies with region. *California Agriculture* 50 (2), 7-12.

Silvopastoral management in temperate and Mediterranean areas. Stakes, practices and socio-economic constraints

M. Etienne

INRA, Unité d'Ecodéveloppement, Site Agroparc, 84914 Avignon, Cedex 9, France. Etienne@avignon.inra.fr

Abstract

The low productivity of many temperate and Mediterranean forests led forest managers to imagine original management plans based on multiple use and diversified practices aiming at summoning up others resources than timber. Among these, the most commonly promoted activities are grazing, professional hunting, landscape or biodiversity conservation, and the prevention of wildfires. To guarantee forest multifunctionality special attention is given to flexible silvopastoral management techniques adapted to the potential multiple uses and the ecological dynamics of the forest. To reduce the threat of fire, silvopastoral management is focused on stimulating dry forage intake and shrub browsing, and is adapted to the structure and spatial organization of fire prevention management plans. To enhance biodiversity silvopastoral management is shifted towards more diversified forest stand structures, and organizing grazing calendars adapted to endangered species' life cycle or to seasonal high grazing pressure on specific ecological targets. These activities are carried out by different stakeholders whose practices can be complementary but sometimes raise up conflicts between the official forest managers and the local societies. The chapter points out the ways silvopastoral management is adapted to such a diversification of woodland productions and to the enhancement of the multifunctionality of peculiar forest ecosystems. A few examples borrowed from several countries under Mediterranean and temperate climates, illustrate how grazing, nature conservation or natural hazard prevention can be integrated into the common management of forests in a sustainable way.

Key words: silvopastoral system, grazing, nature conservation, fire prevention, hunting, multifunctionality

Introduction

Traditionally, Mediterranean and temperate forests have provided a large variety of other products besides wood (Seigue, 1985). Some of them directly come from the trees such as gums and resins extracted from pines, tannins from eucalypti and acacias, cork from *Quercus suber*, and fruits for feeding human beings or animals such as chestnuts from *Castanea sativa*, carob-beans from *Ceratonia siliqua*, pinions from *Pinus pinea* or acorns from many oak species. Other by-products are indirectly linked with the presence of trees such as the mushrooms, among which the truffles are the most appreciated ones, or specific honeys. At the same time, the undergrowth of the forest can provide large food supply for domestic animals or wild game, a wide variety of aromatic and medicinal plants, and some interesting wild food such as berries or asparagus. According to the type of product they are looking for, local societies build up a wide set of management techniques that permit them to enhance some of these functions.

The goal of the vegetation management was to build up and maintain the most appropriate vegetation structure according to the selected main target. If we simplify the description of vegetation structure by means of the combination of the three basic plant types (trees, shrubs and grasses), it is easy to visualize which structure is purchased by each type of operator (Figure 1). Foresters will look for tree stands for timber production or mixed stands for biodiversity enhancement. Livestock farmers will prefer to combine swards with wooded pastures. Meanwhile bee-keepers will look for dense and open shrublands dominated by melliferous plants. Hikers will follow different trails according to season and their impact on blooming or mushrooms springing up. Hunters will move from wooded pastures to close or multi-layered forests according to the type of game they are used to shooting.

A review of silvopastoral management in Mediterranean and temperate forests is proposed on three non-conventional aspects: tree and livestock management, fire prevention and biodiversity enhancement. It is based on examples borrowed from several countries of the Old Continent, and leads to recommendations on how grazing, nature conservation or natural hazard prevention can be integrated into the common management of forests in a sustainable way.

Figure 1. Multistructure and multifunctions of Mediterranean forests according to multiple use. The level of filling of the box is proportional to the area concerned by the activity.

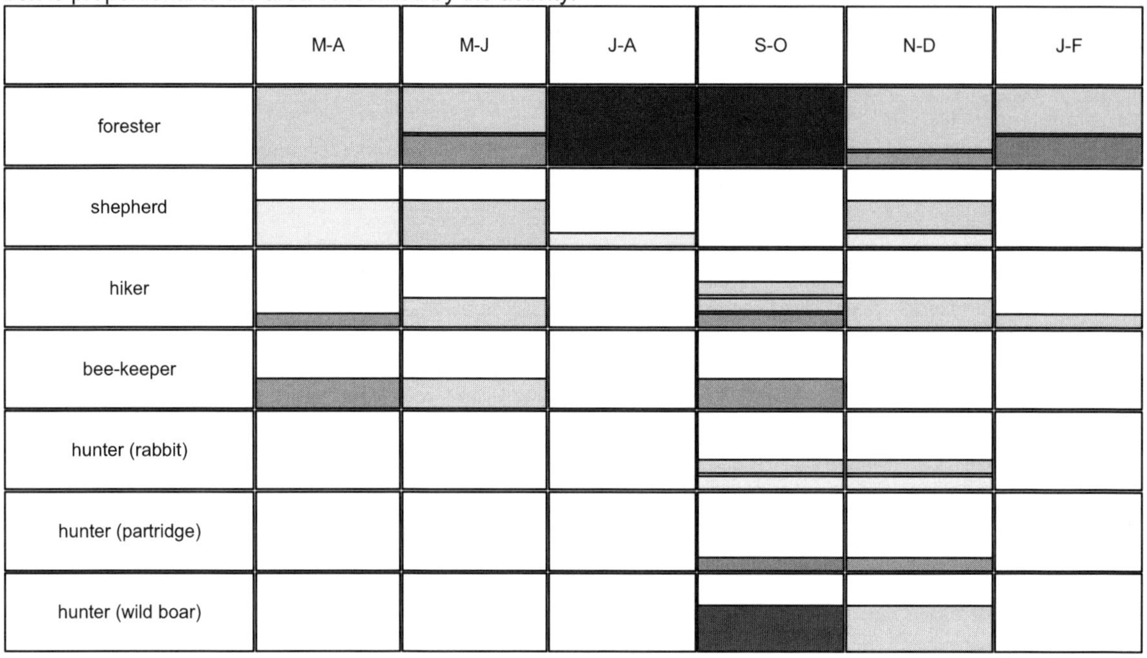

	M-A	M-J	J-A	S-O	N-D	J-F
forester						
shepherd						
hiker						
bee-keeper						
hunter (rabbit)						
hunter (partridge)						
hunter (wild boar)						

Silvopastoral management

Forest grazing has been for a long time a normal activity in Mediterranean forests and under specific temperate conditions but dramatic socio-economic changes in the northern and southern parts of the Mediterranean basin and in temperate mountains or marginal lands during the 20th century led to three contrasting situations. In the southern countries, with rapid population growth, precarious resources and a low standard of living, people's dependence on forest resources has continued and even increased (M'Hirit, 1999). In the north, the transformation of agriculture, together with industrialization and economic growth, has intensified the rural exodus and the abandonment of traditional farming in the hinterland and in the mountains, thus allowing forests to overspread widely (marginal areas of France, Italy and Spain, mountains of Switzerland or UK) or intensive agricultural practices to penetrate into the woodlands (cultivated "dehesas" or "montados" of Spain and Portugal).

In the case of southern countries, silvopastoral management can be assimilated to a mining exploitation of the fodder and firewood resources of the forest. In such a context, silvopastoral planning is necessary to safeguard forest ecosystems but only drastic regulations and a strict supervision of the local farmers' practices can reverse the degradation process (Karmouni, 1997). Presently, in some state forests of Morocco or in the natural reserves of Israel this strategy led to a complete extinction of the traditional silvopastoral systems. In another direction many northern African countries like Tunisia or Algeria did a lot of research on fodder trees and shrubs, mainly based on local species (Tiedeman and Johnson, 1992; Naggar, 1993). But the efforts made to convince the local livestock farmers to establish fodder trees or shrubs in their rangelands, in order to decrease the browsing pressure on woodlands, failed because of the cost, the need of supplemental irrigation during establishment and the need of a temporary exclusion from grazing.

Only highly resilient ecosystems such as *Argania spinosa* woodlands in Morocco (de Ponteves *et al.*, 1990) or *Quercus ilex* coppices in northern Africa (Benabid, 1978; Auclair *et al.*, 1995; Bonin and Loisel, 1995) are able to resist such a high harvesting pressure. Indeed, the small available literature on Mediterranean forest understorey productivity and use provided extremely high output figures compared with the average annual growth (Bourbouze, 1980; Qarro and De Montard, 1989; Tibaoui and Zouaghi, 1991; Madani *et al.*, 2001). To reverse this trend, only drastic socio-economic and political decisions will succeed in shifting the problem of forest vegetation degradation from a curative to a preventive approach in resource use, but this type of decision implies a clear increase of institutional efficiency (di Castri, 1998).

In the case of cultivated woodlands, the structure of the silvopastoral system has been partly conserved but the drastic thinning of the trees and the development of ploughing led to irreparable damage until good sense prevailed again and a compromise was found between traditional management and modern marketing (Gómez-Guttierez and Pérez-Fernández, 1996). Controlled density and pruning treatments prolong the life duration of the oak trees. Density is determined by the need for space for better production of pastures or cereal cropping but also by soil protection requirements. Traditional pruning stimulates acorn production through different cycles: normal pruning every 10 years in order to increase solar radiation at the sward level, lopping every 20 years to form a fruit-producer and harvest some firewood (Carbonero *et al.*, 2003).

The two major tree species used to build up these sustainable agrosilvopastoral systems in Spain and Portugal are *Quercus ilex* and *Q. suber*. The management of the undergrowth consists in a periodic clearing by ploughing and seeding of a cereal for 1 or 2 years and the establishment of a temporary pasture either through natural succession (San Miguel, 1994) or through the sowing of annual legumes (INIA, 1994). Economic studies showed strong differences between commercial net margins according to the type of dominant tree in the silvopastoral system, and the ability of the farmer to combine forage, crop and even hunting productions (Campos *et al.*, 2001).

In the case of abandoned landscapes, traditional practices completely disappeared and are presently considered as an archaic management system. But new social and economic priorities gave a new chance to silvopastoral management as they are deeply concerned with rural employment, landscape management and environment and product quality. This new trend appealed for defining original management rules considering the silvopastoral system as a driving force of a rural area maintenance project (Bland and Auclair, 1996). Some studies compared management techniques of the forest undergrowth, tree canopy or grass layer (San Miguel, 1985; Etienne, 2000) or proposed a modelling approach of the functioning of silvopastoral systems at different scales (Berger *et al.*, 1999; Etienne and Rapey, 1999; Etienne, 2001; Gillet *et al.*, 2002). Other works check the possibilities to combine conventional and silvopastoral feeding resources to build up fodder systems capable of fitting with contrasting livestock systems (Bellon and Guérin, 1992; Talamucci *et al.*, 1996).

Among the most common management techniques, many experiments have been set up on the impact of thinning on forage productivity of the stand. They aimed at defining an optimum tree density in order to find an equilibrium between tree canopy cover and grass cover and quality. This optimum depends on the type of tree and on the regional climate. The type of tree intervenes through phenology, foliage density, growth rate, nitrogen fixation and resistance to grazing. Figure 2 shows how contrasting the canopy cover optimum could be according to the tree species. In Morocco, dry conditions are partly balanced by *Quercus ilex* canopy but the dense foliage of this oak rapidly becomes a limiting factor for light and, over 30% of canopy cover, forage productivity decreases (Qarro *et al.*, 1995).

In France, the shifting between grass growth and tree budding of *Quercus pubescens* generates an intermediate situation with a positive effect of the tree until 60% of canopy cover (Msika, 1993). But, when the tree canopy cover is dominated by *Pinus sylvestris*, only strong thinning reducing stem density from 3000 to 240 or 120/ha permits a sustainable forage production (Dorée et al., 2003).

Another set of experiments dealt with the improvement of the grazing resources by replacing the unpalatable undergrowth with selected forage species. This pastoral improvement may be achieved by a simple fertilization or by introducing legumes such as *Trifolium subterraneum* by overseeding (González and Allué, 1982; Armand and Etienne, 1996) or by sowing (De Zulueta, 1972; Pardini *et al.*, 1987).

Other studies try to improve management techniques in order to stimulate forage production from the trees. Many authors developed thinning strategies that permit the enhancement of the production of sprouts from oak stumps or roots. San Miguel et al. (1983) got a forage productivity from sprout leaves of *Quercus pyrenaica* and *Q. faginea* ranging from 140 to 220 kg DM/ha with a tree density around 750 stems/ha but he reserved this type of thinning to relatively flat soils in order to avoid erosion and top-drying. Msika (1993) got similar results after thinning *Quercus pubescens* stands to the same density but he observed a dramatic decay of this production after 3 years of grazing. Ducrey and Boisserie (1992) found that *Quercus ilex* produced a quantity of stump sprouts proportional to the thinning intensity but they pointed out that the amount of available forage (green leaves) depended also on the age of the stool. A similar positive effect was mentioned on acorns by increasing the average production from 9 to 30 kg DM/ha with *Quercus faginea* (San Miguel *et al.*, 1983) and from 90 to 220 kg DM/ha with *Quercus pubescens* (Msika, 1993).

Figure 2. Impact of tree canopy cover on grass production for three Mediterranean species.

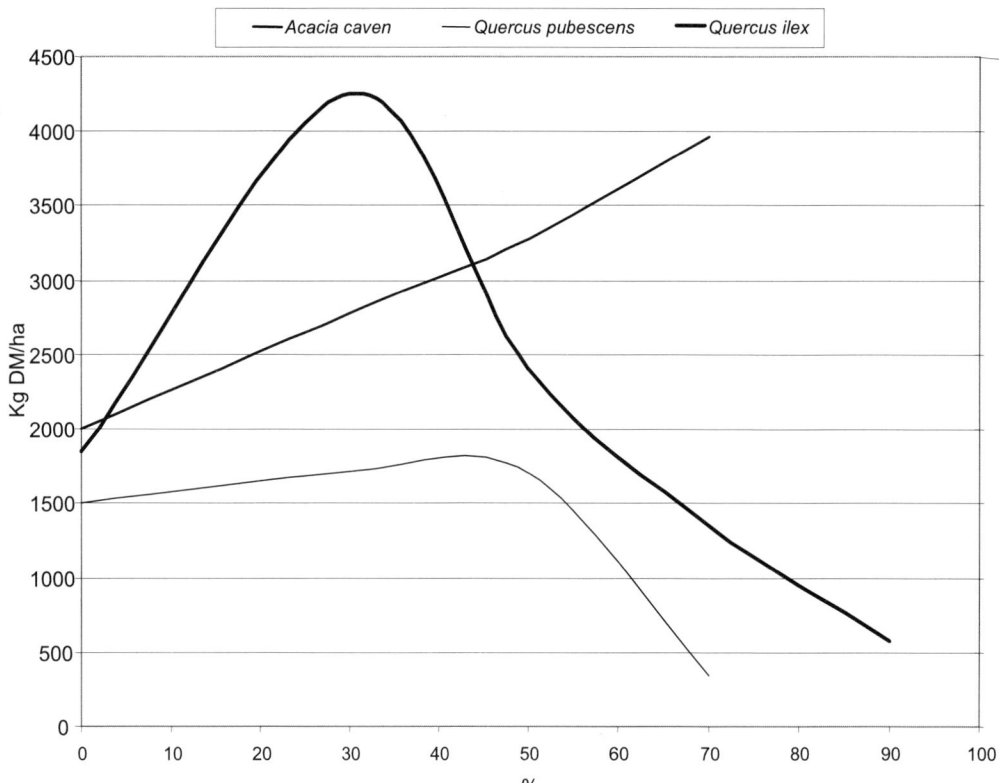

A last group of experiments dealt with widely spaced tree planting in swards in order to reduce livestock production while enhancing quality timber availability and maintaining rural employment and infrastructure (Sibbald, 1996). In order to avoid grazing exclusion during the establishment period, trees are individually protected with tree shelters, plastic net guards or more rarely by a net fence and a spiral guard. In the older experiments, 15 years after planting, there is no significant reduction in carrying capacity and slight differences in tree height between agroforestry plots and control (Sibbald and Dalziel, 2002).

In any case, it is crucial to design systems with silvopastoral plots that fit perfectly into the forage calendar of the livestock farms, by filling gaps in the grazing system of the farm. According to the forage value of the tree leaves and fruits, or the type of interaction between the tree and the grass, the silvopastoral plots can play very diverse roles in the farming system. Silvopastoral plots with *Fraxinus angustifolia* can be reserved for hay in spring and then the sward regrowth can be grazed in summer; meanwhile the leaves of pruned trees can be grazed during autumn (Figure 3).

As another example, *Quercus ilex* silvopastoral plots can be inserted in very different ways into a forage calendar either to feed animals during a critical period or to give more flexibility to the grazing system (Mesón and Montoya, 1987; Bellon and Guérin, 1996).

The economic analysis of silvopastoral systems provided strong variations in annual incomes due both to the impact of climatic variability and to differences in management. For example, in "dehesa" systems, traditional farms get a total income of around 130 €/ha with 100 coming from the animal production and 30 from the trees; meanwhile farms that move to crop intensification get a total income of around 90 €/ha with 80 coming from the animal production and 10 from the crops (Campos and Sesmero, 1987). But the possibility to get incomes from a variety of products and at different periods of time gives flexibility to the system and requires long-term economic assessments (Campos, 1984).

Figure 3. Silvopastoral plots fitting into Mediterranean forage calendars the case of ash and oak "dehesas".

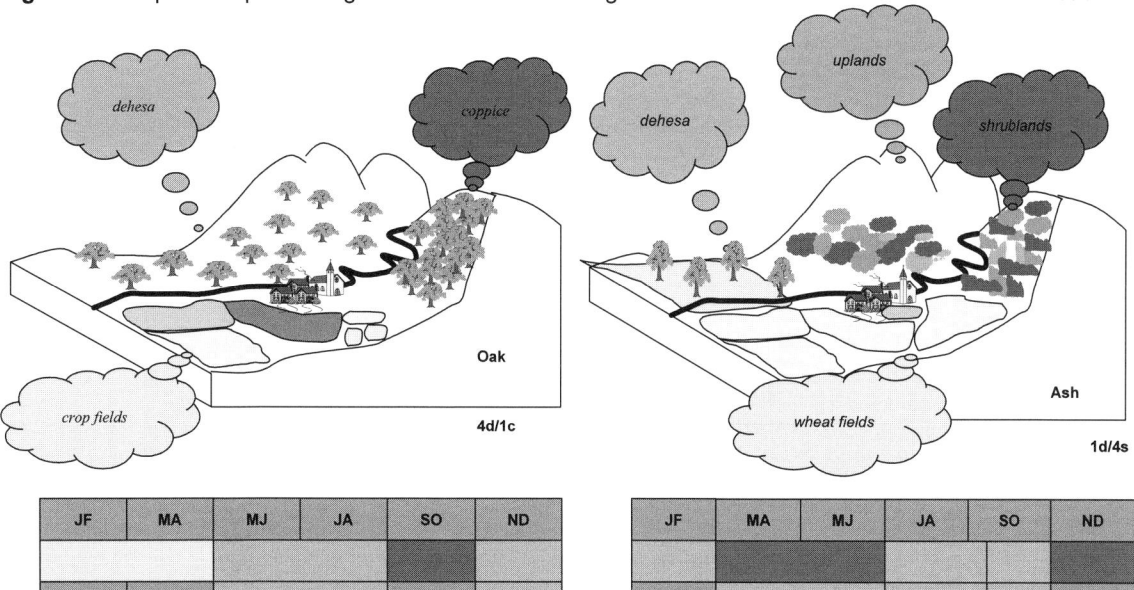

Fire prevention

Only Mediterranean forests are concerned with this goal. In southern France, original management systems have been developed by combining grazing and fuel-break maintenance. The main target is to protect Mediterranean forests from fire, and grazing is part of the management techniques. Pastoral improvements, preferentially with legumes sown under tree canopy cover, increase competition with shrubs, protect well-developed trees, favour wildlife and make forest management easier. A skilled management of these overseedings is required to find the right equilibrium between tree canopy cover, shrub encroachment and climatic variability (Figure 4). Grazing reduces the threat of fire by preventing the build-up of dry forage and of palatable shrubs. Forestry operations such as thinning dense stands, converting coppice into high forest or replacing highly inflammable pioneer species by dense shade species increase the efficiency of the silvopastoral break (Etienne, 1996a).

Figure 4. Changes in silvopastoral plots productivity according to tree canopy cover (30 to 70%) and subterranean clover overseeding (from Etienne, 2000).

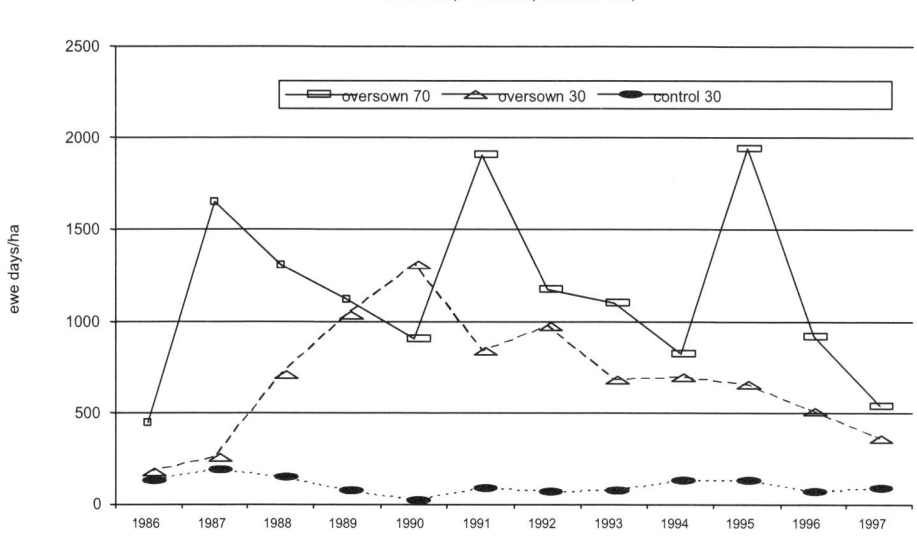

Comparative studies on the five main techniques applied to control shrub encroachment (chopping, burning, uprooting, spraying and grazing) led to a typology based on a balance between advantage and inconvenience according to fuel, labour and environmental aspects (Valette *et al.*, 1993). The main results are summarized in Table 1. As no technique was considered as perfect, more recent studies proposed to combine two or three techniques in order to compensate the negative impact of one by the positive effect of the other. An important work has been developed on the most effective operational sequences to apply in order to establish and maintain fire prevention management plans in a sustainable way (Legrand et al., 1994).

Table 1. Comparative analysis of five fuel management techniques.

	Advantage	Inconvenient
Chopping	Climate free Diversified engines Clean stand Organic matter	Slope limited Rock limited Tree density Chip stocking
Burning	Slope free Rock free Thin fuel eliminated Low cost	Climatic restrictions Vector required Dirty stand Pyrophytes enhancement
Uprooting	Climate free Diversified engines Clean stand Fuel destroyed	Slope limited Rock limited Tree eliminated Loss of organic matter Erosion
Spraying	Slope free Selectivity Long-term effect	Climatic restrictions Cost Standing dry fuel Environment impact
Browsing	Topography free Diversified farming systems Complementary production Yearly grass control Organic matter recycling Trampling effect on litter Browse impact Bucolic stand	Specific equipment Pastoral improvement Selective impact Regeneration damage Dogs Cost Adjustment to a production system

Simultaneously, more sophisticated techniques were developed to enhance browse impact on shrubs either through supplementation or through improved grazing management practices (Etienne et al., 1996; Dumont et al., 2001). Three factors affect particularly browse intensity: the palatability of the dominant shrub species, the grazing management and the type of animal (Table 2).

Table 2. Main factors modifying browse impact (% green leaves and twigs browsed).

Flock management	Herding	Paddock	Camping
	12.4	16.5	48.7
Type of animal	Cattle	Sheep	Goat
	14	22	66
Grazing frequency	2 periods	8 periods	14 periods
	1	15	25
Fertilization	Unfertilized	Fertilized for 5 years	Fertilized for 10 years
	6	12	18
Pastoral improvement	Without overseeding	Sparse overseeding	Dense overseeding
	7	11	17
Dominant shrub	Cistus	Erica	Cytisus
	6	10	37

Grazing management can increase browse impact through feeding management rules (protein-rich supplementation, range fertilization, overseeding) and herding techniques (close herding, high stocking rate, adequate grazing season, appetite stimulation). It also permits partial control of litter fuel stocking by disintegrating the dead material lying on the soil through trampling (Rigolot and Etienne, 1996). As tree regeneration is strongly concerned with browse damage (Figure 5) specific attention must be paid to securing the replacement of tree canopy cover.

Figure 5. Browse impact (% green material browsed) on oak regeneration according to the type of animal and the tree species.

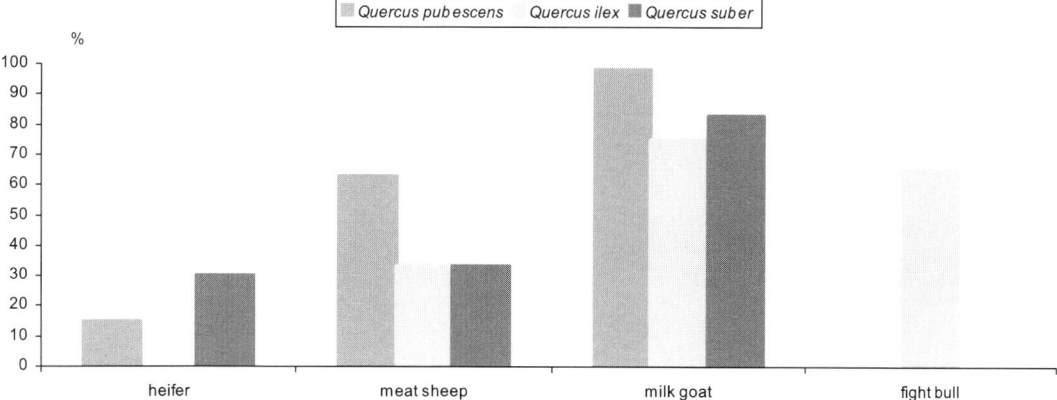

To complete the assessment and evaluate the efficiency of such management plans, a detailed methodology was set up to measure costs and benefits during the many steps of the planning process (Coudour *et al.*, 2000). Land tenure negotiation, plan conception, farmers' involvement, fuel-break building up and functioning, fuel monitoring plus communication and administrative costs summed up from 41,000 to 67,000 € per year. But, if the area covered by the fire prevention plan and the benefits coming from livestock production was taken into account, the net cost averaged 100 €/ha/year and was even negative with milk goat farming for cheese production (Table 3). Many adaptations of such integrated management plans have been set up during the last decade in the French Mediterranean region (Etienne, 1996a).

Table 3. Comparative analysis of the cost (€) of fire prevention management plans in cork-oak forests of southern France. *Net cost includes all the incomes from the livestock production.

	Investment	Functioning	Cost/ha per year	Net cost/ha per year*
Goat farming				
Building up	215,000	47,000	420	395
Current running	93,000	45,000	400	-20
Sheep farming				
Building up	170,000	320,000	315	185
Current running	53,000	314,000	280	150
Cattle farming				
Building up	390,000	47,000	270	130
Current running	107,000	45,000	230	85

According to the type of forest (coppice or high forest, broadleaved trees or conifers), the land tenure, the local livestock farming systems and the other common uses (hunting, cork, hiking,...), the land management plan (Figure 6) will make options between a permanent or temporary grazing, a dense or loose fuel-break network, and global or specific fire-protected areas according to the primary goals of fire prevention (urban interface, heritage forest stands, productive woods, natural reserves,...).

Figure 6. Silvopastoral management plan for fire prevention, structure and benefits (from Etienne, 1996b).

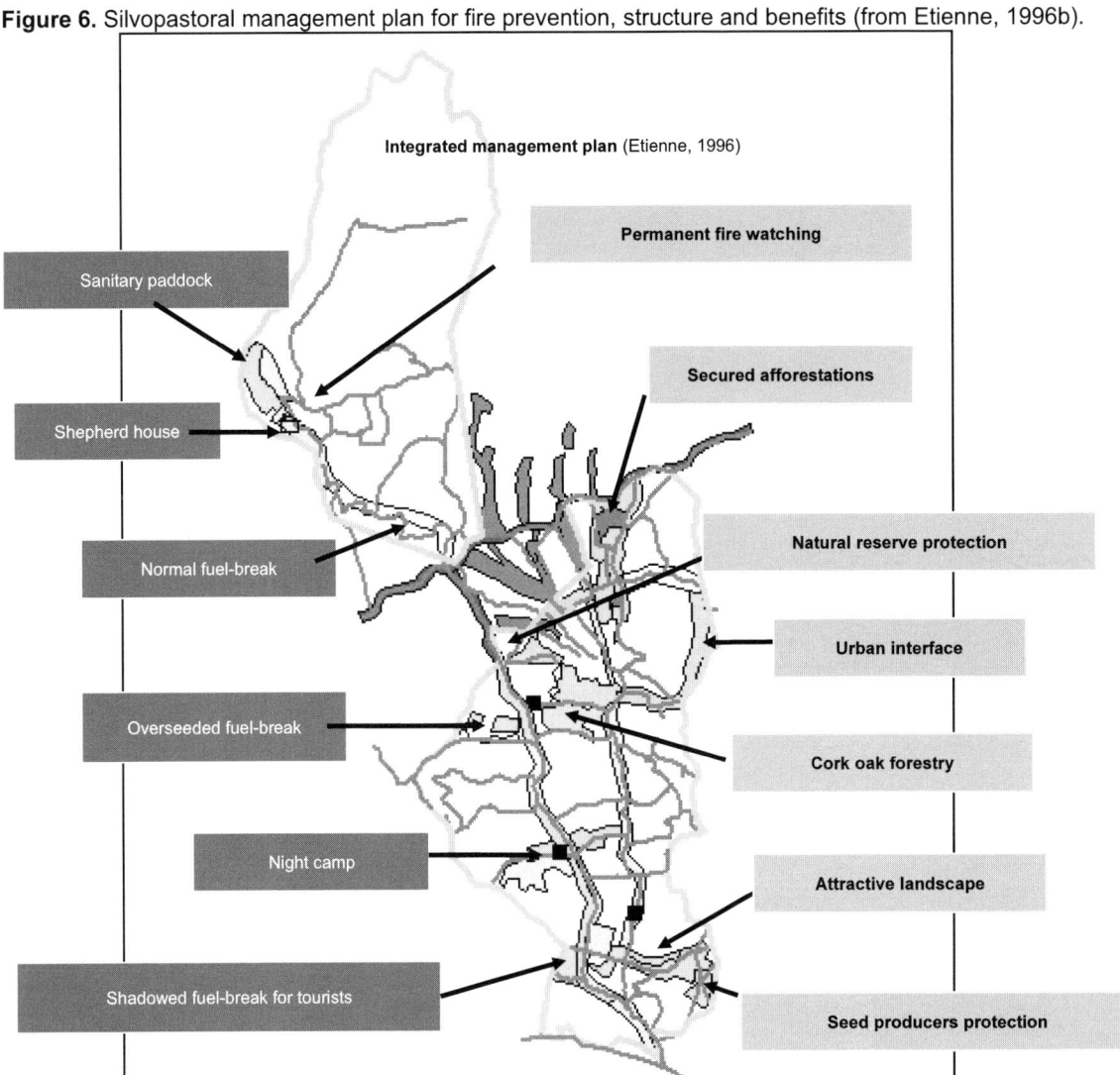

Professional shooting

The economic and social value of hunting can occasionally exceed the value of wood production when a careful management of the wildlife population is applied in order simultaneously to avoid possible damage to the regeneration of the forest and to guarantee an attractive shooting table for hunters who have to pay high hunting rights (Scarasci-Mugnozza *et al.*, 2000). In some regions of Spain the abandonment of the traditional agrosilvopastoral systems called "dehesa" gave the opportunity to test an alternative utilization with deer raising and shooting. New types of management plans have been developed, based on the improvement of feeding resources (San Miguel *et al.*, 1996), on a better balance between the requirements of the animals and the availability of native browse resources, and a rational location of the racks and feeding points in the farmland (Perez-Carral *et al.*, 1995). But the profitability of this management has led to intensification through fencing and artificial feeding, with consequent high stocking rates and the inability to control browsing damage on regeneration and woody vegetation (San Miguel *et al.*, 1999).

Recent studies on grazing and browsing resources for game in the "dehesas" systems (San Miguel *et al.*, 1996) confirmed the importance of acorns and oak leaves in the diet of the deer managed in a traditional way (Figure 7). In this case, the management of the trees is supposed to fill up the gaps in the native grass growth curve by providing fresh leaves during summer through pruning and browse from resprouts during winter through thinning (Perez-Carral *et al.*, 1995).

Figure 7. Acorns, browse and grass ratio in the diet of deer grazing "dehesa" systems all year round.

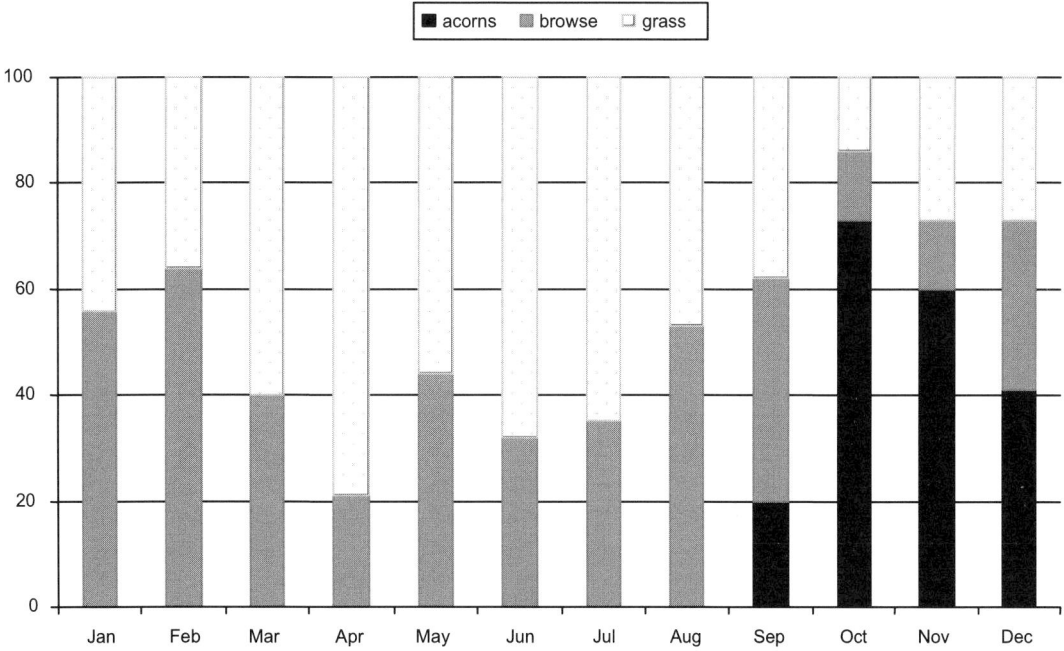

When the system is intensified through clover pastures and cereal crops for green forage, grain and straw, woodlands still play a key role during autumn with the acorn production and during spring with the annual forage growth (Figure 8).

A comparative economic analysis of "dehesa" farms managing deer as a commercial product showed that the incomes from the shooting activity ranged from 60 €/ha for farms mixing cattle and deer rearing, to 220 €/ha for farms exclusively dedicated to deer hunting (Campos et al., 1996). The profitability of such cynegetic farms depends on the ratio between the deer population and the available natural resources. When the farm is fenced and sheltering an excessive population, it may even get negative values.

Figure 8. Forage production of a "dehesa" system grazed by deer all round the year (from Pérez-Carral et al., 1995).

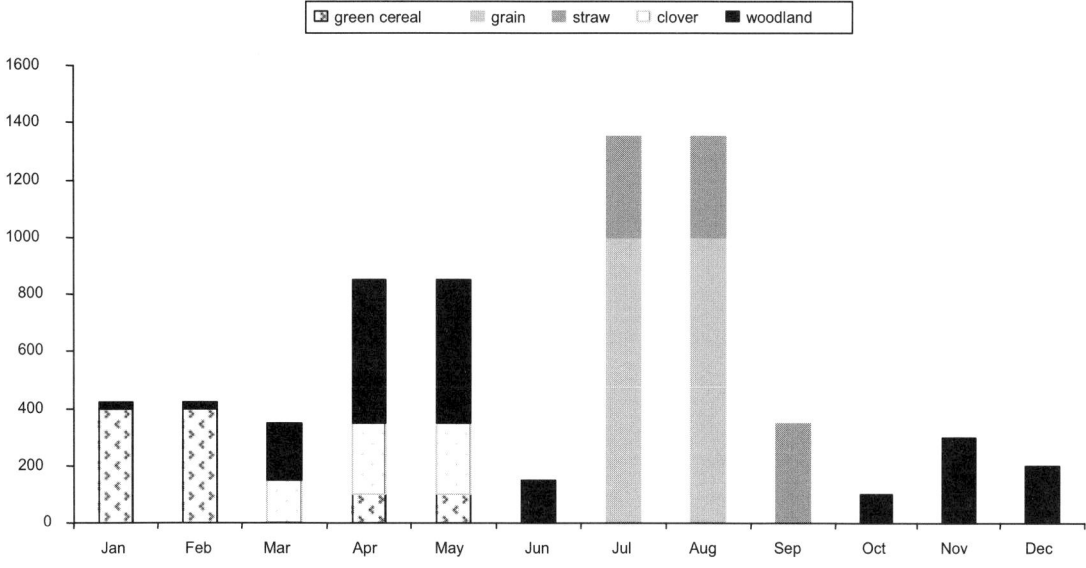

Nature conservation

Mediterranean forest biodiversity is much more marked than in temperate forests because of palaeogeographical and historical factors among which are interactions with human activities for more than 100 centuries (Pons and Quézel, 1985). The higher level of plant richness is particularly clear in the case of woody species, with 247 species for the Mediterranean forests compared with 135 species for the other European ones (Quézel et al., 1999). As Mediterranean forest biodiversity is to a large extent the result of a traditional long use of the environment, many authors argued that repeated disturbances must be guaranteed in order to affect periodically the structure and ecological sequences of Mediterranean forests (Seligman and Perevolotsky, 1994). Particularly in protected zones, human perturbations are required in order to maintain a sustainable mosaic of plant and animal communities that is the only insurance of an optimal biodiversity (Le Floc'h *et al.*, 1998).

When dealing with human perturbations related to an economic activity, many authors confirmed a reduction in plant diversity and richness when grazing pressure decayed (Diez *et al.*, 1991; Seligman and Perevolotsky, 1994) or when tree canopy cover got excessively dense (Puerto and Rico, 1988; Marañón and Bartolome, 1994; Médail *et al.*, 1998). A comparative study on the impact of disturbance levels in three typical French Mediterranean forests (Etienne, 2002) demonstrated clearly how plant richness was positively affected by perturbations, permitting the opening of the undergrowth (shrub clearing) stimulation of sward development (P fertilization and grazing) without getting rid of the tree layer. Whatever the dominant tree was, the higher levels of plant richness were gained by the major perturbation; meanwhile the controls registered the lowest scores (Figure 9). Another comparative study between similar cork-oak forests of southern Spain and northern Morocco showed that an excessive human pressure on woody resources may lead to a loss in biodiversity, mainly because rare or endemic species were replaced by generalist species (Ojeda *et al.*, 1996).

Figure 9. Changes in plant richness according to three types of perturbation in three French Mediterranean forests. T = control, D = roller-chopped, S = silvopastoralism, PA : *Pinus halepensis,* PP : *Pinus pinea,* CL : *Quercus suber.*

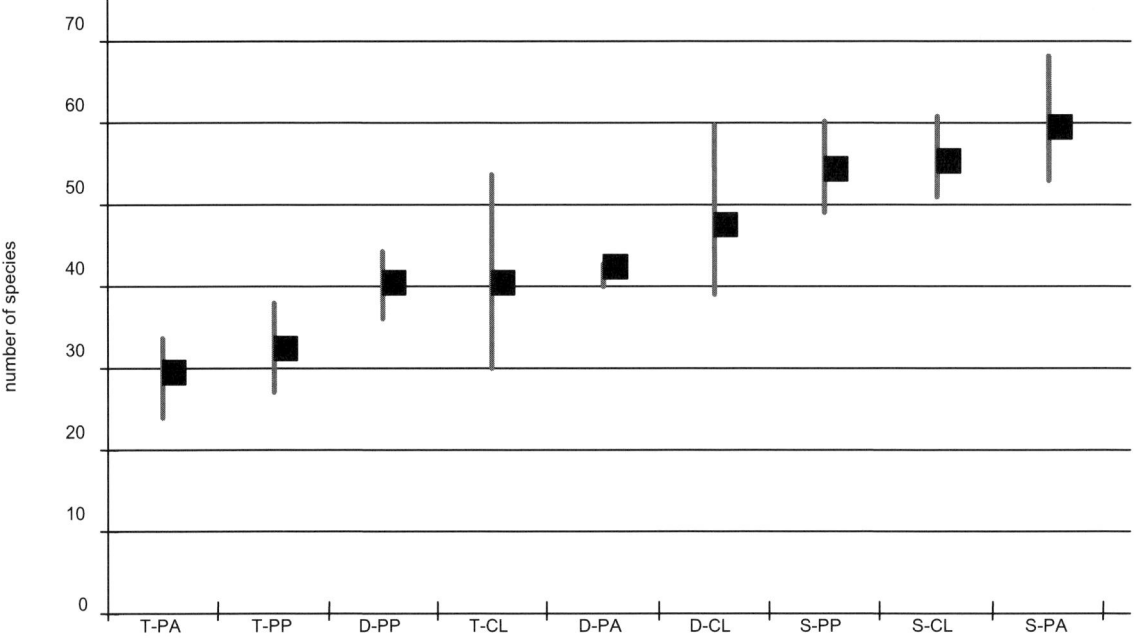

But nature conservation deals with scaling problems and priorities hierarchy because, according to the type of species to protect, the level of perturbation required or the threshold of fragmentation not to be passed over may consistently vary. For instance, the impact of pine encroachment on the endangered species of the National Park of Cévennes had to be monitored and simulated at many different scales according to the home range of the protected species (Figure 10). Consequently specific management units which make sense in relation to the environmental problem, such as ridges covered by mature pine stands and overstanding upon high heritage value catchments, had to be considered in the current management of the forest stands (Etienne, 2003). Similarly, the importance of a sound silvopastoral management was underlined by San Miguel (2003) for conserving adequate habitats for some endangered key species such as the Iberian lynx, the imperial eagle or the black stork.

In order to economically account for the multiple use of the forest, a new methodology was developed to integrate the welfare income into the calculation of the net margin and the profitability rate of Spanish silvopastoral farms (Campos, 1999). Its application to contrasting silvopastoral systems showed the importance of environmental incomes in the complete economic balance of a farm. The integration of environmental services into the evaluation of farming systems managing Mediterranean woodlands led, during the last decade, to a real revaluation of land prices and promoted political efforts to support a better balance between social and private total sustainable incomes (Campos *et al.*, 2001).

Figure 10. Multiscale integration (Etienne, 2002).

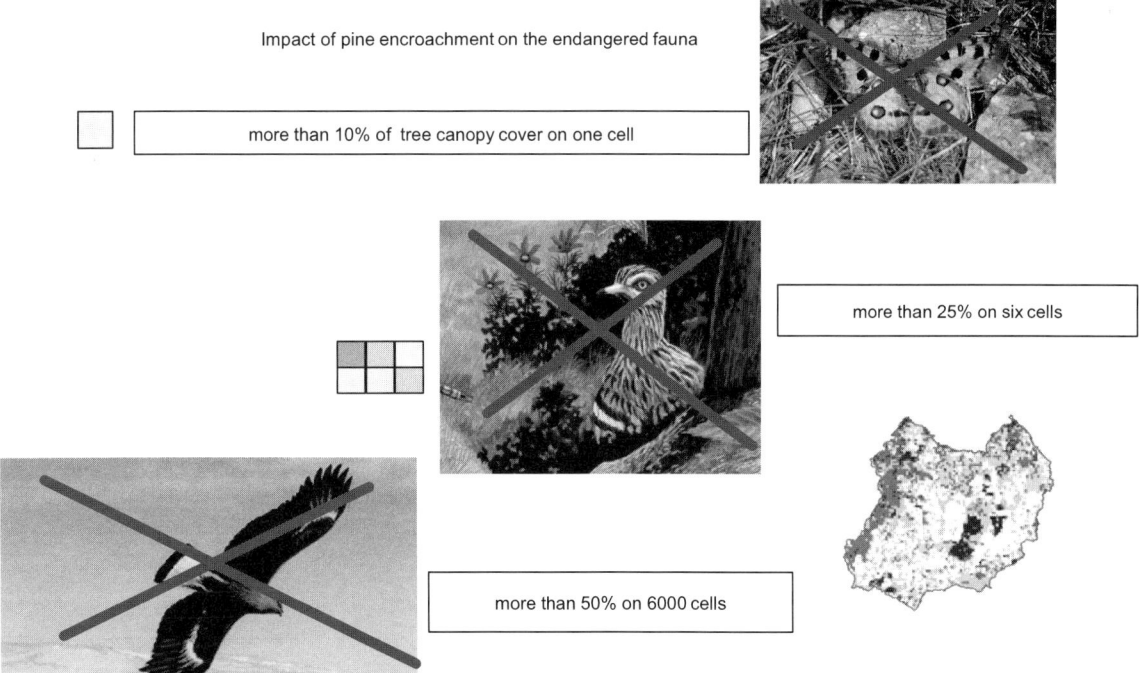

Conclusion

In a rapidly changing socio-economic context, constantly new demands are being made by society on Mediterranean forest managers. Currently, the demand has given priority to multiple use, among which recreation and conservation are taking more and more importance. This imposes the need to develop a scientifically sound conservation strategy, and to implement a locally tailored sustainable management. As sustainability and multifunctionality have to be the two leading principles of a modern Mediterranean forest policy, it is necessary to think about an efficient way to internalize the many positive forest externalities that characterize these ecosystems. It will be the only manner to develop the low-cost management practices that allow sustainable use, care and regeneration of the tree canopy. Particularly, special attention should be paid to flexible and reversible management techniques adapted to the current agricultural conditions and integrating the dynamics of such a complex system.

References

Armand, D. and Etienne, M. (1996) Impact of tree canopy cover on subterranean clover overseeding productivity and use in southeastern France. In: Etienne, M. (ed.) *Western European Silvopastoral Systems*. INRA, Paris, France, pp. 71-81.

Auclair, L., Ben Cheikh, K., Laajili-Ghezal, L. and Pontanier, R. (1995) Use of silvopastoral resources and the production systems in Tunisian Haut Tell (Usage des ressources silvopastorales et systèmes de production dans le Haut Tell tunisien). *Les Cahiers de la Recherche Développement* 41, 7-19. (In French.)

Bellon, S. and Guérin, G. (1992) Old holm-oak coppices... new silvopastoral practices. *Vegetatio* 99-100, 307-316.

Bellon, S. and Guérin, G. (1996) Silvopastoral resource management in the French Mediterranean region. In: Etienne, M. (ed.) *Western European Silvopastoral Systems*. INRA, Paris, France, pp. 167-182.

Benabid, A. (1978) Silvopastoral study of the tetraclinaie from l'Amsittène (Etude silvopastorale de la tetraclinaie de l'Amsittène). *Ecologia Mediterranea* 3, 125-132. (In French.)

Berger, J.E., Etienne, M. and Balandier, P. (1999) ALWAYS: a plot-based biophysical silvopastoral system model. *Ecological Modelling* 115, 1-17.

Bland, F. and Auclair, D. (1996) Silvopastoral aspects of Mediterranean forest management. In: Etienne, M. (ed.) *Western European Silvopastoral Systems*. INRA, Paris, France, pp. 125-142.

Bonin, G. and Loisel, R. (1995) Effects of human impact on forest environment: the Tunisian case. In: Bellan, D., Bonin, G. and Emig, C. (eds) *Functioning and Dynamics of Natural and Perturbed Ecosystems*. Lavoisier, Paris, France, pp. 609-623.

Bourbouze, A. (1980) Goat use of woodland pastures (Utilisation d'un parcours forestier pâturé par des caprins). *Fourrages* 82, 121-145. (In French.)

Campos, P. (1984) *Economía y Energía en la Dehesa Extremeña*. IEAPA, Madrid, Spain.

Campos, P. (1994) Econimic value of Extremeña Dehesa (El valor económico total de los sistemas agroforestales). *Agricultura y Sociedad* 71, 243-256. (In Spanish.)

Campos, P. (1999) Towards welfare income accounting from forestry multiple use (Hacia la medición de la renta de bienestar del uso múltiple de un bosque). *Investigación Agraria de Sistemas y Recursos Forestales* 8, 407-422. (In Spanish.)

Campos, P. and Sesmero, J. (1987) Economic analyses of Portuguese and Spanish Dehesa (Análisis económico de un grupo de dehesas de Extremadura). In: Campos, P. and Martín, M. (eds) *Conservación y Desarrollo de las Dehesas Portuguesa y Española*. Ministerio de Agricultura, Madrid, Spain. pp. 487-534. (In Spanish.)

Campos, P., Vargas, J. and Calvo, J. (1996) Cynegetic resources economy. Reindeer economic management in the Mediterranean. (Economía de los recursos cinegéticos. Gestión económica del ciervo en ambiente mediterráneo). In: *Conservation and Management of Wildlife in Mediterranean Ecosystems*. Instituto de Cultura Juan Gil-Albert, Alicante, Spain, pp. 139-170. (In Spanish.)

Campos, P., Rodríguez, Y. and Caparrós, A. (2001) Towards the dehesa total income accounting: theory and operative Montfragüe study cases. *Investigación Agraria de Sistemas y Recursos Forestales* 1, 43-67.

Carbonero, M., Fernández-Rebollo, A., Blásquez, A. and Navarro, R. (2003) Production and size of *Quercus ilex* acorn after pruning cycle (Evaluación de la producción y del calibre de bellotas de *Quercus ilex* a lo largo de un ciclo de poda). *Reunión Sociedad Española Estudio Pastos* 43, 463-468. (In Spanish.)

Coudour, R., Etienne, M., Millat, C., Beylier, B., Thavaud, P. and Dureau, R. (2000) *Coupures de combustible: le coût des aménagements*. La Cardère, Avignon, France.

de Ponteves, E., Bourbouze, A. and Narjisse, H. (1990) Land use, costume and forestry law of a Septentrional Arganeraie in Morocco. (Occupation de l'espace, droit coutumier et législation forestière dans un terroir de l'arganeraie septentrionale au Maroc). *Les Cahiers de la Recherche Développement* 26, 28-43. (In French.)

De Zulueta, J. (1972) Preliminary assays of *Trifolium subterraneum* sown in areas of difficult etablishments in the extremeñas dehesas (Ensayos preliminares de métodos de siembra de *Trifolium subterraneum* en zonas de implantación dificil de las dehesas extremeñas). *Pastos* 2, 199-211. (In Spanish.)

di Castri, F. (1998) Politics and environment in Mediterranean-climate regions. In: Rundel, P., Montenegro, G. and Jaksic, F. (eds) *Landscape Disturbance and Biodiversity in Mediterranean-Type Ecosystems*. Springer, Berlin, Germany, pp. 407-432.

Diez, C., Luis, E., Tarrega, R. and Valbuena, L. (1991) Degradation process in traditional systems of silvopastoral management in dehesa systems dominated by *Quercus pyrenaica*. In: *Proceedings 4th International Rangeland Congress*. INRA, Montpellier, France, pp. 107-109.

Dorée, A., de Montard, F.X. and Traub, N. (2003) Valorisation and silvopastoral management of natural *Pinus sylvestris* forest stands (Valorisation et gestion silvopastorale de boisements naturels de *Pinus sylvestre*). *Fourrages* 176, 463-478. (In French.)

Ducrey, M. and Boisserie, M. (1992) Natural regrowth after harvesting of *Quercus ilex* in several faros. (Recrû naturel dans des taillis de chêne vert (Quercus ilex) à la suite d'exploitations partielles). *Annual Forestry Science* 49, 91-109. (In French.)

Dumont, B., Meuret, M., Boissy, A. and Petit, M. (2001) Animal grazing: behaviour and breeding (Le pâturage vu par l'animal : mécanismes comportementaux et applications en élevage). *Fourrages* 166, 213-238. (In French.)

Etienne, M. (1996a) Integration of pasture and forage activities in Mediterranean forest to reduce fuel (Intégrer des activités pastorales et fourragères aux espaces forestiers méditerranéens pour les rendre moins combustibles). *Etudes et Recherches sur les Systèmes Agraires et le Développement* 29, 169-182. (In French.)

Etienne, M. (1996b) Research on temperate and tropical silvopastoral systems: a review. In: Etienne, M. (ed.) *Western European Silvopastoral Systems*. INRA, Paris, France, pp. 5-19.

Etienne, M. (2000) Pine agroforestry in the West Mediterranean Basin. In: Ne'eman, G. and Trabaud, L. (eds) *Ecology, Biogeography and Management of Pinus halepensis and P. brutia Forest Ecosystems in the Mediterranean Basin*. Bachuys Publishers, Leiden, The Netherlands, pp. 355-368.

Etienne, M. (2001) Pine trees - invaders or forerunners in Mediterranean-type ecosystems: a controversial point of view. *Journal of Mediterranean Ecology* 2, 221-232.

Etienne, M. (2002) Mediterranean forest management reducing fires and increasing biodiversity (Aménagement de la forêt méditerranéenne contre les incendies et biodiversité). *Revue Forestière Française* 53, 121-126. (In French.)

Etienne, M. (2003) A step-by-step approach to building land management scenarios based on multiple viewpoints on multi-agent system simulations. (accessed May 2003) http://jasss.soc.surrey.ac.uk/6/2/2.html

Etienne, M. and Rapey, H. (1999) Simulating integration of agroforestry into livestock farmers' projects in France. *Agroforestry Systems* 43, 257-272.

Etienne, M., Derzko, M. and Rigolot, E. (1996) Browse impact in silvopastoral systems participating in fire prevention in the French Mediterranean region. In: Etienne, M. (ed.) *Western European Silvopastoral Systems*. INRA, Paris, France, pp. 93-102.

Gillet, F., Besson, O. and Gobat, J.M. (2002) PATUMOD: a compartment model of vegetation dynamics in wooded pastures. *Ecological Modelling* 147 (3), 267-290.

Gómez-Guttierez, J.M. and Pérez-Fernández, M. (1996) The dehesas, silvopastoral systems in semiarid Mediterranean regions with poor soils, seasonal climate and extensive utilisation. In: Etienne, M. (ed.) *Western European Silvopastoral Systems*. INRA, Paris, France. pp. 55-70.

González, A. and Allué, J.L. (1982) Producción, persistencia y otros estudios en una dehesa extremeña. *Anales INIA Serie Forestal* 5, 93-167.

INIA-SEA-ADG (1994) *Improvement of Semiarid Dry Pastures Developed on Acid Soils* (Mejora de pastos en secanos semiáridos de suelos ácidos). Ministerio de Agricultura, Madrid, Spain. (In Spanish.)

Karmouni, A. (1997) Forest grazing: case study of Maghreb countries. In: *Proceedings XI World Forestry Congress*, Antalya, FAO. pp. 321-328.

Le Floc'h, E., Aronson, J., Dhillion, S., Guillerm, J.L., Grossmann, A. and Cunge, E. (1998) Biodiversity and ecosystem trajectories: first results from a new LTER in southern France. *Acta Oecologica-International Journal of Ecology* 19, 285-293.

Legrand, C., Etienne, M. and Rigolot, E. (1994) Methodology of implementation to choose combined techniques for firebreak maintenance (Une méthode d'aide au choix des combinaisons techniques pour l'entretien des coupures de combustible). *Fourrages Méditerranéens* 15, 397-408. (In French.)

Madani, T., Hubert, B., Lasseur, J. and Guérin, G. (2001) Association des bovins, des ovins et des caprins dans les élevages de la suberaie algérienne. *Cahiers Agricultures* 10, 9-18.

Marañón, T. and Bartolome, J. (1994) Coast live oak (*Quercus agrifolia*) effects on grassland biomass and diversity. *Madroño* 41, 39-52.

Médail, F., Roche, P. and Tatoni, T. (1998) Functional groups in phytoecology: an application to the study of isolated plant communities in Mediterranean France. *Acta Oecologica - International Journal of Ecology* 19, 263-274.

Mesón, M.L. and Montoya, J.M. (1987) Three models of integrated agrosilvopastoral systems in Madrid farms (Trois modèles d'intégration agrosylvopastorale sur les parcours communaux de Madrid). In: *Proceedings Colloque Agriculture et Forêt en Région Méditerranéenne Française*. INRA, Toulon, France, pp. 24-31. (In French.)

M'Hirit, O. (1999) Mediterranean forest: ecology, economic richness and social goods (La forêt méditerranéenne: espace écologique, richesse économique et bien social). *Unasylva* 197, 21-28. (In French.)

Msika, B. (1993) Modelisation of herbaceous tree component in *Quercus pubescens* Willd and *Pinus austriaca* Hoss plantations in South Pre-Alps, as a good technique for taking decisions in the silvopastoral management (Modélisation des relations herbe-arbre sous peuplements de *Quercus pubescens* Willd. et *Pinus austriaca* Höss. dans les Préalpes du sud. Un outil d'aide à la décision en aménagement silvopastoral). PhD Ecology, Univ. Aix-Marseille III, France. (In French.)

Naggar, M. (1993) Place des arbustes fourragers dans les aménagements sylvopastoraux. *Fourrages Méditerranéen* 14, 256-264.

Ojeda, F., Marañón, T. and Arroyo, J. (1996) Patterns of ecological, chorological and taxonomic diversity at both sides of the Strait of Gibraltar. *Journal of Vegetation Science* 7, 63-72.

Pardini, A., Talamucci, P. and Margaritelli, L. (1987) Tentativi di incremento dell'offerta di pascolo di un bosco ceduo attraverso concimazione, semina e trasemina. *Monti e Boschi* 38, 47-52.

Perez-Carral, C., San Miguel, A. and Cañellas, I. (1995) Silvopastoral management of a cynegetic farm of Toledo mountains (Gestión silvopastoral de una finca cinegética en los montes de Toledo). *Cuadernos SECF* 1, 259-267. (In Spanish.)

Pons, A. and Quézel, P. (1985) The history of the flora and vegetation, and past and present human disturbance in the Mediterranean region. *Geobotany* 7, 25-43.

Puerto, A. and Rico, M. (1988) Influence of tree canopy on old-field succession in marginal areas of Central-Western Spain. *Oecolica Plantarum* 9, 337-358.

Qarro, M. and De Montard, F.X. (1989) Etude de la productivité des parcours de la zone d'Ain-Leuh. *Agronomie* 9, 477-487.

Qarro, M., Sabir, M., Belghazi, B. and Ezzahiri, M. (1995) Effets des traitements sylvicoles sur le développement des potentialités herbagères dans les taillis de chêne vert. In: *Proceedings Atelier sur le silvopastoralisme*. Centre National de la Recherche Forestière. Rabat, Marocco, pp. 57-69.

Quézel, P., Médail, F., Loisel, F. and Barbero, M. (1999) Biodiversité et conservation des essences forestières du bassin méditerranéen. *Unasylva* 197, 21-28.

Rigolot, E. and Etienne, M. (1996) Litter thickness on tree covered fuel-break maintained by grazing. In: Etienne, M. (ed.) *Western European Silvopastoral Systems*. INRA, Paris, France. pp. 111-122.

San Miguel, A. (1985) Effect of different thinning intensity on a woodlands of *Q. pyrenaica* (Variaciones producidas en un pastizal arbolado con rebollos por claras de distinta intensidad), *Anales INIA Serie Forestal* 9, 97-104. (In Spanish.)

San Miguel, A. (1994) *The Spanish Dehesa* (La Dehesa Española). Fundación Conde del Valle de Salazar, Madrid, Spain. (In Spanish.)

San Miguel, A. (2003) Silvopastoral management and conservation of species and protected areas (Gestión silvopastoral y conservación de especies y espacios protegidos). In: *43 Reunión Sociedad Española Estudio Pastos*. SEEP. LLeida, Spain. pp. 409-421. (In Spanish.)

San Miguel, A., Montero, G. and Montoto, J.L. (1983) Ecological and silvopastoral studies in the *Q. pyrenaica* stand of Guadalajara, preliminor and results (Estudios ecológicos y silvopascicolas en un quejigal de Guadalajara, primeros resultados). In: *23 Reunión Sociedad Española Estudio Pastos*. SEEP, Sevilla, Spain. pp. 1-16. (In Spanish.)

San Miguel, A., Sanz, F., Perez-Carral, C. and Roig, S. (1996) Gestión de recursos alimenticios para la caza mayor en los montes de Toledo. *Pastos* 26, 39-59.

San Miguel, A., Perez-Carral, C. and Roig, S. (1999) Deer and traditional agrosilvopastoral systems of Mediterranean Spain. A new problem of sustainability for a new concept of land use. *Options Méditerranéennes* 39, 261-264.

Scarascia-Mugnozza, G., Oswald, H., Piussi, P. and Radoglou, K. (2000) Forests of the Mediterranean region: gaps in knowledge and research needs. *Forest Ecology and Management* 132, 97-109.

Seigue, A. (1985) *La Forêt Circum-Méditerranéenne et ses Problèmes*. Maisonneuve et Larose, Paris, France.

Seligman, N. and Perevolotsky, A. (1994) Has intensive grazing by domestic livestock degraded Mediterranean Basin rangelands? In: Arianoutsou, M. and Groves, R. (eds) *Plant-Animal Interactions in Mediterranean-Type Ecosystems*. Kluwer, Dordrecht, The Netherlands, pp. 93-103.

Sibbald, A. (1996) Silvopastoral systems on temperate sown pastures: a personal perspective. In: Etienne, M. (ed.) *Western European Silvopastoral Systems*. INRA, Paris, France. pp. 23-36.

Sibbald, A. and Dalziel, A. (2002) Silvopastoral national network experiment. *Agroforestry Forum Newsletter* 2, 9-13.

Talamucci, P., Pardini, A., Argentini, G. and Stagliano, N. (1996) Theoretical silvopastoral systems based on seasonal distribution of diversified resources in an Italian Mediterranean environment. In: Etienne, M. (ed.) *Western European Silvopastoral Systems*. INRA, Paris, France. pp. 183-194.

Tibaoui, G. and Zouaghi, M. (1991) Productivité d'un parcours forestier d'une région subhumide de la Tunisie. In: *Proceedings 4th International Rangeland Congress*. INRA, Montpellier, France, vol. 1, pp. 232-235.

Tiedeman, J. and Johnson, D. (1992) *Acacia cyanophylla* for forage and fuelwood in North Africa. *Agroforestry Systems* 17, 169-180.

Valette, J.C., Rigolot, E. and Etienne, M. (1993) Intégration des techniques de débroussaillement dans l'aménagement de défense de la forêt contre les incendies. *Fourrages Méditerranéen* 14, 141-154.

Conservation "matching funds" from working woodlands in California

L. Huntsinger, A. Sulak, R. Standiford and P. Campos-Palacín
College of Natural Resources, University of California, 145 Mulford Hall, MC 3114, Berkeley, CA 94720-3114, USA. Tenure Research, Institute of Economics and Geography (IEG), Spanish Council for Scientific Research (CSIC), Pinar 25, 28006 Madrid, Spain. buckaroo@nature.berkeley.edu, sulak@nature.berkeley.edu, standifo@nature.berkeley.edu, pcampos@ieg.csic.es

Abstract
Rancher-subsidized oak woodland silvopastoralism is being used to conserve land in California oak woodlands. As part of the Working Landscapes movement, non-governmental organizations like the Nature Conservancy and a proliferation of land trusts are brokering private and public fund transfers in exchange for title restrictions that preclude development on land used by "environmentally friendly" low intensity livestock grazing enterprises. Increasingly, ranchers are willing to trade their option to develop their land for capital, tax benefits and the continued opportunity to enjoy the ranching lifestyle, as a result sharing title with conservation-oriented non-governmental organizations or public agencies. A shift away from the traditional American land acquisition model and from subsidies and other policies for stimulating increased agricultural production, the relatively passive and indirect role of government is attractive to regulation-averse ranchers. Ranchers will invest considerable labour and capital in the ranch in return for lifestyle benefits, including environmental and other non-monetary benefits. Many such benefits are shared by the public. The high value of lifestyle to ranchers is widespread in the USA and California, with more than half of ranchers in various studies holding outside jobs to provide funding to invest in the ranch. Protecting the lifestyle and cultural values that motivate such ranchers is an important consideration for conservationists.

Introduction
Public land management in the United States has become costly and contentious, and does not include sufficient area of important habitats, notably in California's Mediterranean oak woodlands. For this and other reasons interest in private land conservation and management has grown substantially in the last decade. An effort to conserve "working" private agricultural lands in the United States, the Working Landscapes movement is a shift away from traditional American conservation models. In California's oak woodlands, working landscapes equate to conservation of large silvopastoral landownerships.

In this chapter we argue that the tendency of ranchers to forgo opportunity costs in order to enjoy the diverse non-monetary values that accrue from oak woodland ownership augments public conservation investment, providing a form of "matching funds" for conservation of "environmentally friendly" agriculture. The considerable overlap between landowner and public values in the land (shared utility) contributes to the viability of this model for public investment.

Shifting models for conservation
Since the late 1800s, the central approach to land conservation in the United States has been national government land reservation or acquisition (Raymond and Fairfax, 1999). About half of California is public land, owned and managed by the federal government. This is a low proportion compared to the rest of the western United States. The neighbouring state of Nevada is 97% public land. Numerous governmental agencies manage public land for diverse purposes, but most have some mandate for environmental protection.

At the same time, the flaws in the public lands model of conservation have become more apparent. Historic USA land disposition practices resulted in the more productive and well-watered lands being claimed by private landowners, along with critical wildlife habitat (Maestas et al., 2001; Scott et al., 2001). Consequently, some habitat for 95% of all federally threatened and endangered flora and fauna is on private land, and 262 or 19% of these species survive only on private parcels (Wilcove et al., 1996).

One of the richest California habitat types that is largely outside public lands is open oak woodland and savannah, home to more than 300 vertebrate species (Jensen et al., 1990). The woodlands have high biodiversity, with rich acorn mast, mild climate and high productivity, and are perhaps the most significant wildlife habitat in the state on a regional scale. In addition, oak woodlands are valued for their aesthetics and amenities, and are considered part of the "heritage" of California. Historically undervalued for conservation purposes, more than half have been cut and the land cultivated since the mid-nineteenth century (Burcham, 1957). In the 1940s-1960s further oak removal was subsidized by government grants to increase forage production, in an era characterized by the drive to increase commodity production. Oaks were part of the subsequent collateral environmental damage. The woodlands that remain are largely those that have, until the recent advent of hillside vineyard production, been largely non-arable. Such lands have been grazed by livestock as part of silvopastoral production since the Spanish first brought domesticated livestock to California in 1769.

Driven by urban out-migration and a booming second home market, oak woodland area has declined by thousands of hectares per year over the last decade, and it is projected to continue to decline at similar levels through 2040 (CDF-FRAP, 2003). High land costs make government acquisition expensive, but the cost of acquisition pales in comparision to the costs of public management. National agencies are frequently locked into costly legal battles, subject to shifts in political climate, and criticized as insensitive to local needs and conditions. Management initiatives that have been promulgated on a nationwide basis, such as fire suppression, are blamed for fuel accumulations contributing to California's current severe wildfire risk. In addition, complex and costly environmental regulations have alienated many public land users. Budgeting for the major public conservation agencies is politicized, erratic, and has either been reduced or failed to keep up with inflation in recent years. As one example, the agency holding the majority of the public land in the USA has unfunded infrastructural maintenance needs of $190-$300 million and yet underwent a budget reduction in 2003 (GAO, 2003).

In order to conserve the woodlands, American conservationists have had to develop new and, for the United States, unusual tools for conserving them. Operating within the context of an American political system that emphasizes the primacy of individual, private property rights and independence, the Working Landscapes movement, which seeks to preserve private agricultural lands and rural communities through collaboration and incentives, has gained strength in recent years. Non-governmental organizations like the Nature Conservancy and an increasing number of land trusts have taken the lead in this initiative, and broker some of the most common conservation arrangements, with private as well as public funding sources, commonly including tax relief. In California's oak woodlands, Working Landscapes programmes manifest as conservation of silvopastoral grazing enterprises, and draw on the "willingness to pay" of ranchers for a lifestyle.

California's Mediterranean oak woodlands

About 2 out of the approximately 3 million ha of California oak woodlands are open enough to be suitable for grazing (Figure 1), and more than half are owned by livestock producers. Another 25% are leased for grazing (Huntsinger et al., 1997). Most California oak woodland ranches are owned by families who live and work on the ranch. In general, they rely on a base herd of cattle to produce calves each year for market (Table 1). Calves are the major source of agricultural income, followed by firewood, game and lamb.

Table 1. Californian savannah characteristics (Campos-Palacín et al., 2001).

	Californian savannah
Extent	2 million ha savannah out of 3 million ha total oak woodlands and grasslands (Allen-Diaz et al., 1999)
Most common oaks	Blue oak (Q. douglasii), valley oak (Q. lobata), coast live (Q. agrifolia) and interior live oaks (Q. wislizenii)
Ownership	80 %+ private (Ewing et al., 1988)
Owner's primary residence is ranch	80-92% (Huntsinger et al., 1997)
Average ranch size	800-960 ha (Huntsinger et al., 1997; Sulak and Huntsinger, 2002)
Stocking rate of livestock (does not include wild herbivores, nor does it meet total animal demand)	5-10 ha/LU per year (Ewing et al., 1988)
Small stock	Declining
Large stock	Declining; 92% of animal demand is cattle (California Agricultural Statistics, 1990-2001)
Commodity products	Beef, followed far behind by lamb, wool, firewood, game, grazing resources

Woodlands are dominated by each of four common oak species, with mixtures of other oaks (Allen-Diaz et al., 1999). More than 85% of understoreys are dominated by introduced annual grasses. Oak regeneration and/or recruitment is a concern in some areas (Bolsinger, 1988), but there is considerable site to site and study to study variation (McCreary, 2001).

Figure 1. Oak woodlands of California (Allen-Diaz *et al.*, 1999).

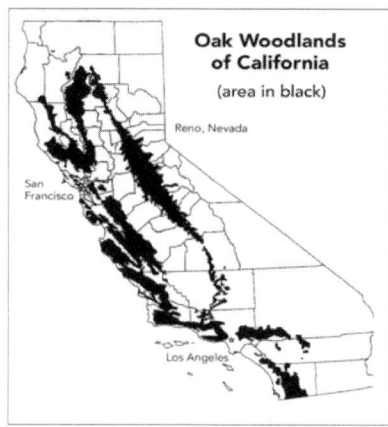

Rancher values and emerging conservation approaches

A review of studies of ranchers throughout the western United States, including the states of California (CA), Arizona (AZ) and Colorado (CO) (Table 2), reveals a consistent pattern of "consumptive" ownership. Most ranchers subsidize their operations with outside income. Smith and Martin (1972), in Arizona in the late 1960s, noted that ranchers were not "economic men", but that what ranching needed from local communities was job opportunities so that ranchers could support their ranches. Torell *et al.* (2001) notes that the lack of a powerful monetary motivation among ranchers grazing federal land is almost universal. Gentner and Tanaka (2002) randomly surveyed 1052 ranchers throughout the west with public land grazing permits. They found that about half of public land ranchers ranch full time and depend on the ranch for over 80% of income, while the other half have the majority of their income and wealth from other sources. But, when asked about their motives for ranching, both groups ranked the lifestyle and way of life above the monetary profits they might make from the ranch.

Table 2. Selected studies of ranchers in the western United States. Oak woodland studies shaded. Bureau of Land Management (BLM); United States Forest Service (USFS).

	Sulak and Huntsinger (2002)	Liffmann *et al.* (2000)	Huntsinger *et al.* (1997)	Rowe *et al.* (2001)	Smith and Martin (1972)	Bartlett *et al.* (1989)	Gentner and Tanaka (2002)
State	CA	CA	CA	CO	AZ	CO	Western US
Sample	Permittees of three forests, similar non-permittees	All ranchers in three CA counties	Oak woodland ranchers statewide	Permittees in CO two counties	All AZ ranch owners	All CO federal permittees	All permittees of USFS and BLM
Off-ranch income	43% not dependent on ranching	44% income is off-ranch	85% have off-ranch income	78% have other source of income	80% hold off-ranch jobs		62% hold off-ranch jobs
Survey type	Interviews	Mail	Mail	Interviews	Interviews	Mail	Mail
Sample size	Small ($n = 37$)	Large ($n = 245$)	Large ($n \approx 200$)	Small ($n = 37$)	Medium ($n = 89$)	Large ($n = 313$)	Very large ($n \approx 1070$)

Such behaviour has been considered "economically irrational", but in fact a closer look shows that rancher behaviour is rational when personal and family goals are considered. Though beyond the scope of this chapter, for the purposes of analysis these goals can be given a monetary value and added to ranch "income", as per the "total economic value accounting method" (Campos-Palacín *et al.*, 2001). In paired landowner surveys in areas as diverse as California and Colorado, ranchers were found to have very similar motivations (Table 3). Enjoying a lifestyle, animal production and living around natural beauty far outstripped the goal of "making money". This does not mean that most ranchers do not need to make money to support their continued ranching, but only that they acknowledge that ranching is not necessarily the way to obtain the greatest monetary returns on investment. Hence ranchers in California's oak woodlands are paying for the opportunity to be ranchers. Of course, there are landowners that own a ranch solely for amenity values, but these are as yet a minority in California.

Table 3. Reasons ranchers in California and Colorado studies gave for continuing to ranch (Rowe et al., 2001; Sulak and Huntsinger, 2002).

Percentage who agree the following is an important to highly important reason to continue to ranch	California $n = 37$	Colorado $n = 34\text{-}37$
Enjoy animal husbandry	95%	97%
Way of life	95%	95%
Family	95%	87%
Tradition	95%	81%
Live near natural beauty	92%	87%
Enjoy the work	89%	89%
It would be difficult to get a job outside the ranch	14%	27%
It's a good way to make money	14%	19%

It appears that there is considerable common ground in the values of ranchers and the public (Table 4), with both seeking environmental and amenity benefits. There are important points of divergence, like public access to the property and the impacts or desirability of certain management practices, but more than half of Californians in a 2001 statewide survey recognized that agriculture provides public benefits. Nearly 60% consider land development and loss of open space to be important problems. The majority approved of the use of public funds to support agriculture as long as conservation benefits were produced (Tarrance Group, 2001). Consequently, the rancher's "willingness to pay" for a lifestyle leverages conservation dollars invested by public and private conservation investors in privately owned working woodlands. The well-documented tendency to forgo opportunity costs in exchange for the lifestyle and other benefits of ranching can be seen as a form of "matching funds" for public conservation dollars when ranchers remain and continue caring for the land.

Table 4. Conceptual table of selected rancher and public values.

Rancher "income"	Public "values"
Natural beauty	Natural beauty
Living on property	Existence and visual appearance
Wildlife and recreation	Wildlife and recreation
Legacy value: heirs	Legacy value: future generations
Production value for agriculture	Healthy agricultural products

Working landscape conservation tools in California

The United States, with its oft-stated premium on "private land rights", is a challenging climate for protection and conservation of private lands. Most ranchers already feel highly over-regulated (Liffmann et al., 2000), and the political and financial costs and unintended consequences of regulation have pushed conservationists into education and incentive-based, and even market-based, modes. Considerable investment has been made in outreach education through university extension services, and this has been shown to have reduced landowner cutting of oaks, among other things (Huntsinger et al., 1997). Approximately 75% of ranchers in California are enrolled in a voluntary tax relief programme under the California Land Conservation Act of 1965, which provides tax benefits to landowners who agree not to sell or develop their land for 10 years. The voluntary and flexible nature of this programme has made it popular with landowners, but studies have shown that it, as well as governmental land-use planning efforts in California, tends to give way when needed most (McClaran et al., 1985; Hart, 1991). Because land-use planning is a responsibility of counties and municipalities in California, zoning regulations are subject to relatively small-scale political influences and often change. In addition, they seldom preclude the subdivision of large properties into small "ranchettes" of 1 to 10 ha, and such small properties do little to protect biodiversity values or traditional agriculture (CDF-FRAP, 2003).

Because of the market for oak woodland properties for housing and other forms of development, and low rangeland calf prices due to competition from government-subsidized grain-based feeds and other forms of intensified production, land values are many times higher than what is justified by earnings from rangeland livestock production. Ranchers historically have only been able to realize the land appreciation by selling all or part of their lands, a practice that is estimated to have resulted in the sale of about 1% of ranch land in California oak woodlands for the years 1985-1992 (Huntsinger et al., 1997). In recent years a market for the "development rights" to privately owned properties has been created through the adaptation of easements, or deed restrictions on land title, for conservation purposes. An agricultural conservation easement on a property title precludes subdivision and development of the land. By selling a conservation easement to a conservation group or agency, the rancher realizes some of the value of land appreciation, but keeps the ranching land and operation intact. Conservation easements have become the most widely used private-sector land conservation method in the United States (Gustanski and Squires, 2000). Land under conservation easements increased by 225% in California in the 1990s (Land Trust Alliance, 2000).

A conservation easement provides financial benefits to the rancher through the marketing of the easement and through various tax benefits, including a reduction in inheritance taxes. Some easements are donated for maximum tax benefit. A prominent example of a private land trust that holds large numbers of easements is the Nature Conservancy.

Though selling or donating an easement is voluntary, once an easement is established it results in permanent sharing of the title with the easement holder. This is transmitted to all future owners and remains a co-ownership. The risks and unintended consequences of this practice are, of course, many (Merenlender *et al.*, 2004), but more than we can take on here.

The value of this right, or easement value, generally is from 35 to 65% of the market value of the property. However, there are many variables and the range of easement values spans from 20% all the way to 90% in rare cases. For example, if land can be sold for $6000/ha to a developer and is valued at $1000/ha for agricultural production, the "easement value" is $5000/ha, which is in fact the average value found in a recent study of 46 programmes in California (Kan-Rice and Sokolow, 2003). This value would be paid to the landowner by a purchaser or can be applied in calculating the tax relief due for easement donation by the owner. In negotiations for easement purchases by a non-governmental organization (NGO), the landowner may accept a payment lower than the market-defined easement value, because of the non-monetary value to the owner of keeping the land for lifestyle, legacy and amenity benefits. The owner accepts the future cost of managing the property and maintaining the agricultural operation.

Each easement and its terms are unique to the property, landowner and funder. By offering tax relief for landowners and sometimes direct grants to NGOs for easement purchase, federal and state governments support the creation of easements. For example, the recently passed federal "Farm Bill" (Public Law No. 107-171 Farm Security and Rural Investment Act of 2002) provides millions of dollars in funding for NGO-brokered conservation easements. Landowners who receive federal funds for farm easements must implement conservation plans developed by the USA Department of Agriculture's Natural Resources Conservation Service. In general, management requirements that are part of easements, if any, are negotiated and influence the easement value.

Easements are the one permanent land conservation method in the USA that encourages the continuation of private agriculture on private land. It has been argued that livestock grazing can be a useful tool for enhancing biodiversity in Mediterranean grasslands and woodlands and, in a more applied sense, for fire hazard control and control of non-native vegetation in California. Grazing can be manipulated at appropriate scales in an effort to change the balance of native and non-native species. Currently, fire and grazing are being used and promoted as a means of enhancing native grassland diversity in different parts of the state (Edwards, 1992; DiTomaso *et al.*, 1999; Meyer and Schiffman, 1999; Reeves, 2001; Fehmi and Bartolome, 2003; Hayes and Holl, 2003; D'Antonio *et al.*, 2004 unpublished results). There are also strong arguments that grazing can degrade ecosystems and harm species (Painter, 1995). Yet the willingness of the public to invest in easements on these livestock enterprises shows a recognition that they produce recognizable public benefit.

Marin County, just north of San Francisco, is one example of a conservation easement programme in California. One of the original non-governmental land trusts, the Marin Agricultural Land Trust, has secured conservation easements on more than 20,000 ha of private ranches and dairies. Easements are coordinated with local land-use planning efforts and with the federally owned National Seashore in the area, so that easements are concentrated in an agricultural zone and buffer the public lands from urban development. Recent surveys show that artisanal and diversified production is more common on land with easements (Gale, 2003). In the north-eastern part of the state, the Nature Conservancy's Lassen Foothills Project was launched in 1997. Before the project started the Conservancy had already purchased about 1000 ha to create the Vina Plains Preserve and had been managing the 15,016 ha Gray Davis Dye Creek Preserve for 10 years. By late 2002 the Lassen Foothills Project had safeguarded an additional 23,200 ha of grasslands, oak woodlands and streamside forests, mainly by purchasing conservation easements. A diverse array of local NGOs, public agencies and private companies are participants in this effort.

Conclusions

Oak woodland ranchers are optimistic about the future of ranching in California, with 77% in one study believing that it was possible to preserve ranching in their community (Sulak and Huntsinger, 2002). Not only are working landscapes "less expensive" than land acquisition, the tendency of landowners to forgo opportunity costs and invest in the enterprise in order to enjoy the diverse non-monetary values that accrue from oak woodland ownership augments public conservation dollars, providing a form of "matching funds" for conservation.

However, it is crucial to remember that ranchers are willing to invest because they enjoy being ranchers (i.e. they receive utility from being ranchers). Conservation easements and other incentive-based programmes are most effective when they allow ranchers to continue their traditions and make an effort to recognize local knowledge, values and interests.

In California's oak woodlands, the Working Landscapes movement seeks to conserve and stabilize large extensive, silvopastoral landownerships. A few landowners owning large amounts of land in a watershed, with similar goals and values, are far easier to work with than numerous small owners with highly diverse motives for owning oak woodland lands. The larger properties also make it easier to preserve intact systems. However, they do raise issues of equity, not only in terms of land distribution, but with respect to accountability - what the public is getting for its money - and other factors. The division of actual costs of various kinds of conservation programmes among the public, the landowner and the non-profit sector is difficult to sort out, and issues of equitability, accountability and the long-term effects of private land conservation on local communities need further study (Merenlender *et al.*, 2004). In his analysis of Oregon's conservation easement program, noted that problems include private control, loss of local revenue, the potential for low-density development, inefficient use of open space, potential exclusionary effects and the destruction of metropolitan vitality.

On the other hand, conservation easements can perhaps optimistically be seen as a way of dividing tenure, a form of "virtual property redistribution". Though the surface of the land remains intact, the easement is a sharing of title with an NGO or public agency. Current easement prices reflect what is generally marketed: the right to develop and to subdivide. Management and monitoring stipulations are negotiated for each property, and each easement contract is therefore unique. Though highly unusual now, in the future easements might include public access and other requirements. The individuality and voluntarism of easements make some of us nervous, as there is little overarching control and a limited ability to plan. On the other hand, such localized efforts may best meet the current conditions for private land conservation in the United States: little overt or direct government role, tailored to very local ecological and social conditions, and a lot of landowner independence and choice. Ironically, however, these choices set up long-lasting restrictions on future landowners though certainly not as consequential as subdivision and development of the property.

References

Allen-Diaz, B.H., Bartolome, J.W. and McClaran, M.P. (1999) California oak savanna. In: Anderson, R.C., Fralish, J.S. and Baskin, J.M. (eds) *Savannas, Barrens, and Rock Outcrop Plant Communities of North America.* Cambridge University Press, Cambridge, United Kingdom, 470 pp.

Bartlett, E.T., Taylor, R.G., McKean, J.R. and Hof, J.G. (1989) Motivation of Colorado ranchers with federal grazing allotments. *Journal of Range Management* 42 (6), 454-457.

Bolsinger, C. (1988) *The Hardwoods of California's Timberlands, Woodlands, and Savannas.* USA Forest Service, Pacific Northwest Forest and Range Experiment Station, PNW-RB-148, Portland, USA.

Burcham, L.T. (1957) *California Range Land: An Historic-Ecological Study of the Range Resource of California.* Division of Forestry, Dept. of Natural Resources, Sacramento, USA.

California Agricultural Statistics, 1990-2001 http://www.nass.usda.gov/ca

Campos-Palacín, P., Huntsinger, L., Standiford, R., Martin-Borroso, D., Mariscal-Lorente, P. and Starrs, P.F. (2001) Working woodlands: public demand, owner management, and government intervention in conserving Mediterranean ranches and dehesas. In: Standiford Richard, B., McCreary, D. and Purcell, K.L. (eds) *Proceedings of the Fifth Symposium on Oak Woodlands: Oaks in California's Changing Landscape.* Forest Service, USA Department of Agriculture, San Diego, USA. pp. 511-528.

CDF-FRAP (California Department of Forestry and Fire Protection, Fire and Resource Assessment Program) (2003) The Changing California: Forest and Range 2003 Assessment. http://www. frap.cdf.ca.gov/assessment2003/index.html.

DiTomaso, J.M., Kyser, G.B. and Hasting, M.S. (1999) Prescribed burning for control of yellow star thistle (*Centaurea solstitialis*) and enhanced native plant diversity. *Weed Science* 47 (2), 233-242.

Edwards, S.W. (1992) Observations on the prehistory and ecology of grazing in California. *Fremontia* 20 (1), 3-11.

Ewing, R.A., Tosta, N., Tuazon, R., Huntsinger, L., Marose, R., Nielson, K., Motroni, R. and Turan, S. (1988) *Growing Conflict Over Changing Uses.* Anchor Press, Sacramento, USA.

Fehmi, J.S. and Bartolome, J.W. (2003) Impacts of livestock and burning on the spatial patterns of the grass Nassella pulchra (Poaceae). *Madrono* 50 (1), 8-14.

Gale, I. (2003) *West Marin Ranchers Increasingly Diversify.* Pt. Reyes Station, Pt. Reyes Light, California, USA.

GAO (General Accounting Office) (2003) *Major Management Challenges and Program Risks: Department of the Interior.* GAO, Washington, DC, USA.

Gentner, B.J. and Tanaka, J.A. (2002) Classifying federal public land grazing permittees. *Journal of Range Management* 55 (1), 2-11.

Gustanski, J.A. and Squires, R.H. (2000) *Protecting the Land: Conservation Easements Past, Present, and Future.* Island Press, Washington, DC, USA.

Hart, J. (1991) *Farming on the Edge: Saving Family Farms in Marin County, California.* University of California Press, Berkeley, USA.

Hayes, G.F. and Holl, K.D. (2003) Cattle grazing impacts on annual forbs and vegetation compostition of mesic grasslands in California. *Conservation Biology* 17 (6), 1694-1702.

Huntsinger, L., Buttolph, L. and Hopkinson, P. (1997) Ownership and management changes on California hardwood rangelands: 1985-1992. *Journal of Range Management* 50 (4), 423-430.

Jensen, D.B., Torn, M.S. and Harte, J. (1990) *In Our Own Hands: A Strategy for Conserving California's Biological Diversity.* University of California, Press. Berkeley, USA.

Kan-Rice, P. and Sokolow, A. (2003) Easements shield suburban farms from development. University of California, http://news.ucanr.org/newsstorymain.cfm?story=515 (accessed October 2004).

Land Trust Alliance (2000) *The Land Trust Census.* http://www.lta.org (accessed October 2004).

Liffmann, R.H., Huntsinger, L. and Forero, L.C. (2000) To ranch or not to ranch: home on the urban range? *Journal of Range Management* 53 (4), 362-370.

McClaran, M.P., Romm, J. and Bartolome, J.W. (1985) Differential farmland assessment and land use planning relationships in Tulare County, California. *Journal of Soil and Water Conservation* 40, 252-255.

McCreary, D.D. (2001) *Regenerating Rangeland Oaks in California.* University of California Agriculture and Natural Resources Communication Services Publication 21601, Oakland, California, USA.

Maestas, J.D., Knight, R.L. and Gilgert, W.C. (2001) Biodiversity and land-use change in the American mountain west. *Geographical Review* 91 (3), 509-524.

Merenlender, A.M., Huntsinger, L., Guthey, G. and Fairfax, S.K. (2004) Land trusts and conservation easements: who is conserving what and for whom? *Conservation Biology* 18 (1), 65-76.

Meyer, M.D. and Schiffman, P.S. (1999) Fire season and mulch reduction in a California grassland: a comparison of restoration strategies. *Madroño* 46 (1), 25-37.

Painter, E.L. (1995) Threats to the California flora: ungulate grazers and browsers. *Madrono* 42 (2), 180-188.

Raymond, L. and Fairfax, S.K. (1999) Fragmentation of public domain law and policy: an alternative to the 'Shift-to-Retention' thesis. *Natural Resources Journal* 39, 649-753.

Reeves, K. (2001) *Holistic Management and Biological Planning in California: You Be the Judge*. California Native Plant Society Association Newsletter Spring 4-7, California, USA.

Rowe, H.I., Bartlett, E.T. and Swanson, L.E., Jr (2001) Ranching motivations in two Colorado counties. *Journal of Range Management* 54 (4), 314-321.

Scott, J.M., Davis, F.W., McGhie, R.G., Wright, R.G., Groves, C. and Estes, J. (2001) Nature reserves: do they capture the full range of America's biological diversity? *Ecological Applications* 11, 999-1007.

Smith, A.H. and Martin, W.E. (1972) Socio-economic behaviour of cattle ranchers, with implications for rural community development in the West. *American Journal of Agricultural Economics* 54, 217-225.

Sulak, A. and Huntsinger, L. (2002) *Sierra Nevada Grazing in Transition: The Role of Forest Service Grazing in the Foothill Ranches of California*. The California Rangeland Trust, Berkeley, California, USA.

Tarrance Group, The (2001) Summary of Findings from a Nationwide Survey. http://www.aftresearch.org/farmbill/report.html (accessed July 2001).

Torell, L.A., Rimbey, N.R. Tanaka, J.A. and Bailey, S. (2001) The lack of profit motive for ranching: implications for policy analysis. In: Torell, L.A., Bartlett, E.T. and Larrañaga, R. (eds) *Proceedings of Annual Meeting of the Society for Range Management: Current Issues in Rangeland Resource Economics*. Kailua-Kona, Hawaii, USA. pp. 47-58.

Wilcove, D., Bean, M., Bonnie, R. and McMillan, M. (1996) *Rebuilding the Ark: Toward a More Effective Endangered Species Act for Private Land*. Environmental Defense Fund, Washington, DC, USA.

Cultural aspects of silvopastoral systems

I. Ispikoudis and K. M. Sioliou
Laboratory of Rangeland Ecology, Aristotle University of Thessaloniki, 54124, Thessaloniki, Greece. ispik@for.auth.gr

Abstract
Silvopastoral systems are multipurpose systems, integrating forest and forage production, meeting a wide variety of ecological and socio-economic conditions all over the world from time immemorial and have been a potential source of fuel, food, medicines and other materials. Attention tended to focus on these characteristics, but cultural aspects were overlooked and they were barely mentioned. Management practices respond to the vagaries of ecological, economic and political circumstances. An important factor influencing silvopastoral landscapes is the cultural attitude of the occupants. Societies view and perceive landscapes in fundamentally different ways, which are translated into unique patterns of activities. Perceptions, practices and customs towards the environment reflect cultural attitudes, spiritual beliefs and resource values. Lifestyles like nomadism, traditional techniques like pollarding and shredding are closely linked with such landscapes. Their symbolic and cultural perception is governed by the local practices, laws and customs. Silvopastoral landscapes represent traditional lifestyles, but they are also imbued with cultural, symbolic and even religious value.

Key words: cultural landscape, traditional lifestyles, traditional techniques

Introduction
In addition to the complex land-use history, numerous types of semi-natural silvopastoral landscapes were created where the impact of humans via grazing, cutting and burning is integral to the structure, composition and dynamics of the vegetation (Bakker, 1989; Sutherland and Hill, 1995; Papanastasis and Peter, 1998).

During the long history of human intervention over the past 10,000 years, equilibrium has been established on those non-cultivated upland ecosystems, which are neither overgrazed and heavily coppiced nor completely protected (Naveh and Dan, 1973; Rackham, 1986). The landscape transformation has resulted in agrosilvopastoral landscapes and the floristic composition and structure of the vegetation communities are determined by local climate, soil and management practices. This man-maintained equilibrium between trees, shrubs, herbs and livestock has contributed much to the biological diversity and attractiveness of these landscapes and is, without doubt, one of its main assets for recreation and tourism. However, in the last 40 years this equilibrium has been more and more distorted by the radical changes in land uses. Due to demographic, social, natural and economic changes, the evolution of these landscapes is rapid and most of the times irreversible, so they should be mapped, registered and evaluated before their complete disappearance. Main reasons for degradation of these landscapes are the temporal changes of management practices.

Origins and types of silvopastoral landscapes
Silvopastoral systems or wood-pasture systems or silvopastoral landscapes of one kind or another exist all over the world and have meshed with the everyday lives of men and women from time immemorial. The dehesas in Spain, the montados in Portugal, the *lovangar* (foliage meadows) of Sweden, the *larchenwisen* (larch meadows) of the eastern Alps, the *kouri* (wood pastures) in Greece, the woodland savannahs of Africa, the American Indian "oak openings" of Michigan, the *pascoli arborati* in Sardinia, the wood pasture in England, etc, are examples of such systems, which account for a high amount of biodiversity, but also represent the tradition and culture of the people of each country (Rackham, 1990). All down the millennia, these systems have been a potential source of fuel, food, medicines and other materials. Silvopastoral systems are intrinsically intersectoral and multipurpose, but attention tended to focus on the economic and ecological components. The cultural aspects were overlooked and they were barely mentioned in the public and scientific discourse, but silvopastoral landscapes represent more than just an aesthetic or ecological value. They represent an identity and cultural value as well, excellent examples of traditional lifestyles and techniques. Silvopastoral landscapes are closely linked with those techniques and/or pastoral management, representing the local know-how. In fact, these landscapes have a strong cultural aspect since all such management practices are based on a body of local or indigenous technical knowledge, which has evolved over time in response to the vagaries of ecological, economic and political circumstances. Undoubtedly one of the most important factors influencing these landscapes is the cultural attitude of the occupants. Societies view and perceive landscapes in fundamentally different ways, which translate into unique patterns of activities. Perceptions, practices and customs towards the environment reflect cultural attitudes, spiritual beliefs and resource values (Tane, 1995). Traditional techniques of livestock husbandry were widespread in universal silvopastoral landscapes, but in most cases are now extinct or highly modified. Silvopastoral systems are the cultural, social, economic and ecological heritage of the people.

The utilization of woody plants in silvopastoral landscapes

Silvopastoral landscapes are linked usually with traditional tree management. There are mainly two techniques, shredding and pollarding. Shredding consists of cutting the lower branches of the tree for fodder, while pollarding involves cutting off the branches of a tree at a height of at least 1.5-2 or 3 m of the trunk in order that the new sprouts are out of the reach of the animals. The trees were first cut at 10-15 years of age or when the stem diameter exceeded 15 cm. Trees were cut every 5-7 years (Rackham, 1986; Austad, 1990) The technique of pollarding was a way of protecting the trees from browsing and it dates back to antiquity, since it is mentioned by Theophrastus (Zaharis, 1977). In this kind of system only trees taller than grazing animals could survive (Rodriguez et al., 1999). The etymology of the word "koura", which means pollarding in Greek, implies exactly the same thing, since it derives from the words "kouros" (young) and "kourizo" (making young) (Dormparakis, 1989).

Leaf and twig fodder cut from trees played a major role in animal husbandry and in many areas stored hay was of critical importance to the survival of livestock housed indoors over winter. Leaf fodder is a major product of trees and woods in many European and Asian countries; sometimes it is the main object of management (Austad, 1990). In addition, leaf and twig fodder harvesting also played a major role in shaping the cultural landscapes and in particular the structure and composition of vegetation. The use of arboreal fodder is both poorly understood and largely ignored by historians, geographers and ecologists.

When people started to keep animals for domestic use, winter fodder had to be provided for livestock. In many countries, the winter is too cold and too long for all livestock for graze outside throughout the year. During the winter, it is necessary for livestock to be stalled in barns for 3 to 6 months from January to April, sometimes from November if the weather is particularly bad (Halstead, 1998).

The practice of collecting twigs and leaves for domestic animals is probably the oldest form of fodder harvesting. In contrast to collecting hay, which requires sharp sickles or scythes, leaf fodder can be collected without tools (Austad, 1990). However, iron tools make such harvesting more efficient. Material finds and illustrations in books, manuscripts and paintings from medieval times back to the Iron Age show that this practice was common in much of Europe (Troels-Smith, 1960; Andersen, 1990), although perhaps not in Britain (Rackham, 1990). Evidence for early Neolithic harvesting of leaf fodder has been found in Switzerland (Guyan, 1981). Pollarding has been common in Scandinavia, whereas shredding has been used widely in central and southern Europe and eastern Asia. This practice, which was common and widespread, affected not only the ecology and scenery of the cultural landscape, but various kind of arts as well. In Denmark, for example, nearly all trees depicted in medieval church fresco paintings have been shredded, while in one case the process itself is depicted (Andersen, 1990).

Both sedentarized populations and nomads practised this tree exploitation. The former were pollarding the trees in order to protect them from the transhumant flocks (a rivalry with the nomads). The latter were pollarding and shredding the trees to feed their animals on their way to the mountains. In Greece, a whole cultural landscape with various forms of at least seven species of deciduous oaks has been created by the people called Koupatsari, the "oak people" (Grove and Rackham, 2001). These "oak people" were sedentarized livestock farmers practising transhumance (Sivignon, 1975).

Tree utilization was a result of experience handed down over generations. All species of deciduous trees were used for animal fodder. Although cutting trees for collecting animal fodder was widely practised in many countries, the choice of species, techniques and utilization varied from country to country and area to area. Among human influences on the landscape some are directly connected with environmental factors. Some characteristics are probably due to the climate, while others result from historical accident (Emanuelson, 1990). The creation of deer parks after the Norman Conquest in the United Kingdom continued the wood pasture tradition. A park was a symbol of minor aristocracy (Rackham, 1978, 1986). Wood pastures are regarded as grasslands with trees (usually pollards) and they are largely responsible for England's special reputation as a land of ancient trees (Rose and James, 1974).

Tree fodder harvesting must have greatly influenced the landscape ecology. Halstead (1998) illustrated the significance of arboreal fodder in the landscape through an example from mountain villages in Greece. Firstly, hay has played a critical role in maintaining and shaping sedentary mixed farming in this agriculturally marginal environment. Occupation by sedentary farmers of the mountain areas in Greece may owe as much to management of woody plants as to the cultivation of crops and animal husbandry. Secondly, along with grazing, burning, logging and cultivation, leaf foddering must be regarded as one of the main cultural activities affecting the historical landscape. Shredding and pollarding of beech and oak trees had a widespread and drastic impact on the landscape of the mountains of Greece. Thirdly, hay collection has affected the landscape of the Greek mountains in the sense that certain areas were set aside, through the restrictions of cutting and grazing, for the development of leafy hay reserves. The selective felling of species other than oak radically altered the composition of the reserves, as a result of which an almost pure oak wood-pasture appears to have been created out of mixed coniferous and broadleaved deciduous forests. The village president authorized cutting periodically.

Perhaps the most interesting is that leaf fodder practices are associated with gender differentiation. The practices women used in exploiting tree resources are often more subtle than men's. Women's role as caretakers and users of the landscape is often crucial because they do most of the work of fodder (FAO, 2000). Women and children usually gathered and cut the branches into shorter pieces (0.75-1 m). The twigs were bunched and tied together with a twisted twig (Haeggstrom, 1983). The number of bunches collected depended on the number of livestock to be fed through the winter and the hay crop for the year. Large quantities of fodder are required for the winter. Domestic animals received at least one feeding with leaves (twigs) a day. Halstead (1998) has extrapolated that when a Greek village collectively owned around 2000 sheep and goats the villagers would have needed between 60,000 and 100,000 bundles of leafy hay per winter, which would have involved the shredding of between 3000 and 10,000 mature oaks, between 30,000 and 70,000 mixed young and mature trees, or 1200 to 2000 large beeches.

The collected bunches had to be dried. In outfields bunches were commonly placed on pollarded trees beyond the reach of animals or on special structures. The secondary branches were used to make short pegs for suspending bunches. These constructions were protected against grazing cattle. In infields bunches were commonly put along stone walls or against the barns, or hung on sticks or fences. It was also common to place bunches on hayricks or stacks. The time required for drying depended on the weather. Bunches were often stored and stacked in outfields. On some farms there were special ventilated barns for storing bunches, often located in the outfields. The farmers had special permanent places for stacking. They also place small semi-permanent stacks in the pollarded trees, beyond the reach of animals. All these stacks, barns and other constructions with specialized architecture were features of special landscapes, sometimes characteristic of an area.

The future of pollarding or shredding today

Comparisons with descriptions from different areas suggest that pollarding or shredding is representative for earlier periods in many European countries (Andersen, 1990; Austad, 1990; Emanuelson, 1990; Rackham, 1990). But, almost all over Europe only elderly people remain in the mountains. Because of the hard work involved in pollarding or shredding, very few farmers still continue this practice. The number of bunches collected is small. Concentrated feedstuffs are available and common everywhere. The reason for still using this kind of fodder, according to the farmers, is to improve the health of their livestock. However, little is known about the value of using tree fodder for controlling sheep parasites. The discontinuation of pollarding or shredding today results in a considerable loss to the heritage because the practice ceases and pollarded or shredded trees disappear from the landscape. Due to their high historical, aesthetic, recreational and ecological value it is essential that existing trees in special areas have to be rejuvenated and that traditional cutting techniques have to be reintroduced for their long-term management (Austad, 1990). Although tree pollarding or shredding for animal fodder was so widespread and intensive, very few people know of the practice or realize its importance, not only for the economy of small farms, but also as an important factor in the ecology of the cultural landscape or in the attractiveness of recreation areas.

Protection of historic buildings and farms should be coupled with protection and management of their associated cultural landscape, in which pollarded and shredded trees are essential elements (Haeggstrom, 1983). In Sweden there are areas where restoration and management have been carried out to preserve the most interesting features of the old cultural landscape. Considering that pollarding is hard work, it is a special national or international responsibility to encourage pollarding by offering farmers subsidies or assistance with labour.

The future of silvopastoral landscapes

Though landscape is a very recent concept it was very quickly incorporated into policy texts, with a very special mention as "cultural landscapes". So far "cultural landscapes" are recognized and protected at the international level and they include silvopastoral landscapes as well. For example the "organically evolved landscape" is one of the three main categories of "cultural landscapes" protected by UNESCO and is defined as: "landscape that results from an initial social, economic, administrative, and/or religious imperative and has developed its present form by association with and in response to its natural environment. Such landscapes reflect that process of evolution in their form and component features" (UNESCO, 2002).

At the European level the Dobriš assessment (*L'Environnement de l'Europe, le rapport de Dobriš*) represents a profound analysis of the European environment. Chapter 8 deals with landscape, recognizing and classifying European agricultural landscapes (Meeus *et al.*, 1990).

The Council of Europe recognized the significance of landscapes and the European Landscape Convention was opened for signature in Florence, Italy, on 20 October 2000 in the framework of the Council of Europe Campaign "Europe, a common heritage" (Council of Europe, 2002). This convention aims for the protection, management and planning of European landscapes. It treats at the same time the cultural and ecological aspects of European landscapes and it is based on two particular principles. Firstly, it is applied to not only the remarkable landscapes but also the ordinary ones. Secondly, it underlines the significance and necessity of the role of citizens in the decision making concerning "their" landscapes. The Council of Europe underlines the necessity to take account of landscape in policy, due to its exceptional value, as an indicator of everyday life. That is why the range of application is very extended; the convention is applied to all the territories of parties who signed it, including natural areas, urban and suburban areas, whether they are water, terrestrial or sea.

The terms used in the convention are defined in Article 1 in order to ensure that everyone concerned with the well being of Europe's landscapes interprets them uniformly:

"Landscape" means an area, as perceived by people, whose character is the result of the action and interaction of natural and/or human factors. The term "landscape" is thus defined as a zone or area as perceived by local people or visitors, whose visual features and character are the results of the action of natural and/or cultural factors. This definition reflects the idea that landscapes evolve through time, as a result of being acted upon by natural forces and human beings. It also underlines that a landscape forms a whole, whose natural and cultural components are taken together, not separately.

"Landscape policy" means an expression by the competent public authorities of general principles, strategies and guidelines that permit the taking of specific measures aimed at the protection, management and planning of landscapes.

"Landscape protection" means actions to conserve and maintain the significant or characteristic features of a landscape, justified by its heritage value derived from its natural configuration and/or from human activity.

"Landscape management" means action, from a perspective of sustainable development, to ensure the regular upkeep of a landscape, so as to guide and harmonize changes which are brought about by social, economic and environmental processes.

Landscape is considered as the main part of the local identity. That is why participation of the public in decision-making regarding 'their' landscapes is considered a requisite parameter of sustainable management (Council of Europe, 2002).

Conclusions

Remnants of the old cultural silvopastoral landscape in many countries can still be found. Traditional techniques of livestock husbandry were widespread in universal silvopastoral landscapes. Although in most cases they are now extinct or highly modified, there are few areas in Europe where such techniques are still practised. In earlier times pollarding and shredding was widespread in Europe. From the Iron Age until the Second World War this form of tree management was common as a means of obtaining winter animal fodder. Pollarded trees are still common. They are an important part of the cultural heritage and the aesthetics of the landscape. A few farmers still cut trees for animal fodder, but trees managed by this practice are rapidly disappearing from the cultural landscape. In some areas, the lack of active pollarding has made it necessary to restore and manage pollarded trees as a means of preserving a valuable aspect of cultural heritage.

References

Andersen, T.S. (1990) Changes in agricultural practices in Holocene indicated in a pollen diagram from a small hollow in Denmark. In: Birks, H.H., Birks, H.J.B., Kaland, P.E. and Moe, D. (eds) *The Cultural Landscape - Past, Present and Future.* Cambridge University Press, Cambridge, United Kingdom, pp. 395-407.

Austad, I. (1990) Tree pollarding in Western Norway. In: Birks, H.H., Birks, H.J.B., Kaland, P.E. and Moe, D. (eds) *The Cultural Landscape - Past, Present and Future.* Cambridge University Press, Cambridge, United Kingdom, pp. 11-29.

Bakker, J.P. (1989) *Nature Management by Grazing and Cutting.* Kluwer, Dordrecht, The Netherlands.

Council of Europe (2002) Council of Europe and the Nature, European Convention on Landscapes (le conseil de l'Europe and la nature, convention européenne sur le paysage). URL http://nature.coe.int/french/main/paysage/conv.htm (accessed February 2002). (In French.)

Dormparakis, X.P. (1989) *Glossary of Classic Greek Language.* I.D. Kollaros and AE, Athens, Greece. (In Greek.)

Emanuelson, U. (1990) A model for describing the development of the cultural landscape. In: Birks, H.H., Birks, H.J.B., Kaland, P.E. and Moe, D. (eds) *The Cultural Landscape - Past, Present and Future.* Cambridge University Press, Cambridge, United Kingdom, pp. 111-121.

FAO (2000) Trees without forests, towards a better awareness, FAO conservation guide 35, http://www.fao.org/DOCREP/005/Y2328E/y2328e07.htm (accessed April 2004).

Grove, A.T. and Rackham, O. (2001) *The Nature of Mediterranean Europe, an Ecological History.* Yale University Press, New Haven and London, United Kingdom.

Guyan, W.U. (1981) Husbandry in Steinzeitdorf Thayngen-Weier II (Zur Viehhaltung im Steinzeitdorf Thayngen-Weier II). *Archaeologie der Scweiz* 4, 112-9. (In German.)

Haeggstrom, C.A. (1983) Vegetation and soil of the wooded meadows in Nato, Aland. *Acta Botanika Fennica* 120, 1-66.

Halstead, P. (1998) Ask the fellows who lop the hay: leaf-fodder in the mountains of northwest Greece. *Rural History* 9 (2), 211-234.

Mecus, J.H.A., Wijermans, M.P. and Vroom, M.J. (1990) Agricultural landscapes in Europe and their transformation. *Landscape and Urban Planning* 18, 289-352.

Naveh, Z. and Dan, J. (1973) The human degradation of Mediterranean landscapes in Israel. In: Di Castri, F. and Mooney, H.A. (eds.) *Ecological Studies 7: Mediterranean Type Ecosystems,* Springer-Velag, Heidelberg, Germany. pp. 273-289.

Papanastasis, V.P. and Peter, D. (1998) *Ecological Basis of Livestock Grazing in Mediterranean Ecosystems.* European Commission, Luxembourg.

Rackham, O. (1978) Archaeology and land-use history. Epping Forest - the natural aspect? *Essex Naturalist N.S.* 2, 16-57.

Rackham, O. (1986) *The History of the Countryside.* J.M. Dent and Sons, London, United Kingdom.

Rackham, O. (1990) Trees and woodland in a crowded landscape - the cultural landscape of the British Isles. In: Birks, H.H., Birks, H.J.B., Kaland, P.E. and Moe, D. (eds) *The Cultural Landscape - Past, Present and Future.* Cambridge University Press, Cambridge, United Kingdom, pp. 53-77.

Rodríguez R., Lopez-Carasco, C. and Zuzuarreguie, J. (1999) Effects on natural regeneration of woodland in a dehesa system grazed by Avilena-Negra Iberica Breed. In: Etienne, M. (ed.) *Dynamics and Sustainability of Mediterranean Pastoral Systems.* CIHEAM, Zaragoza, Spain, pp. 257-254.

Rose, F. and James, P.W. (1974) The corticolous and lignicolous (lichen) species of the New Forest, Hampshire. *Lichenologist* 6, 1-72.

Sivignon, M. (1975) *Thessalia: Geographical Analysis of a Greek Province* (La Thessalie analyse géographique d'une province grecque). Institut des études Rhodaniennes des universités de Lyon, Lyon, France, 572 pp. (In French.)

Sutherland, W.I. and Hill, D.A. (1995) *Managing Habitats for Conservation.* Cambridge University Press, Cambridge, United Kingdom.

Tane, H. (1995) *Ecography: Mapping and Modelling Landscape Ecosystems, Technical Manual.* River Murray Mapping, Murray Darling Basin Commission, Canberra, Australia.

Troels-Smith, J. (1960) Ivy, mistletoe and elm: climatic indicators - fodder plants. Danmarks Geologiske. In: Moe, D. (ed.) *The Cultural Landscape - Past, Present and Future.* Cambridge University Press, Cambridge, United Kingdom. 4p.

UNESCO (2002) Annex 4 of the Convention: guidelines for including specific types into the world heritage sites (Annexe 4 de la convention, *Orientations pour l'Inclusion de Types Spécifiques de Biens Sur la Liste du Patrimoine Mondial).* URL: http://www.unesco.org (accessed February 2002) (In French.)

Zaharis, S.A. (1977) *The Forests of Crete from Antiquity until Today.* Forest Service, no. 39, Aspioti, ELKA, Athens, Greece. (In Greek.)

Comparative analysis of the EAA/EAF and AAS agroforestry accounting systems: theoretical aspects

P. Campos-Palacín [1], P. Ovando-Pol [1] and Y. Rodríguez-Luengo [2]
[1] Institute of Economics and Geography, Spanish Council for Scientific Research, Pinar 25, 28006 Madrid, Spain.
[2] Faculty of Economics and Business Science, Complutense University of Madrid, Spain. pcampos@ieg.csic.es

Abstract

The integration of environmental accounting with a unique social accounting system is arousing growing interest in research circles. There is emerging demand, too, from government, due to European Union (EU) regulations on management of natural resources and damage to the environment. The EU still estimates agroforestry income according to the satellite system 'Manual on Economic Accounts for Agriculture and Forestry', which is based on the 'European System of Accounts' social accounting rules. In this chapter, rather than the conventional national accounting system, an 'Agroforestry Accounts System' is propored that integrates commercial and environmental values in one global social total income indicator. Furthermore, the chapter presents a critical analysis of the theoretical differences between the two accounting systems.

Key words: social accounting, environmental valuation, capital gains, total social income

Introduction

The accounting methodology used by governments in developed countries to measure social income has been criticized by user groups for its notable shortcomings, chiefly from the standpoint of environmental economics (Nordhaus and Kokkelenberg, 1999; Vincent, 1999; Campos-Palacín, 2002a; Perrings and Vincent, 2003; Campos-Palacín and Casado, 2004).

The depletion and degradation of natural resources and the extinction of wildlife species due to economic activity are having an adverse effect on the welfare of ever wider groups of the human population. Therefore, if the economic values of these resources were to have a bearing on how total social income is calculated, society would be better aware of the way in which the use of natural and environmental resources contributes to human economic welfare (Campos-Palacín et al., 2001; Pulido et al., 2003).

In the wake of debate on reforms to the social accounting system in international economic institutions, the Commission and Parliament of the European Union (EU) approved in 1994 a Communication to the effect that the Union should draw up proposals to introduce a system of 'ecological accounting' (Commission, 1994). This recommendation from EU policy-making bodies that environmental values form part of the calculation of national income brought about the creation of working groups in charge of framing technical proposals for reform of the 'European System of Accounts - ESA 95' (hereinafter, "ESA"), which is still in place today (Commission, 1996).

The EU's Directorate-General VI of Agriculture uses the ESA satellite system 'Manual on Economic Accounts for Agriculture and Forestry - EAA/EAF 97' (hereinafter, "EAA/EAF") (Commission, 2000a) to estimate the social income of the primary sector. The key novelty of EAA/EAF is its implied acceptance of the concept of Hicksian income (Hicks, 1946), which sets down that the natural growth of timber forest must be considered in estimating silviculture income.[1] Eurostat has also published pilot proposals for natural resource accounting. In the best of cases, they merely recognize the concept of Hicksian income, but do not suggest how environmental values might be incorporated in measurements of total social income.[2]

It is rare for a measurement of social income to be based on assumptions as dissociated from reality as those underlying agroforestry accounting. That is why the income statistics for multiple use of agrosilvopastoral systems (AFSs) are of key importance to a fair assessment of these systems.

This chapter aims to present a comparative analysis of the theory for measuring the total social income (TSI) of an agroforestry system on the conventional EAA/EAF accounting system and the authors' proposed 'Agroforestry Accounts System' (AAS).

Theoretical comparison of the EAA/EAF and AAS systems

Theory of total economic value

The concept of economic activity
In this chapter, the concept of economic activity extends to any act of processing, consumption and appropriation of scarce new goods and services intended directly or indirectly to meet human needs, whether or not they are normally traded on an institutionalized market.[3]

Private economic values

The total economic value (TEV) of scarce goods and services arises from people's behaviour as consumers.[4] Table 1 shows a classification of people's various potential uses of scarce goods and services that provide the basis for the TEV of those goods and services (Pearce, 1993; Campos-Palacín, 1999).

People in developed countries are accustomed to there being a market in goods and services. This tends to make them think that economic value attaches only to things traded on an institutionalized market. But there are other economic values that people habitually consume without directly buying them on a market, for themselves or through others, e.g. landowners auto-consume an environmental capital income that consists of enjoyment of the recreational and welfare services provided to them by conservation of the habitat of their dehesa estate.[5]

Table 1. Total economic value from AFS. Source: Own elaboration.

Active uses		Passive uses	
Present uses		Future uses	
Direct	Indirect	Option	Existence value
There is often exclusivity or competition in use.	Environmental services considered as an input for the production of other goods and services.	Users' willingness to pay for ensuring future own or third person's goods and services consumption.	Users' willingness to pay for ensuring the future existence of a good or service independently of its active consumption.
E.g. grazing resources, cork, wood, firewood, livestock, crops, hunting, recreation, mushrooms, wild plants, etc.	E.g. habitat sustainability functions, flood damage prevention; carbon fixation; greenhouse gases net emissions saving, etc.	E.g. conservation of biological resources for pharmaceutical uses, etc.	E.g. preservation of unique habitat and endangered wild species, etc.

Public economic values

Recreational visitors enjoying free access consume public environmental services on dehesas, which likewise carry an economic value; such value can be measured by simulating a hypothetical market. The public not visiting dehesas could, as 'passive users', assign an economic conservation value due to option and existence values of these natural areas (Table 1). To date, the conservation value ascribed by non-visiting passive users to dehesas has not yet been measured.[6]

It is difficult to estimate the value of 'indirect environmental services' due to the fact that it is often impossible to define simulated market procedures. 'Indirect active users' and non-visiting passive users tend to be grouped under a single environmental value which is then held to benefit or prejudice 'society as a whole' (Table 1).

The concept of total social income

Academic circles and the public bodies in charge of measuring national income generally accept that the real TSI produced by a territory, in the context of TEV theory, is Hicksian income. This may be stated as the 'maximum possible consumption of the commercial and environmental resources of a territory in an accounting period (one year) without diminishing the value of the original wealth of that territory by the end of the period'[7] (Campos-Palacín 2002b and Caparrós *et al.*, 2003).

TSI, in line with TEV theory and the concept of Hicksian income, should be estimated by aggregating the net value added of raw materials and services consumed in the production process, on one hand, and, on the other, capital gains originated in the wealth of the territory throughout the accounting period.

The European Union's social accounting system

The EAA/EAF system calculates the social income of the territory of a country with reference only to the 'commercial net value added' of raw materials and 'commercial' services consumed in the course of the production process. Hence the value of total output of 'commercial goods and services' in the territory of the nation is the sum of intermediate consumption and commercial gross value added.

The agroforestry accounting system

The AAS structures all information into the 'production account' and the 'capital balance account'. These two accounts are presented for the agroforestry territory under study as a whole and for each disaggregated activity as desired.[8] The production account takes as the value of total output (TO) the sum of intermediate outputs (IO) of raw materials and intra-consumed services[9] and the value of the final output (FO)[10] generated in the accounting period. Total cost (TC) breaks down into intermediate consumption (IC),[11] labour costs (LC) and fixed capital consumption (FCC). The total cost of the agroforestry system comprises the total cost incurred by the landowner and the direct total cost arising from public spending by the environmental authorities on intermediate output of services and internal investment consumed by the multiple activities of the agroforestry system.

The capital balance account presents the value of stocks (initial and final) and changes (entrances and withdrawals) of production-in-progress[12] and fixed capital goods[13] involved during that accounting year in generating the TSI of the agroforestry system.

The two accounts, production and capital balance, enable us to estimate the TSI of the agroforestry system as the sum of net value added (NVA) and capital gains (CG):

$$TSI = NVA + CG; NVA = TO - IC - FCC = LC + NOM; CG = Cr - Cd + FCC;^{14}$$
$$TSI = GVA + Cr - Cd; TSI = LC + NOM + Cr - Cd + FCC; TSI = LC + CI,$$

where Cr is the revaluation of capital goods over the period, Cd is the value of capital goods destroyed, the economic use of which is zero, GVA is gross value added and CI is social capital income.

Comparison of the EAA/EAF and AAS accounting systems

Intermediate output
IO as grazing resources consumed by animals[15] is ignored by the EAA/EAF system. By omitting these in its calculation of forestry income, the EAA/EAF system both underestimates the capital income of the forestry activity by that same value and overestimates the capital income of the animal activity by the value of the grazing resources.

Gross natural growth
Gross natural growth (GNG) of trees and forestry products whose extraction cycle is greater than 1 year is ignored by the EAA/EAF system. Instead, the value of extractions is regarded as output for the given year.[16]

Livestock final output
Livestock final output stock on the EAA/EAF system is the difference between total livestock value at the end and at the beginning of the accounting period. AAS considers the year-end value of livestock production-in-progress only on stating final livestock; also AAS reflects the initial livestock value as part of the intermediate cost.

Private environmental services
The market price of a dehesa significantly depends on its 'private environmental services' (ESc), in the form of recreational enjoyment and habitat conservation, which its landowner auto-consumes exclusively.

ESc are not accounted for in the value of final output under EAA/EAF. This omission is unwarranted by theory and, furthermore, breaches ESA rules, which require a record of all auto-consumed private outputs stemming from private commercial wealth.[17]

Public environmental services
Public 'rights of way' afford visitors free access to consumption of the 'non-controlled public environmental services' (ESnc) of dehesas. Visitors would be willing to pay a given sum of money as consideration for the welfare enjoyed on their visit (recreational value) and a further amount to prevent deterioration of the environmental quality of the dehesa habitat (conservation value). Besides visitors, 'society as a whole', too, even without visiting dehesas, may derive environmental benefits from 'habitat conservation' and the 'indirect ecological functions' of dehesas. EAA/EAF still fails to recognize the value of public environmental services in measuring the total social income of AFSs.

Capital revaluation and destruction
'Capital revaluation' (Cr) of production-in-progress and of durable goods involved in generating the total output of AFSs rightly does not form part of net value added under EAA/EAF, although the figures must be aggregated with net value added to produce a full estimate of TSI.

'Capital destruction' (Cd), like revaluation, is ignored by EAA/EAF, except for livestock. Revaluation of adult breeders or of work is viewed by EAA/EAF as an output for the accounting year. An adult animal, if not fattened or grown, does not generate value added but is a wealth completed since the year opened; like any wealth, however, it has a direct impact on income through capital gains.

Direct expenditures of the government
EAA/EAF does not take account of direct total government's expenditures in activities that affect the production of goods and services in the AFSs in order to calculate its TSI.

Government takes on management or performance of various AFS activities. Government's productive activities are divided into intermediate output of services and internal investments. Government's expenses directly affect output of the public and private goods and services - both commercial and environmental - of AFSs. Direct outputs are added to the costs of other AFS activities, in the amount of the intermediate output of services and depreciation of internal investments. These indirect costs originating with the government's expenditure give rise to 'additional economic values' in total outputs and damages avoided - net of profits - to the capital goods of other multiple uses in an AFS.

The government's total direct expenditures need not be less than the rise in capital income from the aggregate of other activities. Government may allocate expenditure to conservation of unique habitats and endangered species; if so, the guiding principle is not economic efficiency - which in the last analysis simply reflects the preferences of society at the time - but the preservation of a unique natural or man-made resource.

Total social income

Conventional commercial net value added (CNVAc) as estimated under EAA/EAF does not, as pointed out above, represent the whole of commercial net value added (CNVA): there is a group of outputs of private commercial goods and services of AFSs that may be omitted from income figures or, though not omitted, may be an inaccurate reflection of the real output value. The term 'additional commercial net value added' (CNVAa) has been chosen to describe the operating income of omitted or dislocated commercial outputs. The AFSs of at least developed countries have associated ESc that may be measured: in this chapter, a 'controlled environmental net value added' (ENVAc) is ascribed to them. The ESnc for the financial period - consumed by visitors to AFSs and by society as a whole - offer 'non-controlled environmental net value added' (ENVAnc). On aggregate those value added figures provide the 'total social net value added' (NVA) of AFSs as measured on the AAS accounting methodology:

$$NVA = CNVAc + CNVAa + ENVAc + ENVAnc = CNVA + ENVA.$$

Capital gains (CG) may be classified as commercial (CCG) or environmental (ECG). Environmental capital gains are in turn disaggregated into controlled (ECGc) and non-controlled (ECGnc) environmental capital gains. This classification of NVA and CG types enables us to find the comparative differences in calculation of total social income (TSI) under the EAA/EAF and AAS systems:

$$CG = CCG + ECGc + ECGnc = CCG + ECG,$$
$$TSI = CNVAc + CNVAa + ENVA + CG.$$

Recommendations to government for ESA reform

Outlined below, in the form of recommendations to government – in its role as regulator of the ESA social accounting system – are the key requirements that should be met by any future reform to the standard model of social accounting to measure the total sustainable social income of an agroforestry system. All these requirements are met by AAS.
A reformed system should:

1. Apply a broader concept of economic activity, incorporating environmental goods and services.
2. Quantify and place a value on intermediate output originating with direct private and public management of the agroforestry system.
3. Quantify the value of gross natural growth of wood outputs in the estimation of operating income.
4. Consider the value of production in progress used as intermediate cost.
5. Quantify and place a value on production-in-progress and fixed capital balances over 5-year periods in order to present annual estimates of the capital gains of an agroforestry system.
6. Standardize valuation criteria for environmental services so that they may be aggregated with commercial goods and services, thus creating a range of private and social income indicators and a synthetic TSI indicator generated in AFSs.

Acknowledgements

This research was funded in the context of the project Economía y Selvicultura del Alcornocal (Plan Nacional I+D: AGL2000–0936–C02–02) – 'Cork-oak Economics and Silviculture' (National R and D Plan). The authors are solely responsible for the errors or shortcomings that may remain in this work.

Notes

[1] This recognition has had no application whatever to date.
[2] This is the case of forests under the pilot proposal 'The European Framework for Integrated Environmental and Economic Accounting for Forests – IEEAF'– (Commission, 2000b).
[3] Note that the adjective 'environmental' in this chapter refers to an absence of exchange. Thus any scarce goods or services consumed or appropriated that are not usually traded on an institutionalized market are classed as 'environmental value'.
[4] In this regard, investment is conceived of as consumption deferred over time.
[5] For further detail on valuation of environmental services consumed by dehesa landowners, see Campos-Palacín and Mariscal (2003).
[6] This last value may be significant among the European public due to the widespread notice and high regard attained by the public environmental values of the dehesa.
[7] The absence of net donations and new discovery of wealth for the agroforestry territory has been assumed.

[8] An example of a summary classification of territorial activities with final outputs in AFSs: agriculture, forestry, animals, services and infrastructure construction.

[9] The intermediate output of the agroforestry system records the value of intermediate services produced through direct public expenditure by the environmental authority. It has been chosen to record under intermediate output of services the value (cost plus margin) of ordinary work funded by public expenditure and consumed by other uses of the agroforestry system, and under final output of internal investment, the value (cost plus margin) of infrastructure, building works and any other output of durable goods paid for out of direct public expenditure.

[10] Final output (FO) disaggregates into internal investment (IFO), final sales (SFO), existing final output (EFO) and other final outputs (OFO).

[11] Intermediate consumption (IC) is classified into raw materials (RM), services (SS) and production-in-progress used (PPu).

[12] In an agroforestry system, agricultural, forest-related and animal production-in-progress utilized in the accounting period may be a highly significant item in the value of intermediate consumption. Even in a stable price situation, there may be a revaluation of production-in-progress due to the discount effect. Initial cork stocks rise in value at the year-end because the time horizon of the discount applied to the current price assumed for the years in which cork is stripped has diminished by 1 year.

[13] The social fixed capital balance account distinguishes between land values (including the value of non-loggable woodland, forest improvements and agricultural improvements), animal, infrastructure and machinery. It is worth pointing out that knowledge of the environmental capital incomes of the agroforestry system enables us to delimit that part of the social capital value of the land that is owed to environmental services.

[14] Fixed capital consumption has been added to capital gains (CG) to avoid double-accounting, since on calculating capital revaluation (Cr) fixed capital devaluation is discounted implicitly and it is also considered in the calculation of total cost.

[15] The only intermediate output that *is* taken into account by the EAA/EAF model is harvested agricultural output consumed by livestock in the form of supplementary feed or used as seed in the same accounting year as it was harvested.

[16] Even if the quantities of wood and cork extracted and produced (gross annual growth) match, their values differ due to the effect of price discounting in respect of the waiting time for future extraction of the natural growth accruing in the accounting year. The identity that relates both values is: PPu = GNG + PPr, where PPu is the standing-stock value of the quantity of extracted product (production-in-progress used), GNG is the gross natural growth of the forest product in the accounting year, valued at standing-stock prices discounted in accordance with the years remaining to the next round of extraction at the year-end, and PPr is the revaluation of all initial standing stocks of the product at year-end.

[17] The ESc of dehesas are implicitly a commercial value, since the market price of a hectare incorporates them. Omission of the output value of ESc is a breach of ESA rules by the EAA/EAF system, or at least of the underlying economic theory of ESA, although the 'letter' of the rules also ignores ESc.

References

Campos-Palacín, P. (1999) Towards the measurement of the well-being benefit of the forest multifunctionality (Hacia la medición de la renta de bienestar del uso múltiple de un bosque). *Investigación Agraria: Sistemas y Recursos Forestales* 8 (2), 407-422. (In Spanish.)

Campos-Palacín, P. (2002a) Economy of the multiple use of the forest: forests in Jerez de la Frontera (1991-1993) (Economía del uso múltiple del bosque: Montes Propios de Jerez de la Frontera (1991-1993)). *Revista Española de Estudios Agrosociales y Pesqueros* 195, 147-186. (In Spanish.)

Campos-Palacín, P. (2002b) Nature accounting (Los números de la naturaleza). *Ambienta* 15, 3. (In Spanish.)

Campos-Palacín, P., Rodríguez, Y. and Caparrós, A. (2001) Towards the dehesa total income accounting: theory and operative Monfragüe study cases. *Investigación Agraria: Sistemas y Recursos Forestales* 1, 43-67.

Campos-Palacín, P. and Mariscal, P. (2003) Owners' preferences and public intervention: case study in the dehesas in Monfragüe (Preferencias de los propietarios e intervención pública: el caso de las dehesas de la comarca de Monfragüe). *Investigación Agraria: Sistemas y Recursos Forestales* 12 (3), 87-102. (In Spanish.)

Campos-Palacín, P. and Casado, J.M. (2004) (eds) *Environmental Accounting and Economical Activity* (Contabilidad ambiental y actividad económica). Consejo General de Colegios de Economistas of Spain, Madrid, Spain. (In Spanish.)

Caparrós, A., Campos-Palacín, P. and Montero, G. (2003) An Operative framework for total hicksian income measurement: application to a multiple use forest. *Environmental and Resource Economics* 26, 173-198.

Commission of the European Communities (1994) COM (94) 670 final: *Guidelines to be Applied within the EU related to the Environmental Indicators and the National Ecological Accounting. Integration of the Systems of Environmental and Economic Information* (Directrices que debe seguir la UE en relación con los indicadores ambientales y la contabilidad ecológica nacional. Integración de los sistemas de información ambiental y económica). Brussels, Belgium. (In Spanish.)

Commission of the European Communities (1996) *European System of Accounts* – ESA 95. Eurostat, Luxemburg.

Commission of the European Communities (2000a) *Manual on Economic Accounts for Agriculture and Forestry* – EAA/EAF 97 (Rev.1.1). Eurostat, Luxemburg.

Commission of the European Communities (2000b) *The European Framework for Integrated Environmental and Economic Accounting for Forests* – IEEAF. Eurostat, Luxemburg.

Hicks, J.R. (1946) *Value and Capital,* 2nd edn. Oxford University Press, Oxford, United Kingdom.

Nordhaus, W.D. and Kokkelenberg, E.C. (1999) *Nature's Numbers: Expanding the National Economic Accounts to Include the Environment*. National Academic Press, Washington, DC, USA.

Pearce, D.W. (1993) *Economic Values and the Natural World*. Earthscan, London, United Kingdom.

Perrings, C. and Vincent, J.R. (2003) *Natural Resource Accounting and Economic Development*. Edward Elgar, Cheltenham.

Pulido, F.J., Campos-Palacín, P. and Montero, G. (2003) *Forest Management in the Dehesa: History, Ecology, Silviculture and Economy* (La gestión forestal de la dehesa: historia, ecología, selvicultura y economía). Junta de Extremadura/IPROCOR, Mérida, Spain. (In Spanish.)

Vincent, J.R. (1999) A framework for forest accounting. *Forest Science* 45 (4), 552-561.

Comparative analysis of the EAA/EAF and AAS agroforestry accounting systems: application to a dehesa estate

Y. Rodríguez-Luengo,[1] P. Campos-Palacín [2] and P. Ovando-Pol [2]
[1]Contact address: Faculty of Economics and Business Science, Complutense University of Madrid, Campus de Somosaguas, 28023 Pozuelo de Alarcón, Madrid, Spain. [2]Institute of Economics and Geography, Spanish Council for Scientific Research, Madrid, Spain. yrl@ieg.csic.es.

Abstract
This chapter applies the satellite system 'Manual on Economic Accounts for Agriculture and Forestry - EAA/EAF 97' of the 'European System of Accounts' (ESA) and the 'Agroforestry Accounts System' (AAS) to a dehesa estate. Both accounting approaches are applied in order to measure total social income generated by the multiple use of this agrosilvopastoral system. Results show significant differences in capital and total social incomes estimated by EAA/EAF and AAS accounting methodologies.

Keys words: social accounting, total social income, social capital income

Introduction
This chapter compares the income measurements provided by the satellite system 'Manual on Economic Accounts for Agriculture and Forestry - EAA/EAF 97' (Commission, 2000) of the 'European System of Accounts' (ESA) (Commission, 1996) and the 'Agroforestry Accounts System' (AAS) (Campos-Palacín, 1999; Campos-Palacín et al., 2001), both applied to a dehesa estate called Haza de la Concepción.[1]

In Spain, where agrosilvopastoral systems (AFSs) account for about 50% of the surface area of the country (Díaz et al., 1997), it is interesting to see how unlike reality is the picture of total social income drawn by the conventional EAA/EAF methodology in comparison to the measurements produced by the AAS based on Hicksian income theory (Hicks, 1946) and used for this study.

The aims of the chapter are to analyse the shortcomings of EAA/EAF to reflect total social income[2] measurement for agroforestry systems and to show the improvement that could be reached by applying AAS methodology in calculating total social income of agroforestry systems as it is the case of dehesa in Spain.

The Haza estate is a dehesa under public ownership (Cáceres Provincial Authority) in the municipality of Malpartida de Plasencia, Monfragüe shire, Cáceres province.[3] The dehesa encompasses 676 ha of 'useful agricultural land' (UAL), divided into 81% 'forest' land, in turn made up of 65% 'wooded' surface of holm-oak and cork-oak, of an average density of 35 trees per hectare, and 16% unwooded 'pastureland'. Irrigated crop land accounts for the remaining 19% of UAL at Haza, although some of that surface is used for non-irrigated crops (Rodríguez et al., 2004).

The results of the application are set out in disaggregate form in the production account for the following goods and services: cork, forest grazing, cattle, polyphite meadows, other crops, other commercial goods, government-expenditure intermediate services, private environmental services and public environmental services. The data provided in the production account (Table 1) reflect reality objectively when drawn from commercial values. Environmental service values have been estimated on the basis of two contingent valuation surveys carried out in the Monfragüe shire (Campos-Palacín, 1998; Campos-Palacín and Mariscal, 2003).

Comparison of total social income figures provided by EAA/EAF and AAS accounting methodology
Outputs and costs
Table 1 shows the production account developed by the AAS system. Total output involves double accounting by including intermediate output. Total cost also involves double accounting, because government intermediate services are taken into account twice to estimate intermediate consumption costs. The advantage of including double accounting in the production account is that each singular activity considered has its own total output and cost without double accounting. This is the only way to obtain the real income of a singular activity and avoids the omission of output and cost. As has been remarked in the authors' previous chapter (see Campos-Palacín et al., this volume) this is not the case of the EAA/EAF system. The AAS accounting system enables the measurement of real income for a singular activity or for the whole dehesa. The following text of this section is devoted to a description of the major differences in the measurement of incomes between EAA/EAF and AAS approaches.[4]

Due to its omission of forest and cropland grazing resources as intermediate outputs (IO) forming part of the costs of the livestock activity, EAA/EAF overestimates by €122.2/ha of UAL the value added of cattle at Haza, and underestimates by that same amount the value added of the forestry and agriculture activities as a whole (Table 1). In the case of Haza, the estimated value of final sales (SFO) on the EAA/EAF model is overestimated by a little under 10% through inclusion of sales of adult breeders.

Table 1. Social production account by activity in Haza estate. (UAL €/ha, year 2002). [1]ESc: controlled environmental services (private environmental services); ESnc: non-controlled environmental services (public environmental services). Source: Own elaboration.

Class:[1]	Cork	Grazing resources	Cattle	Polyphite meadows	Annual crops	Other goods	GIS	ESc	ESnc	Total
1. Total output (TO)	62.0	92.5	647.1	51.3	174.2	46.0	19.2	96.1	20.8	1209.2
1.1 Intermediate output (IO)	2.5	87.1	39.6	35.1	82.0	0.0	19.2	0.0	0.0	265.6
1.1.1 Private intermediate raw materials (PIRM)	2.5	87.1	39.6	35.1	82.0					246.4
1.1.2 Government intermediate services (GIS)							19.2			19.2
1.2 Final output (FO)	59.4	5.4	607.5	16.2	92.2	46.0	0.0	96.1	20.8	943.6
1.2.1 Internal investment (IFO)	0.8	2.1	60.1	7.1		29.0				99.1
1.2.1.1 Private (PIFO)	0.8	2.1	60.1	7.1		28.5				98.5
1.2.1.2 Government (GIFO)						0.6				0.6
1.2.2 Final sales (SFO)	41.4	0.0	200.4	9.2	55.5					306.4
1.2.3 Existing final output (EFO)	17.2	3.3	347.0	0.0	36.8					404.3
1.2.4 Other final output (OFO)	0.0	0.0	0.0	0.0	0.0	17.0	0.0	96.1	20.8	133.8
2. Total cost (TC)	40.4	72.8	760.8	65.6	212.7	28.6	19.1	0.0	2.9	1203.0
2.1 Intermediate consumption (IC)	33.3	51.8	627.2	21.9	110.3	18.8	1.6	0.0	2.9	867.8
2.1.1 Raw materials (RM)	0.4	42.1	273.4	15.2	63.4	9.6	0.8	0.0	0.0	405.0
2.1.1.1 Private intermediate (PIRM)	0.0	34.1	194.6	3.4	2.2	0.0				234.2
2.1.1.2 Private external (PERM)	0.4	8.0	78.8	11.9	61.2	9.6				169.9
2.1.1.3 Government external (GERM)							0.0	0.8		0.8
2.1.2 Services (SS)	10.2	6.4	6.8	6.7	11.5	9.2	0.7	0.0	2.9	54.3
2.1.2.1 Private external (PES)	0.4	0.6	6.8	6.7	11.5	8.1				34.0
2.1.2.2 Government intermediate (GIS)	9.8	5.8				0.8			2.9	19.2
2.1.2.3 Goverment external (GES)						0.4	0.7			1.1
2.1.3 Production- in-progress used (PPu)	22.8	3.3	347.0	0.0	35.5					408.5
2.2 Labour costs (LC)	7.0	11.9	106.6	28.2	71.5	9.8	16.7	0.0		251.7
2.2.1 Private (PLC)	7.0	11.9	106.6	28.2	71.5	9.6				234.9
2.2.2 Government (GLC)						0.2	16.7			16.9
2.3 Fixed capital consumption (FCC)	0.0	9.2	27.0	15.6	30.8	0.0	0.9	0.0	0.0	83.5
2.3.1 Private (PFCC)	0.0	9.2	27.0	15.6	30.8	0.0				82.6
2.3.2 Government (GFCC)							0.0	0.9		0.9
3. Net operating margin (NOM = 1-2)	21.6	19.7	-113.7	-14.3	-38.5	17.4	0.1	96.1	17.9	6.2

The value of gross natural growth (GNG) of standing cork in Haza is a forestry output that EAA/EAF ignores. The value of annual cork extraction is, on both EAA/EAF and AAS, a final output for the accounting year. AAS estimates final cork output 29% above the figure produced under EAA/EAF. Moreover, unlike EAA/EAF, AAS includes production-in-progress used (PPu) as a component of intermediate consumption cost. At Haza, PPu represents

68% of the intermediate consumption of the cork activity (Table 1). For Haza, EAA/EAF estimates net value added of cork 142% above the figure calculated under AAS.

Table 2. EAA/EAF and AAS accounting systems comparison in Haza estate (UAL €/ha, year 2002). [1]EAA/EAF: *Economic Accounts for Agriculture and Forestry* (Commission, 2000); ESA: *European System of Accounts* (Commission, 1996); AAS: Agroforestry Accounting System (Campos-Palacín, 1999); ESc: controlled environmental services; ESnc: non-controlled environmental services. Source: Own elaboration.

Class: [1]	EAA/EAF	AAS ESA Additional private goods and services Commercial	AAS ESA Additional private goods and services ESc	AAS ESA Total	AAS ESnc	AAS Total
	1	2	3	4 = 1 + 2 + 3	5	6 = 4 + 5
1. Total output (1.1 + 1.2)	506.1	586.3	96.1	1188.4	20.8	1209.2
1.1 Intermediate output	104.0	161.6		265.6		265.6
1.2 Final output	402.1	424.7	96.1	922.8	20.8	943.6
1.2.1 Internal investment	*31.8*	*67.3*		*99.1*		*99.1*
1.2.2 Final sales	*338.6*	*-32.2*		*306.4*		*306.4*
1.2.3 Existing final output	*31.7*	*372.6*		*404.3*		*404.3*
1.2.4 Other final output		*17.0*	*96.1*	*113.0*	*20.8*	*133.8*
2. Intermediate consumption (2.1 + 2.2 + 2.3)	203.9	661.1		864.9	2.9	867.8
2.1 Conventional intermediate consumption	203.9	234.2		438.1		438.1
2.2 Government intermediate consumption		18.3		18.3	2.9	21.2
2.3 Production- in- progress used		408.5		408.5		408.5
3. Labour costs (3.1 + 3.2)	234.9	16.9		251.7	0.0	251.7
3.1 Private labour costs	234.9			234.9		234.9
3.2 Government labour costs		16.9		16.9		16.9
4. Gross operating margin (1 – 2 - 3)	67.34	-91.7	96.1	71.7	17.9	89.6
5. Gross value added (4 + 3)	302.2	-74.8	96.1	323.5	17.9	341.4
6. Fixed capital consumption (6.1 + 6.2)	81.4	2.0	0.0	83.5	0.0	83.5
6.1 Private fixed capital consumption	81.4	1.1		82.6		82.6
6.2 Government fixed capital consumption		0.9		0.9		0.9
7. Net operating margin (4 - 6)	-14.1	-93.7	96.1	-11.7	17.9	6.2
8. Net value added (5 - 6)	220.8	-76.8	96.1	240.0	17.9	257.9
9. Capital current revaluations		28.7	80.7	109.5	24.4	133.8
10. Capital destructions		2.4		2.4		2.4
11. Capital gains (6 + 9 - 10)		109.8	80.7	190.6	24.4	214.9
12. Social capital income (7 + 11)	-14.1	16.2	176.8	178.9	42.3	221.1
13. Social total income (8 + 11)	220.8	33.0	176.8	430.6	42.3	472.9

Omitting the government's direct public expenditure EAA/EAF underestimates, in the case of Haza, direct total cost by slightly more than 3%, while, on the direct total output side, it undervalues intermediate output by 7% and final output investments by 0.6% (Table 1). The conventional EAA/EAF model includes, in the commercial value added of the multiple uses of an agroforestry system, the additional commercial total output originating with public expenditures, but ignores both direct total cost mentioned above and indirect total cost, additional environmental total output and additional capital gains. The application of the AAS approach considers as dehesa outputs the private environmental services (Esc) and the 'non-controlled' public environmental services (ESnc). Both values have been assessed from two respective contingent valuation surveys to dehesa owners and free visitors at the Monfragüe shire. A single visit per hectare and year at the Monfragües shire dehesas is accounted as having a simulated market value of the aggregated conservation and habitat enjoyment consumed by free visitors. Option and existence values that passive users could obtain from dehesa conservation have not been measured in this study.

Capital and total social incomes

Table 1 shows that at Haza the cattle, polyphite meadows and other cultivation activities generate negative net social operating margins[5]. In the case of the cattle activity, the negative social operating margin and the expected negative livestock capital revaluation (Rodríguez et al., 2004) implied its negative social commercial capital income. The negative results for social commercial capital income of the cattle activity must be explained as not only due to net subsidies, which are not considered in its calculation. The presence of the cattle on the dehesa may be favouring the capital incomes of the agricultural, forestry and public and private environmental-service activities. If this hypothesis is accepted, then livestock has part of its income dislocated to the incomes of the above activities; therefore, the dislocated income of livestock may be partly or wholly offsetting its negative commercial capital income[6].

This allows us to surmise that, in the absence of public subsidies, and given current prices for goods and services, livestock output would be derived from breeds and animal numbers different from those currently utilized. In that scenario, the increasingly significant contribution of extensive livestock farming to the revaluation of environmental service output on dehesas would have to offset the property's indifferent and decreasing commercial results.

In the basket of goods and services produced at Haza, total capital income (CI) in 2002 was €221.1/ha. This contrasts with the negative conventional commercial net operating margin (CNOMc) of €-14.1/ha. This last datum is the only capital income figure offered by the EAA/EAF system (Table 2).

Estimation of total social income (TSI) under EAA/EAF adds to the limitations and shortcomings pointed out with regard to its estimation of conventional commercial net value added (CNVAc) and CNOMc, a range of further drawbacks due to omission of capital revaluation (Cr) and destruction (Cd)[7]. At Haza, the omission by EAA/EAF of capital gains means that this capital income, which is 97% of CNVAc and accounts for 45.5% of TSI, is not estimated at all (Table 2).

Capital gains from private environmental services (ECGc) are estimated using a conservative average variation rate for the market price of land; 45% of this value is based on the environmental capital income generated by auto-consumption of private environmental services (Campos-Palacín, 2003). It is estimated that ECGc account for 17% of total social income at Haza. Public environmental services to visitors to Monfragüe have been estimated assuming a rate of variation in visitors' willingness to pay for the aggregate services of habitat enjoyment and conservation in line with inflation for the period elapsed since the contingent valuation survey date. The environmental services consumed by visitors represent only 5% of the estimated social capital income of Haza. A comparison of the total social income calculated by the EAA/EAF and AAS methodologies shows a figure of €220.8/ha under the EAA/EAF model and €472.9/ha for the AAS approach (Table 2).

Conclusions

The compared application of both EAA/EAF and AAS accounting approaches to a dehesa estate allow us to demonstrate the current limitations of the conventional EU rules for measuring agrosilvopastoral income indicators. The conventional approach underestimates both total output and cost involved in the multiple use of a dehesa system. The EAA/EAF account system provides an incomplete measurement of the net value added by omitting important commercial and environmental income figures.

The natural and environmental resources are taken into account by private dehesa owners when private environmental services produce a significant amount of capital income. This could be the case of Monfragüe shire dehesas, where the environmental social capital income accrued from private environmental services represents an important share of the social capital income.

Frequently decisions concerned with natural and environmental resources are taken by the private owners and the government authorities without accurate economic information. In this frame, a global generalization of accounting systems such as AAS is crucial for capturing the complete basket of commercial and environmental incomes accruing to society from agrosilvopastoral systems. In the dehesa study case, a complete measurement for the multiple benefits of active users is developed.

Acknowledgements

The wealth of information required to apply the AAS and EAA/EAF methodologies was obtained with the aid of the managers of the Haza de la Concepción estate, Alfonso Díaz and Micaela Tovar, and of José Antonio Castañares and other dehesa employees. Ángel Rodríguez and Casto Iglesias, of the Management of Monfragüe Nature Reserve, provided the data on public expenditure by the environmental authority in the Monfragüe shire. This research paper was funded in the context of the project Economía y Selvicultura del Alcornocal (Plan Nacional I+D: AGL2000–0936–C02–02) – 'Cork-oak Economics and Silviculture' (National R and D Plan). The authors would like to thank all the above for their invaluable help. The authors are solely responsible, however, for the information and opinions put forth in this chapter and any errors or shortcomings that may remain.

Notes

This chapter follows narrowly the concepts and criteria discussed in the previous paper with the same title and different subtitle by the same authors (see Campos-Palacín et al., this volume).

[2] There is insufficient space to set out in full the measurement of total private income (TPI) from private economic management of AFSs. Campos-Palacín (2002), Campos-Palacín and Rodríguez (2002) and Rodríguez et al. (2004) analyse the private commercial economics of three dehesas, one in Jerez de la Frontera and two in the Monfragüe shire.

[3] In this chapter, the event that an institutional private landowner does not consume private environmental services has been ignored. Hence, on considering such services, the economic results presented are for an individual private landowner. The uniqueness of the ownership of Haza is that its management accepts lower commercial profitability in the present for the sake of higher work and capital income for breeding or conservation of pure cattle breeds (Rodríguez et al., 2004).

[4] The EAA/EAF and AAS comparison in Haza estate has been made assuming an economic steady state. This is the case when the differences in income measurement by the two systems are minimized, because natural growth and harvest are the same physical amount, investment and consumption of fixed capital are the same monetary value too, and prices of year 2002 are presumed constant in the future.

[5] In the case of cattle, considering operating subsidies, private net operating surplus is positive, and contributes very significantly to private capital operating income at Haza (Rodríguez et al., 2004).

[6] This scenario enables us not to rule out the possibility that dehesa owners exercising high auto-consumption of private environmental services maintain extensive livestock operations where negative commercial operating surpluses are offset by the additional private environmental capital income produced by the livestock, as shown in livestock breeders' leisure behaviour in the Doñana shire (Campos-Palacín and López, 1998).

[7] Except the livestock revaluation and destruction, which, as pointed out earlier, *is* taken into account, although in a dislocated way, in the final output of livestock in the estimation of operating income.

References

Campos-Palacín, P. (1998) Contribution of visitors to the conservation of Monfragüe. Public goods, market and natural resources management (Contribución de los visitantes a la conservación de Monfragüe. Bienes públicos, mercado y gestión de los recursos naturales). In: Hernández, C.G. (ed.) *La Dehesa: Aprovechamiento Sostenible de los Recursos Naturales.* Fundación Pedro Arce/Editorial Agrícola Española, Madrid, Spain, pp. 241-263. (In Spanish.)

Campos-Palacín, P. (1999) Towards the measurement of the wellbeing benefit of the forest multifunctionality (Hacia la medición de la renta de bienestar del uso múltiple de un bosque). *Investigación Agraria: Sistemas y Recursos Forestales* 8 (2), 407-422. (In Spanish.)

Campos-Palacín, P. (2002) Economy of the multiple use of the forest: forests in Jerez de la Frontera (1991-1993) (Economía del uso múltiple del bosque: Montes Propios de Jerez de la Frontera (1991-1993)). *Revista Española de Estudios Agrosociales y Pesqueros* 195, 147-186. (In Spanish.)

Campos-Palacín, P. (2003) *Consumption of Private Environmental Services in the Natural Park Los Alcornocales* (Autoconsumo de servicios ambientales privados en el Parque Natural de los Alcornocales). Informe provisional (Internal working paper), IEG-CSIC, Madrid, Spain. (In Spanish.)

Campos-Palacín, P. and López, J. (1998) *Profit and Nature in Doñana. Aiming the Conservation through the Use* (Renta y naturaleza en Doñana. A la búsqueda de la conservación con uso). Icaria Editorial, Barcelona, Spain. (In Spanish.)

Campos-Palacín, P., Rodríguez, Y. and Caparrós, A. (2001) Towards the dehesa total income accounting: theory and operative Monfragüe study cases. *Investigación Agraria: Sistemas y Recursos Forestales, special issue* 1, 43-67.

Campos-Palacín, P. and Rodríguez, Y. (2002) Economic aspects of agroforestry practices. A systems of monetary and biophysical indicators (Aspectos económicos de las prácticas agroforestales. Un sistema de indicadores monetarios y biofísicos). *Cuadernos de la Sociedad Española de Ciencias Forestales* 14, 39-63.

Campos-Palacín, P. and Mariscal, P. (2003) Owners' preferences and public intervention: case study in the dehesas in Monfragüe (Preferencias de los propietarios e intervención pública: el caso de las dehesas de la comarca de Monfragüe). *Investigación Agraria: Sistemas y Recursos Forestales* 12 (3), 87-102.

Commission of the European Communities (1996) *European System of Accounts* – ESA 95. Eurostat, Luxemburg.

Commission of the European Communities (2000) *Manual on Economic Accounts for Agriculture and Forestry* – EAA/EAF 97 (Rev.1.1). Eurostat, Luxemburg.

Díaz, M., Campos-Palacín, P. and Pulido, F. (1997) The Spanish dehesas: a diversity in land-use and wildlife. In: Pain, D.J. and Pienkowski, M.W. (eds) *Farming and Birds in Europe: The Common Agricultural Policy and its Implications for Bird Conservation.* Academic Press, London, United Kingdom, pp.178-209.

Hicks, J.R. (1946) *Value and Capital,* 2nd edn. Oxford University Press, Oxford, United Kingdom.

Rodríguez, Y., Campos-Palacín, P. and Ovando, P. (2004) The commercial economics of a public dehesa in the Monfragüe shire. In: Schnabel, S. and Gonçalves, A. (eds) *Sustainability of Agrosilvopastoral Systems, Dehesas, Montados. Advances of GeoEcology* 37. (In press.)

Preliminary analysis of the impact of payment for environmental services on land-use changes: a case study on livestock farms in Costa Rica

J. Mora [1,2], M. Ibrahim [1], J. Cruz [1], F. Casasola [1], M. Rosales [2] and V. A. Holguin [1]

[1]Grupo Ganadería y Manejo del Medio Ambiente, GAMMA/CATIE 7170, Turrialba, Costa Rica. [2]Livestock, Environment and Development Initiative, FAO, Rome, Italy. LEAD/FAO jmora@catie.ac.cr

Abstract

The project "Integrated Silvopastoral Approaches for the Management of Ecosystems" (GEF-CATIE) promotes the establishment of silvopastoral systems. This is done by paying incentives for environmental services, mainly carbon storage and conservation of biodiversity. The hypothesis of the project is that the payment for environmental services will induce changes in land use, creating beneficial technological change in the well-being of the farmers and natural resources. For the payment of environmental services on each farm, an index of land use that represented the potential carbon storage and/or the conservation of biodiversity was constructed. Primary forest generated a greater index (2.0) and the degraded pastures and annual crops without trees provided a smaller index (0.0). The participatory method was used to identify the problems, necessities for training, planning of land use and technical assistance. In Costa Rica, 137 cattle farms participated in the project, with an average area of 230 ha (range 5.8-175 ha). The lower percentages of land uses were represented by: natural pasture 34%, improved pasture 11%, degraded pasture 21%, fodder bank 0.5%, forests and bushes 28%, infrastructure 1% and other uses 4.5%. The payment for environmental services to 107 farmers at an average of US$172.6 per farm has been granted. In this case we have seen changes from areas with degraded pastures to fodder banks, improved pastures, live fences and/or secondary forests.

Key words: technological change, silvopastoral systems, cattle farming

Introduction

Payment for environmental services (PES) is a novel experience around the world. Landell-Mills and Porras (2002) analysed 287 cases of payment for environmental services, but most of these were related to payments for the conservation or protection of natural resources, mainly in forest systems. Costa Rica has been pioneering PES projects in agroforestry systems including silvopastoral systems; FONAFIFO, which is the national institution responsible for managing PES, has developed incentive schemes for paying farmers for changing land-use practices that will generate environmental services. The linkage of production activities with the marketing of environmental services could constitute a route to reconvert traditional cattle systems towards eco-friendly systems which integrate silvopastoral systems and this could represent one of the best strategies for poverty alleviation while conserving natural resources. This linkage allows the farmer to have the option of continuing to produce food, raw materials and services and at the same time provide benefits for society and the global environment. Motivated by these experiences CATIE in association with CIPAV in Colombia and NITLAPAN in Nicaragua formulated a PES project on silvopastoral systems on cattle farms of the three countries. The project "Integrated Silvopastoral Approaches for the Management of Ecosystems" is financed by GEF (Global Environmental Facility) through the World Bank and coordinated by CATIE. This project is testing two main hypothesis: (1) PES generated on livestock farms will increase the adoption of silvopastoral systems; (2) PES on livestock farms will result in improved farming practices, increased farm income and employment opportunities for rural poor.

The incentive scheme is designed to compensate farmers for changes in land-use systems which will promote conservation of biodiversity and carbon storage. This chapter reports on the decisions that livestock farmers make on land use changes in response to PES. The Silvopastoral Project has a duration of 5 years and results are presented for the first 2 years of the project.

Environmental services

These are services provided by the environment to human beings and those environmental services that man provides to generate and guarantee higher levels of environmental quality. Ultimately, there are a set of activities that are identified to improve the quality of the environment, whilst reducing the negative effects that are generated by the activities of man.

The services that the environment provides to man, include the benefits that nature *per se* provides to the surroundings for its biological balance, among them water balance, greenhouse gases (GHG), the ozone layer, recreational hunting, natural parks and attractive landscape for ecotourism.

Natural ecosystems provide a wide variety of services and agroforestry systems can generate similar services. Among them are hydrological benefits, sediment reduction, prevention of damage to dams and fluvial routes originated from sediments, conservation of biodiversity, GHG reduction, sustainability, prevention of disasters, prevention of floods and earth landslides, conservation of biodiversity, capture and storage of carbon. The importance of these services has caused international agencies such as the World Bank to design and implement strategies to help developing countries adopt innovative solutions to environmental problems (Pagiola and Gunar, 2002). Nevertheless, the GEF Silvopastoral Project represents a *sui generis* experience in Latin America, given the characteristics of the target group (cattle dealers) and the system of indexing of the change in land use as a basis for the payment of environmental services.

Technological changes

Land-use changes bring about an implicit change in production systems and also change the form of management and use of resources, in response to external forces or suitable conditions of the system. To explain these forces economists developed the theory of "induced innovation" (Hayami and Ruttan, 1985), sociologists and communicators perfected the theory of "diffusion of innovations" (Roger, 1983) and other methodological approaches have contributed to a paradigm centred on technological change by participants (Chambers *et al.*, 1985; Engel, 1995). Analysed in the light of these end theories, the experience of the GEF Silvopastoral Project is an attempt to work out a methodology to drive technological change based on the PES and an innovative approach to technical assistance and qualification where the main stakeholders are the farmers and the extension agents.

The GEF Silvopastoral Project is investigating the effect of external stimuli or changes induced by incentives: the (PES) and technical assistance (TA). The project is being conducted in pilot areas of three different countries where the agroecological and socio-economic conditions are different and will analyse how these different conditions affect farmers' decisions on PES. In the pilot areas of Nicaragua, Colombia and Costa Rica 130, 110 and 137 farmers, respectively, are participating in the project.

Experiences of the project in Costa Rica

The GEF Silvopastoral Project commenced in May 2002 in the Central Pacific region of Costa Rica. The zone ranges from 50 to 1000 m asl, in a west-east direction. The annual average temperature is 27°C and annual precipitation ranges between 1500 and 2000 mm, with a relative humidity of 65-80%; the area is classified as a tropical sub-humid zone in transition. The dominant soil orders in the pilot area are Alfisol, Entisol, Inceptisol and Ultisol, with Usticos regimes. The topography varies between flat land and hillsides with slopes greater than 50%. A classification of land-use types was carried out in each of the pilot areas using aerial photographs and ground "survey". A baseline data base on land-use types was developed for each of the pilot areas. The main land-use types were: primary forest, secondary forest, forest plantations, live fencing, native and improved pastures with low- and high- density trees, degraded pastures and forage banks.

An index of potential land use to store carbon and conserve biodiversity was developed for the land-use types and used as a tool to pay PES in consultation with research data and experts. In the payment scheme, the index increases with tree cover on farms. The areas covered with primary forest had the highest index (2) whereas the areas under degraded pastures had the lowest index (0) since this land use results in loss of biodiversity and emissions of GHG.

Collaterally, the change of land use and its associated technology is stimulated by the creation of a systematic process whereby the local capacity integrates the qualification of farmers and technical assistance by extension officers. Technical assistance is delivered under a process of "learning by doing" centred on the situational demands of the participants.

Mechanisms for payment and strategies of induction

Although the payment represents the main inductive motive towards silvopastoral systems, this change of land-use implies a technological development and therefore a system of technical assistance must accompany this process (Figure 1).

Figure 1. Schemes of technical assistance and PSA used in the GEF Silvopastoral Project in Costa Rica.

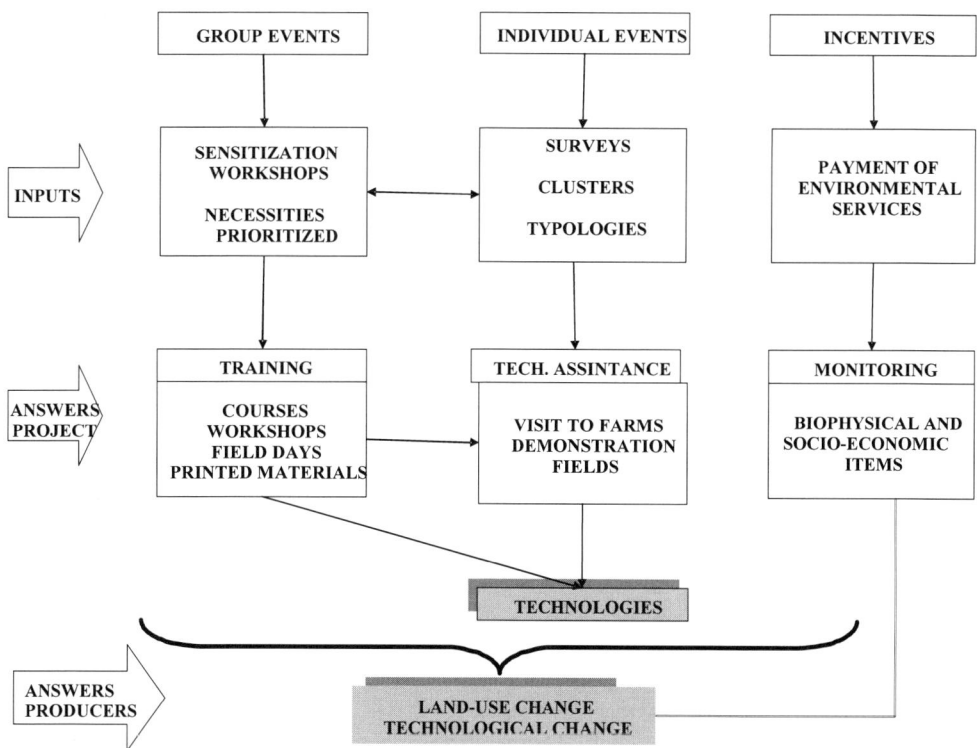

The training of the participants for technical assistance constituted the motive for change. In order to obtain the cooperation of scientists, extension agents and farmers different strategies were adopted, all of them based on a departure from the conventional 'top-down' approach. This has been replaced to fit around the annual cycle of farm work by the stakeholders. Thus, the strategies for training and technical assistance have been organized in two categories: individual events and group events (Table 1), both arranged with the farmers to allow participants a progressive sequence in their levels of communication.

Table 1. Events for training and technical assistance in the GEF Silvopastoral Project in Costa Rica.

Category	Events	No. of events	No. of farmers
Group Events	Sensitization workshops	3	52
	Training courses	5	140
	Field days	3	72
	Institutional meetings	2	0
Individual Events	Farm planning meetings	105	105
	Technical assistance on farm	40	40
	Land-use survey	137	137
	Survey	137	137

The sensitization workshops facilitated the approach between the technicians and the community by means of mutual involvement in the actual roles of each participant. In addition, these events allowed time for the self-diagnosis of technical problems and the necessary time required to implement change. In these workshops the stakeholders shared communication codes, included the objectives of the farmers, understood limitations and learned from each other's experiences.

Sharing experiences with the farmers

Training in different economic, agronomic and social subjects has been a fundamental strategy where the intermediary participants (extension agents) redirect their role in the process. Such redefinition is a new role for extension agents where by "diffusing technologies" they are the facilitators of the shared learning processes. By the end, the final participants (farmers) have acquired a new role as planners and experimenters in their own fields. To achieve this, the accomplishment of the participants' farm plans is centred on the type of technical assistance necessary for the farmers, which will allow them to plan their future land-use scenarios. In practice 105 farm management plans have been worked out between cattle farmer and technician. The dissemination of knowledge by means of technical assistance in the project is characterized by the following: (1) The necessities of the farmers are based on available resources of the agricultural calendar. This means that the technical messages are integrated with farm work, complement field days and individual consultant visits are programmed to follow the farm work based on the rainy and dry seasons. (2) The concept of technical assistance has been re-evaluated, returning to a more basic definition - assisting farmers in the process of learning new technologies or improving what is empirically managed. This is a process of "learning by doing". The concept of "technological menu" or "technological basket" constitutes the basic resource that offers the farmer the best information for the decision-making process.

This system is different from the traditional extension work and the conventional concept of "technological package", which forces the user to implement the technique transferred on the part of the external agent, since what it offers is most appropriate. On the contrary, such assistance centred on the participant stimulates the experimental capacity of the farmer and it encourages decision-making.

Like the farmers the technicians in the project have the capacity to refresh and share their knowledge. The field days are used to develop this creative exercise, where events are organized in two ways: conventional field days, where the experts of the organizations (MAG, CATIE, CACE) show the results of previous experiences and knowledge, and field days in which the farmers share their knowledge and experiences. In these events the participants have the opportunity to learn from their successes and failures, changing their role from one of passive receivers to playing a role as experimenters in the learning of technologies; they become dynamic participants of the technology change process.

Characteristics of the farmers

In Costa Rica, 137 typical small and medium livestock farms have been involved in the projects. The criteria for selection of farmers include: (1) Livestock as the main source of income; (2) small and medium-sized farms; (3) farmers willing to cooperate with the project and accessible. The size of the farms range between 5.8 and 175 ha (mean 23.0 ± 28.37 ha/farm). Eighty per cent of the farms have an area less than 53.3 ha (Figure 2). A large percentage of the farm area is under pastures (> 55%) and a high percentage of the pastures are in a state of degradation (21%). Twenty eight per cent of the land area is under forests (primary, secondary and plantations) and 0.5% under fodder banks.

Figure 2. Distribution of size of farms of participants in the GEF Silvopastoral Project in Costa Rica.

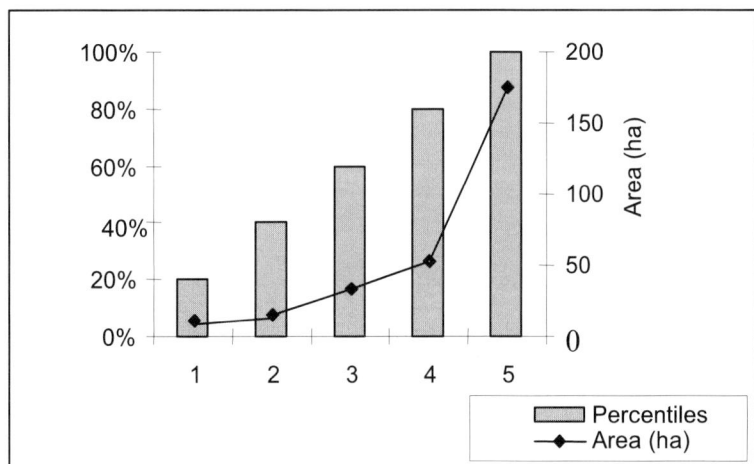

Present and future landscapes

With the use of the GIS and measurements in the field with GPS, maps of the present landscapes were constructed and the baseline data on land use were quantified. This allows the project to quantify the different land uses and to compare the incremental changes over the 4 years of the project. The changes projected for 2004 in the pilot farm are presented (Figure 3). For example, polygons 11 and 20 named as "Tacotal"-which is the common name for secondary successions of abandoned pastures- will change to secondary forest in 2004. However, the most important changes will occur over a relatively large area under natural pasture (33%) and degraded pastures (12%), which will be changed to improved uses, like improved pastures with low-density trees (black polygons) and improved pastures alone (white polygons).

Figure 3. Land-use changes projected in one pilot farm of the Central Pacific of Costa Rica.

a. Present landscape 2003

b. Future landscape 2004

Fifty-five farms were monitored and land use evaluated; approximately 30% of the natural covers are "tacotales" and forests (riparian, primary and secondary) are broken into fragments. The values from Table 2 were obtained from the diagonal of a matrix of vicinity probabilities (Cruz *et al.*, 2004), and they show the degree of fragmentation of each one of the land uses. These values are greater than the rest in the matrix. For example, annual crops (CA) have a probability of 65.8%; this means that, for each ten combinations, 6.58 are with the same land-use class.

Table 2. Matrix of vicinity probabilities between land uses in livestock farms in the Rio Barranca Basin, Pacífico Central of Costa Rica. Source: Cruz *et al.* (2004).

Code	Land uses	Year 1	Year 5
CA	Annual crops	65.8	64.6
PD	Degraded pastures	79.7	75.7
PNSA	Natural pastures without trees	74.4	69.3
PMSA	Improved pastures without trees	63.7	77.7
CS	Semi-perennial crops	45.1	45.9
PNBD	Natural pastures with low density of trees	80.9	77.9
PF-FCf	Timber and fruit plantations	65.7	66.8
BFSSP	Fodder bank and intensive silvopastoral systems	45.6	53.3
PMBD	Improved pastures with low density of trees	70.1	79.6
PNAD	Natural pastures with high density of trees	66.0	59.2
PMAD	Improved pastures with high density of trees	67.7	71.5
TB	Tacotales and natural forest	98.6	98.6
BR	Riparian forest	57.9	57.8

Farmers could adopt silvopastoral systems to sustain production and obtain benefits from environmental services, because vicinity probabilities were increased in the PMSA, PMBD and PMAD land uses (Table 2).

The degraded pastures in the cattle farms in the pilot area basin are relatively important for landscapes and ecosystem management. They are frequently found close to "tacotales" and natural forests (TB) and riparian forests (BR) (29% and 39%, respectively) (Table 3a). However, "tacotales" and forests (TB) and pastures with high density of trees (PNAD and PMAD) are only found together between 2 and 6%. This suggests that the tree cover in the basin is discontinuous and fragmented and the project will have to design strategies to increase tree cover.

Natural grassland (without trees) is most common near the riparian forest and "tacotales" and natural forest, while the pastures improved without trees are found more with natural grasses with low density of trees, riparian forests, "tacotales" and natural forests. Riparian forests are most frequently found closed to degraded pastures and grasses with low density of trees, and are rarely found together with more forested lands (6% and 2%).

The changes likely in land uses in cattle farms by 2007 will probably show changes in the vicinity frequencies between the land uses and compared with those found in 2003 (Table 3b). Degraded pastures will reduce the vicinity frequencies with the forest and "tacotal"; however, this will be increased within the riparian forest, because this combination (degraded pasture-riparian forest) is the one that most often will occur in combinations with degraded pastures (326/970). These changes mean that degraded pastures will decline over all land uses. The improved pastures with high density of trees will increase within forests from 23% to 32%. The forest plantation, fruit plantations and shade coffee plantations will increase their association with natural forests from 28 to 30%. The "tacotales" and natural forests will increase with improved grass pasture with a high density of trees from 2 to 13% and this will result in a landscape with more continuous tree cover. Riparian forests will increase cover from 7% to 38% with improved grasses with low density of trees, and from 16 to 17% with forests and "tacotales".

Table 3. Matrix of likelihood of ocurrences between land uses* on livestock farms in the basin of Rio Barranca in Costa Rica. *Land-use abbreviations presented in Table 2. Source: Cruz et al. (2004).

Land uses	Part a. Class of land uses 2003													Total general
	CA	PD	PNSA	PMSA	CS	PNBD	PFFCf	BFSSP	PMBD	PNAD	PMAD	TB	BR	
CA		19.8	4.36	3.06	0.29	24.56	8.16	1.17	5.11	4.37	1.17	15.09	12.57	100
PD	4.91		2.94	0.58	0.11	13.39	1.71	0.25	5.65	0.61	1.8	2.83	39.22	100
PNSA	3.11	7.93		6.47	0.19	21.01	3.7	0.58	7.78	6.03	1.26	12.79	29.13	100
PMSA	4.32	3.29	13.67		0.41	19.73	1.85	1.85	13.36	1.64	3.49	17.88	18.5	100
CS	5.13	7.69	5.13	5.13		28.21	12.82	0	12.82	5.13	0	7.69	10.26	100
PNBD	5.2	11.5	6.67	2.96	0.34		2.15	0.56	5.68	7.07	1.8	24.75	32.09	100
PF-F-Cf	11.98	10.2	8.13	1.93	1.07	14.87		0.43	10.05	5.13	2.78	28.56	4.92	100
BFSSP	11.39	9.97	8.55	12.82	0	25.64	2.85		1.42	4.27	1.42	13.3	8.36	100
PMBD	2.7	12.1	6.16	5.01	0.39	14.18	3.62	0.08		1.54	2.16	30.97	21.15	100
PNAD	3.68	2.08	7.6	0.98	0.25	28.06	2.94	0.37	2.45		2.57	24.76	24.26	100
PMAD	2.35	14.7	3.82	5	0	10.29	3.82	0.29	8.24	6.18		22.65	22.65	100
TB	3.1	23.9	3.93	2.6	0.09	23.98	3.99	0.28	12.03	6.04	2.3		17.75	100
BR	2.28	28.8	7.93	2.38	0.11	27.51	0.61	0.16	7.27	5.24	2.04	15.71		100
Land uses	Part b. Class of land uses 2007													Total general
	CA	PD	PNSA	PMSA	CS	PNBD	PFFCf	BFSSP	PMBD	PNAD	PMAD	TB	BR	
CA		12.8	2.16	5.99	0.2	8.82	9.32	2.08	2.29	3.3	5.3	17.2	9.5	100
PD	7.93		2.37	4.74	0	2.16	3.55	1.5	18.6	0	4.2	21.2	33.6	100
PNSA	2.38	4.21		15.2	0.2	9.7	3.11	1.3	20.4	2.2	3.3	7.59	30.5	100
PMSA	2.4	3.07	5.54		0.2	4.27	1.87	1.1	27.1	1.6	5.7	19.8	27.3	100
CS	3.03	0	3.03	9.09		0	15.2	0	4.5	0	12	0	12.1	100
PNBD	4.88	1.93	4.88	5.89	0		0.78	0.6	21	5	5.2	21.7	28.1	100
PF-F-Cf	12.6	7.76	3.82	6.3	1.1	1.91		1.6	19.6	0.9	11	30	3.82	100
BFSSP	13.6	12	5.58	13.6	0	5.58	5.58		18.4	0.8	10	9.9	4.73	100
PMBD	3.27	4.29	2.65	9.65	0.4	5.42	2.07	0.5		1	5.9	30.4	34.5	100
PNAD	6.85	0	4.11	8.22	0	18.5	1.37	0.3	14.4		9.6	14	22.6	100
PMAD	2.38	3.04	1.34	6.38	0.3	4.19	3.49	1	18.3	2.1		32.4	25.2	100
TB	3	5.98	1.2	8.59	0	6.84	3.88	0.4	37.2	1.2	13		19.1	100
BR	1.45	8.58	4.38	10.8	0.1	8.03	0.45	0.2	38.1	1.7	8.9	17.3		100

Response of the farmers

The average baseline incentive payment to 107 farmers was US$172.6 per farm and US$8/ha. Most of the incentives were distributed to the farms smaller than 80 ha (Figure 4), following a linear relationship between the size of the farm and the PES at least in farms less than 80 ha (Figure 5). Degraded pastures changed to other eco-friendly uses, mainly improved pasture with trees, live fences, fodder banks, forests and natural regenerations, which may form secondary forest in the future.

Figure 4. Relation between area of farm and baseline payment of environmental service/ha.

Figure 5. Relation between area of farm and baseline payment of environmental service/farm.

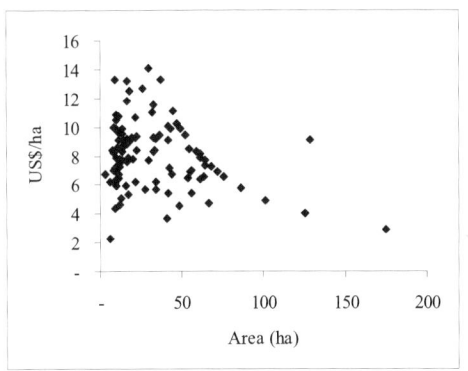

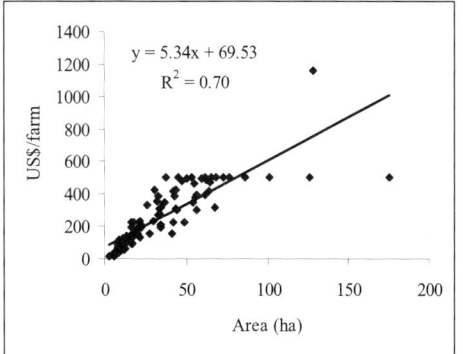

Adoption of eco-friendly technologies

It is hoped that in the future the farmers enrolled in the project would have assimilated and appropriated the technological changes implied by changing from the extensive cattle ranch model towards a silvopastoral system. Thus, the main technologies identified are a high priority in group events and individual technical assistance such as: management of multipurpose breeding grounds of species, management of fodder banks and feeding of cattle with local resources. It is hoped that in the rainy season other necessities arise for resolution, according to the itineraries at that time.

Conclusions

In spite of the project being in the initial stages, differences in the adopted silvopastoral systems have already been observed, based on the natural resources of farms, in such a way that the farms with more land area tend to have more investment. Small farmers are constrained by lack of capital and will establish silvopastoral systems such as live fences and dispersed trees in pastures, which are low-cost technologies. This will permit production while enabling small farmers to gain environmental benefits. Although the PES is the main inducer of change, from a long-term perspective technical assistance also plays an important role in the project. The success of this component, to a certain extent, can guarantee the continuity of technological conversion and the organization of farmers. The predicted changes in land use in cattle farms under PES showed a reduction in degraded pastures, as well as a reduction in the association with forests, "tacotales" and riparian forests.

References

Chambers, R., Pacey A. and Thrupp, L.A. (1985) *Farmer First*. Intermediate Technology Publications, London, United Kingdom. 218 pp.

Cruz, J., Ibrahim, M. and Casasola F. (2004) Effect of farm land use change on the landscape connectivity in Barranca river basin (Efecto del cambio de uso de la tierra en fincas ganaderas en la conectividad del paisaje en la cuenca del río Barranca). In: *VI Semana Científica del CATIE* (Centro Agronómico Tropical de Investigación y Enseñanza). CATIE, Turrialba, Costa Rica, pp. 72-75. (In Spanish.)

Engel, P. (1995) Facilitating Innovation: An Action Oriented Approach and Participatory Methodology to Improve Innovative Social Practice in Agriculture. Ph.D. Dissertation. Agricultural University of Wageningen. Wageningen, The Netherlands.

Hayami, Y. and Ruttan, V.W. (1985) *Agricultural Development, an International Perspective*. Johns Hopkins University Press, Baltimore, London, United Kingdom.

Landell-Mills, N. and Porras, I. (2002) *Silver Bullet or Fools Gold? A Global Review of Markets for Forest Environmental Services and their Impact on the Poor*. IIED (International Institute for Environment and Development), London, United Kingdom.

Pagiola, S. and Gunar, P. (2002) *Payments for Environmental Services* (Pagos por Servicios Ambientales). Environmental Strategy Notes No. 3. The World Bank, Washington. (In Spanish.)

Roger, E.M. (1983) *Diffusion of Innovation,*. 3rd edn. Collier Macmillan, New York, USA.

Adaptation of an agrosilvopastoral system to land-use dynamics: local-level analysis of strategies and practice changes in north-eastern Portugal

J. Alonso [1] and J. Bento [2]

[1] Escola Superior Agrária de Ponte de Lima (ESAPL), Refóios do Lima 4990-706 Ponte de Lima, Portugal.
[2] Universidade de Trás-os-Montes e Alto-Douro (UTAD), Quinta de Prados 5000-911 Vila Real, Portugal; malonso@esapl.pt, j_bento@utad.pt

Abstract

The different components of agrosilvopastoral systems are strongly dependent. Both political and demographic causes and the social and economical consequences of land-use changes affect decision-making and its capacity, calling for an adaptation to new situations and functioning contexts. In this chapter: (a) adopting a GIS, land-use dynamics from 1992 to 2003 were studied for a community and territory of north-eastern Portugal; (b) the causes and impacts of these changes upon the local farms' traditional strategies, namely by animal producers and practices of grazing, were analysed. The observed changes are the outcome of demographic, social and political processes, resulting in the productive specialization of production units and local spaces. Decreases in areas of dry cereal crops and natural pastures ("*lameiros*"), and increases of forest, forages and improved pastures are quite apparent. Concurrently, grazing practices have declined, mainly those of mobile grazing made by small ruminants and native races of cattle, while the stabling regime and fenced pastures have gained importance. The recent land afforestation trend opens interesting possibilities concerning the agroforestry utilization of these territorial systems.

Key words: land-use planning, farming systems, afforestation

Introduction

The orography, geographical location and elevation (700 and 800 m) of Sanhoane village (1169 ha), Miranda Plateau, NE of Portugal, define an Ibero-Mediterranean climate (Leão and Monteiro, 1988) with a great climatic range. Local soils are clearly limited and unbalanced in their chemical, physical and biological elements composition (Martins, 1985). The relative natural isolation and the institutional and political exclusion contributed to a great rural exodus (a 48.3% decrease in resident population between 1960 and 2001), with consequences of ageing and low demographic density (15.1 inhab/km^2) (INE, 2001) and promoting landowner absenteeism. The local economy is agrarian with small family farms, which have very diverse income sources and technical ability, even if integrated in a continuous specialization process.

Traditional farming systems include cereal crops, pastures and forages in dry land areas and extensive livestock production. The local landscape pattern is defined by the stratification of open field and natural pasture areas, alternating with trees and shrub hedgerows, a supplemental feed source in major deficit periods. Historically, this system includes exploited silvopastoral systems, where a wide range of local people use open and community ground, given temporary public access to previously landless shepherds. This cooperative arrangement among local producers and landowners preserves the sustainable yield of the different land uses and agrarian activities.

Materials and Methods

The linkages between the different land-use hierarchies, the socio-economic constraints and the collective and individual decision-making processes determine the organization of the territorial systems and the response to changes at the farm level (Deffontaines, 1996; MacFarlane, 1996). The present study case is based on Sinclair's (1999) methodology and concepts. The adaptation of the agrosilvopastoral system resulting both from the application of a specific rural policy and the particular historical evolution of this territorial and farming system is analysed at the local (village) and individual (farm and plot) levels.

Availability of a detailed cadastral survey for the region (Figure 1) enabled the assignment in a geographical information system (GIS) of agro-ecological and socioeconomical parameters for each parcel, allowing multivariate statistical analysis to be used to study the framework and dynamics of land-use changes between 1992 and 2003. At the same time, extensive fieldwork, including individual queries to landowners and periodic local observations during 3 years (2000-2003), helped us understand the characteristics, organization and management changes of the different component systems and their associated practices.

Figure 1. Land use (1992 and 2003) at village and plot level (space unit analysis).

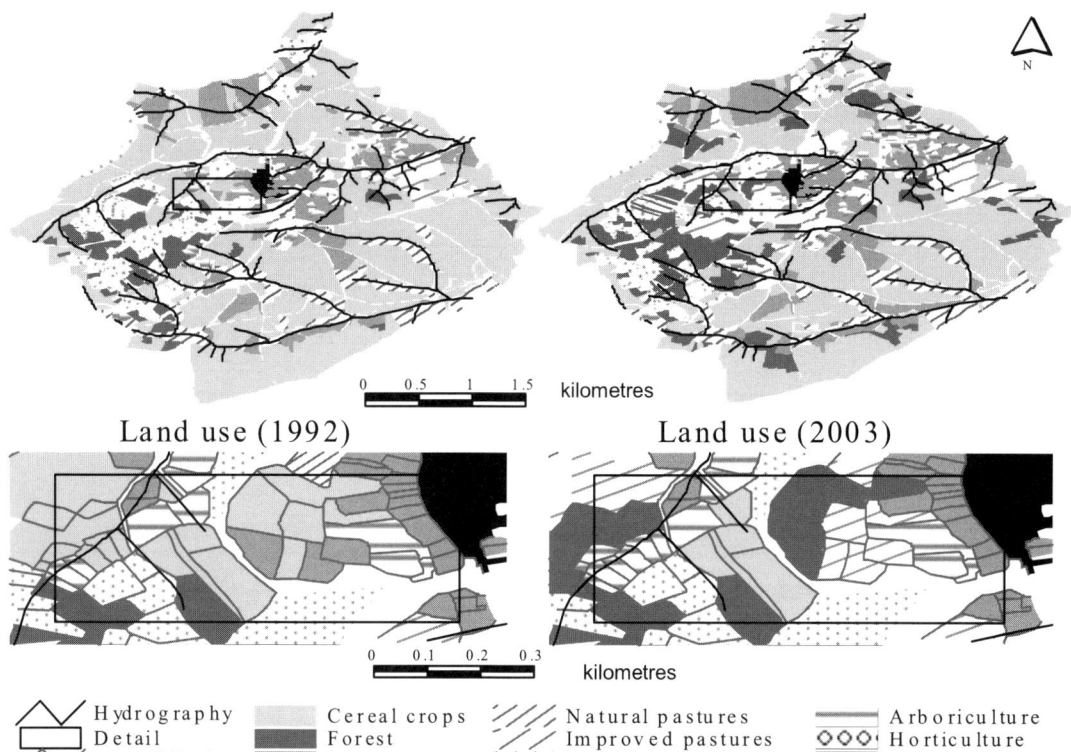

Results and Discussion

A reduction in the area of cereal crops (-24.1%), natural pastures (-4.3%) and shrub lands (-2.5%) was observed during the last decade. There was an increase in forest (17.2%), improved pastures (7.2%) and forage areas (6.9%). These dynamics reflect a strong growth in intensive dairy production by young farmers, and also the afforestation of agricultural areas, mainly by absentee landowners and early retiring farmers, as a direct consequence of the 1992 Common Agriculture Policy (CAP) Reform.

These changes are related to: (a) the historic rural exodus and the differentiated and divergent evolution of agricultural local farm projects; (b) the reduction of available land in parallel with easily acquired new production technology with a higher capacity to impact on the natural environment; and (c) the need to increase production and productivity in order to support the strong investments made by young farmers.

These processes result in a higher degree of specialization of each producer, at the farm and parcel level, responding to rationalization of local resources, accomplished with the internal and external community processes. Fragmentation and diversification, with localized processes of abandonment and extensification, in parallel with intensification and concentration dynamics, are evident at landscape level.

The strong increase in dairy production requires more land for forage production and, consequently, the need to convert traditional natural pasture and cereal crops areas. The lack of family and local labour and less animal grazing as a result of forestation contribute to a less intensive and regular use of local food resources in poor natural pastures in common areas, and reduce the access to temporary private areas and the use of hedgerow food sources.

The installation of improved temporary pastures in fenced spaces implies a simplification of animal production and an increase in total food production as well as of its nutritive value. There has been a reduction in the number of native races of animals, abandonment of spaces with high floristic diversity, an increase of wildfire hazard and destruction of tree hedgerows.

Conclusions

Local land-use dynamics are a component of large territorial processes that continuously operate in space and time, with differential consequences at the village level concerning decisions and actions by the local farmers and landowners. The increased specialization of spatial and socio-economic units makes it difficult to maintain traditional components and their functions as found in local silvopastoral systems. These issues cause productive disintegration processes promoted by different objectives of the European agricultural policy, and place questions of social complementarity and environmental sustainability, perverting the adequacy of the current system to meet natural and human needs. Politicians and planners at the central level should design policies and actions that respect the evolution and functioning of territorial systems, their physical environment and local specific population characteristics. In the former case, the current reality demands integration between activities and users at the community level. The areas afforested by absentee landowners may support animal activities in extensive regimes, utilized by local producers or other external users, with potential benefits for the different stakeholders.

References

Deffontaines, J. (1996) Développement du boisement dans un systéme agraire soumis à des contraintes de qualité de l'eau souterraine. In: Cailliez F., Cavailhes, J., Hubert, B., De Montard, F.X., Guitton, J.L. and Terrasson, D. (eds) *Agriculteurs, Agricultures et Forêts*. Actes de Colloque, Paris, 12-13 décembre 1994, Cemagref-INRA, Cemagref Editions, Paris, France, pp. 83-88.

INE (Instituto Nacional de Estadística) (2001) *XIII Population Statistics* (XIII Recenseamento Geral da População). INE, Lisbon, Postugal. (In Portuguese.)

Leão, C. and Monteiro, L. (1988) *PDRITM – Assessment of the Proposals for Conversion of Mirandês Area* (PDRITM - Avaliação das Propostas de Reconversão para o Planalto Mirandês). University of Tras os Montes, Vila Real, Portugal. (In Portuguese.)

MacFarlane, R. (1996) Modelling the interaction of economic and social behavioral factors in the prediction of farm adjustment. *Journal of Rural Studies* 12 (4), 365-374.

Martins, A. (1985) *Soil Characterization in Trás-os-Montes e Alto Douro and Land Use* (Caracterização Sumária dos Solos de Trás-os-Montes e Alto Douro e sua Ocupação). IUTAD, University of Tras os Montes, Vila Real, Portugal. (In Portuguese.)

Sinclair, F. (1999) A general classification of agroforestry practice. *Agroforestry Systems* 46 (2), 161-180.

Agropastoral systems in Cholistan

A. Farooq, F. Gulzar, S. A. Safdar, F. Sameera and A. Zulfiqar
Dept. of Geography, University of the Punjab, Lahore, Pakistan. drylandpk@yahoo.com

Abstract
Cholistan is an extension of the Great Indian Desert and covers an area of 26,330 km². It lies within the south-east quadrant of Punjab province between 27° 42' and 29° 45' north latitude and 69° 52' and 73° 05' east longitude. The Cholistan desert has extreme summer temperatures (50°C plus) and prolonged droughts; livestock rearing is the only traditional profession of the nomad pastoralists of this desert. The pastoral system is characterized by mass migrations of animals and people throughout the year in search of water and forage. The onset of monsoon and the distribution of rainfall mainly dictate the pattern of movement of nomadic herders. Livestock are the main source of their survival and a number of cultural activites are linked with the animals. The major constraints to the nomadic system are very poor quality of drinking-water and inadequate feed, both of which are acute during summer.

Keywords: nomadic migration, pastoral nomadism, transhumance

Introduction
The silvopastoral system is the oldest system and occurs where forages and/or livestock and trees are cultivated together on the same unit of land. This system is limited today though still found in the Mediterranean region and more widely in the tropics. There are several production concepts that should be considered in understanding silvopastoral system management. These concepts relate to the tree, forage and animal components. Silvopastoral systems are deliberately managed agroecosystems; opinions differ regarding the role of range management and extensive grazing under trees, but grazing under forests has a long history with the production of both animals and tree crops. Systems may seek to introduce or improve forage production and quality under tree plantations, or young trees may be planted into existing pasture. Pastoralists in Cholistan manage their mixed livestock in such a way that milking cows are moved nearby the urban centres, where milk is sold readily, while other animals like camels, goats and sheep are kept in the desert for grazing. Livestock are frequently used for meat, milk and gifts. Communal ceremonies like weddings, funerals and tribal celebrations include slaughtering and exchange of animals. A person's status in the desert nomadic life style is chiefly represented by the size of the herd he owns.

Transhumance
The transhumance system comprises the largest number of immigrating livestock and is characterized by mass movement, including people (Arshad *et al.*, 1999). Patterns of movement (Figure 1) are location specific and dictated by a traditional system of land tenure. The timing of irrigation is determined by the onset of the monsoon and rainfall distribution.

Figure 1. Patterns of movement are location specific and dictated by a traditional system of land tenure.

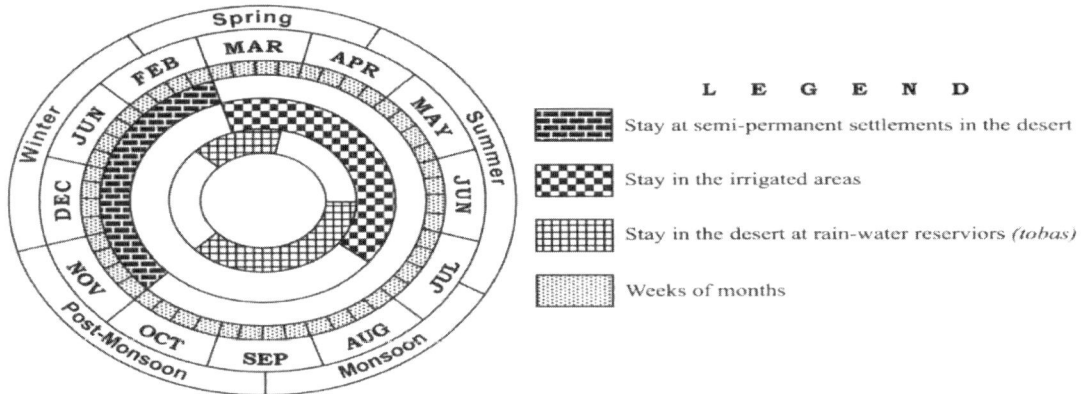

In July/August (monsoon), movement is from the irrigated and riverine areas to traditionally owned *tobas* in Lesser or Greater Cholistan. The distances covered vary from 10 to more than 100 km. Several *tobas* belonging to the same clan may be located within a 1 km radius. At the start of the season, livestock generally graze within a few kilometres of the *toba*, but this distance increases to around 15 km by the end of the season. In October/November, as water or forage is depleted at the *tobas*, migration is to semi-permanent settlements having wells and *kunds*. In March/April, migration is back towards the fringe of the irrigated areas and, after wheat harvest, to the Sutlej River for those with traditional, riverine rights. Irrigation canals are the water sources, but feed supplies are differentiated according to two sub-systems. Pastoral sub-system herds are partly fed on dried forage, on vegetation along canal banks, roadsides, and partly on purchased fodder. Some stubble is available after the wheat harvest in May. Agropastoral system herds are partly fed on dried forage but depend heavily on fodder crops and residues since their owners possess irrigated land. The transhumance system, being heavily dependent on the timing and quantity of rainfall, can be severely disrupted by drought. For example, during a prolonged drought over the last 4 to 6 years preceding this study, most of the herders barely moved south, some staying only a few days or for a few months before being compelled to return. Average herd sizes in the pastoral system were small with a total of 106 sheep units consisting mainly of sheep (46%), cattle (34%) and goats (20%). In the agropastoral system disparity in herd sizes was variable, but the average herd size was much larger at 779 sheep units, with cattle, sheep and camels predominant (Ahmad, 2002). Several constraints to production are identified by the socio-economic study in the transhumance system, all of these being linked to water supply and its balance with forage and fodder. The general constraint is inadequacy of water in the desert. This was compounded by the recent drought when *tobas* became silted in the absence of herders. Some 25 out of 43 *tobas* (Ahmad, 2002) seen in Greater Cholistan were filled with sediment. In the eastern, arid region, *toba* water is of good quality but limited so that feed is still available when thirsty herds are forced to migrate. On the other hand, in the semi-permanent settlements, well water is adequate but of poor, saline quality. The wells are unlined and most have to be re-dug each year because the surrounding sand collapses. In the western hyper-arid region, on the other hand, the quantities of both water and feed are inadequate. Feed is frequently depleted first, so the sheep, whose walking range for pasture is confined to within 3 or 4 km of water, must be moved ahead of the rest of the animals to other *tobas* or to wells. All herds are kept for as long as possible in the well areas, or on the Sutlej floodplain. Many of the wells have brackish water, which, together with the prolonged period of food shortage, results in poor body condition. A major constraint for all landless pastoralists is the scarcity of free grazing during their sojourn on the irrigated fringe or the floodplain even though fresh water is abundant.

Nomadic system

This system applies to the larger herds of camels and goats which remain throughout the year in the desert of Lesser or Greater Cholistan. The size of such camel herds varies from around four to 150 animals, and goat herds are of variable sizes. Depending on the size of the herds to be left in the desert, one or two members of each household will remain behind to tend the herds. In addition, a herdsman will be hired to assist if the herd is particularly large. The other members of the household will follow the normal transhumance system and will return to the irrigated land, taking along one or two camels for transport. Households with only a few surplus camels eg less than five, for their transport needs will leave these behind to be cared for by arrangement with the owners of the larger herds. During winter and summer these nomadic animals drink from wells at the semi-permanent settlements (Jowkar *et al.,* 1996). During the monsoon and post-monsoon they drink from *tobas* like all the other animals. Natural grazing is the exclusive nutritional soure for the nomadic animals living permanently in the desert (Ahmad, 2002). The major constraints to the nomadic system are very poor quality of drinking-water and inadequate feed, both of which are acute during summer. *Haloxylon salicornicum*, an evergreen shrub, provides most of the feed from late winter to summer. Animals must travel long distances (up to 15 km) to search for often insufficient feed. Furthermore, well water salinity increases to very high levels over summer, especially in the western part. The combination of long-distance travel, harsh temperature rising to 50°C or more, undernourishment and highly saline water all contribute to a reported high mortality rate.

Conclusion

Pastoral nomadism is not only an environmentally sustainable way of managing the Cholistan drylands, but it could extend support to national dairy and meat consumption requirements. The likelihood of an increase in the number of livestock, by making feed supplement more accessible and affordable in the dry seasons, could be reduced by increasing offtake through marketing of animals for urban consumption. Support for the livestock sector will automatically increase herders' income and increased offtake through marketing, reducing the likelihood of overgrazing. This reveals that sustainable use of resources with the promotion of indigenous technology will benefit the local people.

References

Arshad, M., Altaf-ur-Rehman Rao and Akbar, G. (1999) Masters of disaster in Cholistan desert, Pakistan: pattern of nomadic migration, *UNEP: Desertification Control Bulletin* 35, 33-38.

Farooq, A. (2002) Socio-economic dimensions and ecological destruction in Cholistan. Ph.D. dissertation, Department of Geography, University of Karachi, Pakistan.

Jowkar, F., Khan, M. and Ajmal Khan M. (1996) *Socio-Economic Dimensions of Resource Management in Cholistan*. Institute for Development Anthropology (IDA), Binghamton, New York, USA.

An evaluation of the effects of forest conservation on reservoir capacity: a case study in the "Cuerda del Pozo" reservoir (Soria)

R. García Díaz [1], F. García Robredo [1] and P. A. Medrano Ceña [2]

[1]Escuela Técnica Superior de Ingenieros de Montes, Ciudad Universitaria s/n, 28040 Madrid, Spain. [2]Asociación de Propietarios Forestales de Soria, Calle Diputación, 1, 42003 Soria, Spain. rgardiaz@montes.upm.es, asfoso@teleline.es

Abstract

The effective life of reservoirs is threatened by erosion and sedimentation in their catchments. When those catchments are covered by forests, erosion and sedimentation are significantly lower. An economic assessment of the maintenance of reservoir capacity due to forest conservation in the basin of Cuerda del Pozo (Soria) has been carried out. Two scenarios were analysed: (i) the current situation with 81% of the basin covered by forest, (ii) total deforestation of the watershed. For each scenario, sediment yield in m^3/year was calculated by using the MUSLE model, and reservoir life was assessed. Once a value was assigned to each m^3 of reservoir capacity, the flow of annual benefits along reservoir life was used to calculate the present net worth and the annual equivalent value of maintaining reservoir capacity.

Key words: protection, sedimentation, reservoir, erosion, economic assessment

Introduction

The objectives of this research are, on the one hand, to state the hydrological benefits generated by the forest and, on the other hand, to quantify the sediment yield coming into a reservoir in order to perform an economic assessment of the contribution of the forests located at its basin headwaters on the maintenance of reservoir capacity.

Materials and Methods

The catchment of the "Cuerda del Pozo" reservoir is located in the north-west area of the Soria province. The main watercourses crossing the basin are the Revinuesa, the Duero and the Ebrillos Rivers, with a set of small streams flowing directly into the reservoir. Two of them (La Herrería and La Dehesa streams) have been studied separately (Foncillas *et al.* unpublished results).

The dominant vegetation in the basin, accounting for 81% of the total basin area, is forest with a well-developed tree layer, with or without a shrub and grass layer. This vegetation community works as a true silvopastoral system.

The forest in the basin consists mainly of *Pinus sylvestris*, *Pinus pinaster*, *Pinus uncinata*, *Quercus pyrenaica*, *Fagus sylvatica*, *Quercus faginea* and other tree species. About 13% of the total basin area is covered by shrub and pasture including species such as *Erica* sp, *Thymus* sp, *Cistus laurifolius*, *Calluna vulgaris*, *Rosa* sp, *Corylus avellana*, *Genista* sp and *Cytisus scoparius* (Peña, unpublished results). Land-use distribution in the basin is shown in Table 1.

Table 1. Vegetation categories in the first scenario.

Basin	Forest (ha)	Shrub/pasture (ha)	Agriculture (ha)	Other (ha)	Total (ha)
Revinuesa	10,625	610,000		330	11,565
Duero	11,116	2,082		452	13,650
Ebrillos	14,611	851,000			15,462
Direct slopes	6,392	3,260	2,524		12,176
Total	42,744	6,803	2,524	782	52,853

Under the tree layer there are also shrubs and grasses, which have usually been grazed by livestock, especially cattle. The use of the grass under the canopy cover is common and more intense in the older pine stands. However, the grass yield in these areas is lower than in the areas denoted as shrub/pastureland.

Soil and water conservation in a watershed are highly dependent upon vegetation density and structure. When forest cover is significant, part of the rainfall is intercepted by leaves, branches and stems and released gradually, flowing down the stems to the ground. The existence of roots and surface litter also improves infiltration, thus regulating runoff, cutting down flood peak discharge and temporarily reducing water volume in the reservoir.

This protective effect of vegetation is quantified by means of the C factor of the universal soil loss equation (USLE), ranging from 0.004 for forests with a canopy cover $\geq$ 0.75, to 0.25 in agricultural areas (Mintegui Aguirre, 1985). In the study area, C factor values range from 0.015 to 0.15. The effect of the forest cover on the increase in relative atmospheric moisture and the length of the grass growing period has also been assessed by direct observation.

The effects mentioned above lead to a lower runoff volume and a lower flood peak discharge. The change in runoff is quantified by means of the curve number method, taking into account the soil water content conditions before rainfall: average previous humidity (condition II) or high previous humidity (condition III). The current situation within the study area shows that the curve number in condition II is between 61 and 66, whereas it goes from 78 to 83 for condition III.

In order to evaluate the effect of the vegetation on erosion and sedimentation, two scenarios have been considered: (i) the current situation with the watershed covered by forest, (ii) the deforestation of the basin. In this second scenario the forest is assumed to be replaced by agricultural crops where the slope is lower than 12% and by grassland or shrubland where the slope is higher than 12%. Since grassland and shrubland are intimately mixed, their different effects on erosion have been accounted for by assuming a particular combination of both vegetation categories based on the usual land-use distribution in non-forested areas of the watershed.

Table 2. Land-use distribution and hydrological parameters in the second scenario.

Basin	Shrub/pasture (ha)	Agriculture (ha)	Total (ha)	C	NH (II)	NH (III)
Revinuesa	10,765	800,000	11,565	0.132	74.46	87.02
Duero	13,650		13,650	0.122	74.00	86.75
Ebrillos	12,962	2,500	15,462	0.143	74.97	87.32
La dehesa	1,377	700,000	2,077	0.206	77.95	89.05
La Herrería	260,000	500,000	760,000	0.165	76.02	87.94
Other direct slopes	5,339	4,000	9,339	0.177	76.57	88.26
Total	44,353	8,500	52,853			

Table 2 shows the land use distribution and the values of the C factor and the curve numbers (NH) in the second scenario.

The MUSLE model (Williams and Berndt, 1977) has been used to estimate sediment yield values for different return periods in order to calculate the average sediment yield over the 100-year horizon. The figure obtained has been increased by 20% in order to account for bed-load solid discharge. Based on sediment yield estimates, the effective life of the reservoir for the two scenarios has been assessed.

In order to carry out this assessment, it would be interesting to have a reference value of reservoir capacity, which depends not only on the value of water, but also on water reuse. For the purpose of this research, the value of a cubic metre of reservoir capacity (p) has been assumed to be 0.06 €, as a result of updating the value reported by Simón Navarrete *et al.* (1993).

The difference in the annual sedimentation calculated for both scenarios (q) will be the saving in reservoir capacity loss. At the end of the first year after reservoir construction, the value of this saving is pq; after the second year, that value will go up pq, that is, the saving in capacity loss after the second year will be valued $2pq$, and so on.

The present value of the benefit generated by the forest can be calculated as:

$$PV = \frac{pq}{1+i} + \frac{2pq}{(1+i)^2} + \ldots\ldots + \frac{T \times pq}{(1+i)^T} = \frac{(1+i)^{T+1} - (1+(T+1) \times i)}{i^2 \times (1+i)^T} \times pq \qquad (1)$$

Where:
p = value of a cubic metre of reservoir capacity.
q = difference in sedimentation for both scenarios.
i = discount rater.
T = effective life of the reservoir.

The annual equivalent value (a) of the present value is given by:

$$a = \frac{PV \times i \times (1+i)^T}{(1+i)^T - 1} \cong i \times PV \qquad (2)$$

Results and Discussion

The results obtained confirm the effects of forest cover on rainfall interception, infiltration improvement, runoff control and flood peak discharge reduction. It has also been observed that the increase in relative atmospheric moisture, the reduction of the drought period (Montoya Oliver and Mesón García, 1981) and the decrease in water runoff speed are also effects of a dense vegetation cover.

In scenario 1 (forested watershed), the average sediment yield in the basin attains the figure of 45,760.48 t/year, equivalent to 28,600.30 m^3/year, or 0.866 t/ha per year. In these conditions, the effective life of the reservoir could have a theoretical value of 6000 years, assuming that the construction materials could last.

However, in scenario 2, the result obtained for sediment yield were 421,514.54 t/year, equivalent to 263,446.59 m^3/year. It can be noticed that, in this scenario (deforestation conditions), sediment yield values are ten times higher than those obtained for scenario 1. The effective life of the reservoir in the second scenario is 600 years.

By substituting these values in formulae (1) and (2), the following results have been obtained:

$$PV = \frac{(1+i)^{T+1} - (1+(T+1)\times i)}{i^2 \times (1+i)^T} \times pq = \frac{1.03^{601} - (1+601\times 0.03)}{0.03^2 \times 1.03^{600}} \times 0.06 \times 234{,}846.29 = 16{,}126{,}106 \text{ €}$$

For a 3% discount rate, $PV = 16{,}126{,}106$ €.

$$a = \frac{PV \times i \times (1+i)^T}{(1+i)^T - 1} = \frac{16{,}126{,}106 \times 0.03 \times 1.03^{600}}{1.03^{600} - 1} = 483{,}783.19 \text{ €}$$

For a 3% discount rate, $a = 483{,}783.19$ €/year, equivalent to 9.15 €/ha per year.

Conclusions

The annual sediment yield for the forested watershed is 0.866 t/ha per year, resulting in a volume of 28,600 m^3 in the whole basin, whereas for a hypothetical vegetation of shrub/grass and agricultural crops the figure obtained would be 7.98 t/ha per year, that is, 4.98 m^3/ha per year.

The forest cover of the basin is responsible for reducing by nine-tenths the sediment yield arriving at the reservoir, and therefore it is responsible for multiplying by ten (from 600 to 6000 years) its effective life.

According to the methodology used, the economic benefit of keeping reservoir capacity through forest conservation is 16×10^6 € along the reservoir lifespan, equivalent to 0.48×10^6 € per year. The benefit per hectare in most of Spain is of the same order of magnitude as that obtained from timber production.

References

Mintegui Aguirre, J.A. (1985) *Forest Hydrological Restoration and Dominated Lands* (La Restauración Hidrológica-Forestal y las áreas dominadas). Fundación Conde del valle de Salazar, Madrid, Spain. (In Spanish.)

Montoya Oliver, J.M. and Mesón García, M.L. (1981) Influence intensity and effects of dehesa trees on the phenology and species composition of the understorey (Intensidad y efectos de la influencia del arbolado de las dehesas sobre la fenología y composición específica del sotobosque). *Anales del INIA Serie Forestal*, no. 5, INIA, Madrid, Sapin. (In Spanish.)

Simón Navarrete, E. de, Mintegui Aguirre, J.A., García, J.L. and Robredo, J.C. (1993) *Forest Hydrological Restoration in Mediterranean Hydrological Basins* (La restauración Hidrológica-Forestal en las cuencas hidrográficas de la vertiente mediterránea). Junta de Andalucía, Consejería de Agricultura y Pesca, Granada, Madrid, Spain. (In Spanish.)

Williams, J.R. and Berndt, H.D. (1977) *Sediment Yield Prediction Based on Watershed Hydrology*. Transactions of the ASAE, 20 (6), 1100-1104.

Characterization of tree species in silvopastoral systems in the mountain region of Tabasco, Mexico

D. Grande [1], G. Pérez [2], H. Losada [1], M. Maldonado [3], J. Nahed [4] and F. Pérez-Gil [5]

[1] Área de Desarrollo Agropecuario Sustentable, División de Ciencias Biológicas y de la Salud, Universidad Autónoma Metropolitana Iztapalapa, Mexico, D.F., Mexico. [2] Centro Regional Universitario del Sureste, Universidad Autónoma Chapingo, Teapa, Tabasco, Mexico. [3] División de Ciencias Agropecuarias, Universidad Juárez Autónoma de Tabasco, Teapa, Tabasco, Mexico. [4] División de Sistemas de Producción Alternativos, El Colegio de la Frontera Sur, San Cristóbal de Las Casas, Chiapas, Mexico. [5] Dept. de Nutrición Animal, Instituto Nacional de Ciencias Médicas y Nutrición "Salvador Zubirán", Mexico, D.F., Mexico. ifig@xanum.uam.mx, hrlc@xanum.uam.mx, freyes@taurus1.chapingo.mx, noelmauricio@yahoo.com, jnahed@sclc.ecosur.mx, fernandoperezgil@hotmail.com

Abstract

The mountain region of Tabasco, in the south-east of Mexico, is a tropical zone where diverse silvopastoral systems (SPS) are practised; the principal SPS in the region are scattered trees on grasslands and living fences, which include native or introduced grasses associated with diverse trees. The more important trees in the SPS of the mountain region include native species *Brosimum alicastrum, Bursera simaruba, Byrsonima crassifolia, Castilla elastica, Cedrela odorata, Ceiba pentandra, Cordia alliodora, Enterolobium cyclocarpum, Erythrina* sp *Gliricidia sepium, Guazuma ulmifolia, Haematoxylum campechianum, Pachira aquatica, Swietenia macrophylla* and *Tabebuia rosea*. These species offer diverse products, uses and environmental benefits, derived from their use as hedges, shadow, human food, ornamental, medicinal, firewood, housing construction materials, wood, fodder or soil fertility conservation purposes. Multipurpose native trees of the SPS in the mountain region of Tabasco show the importance and multiple benefits that could be obtained from well-known and adapted species, which contribute to a better use of the available resources and to a higher sustainability of the system.

Key words: multipurpose trees

Introduction

During the last four decades, around 90% of the natural tree vegetation has been lost through agricultural activity in the state of Tabasco, Mexico (López, 1980). This has adversely affected the environment and availability of natural resources. For example, soil organic matter and fertility were reduced and degradation by water erosion and mechanized till accelerated (Larios and Hernández, 1992). On the other hand, in different regions of the state, and particularly in the mountain region, the producers practise diverse silvopastoral systems (SPS). Based on these considerations, the objective of this study was to characterize important tree species of the SPS practised in the mountain region of Tabasco, Mexico.

Materials and Methods

In the mountain region of Tabasco state, in the south-east of Mexico, there are four state municipalities - Jalapa, Tacotalpa, Macuspana and Teapa. The mountain region of Tabasco is a tropical zone with a humid warm climate, annual rainfall of 3500 to 4000 mm and annual mean temperature of 26°C; the maximum altitude in the region is below 800 m asl and the original vegetation was tropical evergreen rainforest. The principal agricultural activities include livestock production under extensive grazing with introduced and native grasses, commonly under silvopastoral management, and sown tropical crops such as sugar cane, coffee, cocoa, bananas or rubber. Information about characteristics of the silvopastoral systems and tree species was obtained by mean of interviews with producers, field journeys and direct evaluation of trees in diverse sites of the region.

Results and Discussion

The principal SPS in the mountain region of Tabasco, Mexico, are scattered trees (ST) on grasslands and living fences (LF). The SPS of the mountain region include native or introduced grasses associated with diverse trees, principally for beef and milk production or for fattening sheep. Initially a list of more than 40 tree species for the SPS was obtained. From this some important native species were selected, assuming that these species would be well adapted to the zone and offer diverse benefits (Table 1).

All the listed trees in the SPS are native multipurpose species and important components of the original vegetation of the area. Most are very well adapted to the predominant environmental conditions, and their propagation or establishment is widely known by the producers. Several species are nurtured, protected or planted by the peasants of the region, mainly for the use or benefits that they get from their presence in SPS.

The main characteristic that the producers of the mountain region of Tabasco value from the trees in the SPS is their use as wood or as a firewood source for different purposes (for example fine wood with commercial value obtained from species such as *Swietenia macrophylla, Cedrela odorata, Cordia alliodora* and *Tabebuia rosea*); another remarkable example is the "tinto" (*Haematoxylum campechianum*), a coloured species with a high content of tannins. This species is of very little interest to these users, but they are valuable to the producers for poles (for self-use or sale) and used for fencing. The species has great durability in the low flooded soils of the area. Some wood trees such as *C. odorata* and *S. macrophylla* are native species with great potential for reforestation in the zone, which is very important.

Some scattered trees on grasslands (*Ceiba pentandra*, *Enterolobium cyclocarpum*) provide shade for animals, while other species (for example *Gliricidia sepium*) are mainly used by the producers in live fences. The producers use several tree species adapted to the area, such as *H. campechianum*, *Pachira aquatica* and *T. rosea*, which are used directly as live fences or as scattered trees on grasslands in flooded areas. At the moment, most of the producers give little importance to the tree species as forage or fruit sources for animal intake. Valuable fodder trees widely distributed and available in the SPS of the area such as *G. sepium* or *Erythrina* sp are not considered by the producers as fodder resources for the animals, although their potential in animal feeding has been pointed out and promoted in agroforestry systems in other tropical regions of the world (Kass, 1994; Simons and Stewart, 1994). Producers are aware of the value to animals of the forage and/or fruit intake of "cuajilote", *Parmentiera edulis*, "guácimo", *Guazuma ulmifolia*, "ramón", *Brosimum alicastrum*, or "guanacaste", *Enterolobium cyclocarpum*, but they only marginally use them.

Besides forage, a lot of producers are not aware of the other benefits from trees in the system. Although the flowers of many trees are important for honey production, the producers do not consider this function important. Likewise, the ecological or environmental benefits of tree legume species *E. cyclocarpum*, *Erythrina* sp, *G. sepium* and *H. campechianum*, which improve the nutrient and organic matter content of the soil, through nitrogen fixation or the natural mulch incorporated into the soil, are not considered by the producers as important in the system. The benefits of some species (*Castilla elastica*, *G. sepium*), which serve as wildlife or insect shelters, or as host organisms for the growth of diverse epiphyte plants (included valuable orchids), are not considered important by the producers.

Table 1. Main attributes, establishment methods and characteristics of environmental adaptation of tree species in the SPS of the mountain region of Tabasco, Mexico. Shade: S, edible fruit: Efi, firewood: FW, fodder: F, wood: W, hedge: H, live fence: LF, medicinal: M, ornamental: O, poles: P.

Species	Considered attributes	Main attribute	Establishment or propagation method	Relevant characteristic of environmental adaptation
Brosimum alicastrum	W, S, F, LF	W	Natural regeneration; partially protected	---
Bursera simaruba	W, M, FW	W	Sown (stakes)	---
Byrsonima crassifolia	S, Efi, FW, F	Efi	Sown, protected	Degraded and wide soil drainage conditions
Castilla elastica	S, L	S	Protected	Wide range of soils
Cedrela odorata	W, H	W	Sown, protected	
Ceiba pentandra	S, W	S	Protected	Wide range of soils
Cordia alliodora	W, S, FW	W	Natural regeneration	---
Enterolobium cyclocarpum	S, FW, W, F	S	Protected	---
Erythrina sp	S, LF, Efl	S	Sown (stakes)	---
Gliricidia sepium	LF, FW, W	LF	Sown (stakes)	---
Guazuma ulmifolia	S, FW, W	S	Natural regeneration	Wide range of soils
Haematoxylum campechianum	P, LF, S	P	Sown, protected	Flooded soils
Pachira aquatica	LF, S	LF	Sown	Flooded and wide range of soils
Swietenia macrophylla	W	W	Sown	Soils with drainage problems
Tabebuia rosea	W, O, FW	W	Natural regeneration	Flooded soils

Table 1 shows a summary of the main attributes that the producers give to the trees of the SPS in the mountain region of Tabasco; it also gives the form in which they favour their propagation or establishment, and the adaptation of some species to particular environmental conditions of the region. Based on their density and frequency, *C. alliodora* and *C. odorata* are the main scattered trees in grasslands, while *G. sepium*, *Bursera simaruba*, *Erythrina* sp and *P. aquatica* are the principal tree species in the live fences of the area; *G. sepium* is the species with the highest density and frequency, located in all the evaluated sites, and for that reason is the most important tree in live fences. *T. rosea* and *H. campechianum* are tree species most adapted to flooded soils in the two main SPS of the mountain region of Tabasco.

Some isolated trees, such as *B. alicastrum*, *C. elastica*, *C. pentandra*, *E. cyclocarpum* or *S. macrophylla*, can grow to 20 m or higher; others, such as *C. alliodora*, *C. odorata*, *G. ulmifolia* and *T. rosea*, grow to 8 m. Some species such as *B. simaruba* and *C. alliodora* grow in live fences to 17 and 14 m, respectively. A live fence species such as *G. sepium* is not very high (mean 5 m), due to the browsing in the SPS. The most outstanding characteristics of the multipurpose trees in the SPS of the mountain region of Tabasco are presented in Table 2.

There are tree-grass, tree-animal and animal-tree interactions in the Tobasco SPS. Six important factors are taken into account by the producers. The three main effects are: (1) the effect of the trees on grass yield; (2) the animal welfare, from the tree shade in the SPS; and (3) other products, services or benefits that the animals receive from trees in the SPS, such as the construction of cattle facilities or the use of trees in live fences. The producers also consider yield of meat or milk and the damage that the animals cause to mature or young trees as important.

Table 2. Characteristics of the multipurpose trees in the two silvopastoral systems of the mountain region of Tabasco, Mexico. [1] Percentage of its presence in the evaluated sites. [2] Diameter at breast height.

	Scattered trees on grasslands				Live fences			
Species	Density Trees/ha	Frequency[1] %	Height m	DBH[2] cm	Density Trees/100 lineal m	Frequency[1] %	Height m	DBH[2] cm
Brosimum alicastrum	---	---	---	---	isolated	isolated	25	88
Bursera simaruba	17	3	7	33	4	71	17	53
Byrsonima crassifolia	isolated	isolated	6	20	---	---	---	---
Castilla elastica	isolated	isolated	18	50	---	---	---	---
Cedrela odorata	48	44	8	33	---	---	---	---
Ceiba pentandra	12	3	16	92	---	---	---	---
Cordia alliodora	52	50	9	9	2-3	28	14	27
Enterolobium cyclocarpum	isolated	isolated	15	90	---	---	---	---
Erythrina sp	5	3	5	30	33	57	6	24
Gliricidia sepium	---	---	---	---	85	100	5	29
Guazuma ulmifolia	24	14	8	31	2	14	2	8
Haematoxylum campechianum	66	5	7	12	26	42	9	16
Pachira aquatica	7	8	7	36	8	57	9	24
Swietenia macrophylla	isolated	isolated	18	40	---	---	---	---
Tabebuia rosea	50	25	8	12	6-14	57	5	11

All the interactions mentioned are positively modified by the management practices used in the SPS. The management and utilization of several native or introduced grasses totally or partially adapted to tree shade, such as *Axonopus compresus*, *Paspalum conjugatum*, *P. notatum*, *Brachiaria brizantha*, *B. decumbens*, *B. humidicola*, *Cynodon nlemfuensis*, *Panicum maximum* and *Pennisetum purpureum*, are associated with diverse tree species in the SPS of the region. Grassland rotation and grazing management, as well as other regional management practices commonly applied to grasses (like grassland burning, hand weeding with "machete" or the grassland "cleanness"), are also considered by the main interactions in the SPS.

Producer management can negatively affect some aspects of the SPS. Tree density, grazing management, chemical weed control, hand weeding and burning all effect the botanical composition of the grassland.

The importance and the ways in which producers modify the effects of the main interactions in the SPS are presented in Table 3.

Table 3. Effects, importance and the methods by which the producers modify the main interactions in the SPS of the mountain region of Tabasco, Mexico. [1] Based on the importance that the producer gives to the effects of interaction in the SPS: *null or little importance; ** important; *** very important.

Interaction type	Effect of interaction	Importance of the effect in the SPS[1]	Method by which the producer modifies the effect of interaction
(1) Tree-grass	Grass yield	***	Diversity, density and distribution management of tree species; management of grass species adapted to tree shadow; other grassland management practices (for instance grassland burning, the hand weeding with "machete" or the grassland "cleanness") or of trees (pruning)
(2) Tree-animal	Tree shade (animal welfare)	***	Diversity, density and distribution management of tree species; other management practices
	Other services or benefits for the animals derived from trees (for instance live fences or other construction of cattle facilities)	***	Diversity, density and distribution management of tree species; other management practices
	Product yield (meat, milk, etc.)	**	Diversity, density and distribution management of tree species; other management practices
(3) Animal-tree	Damage to adult trees (mechanical or in some other way)	**	Diverse management practices
	Damage to young trees (seedlings) (cattle trampling, intake or others)	**	Diverse management practices

Conclusions

The principal SPS in the mountain region of Tabasco, Mexico, are scattered trees on grasslands and living fences. The most important trees in the SPS are multipurpose native species, most of them very well adapted to the predominant environmental conditions, whose propagation and establishment are widely known by the producers. The main characteristic that the producers of the mountain region of Tabasco value in most of the present trees in the SPS is their use as wood or firewood sources. However, multipurpose native trees of the SPS in the mountain region of Tabasco, Mexico, show the importance and multiple benefits that could be obtained from well-known and adapted species, which make better use of the available resources and ensure a more sustainable system.

References

Kass, D.L. (1994). *Erythrina* species – pantropical multipurpose tree legumes. In: Gutteridge, R.C. and Shelton, H.M. (eds) *Forage Tree Legumes in Tropical Agriculture.* CAB International, Wallingford, United Kingdom, pp. 84-96.

Larios, J. and Hernández, J. (1992) *Physiography, Environment and Agricultural Land Use in Tabasco, Mexico* (Fisiografía, ambientes y uso agrícola de la tierra en Tabasco, Mexico). Universidad Autónoma Chapingo, Chapingo, Mexico. (In Spanish.)

López, R. (1980) *Vegetation Types and their Distribution in Tabasco and Northen Chiapas* (Tipos de vegetación y su distribución en el estado de Tabasco y norte de Chiapas). Universidad Autónoma Chapingo, Chapingo, Mexico. (In Spanish.)

Simons, A.J. and Stewart, J.L. (1994) *Gliricidia sepium* – a multipurpose forage tree legume. In: Gutteridge, R.C. and Shelton, H.M. (eds) *Forage Tree Legumes in Tropical Agriculture.* CAB International, Wallingford, United Kingdom, pp. 30-48.

Non-wood products in Russia

A. V. Griazkin and T. D. Smelkova
Forestry Dept., Saint Petersburg Forest Academy, 195021, Saint-Petersburg, Institutskiy per. 5, FTA, Russia. anatoliy@griazkin.spb.ru

Abstract
Russia has the largest global forest reserve. These can be used not only for wood production but also for medicinal and food plants, gum, other chemicals from trees, hay making, mushrooms and bee-keeping. The economic exploitation of these resources helps increase the efficiency and profitability of the forest. Scientific research in forestry started with a decree by Tsar Paul I in 1798 to create the Forest Department of Russia. Initially secondary forest products were charcoal for the foundry industry, potash, tar distillation, and honey. Forests also yielded wild animals for fur, berries, mushrooms and birds. More recently forests have been used to grow medicinal plants and secondary chemical products such as tannins, tar from birch, fir balsam, a coniferous vitamin flour, food plants, mushrooms and resin. This resin, from which rosin and turpentine are distilled, is one of the most valuable products. In some years resin production in Russia was 168,000 t from a world production of 976,000 t.

Key words: chemical products, resin, mushrooms, multiproduct forestry

Introduction
Russia has about 1180 million ha of forest, the most of any country in the world. The history of scientific forestry in Russia goes back more than 200 years. In the early years forests were exploited for charcoal, potash manufacture, tar distillation and honey. Economic activities in the forest are carried out by 1680 forestry enterprises, most of which process non-wood products. Forests produce not only wood, but medicinal and food plants, resin and other dendrochemistry material, the products of bee-keeping and mushrooms. In addition there are huge areas of forest pastures and haymaking is an essential component of management. More recently greater priority has been given to the production of medicinal and timber-chemical material, berries and mushrooms (Griazkin *et al.*, 1993; Kozubov and Taskaev, 2000; Tutygin *et al.*, 2000).

Materials and Methods
The use of non-wood forest resources in Russia is regulated by federal rules and standards OST 53-83-85, State Forest Service (1986, 1987, 1995), OST 61-6-1-91 and SR 2.3.4.009-93. Information and records have been confirmed in official reports from the enterprises located mainly in the north-west of Russia. Information on the historical part is taken from statistical yearbooks across the former USSR, Mendeleeva's survey work *The Explanatory Tariff*, issued in 1892, and V.E. Tishchenko *Rosin and Turpentine* (1895). Modern data are gathered from publications over the last decade (Griazkin *et al.*, 1993; Coppen and Hone, 1995; Niskanen and Demidova, 1999; Kozubov and Taskaev, 2000; Tutygin *et al.*, 2000; Wong *et al.*, 2001).

The account of resources and productivity of the basic kinds of food, medicinal and technical plants in Russia is carried out at forest inventory by a unified method. According to the working rules all pine forest stands of high productivity site index (I-III classes of Bonitet) up to final felling should be given for tapping.

Results and Discussion
The use of non-wood production is one potential way of increasing the efficiency of forestry. It is known that the value of forest berries and mushrooms per unit area can exceed the value of timber by several times. There is general concern over the low productivity of forest stands (Griazkin *et al.*, 1993; Kozubov and Taskaev, 2000; Tutygin *et al.*, 2000). In the near future it is predicted that the value of non-wood resources will grow, and their industrial development will enable the profitability of forestry to improve, employment to increase and people to be provided with natural raw materials.

Table 1. Average productivity of berries in woodland in the Republic of Komi, kg/ha.

Type of wood	Cranberry *Oxycoccus palustris* L.	Crowberry *Vaccinium vitis idaea* L.	Bilberry *Vaccinium myrtillus* L.	Blueberry *Vaccinium uliginosum* L.	Cloudberries *Rubus chamaemorus* L.
Pinetum vaccinosum	-	60	140	-	-
Pinetum myrtillosum	-	240	70	-	-
Pinetum polytricosum	-	120	90	50	-
Pinetum equisetumosum	-	-	-	80	40
Pinetum sphagnosum	270	-	-	80	60
Sphagnosums bog	320	-	-	50	50
Piceetum vaccinosum	-	50	140	-	-
Piceetum myrtillosum	-	170	70	-	-
Piceetum polytricosum	-	130	90	-	40

The volume of production of food and herbs in many respects is defined by the productivity of forests and the natural climatic conditions. The productivity of berries, mushrooms and different kinds of raw material depends on forest type (Griazkin *et al.*, 1993), with a potential fivefold difference across types (Table 1).

In general the crop of berries is dispersed over a huge area and is non-uniform. Most of the production is now inaccessible due to rural abandonment, absence of roads and uneven population distribution. The effect of structure of stocks of non-wood forest products on availability can be seen in an example from the north-west region of Russia (Table 2).

Table 2. Non-wood forests products in the north-west region of Russia.

Type of production	Producing area (10^3 ha)	Stocks ($\times 10^3$ t)	
		Total Biological	Harvested
Cranberry *Oxycoccus palustris* L.	201.8	27.2	13.6
Cowberry *Vaccinium vitis idaea* L.	112.1	16.8	8.3
Blueberry *Vaccinium uliginosum* L.	47.0	3.2	1.6
Raspberry *Rubus idaeus* L.	11.1	1.7	0.8
Bilberry *Vaccinium myrtillus* L.	212.0	31.9	15.9
Cloudberries *Rubus chamaemorus* L.	23.0	1.2	0.5
Mushrooms	918.4	35.9	17.9

Under Russian Sanitary Rules (1993) gathering of 57 species of mushrooms for commercial purposes (from 200 growing in Russia) is allowed. The total Russian stocks of mushrooms in forests are estimated to be 4.3 million tons, of which 10-12% are accessible to gathering (State Forest Service, 1987).

One of the most valuable products from the forest is resin, from which rosin and turpentine and their derivatives (more than 800 kinds of products) are extracted. In some years the volume of resin production in Russia reached 168,000 t (Griazkin *et al.*, 1993; Tutygin *et al.*, 2000).

Conclusions

About 74% of territory of Russia has more than 30% forest cover and 59% of the population lives here. In these areas there are many opportunities for industrial and sustainable development of non-wood forest production. The limiting factor is the absence of preferential crediting, insolvency of consumers, absence of investment for development by the manufacturers of processed edible forest products.

In the densely populated areas of Russia there are other problems in producing non-wood forest products - increased recreational pressure, loss of biodiversity, degradation of natural ecological systems and decreasing of their stability. In these areas strict regulation of wildlife management is required.

References

Coppen, J.J.W. and Hone, G.A. (1995) *Gum Naval Stores: Turpentine and Rosin from Pine Resin*. 2. Non-wood Forest Products, Natural Resources Institute, FAO of the United Nations, Rome, Italy.
Griazkin, A.V., Evdokimov, A.M., Egorenkov, M.A. and Konovalenko, V.M. (1993) *The Tapping and Harvesting of Minor Forest Products*. Ecology Press, Moscow, Russia.
Kozubov, G.M. and Taskaev, A.I. (2000) (eds) *Forestry and Wood Resources of Republic Komi*. The Information. Cartography, Moscow, Russia.
Mendeleev, D. I. (1892) The explanatory tariff, or research about development of the inductry of Russia in connection with its general custom duties. S. Petersburg, Russia. pp 243-770.
Niskanen, A. and Demidova, N. (1999) *Research Approaches to Support Non-Wood Forest Products Sector Development*. EFI Proceedings 29, Joensuu, Finland.
OST 53-83-85 *Wild Berries, Fruits and Nuts. Methods of Definition of a Crop and Resources*. Standart, Moscow, Russia.
OST 61-6-1-91 *Dried Mushrooms*. Standart, Moscow, Russia.
SR 2.3.4.009-93 *Sanitary Rules on Preparation to Processing and Sale of Mushrooms*. Moscow, Russia.
State Forest Service of the USSR (1986) *Technique of Definition of Stocks of Herbs*. Moscow, Russia.
State Forest Service of the USSR (1987) *Technique of Revealing of a Wild-Growing Source of Raw Materials at Forest Inventory*. Moscow, Russia.
State Forest Service (1995) *Rules of Tapping in the Forests of the Russian Federation*. Moscow, Russia.
Tishchenko, V. E. (1895) Rosin and turpentine. Boxing coniferous and getting resin in the North American United States and other states. Getting turpentine, turpentine and rosin, their property, a chemical compound and application. Department of trade of the Ministry of Finance, S. Petersburg, Russia. 246 pp.
Tutygin, G.S., Gaevsky, N.P. and Petrik, V.V. (2000) *Non-Wood Products*. University Press, Archangel, Russia.
Wong, J.L.G, Thornber, K. and Baker, N. (2001) *Resource Assessment of Non-Wood Forest Products - Experience and Biometric Principles*. FAO Technical Papers, Rome, Italy.

Transhumance and silvopastoral dependence in the Great Himalayan National Park Conservation Area – a landscape-level assessment

P. K. Mathur [1] and B. S. Mehra [2]

[1]Dept. Landscape Level Planning and Management, Wildlife Institute of India, Post Box 18, Chandrabani, Dehra Dun – 248 001, Uttaranchal, India. [2]WWF-India, 172-Lodi Estate, New Delhi – 110 003, India. mathurpk@wii.gov.in, badrishm@yahoo.com

Abstract

The implications for conservation of transhumance in a mountainous landscape in the north-west Himalayas with 35,000-38,000 livestock under silvopasture was studied. A hierarchical approach to understanding the physical, biological and sociocultural environments was used. Assessments of plant diversity and biotic pressure from four types of grazing resources showed that environmental forces and human actions over a long time span have shaped the composition and structure of forests and alpine pastures along an altitudinal gradient. 161 pastures and different migratory routes were mapped. There is a need for a careful delineation of protected area boundaries in such landscapes based on the physical characteristics and presence of natural resources. Adopting practices based on sound principles of spatio-temporal use of silvopastoral resources through appropriate distribution of livestock pressure across different migratory routes at any point in time is the only way to ensure continuance of transhumance along with the sustainable management of forests and alpine pastures.

Key words: biodiversity, protected areas, alpine pastures

Introduction

Traditionally, mountain people have relied on forests, grazing lands and livestock for their livelihoods. The Great Himalayan National Park Conservation Area (GHNPCA) has great regional conservation significance and presents a rich and diverse landscape. Transhumance has historically been practised in the landscape. However, recently it appears that the livestock population and herd size have increased considerably and practices of transhumance and silvopastoral dependence are not compatible with long-term conservation objectives. Hence, a comprehensive study of the compatibility of the current practices of transhumance and silvopastoral dependence with conservation objectives was carried out to study how best the diversity and productivity of existing silvopastoral resources can be maintained along with sustainable livestock grazing.

Materials and Methods

The study was carried out in the GHNPCA, Kullu district, Himachal Pradesh. The landscape (1171 km^2) covered a national park, two contiguous wildlife sanctuaries (Tirthan and Sainj) and a multiple use area or the ecodevelopment zone (EZ). The country's wildlife law prohibits resource use in a park while allowing regulated livestock grazing in a sanctuary. Traditionally the GHNP, prior to its final notification as a park in 1999, used to maintain ca. 35,000 migratory sheep and goats during summer and they extensively utilized its silvopastoral resources. Subsequently, livestock entry was banned in the park. A study undertaken during 1995-1999 used a combination of standard field methods and modern techniques, viz. remote sensing and a geographical information system employing the hierarchical approach, to have a better understanding of the physical, biological, ecological and socio-economic environments. Plant diversity, wildlife use, biotic pressure and silvopastoral dependence were assessed in four types of grazing resources vis-à-vis practices of transhumance. These resources were: in village surrounds; along migratory routes; transitory forest camping sites; alpine pastures. Mathur and Mehra (1999) provide details on area description and methodology for field data collection and analyses.

Results and Discussion

The study distinctively highlighted the disproportionate distribution of resources (forests, alpine pastures, agriculture and horticulture areas) in the four constituent areas (Mehra and Mathur, 2001). Vertical and horizontal complexity has resulted in a high spatial heterogeneity. Seasonal and staggered use of silvopastoral resources has kept grazing pressure localized and confined to smaller areas (e.g. proximity to village surrounds, camping sites along forest migratory routes and alpine pastures) and thus insignificant at the overall landscape level. Pastoralism has been practised on sound principles of spatio-temporal use, allowing a high species diversity. Resource mapping and ground validation allowed identification of 161 alpine and subalpine pastures along different grazing routes. Villagers respected grazing rules as they adhered to predetermined migratory routes, ultimately avoided conflict among themselves (Baviskar, 1998). The cyclic movement to more or less permanent seasonal bases allowed regeneration and conservation of living resources. However, a concurrent study short-listed ten tree, seven shrub and 16 herb species under some direct or indirect threat as they are being used by villagers, pastoralists and herb collectors for long-term ecological monitoring in GHNPCA (Mathur and Uniyal, 1999). Staggered flowering and spatial and temporal dispersal of livestock in different pastures provided favourable conditions for plant growth. The study provided an insight into the ecology, socio-economics and overall conservation of GHNPCA in relation to transhumance and also highlighted those natural factors, market forces, forest management practices, resource use and other illegal activities (poaching, timber extraction, medicinal plant collection, etc.). The latter probably have a greater, compounding and permanent influence on silvopastoral resources.

Conclusions

The study found that (i) mountain people rely on the whole landscape for their livelihoods; consequently, policies and institutions for mountain forests and agroforestry must recognize interactions between agricultural land uses, forests and livestock; (ii) every strategy for ensuring that mountain people derive sustainable livelihoods from their forests must be tailor-made for the local physical, biological, cultural and political environment – and ways of responding to change must always be included. The exclusion of traditional silvopastoral use in GHNP and subsequent active protection may lead to an overall recovery of forests and pastures. This way, the legal role of a national park would be fulfilled. However, it is difficult to predict how this overall ecological recovery would affect individual plant and animal species or the overall species diversity. Further, the restriction of access to nearly 65% of the total area of GHNPCA is likely to increase livestock pressure on the remaining area and this is expected to accelerate the degradation of those remaining silvopastoral resources. The study calls for (i) a careful delineation of wildlife protected area (PA) boundaries in high altitude landscapes, based on physical characteristics and the availability of natural resources; (ii) a development of local and cultural participatory forest and range resource management programs, rather than policies based on exclusion of grazing; and (iii) distribution of uniform livestock pressure as far as possible across different grazing routes and silvopastoral resources.

References

Baviskar, A. (1998) Intensive micro-study to assess socio-economic conditions of people using the great Himalayan National Park. In: *Final Report, FREEP-GHNP Research Project*. Wildlife Institute of India, Dehra Dun, India. pp. 1-19.

Mathur, P.K. and Mehra, B.S. (1999*)* Livestock grazing and conservation of biodiversity in the high altitude ecosystem – an integrated landscape management approach. In: *Final Report, FREEP-GHNP Research Project*. Wildlife Institute of India, Dehra Dun, India. pp. 1-192.

Mathur, P.K. and Uniyal, V.P. (1999) *Long Term Ecological Design - Great Himalayan National Park Conservation Area, Forestry Research Education and Extension Project (FREEP)-GHNP Research Project, Final Project Report*. Wildlife Institute of India, Dehradun, India. 118 p.

Mehra, B.S. and Mathur, P.K. (2001) Livestock grazing in the Great Himalayan National Park Conservation Area - a landscape level assessment. *Himalayan Research Bulletin* XXI (2), 89-96.

Economic, social and cultural benefits of silvopastoral systems in Nigeria

I. O. Oladele
Japan International Research Center for Agricultural Science, 1-1Ohawashi, Tsukuba City, Ibaraki 305-8686, Japan.
oladele@jircas.affrc.go.jp

Abstract
Three out of the five agricultural zones in Nigeria are prominent and have a comparative advantage for livestock production. A major constraint is the availability of feeds due to the serious weather fluctuations. The introduction of silvopastoral systems in the National Fadama Development Programme sponsored by the World Bank in Nigeria has left serious consequences for the farmers economically, socially and culturally. This chapter examines these different benefits from a survey of farmers in northern Nigeria. Prominent economic benefit is the reduction in production cost, while the resultant conflict between the farmers and pastoralists is the overriding social effect while the preservation of values and norms was indicated as the cultural reason. The chapter concludes with the pragmatic steps in which these benefits could be harnessed for sustainable development of the livestock industry in northern Nigeria.

Introduction
Nigeria has a land area of 923,768 km^2 with a great climatic diversity that permits the production of different classes of livestock. In 1987 the country was zoned into five agricultural zones to cover the different ecological environments. This is predicated on the need to strengthen the farming system research and extension links. Of the five zones, livestock production is prominent in three zones namely, north-west, north-east and central zones. These areas represents 79.1% of the total land area of the country with the central zone located on latitude 6° 30' to 11° 20', north-east from 8° to 14° and north-west from 9° to 14°. The average growing period for these area is 110 days due to low rainfall. The three zones produce 90% of cattle, 66% of sheep and 56% of goats in the nation (FMNAR, 1997).

To improve livestock production that is often constrained by adequate feeds for the animal, agroforestry practices were introduced to the farmers by the joint effort of International Institute of Tropical Agriculture (IITA), International Livestock Research Institute (ILRI) and local afforestation projects. The acute shortage of feed in the long dry periods encouraged the adoption of silvopasture. The practice is based on the combination of tree and shrub species with animal production. The species used in northern Nigeria include *Parkia* species, Gum Arabic, *Leucaena leucocephala*, *Faidherbia albida*, *Gliricidia sepium* and *Dacryodes edulis* (NARP, 1996).

The rapid adoption of silvopasture depends on the combination of economic, social and cultural benefits that the combination of tree species and animal production offers to the farmers. Many of these benefits are based on the indigenous knowledge system of the people. The benefits derived by the farmers would enhance the continuous practice of silvopastoral systems; otherwise, the system could be discontinued like other agricultural innovations. This chapter therefore examines the different types of benefit derived by farmers through silvopasture.

Materials and Methods
A rapid rural appraisal was used to determine the benefits of silvopastoralism among 300 farmers. A snowballing sampling technique was used to locate the respondents. From a checklist of 25 items on the benefits of silvopasture, respondents were asked to indicate the economic, social and cultural benefits derived. These were later sorted and subjected to frequency counts and percentage as well graphical representations.

Results and Discussion
Of the 25 items listed, eight were indicated as economic benefits. Their response was based on their perception of these factors in economic terms relative to their circumstances. The factors are livestock–tree integration (25%), availability of non-timber forest products (50%), provision of shade and shelter for animals (20%), use of trees as windbreaks (22%), use for animal training for traction (18%), reduced cost of feeds and production (60%), improved pasture feeding (15%), enhanced record keeping (10%), provision of fuelwood (14%) and reduced commercialization (30%).

Figure 1 shows the social benefits derived by the respondents with the prevention of conflict as the most prominent. This is due to the fact that grazing has traditionally been on a communal basis, with no individual possessing sole right to any grazing land. Farmers usually used uncultivated bush, fallow farms, forest and grazing reserves, and grass and water were considered free resources available to the stock that got to them first. Sometimes, at the invitation of the farmers, pastoralists would kraal their animals on farms to utilize crop residues and, in the process, the animals would provide dung to enrich the soil. However, over the years, there has been a progressive deterioration in the symbiotic relationships, and conflicts between the farmers and pastoralists have become routine in Nigeria, which tend to be violent and fatal and therefore threatening to the peace and stability of the states where they occur.

Figure 1. Social benefits of silvopasture among farmers in northern Nigeria. Socio-economic status (SES), wealth indicator (WIDD), multiple Income generating activities (MIGA), gender influence (GI), manure exchange (ME), reduced transhumance (RT), conflict prevention (PCONFLICT) and changed migration pattern (CMP).

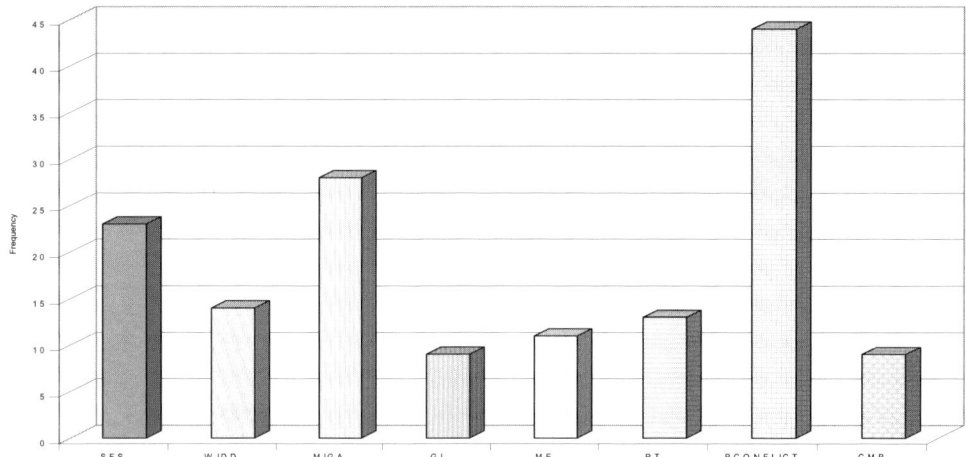

The factors on the social benefits derived by respondents are represented in Figure 1. Figure 2 presents the cultural benefits among farmers with the enhancement of the values and norms of the respondent being the most prominent. This may be attributable to the prevailing sociocultural milieu rooted in the beliefs and systems of production of the respondents. These cultural benefits are enhancement of seclusion (ES) as women in Muslim communities are not allowed to mix freely with men in public places, preservation of indigenous trees (PIT) - some indigenous beliefs are associated with these trees - supply of staking materials (SSM), which are used for other crops such as tomatoes and twining vegetables, forage palatability (FP), enhancement of ethnoveterinary practice (EEP) - dried leaves and soaked roots and other parts of the trees used in silvopasture are used in treating animal pests and diseases reduction of bush burning (RBB) - the act of burning bush in order to kill wild animals is not practised around silvopasture - and the enhancement of values and norms (EVN) - there is the encouragement of the pastoralist living style and gender roles division such that younger generation are exposed to these practices.

Figure 2. Cultural benefits of silvopasture among agropastoralists in northern Nigeria.

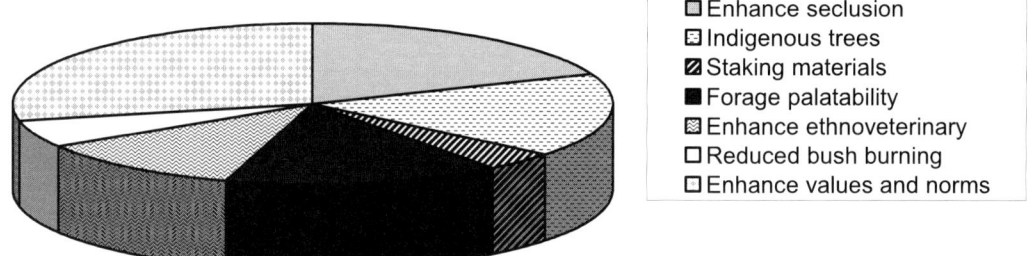

Conclusions

The chapter has clearly shown that the reasons for the adoption and practice of silvopastoralism among farmers are embedded in the economic, social and cultural compatibility and enhancement of this technology. This will serve to ensure the sustained use of the technology and its wider spread. Also, the realization that the economic, social and cultural needs of the pastoralists can be met with the silvopasture system would ensure that more silvopasture systems are established, and the existing ones maintained.

References

FMNAR (1997) Nigeria: National Agricultural Research Strategy Plan 1996-2010. In: Shaib, B., Aliyu, A. and Bakshi, J.S (eds) Department of Agricultural Sciences, Federal Ministry of Agriculture and Natural Resources, Intec Printer, Ibdan, Nigeria. pp. 41-150.

NARP (1996) Evolving the Nigerian Agricultural Research Strategy Plan: agro-ecological inputs. In: Adedipe, N.O, Bakshi, J.S, Odegbaro, O.A. and Aliyu, A. (eds) *National Agricultural Research Project*. Oluseyi Press, Ibadan, Nigeria. 486 pp.

Ancient wood-pastures in Scotland

M. A. Smith
Scottish Natural Heritage, Anderson Place, Edinburgh EH6 5NP, UK. Mike.Smith@snh.gov.uk

Abstract
The aim of the project was to try and identify potential wood-pasture sites through remote sensing, carried out by undertaking an analysis of existing maps and map-based inventories. The object was not to identify every possible site but to find a practical methodology that would give an estimate of their extent and distribution. From this it was estimated that there are between 8500 and 17,000 ha of ancient wood pastures in Scotland. The next challenge is to develop best management practices that can maintain this habitat and contribute to the development of ecological networks within a sustainable multipurpose land use.

Key words: veteran trees, silvopasture, cultural landscape

Introduction
Ancient wood pastures (AWP) are found where the woodland and grazing history have evolved together to produce a grazing-maintained habitat with elements of both woodland and pasture. These dynamic processes have resulted in woodland structure composed of biologically important open grown trees (Rotheray, 1997; Denton et al., 2001), which can attain a great age, over a range of semi-natural ground floras. These wood-pastures have developed in both lowland and upland situations (Quelch, 2000) but their presence can often be masked by more recent management practices. As a result the development of a methodology that could help identify these sites has been undertaken.

Materials and Methods
AWP sites were identified by comparing sites with open woodland cover, as recorded by the Scottish Semi-Natural Woodland Inventory (SSNWI), with those that are mapped as open woodland on the Ordinance Survey Maps of 1860, the premise being that sites with open woodland on both are potentially ancient wood-pastures. Ascertaining wood-pasture status was only possible by field survey to establish whether the site still maintains populations of trees that were present in both 1860 and the present day. By combining data from the SSNWI and the maps with the field survey exercise it is possible to derive an estimate of the number of SSNWI polygons that contain AWPs and, from this, to estimate the area. A statistically useful sample of the sites was selected and these were visited to confirm their AWP status. The results of the field survey are then combined with the remote sensing part of the project to estimate the proportion of the SSNWI polygons that are AWPs, using ArcView GIS software.

Results and Discussion
Of the sites visited, 14 of the smaller (1-10 ha) 25 were confirmed as AWP and 38 of the larger (> 10 ha) 55 sites were confirmed as AWP. The calculations in Table 1 give an estimate for the total number of wood-pasture patches in Scotland.

Table 1. Estimated numbers of wood-pasture SSNWI polygons in Scotland.

	1-10 ha	> 10 ha	Total
Number of potential patches (from LCS88) (N)	7,600	3,000	
Initial sample size (n1)	1,000	1,271	
Number of these that match	116	187	
Survey sample size (n2)	25	55	
Number of these that are verified	14	38	
Proportion of initial sample that match OS 1860 map (p1)	0.116	0.147	
Proportion of matching sample (p2)	0.560	0.691	
Estimated number of real patches (Nw)	494	305	799
Estimated variance of Nw	9,273	1,001	10,274
Standard deviation of Nw	96	32	101
% Coefficient of variation of Nw	19.5	10.4	12.7
Multiplier	1.46	1.22	1.28
Lower 95% confidence limit	338	249	623
Upper 95% confidence limit	721	374	1,023

The estimated number of real patches is calculated as $N \times p1 \times p2$ (see table). Its variance estimated as $Nw \times Nw \times ((1 - f) \times (1 - p1)/(n1 \times p1) + (1 - p2)/(n2 \times p2))$ where f is the sampling fraction for the initial sample (n1. Calculation of the 95% confidence limits assumes a log-normal distribution for the number of patches (the 'multiplier' is just an intermediate step). The calculations in Table 2 develop an estimate for the total area of wood-pasture patches in Scotland.

Table 2. Estimated area of wood-pasture in Scotland. The estimated area Aw is calculated as Nw x sum (verified survey areas)/number patches, i.e. Nw x abar where 'abar' is the mean of the OK truthed patches. Its variance estimated as Aw x Aw x (var (Nw)/Nw2 + var (abar)/abar2). Calculation of the 95% confidence limits again assumes a log-normal distribution for the area of patches.

	1-10 ha	> 10ha	Total
Mean of the verified sample survey areas (abar)	3.845	33.489	
Variance of abar	0.444536	34.01531	
Variance of Aw	245,440.2	4,285.922	4,531.363
Standard deviation of Aw	495	2,070	2,129
% Coefficient of variation of Aw	26.1	20.3	17.6
Multiplier	1.65	1.48	1.41
Lower 95% confidence limit	1,148	6,892	8,604
Upper 95% confidence limit	3,139	15,134	17,047

The inventory of ancient wood-pastures has been set up on ArcView GIS software, which indicates sites as polygon themes. These can be displayed over OS map data at an appropriate scale or alongside other Scottish Natural Heritage GIS-based inventories. This will allow AWP to be shown as part of forest habitat networks or wider ecological networks as they are developed (Smith and Holl, 2002). Understanding the resource of AWP in Scotland is still preliminary and the inventory gives the first estimate of the extent of the resource, which will become more statistically accurate as new sites are added.

Conclusions

So far attention to wood-pasture in Scotland has been limited to land-use historians, naturalists, conservation bodies, the government agencies and a handful of interested individuals and enlightened landowners (Quelch, 2001). This work by SNH to develop the inventory of existing sites helps to increase knowledge of the habitat and draw it to the attention of the owners of remnant AWP. Some are on public land: in forests, country parks and nature reserves. Government agencies and local authorities should lead the way in establishing best practice for restoration and management. The development of ancient and new wood-pastures as a sustainable multipurpose land use will require continued research and also the interest and cooperation of innovative landowners who would be willing to put theory and research into practice, marrying the benefits of trees and livestock farming. Both are vital to the long-term management of the Scottish countryside. Grazing regimes suitable for the enhancement and maintenance of ancient wood-pasture are likely to be consistent with those for other important habitats that made up the pre-clearance Scottish cultural landscape.

References

Denton, M.L., Godfrey, A., Hemingway, D.G. and Skidmore, P. (2001) *Saproxylic Invertebrate Survey Hamilton High Parks SSSI 1999-2000*. Scottish Natural Heritage Commissioned Report BAT/LI02/99/00/48 (unpublished report).

Quelch, P. (2000) Upland Pasture Woodlands in Scotland part I. *Scottish Forestry* 54 (4), 209-214.

Rotheray, G. (1997) *The Conservation of Saproxylic Diptera in Scotland*. Scottish Natural Heritage Commissioned Report F96AC306 (Unpublished report).

Smith, M. and Holl, K. (2002) *Wood-pasture in Scotland: Classification, Inventory and Management*. Scottish Natural Heritage Research Survey and Monitoring Report.

Environmental and resource economics: would the Delphi method follow a non-iterative process of surveys?

M. Soliño

Research Group: Environmental Economics Applied to Natural Spaces and Rural World. University of Vigo. Faculty of Economics, Lagoas Marcosende s/n. 36310, Vigo, Spain. mario@uvigo.es

Abstract

There are still few applications in Spain of the Delphi method, an economic valuation technique based on the analysis of collective preferences. This method allows the estimation of the minimum compensation required to accept a measure or policy that improves some public good, in this case the environment, but imposes some cost on some individuals or groups. Also, the method provides valuable information to arrange the high-priority performance lines in the design of agri-environmental policies. In this chapter, two recent Delphi applications in Spanish forestry and agricultural systems are critically compared. Taking into account some of their results, one of the principles of the method - the iterative survey process - is discussed. To achieve this, a quantitative analysis of stability and consensus of the answers is carried out. Finally, it is argued that it would not be reckless to carry out a Delphi round only in terms of cost-efficiency.

Key words: compensations demand, agri-environmental policy, Common Agricultural Policy, silvopasture

Introduction

The design of agri-environmental policies usually implies a series of legislative modifications affecting some target groups. In order to efficiently reach the objectives, the collaboration of the affected population is essential, and this is not always easy to obtain. Sometimes the appropriate information about the related consequences is not provided; other times decisions are taken without negotiation and, when the measures are applied, they are not succesful because the population rejects them. Certainly, if policies want to achieve their objectives, cooperation in the expenditure of public policy is necessary, both with the affected population and with the agents that would apply them. Participation has some important costs (transaction costs) and therefore it is necessary to carry out a correct design of the consultations to be taken. It is then useful to know the public assessment of the policies and the proposed changes, so that the final policy could reflect the opinion of the general public as well as the experts and that of the agents of local development and technicians.

Materials and Methods

The Delphi method (Dalkey and Helmer, 1963) could be a useful tool for the collective decision-makers themselves to provide information to help adapt the existing policies to environmental objectives and to rural development, improving their effectiveness or social acceptance. In such a case, the application of this methodology will help us to know which is the most acceptable and effective arrangement and design of intervention instruments by the policy maker and the minimum compensation that the affected agents would demand to carry out agri-environmental measures. The application of the Delphi method to environmental problems or policies is currently under way in research projects in Spain. Recently three studies have been carried out (Colino *et al.*, 1999; Mariscal and Campos, 2000; Soliño, 2003a,b) with a common connection: the primary sector (agriculture, forest and grazing systems) and Common Agricultural Policy (CAP) Reform. In all of them a forecast exercise has been carried out that could be useful to policy makers in the future when executing the imminent Agenda 2000. Based on some of the results of these studies, the necessity of an iterative process in a Delphi analysis will be discussed and, a constructive approach adopted throughout the discussion on the theory underlying the procedure.

Results and Discussion

The recommendation that the Delphi method should be an iterative (repeatable) process of two surveys would be debatable in view of the quantitative and qualitative results presented (Table 1). At least two surveys applied to the same individual are necessary to have controlled feedback, to obtain a consensus and diminish the dispersion of answers. Although this last point may be unquestionable, the consensus objective is questionable. In Table 1 we can observe that the central measure - reference when settling down if a consensus has been arrived at or not - is similar between the first and the second round of the analysed studies.

However, the median remains constant between the first and the second round, although the dispersion of the answers significantly diminishes between both rounds. To measure the degree of dispersion of the answers we use the interquartilic relative range (IRR). This shows that indeed, in the second round, the dispersion is smaller than in the first one. Continuing with the discussion about the necessity of following an iterative process of surveys, we have estimated a prediction model of the answers' dispersion for the second Delphi round. The results show how a unit increase in IRR1 would cause a 0.433 increase in the IRR2. Estimated parameters are highly significant (100%) and the goodness of fit of the model is about 90%. In spite of treating this valuation with caution because of the low number of observations, the results of the analysis show a clear line of behaviour.

Table 1. Compensations demanded and consensus in the answers * Variation calculations based on the data corresponding to the flat land. Source: own elaboration on Soliño (2003a,b) and Mariscal and Campos (2000). =: no variation; ∇: reduction; Δ: increment.

Delphi analysis		First round			Second round			RV(Me)
		Percentile 25	Median	Percentile 75	Percentile 25	Median	Percentile 75	
Natural Network 2000 - Minimum grant demanded (€) for hectare of performance								
Initial reforestation costs	Conifer	1547.61	1953.29	2524.25	1840.60	1953.29	2065.98	=
	Decidous	2404.05	3005.06	3268.00	2404.05	3005.06	3005.06	=
Maintenance premium	Conifer	300.51	300.51	360.61	300.51	300.51	300.51	=
	Decidous	312.53	465.78	601.01	360.61	464.28	465.78	∇ 0.3%
Compensatory premium	Conifer	21.04	33.06	60.10	33.06	33.06	46.58	=
	Decidous	21.04	33.06	60.10	33.06	34.56	52.59	Δ 4.5%
Consolidation /restocking of forests	Conifer	901.52	901.52	1202.02	901.52	901.52	901.52	=
	Decidous	1089.33	1202.02	1803.04	1202.02	1202.02	1202.02	=
Dehesas - compensations for the reforestation of oaks or cork-oaks (€/ha per year)								
Maintenance	Flat land	135.23	240.40	330.56	150.25	240.40	270.46	=
	Land with low slope	-	-	-	180.30	240.40	270.46	=*
Income loss for pasturing	Flat land	102.62	135.23	277.97	60.10	135.23	157.77	=
	Land with low slope	-	-	-	60.10	120.20	180.30	∇11.1%*

Conclusions

Although little work has been carried out in Spain applying the Delphi method to the field of environmental and natural resource economics, the good quality research that has been carried out in the last years and the application of this technique to an analysis of the primary sector make it likely that further research will be carried out using this methodology in the future. The analysis can be applied to silvopastoral systems as a land-use scenario within a reformed CAP. Improvement of the method has been discussed. The second round does not seem to be cost-efficient and omitting it results in considerable savings of money and time. In addition, a single round would diminish the risk of desertions in the iterations because of the time delay and the decrease in psychic fatigue of the experts. However, given the low number of studies available, at least it has been proven that it makes sense to think deeply about why an iterative process should be used. Some researchers may see the Delphi method as a useful tool for their analyses and this would help reform the methodology, as discussed.

References

Colino, J., Noguera, P., Riquelme, P.J., Carreño, F. and Martínez-Carrasco, F. (1999) *Informe sobre La Reforma de la PAC y el Sector Agrario de la Región de Murcia*. Consejo Económico y Social de la Región de Murcia, Murcia, Spain.

Dalkey, N.C. and Helmer, O. (1963) An experimental application of the Delphi method to the use of experts. *Management Science* 9, 295-310.

Mariscal, P.J. and Campos, P. (2000) *Aplicación del Método Delphi a un Grupo de Propietarios de Dehesas de la Comarca de Monfragüe (Cáceres)*. Informe final CSIC, Madrid, Spain.

Soliño, M. (2003a) Programas Forestales en las Comunidades de Montes Vecinales en Mano Común de la Red Natura 2000: Un Análisis Delphi. *Revista Galega de Economía* 12 (1), 225-246.

Soliño, M. (2003b) Nuevas Políticas Silvo-Ambientales en Espacios Rurales de la Red Natura 2000: Una Aplicación a la Región Atlántica de la Península Ibérica. *Revista Investigación Agraria: Sistemas y Recursos Forestales* 12 (3), 57-72.

The influence of goat grazing on ground vegetation and trees in a forest stand

A. Zingg [1] and P. Kull [2]

[1] Swiss Federal Institute for Forest, Snow and Landscape Research WSL, Birmensdorf, Switzerland. [2] Office for Nature Conservation and Landscape Management, Lucerne, Switzerland. andreas.zingg@wsl.ch, peter.kull@lu.ch

Abstract

Grazing of domestic animals in forests is prohibited by law in Switzerland. The historical reasons which led to this legislation no longer apply. Nonetheless the regulation persists and is defended by foresters even if they have usually had no practical experience. As part of a demonstration of former land-use practices in an open-air museum in 1998, it was possible to establish a goat pasture in a closed forest. The growth and yield research group of the Swiss Federal Institute for Forest, Snow and Landscape Research WSL, together with the vegetation science group, studied how grazing goats affected trees and ground vegetation. The whole forest stand was recorded, i.e. all trees taller than 1.3 m, by measuring their diameters and their coordinates. In addition, the ground vegetation was recorded on sample plots. This survey was carried out before the goats were allowed to enter the forest and repeated twice in the following years. During the pasture season three adult goats and two kids could circulate freely between the forest and the meadow for a restricted time. The result is that goats destroy the regeneration almost completely and alter the herb layer. Goats also damage larger trees, but there was no damage to the dominant trees of the stand.

Key words: goat, forest pasture, forest stand, grazing, browsing

Introduction

Until the first Swiss Forest Act of 1876 came into force, traditional land use included pasturing of all kinds of livestock in the forests. Open agricultural land was intensively used to produce crops; therefore forests were used as pastureland for livestock. Damage from grazing was frequent and as a consequence forest structures declined. Additional use of these forests as a source of winter fodder for the animals led to the situation where a number of forests with important protective functions were unable to fulfil these anymore. This problem was already recognized in the 18th century and many historical documents deal with the problem of animal husbandry in the forest. Landolt (1862) showed that many of the inherent problems were the consequence of over-utilization, with grazing as a major component. As a consequence grazing in forests with protective functions was prohibited by law. This was extended to all other forests within the following 50 years. In 2004 grazing is prohibited in almost all forests in Switzerland and only possible in special cases with a permit from the forestry administration or even the cantonal governments.

At lower altitudes grazing in forests as a form of forest utilization has disappeared almost completely in the last 150 years. Today new ideas arise: using goat grazing to improve the habitat for rare plant species, cattle grazing to influence forest structures (Bebi et al., 2001). Often the initiators of the new ideas are not aware of possible problems already known from earlier times (Gotsch et al., 2002).

The Swiss Forest Museum planned to demonstrate to the public the different effects of traditional forest utilization. One of these themes was the impact of goat and cattle grazing on forest stands. Once the necessary permit had been given, those responsible in the Forest Museum decided to document the effects scientifically. The Ballenberg Open-air Museum provided the animals and the necessary fencing. The first season of forest pasturing was in summer 1998. The main research questions were: What are the effects of goat grazing on an existing stand? What is the influence of goat grazing on the ground vegetation (diversity, abundance, composition) and how is it influenced in the long term?

Materials and Methods

The research and demonstration plot was established at the Ballenberg Open-air Museum in Hofstetten near Brienz in the Bernese Oberland, at an altitude of 680 m asl, adjacent to a meadow usually grazed by the goats. The vegetation type is a beech forest with sweet woodruff (*Galium odoratum* L.), i.e. a *Galio-odorati Fagetum* typicum phytosociologic association (Keller et al., 1998) with a tendency to a mixed lime forest. The grazing area inside the forest measured 941 m^2. A case study approach was used.

The forest stand consists of 47% beech (*Fagus sylvatica* L.) (percentage of the total basal area of trees with a dbh $d_{1.3} \geq 8$ cm), 3% sycamore maple (*Acer pseudoplatanus* L.) and Norway maple (*Acer platanoides* L.), 35% small-leaved lime (*Tilia cordata* L.), (*Quercus* L.) oak (12%) and 0.2% Norway spruce (*Picea abies* (L.) Karst.).

In 1998 all trees and shrubs taller than 1.3 m ($d_{1.3} \geq 0$) were measured and mapped. Damage to the plants was recorded. The ground vegetation was surveyed in three concentric circular plots of 30, 200 and 500 m^2, using the method of Braun-Blanquet to estimate the abundance (Mueller-Dombois and Ellenberg, 1974). The plots were permanently marked. The first tree and vegetation survey was taken in May 1998 before grazing started.

During the summer and autumn of 1998 a group of three adult goats with two kids had access to the forest for a total of 8 weeks. Over the whole period, the goats could circulate freely between the meadow and the forest. It was observed that the goats seemed to prefer grazing in the forest. In spring 1999 all the recorded trees were checked for damage and the vegetation survey was repeated. In 1999, 2000 and 2001 the forest was again accessible to the goats and the survey of damage and vegetation was repeated in 2001. Diameter measurement was not repeated until autumn 2003.

Results and Discussion

The forest stand is a mixed broadleaved forest where the upper canopy consists mainly of beech and lime interspersed with some oak, maple and Norway spruce. In the middle layer only a few trees were present. The lower canopy (regeneration) consisted of a good natural regeneration of beech and Norway spruce together with the other species and some hazel nut (*Corylus avellana* L.) and hawthorn (*Crataegus* L.) along the forest edge. The canopy coverage was 90 to 100%, the regeneration coverage was approximately 50 to 70%.

Figure 1. Reduction in number of living trees (M) of individual species between 1998 and 2003.

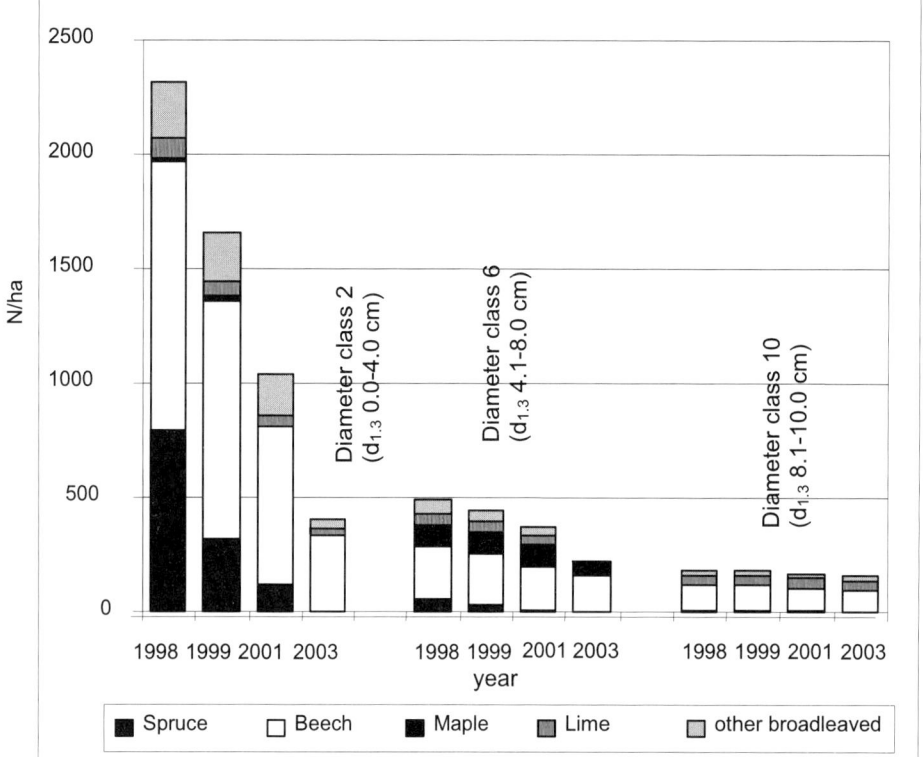

A reduction in the number of stems per diameter class (Figure 1) occurred only in classes 2 (0–4 cm) and 6 (4–8 cm). In the larger diameter classes the changes were minor. With the exception of a few surviving trees in the lowest diameter class, the regeneration disappeared. Spruce disappeared completely whereas beech, maple, lime and other broadleaved species were drastically reduced. Little damage by goats was recorded to trees with a diameter greater than 8 cm, and the affected trees were mainly lime and maple. Large spruce, beech and oak were left untouched. The main damage caused was peeling of the stems and in the case of spruce even the twigs. Browsing damage was recorded on many trees; on some plants the terminal bud 2.5 m high was browsed.

The results from the vegetation surveys are compiled in Table 1. There was no change in the tree layer, but a major change in the shrub layer occurred, which reflects the peeling and browsing already shown in the tree data. The change to the herbal layer seems to be relatively moderate. However, knowing that ten species disappeared and 13 new species appeared - mainly species typical of meadows - it can be concluded that the probable change caused by grazing activity is considerable.

Table 1. Vegetation survey of goat-grazed forest.

Plot size: 500 m^2		1998	1999	2001
Tree layer: height	m	27	27	28
Canopy cover	%	95	95	95
Number of species	N	7	7	7
Shannon-Weaver index		1.46	1.46	1.46
Shrub layer: height	m	7	8	8
Canopy cover	%	40	30	25
Number of species	N	10	7	5
Shannon-Weaver index		0.38	0.24	0.21
Herbal layer: height	m	50	50	50
Canopy cover	%	15	20	20
Number of species	N	53	49	56
Shannon-Weaver index		0.62	0.47	0.51
Mosses: Canopy cover	%	0	0	0

Conclusions

Goat grazing in a forest stand may not harm the stand itself, but it makes regeneration impossible. This means that young stands or stands to be regenerated do not tolerate goat grazing. An active management and a control of goat grazing in forests are necessary. The change of the herbal layer is considerable but no measures need to be taken in favour of certain plants. There were no endangered species in the stand. If this were the case, it might be necessary to protect such plants against goat grazing. In this case study length and intensity of goat activity have to be considered. It would therefore be useful to establish a controlled experiment to obtain more reliable data on the influence of goat grazing on a forest.

References

Bebi, P., Kienast, F. and Schönenberger, W. (2001) Assessing structures in mountain forests as a basis for investigating the forests' dynamics and protective function. *Forest Ecology and Management* 145, 3–14.

Gotsch, N., Finkenzeller, N., Beck, J., Bollier, D., Buser, B. and Zingg, A. (2002) *Bedeutung und Zukunft von Waldweiden im Schweizer Alpenraum: Auswertung von Daten des Landesforstinventars und einer Befragung. von Förstern.* Ergebnisse des Komponentenprojektes H. Polyprojekt Primalp c/o Institut für Agrarwirtschaft, Zürich, (Importance and future of pasturing in forest in the Swiss Alps.) 32 pp.

Keller, W., Wohlgemuth, T., Kuhn, N., Schütz, M. and Wildi, O. (1998) Waldgesellschaften der Schweiz auf floristischer Grundlage. Statistisch überarbeitete Fassung der «Waldgesellschaften und Waldstandorte der Schweiz» von Heinz Ellenberg und Frank Klötzli . Mitteilungen der Eidgenössischen Forschungsanstalt für Wald, Schnee und Landschaft. 73 (2) 91-357.

Landolt, E. (1862) *Bericht an den hohen schweizerischen Bundesrat über die Untersuchung der schweiz. Hochgebirgswaldungen, vorgenommen in den Jahren 1858, 1859 und 1860.* Weingart, Bern, 367 pp. (Report to the Swiss Federal Council about the survey of the forest in the high mountains of Switzerland, conducted in the years 1858, 1859 and 1860.)

Mueller-Dombois, D. and Ellenberg, H. (1974) *Aims and Methods of Vegetation Ecology.* John Wiley and Sons, New York, USA.

Session 5.
Future perspectives of silvopastoral systems in a global context

Silvopastoral systems for rural development on a global perspective

A. Pardini
DiSAT – University of Florence, Italy. andrea.pardini@unifi.it

Abstract
Rural areas of developing countries are home to about 2800 billion people that get food, energy, medicines and other products from forestry, agriculture and animal rearing. The monthly share of one of these inhabitants is very little on a world scale; the search for wealth by such a large number of people can cause excessive exploitation and irrational management of natural and agricultural systems which, because of this, experience reduced productivity and resilience. Even if land productivity could be increased tenfold by introducing modern technologies and scientific knowledge, this would not be enough to avoid further environmental degradation and land abandonment. Sustainable management of the ecosystems together with an increase in wealth needs a link between traditional land uses and management with modern income sources and the integration of local and regional rural economies within the economies of the other areas of the world. This link is favoured by the multiple use of the territory and the passage from simplified farming systems to complex agrosilvopastoral systems.

Introduction
The low population density and the primitive technologies available in the past had little impact on the natural environment. The demographic growth and powerful modern technologies developed in the last century have had a greater impact on the land than that of the entire previous history. Nowadays over 50% of the present world population, lives in rural areas. This is about 3 billion people (FAOStat, 2003) who, on a world average, share very low incomes and overuse natural resources in an attempt to increase their wealth through immediate incomes (Eldridge and Freudenberger 1999).

Decreased productivity and degradation of the quality and resilience of natural ecosystems have been the result in large areas. Modern technologies and available knowledge cannot compensate in large areas for the diverging trends of a worsening environment and the search for wealth (Baker, 1993).

The natural ecosystems have been simplified everywhere to create easy management and to maximize short-term productivity. Many agronomic actions are possible to restore, maintain or increase the productivity of agricultural ecosystems; however, it is understood that not even the best increase possible in farm production would be sufficient to reduce consistently the existing disadvantage of most rural communities (Pardini *et al.*, 2002a).

The improvements suggested by modern agronomic science comprise: sowings and oversowings of pastures and rangelands, soil cultivation and fertilization, sowing of selected cultivars in developed areas and valorization of plant ecotypes and native animal breeds, especially in marginal areas, tree plantations and windbreaks, limitation of tree and shrub felling in natural areas, removal of encroaching shrubs from pastures and rangelands, livestock rotation in grazing sectors and balanced animal stocking rates.

Exhaustive discussion of these management tools is already available in the literature (Heady and Heady, 1982; Hodgson and Illius, 1996; Pardini, 2000) and will not be further discussed here. These actions can optimize plant and animal productivity by standard modern ordinary management; however, solutions are only temporary if the number of people living on the primary production of the land increases and they want to improve their wealth further.

In fact, the natural productivity of the ecosystem is a very difficult limit to overcome and when land productivity has been maximized no further improvements are possible unless effective new scientific advances and new technologies are available (Leith and Whittaker, 1975). At this stage, the economic diversification of the area or country is necessary, including the integration of local economies within a global perspective (Pardini *et al.*, 2003). New sources of energy, food and incomes are necessary for rural people and must also be derived from the diversification of the entire local and regional economy; this is easier if natural resources of the farm and the territory are also diversified (Pardini, 2002b).

The discussion of seven case studies in this chapter suggests some opportunities for this purpose.

Utility of the diversification of resources
Economic diversification is favoured when the resources of the territory are also diversified, so these two aspects ought to be integrated. Unfortunately, this is often not the case: scant availability of technologies results in the suboptimal utilization of the territory in the developing countries, while capital is used to maximize profits regardless of the possible long-term possible damage to the ecosystems in the developed areas.

A good level of resource diversification exists in limited areas of the world. Unfortunately management difficulty and reduced incomes are part of the diversification, and they are not proposed as a model for all agricultural systems. However, it is notable that the modern crop systems based on very limited diversity became unstable over a single century, while land productivity has been maintained better where higher biological diversity was left within farms and territories. This comparison confirms that the opportunistic simplification of the resources can maximize profits in the short term but can lead to severe crisis in the long term. In a case study that investigated 40 pastoral systems of all the countries of the Mediterranean basin, up to eight different forage resources were present in a single case, with an average of five (Argenti *et al.,* 1999). Nowhere else in the world is there such a diversification of pastoral, agricultural and forestry resources that integrate rangeland, native and sown pastures, forage crops, shrubs and tree plantations, forests, irrigated areas with intensive crops, within the agrosilvopastoral systems of single farms. Contemporarily there is a great diversification of peoples, traditions, languages and cultures and together they all contribute to the maintenance of the biological and cultural value of the area (Argenti *et al.,* 1996; Papanastasis and Mansat, 1996; Etienne, 1996; Ferchichi, 2000).

Sustainability is intrinsically linked to the maintenance of elevated biodiversity (West, 1993). This is a concept which will be of increasing importance in the future if climate modifications come to exacerbate the instability of production.

Land-use diversification contributes to biodiversity and both, in turn, will favour the integration of pastoralism within the whole national economies. From this point of view, plans of action aimed at the improvement of farming systems should not be limited to plant and animal production. Of course, the demand on primary resources remains important; (Skerman *et al.,* 1988) however, all future interventions should also be planned with a view to the multi-use of the territory and the integration of pastoral communities within the national and regional perspectives (Vallentine, 1989). The major possibilities of integration comprise: the production, certification and sale of typical and quality foods; the development of new plant industries; the development of cultural and recreational uses of the territory. Finally, these three kinds of development should allow the linkage of pasture management to the rest of the economy (Pardini, 2002a).

Seven case studies discussed in detail

A few examples from the five continents are discussed below, including seven systems located in areas with different climates: Mediterranean (central Italy, south-western Australia), arid tropical (Sultanate of Oman, northern Somalia, coastal Peru), tropical with seasonal rainfall (Bolivia), tropical with rainfall throughout the year (Republic of São Tomé and Principe).

Mediterranean area of central Italy

The variability of the Italian farming systems is favoured by differences in climate and soil differences. The variability of resources makes the management difficult and contributes to the maintenance of productivity below the maximum possible in specialized systems, but also contributes to ecological sustainability and to the beauty of the landscape (Eldridge and Freudenberger, 1999). Agritourism and educational farms are new enterprises which already exist in the territory and further development of this economic sector is sought.

In the case-study farm there is a small number of cattle and some horses that are also kept to the advantage of tourists, who practise horseback riding in park forests. The financial income from per-hour tourist or educational activities is much higher than the potential income from any agricultural or pastoral activity, and the revenue is then reinvested in activities directed at maintaining the beauty of the natural landscape, and great care is taken to preserve both diversity and quality of the ecosystems since it is their beauty which constitutes the most interesting source of revenue (Pardini *et al.,* 2002b).

Improvements of this system are possible by the production of typical animal and plant foods that have not yet saturated the internal market and that are also interesting for trade. Added incomes can come from certified organic foods (Faiello *et al.,* 2003). On-farm sales to agritourism customers can be further developed.

Ley farming system in the Mediterranean south-western Australia

Over 19 million hectares of the native *Eucalyptus* forest in the southern part of Western Australia were destroyed and converted to ley farming. Unfortunately, tree felling and replacement with annual plants have dramatically changed the level of the water table, which has brought to the soil surface the salts deposited in past ages and caused progressive land salinization. About 9.4% was already salinized in 1994 and this percentage will probably rise to 17.1% in the next 10-20 year period after 2000. Moreover, the use of a small number of plant species on large areas has diffused plant phytopathologies, produced high dependence on pesticides and other chemicals, and accumulated oestrogens in animal meat.

The actions proposed by Australian scientists comprise the protection of remnants of native forest, the reintroduction of trees and shrubs, the pasage shift from ley to phase farming with the introduction of perennial pasture plants on at least 4 million ha. These actions are useful; however, they can be further integrated by the reintroduction of trees on cropped fields and pasture hedges.

The introduction of new crops including olive oil, timber and chipped wood from tree plantations, cork-oak to supply the emerging wine industry and new annual crops, and also different animal products like sheep cheese and, the adoption of low-impact agriculture, will all provide a benefit for the land. If such solutions are adopted, the Western Australian landscape of the future will be more a mosaic of different resources than it is at present as it will partially shift to agrosilvopastoral systems (Pardini *et al.,* 2001).

Arid rangelands of the southern Sultanate of Oman
The productivity of the rangeland is very low and concentrated within the only 2 months when the monsoon causes rainfall. Grazing animals are camels and local sheep, goats and cattle, and the present animal stocking rate is about ten times higher than pasture carrying capacity. In 20 years the open forest that once covered the area has been completely destroyed.

An improved management strategy should include: reduction of the animal stocking rate, grazing exclusion from the more degraded areas, overseeding of pasture species and new planting of shrubs and trees (Longhi *et al.*, 1999).

Nevertheless, the arid climate will restrict the positive effects of any action and the productivity of this land will remain very low; moreover, restoration of the degraded rangeland will not be possible unless the number of livestock is dramatically reduced. Fortunately the wealth of the population of the Sultanate is maintained by oil extraction, which constitutes up to 95% of the entire revenue. Animal husbandry is nowadays just a tradition that adds only a little supplementary income, and the resources of the territory can be better valorized with the conservation of the natural resources, which are also the base for the very young tourist industry, which has just begun to cash in on the beauty of the territory by generating monetary incomes (Pardini *et al.*, 2000).

Arid rangeland in northern Somalia
The previous vegetation was an open forest of mainly *Acacia* species, nowadays dramatically reduced by the collection of firewood and timber and by overgrazing. The present animal stocking rate is much higher than pasture carrying capacity. The rangeland is now very degraded and the current management is not sustainable (Pardini and Angeloni, 2002).

Some improvements are possible: grazing exclusion from the more degraded areas; reduction of the animal stocking rate; modernization of sorghum cropping techniques; overseeding of pasture species; introduction of new trees and shrubs to make green fences and windbreaks; low-input mechanization including draught animals; the unification of small fields into larger cropped areas.

In spite of this, the arid climate, which results in low productivity, will limit the positive effect of the suggested improvements. Consequently, the human impact on the rangeland must be reduced by providing new building materials and renewable energy resources (Cavestro *et al.*, 2001). Non-traditional activities should be introduced or further developed to provide monetary incomes and these must be environmentally friendly: the fishing industry, the mining industry, the production of perfumes from the *Aloe* species, the handcrafting of animal leather and wood carving are just a few of the potential developments. These last proposals will benefit from the integration of tree and pasture plants in agrosilvopastoral systems and by organizing the sale of local produce within national and international markets.

Arid rangeland in coastal Peru
Rainfall is concentrated in about 2 months each year, resulting in low annual production. In the short rainy season the animals are moved from the irrigated valleys to the mountain rangelands. The actual stocking rate is very variable but reasonable during most years, and rangeland degradation is far greater than could be attributed merely to overstocking and is probably due to tree felling.

In the past, an open forest of *Caesalpinia tinctorea* obstructed the movement of the air masses from the Pacific Ocean, causing the condensation of the humid air. Part of the water collected by the tree crowns was also available at ground level in the pasture. The forest was cut down during the last century to obtain firewood (for cooking and heating) and timber (houses and fences). The construction of the Pan-American railway had the worst impact, since most of the timber used was taken from these forests. More recently, the trucks that carried sheep to graze on the mountain rangeland during the rainy season also loaded up with wood to sell in other regions on their return journey. The number of livestock has also recently increased and has hindered the renovation of the tree species and resulted in a further reduction of the herbaceous cover.

A restoration project has proved that artificial nets, used to obstruct the movements of the air masses, can condense enough water from the fog to enable the growth of tree plantations. Once the crowns have grown tall and broad the water supply becomes autonomous and also supplies sufficient water for the pasture if the number of grazing livestock is controlled.

Above all, it is essential that the people have renewable energy resources and higher monetary incomes, thus also enabling them to pay for the presence of the animals in the irrigated swards for longer and even during the rainy season.

Pastures in the seasonal rainfall areas of Bolivia
The favourable seasonal distribution of the rainfall allows the plant-growing season to continue almost all through the year. Productivity is very high for pasture, cropped fields and orchards. Some examples of modern and good management are already present beside areas which are still traditionally managed. This includes windbreaks on crop fields and pastures, rotations of pasture and crops, agroforestry techniques especially in family-run farms, and progressive mechanization.

These techniques have already achieved a good level of rationality and enable good productivity all the year round and they should be extended to more farms.

However, the commercial value of agricultural produce is very low and people prefer to crop coca plants, which give ten times more than the best normal crop. It is understood that a closed agricultural development of the area will have only slow effect on people's wealth and that economic differentiation is also necessary. This is already present with tourism, especially in conservation areas and parks, and will benefit from resource diversification at farm and territory scale.

Rangelands in the Republic of São Tomé and Principe
The two islands include a major proportion of virgin forest and both provide shelter for a number of endemic plant and animal species. The ecological value of these islands is very high, but parts of the tropical rainforest are burned to get cropping land and pasture. Activities of agriculture and animal husbandry are at a minimum level; nonetheless, their impact on the environment is extremely negative, even if limited to small areas because of the low population density. However, the population growth is 5.6% per year and could cause major changes in the vegetation in the course of a few decades.

Agroforestry techniques adopted in small areas close to the villages could achieve more than forest clearing (Pardini et al., 1996). Fruit trees and vegetables could be grown alongside and small animals such as goats and geese could become part of the food production. The diversification of plant products could sustain the diversification of the economy: coconut balsam, bamboo furniture and wood carving are just a few of the very new activities. These products are destined for tourists rather than for local people; in fact, the money spent by one tourist in one day is more than the monthly pay of an agricultural worker employed by the government. Nevertheless, the development of a tourist-related industry demands the kind of facilities which are very inadequate at present and should be developed because the use of the forest for tourism and cultural uses could contribute more to the wealth of the people than the present agriculture and animal husbandry, and would be based on the conservation of the natural ecosystems.

Diversification of farming systems and integration within the global economy
From this point of view, plans of action aimed at the improvement of rangelands should not be limited to the forage production. Naturally, the demand for food will, for a long time, continue to be the primary need in the developing countries; it is, however, essential that all future intervention is planned with a view to the diversification of the local pastoral economies and their integration within the global perspective.

If the future development of the rangelands is to be based on diversities, it is important to remember that environmental diversity also comprises the differences between peoples, their knowledge, languages and traditions. All these aspects are part of the relationship between man and the use of the territory, and could now be used to generate monetary incomes through the genetic exploitation of biodiversity, recreational activities, tourism and culture.

There are sectors of the population which already live on what the rangelands can offer, in both developed and developing countries. This potential offer is enhanced when a balanced number of domestic and wild grazers and browsers is present. What we are aiming at, therefore, is not a disruptive change from the traditional uses of the rangeland to a completely different kind - or still less the abandonment of the ecosystem - but merely a rationalization of the traditional activities and their integration within the economies of the other areas of the world.

References

Argenti, G., Talamucci, P., Stagliano, N. and Gusmeroli, F. (1996) Primi risultati sulla possibilità di riequilibrio di sistemi pastorali attraverso la transumanza tra ambienti mediterranei e alpini. *Rivista di Agronomia*, n 3 Suppl., pp. 408-413.
Argenti, G., Bianchetto, E., Londardi, P., Sabatini, S. and Stagliano, N. (1999) Some examples of Mediterranean pastoral systems. DISAT, University of Florence, Florence, Italy 62 p.
Baker, M.J. (1993) *Grasslands for Our World*. SIR Publishing, Wellington, New Zealand.
Cavestro, L., Pardini, A., Angeloni, A. and Innocente, S. (2001) An outlook of pastoralism in Northern Somalia after the civil war. *Journal of Agriculture and Environment for International Development* 2-3, 149-186.
Eldridge, D. and Freudenberger, D. (1999) People and rangeland building the future. In: *Proceedings of the VI International Rangeland Congress, Queensland, Australia*, 2 volumes. Elect Printing, Fyshwick, Australia.
Etienne, M. (1996) Research on temperate and tropical silvopastoral systems: a review. In: Etienne, M. (ed.) *Western European Silvopastoral Systems*. INRA, Paris, France, pp. 5-22.
Faiello, C., Pardini, A. and Litjens, L.C. (2003) Current socio-economic trends of organic plant and animal produce in Western Australia. In: *Proceedings of the SAFO (Sustainable Animal Health and Food Safety in Organic Farming Association) Meeting in Florence (I)*, University of Florence, Florence, Italy. pp. 4-7. (In press.)
FAOStat (2003) Statistical Databases http//www.fao.org.
Ferchichi, A. (2000) Rangeland biodiversity in pre-Saharían Tunisia. Options Méditerranéemes, 45, 69-72
Heady, H.F. and Heady, E.B. (1982) *Range and Wildlife Management in the Tropics*. Longman, New York, USA.
Hodgson, J. and Illius, A.W. (1996) The ecology and management of grazing systems. CAB International, Oxford, UK. 466 p.
Leith, H. and Whittaker, R.H. (1975) Primary productivity of the biosphere. Springler Verlag, Berliun.
Longhi, F., Pardini, A. and Grammatico, V. (1999) Biodiversity and productivity modifications in the Dhofar rangelands (Southern Sultanate of Oman) due to overgrazing. In: *Proceedings of VI International Rangeland Congress*, Townsville, Queensland, Australia. (In press.)
Papanastasis, V.P. and Mansat, P. (1996) Grasslands and related forage resources in Mediterranean areas. In: *Proceedings XVI EGF Meeting*, Grado I, pp. 47-57.
Pardini, A. (2000) *Pascoli e Foraggere Tropicali e Subtropicali*. Ipertext on CD-Rom, EuroPlanet informatica S.R.L., Florence, Italy.
Pardini, A. (2002a) Mediterranean pastoral systems and the threat of globalization. In: *XI Meeting Mediterranean Pastures Network*. FAO-CIHEAM, Djerba, Tunisia.

Pardini, A. (2002b) Rangeland management. In: Biodiversity, Conservation and Habitat Management, in: *Encyclopedia of Life Support Systems* (EOLSS). EOLSS Publishers, Oxford, United Kingdom [http://www.eolss.net], p. 67.

Pardini, A. and Angeloni, A. (2002) Economia e società nella Somalia Settentrionale dopo la guerra civile. *Bollettino Società Geografica Italiana* 3, 651-666.

Pardini, A., Longhi, F. and Zari A. (1996) Possibilità di sviluppo economico eco-compatibile nella Repubblica Democratica di Sao Tomé e Principe. *Rivista di Agronomia Subtropicale e Tropicale* 90 (4), 497-514.

Pardini, A., Lombardi, P. and Longhi, F. (2000) Impatto del pastoralismo sulle risorse naturali del Sultanato di Oman. *Bollettino della Società Geografica Italiana* XI (11), 237-248.

Pardini, A., Gremigni, P., Longhi, F. and Lombardi, P. (2001) Impatto del "ley farming system" sulle risorse ambientali del Western Australia. *Bollettino della Società Geografica Italiana* VI, 651-672.

Pardini, A., Borneo, A. and Angeloni, A. (2002a) A participatory research approach to investigate social changes in pastoral communities. *Herba* 13, 65-72.

Pardini, A., Mosquera, M.R. and Rigueiro, A. (2002b) Land management to develop naturalistic tourism. In: *Proceedings V International IFSA (International Farming Systems Association) Symposium*. Florence, Italy.

Pardini, A., Longhi, F., Orlandini, S. and Dalla Marta, A. (2003) Integration of pastoral communities in the global economy. In: *Proceedings of the International Conference: "Reinventing Regions in a Global Economy"* (www.regional-studies-assoc.ac.uk).

Skerman, P.J., Cameron, D.G. and Riveros, F. (1988) *Tropical Forage Legumes.* Plant Production and Protection Series, no. 2, FAO, Rome, Italy.

Vallentine, J.F. (1989) *Range Development and Improvements.* Academic Press, San Diego, USA, 524 pp.

West, N. (1993) Biodiversity of rangelands. *Journal of Range Management* 46, 2-13.

A silvopastoral system for Eastern Europe – based on the example of Poland

K. Boron
Dept of Soil Reclamation and Peat Bogs Protection, Agricultural University in Kraków, Poland. rmboron@cyf-kr.edu.pl

Abstract
Silvopastoral systems have been widely introduced since the 6th century in many countries of the temperate climatic zone, except Poland. Political and economic changes and accession to the EU create the opportunity to introduce modern forestry and agriculture systems in Poland as a form of alternative land use. Silvopastoralism is an extensive land utilization method; therefore a basic question is how such a system can be integrated into a specific agricultural and forestry tradition in a country such as Poland. At present only private landowners have an interest in silvopasture and are able to practise it. Lack of a scientific base and experimental data on silvopastoral systems applicable to Poland is the most serious factor limiting its uptake. This chapter reviews the weakness and advantages of the system in different situations in Poland. The aim of the work was to investigate the possibility of application, modification in land-use planning, landscape management and protection.

Introduction
Extensive silvopastoral systems are managed to improve environmental conditions. After the Second World War private farm and collective farm owners attempted to maximize crop yield by intensifying crop production and, following government requirements, cropping was obligatory. These measures had only very moderate success. Although small areas of scattered private arable lands and grasslands resulted in a vibrant mosaic of land-use types, it was a poor utilization of land. Big collective farms were difficult to manage and in many cases improperly administered. A culture of low outputs was a deterrent to productive working.

Increases in output have had a serious negative effect on the environment. Ineffective drainage systems have had dramatic effects on water regimes in a country with poor soil water retention - in general - and such poor water management exacerbates a naturally adverse water balance. Increased water loss from soil has been especially destructive for wetlands and resulted in loss of organic matter from soil. Heavy industrialization caused, often irreversible, soil degradation and water and soil contamination.

Materials and Methods
Data were extracted from the *Statistical Year Book of the Republic of Poland* for 2003 (Central Statistics Office, 2003a, b) and different maps (Siuta *et al.*, 1996 and Gasiorek, 1999). Five places in the southern part of Poland were chosen to represent a range of mainly geographical and climatic conditions. The data were selected to consider the possibility of introducing silvopastoral systems which might be practically suitable for the southern part of Poland.

Results and Discussion
Knowledge of the climate is necessary to determine the suitability of an area for silvopasture. Some basic climatic data are presented (Table 1).

Table 1. Climatic data for selected locations in Poland (1971-2000). * For the year 2002.

Place	Elevation (m asl)	Precipitation (mm)	Temperature (°C) Average	Max	Min	Sunshine* (hours/year)	Wind speed* (m/s)
Poznań	87	507	8.5	37.0	28.5	1831	3.8
Bielsko-Biała	398	942	8.1	34.2	27.4	–	5.1
Nowy Sącz	694	1576	8.2	36.1	29.2	1576	4.9
Zakopane	855	1107	5.4	31.8	27.1	1437	5.3
Śnieżka	1603	1150	0.6	23.6	32.1	1318	5.9

The average elevation in Poland is 173 m asl and 8.7% of the total area is above 300 m asl. Mild climatic conditions can be found in Poznań (Poznań Lake District) and more severe conditions at the summit of Śnieżka (Karkonosze mountain range). The Nowy Sącz (Nowy Sącz Valley) area has the highest rainfall of all sites analysed. On the basis of these climatic data it can be concluded that there is no serious limitation on the introduction of silvopastoral systems within Poland.

Land utilization Agricultural land in Poland covers 168,990 km^2, of which private farms occupy 148,580 km^2 (2002). Meadows cover an area of 25,310 km^2 (private – 22,450 km^2) and pastures 10,300 km^2 (private – 8640 km^2). Mean farm size in Poland was 7.4 ha in 2002.

Forest lands In 1995 forestlands covered an area of 89,460 km^2. This had increased to 91,130 km^2 by 2002 (Table 2).

Table 2. Statistical information on forests and land use in Poland, 1995 and 2002.

	1995	2002
Total forest area (km^2)	87,560	89,180
Public forest area (km^2)	72,620	73,630
Private forest area (km^2)	14,940	15,550
Forest cover (%)	28.0	28.5
Afforestation (%)	27.7	28.2
Renewals and afforestation (km^2)	778	568
Private afforestation (km^2)	68	131
Bare ground and irregularly stocked open stands (km^2)	110	19
Agricultural land and wasteland (km^2)	156	203

Pastoral Information on cattle and sheep numbers on farmland are presented in Table 3.

Table 3. Cattle and sheep numbers on private farms exceeding 1 ha of agricultural land by area groups.

Area of land (ha):	1.00-4.99	5.00-14.99	15.00-19.99	20.00-49.99	> 50
% of farmland	58.7	31.2	4.3	4.9	0.9
Cattle (head)	641,000	2,113.000	752,000	1,293.000	368,000
Sheep (head)	72,000	86,000	28,000	72,000	38,000

The total sheep population in Poland was estimated as 296,000 in 2002, a substantial decline from a peak of 5 million in 1996. The main reasons for this decline include very low sheep wool and skin prices and low efficiency of meat production (Wierzchoś, 2004).

The aims of silvopastoral systems are summarized in Table 4.

Table 4. Summary of the key aims of silvopastoral systems.

Environmental	Economical	Other
1. Support of biodiversity	1. Animal production and products	1. Pastoral culture
2. Soil erosion limitation	2. Timber production	2. Support of pastoral tradition
3. Climate attenuation	3. Unemployment decrease	3. Landscape formation increase
4. Soil-water retention	4. Tourism development	4. Protection of local sheep race
5. Soil fertilization increase		
6. Soil and water protection		

Conclusions

Silvopastoral systems are not currently practised in Poland. Climate, soil and area configuration will not seriously constrain silvopastoral systems in the country. Silvopastoral systems should be located mainly in hilly and mountainous parts of the country because of the strong pastoral tradition in those regions. Silvopastoral systems should be closely connected with sheep grazing because of its extensive nature.

References

Central Statistical Office (2003a) *Statistical Yearbook of the Republic of Poland.* CSO, Warsaw, Poland.

Central Statistical Office (2003b) *Environment.* CSO, Warsaw, Poland.

Gąsiorek, S. (1999) Energetyczne aspekty zintegrowanego użytkowania pastwisk górskich Zeszyty Nauk. Akademii Rolniczej im H.Kołłątaja w Krakowie Rozprawy nr.249, Kraków 3. Siuta J., Zielińska A., Makowiecki K., Sroka L Instytut Ochrony Środowiska "Polska, Potrzeby dolesień" (Mapa w skali 1:1000 000).

Siuta, J., Kucharska, A. and Sienkiewicz, R. (1996) Polska; Wieloczynnikowa degradacja środowiska (Mapa w skali 1:750 000) Państwowa Inspekcja Ochrony Środowiska - Instytut Ochrony Środowiska, Warsaw.

Wierzchoś, E. (2004) Personal contact Kraków. Agricultural University of Kraków, Department of Sheep and Goat Husbandry.

From silvopastoral to silvoarable systems in Europe: sharing concepts, unifying policies

C. Dupraz
INRA, UMR-SYSTEM, Equipe d'agroforesterie, 2, place Viala, 34060 Montpellier, France
Dupraz@ensam.inra.fr, http://www.montpellier.inra.fr/safe/

Abstract
Silvopastoral and silvoarable systems are often considered separately, with distinct management schemes, research approaches and policy-making perspectives. However, both are multi-species agroecosystems with a tree component. While the motto in agriculture and forestry has been for decades to reduce complexity for simplifying management and efficiency, multi-species systems are now sometimes considered as more dependable in production and more sustainable in terms of resource conservation than simple ones. In spite of many enthusiastic papers, evidence is, however, not yet that conclusive. Recent findings on silvoarable systems help us to understand how some functional aspects of trees like their root opportunism or phenology may result in very efficient tree-crop systems in some climate and soil conditions. We present here the approach by the SAFE (Silvoarable Agroforestry For Europe) consortium to biophysical modelling and policy-making for silvoarable systems. Integrated models of tree-crop interactions are linked to an economic model and predict the efficiency of tree-based systems. Some concepts of precision agriculture may be applied to silvopastoral and silvoarable systems in the near future. The integration of such findings up to farm and regional levels and the consequences for policy-making within the current reform of the Common Agricultural Policy are discussed. We advocate a unified policy scheme for both silvopastoral and silvoarable systems.

Key words: tree root management, policy-making, Common Agricultural Policy

Introduction
Silvoarable and silvopastoral systems share many common features, both for biophysical (low-density tree stands and herbaceous plants interactions) and socio-economic aspects (combining short-term and long-term revenues). Silvoarable agroforestry (SAF) comprises widely spaced trees intercropped with arable crops. Most research on agroforestry systems was concentrated in the tropics until the early 1990s (Van Noordwijk et al., 2003), but recent findings (Zhu, 1991; Burgess et al., 2003; Dupraz et al., 2004; Gillespie et al., 2000) indicate that modern temperate silvoarable production systems are very efficient in terms of resource use, and can be compatible with modern machinery. Dupraz and Newman (1997) suggested that agroforestry could be an innovative agricultural production system that will be both environment-friendly and economically profitable for Europe. Growing high-quality trees in association with arable crops in European fields may improve the sustainability of farming systems, diversify farmers' incomes, provide new products to the wood industry, and create novel landscapes of high value (SAFE, 2003).

However, during the late twentieth century, the advantages of agroforestry systems have been overlooked. Even traditional silvoarable landscapes, whose benefits are widely recognized, have received little attention from policy makers and research organizations. Across Europe the integration of trees and arable agriculture is currently unattractive to farmers, simply because the available grant or subsidy schemes are designed for forestry or agriculture and don't permit agroforestry. In some countries, agroforestry systems can actually be declared illegal, because they are a category that is not recognized for taxation purposes. This preposterous situation has had unfortunate consequences with some EU-funded silvoarable agroforestry experiments being closed prematurely because local agencies deemed that they were not eligible for agricultural or forestry grants. A mixed or combined status of agroforestry plots is currently not available, either at the European level or at the national level, preventing both forest and agricultural grant policies from being applied to agroforestry systems.

What future for agroforestry systems in Europe?
In recent years, the European Union has introduced a series of measures to promote the integration of trees within existing farm businesses. The 'Silvoarable Agroforestry for Europe' (SAFE) project, funded under Key Action 5[1] of the Fifth Framework Programme,[2] is one example. The SAFE project will provide up-to-date appraisals of the productivity of tree-crop mixtures, based on extant experiments and progress in modelling of tree-crop interactions at the field scale (Dupraz et al., 2004). It will develop a computer model to compare the economics of silvoarable agroforestry with arable and forestry systems, so that the financial implications of silvoarable agroforestry for European farmers can be examined (Graves et al., 2004). The Hi-SAFE process-based biophysical model includes unique features such as a 3D dynamic simulator of tree-crop root interactions. Most research has so far focused on above-ground tree management options, but recent findings indicate that below-ground tree management options such as root control or precision fertilization should now be considered for managing tree-based systems. Annual and perennial crops, including fodder

crops, can be powerful tools to monitor tree roots in agroforestry systems. A major finding so far is that the productivity of such silvoarable systems, as measured by their land equivalent ratio (Willey and Rao, 1980, adapted to agroforestry systems by Dupraz, 1998), is very attractive: silvoarable systems with winter cereals and deciduous tree species would be above 1.3 (Figure 1), which means that the overall productivity of agroforestry exceeds that of separated farm and forest systems by 30% (Dupraz *et al.*, 2004).

Figure 1. The first estimates of the land equivalent ratio of this poplar-wheat silvoarable system in southern France are above 1.3.

The environmental aspects of silvoarable systems in Europe are only starting to be explored. Such impacts may further favour the adoption of SAF if their financial value is recognized through 'shadow' payments to farmers for environmental services. However, even if silvoarable systems are more productive than monocultures, farmers will not adopt them if subsidy schemes totally distort the 'level playing-field'.

Incorporating agroforestry in the regulations: the French case

Agroforestry has been permitted since 2002 as a standard practice for French landowners and farmers. Agroforestry is advertised by the French Ministry of Agriculture to French farmers and landowners in a four - page pamphlet. Both silvoarable and silvopastoral options are included. Grants are available for planting the trees, and the usual crop payments are available for intercrops, on a "cropped area basis". The policies are valid throughout France (including overseas tropical departments), but some provinces may add additional grants for planting trees.

For planting new agroforestry plots, the landowner can apply for a grant for planting the trees. This grant is the same as for a forest plantation: it is a percentage of the total cost of planting and tending the trees during the first 3 years (the usual rate is 40%). Crops planted between the trees are eligible for CAP payments, but it is not possible to get these on a silvoarable plot obtained from clearing a forest, or planted on a parcel that was not eligible for CAP payments prior to tree planting.

In addition, a farmer who manages an agroforestry plot may apply to a specific agri-environmental scheme (second pillar of the CAP). The reason for this scheme is to promote agroforestry by compensating additional costs compared to a standard agriculture plot. The measure is contracted on a 5 - year term, and two options are available: one for creating a new agroforestry plot (240 to 360 €/ha per year during 5 years), and one for tending an existing silvoarable plot (100 to 140 €/ha per year during 5 years). The value depends on the agricultural activity on the plot, because tree protection costs are different for annual crops, grazing by small animals or grazing by cattle.

However, the current reform of the CAP will induce large changes to this regulation. Will agroforestry still be allowed to farmers after 1 January 2005?

Agroforestry and the new European CAP

A crucial debate for the future of agroforestry in Europe is now on. Recent European regulations define the conditions for the Single Payment Scheme (SPS) that will be in force in the European Union as of 1 January 2005. Will agroforestry finally be an option for European farmers with the reformed CAP, after January 2005?

How to incorporate agroforestry in the CAP

Two possibilities are explored: either AF could be considered as a normal agricultural practice, and AF plots could be fully included in the SPS (provided they meet some criteria) or AF plots could be considered as a mixture of agriculture and forestry, and only the agricultural part would be included in the SPS. A third possibility would be that AF is not included in the SPS at all, which would mean that AF has no future in Europe for the next decades.

Regulation 1782/2003 includes a provision that areas of 'woodland' should be excluded from the area of the farm eligible for SPS. This may incite farmers to remove trees from farmed landscapes, which would not only induce landscape and environmental damage, but also prevent farmers from investing in new agroforestry systems. In addition, a recent Guidance Document (AGRI/2254/2003) recommends that the threshold of 'woodland' is 50 stems per ha, which would classify most agroforestry plots as woodlands, and exclude them from the SPS. This would prevent any agroforestry system with scattered trees from existing in Europe. However, Article 5 of Regulation 2419/2001 indicates that: 'a parcel that both contains trees and is used for crop production shall be considered an agricultural parcel *provided that the production envisaged can be carried out in a similar way as on parcels without trees in the same area'*. This is perfectly suited for agroforestry: in a carefully designed and managed agroforestry system, crop production can effectively be carried out in good conditions. Fortunately, this wording was retained in the most recent regulation (R796-2004 of 21 April 2004, Article 8), which details how the SPS will be enforced.

It is interesting to observe that Regulation 2254-2003 considers that border hedges of up to 4 m wide that serve as boundaries between agricultural parcels and are traditionally part of good agricultural practices in the region concerned will be considered as being included in the SPS area. A 2 m width will be attributed to each adjacent agricultural parcel. Internal features will be, under the same conditions, accepted as forming part of the agricultural parcel where their width is less than or equal to 2 m. Member States may, however, after prior notification to the Commission, allow a width greater than 2 metres if those areas were taken into account for the fixing of the yields of the regions concerned. This could be a circuitous way of including agroforestry in the SPS (tree lines are often less than 2 m wide), but a more direct approach would be more appropriate.

The project of a new European Regulation (http://europa.eu.int/comm/agriculture/capreform/rurdevprop_en.pdf) on support for rural development was published in July 2004 and includes for the first time in history a full article on agroforestry systems. The introduction section of the document states that "Agri-forestry systems have a high ecological and social value by combining extensive agriculture and forestry systems, aimed at the production of high-quality wood and other forest products. Their establishment should be supported." A new era for agroforestry systems in Europe is about to start.

Defining agroforestry systems at the European level

When designing regulations, definitions of forest or forestland are often required, to circumvent the domain of each regulation. There are internationally accepted definitions of 'forest' or 'forestland' used by the UN-ECE/FAO and the UNFCCC, which use threshold values of crown cover, tree height at maturity, minimum area and bounding areas. However 'woodland', as used in EU Regulation 1782/03, is less well defined. If agroforestry is to be recognized as an accepted land use, a clear definition of an agroforestry plot should be introduced. The SAFE project is currently suggesting that agroforestry systems could be defined by a tree plantation design and management that allow significant crop or grass production (at least 50% of the reference yield without trees), and with a tree density of less than 200 trees/ha (only trees with a diameter at breast height above 15 cm are included). Hedge trees should be included in the calculation. This suggestion conflicts with Guidance Document AGRI/2254/03, which states that any plot with more than 50 trees/ha (irrespective of their size) would be 'woodland' and therefore excluded from the SPS system. But the Guidance Document also states "the Commission services take the view that wood within this meaning should be interpreted as meaning areas within an agricultural parcel with tree-cover (including bushes etc.) preventing growth of vegetative under-storey suitable for grazing". If this approach is extended to silvoarable systems, agroforestry is clearly not 'woodland'.

The Integrated Control System (IACS) and the Land Parcel Identification System (LPIS) should be designed to allow agroforestry systems to operate. Some countries (such as France) declare for taxation purposes parcels as partially covered by one 'activity' (e.g. farming) and partially covered by another 'activity' (e.g. tree growing). If agroforestry is not recognized as an agricultural system in the SPS, the EU should make clear to all EU countries that they have the flexibility to allow multiple activities within parcels in their national IACS systems (e.g. for agroforestry, 'forestry' and 'cropping' in the same parcel), and define fair rules for these plots in the SPS scheme.

Finally, agroforestry should now be recognized as a distinct land-use system, and all European countries should be allowed to define a niche for agroforestry in their own taxation system. This would probably favour the mixed activity status of the plot, as land taxes paid on woodland or agricultural land are often very different.

Conclusion

There is a need for a special 'agroforestry status' to be designed for the countries where tax policy and grant availability are dictated by land-use classes. Policies for agriculture and forestry grants should recognize that both silvoarable and silvopastoral systems are 'legal' forms of land-use which should be permitted and placed on a 'level playing-field' with conventional agriculture or forestry.

From these elements, we can conclude that the adoption of a very new system of land use like modern agroforestry can simply be impossible if the rules for agriculture support ignore intercrops. This is true even if their productivity or environmental advantages are demonstrated and significant. In a world were subsidies may represent half the revenue of the farmer, agroforestry has no future if the crops are not included in the subsidy schemes, even if agroforestry is an excellent system for productivity and environmental benefits. On the contrary, in a world where the crops get no subsidy, agroforestry would be a very attractive system to all farmers. This is a clear example of legal and policy regulations exercising an unforeseen constraint on the implementation of innovative options for rural development and landscape enhancement.

Acknowledgements

This research was carried out as part of the SAFE (Silvoarable Agroforestry for Europe) collaborative research project. The EU under its Quality of Life programme funds SAFE, contract number QLK5-CT-2001-00560, and the support is gratefully acknowledged.

Notes

[1] Key Action 5: Sustainable agriculture, fisheries and forestry, including integrated development of rural areas
[2] Fifth Framework Programme: http://europa.eu.int/comm/research/quality-of-life/leaflets/en/whatisthe5th.html

References

Burgess, P.J., Incoll, L.D., Hart, B.J., Beaton, A., Piper, R.W., Seymour, I., Reynolds, F.H., Wright, C., Pilbeam, D.J. and Graves, A.R. (2003) *The Impact of Silvoarable Agroforestry with Poplar on Farm Profitability and Biological Diversity*. Final Report to DEFRA. London.

Dupraz, C. (1998) Adequate design of control treatments in long term agroforestry experiments with multiple objectives. *Agroforestry Systems* 43 (1/3), 35-48.

Dupraz, C. and Newman, S. (1997) Temperate agroforestry: the European way. In: Gordon, A.M. and Newman, S.M. (eds), *Temperate Agroforestry Systems*. CAB International, Wallingford, United Kingdom/New York. pp.181-236.

Dupraz, C., Vincent, G., Lecomte, I., Jackson, N., Van Noordwijk, M., Mayus, M. and Mulia, R. (2004) Integrating tree-crop dynamic interactions with the Hi-SAFE model. Keynote paper. In: *First Congress of Agroforestry*. IFAS, Orlando, Florida, USA. 98 pp.

Gillespie, A.R., Jose, S., Mengel, D.B., Hoover, W.L., Pope, P.E., Seifert, J.R., Biehle, D.J., Stall, T. and Benjamin, T.J. (2000) Defining competition vectors in a temperate alley cropping system in the Midwestern USA; 3. Production physiology. *Agrofororestry Systems* 48, 25-40.

Graves, A.R., Burgess, P.J., Liagre, F., Dupraz, C., and Terreaux, J.P. (2004) The development of an economic model of arable, agroforestry and forestry systems. In: *Book of Abstracts, 1st World Congress of Agroforestry*. University of Florida, Florida, USA. 242 pp.

SAFE (2003) Second annual report of SAFE project. (http://www.montpellier.inra.fr/safe/)

Van Noordwijk, M., Roshetko, J.M., Ruark, G.A., Murniati, Delos Angeles, M., Suyanto, Chip Fay and Tomich, T.P. (2003) *Agroforestry is a form of Sustainable Forest Management*. UNFF intersessional expert meeting on the role of planted forests in sustainable forest management. Wellington, New Zealand, 24-30.

Willey, R.W. and Rao, M.R. (1980). A competitive ratio for quantifying competition between intercrops. *Experimental Agriculture* 16, 117-125.

Zhu, Zhaohua (1991) *Agroforestry Systems in China*. The Chinese Academy of Forestry and International Development Research Centre (IDRC), Singapore, 216 pp.

Silvopastoral systems as a forest fire prevention technique

A. Rigueiro-Rodríguez, M. R. Mosquera Losada, R. Romero Franco, M. P. González Hernández and J. J. Villarino Urtiaga

Dept. of Plant Production, Escuela Politécnica Superior, Universidad de Santiago de Compostela, Campus Universitario, 27002-Lugo, Spain. anriro@lugo.usc.es

Abstract

Silvopastoral systems are a form of multiple purpose land use, one function of which can be to reduce the risk of forest fires, which continue to be a threat to forests. In Spain, from 1991-2000 more than 175,000 ha were destroyed by fire annually, a third of this being woodland. Control of the undergrowth, which acts as a plant fuel for fire, should be made through prevention techniques that include cutting, controlled burning and controlled grazing. Grazing can also be a cause of fire, but when a suitable livestock breed is chosen and correctly managed, with animals which will reduce the understorey, it can prevent fires. These animals can also create income from meat production, as well as bringing other benefits, such as landscape enhancement, improved access and production of secondary products such as mushrooms. In long-term research in Galicia on this technique over four decades, encouraging trials have been conducted with goat, horse, sheep and pig grazing in *Eucalyptus globulus*, *Pinus pinaster*, *Pinus sylvestris* and *Pinus radiata* plantations. In this chapter, the key findings of this research will be presented alongside research on the same topic from elsewhere. Current research on grazing management of horses in a *Pinus radiata* plantation showed that, in the medium term, control of understorey is important and there was no difference between continuous and rotational grazing.

Key words: plant fuel, fire, grazing, silvopastoral systems

Introduction
Forest fires in Europe

Forest fires are one an important problem in many countries. In Mediterranean Europe, particularly in the north-west of the Iberian peninsula, the incidence of fire is very important and worrying.

In EU Mediterranean countries the surface area affected by forest fires has tended to decrease in recent years (Figure 1), although it is still significant, whereas the number of fires has shown a growing upward trend, decreasing somewhat between 2000 and 2002.

Figure 1. Number of fires and burnt areas from 1980 to 2002 in the five EU Mediterranean Member States (France, Greece, Italy, Portugal and Spain).

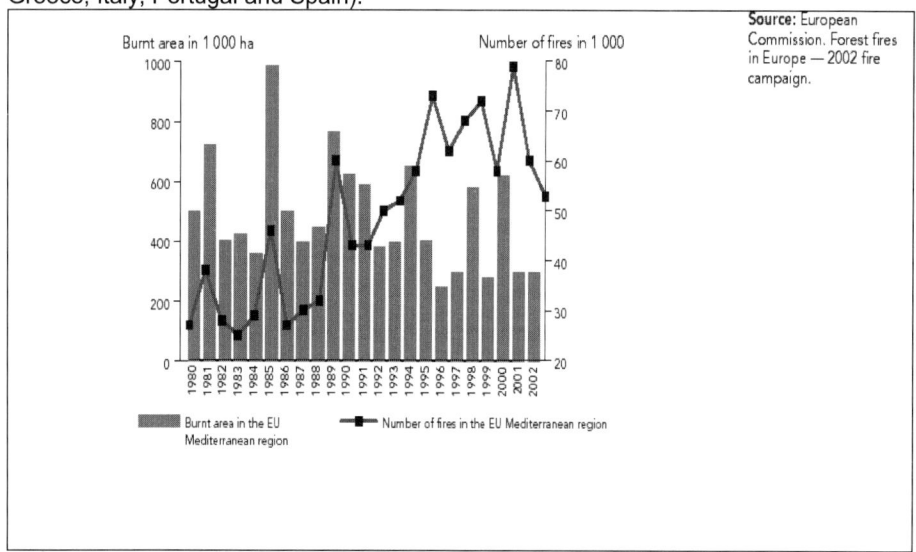

In Spain, in the last 10 years between 1993 and 2002, 202,248 forest fires were registered, burning a surface area of 1,433,014 ha, that is, 5.5% of the surface area of Spain. Of this, 515,578 ha was forest and the rest scrubland, pasture and open woodland. During this period, 285 of the fires were considered extremely serious, having more than 500 ha of surface area affected.

Two of the most important problems facing the forestry sector in Galicia are the fragmentation of property into small plots, causing a large number of smallholdings, and forest fires, which for decades now have been causing economic and ecological damage to the Galician countryside. In 2003, in Galicia, which has a surface area of just 3,000,000 ha, there were 8553 fires, which destroyed 19,835 ha, 4,896 ha with tree cover and 14,939 ha which was scrubland. In recent years efficiency in extinguishing fires in Galicia has increased considerably through improved organization and a significant increase in the economic resources provided for this purpose. However, the problem continues to be extremely worrying; the annual number of fire locations has stabilized or grown slightly, and the attention given to prevention methods involving control of the understorey (which provides plant material to fuel fires) is insufficient.

There are many causes of forest fires, common causes being fire through agricultural activities, the existence of large masses of combustible tree species and high living or dead biomass levels. However, in some regions deliberate fires are frequent. Thus, for example, of the fires which occurred in 1996 in Galicia, 86.6% were deliberately started, compared to an average for Spain as a whole of 65.6% (Ministerio de Agricultura, 1996).

Silviculture as a means of preventing forest fires in reafforested areas

At present, when reafforestation is being planned, the tendency is to use wide spacings to ease machinery access between the trees so that clearing, etc., can take place to reduce vegetation competition with the tree seedlings. This will also reduce the combustible plant material in the undergrowth, mitigate the danger of fire and represent good silvicultural practice. Other issues, apart from spatial orientation of the trees, are the diversification of the tree sizes and pruning regime to reduce crown fire risks. Clearing can be specific, total or selective. The most commonly used methods are as follows:

(i) Manual: Prohibitive for extensive areas in many countries as the labour costs are high.
(ii) Mechanical: Tractor-driven manual brush mowers or power brush mowers (vertical axis and horizontal axis, chains or drills) can be used. Heavy brush mowers shred the scrub, which facilitates the mineralization of the organic matter. When the steepness of the slope permits, surface clearing can also take place using a cutter or disc harrow, which pulls out the herbaceous species and scrub. On steep slopes, the land can be terraced, planting taking place at the edge of the terrace and keeping the platform free of scrub by clearing pathways using the brush mowers, cutters or disc harrows. However, this technique is costly and environmentally degrading as large amounts of earth are moved.

In the same way as the rest of the pruned or cut material, the biomass undergrowth can also be used as fuel, either directly or pyrolysed to give liquid or gaseous fuel.
(iii) Fire: Controlled burning or prescribed fires are effective means of preventing forest fires, although great care must be taken in controlling them.
(iv) Chemical: Chemical clearing is expensive and causes concern from an ecological point of view. Therefore, we only consider the use of viable herbicides when a specific and small area has to be cleared for some reason. Sulfosate, hexazinone, triazine, piclorum, asulum, 2,4- D, triclopir, etc. are recommended. The last four are for selective action only, and will not damage plants with thin leaves (Gramineae, for example). Today, herbicides are commonly used along plantation lines (strips approximately 1 m wide) to reduce competition from naturally regenerating herbage in the first few years of reafforestation.
(v) Grazing: Using livestock compatible with the type of forest and which are capable of consuming the understorey herbage. This technique can be used in combination with those mentioned above.

Under some of the clearing systems mentioned the plant remains can stay on the ground to break down and be incorporated into the soil, but under other systems (exploitation of the biomass as energy, for example) the plant matter is extracted and can have a negative impact on the soil organic matter biomass in the medium and/or long term. By burning, the organic matter is mineralized quickly and the nitrogen can be volatilized to a greater or lesser extent. By grazing, the vegetation is extracted, but at least it is returned to the earth in the form of excrement (Baker, 1979; García Salmerón, 1991; Valette *et al.*, 1993).

Silvopastoral systems

Agroforestry is an age-old practice throughout the world. It consists of combining trees with crops and/or animals in the same management unit, whereby both ecological and economic interactions are established between the components (Nair, 1989). The advantages of agroforestry activities are the consequence of their defining characteristics, which, according to Anderson and Sinclair (1993), can be summed up as productivity, stability and sustainability.

The productivity of these systems is primarily in the multiplicity of the products obtained. From the same area wood, food, meat, milk, wool, forage, firewood, honey, medicinal and pharmaceutical products, basketry, resin, flowers and acorns can be obtained. These systems are of economic importance, in that highly valuable non-productive functions take place, such as the prevention of wind and rain erosion and forest fires; the microclimate, soil fertility and quality of the landscape are all improved; habitats for native flora and fauna are created; water is regulated and purified; the recreational use of forest ecosystems is enhanced; they contribute towards controlling pests and disease; they increase biodiversity; and they reduce pollution (Hislop and Sinclair, 2000; McAdam and Sibbald, 2000; Sinclair *et al.*, 2000).

The greater stability of agroforestry systems is related to the short-term economic return of agricultural and/or livestock products which are compatible and complementary to the wood-producing functions of the forest. This provides owners with a steady form of income over the useful life of the farm, compared to traditional forestry management (Sharrow, 1999), while at the same time the risk of losses is reduced as a result of possible unfavourable market developments, climate, or even political decisions. These can imply a reduction in the vulnerability of the systems in the short, medium and long terms (Anderson and Sinclair, 1993; Sharrow, 1999).

These systems improve the socio-economic conditions of rural areas, especially the most depressed areas, by creating employment, increasing income and reducing risks. They are land management systems in which modern technology combines with traditional techniques. This makes them compatible with the sociocultural characteristics of local populations, helps their establishment and facilitates their stable integration into rural society (Nair, 1991).

Agroforestry systems are many and varied, as are their functions and production. In this study we focus on silvopastoral systems, which are the most highly developed and oldest agroforestry practices in temperate and industrialized regions (Nair, 1991). This study concerns grazing in forests and woodland which have been cleared to a greater or lesser extent and where the grazed undergrowth is shrubs and grasses. This type of system can be used to reduce the amount of combustible material and hence the risk of forest fires (Rigueiro-Rodríguez, 1992, 2002). One of the best examples is to be found in Mediterranean Iberia in the so-called Spanish "dehesa" or Portuguese "montado" system (Campos and Martin, 1986).

Woodland
The tree is an important component of silvopastoral systems, constituting the plant roof of the system and carrying out various functions: (1) To produce wood. (2) To provide fresh and preserved food directly for livestock (browse, fruit, etc.). This is frequent in tropical agroselviculture and in the Mediterranean region, where the "dehesa" is the greatest example of a silvopastoral system. The trees are pruned during the years when pasture production is low to use the leaves and twigs as forage (Joffre *et al.*, 1989). (3) To reduce fertilization inputs, by using leguminous tree species, which fix atmospheric nitrogen, improve forage production and, therefore, animal production. (4) To provide shade or refuge for the animals.

It is desirable for the woodland in silvopastoral systems to meet certain characteristics (King, 1980; Rigueiro-Rodríguez, 2000): Must have apical dominance and good natural pruning, or tolerate intense pruning (Beaton and Hislop, 2000).

Species with a low top diameter-trunk diameter ratio, with clear tops, which let light through to the ground and do not intercept rain to any great extent, are advisable. Furthermore, it is better that they do not release allelopathic substances which could affect the undergrowth species.

They must be efficient nutrient pumps with deep root systems to reduce competition with the shrub and herbaceous strata and thus obtain greater productivity from the tree and forage components.

The trees must be compatible with the type of livestock used. Species such as *Pinus palustris*, *Pinus elliotii*, *Pinus radiata*, *Pinus pinaster* and *Pinus sylvestris* and other types of pine meet most of the characteristics mentioned above, such as natural pruning or tolerance of artificial pruning, capacity to grow in a wide frame, a high level of apical dominance, as well as acting as nutrient pumps. In the case of the eucalypts, there are some disadvantages, such as its allelopathy and tendency of their tops to open excessively if they have too much space, although this does have the advantage of allowing light to penetrate through to the lower strata of the undergrowth (Silva-Pando, 1988; Rigueiro-Rodríguez, 1992, 2000).

In the United Kingdom, positive results have been obtained for *Acer pseudoplatanus*, *Fraxinus excelsior*, *Prunus avium*, *Pinus sylvestris* and various different species of the *Populus* genus (McAdam and Hoppe, 1996; Beaton and Hislop, 2000; McAdam and Sibbald, 2000) in silvopastoral systems. In Greece, silvopastoral systems with tree species such as *Quercus petraea*, *Quercus frainetto*, *Castanea sativa*, *Pinus halepensis*, *Pinus brutia*, *Pinus pinaster*, *Pinus nigra*, *Quercus ilex*, *Quercus coccifera* and *Quercus suber* (Papanastasis, 1996) are quite common. In France, these systems usually include *Pinus pinea*, *Pinus halepensis*, *Quercus ilex*, *Quercus suber* or *Quercus humilis* (Étienne, 1996). In Spain, as well as the Mediterranean "dehesas", in the high fire-risk areas of the north-west, an area particularly punished by forest fires, as we have mentioned, *Pinus pinaster*, *Pinus radiata*, *Pinus sylvestris*, *Eucalyptus globulus* and *Castanea sativa* are common in silvopastoral systems, and the potential, other tree species such as *Quercus rubra*, *Eucalyptus nitens*, *Castanea* × *coudercii*, *Pseudotsuga menziesii*, *Fraxinus excelsior*, *Betula alba* or *Populus* × *canadensis* (Rigueiro-Rodríguez, 2002; Ibarra *et al.*, 2000) is currently being investigated.

In New Zealand and Chile, the most commonly used tree species in silvopastoral systems is *Pinus radiata*. This species has been planted on already established pastureland, with an increase in the overall profits of up to 12% (Hawke, 1991; Knowles, 1991; Hawke and Knowles, 1997). In Australia, as well as *Pinus radiata*, other species of the *Eucalyptus* genus are used, such as *E. saligna, E. maculata, E. camaldulensis* and *E. globulus*. In this case, the silvopastoral systems are of a protective rather than productive nature. Their main contribution is to prevent soil erosion and control the acidity and salinity of the soil.

In the north-western region of the United States the most common silvopastoral systems include the plantations of *Pinus ponderosa* and *Pinus contorta* in the natural extensive pastures called "rangelands" (Williams *et al.*, 1997). On the other hand, in the south-west of the US, much more intensive systems have been developed using *Pinus elliottii* or *Pinus palustris* (Lewis and Pearson, 1987; Williams *et al.*, 1997).

Livestock
The livestock used must be as compatible as possible with the woodland, and capable of utilizing the undergrowth (Silva-Pando, 1988; Rigueiro-Rodríguez, 1992, 1998, 2000, 2002). It will be necessary to continually adjust the stocking rate and the composition of the herd to match the productivity and botanical composition of the pasture. On the other hand, the initial stocking rates will depend on the function of the livestock; if the aim is to bring the combustible plant fuel in the undergrowth under control quickly, the stocking rates will be high for a short period of time, while, if the aim is to achieve a sustainable system over a considerable period of time, the stocking rates will be lower.

Goats and horses, especially native breeds, are suitable animals for controlling living woody plant fuel in the undergrowth, thus reducing the risk of forest fires. The trampling effect also helps crumble the dead plant fuel, accelerating decomposition and mineralization. This is further enhanced by the fertilization provided by excrement (Rigueiro-Rodríguez, 1992). Sheep and cows will eat the herbaceous pasture, so it is advisable to introduce these animals when the undergrowth is mainly grass through the previous grazing of goats and horses or when pasture has been reseeded under the trees (Rigueiro-Rodríguez, 1992).

In reafforested areas in artificial pastureland, to prevent possible damage the animals could cause in the first years after planting, larger-sized seedlings could be used or the trees protected artificially by fencing them off either individually or in groups (McAdam, 1991; Fletcher *et al.*, 1993; Mosquera *et al.*, 2001). Another option would be to utilize the pasture by cutting it while the risk of the animals damaging the trees exists (Sharrow, 1983).

Silvopastoral systems to prevent forest fires (experiments in Galicia, NW Spain)

Grazing has been a common practice in most Iberian regions, and in some areas it continues to be important. Natural scrubland is used, either improved, unimproved or reseeded. In the north of the Iberian peninsula there is a history of experiments, on the grazing of scrub, its improvement and conversion to pastureland (Sineiro, 1982; Osoro *et al.*, 2002; Sineiro and Díaz, 2002).

Making livestock compatible with woodland has traditionally been difficult and the administration has often prohibited the entry of livestock in wooded areas and forests, fearing that the trees themselves would be damaged by the animals or that they would not regenerate. This prohibition, often imposed without consultation, has caused conflict and has led some owners to burn their own land deliberately. Grazing in woodland is still a matter of conflict in regions such as Galicia, which is why it is a problem that needs to be tackled and regulated. An answer to the problem could lie in the introduction of silvopastoral systems.

Research has been carried out on establishing silvopastoral systems in Galicia for the last 20 years. Work began at the Lourizán Centre for Forestry Research (Pontevedra), and was later supplemented by teams from the Mabegondo Agrarian Research Centre (A Coruña) and the Department of Crop Production Lugo Engineering School. Techniques investigated have successfully reduced combustible plant material in the undergrowth and, consequently, the risk of forest fires (Rigueiro-Rodríguez, 1985, 1986, 1992, 2002; Silva-Pando, 1988, 1991, 1993). Reseeding of the understorey scrubland with more productive, nutritious, digestible and palatable species creates wooded pastureland which, as well as reducing the risk of forest fires, provides a more valuable herbage resource (Piñeiro and Pérez Fernández, 1988; Rigueiro-Rodríguez, 1985, 1992; Silva-Pando, 1993). Tree species used have been maritime pine (*Pinus pinaster*), Scots pine (*Pinus sylvestris*), radiata pine (*Pinus radiata*), birch (*Betula alba*) and white eucalytpus (*Eucalyptus globulus*). Currently, experiments are being carried out with other tree species and trials with horses are taking place in a *Pinus radiata* plantation in the province of Lugo, aimed at comparing the effect of continuous and rotational grazing on reducing the plant fuel in the undergrowth (Rigueiro-Rodríguez *et al.*, 2001).

The most significant results from these Galician experiments will be presented below. These will focus on the use of livestock grazing the combustible understorey to reduce the risk of forest fires.

Trees

Most (70-80%) of the planted forest area of Galicia is pine and eucalyptus. The density of the woodland (number of trees per ha, ground cover in vertical projection of crown area) is inversely related to the productivity of the undergrowth (Dodd *et al.*, 1972).

Natural undergrowth pasture

The herbaceous-shrub stratum in the Galician pine and eucalyptus groves is predominantly made up of heliophilic fruticose species and nemoral and heliophilic herbaceous species. Hence it is important to know the quality of the main herbaceous, shrub and tree species growing under shaded conditions in Galicia. These include some herbaceous monocots (such as *Dactylis*, *Molinia*, *Holcus*, *Agrostis*, *Lolium*, *Briza* and *Pseudoarrhenatherum* genera), herbaceous dicots (such as *Achillea*, *Erodium*, *Lamium*, *Plantago*, *Rumex*, *Trifolium*, *Senecio*, *Stellaria*, *Urtica*, *Capsella*, *Mentha*, *Taraxacum* and *Daucus* genera); shrubs such as *Cytisus*, *Ulex*, *Erica*, *Lonicera*, *Pterospartum*, *Daboecia*, *Rubus*, *Genista* and *Calluna* genera and tree leaves and twigs under 0.5 cm from genera *Alnus*, *Betula*, *Fraxinus*, *Quercus*, *Pinus*, *Fagus*, *Salix*, *Populus*. Herbs had higher nutrient contents than shrubs, and are better for extensive silvopastoral systems (Rigueiro-Rodríguez *et al.*, 2002). Dicots had a higher mineral content than monocots (Pinto *et al.*, 2002; Rigueiro-Rodríguez *et al.*, 2002). Species from the *Cytisus*, *Rubus* and *Ulex* genera had a higher forage potential than *Erica* or *Calluna*, and *Pterospartium* had a lower protein content even though it was a legume. Trees were of a higher quality than the shrubs, and during the summer they had similar protein, phosphorus and mineral contents to the herbaceous species. This makes them suitable for use in regions where pasture production is low during the summer drought period.

Maritime pine and white eucalyptus have open crowns which let sufficient light through to the undergrowths or understorey growth, even when the plantations are a normal density. Consequently, these formations have a herbaceous sub-shrub stratum dominated by heliophilic fruticose species and nemoral and heliophilic herbaceous species. The productivity of the undergrowth is between 2500 and 3200 kg DM/ha per year. In the Scotch and Monterey pines, at normal plantation densities, less sun reaches the undergrowth through the tree cover and therefore the heliophilic scrub finds growing conditions less favourable. However, the herbaceous and woody species are more shade tolerant, although productivity is slightly lower, between 1400 and 2800 kg DM/ha per year (Silva-Pando, 1993).

In several experiments, the woody undergrowth in ungrazed experimental plots accumulated between 25 and 50 t DM biomass. To control the scrub by livestock, the animals must graze on the young shoots in a herbaceous state. This is when they are most nutritious and palatable and, consequently, better grazed. Therefore, before letting the animals onto the land to graze it is advisable to treat the undergrowth by crushing, burning or wearing it down (manually or mechanically) (Rigueiro-Rodríguez *et al.*, 2002).

Livestock and its management

The livestock must be compatible with the woodland and must be native, traditional breeds, capable of obtaining adequate nutrition from the natural pasture that grows under the trees. In a first phase, when the woody pasture is abundant, it is advisable to introduce animals that consume a high proportion of woody plants in their diet (lignivores), such as goats and horses. Grazing will modify the undergrowth, reduce the cover of woody species and increase the cover of herbaceous species. During this process it is advisable to substitute the lignivores for herbivores (such as sheep and cows). However, goat and horse grazing should not be suppressed completely in order to prevent the scrub from reappearing (Rigueiro-Rodríguez *et al.*, 2002).

Horses are compatible with eucalyptus and pine, even when the trees are young, as the animals do not eat them and they control the gorse, broom and hard grasses. They are only compatible with leafy vegetation when they cannot reach the crowns. Goats are compatible with the eucalyptus, and will not harm young trees, although they will browse the accessible tops and any weak bark of pine and other leafy species. They will eat the shoots of the gorse, broom, bramble, small briar and herbaceous species. Sheep and cows graze the herbaceous pasture and, if they are native breeds, the young shoots of the woody species. They are considered compatible with pine, eucalyptus and other leafy species, as long as they cannot reach their tops (Rigueiro-Rodríguez, 1992; Rigueiro-Rodríguez *et al.*, 2002).

In the experiments carried out on Marco da Curra public land (Monfero, A Coruña), 550 m asl and on a schist soil, with a mean annual precipitation of 1593 mm and a mean annual temperature of 10.6ºC, in a 30-year old *Pinus pinaster* stand with a density of 450-700 trees/ha and in a *Pinus sylvestris* stand of the same age and with a density of 500-800 trees/ha, understorey vegetation was effectively controlled with an initial stocking rate of 2 goats/ha. This varied as the undergrowth grew thicker and after the third year the stocking rate was 1 goat plus 3 sheep/ha. Grazing was described as "rotational-extensive" (1 month grazed, 3 months rested) to have high early stocking rates, which increased the effectiveness of vegetation control and hence reduced fire risk rapidly (Rigueiro-Rodríguez, 1992, 2002).

In the experimental plots with white eucalyptus (*Eucalyptus globulus*) in the Coto de Muiño (Zas, A Coruña), which belongs to the Empresa Nacional de Celulosas, situated 420 m asl and also on a schist soil, the mean average precipitation was 1640 mm and the mean average temperature 11.9ºC. The tree density was 2000 trees/ha and the owners have used livestock for more than 30 years, grazing freely or continuously. The normal stocking rate is 1 goat/2 ha plus 1 mare/4 ha. In the spring, there is an excess of pasture and the owners allow the neighbours' cows onto the land, at a stocking rate of approximately 1 cow/ha (Rigueiro-Rodríguez, 1992, 2002).

A trial is being conducted with indigenous Galician horses in 25-year-old Monterey pine at 500 m asl on land belonging to Sambreixo (Parga-Guitiriz-Lugo). Initial tree density was 800 trees/ha, reduced to 400 after recent clearing. The mean average temperature is 10.9ºC and the mean average precipitation 1477 mm. The overall stocking rate is 0.5 horses/ha and two grazing systems are being compared: continuous (two replicas in two 6 ha plots) and rotational (two replicas in two 6 ha plots divided into four sub-plots of 1.5 ha; the land is occupied for 1 month and rested for 3). Vegetation understorey is mainly: gorse (*Ulex europaeus*, *Ulex gallii*), bramble (*Rubus* sp.), small briar or heather (*Erica umbellata*, *Erica cinerea*, *Erica ciliaris*, *Calluna vulgaris*), Portuguese broom (*Cytisus striatus*), Scotch broom (*Cytisus scoparius*), Spanish broom (*Genista florida*), carqueixa (*Pterospartum tridentatum*), fern (*Pteridium aquilinum*), *Daboecia cantabrica*, *Halimium lasianthum*, *Pseudoarrhenatherum longifolium*, *Agrostis curtisii*, *Agrostis capillaris*, *Holcus lanatus*, *Holcus mollis*, *Avenula marginata*, *Molinia caerulea*, etc. Saplings of other native trees present were: oak (*Quercus robur*), chestnut (*Castanea sativa*) and birch (*Betula alba*). The pine was not naturally regenerating.

Control of combustible plant material

Control of the living plant fuel by livestock in the eucalyptus grove in Coto do Muiño (Zas, A Coruña) is extremely important. In a plot where the trees were felled 3 years ago and where, subsequently, the rest of the trees and scrub were burned and the livestock allowed onto one area of land, the other fenced off, the biomass of the undergrowth is 80% less than the areas where the livestock were prevented from grazing. This treatment has been successful in preventing forest fires (Rigueiro-Rodríguez, 1992).

In the pines of Marco da Curra (Monfero, A Coruña), before beginning the experiment, the undergrowth scrub had a biomass of 40-50 t DM/ha and an average height of over 2 m. Vegetation control has been very effective, the cover now being herbaceous species with a maximum height of 10-15 cm and a biomass stabilised at 0.5-2 t DM/ha (Silva-Pando, 1988). On this land, the aerial biomass of the undergrowth recovers at a rate of 5 t DM/ha per year in ungrazed plots (Rigueiro-Rodríguez, 1992).

Figure 2. *Ulex* sp. pasture (biomass) on offer at the beginning of each rotation in the first five rotations (20 months) in the experiment in Sambreixo.

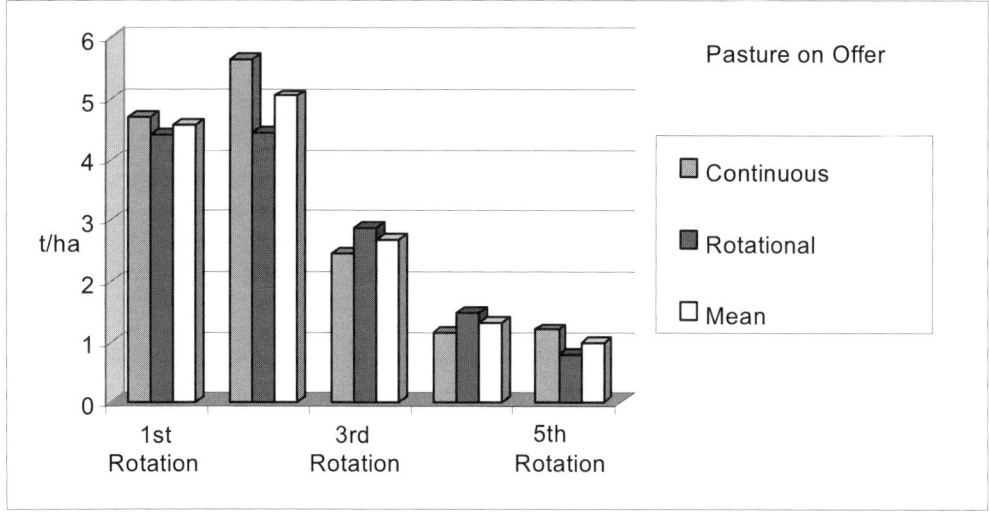

In the Sambreixo horse grazing trial, the pasture (biomass) on offer during each rotation (when the animals enter each sub-plot under rotational grazing, and the simultaneous estimation of continuous grazing) for the gorse, the dominant species in the undergrowth, is presented (Figure 2). The pasture on offer is initially (first two rotations) higher in the plots subjected to continuous grazing than other grazing but the trend changes, up until the fifth rotation, where, under the greater pressure applied by rotational grazing, there is more pasture on offer in the continuous grazing plots. The residual pasture (biomass) (when the animals leave each sub-plot under rotational grazing and the simultaneous estimation of continuous grazing) for the same scrub species is shown in Figure 3. In the two first rotations, the residual pasture is greater in the continuous grazing plots than the rotational, a trend that is maintained, although to a lesser extent, in the remaining rotations. After 20 months of grazing over all systems the pasture on offer is reduced by 66% and the residual pasture by 87.5%. This demonstrates the effectiveness of grazing in reducing the plant fuel in the undergrowth. The clearing effect is initially higher under rotational grazing than continuous grazing but differences vanish over time. Horses control the shrub stratum dominated by gorse, showing their preference for leguminous plants; when these have been grazed out, the animals eat and control other, less palatable species, such as *Rubus* sp.

Figure 3. *Ulex* sp residual pasture (biomass) at the end of each rotation in the first five rotations (20 months) in the experiment in Sambreixo.

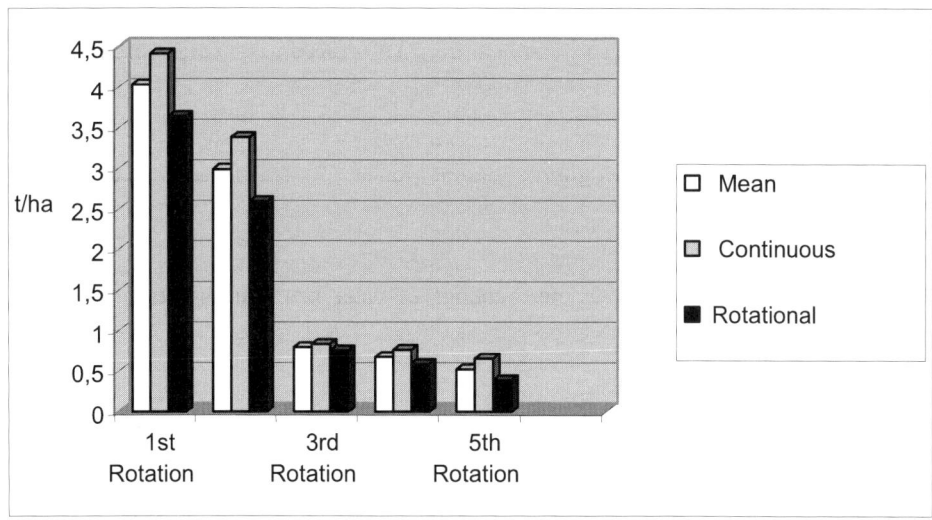

Conclusions

Silvopastoral systems are an appropriate technique to control the understorey vegetal fuel from different forestry stand types, reducing fire risk. On the other hand, silvopastoral systems allow people to organize grazing in the forest, which reduces social conflicts and therefore one of the main causes of fire. Goats and horses are a good tool to reduce understorey biomass when it is mainly constituted by woody pasture. There are no differences between rotational and continuous grazing systems with horses in the short term regarding the control of vegetal fuel of the understorey of Monterey pine.

References

Anderson, L.S. and Sinclair, F.L. (1993) Ecological interactions in agroforestry systems. *Forestry Abstracts* 54 (6), 489-523.
Baker, F.S. (1979) *Principles of Silviculture*, 2nd edn. Mc Graw-Hill, New York, USA.
Beaton, A. and Hislop, M. (2000) Trees in agroforestry systems. In: Hislop, M.and Claridge, J. (eds) *Agroforestry in the United Kingdom*. FC Bulletin 122, Edinburgh: Forestry Commission, pp. 31-43.
Campos, P. and Martin, M. (1986) *Development and Conservation of the Portuguese and Spanish Dehesas* (Conservación y Desarrollo de las Dehesas Portuguesas y Españolas). Ministerio de Agricultura, Madrid, Spain. (In Spanish.)
Dodd, C.J.H., McLean, A. and Brink, V.C. (1972) Grazing values as related to tree-crown covers. *Canadian Journal of Forest Reseach* 2 (3), 185-189.
Étienne, M. (1996) Browse impact in silvopastoral systems participating in fire prevention in the French Mediterranean region. In: Étienne, M. (ed.) *Western European Silvopastoral Systems.* INRA, Paris, France, pp 93-102.
Fletcher, R., Logan, R., Monroe, J., Stephenson, G. and Withrow-Robinson, B. (1993) *Agroforestry in Western Oregon.* Benton County Extension Service, Corvallis, Oregon, USA.
Garcia Salmerón, (1991) *Forest Nursery Practices* (Manual de repoblaciones forestales). E.T.S. Ingenieros de Montes, Fundación Conde del Valle de Salazar. Madrid, Spain.
Hawke, M.F. (1991) Pasture production and animal performance under pine agroforestry in New Zealand. *Forest Ecology and Management* 45, 109-118.
Hawke, M.F. and Knowles, R.L. (1997) Temperate agroforestry systems in New Zealand. In: Gordon, M. and Newman, S.M. (eds) *Temperate Agroforestry Systems.* CAB International, Wallingford, United Kingdom, pp. 85-118.
Hislop, M. and Sinclair, F., (2000) Introduction. In: Hislop, M. and Claridge, J. (eds) *Agroforestry in the United Kingdom*. Forestry Comission. Bulletin 122. London, United Kingdom, pp. 1-6.
Ibarra, A., Albizu, I. and Besga, G. (2000) Silvopastoral option in the Basque Country (La opción del silvopastoralismo en el País Vasco). *Sustrai* 56, 40-43.
Joffre, R., Vacher, J., Llanos, C. and Long, G. (1989) The dehesa: an agrosilvopastoral system of the Mediterranean region with special reference to the Sierra Morena area of Spain. In: Nair, P.K.R. (ed.) *Agroforestry Systems in the Tropics.* Kluwer Academic Publishers, Dordrecht, The Netherlands, pp. 427-456.
King, K.F.S. (1980) Multiple-use research. In: *IUFRO/MAB Conference: Research on Multiple Use of Forest Resources.* USDA – Forest Service. GTR WO-25, WT, Doolittle, USA.
Knowles, R.L., (1991) New Zealand experience with silvopastoral systems: a review. *Forest Ecology and Management* 45, 251-267.
Lewis, C.E. and Pearson, H.A. (1987) Agroforestry using tame pastures under planted pines in the southeastern United States. In: Ghotz, H.L. (ed.) *Agroforestry: Realities, Possibilities, and Potentials.* Martinus Nijhoff, Boston, USA, pp. 195-212.
McAdam, J. (1991) An evaluation of tree protection methods against Scottish Blackface sheep in an upland agroforestry system. In: Jarvis, P.G. (ed.) *Agroforestry: Principles and Practice.* Elsevier, Amsterdam, The Netherlands, pp. 119-126.
McAdam, J. and Hoppe, G.M. (1996) Pasture production between trees in a silvopastoral agroforestry system. *Grassland Science in Europe* 1, 119-122.
McAdam, J. and Sibbald, A. (2000). Grazing livestock management. In: Hislop, M. and Claridge, J. (eds) *Agroforestry in the United Kingdom*. Forestry Comission, Bulletin 122, London, United Kingdom, pp. 44-57.
Ministerio de Agricultura (1996) *Agriculture Statistics Yearbook* (Anuario de Estadística Agraria). Madrid, Spain. (In Spanish.)
Mosquera, M. R., Rigueiro-Rodríguez, A. and Villarino, J. J., (2001) *Silvopastoral Systems Establishment.* (Establecemento de Sistemas Silvopastorais). Consellería de Agricultura, Gandeiría e Política Agroalimentaria, Xunta de Galicia, Santiago de Compostela, Spain. (In Spanish.)
Nair, P.K.R. (1989) Classification of agroforestry systems. In: Nair, P.K.R. (ed.) *Agroforestry Systems in the Tropics.* Kluwer Academic Publishers, Dordrecht, The Netherlands, pp. 39-52.
Nair, P.K.R. (1991) State of the art of agroforestry systems. *Forest Ecology and Management* 45, 5-29.
Osoro, K., Celaya, R. and Martínez, A. (2002) Basic knowledge for the pasture resources management of Cantabrian Mountains (Conocimientos básicos para la gestión de los recursos pastables de la Cordillera Cantábrica). In: *Seminario sobre Producción de Pastos Extensivos.* Centro de Investigaciones Agrarias de Mabegondo CIAM, Xunta de Galicia, A Coruña, Spain, pp. 51-86.
Papanastasis, V. (1996) Silvopastoral systems and range management in the Mediterranean region. In: Étienne, M. (ed.) *Western European Silvopastoral Systems.* INRA, Paris, France, pp. 143-156.
Piñeiro, J. and Pérez Fernandez, M. (1988) Producción de pastos entre pinos. *Agricultura* 672, 480-484.
Pinto, M., Besga, G. and Rodríguez, M. (2002) Chemical composition of species-rich pastures. *FAO REU Technical Series* 64, 233-237.
Rigueiro-Rodríguez, A. (1985) Animal use in Galician woodland pasture, towards the integrated woodland use (La utilización del ganado en el monte arbolado gallego, un paso hacia el uso integral del monte). In Vélez, R. and Vega, J.A. (eds) *Estudios sobre Prevención y Efectos Ecológicos de los Incendios Forestales***.** ICONA (MAPA), Madrid, Spain, pp. 61-78. (In Spanish.)
Rigueiro-Rodríguez, A. (1986) Silvopastoral treatments for fire risk prevention in Galicia. (Tratamientos silvopastorales para la prevención de incendios en Galicia). In: Castelló, J.I. and Terradas, J. (eds) *Bases Ecologiques per la Gestió Ambiental.* Diputación de Barcelona, Barcelona, Spain, pp. 25-27. (In Spanish.)
Rigueiro-Rodríguez, A. (1992) Controlled grazing in Galician forests (Pastoreo controlado en los bosques gallegos). *El Campo* 124, 29-33. (In Spanish.)
Rigueiro-Rodríguez, A. (2000) Silvopastoral systems in the Atlantic Iberia (Sistemas silvopastorales en la Iberia Atlántica). In: *XL Reunión Científica de la Sociedad Española para el Estudio de los Pastos.* SEEP, Coruña, Spain, pp. 649-658. (In Spanish.)
Rigueiro-Rodríguez, A. (2002) Silvopastoral systems in Galicia. (Sistemas Silvopastorales en Galicia). In: *Seminario sobre Producción de Pastos Extensivos.* Centro de Investigaciones Agrarias de Mabegondo CIAM, Xunta de Galicia, A Coruña, Spain, pp.43-50. (In Spanish.)
Rigueiro-Rodríguez,A., Mosquera-Losada, M.R., Silva, F.J., Rodríguez, R., Castillón, P.A., Álvarez, P., M.R. Mosquera, M.R. and González, M.P. (1998) *Silvopastoral Systems Handbook* (Manual de sistemas silvopastorales). Universidad de Santiago de Compostela. Santiago de Compostela, Spain. (In Spanish.)

Rigueiro-Rodríguez, A., Mosquera, M.R., López Díaz, M.L., Pastor, J.C., González Hernández, M.P., Romero, R. and Villarino, J.J. (2001) Forest fire risk reduction through the use of Galician mountain horse (Reducción do risco de incendios forestais mediante o pastoreo do cabalo galego de monte). *O Común dos Veciños* 3, 12-14. (In Spanish.)

Rigueiro-Rodríguez, A., López-Díaz, M.L., Iglesias-Rego, R., Fernández-Núñez, E., Fernández-Gómez, S., Jardón-Bouzas B. and Mosquera-Losada, M.R. (2002) Macronutrient concentration of main natural herbs, shrubs and forage trees in NW Spain. *Grassland Science in Europe* 7, 90-91.

Sharrow, S.H. (1983) *Agroforestry: Growing trees, Forage and Livestock together. The Woodland Workbook.* Extension circular 1114, Oregon State University Extension Service, Oresa, USA.

Sharrow, S.H. (1999) Silvopastoralism: competition and facilitation between trees, livestock and improved grass-clover pastures on temperate lands. In: Buck, L.E., Lassoie, J.P. and Fernández, E.C.M. (eds) *Agroforestry in Sustainable Agricultural Systems.* Lewis, Boca Raton, USA, pp. 111-130.

Silva-Pando, F.J. (1988) *Aprovechamientos Silvopastorales.* Actas curs d'Estudis Pirenencs 1-21, Seo de Urgel.

Silva-Pando, F.J. (1991) Ecological effects of agroforestry on pinewoods and eucalypts woods in Galicia (NW of the Iberian Peninsula). In: *Proceedings of the First European Symposium in Terrestrial Ecosystems: Forest and Woodlands.* Elsevier Applied Sciences, Amsterdam, The Netherlands, pp. 874-985.

Silva-Pando, F.J. (1993) Agroforestry practices in eucalypts stands and pinewoods: undergrowth production (Prácticas agroforestales en pinares y eucaliptales atlánticos) I: Producción del sotobosque. In: *Congreso Forestal Español.* Lourizán (Pontevedra), Ponencias y Comunicaciones, Tomo II, SEEP, Pontevedra, Spain. (In Spanish.)

Sinclair, F., Eason, B. and Hooker, J. (2000) Understanding and management of interactions. In: Hislop, M. and Claridge, J. (eds) *Agroforestry in the United Kingdom.* Forestry Commission, London, UK, Bulletin 122, 17-30.

Sineiro, F. (1982) Aspectos del uso ganadero del monte en Galicia. *Pastos* 12 (1), 1-39.

Sineiro, F. and Díaz, N. (2002) Sistemas de producción sobre pastos establecidos en terras a monte: síntese de 20 anos de investigación en Galicia. In: *Seminario sobre Producción de Pastos Extensivos.* Centro de Investigaciones Agrarias de Mabegondo CIAM, Xunta de Galicia, A Coruña, Spain, pp. 11-42. (In Spanish.)

Valette, J.C., Rigolot, E. and Etienne, M. (1993) Integration des techniques de débroussaillement dans l'aménagement de défense de la forêt contre les incendies. *Forêt Méditerranéenne* 14 (2), 141-154.

Williams, P.A., Gordon, A.M., Garrett, H.E. and Buck, L. (1997) Agroforestry in North America and its role in farming systems. In: Gordon, A.M. and Newman, S.M. (eds) *Temperate Agroforestry Systems.* CAB International, Wallingford, pp. 9-84.

Forestry, pastoral systems and multiple use woodland

F. J. Silva-Pando
Forestry and Environmental Research Centre of Lourizán, P.O. Box 127, 36080 Pontevedra, Spain, and Plant Production Dept., Polytechnic High School University of Santiago de Compostela, Univ. Campus, 27002 Lugo, Spain. silva@inia.es

Abstract
The multiple use of forests could be understood as the deliberate and conscious use of forest lands for simultaneous production of several products or services. Silvopastoral systems can be included in a kind of multiple use because of the concurrent production of wood and meat, furthermore, other products can be obtained. In this chapter, are presented several examples of silvopastoral systems, mainly from the Iberian peninsula, with production of mushrooms, cork, hunting and medicinal plants. Several products produced by this kind of systems under a philosophy of sustainable forest management are described.

Key words: silvopastoralism, Iberian peninsula, production, added-value services

Introduction
King (1980) defines forest and grazing land systems as "ground management systems, where forests are used for wood production, and for tame animals breeding. It points out that in this system animals live and pasturage is permitted in the forest." The above-mentioned systems include as components trees, animals and shrub or herbaceous-like plants. They have the dual aim of conservation and improving the land type as well as optimizing crop production from the forest and the animals. Gregory in King (1980) defines multiple use as "the conscious and deliberate use of land for the concurrent production of more than one good or service". Applying this for the forest and considering the multiple functions and products that can be obtained, the result will be the conscious and deliberate use of the forest.

Taking into consideration these definitions and that of agroforestry (Lundgren and Raintree, 1982 in Nair, 1993) forest and grazing land systems are multiple use forms, but they are not restricted to the delivery of only wood and meat. Other products, services and benefits can be obtained concurrently from the same area. This is defined as the "value of use" (Alvarez and Field, 1997; Campos Palacín, 1999), which can be active or passive, or non-market.

Researchers could learn lessons in how to manage forests sustainably without exhausting the natural resources or impoverishing the ecosystem by observing the sophisticated management carried out by the indigenous people.

The production and extraction of traditional products in addition to wood from the forest are important for the maintenance of biodiversity and the accumulation of carbon and the viability of rural populations (Lamb, 1993; Sánchez, 1995).

Lamb (1993) defines the non-timber-yielding forest products (NTFP) as "all products for commercial, industrial or subsistence use" derived from the forest and its total plant and animal biomass, obtainable continuously from the forest ecosystem in quantities and ways that do not damage the basic reproductives functions of the plant community. Many products can be obtained from forests (Lamb, 1993). In India more than 5000 products of almost 3000 species have been obtained from forests (Uma Shankar *et al.*, 2003). Reid and Wilson (1986) and Nair (1993) describe products which can be obtained from several arboreal species. The use of the NTFP can help develop small companies, encourage new sources of income and new markets as well as helping survival and subsistence of local people.

The extraction of the NTFP depends on intrinsic and extrinsic factors. Intrinsic factors include the existence of the product, its value and capacity to recover. Extrinsic factors are market demand, product price, harvest cost and product management (Uma Shankar *et al.*, 2003). These natural products and resources can coexist with grazing and wood extraction. This multifunctional extraction has gone on for many years, often at variable intensities and motivated by the structural and functional changes imposed on the woodland.

In some areas, such as in developing countries, the NTFP are starting to capture a part of the market share. However, in some developed countries, such as the United States of America, the industry associated with NTFP is growing whereas conventional wood production is declining. In these countries, industrial use (wood pulp, chipboard wood, etc.) is an important part of production.

In this chapter some of the non-timber products that can be obtained in the Iberian peninsula and the occidental part of the Mediterranean region are reviewed. Some examples of multiple-product output from forest and grazing land systems will be given. The chapter will not cover services from or benefits accrued from the forest.

Forest products
Products with market value which can be obtained from the forest are:

Products derived from wood and wood-forest products (WFP) e.g. saw wood, boxwood, poles, paper fibre, firewood, charcoal, crude chemical products.

NTFP such as pasture, meat, resin, oils, latex, tannins, medicines, poisons, honey, fruits, seeds, cosmetics, scents, bark, cork, spice, food, colouring. These include products from the non-tree components of the forest.

Wood-derived products
There are many tree species which can be used for the production of wood, mainly saw wood, boxwood and wood pulp, and there are others where wood is not the main product (Table 1).

The forest productivity is a function of the species and environmental characteristics, the principal one being light and water availability. The woodland canopy competes with the rest of the vegetation by absorbing solar radiation (Nair, 1993; Silva-Pando et al., 2002). Depending on the age and structure of the tree, the trunk and branches can be put to one of the above-mentioned uses. In developed countries the production of firewood is decreasing, while in the underdeveloped countries approximately 80% of wood production is designated for that use.

NTFP
These products are not extractable from all forest areas of the world. In the case of the Iberian peninsula and an important part of the Mediterranean region the following products can be exploited:

(i) *Meat.* In grazed forests, livestock and wild animals for meat production are an important output (e.g. Rigueiro Rodríguez, 1985; Reid and Wilson, 1986; Nair, 1993; Montero et al., 2003; Silva-Pando et al., 2002).

(ii) *Cork.* Cork is an exclusive product from the cork oak (*Quercus suber*) with applications in construction, bottle cork and several industries. The Iberian peninsula is responsible for approximately 80% of world production. Extraction rates vary between 250 kg/ha per year in woodland and 4500 kg/ha per year in coppice in 9-10-11 year rotations (Riesco Muñoz and Amurrio Ordónez, 1997; Montero et al., 2003). Density of cork-oaks varies between 100 and 400 trees/ha having a basal area of 15-25 m^2/ha and 0.6-0.7 canopy cover (Riesco Muñoz and Amurrio Ordónez, 1997; Montero et al., 2003). At this level cork production is compatible with the livestock production. However, the Iberian cork-oak system is a very old land use and current levels and quality of cork production are low, largely because of inappropriate shrub cutting and a high density of cattle.

(iii) *Tannin* is used for leather tanning and for extraction of several products. Traditionally tannery products from the Iberian peninsula come from chestnut (most used) oak, holm-oak, cork-oak, pine and lentisco (*Pistacia lentiscus*). Nowadays, natural tannery products have been largely replaced by synthetic ones. In 1935 Spain produced 35,106 kg of vegetal tannery products, but production gradually declined from the 1960s. Young trees must be used; they have a maximum of 13.5% of tannin, principally in the bark, and in the case of the chestnut tree they must be between 20 and 30 years old (Berrocal del Brío et al., 1998).

(iv) *Liga* is a product of holly tree (*Ilex aquifolium*) bark. This is a gum that can be obtained by cooking after fermentation of the bark. It is used for bird hunting (Ceballos and Ruiz de la Torre, 1979; García González, 2002).

(v) *Resin* can be obtained from some pines, principally *Pinus pinaster* and to a lesser extent *P. halepensis* and *P. nigra*, but in the Iberian peninsula an incision is made in the bark or the cambium is treated with acids, to distil turpentine (20%), colophony (70%) and other products such as varnish (10%). These products rely on the industrial production for high consumption (Riesco Muñoz and Amurrio Ordónez, 1997). Due to the irregularity of world market prices and the cost of resin collection, production has declined. Production varies between 2 and 4 kg/tree per year, i.e. equivalent to 100-300 kg/ha per year depending on the method used to collect resin (Riesco Muñoz and Amurrio Ordónez, 1997). The exploitation of resin is compatible with timber extraction according to Woodland Regulations since 1971.

(vi) *Fruits.* The principal forest species of fruits are from the chestnut tree, cherry tree, holm-oak, walnut tree and pine tree. Production from walnut and cherry trees and strawberry shrubs are for human consumption and production from holly, rowan, wild cherry and blackthorn are for wild animals. The production of chestnuts varies between 20-200 kg per tree (with mean value of 40) depending on the climate. This can represent between 1-3.5 t/ha per year with a density of 80-350 trees/ha (Berrocal del Brío et al., 1998; Fernández de Ana Magán et al., 1998). This is used principally for human consumption, although in some cases it will be left for animal consumption. The production of chestnuts (*Castanea sativa*) is related to wood production through the cycle of branch cutting between 20 and 30 years old.

Acorn production from the holm-oak can vary between 300 and 1000 kg/ha per year (Espárrago et al., 1993; San Miguel Ayanz, 1994). These acorns are used to fatten pigs for high-quality ham. Normal grazing densities are 0.68 pigs per ha (Espárrago et al., 1993). In some places in central and southern Spain these acorns will be used to produce preserves, sweets, jams and pastry for human consumption (García Gómez et al., 2002). The nutritional value of 2 kg of acorn is equivalent to 1 kg of medium-quality barley (San Miguel Ayanz, 1994).

The pine tree (*Pinus pinea*) can produce between 100 and 300 kg of fruits/ha/year at densities between 125 and 140 trees per hectare, but at this level of production wood production is reduced (Yagüe Bosch, 1993). These pine nuts are used mainly for human consumption.

Others are also for the consumption of the wild animals like the holly tree, rowan, wild cherry tree, blackthorn, etc.

Several understorey shrubs, like blackberry (*Rubus idaeus*, *Ribes* sp), redcurrants, bilberry (*Vaccinium myrtillus*), produce edible fruits for jam, wine, ice cream, preserves, juice or medicines and other products. Levels of production vary between 800 kg fruit/ha for bilberry and 3000-7000 kg/ha for redcurrant and raspberry cultivation respectively (Rigueiro et al., 1988). Matching clonal selection to the ecological conditions and canopy cover is an important aspect of growth.

(vii) *Seeds.* Olive tree seeds are used for oil extraction and for human consumption. During the 1990s there was a reduction in demand and consequently production, but the more recent high interest in olive oil consumption for health reasons has led to a demand and price increase. Carob tree seeds are used in the making of a livestock feed "garrofín" (Ceballos and Ruiz de la Torre, 1979).

(viii) *Honey*. Many forest trees and shrubs are used by bees in the production of honey. Introduced species such as the eucalyptus are also used to produce honey from the low altitudes of the Galician region. This exploitation occurs in places where beehives are in a clearing surrounded by woodland or scrubland. Normally, production levels are low, between 5 and 20 kg per beehive, with a high economic value.

(ix) *Medicines and aromatic plants*. Since ancient times medicines have been extracted from forest resources; however, since the 19th century synthetic medicines have been used. Medicinal products can be obtained from the trees (Table 1) and from woody shrubs and herbaceous plants from undergrowth (e.g. gentian, false hellebore, arnica, garlic). The Mediterranean region is very rich in aromatic plants (e.g. thyme, lavender, "cantueso" mint) and these can be obtained from the forest-pasture system.

(x) *Branches*. Leaves and branches used to be used alongside other products. This contributed to animal nutrition when the leaves and fruits were eaten except when the resouce was reduced by overuse or adverse conditions (San Miguel Ayanz, 1994). Production ranges between 50 and 900 kg of twigs/ha plus 90-140 kg of acorns (San Miguel Ayanz, 1994). The most interesting species are from the Leguminosae, such as the carob tree, although most trees, such as holm-oak, cork-oak, gall oak or ash tree, are used for this purpose.

(xi) *Mushrooms*. Mushrooms and truffles are forest products which are highly valued not only in Europe, but in other parts of the world (Zamora-Martínez and Nieto de Pascual-Pola, 1995). They have an economic value but are of value ecologically because of their interaction with forest fires (Fernández de Ana Magán *et al.*, 1998). Growing season climate and arboreal species determine the composition and production of the mushroom flora. More than 50 species can be harvested, although only some of them are useful. Production varies between 15 and >100 kg/ha per year (Fernández de Ana Magán *et al.*, 1989, 1998; Agreda and Fernández, 2001; Zamora-Martínez and Nieto de Pascual Pola, 1995; Martínez Peña and Fernández Toirán, 1997).

The truffle (*Tuber melanosporum*) grows principally in oak wood, kermes oak wood and gall-oak groves either as a mycorrhizal species naturally or as mycorrhized plants. Production varies between 15 and 200 g/ha per year, but the market prices that are reached can be very high (12-540 €/kg in 2001) (Reyna *et al.*, 2002).

(xii) *Fertilizer*. In some places on the Iberian peninsula, plants or remains of the leaves and branches are used as cattle fertilizer. In Galicia it is common to cut the gorse to bed cattle, and it will mix with solid and liquid to make cattle waste. Once sufficient material has accumulated, it will be spread over the cultivation ground.

(xiii) *Decoration*. Species such as holly tree branches (García González, 2002), mistletoe, *Ruscus aculeatus*, cyclamen, heather twigs and others can be used for ornament.

(xiv) *Firewood*. Products from pruning can be burnt. Often these are used by the owner of a second home as well as in the countryside. Production varies according to the region. In the Mediterranean zone it reaches between 500 and 1000 kg/ha per year, depending on mass, age and treatments (San Miguel Ayanz, 1994; Montero *et al.*, 2003). In the Atlantic zone more production can be expected due to the higher forest productivity.

(xv) Other products such as latex, poison, cosmetics, scent, spices and colouring are not produced in significant quantities on the Iberian peninsula.

Use of all these resources is not always possible due to problems of compatability. However, systems such as mushrooms and wood production or the windbreak effect in the crops are complementary type (Sibbald, 1988, in Silva-Pando and Rozados Lorenzo, 2002). Pasture production under trees in the first few years is supplementary type. Competition is the most common ecological type, such as the case between grazing and forest for the light or mushroom and scrubland for the space (Pilz and Molina, 2002) or fruit and wood (Yagüe Bosch, 1993). A relationship model (based on Bergez and Msika, 1995) is presented (Table 1), where the influence of the different components on the ecosystem can be seen.

Table 1. Agroforestry characteristics of Iberian forestry species. Y = yes; N = no. Climate: A = Atlantic; S = sub-Mediterranean; M = Mediterranean. Precipitation: a = < 600 mm; b = 600 - < 1000 mm; c = 1000 - < 1500 mm; d = > 1500. Saw wood: N = no suitable; S = low quality; SS = medium quality; SSS = high quality. Fruits: SH = human consumption; SA = animal consumption.

Species	Climate	Precipitation	Saw wood	Industrial use	Fuelwood	Resin	Fruits	Seeds	Honey	Medicinal plants	Browsing	Manure	Ornamental	Tannins	Notes
Abies alba	S	b,c	SSS				N						Y		
Acacia dealbata	A	c	N				N	Y					Y		
Acacia melanoxylon	A	b,c,d	S				N	Y						Y	
Alnus glutinosa	A	c,d	S		Y		N			Y	Y	Y		Y	
Arbutus unedo	S	b,c	N				SH			Y	Y		Y	Y	Charcoal
Betula celtiberica	A	c,d	S	Y			N			Y		Y		Y	Drinks
Castanea sativa	S	b,c	SS				SH	Y				Y		Y	
Cedrus atlantica	M	b,c	SS				N						Y		
Corylus avellana	A	c,d	N				SH			Y		Y	Y		
Eucalyptus camaldulensis	M	b	S	Y						Y	Y				
Eucalyptus globulus	A	c,d	S	Y						Y	Y				
Fagus sylvatica	A	c,d	SSS				SA					Y	Y		
Fraxinus angustifolia	M	b	S		Y					Y	Y	Y			
Fraxinus excelsior	A	c,d	SS		Y					Y	Y	Y	Y		
Ilex aquifolium	A	c,d	N				SA			Y			Y		
Juglans nigra	S	b,c	SSS				SH						Y		
Juglans regia	S	b,c	SSS				SH						Y		
Juniperus communis	M	b	SS		Y		SH	Y		Y	Y		Y		Ginger
Laurus nobilis	S	c	N							Y			Y		
Olea europaea	M	a,b	N				SH	Y					Y		Olives
Pinus halepensis	M	a,b	N			Y									
Pinus nigra	M	b	S												
Pinus pinaster	A	c,d	S	Y	Y	Y				Y		Y			
Pinus pinea	M	b	S		Y		SH						Y		
Pinus radiata	A	c,d	S	Y								Y			
Pinus sylvestris	M	c	SS		Y							Y			
Populus alba	M	b,c	S	Y									Y	Y	
Populus nigra	M	b,c	S	Y									Y	Y	
Prunus avium	S	c	SSS				SH	Y		Y		Y			Cherries
Quercus ilex	M	a,b	S		Y		SA							Y	Charcoal
Quercus petraea	A	d	SS				SA				Y				
Quercus pyrenaica	S	c	S		Y		SH							Y	
Quercus robur	A	d	SS		Y		SA				Y			Y	
Quercus rotundifolia	M	b	S		Y		SH							Y	Charcoal
Quercus suber	S	b,c	S				SA							Y	Cork
Robinia pseudacacia	A	c,d	S		Y			Y	Y		Y	Y	Y		
Taxus baccata	A	c,d	S				S			Y			Y		

Iberian forest and grazing land system

San Miguel Ayanz *et al.* (2002) describe the principal Iberian agroforestry systems, in which forest grazing areas are included. This practice is a very ancient one, although in some cases it has been introduced to align with environmental and economic criteria.

(i) Pine forest. Grazing in the pine forests has been described by Rigueiro Rodríguez (1985) and Silva-Pando *et al.* (1998). The principal arboreal species used are *Pinus pinaster, P. radiata, P. sylvestris, P. pinea* and *P. halepensis* (the last two in the Mediterranean area). These are all used by cows, sheeps, goats, horses and pigs. The undergrowth can have characteristics of scrubland or grazing land. Wood, meat and mushroom can be harvested at the same time (Silva-Pando *et al.*, 1998). Fallen pine needles can be used used to bed cattle and afterwards can be used as fertilizer for herbaceous cultivation (maize, rye, potato, etc.). Resin may also be exploited. Mushrooms can be produced with other products but grazing must be excluded during autumn (Silva-Pando *et al.,* 1998).

Liquid waste from urban sewage treatment plants and milk treatment plants (Mosquera-Losada *et al.,* 2001) has also been successfully used as fertilizer in this agroforestry system.

(ii) Eucalyptus. The most planted species in Galicia is *Eucalyptus globulus*. It is compatible with grazing by the same types of animals as are found for the pine trees (Rigueiro Rodríguez, 1985; Silva-Pando *et al.,* 1998). It can be combined with collecting gorse for cattle bedding and setting up beehives for honey production. Another issue, not dealt with in this chapter, is the biomass reduction of the undergrowth scrubland and the resulting reduction in risk of damage by forest fire (Silva-Pando and González Hernández, 1992).

(iii) Dehesa is an agroforestry system existing since at least 1000 AD. Its structure and functioning has been analysed in depth (see San Miguel Ayanz, 1994). The most used common tree species are the holm-oak, cork-oak, Pyrenean oak, gall oak, *Pinus pinea* and ash, often with other species such as the olive, juniper, white juniper and carob. The grassland types are called "majadales", "bonales", "fenalares" and rushies, the second one being the most productive. The products that can be obtained are wood (not the principal one), grazing for cattle, fruits (for pig feeding), cork, branches for cattle, firewood, mushrooms (e.g. various Boletaceae, *Auruncula*, earthstars, thistle mushroom), hunting, oats cultivation, barley, rye, wheat (generally in small areas of land and not very important, except as a control measure of the woody vegetation or as a nutritional complement for hunted or domesticated animals) (San Miguel Ayanz, 1994).

A special type of exploitation is the case of the collecting of fruits in limy oak wood, kermes oak wood and gall oak groves, where there is close association with sheep grazing, the use of firewood and hunting (Reyna *et al.,* 2002).

(iv) Holly tree wood. Nowadays, the principal use of *Ilex aquifolium* is for stockbreeding and for decorative plants. The stockbreeding use can be for cows, sheep, horses or goats through a rotational grazing system up to a field capacity of 0.2-1.0 LU/ha. Heavy use can be made in spring and branches and leaves cut in autumn for fodder (García González, 2002). In some cases, overgrazing of the scrubland can reduce regeneration. It is very common to charge a small amount of money, for grazing in communal forestland. Hunting takes place in private game preserves by local hunters. Thrushes and doves are traditionally hunted (García González, 2002).

Some decorative species are protected and, if their foliage is used for decoration, it is often disguised as a forestry treatment. Typically foliage is used as a Chrismas decoration. The level of use depends more on the market demand than availability but over-harvesting and loss of the resource tends not to occur, partially because the holly tree is a protected species in many areas of the Iberian peninsula. This wood is only used in small pieces for cabinetwork, lathe work, marquetry, piano keys or construction (García González, 2002).

(v) Chestnut tree wood. *Castanea sativa* wood will be used for two principal objectives: the first one is wood exploitation, usually as coppice or plantations, and the second one is for fruit exploitation, combined with wood production and other possible uses (Berrocal del Brío *et al.,* 1998). High-density sheep grazing can be carried out over short periods of time at 20-100 sheep/ha during the wintertime for the exploitation of the grazing and in spring the pasturage is used at a time of maximum production and digestibility. The nutritional value of 2 kg of chestnut is equivalent to 3 kg of barley.

The exploitation of mushrooms has a high economic value, including species from the genus of *Amanita*, *Boletus*, *Russula* and *Cantharellus cibarius* (Fernández de Ana Magán *et al.,* 1998).

(v) Birch, chestnut and several oak-type forests. These forests, which are principally in mountain areas, have been exploited to collect wood and forage for cattle. The exploitation of mushrooms, some medicinal plants and small fruits from the undergrowth is possible. Hunting is a very important resource across the entire Atlantic region (San Miguel Ayanz *el al.,* 2002).

(vii) Vineyards. In the north of Portugal, *Quercus lusitanica*, *Ulmus* sp and *Prunus* sp are used to support the vineyards for the production of green wine from vine stems of the vine (*Vitis vinifera*). In the intermediate spaces between the grape rows, maize, pulses and vegetables are grown for cattle food. Sometimes gorse can be cultivated for animal bedding or even for fodder (Altieri and Nicholls, 2002). Other products used are wood, bark and fruits of some of the exploited trees.

Conclusions

There is a wide diversity of resources available in forest ecosystems. It is necessary to compile a catalogue of the available resources for the south-west mountains of Europe. In general, the exploitation compatibility of these resources with the forage resource is very high.

Management of the forest has to take into account the different requirements and the interactions between the species. The ecological interactions between processes are important as they affect the light, space, nutrients, water and utilization by the herbivores. The achievement of optimum output will depend on product demand, production and prevailing socio-economic conditions.

References

Agreda, T. and Fernández, M. (2001) South-eastern mycological profits in *Pinus pinaster* stand in de Soria province (Rendimiento micológico de una masa de *Pinus pinaster* Ait. del Sudeste de la Provincia de Soria). In: *Actas III Congreso Forestal*

Español-Sierra Nevada 2001. Junta de Andalucía, Sociedad Española de Ciencias Forestales, Coria Gráficas, S.A. Sevilla, Spain. (In Spanish.)

Altieri, M.A. and Nicholls, C.I. (2002) The simplification of traditonal vineyard based agroforests in northwestern Portugal: some ecological implications. *Agroforestry Systems* 56, 185-191.

Alvarez, M. and Field, D. (1997) Social benefits of multiple use management of natural parks (Beneficio social de la gestión de uso múltiple de los espacios naturales). In: Rojo Alboreca, A., Díaz-Maroto, I., Álvarez González, J.G., Barrio, M., Castedo, F., Riesco, G. and Rigueiro, A. (eds) *Actas del Congreso de Ordenación y Gestión Sostenible de Montes*. Consellería de Medio Ambiente, Santiago de Compostela, Spain. (In Spanish.)

Bergez, J.-E. and Msika, B. (1995) A silvopastoral model for the EU. *Cahiers Options méditerranéennes* 12, 231-238.

Berrocal del Brío, M., Gallardo Lancho, J.F. and Cardeñoso Herrero, J.M. (1998) *Castanea sativa. Fruit and Timber Production.* (*El castaño*. Productor de fruto y madera). Creador de paisaje y protector, Mundi-Prensa, Madrid, Spain. (In Spanish.)

Campos Palacin, P. (1999) Hacia la medición de la Renta de bienestar del uso múltiple de un bosque. *Investigación Agraria, Sistemas de Recusos Forestales* 8 (2), 407-422.

Ceballos, L. and Ruiz de la Torre, J. (1979) *Árboles y Arbustos de la Península Ibérica*. E.T.S. Ingenieros de Montes-Fundación Conde del Valle de Salazar, Madrid, Spain.

Espárrago, F., Vázquez, F.M., Buzarco, A. and Pérez, M.C. (1993) Acorn producrion of *Quercus rotundifolia* Lam.: annual variability and economic importance (Producción de bellota en *Quercus rotundifolia* Lam.: Variabilidad anual e importancia económica). In: Silva-Pando, F.J. and Vega, G. (eds) *Ponencias y Comunicaciones del I Congreso Forestal Español-Lourizán*. Grafol S.L., Vigo, Spain. (In Spanish.)

Fernández de Ana Magán, F.J., Rodríguez, A. and Rodríguez Fernández, R.J. (1989) Relationship between mycology productivity and silvicultural treatments of *Pinus pinaster* (Relación entre a productividade dos fungos micorrízicos e os tratamentos selvícolas en *Pinus pinaster* Ait.). In: *VI Xornadas Agrarias Galegas*. Xunta de Galicia, Sergude. (In Spanish.)

Fernández de Ana Magán, F.J., Verde Figueiras, M.C. and Rodríguez Fernández, A. (1998) *Castanea Sativa Stands. Ecosystems in Danger* (O souto, un Ecosítema en Perigo). Xunta de Galicia, Santiago de Compostela, Spain. (In Spanish.)

García Gómez, E., Pereira Sieso, J. and Ruiz Taboada, A. (2002) Acorn use as food by farms (Aportaciones al uso de la bellota como recurso alimenticio por las comunidades campesinas). *Cuadernos Sociedad Española de Ciencias Forestales* 14, 65-70. (In Spanish.)

García González, M.D. (2002) *Ilex aquifolium* in the north Iberian mountains: current and traditional use (Las acebedas del Sistema Ibérico Norte: Sus aprovechamientos tradicionales y actuales). *Cuadernos Sociedad Española de Ciencias Forestales* 14, 71-76. (In Spanish.)

King, K.F.S. (1980) Multiple-use research. In: Doolittle, W.T. (ed.) *IUFRO/MAB Conference: Research on Multiple Use of Forest Resources*. USDA-Forest Service Flagstaff, GTR WO-25, Washington DC, USA. pp. 3-9.

Lamb R. (1993) More than wood. *Forestry Topics Report* 4, 6-52. FAO Forestry Department.

Martínez Peña, F. and Fernández Toirán, M. (1997) Mycology species production in *Pinus sylvestris* stands of different ages (Producción de especies fúngicas en masas de *Pinus sylvestris* L. de diferentes edades). In: Puertas, F. and Rivas, M. (eds) *Actas I Congreso Forestal Hispano-Luso. II Congreso Forestal Español. Irati 97*. Gráficas IRATI, Pamplona, Spain. Vol. IV, pp. 405-410. (In Spanish.)

Montero, G., Martín, D., Cañellas, I. and Campos, P. (2003) Silviculture and production of *Q. suber* (Selvicultura y producción del alcornocal). In: Pulido, F., Campos, P. and Montero, G. (eds) *La Gestión Forestal de las Dehesas*. Instituto de Promoción del Corcho, La Madera y Carbón (IPROCOR). Mérida, Spain. pp. 49-94. (In Spanish.)

Mosquera-Losada, M.R., López-Díaz, L. and Rigueiro Rodríguez, A. (2001) Sewage sludge fertilization of a silvopastoral system with pines in northwestern Spain. *Agroforestry Systems* 53, 1-10.

Nair, P.K.R. (1993) *An Introduction to Agroforestry*. Kluwer Academic Publishers-ICRAF, Dordrecht, The Netherlands.

Pilz, D. and Molina, R. (2002) Commercial harvests of edible mushrooms from the forests of the Pacific Northwest United States: issues, management, and monitoring for sustainability. *Forest Ecology Management* 155, 3-16.

Reid, R. and Wilson, G. (1986) *Agroforestry in Australia and New Zealand. The Growing of Productive Trees on Farms*. Goddard and Dobson, Box Hill.

Reyna, S., Folch, L. and Alloza, J.A. (2002) Truffle culture: profitable dehesa in lime soils (La Truficultura: una dehesa rentable para los encinares en suelos calizos). *Cuadernos Sociedad Española de Ciencias Forestales* 14, 95-101. (In Spanish.)

Riesco Muñóz, G. and Amurrio Ordóñez, M. (1997) *Management of No Timber Forest Resources: Resins, Cork, Pasture and River Fish*. (Ordenación de recursos forestales no madereros: resinas, corcho, pastos y pesca fluvial). Unicopia, Lugo, Spain. (In Spanish.)

Rigueiro Rodríguez, A. (1985) Livestock in Galician forests: towards and integrated-use of the foresland (La utilización del ganado en el monte arbolado gallego, un paso hacía el uso integral del monte). In: Vélez Muñoz, R. and Vega Hidalgo, J.A. (eds) *Estudios sobre Prevención y Efectos Ecológicos de los Incendios Forestales*. ICONA (MAPA), Madrid, Spain. pp. 61-78 (In Spanish.)

Rigueiro, A., Silva-Pando, F.J. and Salinero, M.C. (1988) Multifunctional use of the forestland (O aproveitamento múltiple do monte). *Revista Gallega Estudios Agrarios* 10, 63-79. (In Galician.)

Sánchez, P.A. (1995) Science of agroforestry. *Agroforestry Systems* 30, 5-55.

San Miguel Ayanz, A. (1994) *Spanish Dehesa. Origin Type, Characteristics and Management* (La Dehesa Española. Origen, tipología, características y gestión). E.T.S. Ingenieros de Montes de Madrid-Fundación Conde del Valle Salazar, Madrid, Spain. (In Spanish.)

San Miguel Ayanz, A., Roig, S. and Cañellas, I. (2002). Agroforestry practices in the Iberian peninsula (Las prácticas agroforestales en la Península Ibérica). *Cuadernos Sociedad Española de Ciencias Forestales* 14, 33-38. (In Spanish.)

Silva-Pando, F.J. and González Hernández, M.P. (1992) Agroforestry helps to prevent forest-fires. *Agroforestry Today* 4 (4), 7-9.

Silva-Pando, F.J. and Rozados Lorenzo, M.J. (2002) Agrosilviculture, agroforestería, agroforestry practices, multiple uses: definitions and concepts. (Agroselvicultura, Agroforestería, Prácticas Agroforestales, Uso Múltiple: Una definición y un concepto). *Cuadernos Sociedad Española de Ciencias Forestales* 14, 9-21. (In Spanish.)

Silva-Pando, F.J., González Hernández, M.P., Rigueiro Rodríguez, A., Rozados Lorenzo, M.J. and Prunel Tuduri, A. (1998) Grazing livestock under pinewood and eucalyptus forest in Northwest Spain. *Agroforestry Forum* 9 (1), 36-43.

Silva-Pando, F.J., Rozados Lorenzo, M.J. and González Hernández, M.P. (2002) Pasture production in a silvopastoral system in relation with microclimate variables in the Atlantic coast of Spain. *Agroforestry Systems* 56, 203-211.

Uma, S., Lama, S.D. and Bawa, K.S. (2003) Temporal patterns of extraction of non-timber forest products in Chel Range of Darjeeling Himalaya. *Forests, Trees and Livelihood* 13, 114-133.

Yagüe Bosch, S. (1993) Silviculture and *Pinus pinea* L. production in Avila province (Silvicultura y producción del pino piñonero (*Pinus pinea* L.) en la provincia de Ávila). In: Silva-Pando, F.J. and Vega, G. (eds) *Ponencias y Comunicaciones del I Congreso Forestal Español-Lourizán 1993*. Grafol, Pontevedra, Spain. Vol. 2, pp. 479-484. (In Spanish.)

Zamora-Martínez, M.C. and Nieto de Pascual-Pola, C. (1995) Natural production in the southwestern rural territory of Mexico City, Mexico. *Forest Ecology Management* 72, 13-20.

An assessment of the role of grazing in European habitats

R. G. H. Bunce, M. Pérez-Soba, B. S. Elbersen, W. K. R. E. van Wingerden

Alterra, P.O. Box 47, 6700 AA Wageningen, The Netherlands. marta.perezsoba@wur.nl

Abstract

Policy responses are eventually needed to maintain traditional grazing or to ameliorate its impacts. The driving forces, pressures, states, impact and responses (DPSIR) framework is used to address the environmental implications of extensive grazing and how policy responses may be developed, using as an example the dehesas/montado silvopastoral systems. The impact of such grazing depends on the inherent environmental character of a given region, and the DPSIR framework allows a rapid assessment of silvopastoral habitats and their linked extensive livestock systems using consistent procedures, so that the threats can be determined and appropriate policies set up to maintain them. Otherwise they will disappear - as has already taken place in much of northern Europe.

Key words: DPSIR framework, extensive grazing, agricultural biodiversity, cultural landscapes

Introduction

The present composition of vegetation is highly correlated with the initial environment and has become integrated with human activity. It is therefore essential to consider the inherent environmental character of a given region, before assessing the role of grazing in the maintenance of systems with a high nature conservation value. Most of the animals in these systems are dependent on the semi-natural vegetation, with minimal supplements, which are often linked to silvopastoral systems in the Mediterranean region.

Once the delicate balance between grazing animals and the vegetation is altered, it is invariably difficult to return to the original ecological condition because often with cessation of grazing a few aggressive species take over. In a short time the other species disappear from the system and cannot return because they are not present in the seed bank and have insufficient populations to disperse effectively. Many such changes are therefore irreversible except by expensive restoration procedures. The driving forces, pressures, states, impacts and responses (DPSIR) NERC framework (1995) enables a rapid assessment of such linkages to be made and, whilst it can never replace detailed scientific surveys, it enables expert judgement to be formalized. The DPSIR framework is currently being adapted for the TRANSHUMOUNT project (EVK2-2002-00608), an Accompanying Measure of the Fifth Framework. This project is summarizing the role of transhumance in the mountain habitats in Europe in order to identify policies for their maintenance and conservation. The principal habitats used by transhumant animals have been identified and the framework is being used to define the pressures, state and impact of the grazing animals.

Materials and Methods

This chapter uses the DPSIR framework, which was used previously by Petit *et al.* (2001) for assessing the consequences of environmental change for biodiversity and allows expert judgement to be related to present and potential future states of habitats, using known relationships. The model segregates information on the pressures acting on the environment, the state of the individual environmental components and the anthropogenic responses. In this case, the driving force is agriculture. The pressures are exerted by (i) grazing animals i.e. sheep, cattle, goats and horses; (ii) changes in the species of grazing animals, specifying the influence of a shift from one species to another; (iii) the effect of intensive grazing/overgrazing and links to erosion; (iv) the application of artificial fertilizers; and (v) abandonment, including removal of grazing and its influence on vegetation succession. The state is reflected by the composition of habitats and the impacts follow from their interactions. A policy response is then needed in order to either maintain traditional grazing or to ameliorate its impacts. It must therefore be evaluated for each new set of conditions.

Results and Discussion

An example of the current pressure, state and impact of grazing on the dehesas/montado habitats, one of the major silvopastoral Mediterranean habitats, is summarized in Table 1.

Table 1. DPSIR Framework applied to dehesa/montado habitats.

Definition	Open forest or scattered trees, mainly of *Quercus ilex* and *Q. suber*, but also *Pinus pinea* and *Q. pyrenaica*
	Other land covers within these areas may be:
	Cereals, involving ploughing
	Fallow land left for several seasons
	Grassland of many categories
	Scrub of all three categories
	Mosaics of all types
Distribution and extent	Principally in Spain and Portugal, but also elsewhere in the Mediterranean
	Widespread, covering several million ha
Pressures	Grazing by all animals
	Abandonment
	Overgrazing
State	Highly variable according to local practices and the land cover as described above
	Structure is directly dependent upon grazing intensity
	Depends on the land cover types between the trees
	Scrub is controlled by local practices
Impact	Changes in the pattern of use lead to rapid loss of biodiversity
	Abandonment leads to expansion of small numbers of competitive species
	Depends on the state of the land cover
Conclusion	Transhumance is an integral feature of many sub-systems in dehesas/montados, and is essential for the maintenance of biodiversity

This example shows that silvopastoral systems are very sensitive to changes in management procedures and the biodiversity in the grasslands within them can therefore change rapidly. Whilst this study is not definitive and needs further amplification and the inclusion of other expert views, it does show how the DPSIR framework can be used to assess the current situation. The large literature on extensive grazing concentrates on semi-natural grasslands where its role is undoubtedly beneficial, but may not be providing a balanced overall view. From the vegetation viewpoint, the cultural and landscape benefits of extensive livestock systems could have greater significance at a strategic level than for biodiversity in regions where tourism is important. It is therefore essential to make a review of the role of silvopastoral systems in order to produce a sound assessment of their overall significance.

Conclusion

Overall, the balance of available information indicates that traditional extensive grazing and silvopastoral systems positively contribute to biodiversity, landscape value and rural development in semi-natural grasslands, even when scattered trees are present. However, in true forests and some other habitats its role needs careful assessment. The strength of these extensive systems is that they are part of a wider historical system in which agriculture has gradually adapted to the specific natural circumstances. However, under the influence of productivity increase, regional and on-farm specialization and rationalization driven by the CAP, many extensive grazing systems have disappeared. Different policy instruments are therefore required at local, regional and national levels, not only because of financial considerations but also due to the widely differing targets involved, i.e. biodiversity, cultural systems, landscape, conservation and rural development. The evidence for the effectiveness of current policies is fragmented and inadequate to really determine optimal maintenance. Therefore it is important to carefully review current policies to see if more can be learned. Consequently, silvopastoral habitats and their linked extensive livestock systems need to be identified using consistent procedures, so that the threats can be determined and appropriate policies set up to maintain them. Otherwise they will disappear - as has already taken place in much of northern Europe.

References

National Environmental Research Institute (NERC) (1995) *Recommendations and Strategies for Integrated Assessment of Broad Environmental Problems*. Report to the European Environmental Agency, Copenhagen, Denmark.

Petit, S., Firbank, L., Wyatt, B. and Howard, D. (2001) MIRABEL: models for integrated review and assessment of biodiversity in European landscapes. *Ambio* 30 (2), 81-88.

Situation and perspectives of silvopastoral systems in Germany

P. Finck [1], U. Riecken [1] and F. Glaser [2]

[1] Federal Agency for Nature Conservation, Konstantinstr. 110, D-53179 Bonn, Germany. [2] Luftbild Brandenburg GmbH, Karl-Liebknecht-Str. 1, D-15711 Königs Wusterhausen, Germany. FinckP@bfn.de, RieckenU@bfn.de; Luftbild.Brandenburg@t-online.de

Abstract

Until the 18th century, silvopastoral systems had been one of the main forms of forest utilization in Germany. However, in many cases the excessive use of woodlands resulted in their loss. To counter this, domestic animals were widely banned from woodlands and a well-organized forestry programme was implemented. This and the widespread disbandment of common lands caused a dramatic decline in silvopastoral systems.

The remaining silvopastoral systems in Germany nowadays centre in the Alpine area. Beyond this region only small-sized traditional wood-pasture stands remain, the majority of which are no longer grazed by domestic animals.

Current attempts to preserve the unique character of these wood-pastures focus on conventional silvicultural management methods. In a modelling project the feasibility of the restoration of wood-pastures by grazing is tested and scientifically monitored.

Key words: wood-pastures

Introduction

Until the 18th century, silvopastoralism had been a widespread use of land in Germany. Livestock was tended by herdsmen on common lands. As a rule the surrounding woodlands were included in these grazing activities. However, the excessive use of woodlands in the 16th to the 18th century resulted in their devastation and loss. This was caused not only by intensive grazing of domestic animals but also by other forms of forest utilization. Consequently the woodland area declined dramatically in most areas in Germany and depleted heathlands were widespread. To counter this development domestic animals were widely banned from woodlands in the 18th and 19th century. Furthermore a well-organized forestry programme was implemented, resulting in the conversion of woodlands into timber-producing forests. Hence the characteristic structures and appearance of wood-pastures as well as numerous species depending on these systems declined dramatically in Germany during the last 200 years (Küster, 1995).

Present situation of silvopastoral systems and wood-pastures

A recent study by Luftbild Brandenburg (Glaser and Hauke, 2004) on traditional wood-pastures found that the centre of distribution for active silvopastoral systems nowadays is within the Alpine biogeographical region. In this region rights for grazing of mountain forests exist for up to 55,000 ha. However, these rights are made use of to a varying extent. Due to the low stocking with cattle and their seasonal use, these Alpine silvopastoral areas do not show many of the characteristics of traditional wood-pastures. Furthermore, coniferous trees dominate these forests. Beyond the Alpine region no more than about 200 sites larger than 5 ha of traditional wood-pasture stands (amounting to less than 5000 ha) remain. More than 50% of these sites have not been grazed for several years or decades and therefore have lost their sparse character. The tree species composition changed from oaks and hornbeams to beeches. Especially, species that depend on dead and decaying wood as well as species adapted to open woodland have dramatically declined. These sites are concentrated in Bavaria, Baden-Württemberg and Lower Saxony (Figure 1). Furthermore some of these sites do not show all the characteristic features of wood-pastures since they have not developed from woods but have emerged from extensively used pastures. This means particularly that those coenoses adapted to ancient oak trees in sparse stands do not prosper at these sites.

At present in Germany there are several attempts by nature conservationists as well as public forestry (Niedersächsische Landesregierung, 1991) to preserve the characteristic appearance of wood-pastures. However, forestry officials favour conventional silvicultural management methods to the idea of reintroducing domestic animals to the forests. In these circles the fear of renewed damage to the forests is widespread. It must be remembered, however, that this damage had been caused by overgrazers combined with other ways in which the forests can be utilized. In the "Rheinhardswald", for example, in a forest area in the north of Hesse, 35,000 livestock used to graze on 120 km^2 during the middle of the 18th century (Gerken and Sonnenburg, 2003).

Figure 1. Present situation of wood-pastures in Germany.

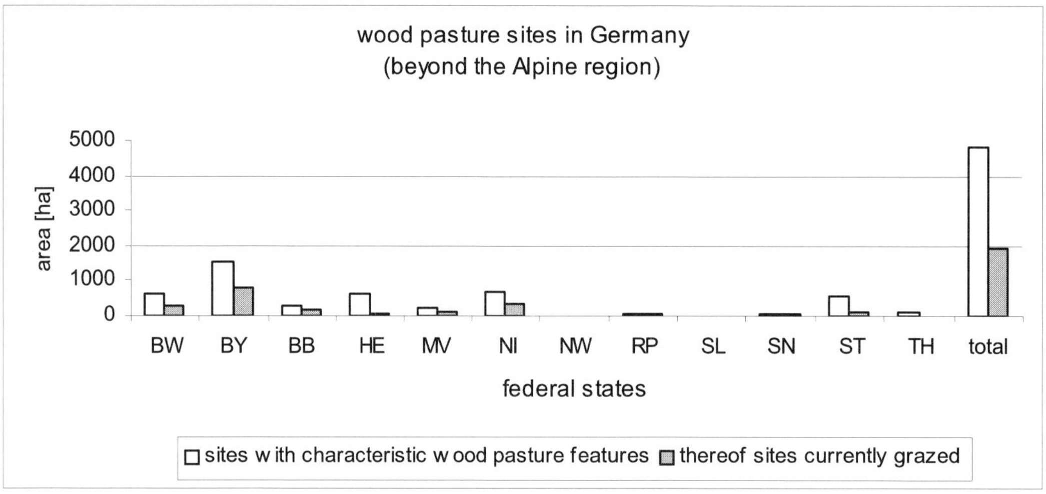

A model project for the restoration of a wood-pasture

The feasibility of restoring a former wood-pasture site by grazing is currently being tested in a model project in the Solling mountain range (Lower Saxony). This 5-year project is funded mainly by the German Federal Agency for Nature Conservation, together with the local authority of the Solling-Vogler nature park and the local forestry commission. Heck cattle and Exmoor ponies have been introduced to a 170 ha area that partially used to be a wood-pasture. The objective of the scientific monitoring (conducted by the Technical University of Lippe-Höxter) was the establishment of causal links between the behaviour of the cattle and horses and ecological as well as economical parameters. These included: (1) documentation of the wood stand conditions and abiotic parameters, (2) research on selective taxa of plants and animals, (3) ethno-ecological studies of the Exmoor ponies and Heck cattle, and (4) documentation and analysis of the economics of the project.

Since 1999 more than 2700 species have been recorded in the project area, of which more than 300 are listed as endangered in the German Red Data Book (Gerken and Sonnenburg, 2003). The number of species with a characteristic adaptation to sparse oak woods and ancient woodland sites is high. The following preliminary results can be deduced from the project: (1) Exmoor ponies and Heck cattle are principally suited for wood pasturing; (2) a considerable number of dynamic processes have been initiated in the project area by grazing; (3) by introducing cattle and horses food chains with new qualities emerge (e.g. dung → coprophagous insects → insectivorous bats → birds); (4) beeches of less than 2 m height are clearly impaired in their growth by the cattle and horses, whereas trees exceeding this height are not seriously damaged and therefore require supplementary silvicultural management to restore the sparse character of a wood-pasture.

Outlook

The preservation of the typical coenoses of silvopastoral systems, especially wood-pastures, is of high importance in Germany. However, in the long term it can only be achieved beyond the Alpine region if innovative systems of land use are established in large areas. Especially the existing rigid boundaries between the utilization of woodlands and open landscape must be broken down. For the open landscape this development can be faciltated by the establishment of "semi-open pasture landscapes" (Gerken and Görner, 2001; Finck *et al.*, 2002; Riecken *et al.*, 2002). Furthermore, taking into account that the production of timber in many state-owned forests is lacking profitability, at least the state-run forestry has to consider the multifunctionality of forests for society. Therefore the preservation and restoration of wood-pastures by means of appropriate grazing by livestock should increasingly be considered as a management objective.

References

Finck, P., Riecken, U. and Schröder, E. (2002) Pasture landscapes and nature conservation - new strategies for the preservation of open landscapes in Europe. In: Redecker, B., Finck, P., Härdtle, W., Riecken, U. and Schröder, E. (eds) *Pasture Landscapes and Nature Conservation.* Springer, Heidelberg, Berlin, New York, USA, pp. 1-13.

Gerken, B. and Görner, M. (2001) Über große Weidetiere und die künftige Landschaftsentwicklung in Europa. *Natur- und Kulturlandschaft* 4, 11-18.

Gerken, B. and Sonnenburg, H. (2003) *Das Hutewaldprojekt im Solling.* Verlag Huxaria, Höxter, Germany.

Glaser, F.F. and Hauke, U. (2004) Historisch alte Waldstandorte und Hudewälder in Deutschland - Ergebnisse bundesweiter Auswertungen. *Angewandte Landschaftsökologie* 61. 194 pp.

Küster, H. (1995) *Geschichte der Landschaft in Mitteleuropa von der Eiszeit bis zur Gegenwart.* C.H. Beck'sche Verlagsbuchhandlung, Munich, Germany.

Niedersächsische Landesregierung (1991) *Niedersächsisches Programm zur langfristigen ökologischen Waldentwicklung in den Landesforsten (LÖWE).* NL. Hanover, Germany.

Riecken, U., Finck, P. and Schröder, E. (2002) Significance of pasture landscapes for nature conservation and extensive agriculture. In: Redecker, B., Finck, P., Härdtle, W., Riecken, U. and Schröder, E. (eds) *Pasture Landscapes and Nature Conservation*. Springer, Heidelberg, Berlin, New York, USA, pp. 423-435.

Wood-pasture and parkland: overlooked jewels of the English countryside

R. Isted
English Nature, Northminster House, Peterborough PE 1 1UA, UK. rebecca.isted@english-nature.org.uk

Abstract
England is home to some of the oldest and widest trees in Europe, frequently concentrated in wood-pasture. Until recently the value of this habitat for wildlife had been largely overlooked. Many wood-pastures are important for their veteran trees and the species within them; however, wood-pastures are frequently rich in other wildlife such as butterflies and birds because they are mosaics of scrub, woodland and grassland. Grazing is a vital part of the conservation management to maintain the habitat. Actions and targets have been identified to promote the conservation of this habitat as part of the UK government's biodiversity action plan. Significant progress is being made towards these targets, including restoration of neglected parkland. This is being achieved through partnership with over 50 government and non-government organizations.

Key words: veteran trees, habitat action plan, grazing

Introduction
England is home to some of the oldest and widest trees in Europe, frequently concentrated in old parkland and wood-pasture. These trees and the landscapes in which they sit are of immense wildlife value, but also of great historical and cultural value. Until recently they were perhaps better celebrated in paintings and literature than in natural history accounts. However, in the last decade, this was changed through the efforts of the Ancient Tree Forum, the Veteran Tree Initiative and now the Wood-pasture and Parkland Habitat Action Plan led by English Nature (Kirby, 1999; Read, 2000; Goldberg, 2003). These jewels of the English countryside are undergoing a renaissance.

Discussion
Wood-pasture is a general term used to describe situations where trees and shrubs exist, usually at lower density than in conventional woodland, in conjunction with extensive grazing by domestic livestock or deer. They evolved as a way of combining some wood production, or the production of foliage for winter browse, with grazing because the regrowth from lopped trees was beyond the reach of the animals. They include parks such as Moccas in Herefordshire, former wooded commons such as Ebernoe in Sussex and some of the Royal Hunting Forests at Epping, the New Forest or Hatfield Forest. Wood-pastures were also formerly widespread in the uplands.

Most wood-pastures are important for their veteran trees and the wildlife within them: 56 Biodiversity Action Plan (BAP) priority species either depend on or are closely associated with parkland and wood-pasture, including dead wood beetles, moths, lichens and fungi. Another 38 lower plant species rely on the unique habitats provided by the old trees and the landscape they are found in (Scarborough and Smith, 2001).

Wood-pastures are frequently mosaics of scrub, woodland and grassland, with many edges and transition zones. Many butterflies, birds and other species seem to do best in such mosaic conditions. Recent work (Vera, 2000; Kirby, 2003) suggests that these conditions may have been more common in the original forests than had once been thought.

The trees in wood-pastures are some of the oldest living organisms in the UK. Individual trees are named on maps and may be associated with local traditions. In landscaped parks they provide the setting for some of the great houses and may be older than the houses themselves. English Heritage has over 1500 sites listed on its register of parks and gardens (Isted, 2004), but wood-pastures in the shape of commons were also part of the landscape for all people to enjoy. Place names and records in the Domesday Book indicate that wood-pasture was a dominant land use in parts of the country, certainly until medieval times (Hooke, 2003; Jones, 2003).

Conclusions
The current state of wood-pasture/silvopastoralism in the UK
Except for conservation/biodiversity benefits, little grazing land now has to be managed as wood-pasture; there is no longer a need to manage trees alongside pasture to provide winter fodder and shelter. The habitat has been identified as a priority for conservation within the UK as part of the Biodiversity Action Plan (Kirby, 1999). Preliminary research (Smith and Bunce, unpublished) shows that veteran trees do occur in the rest of Europe, but Great Britain tends to have more clusters of trees, in wood-pasture/parkland, than elsewhere, where veteran trees tend to be found as boundary features. Britain also seems to have relatively more really big trees (those over 150 cm dbh) compared to the rest of Europe. So conservation of this habitat is important on the European scale, not just within England or GB.

Significant progress is being made towards the targets in the Habitat Action Plan (Goldberg 2003; Isted, 2004; UK BAP website), including restoration of neglected parkland and reintroduction of grazing to sites which have not been grazed by livestock for 50 years or more.

This is being achieved through partnership with over 50 government and non-government organizations. The National Trust at Fellbrigg Hall near Norwich are part-way into a restoration programme that is reintroducing grazing to what was once a deer park (until the 1770s) and was grazed by cattle until the late 1930s. Approximately 25 hectares of parkland have so far been reinstated. The existing veteran and mature native trees have been released from the closed canopy woodland, so, as well as scattered trees over grassland, there are glades and groves of trees. Once the canopy has been opened up it has been relatively easy to establish a grazing sward and cattle have been reintroduced to two restored areas.

Castle Hill SSSI in North Yorkshire is a relict wood-pasture, growing on a limestone knoll. The abundant dead and fallen timber is noted for its rare beetles. It is surrounded by a Forest Enterprise (FE) plantation called Deer Park. Within the conifer plantation, foresters have discovered a large number of veteran oak trees, many of which were being shaded out and were dying due to the competition from the conifers. FE are now removing the conifers from around the veteran trees to give them the space they need to survive. Full restoration, including returning grazing animals to the wood, is under discussion.

References

Goldberg, E.A. (2003) *Wood-Pasture and Parkland Habitat Action Plan. Third Advisory Group Annual Meeting 31 October 2002.* English Nature Research Report 539. English Nature. Peterborough.

Hooke, D. (2003) English woodlands: historical landscapes and archaeology. In: Rotherham, I.D. and Handley, C. (eds) *Working and Walking in the Footsteps of Ghosts.* The Landscape Conservation Forum, Sheffield, United Kingdom, p. 10.

Isted, R. (2004) *Wood-Pasture and Parkland Habitat Action Plan. Fourth Advisory Group Annual Meeting 10 July 2003.* English Nature Research Report. No. 582. English Nature. Peterborough.

Jones, M. (2003) South Yorkshire's ancient woodlands: past, present and future. In: Rotherham, I.D. and Handley, C. (eds) *Working and Walking in the Footsteps of Ghosts.* The Landscape Conservation Forum, Sheffield, United Kingdom, pp. 11-24.

Kirby, K.J. (1999) Trees, people and profits – into the next millennium: biodiversity and forestry. *Quarterly Journal of Forestry e* 93, 221-226.

Kirby, K.J. (2003) *What Might a British Forest-Landscape Driven by Large Herbivores Look Like?* English Nature Research Report 530. English Nature. Peterborough.

Read, H. (2000) *Veteran Trees: a Guide to Good Management.* English Nature. Peterborough.

Scarborough, H. and Smith, T. (2001) *Biodiversity: Linking the Habitat Action Plan for Wood Pasture and Parkland with the Requirements of Priority and Other Species.* English Nature Research Report 432. Peterborough.

Smith, M. and Bunce, R.G.H. (unpublished) A preliminary assessment of the distribution and abundance of veteran trees in North West Europe. English Nature, Peterborough, UK. BAP Website: www.ukbap.org.uk.

Vera, F.W.M. (2000) *Grazing Ecology and Forest History.* CABI, Wallingford, United Kingdom.

Local capabilities development and silvopastoral intervention in Chiapas, Mexico

G. Jiménez-Ferrer [1], L. Soto-Pinto [1], J. Nahed-Toral [1], T. Aleman [1], B. Ferguson [1], M. Ibrahim [2] and F. Sinclair [3]

[1]ECOSUR (El Colegio de la Frontera Sur), Ganaderia y Ambiente, Carr. Panamericana s/n, 29290, Barrio Maria Auxiliadora, San Cristóbal de las Casas, Chiapas, Mexico. [2]CATIE (Centro Agronomico Tropical de Investigacion y Enseñanza), Apdo. Postal 7170. Turrialba, Costa Rica. [3]School of Agriculture and Forest Science, University of Wales, Bangor, LL57 2UW Bangor, Gwynedd, UK. gferrer@sclc.ecosur.mx, mibrahim@catie.ac.cr, f.l.sinclair@bangor.ac.uk

Abstract

The objective of this chapter is to present the methods followed by researchers, farmers and development institutions to design silvopastoral alternatives in Chiapas, Mexico. This work emphasizes the importance of the following steps for silvopastoral systems research and participatory gestation: diagnosis and design, training, on-farm establishment, monitoring, evaluation and public policy proposals.

Key words: agroforestry, local knowledge, network

Introduction

During the last years, there has been a debate in academic and development forums regarding the necessity of shifting from conventional livestock systems towards more robust production systems able to improve yield in the context of natural resources conservancy (Sánchez and Rosales, 1999). Moreover, technicians, farmers and researchers have underlined the necessity to generate social and technological alternatives, with local farmers' knowledge and experience as a base in joint collaboration with the research and development centres (Sinclair and Walker, 1999).

Silvopastoral systems integrate trees into livestock systems for multiple purposes (fodder, wood, fruit and environmental services); however, despite the abundance of research results on several silvopastoral topics in the south of Mexico, technological adoption by farmers is relatively low. Similarly, specific policies for scaling silvopastoral systems towards a conservative and productive as well as a commercial speciality activity are scarce. The implementation of silvopastoral systems, development of local capabilities and improvement of cooperation mechanisms between farmers, researchers and development agencies are urgent. This chapter presents a description of the methodology and development experience for designing silvopastoral systems in Chiapas, Mexico.

Materials and Methods

The current work was carried out with the participation of the following partnerships: farmer groups of three region of Chiapas (CIOAC (Central Independiente de Obreros Agricolas y Campesinos), FIPI (Frente Independiente de Pueblos Indios) and ARIC Selva (Asociacion Rural de Interes Colectivo de La Selva); research institutes (ECOSUR, UNACH-University of Chiapas), INNSZ (Instituto Nacional para la Nutricion "Salvador Zubiran"), CATIE (Centro Agronomico Tropical de Investigacion y Enseñanza); and governmental agencies in Chiapas (SDR (Secretaria de desarrollo Rural)). The main thrust of this integrated development was to adopt a different focus from the conventional agricultural research. This research considered the necessity to approach diverse ecological, technological and social aspects of production systems, using a participatory methodology, in which several research stages were jointly covered by researchers and farmers to design agroforestry interventions in order to improve the local production systems.

Results

Methodological phases were defined as follows: (a) social, entailing (b) diagnosis of livestock production systems; (c) training courses (d) survey of local knowledge on fodder trees and silvopastoral practices; (e) design and on-farm establishment of silvopastoral interventions through prototypes; (f) negotiation of silvopastoral alternatives; (g) launching of the Silvopastoral Network; (h) monitoring and evaluation; (i) promotion of public policies. Table 1 shows a synthesis of the methodological phases, tools and products. The central contributions of this process were: (1) planning actions in the community and regional scales with national and international collaboration, (2) characterization and identification of the main constraints and potentialities of several livestock production systems, (3) ethnobotanical and chemical studies of fodder trees and shrubs incorporating local knowledge, (4) design of silvopastoral prototypes for several regions of Chiapas, (5) promotion of silvopastoral projects on local scale, (6) launch of the Silvopastoral Network between producers, researchers and agencies of development, and (7) baseline document for the elaboration of the Silvopastoral Plan for Chiapas (Aleman and Ferguson, unpublished).

Table 1. Methodology outline for research and development of silvopastoral systems in Chiapas, Mexico.

Phase 1. Actions on planning
(A) Information exchange between partnerships.
(B) Definition of the objectives.
(C) Collaboration between researchers, farmers, development agencies and financing institutions.
Tools Participatory workshops and collaboration agreement signature
Phase 2. Diagnosis and design of agroforestry alternatives for the livestock systems, training courses.
(A) Diagnosis of the livestock systems in Chiapas State.
(B) Training courses on agroforestry and silvopastoral potential, prototypes and design.
(C) Ethnobotanical study of fodder trees and shrubs. Evaluation of the fodder species potential: (i) local knowledge, (ii) determination of the chemical composition and studies on tree foliage digestibility and palatability.
(D) Design and on-farm establishment of silvopastoral prototypes for three livestock systems: (i) livestock-forest for highlands; (ii) livestock system for lowlands; (iii) sheep system for highlands.
Tools Training, diagnostics, workshops, strategic research (on-farm)
Phase 3. Alternatives and negotiation
(A) Collective consensus (between indigenous) for the development of silvopastoral projects.
(B) Development of projects for cattle and sheep in highlands and lowlands
(C) Negotiation of projects between development agencies, farmers and research institutes
Tools Project formulation and workshops
Phase 4. Monitoring and evaluation
(A) Analysis of social and technological restrictions.
(B) Measurement of social variables: empowerment, organization, participation, new knowledge and capabilities, new expectations, inter-institutional collaboration.
Tools Workshop, fieldwork, questionnaires
Phase 5. Silvopastoral network and public policies
(A) Improvement of network between farmers and development agencies (Northern Alliance for Silvopastoral Systems) (La Selva Livestock Alliance)
(B) Launch of the Network between research institutes in Chiapas
(C) Management plan for livestock production in Chiapas, Mexico (producers, researchers, governmental institutes and international centres)
Tools Participatory workshop, agreement collaboration and state plan

Conclusions

This experience has helped generate new capabilities at different levels. It creates an opportunity in southern Mexico to realize the introduction of silvopastoral systems. Farmers have new expectations, space for inter-institutional interactions, participatory roles and empowerment to negotiate further projects. It has allowed research centres to play new social roles in solving some aspects of the current ecological and socio-economic crises in the region. Governmental institutions can now incorporate the views of farmers and researchers in the public policy and planning process.

References

Aleman, T. and Ferguson, B. (Unpublished) *Working Document for Farming and Environment Planning with Focus on Silvopastoral Systems in Chiapas* (Documento base para el Plan Rector de Ganaderia y Medio Ambiente con enfasis en sistemas silvopastoriles en el Estado de Chiapas). El Colegio de la Frontera Sur, Universidad Autonoma de Chiapas, Instituto Nacional de Investigaciones Agricolas y Forestales, San Cristobal, Chiapas Tuxtla Gtz., Chiapas, Mexico. (In Spanish.)

Sinclair, F.L. and Walker, D.H. (1999) A utilitarian approach to the incorporation of local knowledge in agroforestry research and extension. In: Luck, L.E., Lassoie, J.P. and Fernandes, E.C.M. (eds) *Agroforestry in Sustainable Agricultural Systems*. CRC Press, New York, USA, pp. 245-275.

Sánchez, M.D. and Rosales, M. (1999) Agroforestería Pecuaria en América Latina. In: *2ª Conferencia Electrónica sobre Agroforestería Pecuaria*. FAO, Rome, Italy. 515 p.

Silvopastoralism as a land-use option for sustainable development on grassland farms in Northern Ireland

J. H. McAdam
Dept. of Applied Plant Science, Queens University Belfast and Applied Plant Science Division, Dept. of Agriculture and Rural Development, Newforge Lane, Belfast BT9 5PX, UK. jim.mcadam@dardni.gov.uk

Abstract
In Northern Ireland, most (78%) of land is in pastoral agriculture and only 6% (lowest in the EU) in forestry. Grassland intensification as a result of reseeding, fertilizer application and increased stocking has resulted in pasture with low biodiversity and nutrient leakage into watercourses. Farms are family owned (mean 35 ha) and there is a need for diversified income sources. Within such a land-use scenario there is a need to increase tree cover, especially in a sustainable fashion, on grassland farms and create multifunctional land-use options. Silvopastoral systems where protected *Fraxinus excelsior* trees have been planted at 400/ha into intensively managed pasture grazed by sheep were established. Such systems have proved to be sustainable, have permitted farming to continue within the silvopasture and have generated income from the forest products.

Silvopasture will not replace farm woodlands and little has been planted but when farmers are shown silvopasture in practice uptake is enhanced. The way ahead is to offer options which highlight short- and medium-term outputs, exploit environmental and welfare benefits and are attractive to the organic and rural community sectors.

Key words: grassland intensification, *Fraxinus excelsior*, rural development

Introduction
In Northern Ireland, 78% of the land is in agriculture and of this only 5% is cropped, the remainder being pastoral. Tree cover is low (6%) and broadleaved woodland is 1-2%. Pasture is largely managed intensively, biodiversity is low and nutrient leakage into watercourses is a major problem. Farms are small (35 ha) and largely stock-rearing. It is widely recognized that part of the solution to current income problems must come from farmers' willingness to diversify into other land-use types which can deliver the diversity of products under the necessary controls which the public and government wish to pay for. EU policy is to decrease levels of livestock output, tighten nutrient management on farms, increase tree cover, stabilize rural communities and enhance biodiversity through a more sustainable and lower input agriculture. The industry might develop in two groups of agricultural production: 'competitive pillar' - a relatively intensive agriculture industry competing on world markets in a strictly business-orientated method of raw material/food production; 're-creational pillar' - conservation, amenity, recreation and environment (CARE goods) - state subsidy aimed at producing CARE goods through funding to farmers/landowners. The 'bridge' between these two pillars is rural development policy, which can provide benefits in both areas. The integration of trees onto farms and into livestock systems (silvopasture) at a range of scales and levels offers a strategic policy option to realize some of these goals.

Trees on farms in Northern Ireland
Research on silvopastoral systems to date has concentrated on quantifying production, and fewer resources have been directed towards the investigation of ecological interactions (Crowe and McAdam, 1999). In 12-year-old silvopastoral systems, output was only marginally reduced 11 years after planting ash at 400 stems/ha. Overall compatibility with farming systems was better than predicted. Trees in farmland have been shown to create systems which are sustainable in the long term, and produce a type of habitat that enhances biodiversity, particularly of invertebrates, birds and plants. Research on livestock in silvopasture has shown that stock also take advantage of the shade and shelter (Hislop and Claridge, 2000). By using flexible, sensitive design, a range of tree species and varying planting densities, silvopastoral systems can be created that will have positive landscape and environmental benefits. There are strong indications that such systems may allow less nutrient runoff and require fewer pesticides than conventional grassland and arable systems.

Agriculture has come through a series of crises recently and farm incomes are currently severely depressed. In difficult times farmers generally concentrate on short-term goals and needs, longer term needs being much less attractive. This tends to severely limit the opportunity for innovative long-term planning. The needs which can be justifiably met by planting trees tend to be longer term. However, currently farmers are by necessity concentrating on the short- to medium-term goals. Although this fact has always been recognized as major drawback to farmer investment in woodland-related enterprises, it would appear that this limitation is particularly strong at present. The position of trees in silvopasture becomes even more difficult as it is viewed as an unproven technology in a range of woodland options which are already considered as limited in achieving short- to medium-term goals. Speculating on the potential for agroforestry planting in Northern Ireland, given the current state of the industry, it is likely silvopasture will substitute for, rather than complement, current or proposed woodland planting (McAdam *et al.*, 1999). If the potential area of tree planting in Northern Ireland arises only through substitution of conventional broadleaved woodland planting, then, dependent, on the level of substitution, only a very small area (50 ha) of silvopasture would potentially be planted per annum. Hence substitution of candidate land for farm woodland is clearly not the route to follow to encourage adoption of silvopasture.

The target audience for silvopastoral development in Northern Ireland would appear to be those farmers and landowners included in the 're-creational pillar' category, some organic farms and certain conservation organizations. It may be a waste of time targeting large, intensive farming units, which are unlikely to take up the system, apart from where nutrient management may be a key issue. The challenge remains to develop systems which yield a short- to medium-term product from the woodland component without excessively compromising the agricultural component of the system. The potential of woody crops for energy and fibre should influence these considerations. The markets for fruit, foliage, trees which produce medicinal and cosmetic products and landscape trees may represent areas for business development but are highly niche opportunities which are difficult to promote in a generalized way. Opportunity for bioremediation may exist in tree/crop or tree/pasture systems. Hence the potential target audience would be: farmers and landowners in the 're-creational pillar' (of the CAP) category; all farmers interested in agri-environment measures; conservation bodies and community groups; farmers with specific nutrient management problems, e.g. riverside and general bioremediation scenarios (dirty water, sewage sludge, slurry, etc.); the organic sector (McAdam and Crowe, 2002).

Conclusions

In the light of current policy directions in areas such as part-time farming, environmental issues, need to diversify farming systems in sympathy with the environment, decoupling and the mid-term review of the CAP, there is a need to increase the utilization of trees within the rural landscape generally and within the farming industry specifically. Timber produced might be utilized for high value-added products and for farmland enhancement. It is likely that silvopastoral systems will be able to offer added value in terms of sustainability, environmental benefit or CARE goods. In the current climate, planting trees at a range of levels will not appeal to large, intensive farming units. Any tree planting strategy should include a range of options which highlight the short-medium-term outputs possible, should highlight the environmental benefits and animal welfare generated, must align with requirements of agri-environment schemes and be attractive to rural community groups and the organic farming sector.

References

Crowe, S.R. and McAdam, J.H. (1999) Silvopastoral practice – on farm agroforestry in N. Ireland. *Scottish Forestry* 53 (1), 33-36.

Hislop, M, and Claridge, J. (2000) *Agroforestry in the United Kingdom.* Bulletin 122, Forestry Commission, Edinburgh, United Kingdom

McAdam, J.H., Thomas, T.H. and Willis, R.W. (1999) The economics of agroforestry systems in the United Kingdom and their future prospects. *Scottish Forestry* 53 (1), 37-41.

McAdam, J.H. and Crowe, S.R. (2002) *An Agroforestry Strategy for Northern Ireland. Farming with trees: New Optios for Short-Term Profit.* United Kingdom. Agroforestry Forum, Annual Meeting, Royal Agricultural College. Cirencester, United Kingdom.

The potential for agroforestry in the Falkland Islands

J. H. McAdam
Dept. of Applied Plant Science, Queens University Belfast and United Kingdom Falkland Islands Trust, Newforge Lane, Belfast BT9 5PX, Northern Ireland, UK. jim.mcadam@dardni.gov.uk

Abstract
In the Falkland Islands (UK dependent territory; lat. 51-52°S, long. 57-61°W) land use is almost exclusively sheep grazing and the vegetation is acid moorland with no indigenous tree cover. Shelter from the strong cold winds would be extremely advantageous to sheep flocks and might allow crops to be grown in a diversified rural economy. Such trees might be integrated in silvopastoral systems. Previous experience with more widespread tree planting was largely unsuccessful and pessimistic and in 1989 the UK Falkland Islands Trust (UKFIT) commenced a programme of research into factors affecting the establishment of conifers to act as initial windbreak protection in shelterbelts. A trial was established on a grassland and a heath site comparing the effect of ground preparation and nutrition on establishment and early growth. Since then the Department of Agriculture (DoA) has established five shelterbelts. There is a need for high-quality, wind-firm seedlings and work has commenced on a research programme. The trials have demonstrated that trees can be successfully established into native grassland in the Falkland Islands and that shelter is likely to have a significant impact on pasture production. The resultant soil improvement is likely to improve the chance of growing improved forage species and impact significantly on land development. The original trees planted in 1989 are now at a stage where stock might be introduced in a silvopastoral system where shelter and nutrition could be enhanced.

Key words: shelterbelts, *Pinus contorta*, pasture production, sheep grazing

Introduction
The Falkland Islands have a cool (2°C-9°C temp. range), dry (max 750 mm/year) climate with exposure to salt-laden winds, high levels of incident sunshine and periodic dry spells. The soils are acid peats with low levels of mineralization and poor drainage in many areas. Vegetation is dwarf shrub heath or Magellanic moorland and trees are virtually absent (Summers and McAdam, 1993). The flat or gently undulating landscape results in little opportunity for shelter from the cold winds (McAdam, 1985). The inherently low soil fertility and the poor quality and productivity of the natural pastures result in levels of sheep production which are low due to the poor nutritional base and adverse climatic exposure (McAdam, 1985). Agriculture has been traditionally dominated by sheep from large wool-producing farms operating at a low level of input and output and farm ownership has been restructured towards smaller, family operated units since the 1980s. There is a need to diversify the range of agricultural production and to consider wider land-use issues in relation to a diversified rural economy. Trees, by enhancing soil quality, providing shelter and producing an industrial product for local use are an essential prerequisite to this view of an expanded rural economy. There are currently only a few hectares of established woodland on the Islands, largely as a result of soil and climate limitations, unenthusiastic advice and the availability of planting material of suitable provenance (Low, 1986). Strategically placed shelter around settlement farms or in breeding stock camps could be used over the critical periods of lambing and shearing to make a very significant impact on lamb survival and on sheep recovery after stress. The production of beef cattle is to be promoted following the decision to build an abattoir, and improved pasture and cereal or fodder cropping will be needed to sustain this stock.

There have been sporadic attempts to establish trees over the past 80 years or so (McAdam, 1982; Low, 1986). These have been largely unsuccessful (with a few exceptions) up until a programme was initiated by the UK Falkland Islands Trust in 1989 (McAdam, 1996). This chapter summarizes trials to establish introduced conifers into pasture as shelterbelts and an ongoing programme of tree improvement and integration into silvopastoral systems.

Materials and Methods
In an initial pilot trial, tree planting and establishment were investigated (by UKFIT) on two sites, a shrub grassland and a shrub heath. Lodgepole pine, *Pinus contorta*, was either slit (or notch) planted or pit planted (small pit dug to disturb the soil). Small plots of trees received either no fertilizer or phosphate only, or kelp compost (in the pit).

Subsequent to the above trial, in 1995 the DoA planted five shelterbelts on farms at a range of locations. These were approximately 2 ha each in mainly coastal provinces, using lodgepole pine (Low and Kerr, in prep.). In light of the success of the above trials, a programme to investigate tree seedling production and species selection was carried out. Trees (*Salix, Eucalyptus, Pinus contorta, P. radiata,* and *Nothofagus*) were grown either bare root or in a range of containers and planted out in the Falklands in spring 2003.

Results and Discussion
Pilot trial. Growth on the grassland site was better than on the heath site and overall tree growth was significantly greater from pit-planted trees than slit-planted trees (McAdam, 1999). Poorer sites responded to phosphate but the effects had gone after 5 years. Composted seaweed in the pit at planting enhanced tree growth by approximately 5 cm

over the critical establishment period (McAdam, 1999). Shelterbelts were successfully established on all sites, and confirmed the importance of provenance selection (Low and Kerr, in prep.). A booklet on shelterbelt establishment resulted (Low and McAdam, 2001). *Seedling production.* By containerizing willow cuttings on transport from Europe they were able to be successfully established (Olave *et al.*, 2003). Containerized trees always established and grew better than bare root trees. Although it will take some time for the real benefits of shelter to be demonstrated, and the improvement in soil quality which will allow improved forage grasses, legumes and cereals to be grown, the indicators are that trees can be successfully grown in the shrub grasslands of the Falkland Islands and have the potential to make a significant impact on land use. The small pilot programme was greatly expanded by the Department of Agriculture's demonstration shelterbelt trial throughout the Islands. If tree planting is to be more widespread in the future, it is essential to locally produce high-quality seedlings at an economic cost. Now that cuttings can be successfully relocated from the northern to the southern hemisphere and more is known about containerization, advances in local tree seedling production can be made. The integration of the established pines into farming systems can be envisaged at several scales: (a) stock could be allowed controlled access to the original (1989) planting to create close-contact silvopasture or woodland grazing; (b) an area of wide-spaced and protected trees from fast-growing, high-quality stock could be planted in the lee of the shelterbelts and grazed (silvopasture) to create a gradual transition from open pasture to shelterbelt; (c) the shelterbelt could remain as a stock-free area for the foreseeable future with stock/crops utilizing the land up to its boundary (agroforestry). It is recommended that option (c) is the best course of action to follow at present with a strategy to adopt (b) and (a) in the medium and long term, respectively.

References

Low, A.J. (1986) Tree planting in the Falkland Islands. *Forestry* 59 (1), 59-84.
McAdam, J.H. (1999) The potential for trees to improve the shrubby grasslands of the Falkland Islands. *Grasslands Science in Europe* 4, 205-209.
McAdam, J.H. (1982) Recent tree planting trials and the status of forestry in the Falkland Islands. *Commonwealth Forestry Review* 61 (4), 259-267.
McAdam, J.H. (1985) The effect of climate on plant growth and agriculture in the Falkland Islands. *Progress in Biometeorology* 2, 155-176.
McAdam, J.H. (1996) The potential role for trees and forestry in the Falkland Islands. In: *A Report to the United Kingdom Falkland Islands Trust.* United Kingdom Flakland Islands Trust, Westminster, United Kingdom. p. 49.
Low, A.J. and McAdam, J.H. (2001) *Guide to Shelterbelt Establishment in the Falkland Islands.* UKFIT and DoA, Westminster, United Kingdom.
Low, A.J. and Kerr, A.J. (in prep.) Shelterbelt trials in the Falkland Islands.
Olave, R., McAdam, J.H., Dawson, W.M. and Lennie, G.J. (2003) Potential use of containerised willow transplants in the Falkland Islands. In: Dumroese, R.K., Riley, L.E. and Landis, T.D. *Proceedings of Western Forestry and Conservation Conference. Forest amd Conservation Nursery Associations.* USDA Forest Service, Rocky Mountain Research Station. Coeur d'Alene, Idaho, USA.
Summers, R.W. and McAdam, J.H. (1993) *The Upland Goose.* Bluntisham Books, Bluntisham, United Kingdom.

Future perspectives for silvopastoral systems in NW Spain

A. Rigueiro-Rodríguez [1], M. Rois-Díaz [1], M. Pinto [2], A. Oliveira [3] and M. R. Mosquera-Losada [1]

[1]Departamento de Producción Vegetal, Escuela Politécnica Superior, Universidad de Santiago de Compostela, 27002-Lugo, Spain. [2]NEIKER, Basque Country, Spain. [3]Universidad de Oviedo, Asturias, Spain. anriro@lugo.usc.es

Abstract

Spain is divided in two main regions based on climatic parameters: Atlantic and Mediterranean. While silvopastoral systems are well established in Mediterranean areas, they are not widely used in the Atlantic area. The objective of this chapter is to evaluate the possibility of using silvopastoral systems in the Atlantic-influenced area of northern Spain, mainly in new plantations but also in old stands of coniferous and broadleaved species. The chapter describes the potential of silvopastoral systems using a range of species.

Key words: Atlantic mountain region, livestock

Introduction

Spain has a temperate climate, with two different contrasting temperature and precipitation regimes, Atlantic and Mediterranean. The Atlantic area of Spain comprises four northern regions from the west to the east: Galicia, Asturias, Cantabria and Basque Country, which are characterized by a short dry summer and high relative humidity with annual precipitation above 600 mm and temperatures with low contrast between seasons due to the maritime influence. Galicia and the Basque Country have around 32% and 5% of their area above 600 and 1000 m asl, respectively, but this percentage reaches 50% for the other two regions, where about 20% is higher than 1000 m asl. Atlantic Spanish regions account for 15% of the population, 10% of the national land area and around 16% of the total forest and other woodland areas of Spain.

From the coast up to the mountains natural tree cover is mainly mixed stands of different *Quercus* species, *Betula alba* and *Castanea sativa*. *Fagus sylvatica* appears in the upland forests. However, this area was dramatically deforested in the last thousand years and an important effort was made during the last century to afforest mountain and another agricultural land, mainly with fast growing species, like *Pinus radiata*, *Pinus pinaster* or *Eucalyptus globulus*.

Materials and Methods

The objective of this study was to evaluate the current state of the main forest and farm systems found in the Atlantic-influenced area of Spain, taking into account the national (INE, 2004) and regional inventories (Xunta de Galicia, 2001 and 2002). The forest and its associated livestock farms will be described, a silviculture description for the main groups of species will be made and the role of silvopastoral systems in the socio-economic context of the north Atlantic-influenced area of Spain will be discussed.

Results and Discussion

In all the regions studied, an important increase in forestland had occurred by the end of the 20th century, and in 1996 more than 35% of the area of these regions was occupied by forest (Table 1). In Galicia around 17% of the forest area (195,000 ha) was burnt in 1989. The importance of the forest in northern Spain can be highlighted by the fact that timber and other forestry products from this area account for around 70% of output of the whole country.

Table 1. Area (ha; percentage over total region area in brackets) in 1996 of coniferous (C), broadleaved (Br) and mixed tree species in the different regions of North Spain (Second National Inventory).

1996 Species group	Region			
	Galicia	Asturias	Cantabria	Basque Country
Br	103,607 (3%)	208,509 (20%)	133,348 (25%)	153,569 (21%)
C	378,380 (13%)	97,933 (9%)	23,183 (4%)	173,419 (24%)
Mixed	563,389 (19%)	61,687 (6%)	9,013 (3%)	63,017 (9%)
Total	1,045,376 (35%)	368,129 (35%)	165,544 (32%)	390,005 (54%)

The evaluated area produces almost 80% of Spanish broadleaved forest products (76% *Eucalyptus globulus*) and over 60% of Spanish coniferous forest products (mainly *Pinus pinaster* in Galicia and *Pinus radiata* in the Basque Country) (Figure 1). The main coniferous species, taking into account the occupied area in the evaluated regions are *Pinus pinaster* (56%), *Pinus radiata* (25%) and *Pinus sylvestris* (9%); and main broadleaved species are *Quercus* sp. (26%), *Eucalyptus* sp. (20%), *Fagus sylvatica* (15%) *Castanea sativa* (8%) and other broadleaved species (30%) like *Betula alba*, *Prunus* sp., etc.

One of the main environmental risks of this region is forest fires. In the last decade, Spain was the European country most affected by fire, with one-third of the forest area burnt (UNECE/FAO, 2002). In the north, mainly in

Galicia, burnt forest area accounts also for 40% of Spain. This is mainly due to the annual rainfall and temperature distribution, which allows very good shrub growth during the spring, and there is a strong drought during the summer, which converts the vegetation into fuel that burns easily.

Figure 1. Forest production (total m³ wood expressed as percentage of total production of Spain) in Atlantic Spanish regions per forest type.

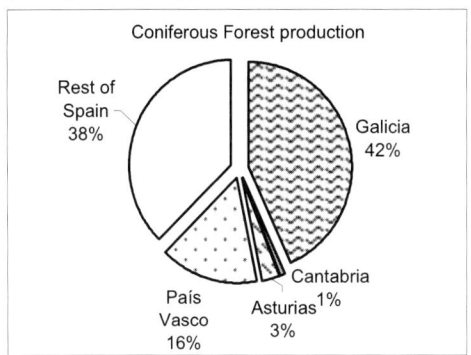
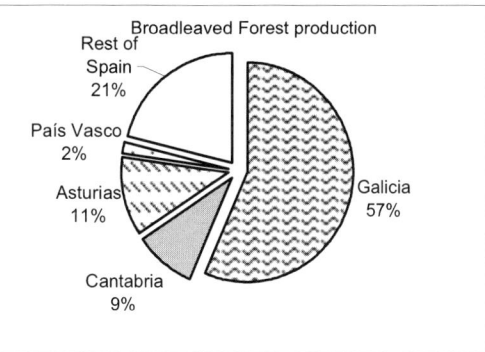

From an animal point of view, 26% of the cows and horses of Spain are in the Atlantic area, the main products being milk (50% of the whole of Spain) and meat (23 and 18% of beef and horse meat, respectively, of those produced in Spanish production). It can be seen that Atlantic northern Spain is an important provider of forest and animal products, the latter mainly from grazing.

In the past, an important part of forest area was used for silvopasture, through thinning of mixed oak, birch, chestnut and beech areas. Forest species composition has broadly changed over the last century. In Galicia and the Basque Country the forest consists of eucalypts and coniferous species and in Cantabria and Asturias broadleaves (with a very low proportion of eucalypts on level areas). These have been protected by the government mainly in the mountain areas (more than 10% of the land area).

Coniferous and eucalyptus stands are very susceptible to fire. Hence it is necessary to reduce the fuel load from the understorey vegetation in these areas, and a high proportion of the budget of the regional governments is dedicated to fire prevention (mainly clearings, firebreaks, fire detection) and control. Initial planting density in these stands is usually high. Management of eucalyptus stands does not include thinnings (crop forest product) and animals like horses (which will eat woody plants) are currently used to prevent fire risk. Thinning is usually planned for *Pinus* species, but unfortunately it is not carried out in most cases because of the high price of the thinning. Hence, in many cases stands are too dense or the timber product quality is reduced due to the harvest of the best trees to increase the thinning profits. Thinning will increase light input into the forest and it can increase understorey pasture production, and therefore animal production, which can increase land profitability, improving land return from a forestry (individual tree growth is enhanced) and agronomic (pasture production is enhanced compared to unthinned stands) point of view. During the regeneration period special care should be taken to protect the trees or to control access of animals to the new area. The stand age when animals are allowed to graze in young stands depends on the tree and animal species. For example, goats should be introduced much later than sheep, cows or horses. It also depends on the growth rate of tree species: *Pinus sylvestris* grows much more slowly than the other pines which are found in the region.

Most of the broadleaved tree species (with the exception of eucalypts) are in a poor silvicultural state. They need thinning, but with low intensity, which would not allow too much pasture growth in the understorey. However, grazing with wildlife animals in these stands at a low stocking rate will increase their biodiversity, and also allow the preservation of some native livestock breeds like pigs, goats, cows, sheep or chickens. These can also be fed with fruits from the trees. In this sense, silvopastoralism could be used as a way of increasing cover of broadleaved species through new forest establishment. Some species can be planted at a very low density. This will not affect pasture production for many years, will increase it during the summer, but will allow animal husbandry to proceed and therefore a return to be made from the land in the short term, enhance animal welfare and promote biodiversity. Genetically improved forest plant varieties should be made available to guarantee well-formed trees for harvesting. More research should be carried out to develop adequate planting densities and management plans for this type of silvopastoral system in Spain.

Conclusions

The Atlantic regions of northern Spain are lands that can enhanced the use of silvopastoral systems which utilize fast and slow growing tree species. Such systems will enhance the stability of rural populations and increase profits from the land.

References

Xunta de Galicia (2001). Galician forests in figures (O monte galego en cifras). Dirección Xeral de Montes e Medio Ambiente Natural. (In Galician). http://www.xunta.es/conselle/cma/CMA07g/CMA07ga/O%20Monte%20Galego%20en%20Cifras.PDF

Xunta de Galicia (2002) Agriculture Statistics (Anuario de Estadística Agraria). Consellería de Agricultura. http://www.xunta.es/conselle/ag/public/aea2000/indice.htm

INE (2004) INE base. http://www.ine.es/inebase/menu4.htm#18

UNECE/FAO (2002) Forest Fires Statistics. http://www.unece.org/trade/timber/ff-stats.html (accessed April 2004)

Outdoor pig production in the Basque Country (Spain)

R. Ruiz [1], A. Domingo [2] and L. M. Oregui [1]

[1]NEIKER A.B., Apdo. 48, 01080 Vitoria-Gasteiz, Spain. [2]Txerrizaleok Elkartea, Pza. Simon Labayen s/n, 20490 Albiztur, Guipúzcoa, Spain. rruiz@neiker.net, txerrizaleok@yahoo.es

Abstract

In 1999 a group of butchers and farmers with limited experience in home-grown pig fattening became interested in the production of pig meat from an outdoors system. The goal was to provide the Basque market with higher-quality products. An R+D project was carried out to design an outdoor pig production system adapted to the local conditions. This involved taking into account the interests and opinions of farmers to assess production parameters, and butchers and final consumers in terms of quality. The cross of Large White-Landrace/Duroc (LWLD) offered the best compromise between the productive objectives of the farmers, and the higher-quality meat requirements of both butchers and consumers. Hence LWLD was chosen to develop an outdoor fattening system and a farmers association, "Txerrizaleok elkartea", was formed. The present chapter describes the current situation of the sector, and its implications.

Key words: productivity, meat quality, implications, silvopasture, forest grazing

Introduction

Livestock farming in the Basque Country is mainly based on systems whereby dairy cows, sheep and beef cattle make use of natural and improved grasslands to a great extent. However, around 50% of the land is covered by forest, either natural (*Fagus, Quercus*, etc.) or for commercial purposes (pine woods). Nowadays, livestock make little use of them, although they might be a potential resource for animal feeding.

On the other hand, the demand for higher-quality and safety meat has been notable during the last decade, especially for beef meat. This fact was perceived by some farmers and local butchers as an opportunity, and they expressed their interest in developing outdoor pig production systems. Hence, a project was carried out in 2000 to assess the possibilities of an outdoor fattening system (Ellis and McKeith, 1993; López-Bote *et al.*, 2000), and its implications for carcass and meat quality (Gispert and Diestre, 1996; Serra *et al.*, 1998).

As a consequence of the project, a cross-breed coming from Large White, Landrace and Duroc offered the best compromise between the productive objectives of the farmer and the higher-quality meat requirements of both butchers and consumers (Ruiz *et al.*, 2003). So it was decided to develop an outdoor fattening system for the conditions found in the Basque Country. To promote this activity as a complementary alternative, an association called "Txerrizaleok elkartea" was formed in October 2001.

Current situation

Nowadays, "Txerrizaleok elkartea" encompasses 16 farmers (nine in Gipuzkoa and seven in Bizkaia) and two technicians responsible for the suitable management of the association. They produce and commercialize around 3000 pigs and 340,000 kg meat per year (55 pigs per week), having an annual budget of around 1 million €. The objective of the association is to obtain high-quality meat and derived products and the joint commercialization of the pigs reared in small or medium-sized farms under outdoor conditions.

Before joining the association, each farm has to be evaluated and approved by the technical staff, who will later exert a strict control throughout the productive cycle in order to guarantee the fulfilment of the regulations previously accepted by the farmer. Farmers must be able to produce every year at least two cycles of two lots of 50-60 animals each. They have to follow the "Regulations of production conditions" and "Regulations to achieve high-quality meat" agreed by the association. Basically they refer to the set of measures related to basic instalments and stocking rates, origin and genetic features of the piglets fattened, land management, characteristics of the feedstuff used, health issues, etc.

The land must be parcelled and adequately fenced to allow producing differentiated lots, offering a minimum availability of 300 m^2 land per animal. Vegetation ranges from natural grasslands in quite flat areas, to a mixture of stony patches and forest (*Fagus, Quercus, Acaeciae, Pinus,* etc.) with a high degree of shrubby cover on steep slopes. Once a lot is slaughtered, patches will remain empty until optimal conditions are restored.

Certain standard instalments were agreed to provide better conditions for the pigs. Basically they consist of an open shelter, well ventilated, at least 2 m high, offering a minimum of 0.8 m^2 per animal. Concrete floor must be placed in and out, overpassing at least by 3 m the feeding and watering areas to avoid mud formation.

Piglets are transferred to the farms with 20-25 kg live weight (LW); they are individually identified with ear tags and ringed in the nose. Then they are managed under a free-range system and a cereal-based diet is provided *ad libitum* until they reach around 150 kg LW, after about 5 months.

After slaughtering pigs at around 150 kg live weight, every carcass is assessed by the technical staff of the association, who will determine which ones fulfil the minimum quality standards fixed within the regulations of the Basatxerri label, and will adequately seal and market them. The remaining ones will be sold without the label.

Txerrizaleok jointly sells the fresh meat coming from the pigs, as well as some derived products (cured ham or foreleg, cooked ham, bacon, spicy sausage, salami sausage, paté, cured loin, etc.), with the same label, Basatxerri, in the butcheries ascribed to Eusko Label.

Benefits of the system

This outdoor system provides the local market with a differentiated meat product (Lázaro *et al.*, 2002; Ruiz *et al.*, 2003). These systems are usually much better perceived by consumers in terms of animal welfare (Gentry *et al.*, 2001). It allows the use of natural resources growing in forested, sloping and/or underused areas. It is regarded as a part-time alternative for less-favoured areas, with low requirements in terms of labour (on average less than 30 min/day) and interesting economic profitability when suitably managed. In a particular farm, the added income coming from this activity allowed one member of the family (the wife) to stop working outside the farm. In the future, this activity can play a key role in the utilization of forest areas in the Basque Country.

Constraints and requirements for research

The association has developed a "Plan for the Management of Pig Slurry" and a "Plan for Rotation Management of the Land", which are personalized for each farm. However, they must be improved, as there is still limited knowledge available in relation to some issues: importance of grazing in the fulfilment of nutritive requirements in pigs managed under these conditions, effects on natural vegetation, optimum reseeding strategies, N excretion, determination of optimal stocking rates, definition or improvement of the rotation practices, possibilities of erosion in sensitive areas (access from feeding areas to patches), etc.

References

Ellis, M. and McKeith, F.K. (1993) Factors affecting the eating quality of pork. In: Hollis, G.R. (ed.) *Growth of the Pig*. CAB International, Wallingford, United Kingdom, pp. 215-239.

Gentry, J.G., Miller, F.M. and McGlone, J.J. (2001) Alternative production systems: influence on pig growth and pork quality. In: *2nd International Virtual Conference on Pork Quality via Internet*, 5 November to 6 December. http:\\www.conferencia.uncnet.br/pork/seg/ pal/anais01p2_mcglone_en.pdf

Gispert, M. and Diestre, A. (1996) Harmonization of the classification methods of pig products in the EU (Harmonización de los métodos de clasificación de canales porcinas en la UE). *Eurocarne* 51, 45-50. (In Spanish.)

Lázaro, R., Toldra, F., Ferrer, J.M., Silió, L., Rodríguez, M.C. and López-Bote, C.J. (2002) Effect of exercise on the activity of proteolytic enzymes in skeletal muscle and carcass quality of Iberian pigs. *Journal of Agriculture Science* 80 (1), 214.

López-Bote, C., Fructuoso, G. and Mateos, G.G. (2000) Pork production systems and meat quality: the Iberian pig (Sistemas de producción porcina y calidad de la carne. El cerdo ibérico). In: *XVI Curso de Especialización FEDNA (Fundación Española para el Desarrollo de la Nutrición Animal)*. Escuela Técnica Superior de Ingenieros Agrónomos, Madrid, Spain. (In Spanish.)

Ruiz, R., Oregui, L.M. and Domingo, A. (2003) *Design of an Outdoor Pork Production System in the Basque Country Region* (Diseño de un sistema de producción de porcino al aire libre en la Comunidad Autónoma del País Vasco). ITEA, Zaragoza, Spain. (In Spanish.)

Serra, X., Gil, F., Perez-Enciso, M., Oliver, M.A., Vázquez, J.M., Gispert, M., Díaz, I., Moreno, F., Latorre, R. and Noguera, J.L. (1998) A comparison of carcass, meat quality and histochemical characteristics of Iberian (Guayerbas line) and Landrace pigs. *Livestock Production Science* 56, 215-223.

Technology transfer in silvopastoral agroforestry: a toolbox for UK farmers

A. R. Sibbald
The Macaulay Institute, Craigiebuckler, Aberdeen AB15 8QH, United Kingdom. a.sibbald@macaulay.ac.uk

Abstract
Silvopastoral agroforestry has not been taken up by many farmers in the UK, partly because government funding has not been available and partly because many UK farmers have no experience of such systems or of managing trees. Changes in the EU's Common Agricultural Policy towards more biodiverse, environmentally friendly and sustainable land-use systems should result in future funding for silvopastoral agroforestry. This will still leave the problem of the lack of experience of UK farmers. Consequently, a browser-based, one-stop shop of information about silvopastoral agroforestry and managing silvopastoral systems has been constructed. The content of the toolbox was developed at a workshop using a soft-systems approach involving farmers, advisers and researchers. The workshop produced the information that farmers would require to assess if silvopastoral agroforestry was suitable for their farm and to set up and manage a silvopastoral system. The browser-based toolbox allows farmers and their advisers to easily access this information.

Key words: silvopastoral, agroforestry, technology transfer, toolbox

Introduction
Silvopastoral agroforestry has not been taken up by many farmers in the UK, partly because government funding has not been available, partly because many UK farmers regard agriculture and forestry as mutually exclusive enterprises (Crowe and McAdam, 1999), meaning that few farmers have experience of managing trees on their land, and partly because advice on the management of silvopastoral systems has not been readily available.

Silvopastoral systems have been shown to be more biodiverse (Burgess, 1999) and more environmentally friendly (Lehmann et al., 1997) than conventional livestock systems and to be sustainable in the widest sense of the term (Sibbald, 1999). Changes in the EU's Common Agricultural Policy towards more biodiverse, environmentally friendly and sustainable land-use systems and the decoupling of production support should result in future financial support for silvopastoral agroforestry (see McAdam, in this volume chapter 2). However, this will still leave the problem of the lack of experience of UK farmers, highlighting the need for a readily available silvopastoral information system for farmers.

Materials and Methods
The content of the information system was developed at a workshop using the soft-systems, rapid appraisal method (Van Beek and Nunn, 1995). The participants included: farmers who had silvopastoral systems on their farms; owner-occupier farmers who had no experience of silvopastoral systems; land managers who had no experience of silvopastoral systems; advisers for both agroforestry and farm woodlands; a professional forester and silvopastoral researchers.

The workshop was held at the Glensaugh Research Station of the Macaulay Institute, where one of the sites of a UK silvopastoral network experiment (Sibbald et al., 2001) is located.

The background information provided to the participants included the experiences of the farmers who had established silvopastoral systems on their farms and a summary of the results from the UK silvopastoral experiment covering agricultural production, tree development, environmental impacts (Burgess, 1999) and economics (McAdam et al., 1999).

The participants visited the silvopastoral research site, which has three different tree species, planted at different densities in grazed pasture 14 years previously. The practical issues relating to tree establishment (site selection, tree protection, use of herbicides) in the presence of grazing sheep, tree management (tree selection/thinning, form pruning) and pasture and grazing management under trees were discussed.

The participants were then divided into smaller groups, facilitated and recorded by researchers, to discuss the following questions:

What information will farmers require in order to decide if silvopastoral agroforestry is an appropriate land use on their land?

What information will farmers require in order to set up and manage a silvopastoral agroforestry system on their land should they decide to go ahead with it?

What is the best format for the information and how should it be made available?

A plenary session was held after each discussion with each group summarizing its views. The issues raised in the discussions were used as the framework for the information system.

Results and Discussion

The following issues arose from the discussions on Question 1:

> What subsidies are available and how are they accessed?
> What are the establishment costs?
> What markets are available for the timber?
> What will be the positive and negative impacts on the current farming system?
> What impacts on farm resources - labour, machinery, cash flow?
> What tree species are appropriate to my site?
> Can I expect agroforestry to provide environmental benefits?

The following issues arose from the discussions on Question 2:

> What species/provenance/size of tree seedling?
> What tree protection system?
> When to plant? How to plant?
> What changes in livestock management will result?
> Tree management during establishment?
> Tree management during later phases - pruning/thinning?
> What training is available on tree selection, management and maintenance?

In discussing Question 3, the participants agreed that most farmers now have personal computers and many are connected to the Internet. However, there are farmers not familiar with information technology who would require access to the information through advisers. The preferred format was, therefore, an Internet- or CD-ROM-based system which would be freely accessible to advisers (farming, forestry and environmental) as well as to farmers.

The answers to the issues from Questions 1 and 2 are, therefore, covered in a browser-based information system, called the Silvopastoral Agroforestry Toolbox. The information in the Toolbox was derived from the outcomes and experiences of the farmers who had practised silvopastoral agroforestry and from the UK silvopastoral network experiment. The Toolbox also draws upon published agroforestry information (Hislop and Claridge, 2000) and information on, for example, site selection for trees (Ray, 2001) and form pruning (Kerr and Evans, 1993). The Toolbox allows farmers and others to get answers to their questions through a linked structure, an example of which, starting at the top level, is shown below:

> How easy is it to manage? > Tree management > Post-establishment management > Pruning for quality timber > What to prune > When to prune > How to prune > Tools for pruning

Conclusions

The Toolbox has been launched as a web site http://www.macaulay.ac.uk/agfor_toolbox/ and as a CD-ROM. The CD-ROM has been widely distributed in the UK and abroad to farmers, advisers, researchers and students of agroforestry systems.

References

Burgess, P.J. (1999) Effects of agroforestry on farm biodiversity in the United Kingdom. *Scottish Forestry* 53 (1), 24-27.
Crowe, S.R. and McAdam, J.H. (1999) Silvopastoral practice on farm agroforestry in Northern Ireland. *Scottish Forestry* 53 (1), 33-36.
Hislop, A. and Claridge, J.N. (2000) *Agroforestry in the United Kingdom.* Bulletin 122, Forestry Commission, Edinburgh, United Kingdom.
Kerr, G. and Evans, J. (1993) *Growing Broadleaves for Timber.* Forestry Commission Handbook No. 9, HMSO, London, United Kingdom.
Lehmann, J., Wulf, S. and Zech, W. (1997) Can trees recover nutrients under high leaching conditions? In: *Agroforestry for Sustainable Land Use.* INRA, Montpellier, France, pp. 155-158.
McAdam, J.H., Thomas, T.H. and Willis, R.W. (1999) The economics of agroforestry systems in the United Kingdom and their future prospects. *Scottish Forestry* 53 (1), 37-41.
Ray, D. (2001) *Ecological Site Classification - User's Guide.* Forestry Commission, Roslin, Midlothian.
Sibbald, A.R. (1999) Agroforestry principles - sustainable productivity? *Scottish Forestry* 53 (1), 18-23.
Sibbald, A.R., Eason, W.R., McAdam, J.H. and Hislop, A.M. (2001) The establishment phase of a silvopastoral national network experiment in the United Kingdom. *Agroforestry Systems* 39, 39-53.
Van Beek, P. and Nunn, J. (1995) *Introduction to Rapid Appraisals.* Systems Studies Series, SyTREC Pty Ltd., Karralee, Queensland, Australia. 36 pp.

Model and procedures for decision-making in management of diffusion, adoption and improvement process of agroforestry technology

M. J. Suárez Hernández [1], G. Hernández Pérez [2] and R. Suárez Mella [3]

[1] Grasses and Forages Research Station "Indio Hatuey", Matanzas, Cuba. [2] Dept. of Industrial Engineering, Central University of Las Villas, Santa Clara, Cuba. [3] Dept. of Industrial Engineering, Matanzas University, Matanzas, Cuba.
chuchy@indio.atenas.inf.cu, ghdez@rectorado.uclv.edu.cu, rogelio.suarez@umcc.cu

Abstract

The study consists of proposing and validating a model and its procedures to back up decision-making, from a strategic perspective, in order to manage the process that involves making known, adopting and improving agroforestry technologies in stockbreeding businesses. The model is made up of six procedures: inventory, linked to an in-company technological audit; security, in relation to the design and putting into practice of a security system; evaluation, assessing the technological wealth of the company and opportunities, which makes it possible to design a technological strategy; increase in wealth, linked to the choice of agroforestry technologies and the valuation of such wealth; optimization of this wealth, on the basis of organizational decisions; and protection of the above through industrial property activities.

Key words: stockbreeding businesses, technology, innovation management

Introduction

The Cuban livestock sector plays a key role in human food production. However, it has not been possible to satisfy the demand, due to insufficient production levels. This demands a paradigm change, in which sustainable technologies and production systems, especially based on an agroforestry approach, are replacing the intensive systems using high external inputs. This situation has been influenced by a lack of training of management and technical staff, scarce funding, incipient mechanisms of technology transfer, low adoption rate, insufficient vision and low product diversification. Another problem is the lack of communication between the scientific and the production sectors, necessary for the development of the country. Many of these problems are caused by the lack of experience about technology and innovation management (TIM), which is considered a key issue for the future of the country. These problems are worsened by the lack of efficient management procedures. These include diagnosis, evaluation and monitoring of technological processes. Technological development plans are needed within a strategy to enrich and improve its technological patrimony and to protect its products and technologies (Suárez, 2002). Thus there is an insufficient innovation performance. That is why it is necessary to develop and validate a general model for decision-making to encourage TIM in the livestock enterprises. Thus the competitiveness of the livestock enterprise could be improved. The general objective of the research was to develop and validate a general application model to support decision-making in Cuban livestock enterprises. This is the first time that such a model has been developed for livestock enterprises.

Materials and Methods

In the past, different models have been developed from the prevailing linear model in the 1960s (the technology-push and demand-pull models) to the currently prevailing models in networks. However, all these models have faults in that they are mainly theoretical-conceptual, although they explain the innovation phenomenon and its causes. However, they did not go beyond a general conceptual level and specific sector solutions were missing. Nevertheless, Morin (1985) made an important contribution by separating TIM into six basic functions. So there is not an absence of models and procedures, but there are no appropriate or pertinent ones for the livestock sector and they do not completely comprise TIM. The general model proposed, as the basis of specific procedures and tools developed from the TIM functions defined by Morin, constitutes the basis for the formulation and implementation of the technological strategy of livestock enterprises. These should be integrated with the process of strategic planning. This model integrates the six functions for managing the technological resources of the enterprise with information feedback between these functions that have been outlined above. The objective is to formulate a technological strategy and technological development plan (TDP). These specific procedures comprised the following aspects:

The inventory function, which is a technological audit and the determination of the innovation intensity in the livestock enterprise (IILF), the evaluation of the excellence level in the livestock enterprise (ELLF) and the satisfaction level of the workers (SLW), as well as the determination of training needs of the management and technical staff.

The monitoring function comprises the design of the technological and competitive monitoring system, which constitutes the key areas of monitoring associated with agroforestry and the information sources.

The evaluating function is linked to the technological capacity (TCLF). This involves the selection of the most appropriate technological options to work with an agroforestry orientation through different matrices and the definition of the technological objectives that should be fulfilled, as well as the formulation of the technological strategy and the elaboration of the TDP.

The enriching function allows selection of the most appropriate agroforestry technologies from the combined use of a multicriteria method developed to evaluate the excellence level of the technologies that are used in livestock enterprises and an adaptation of the factor-location method.

The optimizing function is focused on the fact that the enterprise must have an organizational structure for TIM, but it must be a case decision, which is difficult to classify. However, it must include the definition of the most adequate structure in each case, the centralization level of the technological activities and the hierarchical level of the decision-making associated with TIM.

The protecting function is associated with the fact that the enterprise must design an active policy regarding the rights of the industrial and intellectual property that limit or prevent the uncontrolled use and commercialization by third parties of the innovation generated by the enterprise.

Results and Discussion

The "Hermanos Sartorio" Livestock Enterprise was selected to validate this research and promote the adoption and improvement of agroforestry technologies. There were no functions, responsibilities and, least of all, a structural unit to take up the processes linked to TIM. The function of monitoring was not defined and the legislation about intellectual property was not well known at the beginning of this research (year 2000). There was no inventory of the enterprise technological resources or a technological strategy and there was no systematic follow-up of its innovating performance.

With the application of the model and its procedures, five important agroforestry technologies were selected. The strategic choice of the enterprise was within the area of expertise. However, the strategy of the technological niche was selected as more appropriate, based on the characterization and definition of the technological and competitive position of the enterprise. Another reason for this selection was the attractiveness of a technology that reduces costs and environmental impact, creates higher returns on investments and increases forage production, according to Escorsa and Puerta (1991), based on Arthur D. Little (1981). This strategy has been developed by the enterprise since 2000. The strategic decisions taken were to invest in internal R & D in order to improve and diffuse special management options that were developed by the enterprise in cooperation with the GFRS "Indio Hatuey", which have a high development potential, as well as potential for the improvement of silvopastoral systems. The selection of these technologies resulted in technology excellence levels (TEL) ranging from 69 to 85 (Suárez Mella, 1996). The creative adaptation of the factor-location method, proposed from 17 indicators, also allowed the evaluation of the technological attractiveness of the above-mentioned technologies (achieving values between 847 and 917 out of 1000 points). Considering the joint application of both methods, the adequacy of the decision of including these agroforestry technologies could be proved because they achieved high values in the TEL and in the application of factor-location.

Table 1. Changes in the use of technological resources at the "Hermanos Sartorio" Livestock Enterprise. NE: not evaluated that year.

Use of technological resources	1998	2000	2002	2003
Satisfaction level of the workers (%)	NE	80.58	88.35	NE
Innovating performance	NE	Innovating	NE	Highly innovating
Excellence level in livestock enterprises (%)	37.7	NE	75.8	79.48
Innovating intensity of the enterprise (%)	NE	73.4	NE	88.4
Technological capacity of the enterprise (%)	NE	62 (medium)	NE	86 (high)
Technologies used in the enterprise	14	NE	NE	18
Development of innovations in the enterprise	29	NE	NE	66

On the other hand, other key intangible results were achieved in the enterprise. These include an increase of the R & D capacity and its management, as well as the definition of a technological strategy and the identification and development of essential competences linked to R & D and technological innovation. Furthermore, the generation of three technologies that differentiate it from the rest of the enterprises in the sector and the inventory of the most relevant technologies and the establishment of a monitoring system were made possible. A technological alliance with the GFRS "Indio Hatuey" also contributed, as well as the creation of the position of technology and innovation manager and of a technological committee. These can be regarded as the most adequate organizational forms for TIM, as well as the existence of a programme of moral and material incentives that stimulate innovation.

The application of the model and specific procedures in the enterprise under study are shown in Tables 1, 2 and 3.

Table 2. Improvement of economics indicators selected in the "Hermanos Sartorio" Livestock Enterprise.

Economic indicators selected	1998	2000	2001	2002
Incomes (thousand pesos/year)	6,542	7,409	8,804	10,380
Gains or losses (thousand pesos/year)	(906)	17	439	20[1]
Cost/peso	1.13	0.99	0.95	0.99[1]
Milk production cost (pesos/litre)	0.89	0.88	0.79	0.67
Sale price of milk (pesos/litre)	0.83	0.83	0.98[2]	0.98[2]
Meat cost (pesos/kg)	1.52	1.10	1.47[3]	1.40[3]

[1] The gain decreased as compared to the previous year, as well as the cost, because in 2002 the "Hermanos Sartorio" enterprise assimilated with another enterprise and started to invest in its development.
[2] The sale price increases due to a quality increase, because of the high protein level in the forage of trees.
[3] The cost increased because the animals are fattened in an intensive system, which also increases meat quality and production.

Table 3. Improvement of productive indicators selected in the "Hermanos Sartorio" Livestock Enterprise.

Productive indicators selected.	1998	2003
Daily production/milking cow (litres)	4.4	7.0
Annual production/cow (litres)	947	1,507
Annual production/hectare (litres)	121	192
Total mortality rate (%)	5.1	1.3
Average daily live weight gain (grams)	293	500-600

Through the application of the general model and the specific procedures and tools in the "Hermanos Sartorio" Livestock Enterprise, it was possible to validate them in the decision-making processes of the enterprise associated with the development of agroforestry, as well as to appreciate the tangible and intangible results that allow the transition towards highly innovating and competitive enterprises.

References

Arthur D. Little (1981) *The Strategic Management of Technology.* Arthur D. Little, Cambridge, United Kingdom.
Escorsa, P. and Puerta, E. (1991) La Estrategia Tecnológica de la Empresa. *Economía Industrial* 281, 93-107.
Morin, J. (1985) *L'Excellence Technologique.* Publi Union, Paris, France.
Suárez, J. (2002) El papel clave de la tecnología y la innovación en la ganadería cubana y su efecto en el desarrollo del sector. In: *III Convención Internacional de Educación Superior.* Ministerio de asuntos exteriores, Ciudad de la Habana, La Habana, Cuba.
Suárez Mella, R. (1996) Modelo de evaluación del nivel de organización de la producción en empresas de la industria mecánica. TMSc thesis, ISPJAE, Ciudad de la Habana, Cuba.

DECLARATION FOR SILVOPASTORALISM

Silvopastoralism is an ancient way of managing forestland and a recent way of managing pastureland. It is a type of agroforestry system, fully in line with the global action plan for areas of sustainable development: Agenda 21.

Silvopastoralism can be applied in forest, grazing and arable lands in a sustainable way for diversification and multipurpose land use. It can increase biodiversity, protect the environment, combat and prevent desertification, promote the landscape, improve health and increase rent income in the short, medium and long term for managers when the social, cultural and economic benefits are accounted for. Silvopastoralism promotes land sustainability, integrated land-use management and offers benefits to managers, local communities and the public.

Forest areas: Agenda 21 promotes the maintenance of sustainable forests through conservation and management, by maintaining or restoring the ecological balance and expanding the contribution of forest to human needs and welfare through the utilization of non-timber products. Silvopastoralism can be a way of conserving and enhancing forest protection from fires, and increasing rent incomes and biodiversity. At the same time, it promotes the better use and development of some types of forests and woodlands, including planted forests, through appropriate, environmentally sound and economically viable activities and management of plant and animal species (including autochthonous breeds). It promotes management of wildlife, attractive landscapes, as well as ecotourism, which ensures the adequate participation of the private sector, rural communities, indigenous people, youth and other user groups in sustainable forest management.

Agricultural areas: Agenda 21 promotes the expansion of areas under forest and tree cover, which unfortunately in many cases encourages rural abandonment and necessitates increased public investment for their protection. Promotion of silvopastoralism will increase rent income in the short, medium and long term for managers, therefore taking into account social, cultural and economic aspects which promote land sustainability. Silvopastoralism prevents rural abandonment, enhances more viable rural communities, promotes rural tourism and enjoyment of land for different population sectors, and enhances sustainable economic management of the natural resources. Therefore it is a land user-centred approach instrumental to the attainment of sustainability.

Multipurpose land use is fulfilled through the increase of resource use efficiency at spatial and temporal scales, the reduction of hazards and risks, the enhancement of system stability (multiple species) and the promotion of the social and recreational use of rural land.

Biodiversity increase is attained through patchiness and heterogeneity, which promotes efficient use of land resources and therefore plant species and structural richness of habitat. This system can be used for preserving autochthonous animal breeds and wildlife preservation and for prevention of hazards (erosion, fires).

Protection of the environment is reached through the maintenance of soil fertility, the enhancement of buffers against pollution (N, P…) and the reduction of fertilizer needs, combating desertification, carbon sequestration and carbon reservoirs, the improvement of water and soil quality and the prevention of climate warming.

Health protection is attained through the improvement of community and animal welfare and reduction of pollution effects.

Finally, silvopastoralism **promotes landscape** as it enhances its conservation and amenity value.

Therefore, silvopastoral systems should be promoted for preserving and enhancing productivity in marginal areas (mountains) in a sustainable way as well as for increasing stability, profitability, biodiversity and multipurpose uses of more productive areas since they improve traditional farm and animal husbandry activities and have the potential to increase the participation of diverse groups of people and offer more employment of people.

The inherent complexity of silvopastoral systems requires adequate management plans. This complexity makes research necessary at different levels: traditional knowledge evaluation, surveys, identification of management problems at a local level, research, training of all relevant sectors and communities and technology transfer.

In the European context, silvopastoralism should be considered a viable alternative land use in agricultural lands and included in the reformed CAP as a way of strengthening rural development and decoupling subsidies from production and expansion of the European forest strategy.

SUMMARY - SILVOPASTORAL SYSTEMS CONCLUSIONS

Silvopastoralism is an ancient way of managing forestland and more recently a way of managing pastureland. It leads to increases in production from forestry in the short, medium and long term and, as a structurally diverse ecological system and a multiple product system, often is a sustainable use of land.

If managed in a sustainable way, silvopastoralism can enhance biodiversity and contribute to the preservation of many endangered species that depend on ecotones between woodlands and open landscapes. Areas managed for silvopastoralism can reduce fire and erosion risk in forests, creating the necessary balance between full ground cover (erosion control) and shrub control (fire risk). In this context, manipulation of the type and density of grazing animals can greatly influence the biological and socio-economic outcomes from the system. Silvopastoral systems can favourably impact biodiversity, landscape and rural welfare issues which underpin agri-environment objectives through a number of attributes, including efficient nutrient cycling, buffering against non-point source pollution, fulfilling animal welfare criteria, employment generation and income enhancement, reversal of rural abandonment and creation of viable rural communities. Also, the silvopastorism production improves the landscape of the countryside and it could be used for ecotourism projects.

In the policy arena, silvopastoral systems can contribute to reduction of livestock grazing pressure, better nutrient management and amelioration of animal living conditions if introduced to intensive or semi-intensive farming systems. Silvopastoralism aligns closely with current key EU policy for intensively managed pasturelands which will be decoupled from subsidies for production to a more area-based system of payment with strict environmental and other cross-compliance measures attached. In a more worldwide context, silvopastoralism can be a mechanism to create land-use systems with levels of carbon sequestration which are higher than those from pastureland and which can buffer the more adverse effects of climate change.

Silvopastoralism can create multifunctional landscapes which can be used for realizing a range of purposes, including nutrient-enriched waste disposal systems that meet crop fertilizer requirements; a range of plant, livestock and socio-economic products; key recreational educational, conservation and landscape resources; and conserving animal autochthonous breeds. The general public expects to see all these benefits in return for public investment and as a cultural right. In the latter context, the re-creation and maintenance of traditional patterns of land use, including transhumance and forest grazing, are of fundamental importance in the preservation of national heritage and respect for the countryside.

FUTURE RESEARCH PERSPECTIVES

Whilst the above benefits of silvopastoral systems are more or less known, the potential to integrate silvopastorism into current land-use systems is poorly researched. In this context, valuable lessons can be learned from global experiences in similar land-use systems. Key research issues that emerged can be grouped into six categories:

a. Inventory of silvopastoral resources and current research.
b. Component evaluation.
c. Ecological interaction.
d. System management.
e. Socio-economic, human and cultural aspects.
f. Dissemination of information to land-users.

a. *Inventory of silvopastoral resources and available research*

There is a need to consolidate the knowledge base, summarize recent progress and make this available to a wide range of countries, across Europe especially, including the ten new accession countries.

Key outcomes could be: (i) *dissemination of the findings of the congress;* (ii) *creating an inventory of the systems already practised or researched in a way that can be easily interrogated;* (iii) *establishment of research groups and networks (particularly benefiting from global expertise) to discuss and share knowledge.*

b. *Component evaluation*

(i) Modelling of the growth of trees in a range of cases to enable them to be integrated into novel and traditional systems. The potential of new forage trees such as *Morus alba* and *Robinia pseudoacacia* between others and legume trees. Implications of climate change for tree selection.
(ii) Influence of trees on pasture, especially within reduced fertilizer inputs or as a sink for sludge disposal. Quality of pasture for animal production.
(iii) Animal feeding, health, stocking rate and welfare issues.

c. *Ecological interactions*

The impact of a range of climates and soils on factors which are specifically related to sustainable management and products of multiple outputs. Concentrate especially on nutrient management to meet policy objectives.
(i) Experience in North America, where woody plants have been successfully used as riparian buffers, and in reduction of non-point source pollution through silvopastoral practices should be used in the European environment.

(ii) The ecophysiology of trees and pasture as systems and their subsequent response to disturbance, whether by grazing, fire, etc.
(iii) The carbon economy of silvopastoral systems.
(iv) The mitigation of climate warming.
(v) The role played by silvopastoralism in the conservation of biodiversity.

d. *Management*

The integration of biological and ecological components and their interaction into sustainable management systems to meet a range of objectives. These may need to consider economic, cultural and social needs of farmers.
(i) Modelling as a predictive tool in vegetation management.
(ii) Ecological and landscape assessments for land-use options.
(iii) Management for specific issues such as fire and erosion control and nutrient conservation.
(iv) Silvopastoralism in organic and integrated farming systems.

e. *Socio-economic human and cultural aspects*

Adaptation of systems to the particular socio-economic needs of the EU member states.
(i) Preservation and encouragement of traditional versus new systems (reduction of conflict between wood preservation and conservation of traditional silvopastoralism activities).
(ii) Transhumance and silvopastoralism.
(iii) Quantification of non-market values from silvopastoral systems and their integration into viable farm businesses.
(iv) Synthesis of agricultural, forest and environmental policy drivers within silvopastoral systems.
(v) Increased understanding of the motivations and potential and actual responses to policy and economic alternatives of landowners and silvopastoral enterprises.
(vi) Inventory and assessment of traditional ecological knowledge, as well information needs, of silvopastoral practitioners.

f. *Dissemination to land-users*

As silvopastoral systems cut cross a range of disciplinary boundaries, there is a need to develop broad-spectrum technology transfer mechanisms and train extension workers to disseminate this information.
(i) Development of knowledge-based systems for technology transfer.
(ii) Dissemination to meet policy objectives.
(iii) Use of networks to target specific issue scenarios.
(iv) Assessment of the effectiveness of technology transfer and adaptive use of feedback from practitioners.

RECAPITULATION

1. A series of research issues need to be addressed to underpin policy directives and technology transfer. These can be grouped under: inventory of resources and research; component evaluation: ecological interaction, management, socio-economic, human and cultural; dissemination.
2. There is a need to invest in technology transfer through training of extension workers and develop new dissemination mechanisms involving the farmer, researcher and adviser. The importance of farmers in these systems will definitively promote sustainable land use.
3. There is currently no policy on silvopastoralism in relation to its position among:

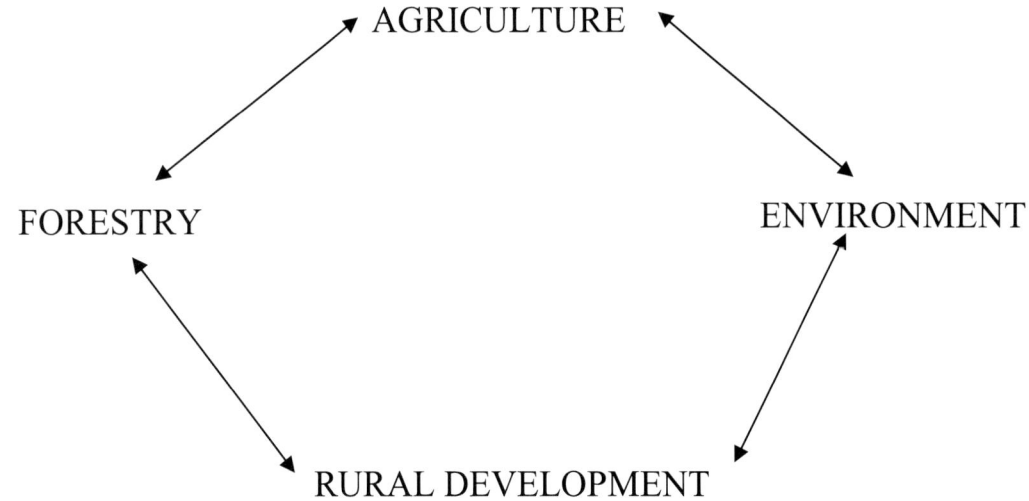

4. There is a need for a silvopastoralism policy in the context of:

- The reformed CAP. Inclusion of silvopastoralism in the Single Payment Scheme (SPS).
- The strengthening of rural development as the second pillar of the CAP.
- Decoupling of subsidy from production.
- Expansion of the EU.
- The comments made by the members of the SAFE Project in Toulouse, April 2004, on the "Implementation of the mid-term review of CAP related to agroforestry" should be seen alongside this Statement.

The most notable outcomes are:

1. The definition of 'woodland' should be clarified to ensure that it does not lead to the removal of trees from farmed landscapes.
2. In Guidance Document AGRI/2254/2003 of the Agriculture Directorate General of the European Commission, the threshold recommended for 'woodland' (and hence exclusion from payment under the SPS) is > 50 trees/ha. Exceptions listed should cover 'agroforestry'.
3. Agroforestry should remain classified as agriculture in the wording of Article 5 of Regulation 24/9/01.
4. Fifty trees trees/ha is an acceptable definition of woodland for the purpose of 1782/03 but should be classified to say "50 trees/ha of more than 15 cm diameter at breast height".
5. For silvopastoral systems, SPS can be maintained provided that no more than 50% of the non-shaded pasture production is maintained.
6. The EU should clarify that they have the flexibility to allow multiple activities within parcels in their national IACS systems (e.g. forestry and cropping in the same parcel).
7. Trees planted at agroforestry spacings do not constitute a perennial crop.
8. National definition of 'good agricultural and environmental condition' could include the phrase 'well-managed agroforestry is recognized as a mechanism of improving landscape and environmental diversity'.

INDEX

accounting
   EAA/EAF 97 (European System of Accounts)
      compared with AAS (Agroforestry
      Accounts System) 324–327
      applied to a dehesa 330–333
acidity, soil 96, 104(tab), 165–167
acorns 101–102, 110–111, 174–175, 180–181, 389
adaptation: to the environment 231, 352(tab)
adoption: of silvopastoral systems by farmers 16–17, 21–22, 75–76, 341
aeration, soil 225
afforestation 68–69, 197–199
   fire prevention in reafforested areas 381
Africa 106–107, 137–138, 150–151, 300, 359–360
age: of trees 96
Agenda 2000 (EU) 363
aggregates, water-stable soil 226–227
Agroforestry Accounts Systems (AAS) 324–326, 330–333
Agrostis tenuis: establishment of pasture 266–268
alkanes: in diet estimation 124–126
   method validation 127–128
alley pasture systems 15–16, 27, 73, 76, 197–199
Alps: cattle grazing 259–260
amelioration potential, environmental 73–74, 129–131
Apennines: oversowing of ski lanes 152–153
arbuscular mycorrhizae 192–193
ArcView GIS software 362
aridity, regions of 371
assistance, technical 337
astringency, tannin 90
Atlantic regions
   changes in agricultural production 231
   establishment of herbaceous strata 96–98
   experiments with Rubia Gallega cattle 232–234
   pasture establishment for extensive systems 266–268
   perspectives for silvopastoralism 408–409
   pig production 411–412
   post-fire characterization of shrubland 269–270
   production in coniferous and broadleaved forests 409
   tree and shrub characteristics 391(tab)
   tree-pasture interactions 94–96, 188–190
   types of silvopastoral systems 93–94, 248–250
Australia 103–104, 255–256, 370

Balaninus elephas 175
Bangladesh: improvement of smallholdings 41–42
Basque Country 411–412
beech woods 55–57, 288–289
Belize: constraints on adoption of silvopastoral systems 16–17
berries 355(tab), 356(tab), 389
biodiversity
   of Cantabrian beech woods 55–57
   changes in abandoned dehesa 290–292
   effects of breed and stocking rate 246–247
   effects of burning 269–270
   effects of cutting and fertilization 241–243
   effects of vineyard cover crops 255–256
   enhanced by grazing 129–131, 163–164
   helped by land-use diversification 370
   indices 245
   Mediterranean swards 38
   required for sustainability 370
   of silvopastoral systems in Latin America 220–221
   of traditional Galician systems 248–250
   of windbreaks in Costa Rica 220
Biodiversity Action Plan: priority species 400–401
biomass
   estimation above and below ground 211–212
   fuel biomass
      effects of grazing 112–113
   microbial biomass 227
Bolivia 177–179, 371–372
Brazil: grazing within tree plantations 26–27
breeds
   effect of breed on goat weight 132–133
   effect of breed on vegetation dynamics and biodiversity 246–247
   indigenous cattle 232–234
broilers: litter 195–196
broom 248–250
browse impact 304–305
Bryant and May (UK) 20
Buddleia cordata 108–109
burning, controlled
   and biodiversity 269–270
   improvement of fallows in slash and burn systems 44–45
   as a method of fire prevention 381
   post-fire characterization of shrubland using structural variables 283–284
   and vegetation dynamics 142–144
   weight changes in goats and sheep on burned shrubland 257–258

Caesalpina yucatenenses 44–45
calabash trees 25
calendars, forage 302, 303(fig)
Calluna vulgaris 241–243
calves 140–141, 233(tab)
Canada
   carbon sequestration 210–217
   sheep methane emissions 216–217
Cantabrian mountains, Spain 55–57, 241–243, 246–247
Capra pyrenaica 160–161
Capsella bursa-pastoris 267–268
carbohydrates: in forage 88–89
carbon
   C budget for monoculture pasture 216(fig)
   C budget for poplar-based system 215(fig)
   content of ryegrass 213(tab)
   content of tree components 213(tabs)
   in and leached from soil 212, 214(tabs)
   quantification in trees 211
   sequestration 15–16, 74–75, 210–217, 220
Cashmere goats 132–133, 246–247
cattle farming and pastures
   alley pasture systems 15–16

assessment of contribution to income 333
bovine/*Quercus* experimental system 197–199
browse impact 304(fig)
cattle population in Poland 375
compatibility with woodland 384
dense oak forests 129–131
dispersed trees on Costa Rican farms 119–120
effects and management of Alpine grazing 259–260
effects on environment 24
experiments with Rubia Gallega breed 232–234
fodder banks 15
gorse and heather 157–158
grazing behaviour of calves 140–141
impact of grazing in protected mountain area 285–286
importance of trees in pastures 13, 14
Leucaena leucocephala promotes weight gain 221
multi-strata systems 16
net margin index per cow 234(tab)
pine-forest grazing 72–73, 75
recommendations for sustainable systems in Cuba 237
reforestation enterprises 26–27
seasonal nature of *Leucaena leucocephala* ingestion 168–169
spatial dependence and seasonal patterns of animal activity 239–240
transition of pastoral landscape to "new wilderness" 148–149
tree protection with unpalatable plants 202–203
Celtiberian goats 132–133, 246–247
Chaqueño Mountains, Bolivia 177–179
chestnuts 389, 392
chickens: litter 195–196
children: exploitation of tree resources 320–321
Cholistan Desert 346–347
classification: grazed forest stands 51–52
cocksfoot 172–173, 182–183
cocoa: effect of duck grazing on yields 62–63
coffee plants 117–118, 204–205
Colombia 16, 26
Common Agricultural Policy (EU) 19–20, 377–378
compaction, soil 225
composition, chemical
    of grasses 141(tab)
    of trees and shrubs 83–84, 138(tab)
    of tropical fruits 84
conflicts 359, 359(fig)
conservation
    easements and land trusts 297, 315–316
    effects on reservoir capacity 348–350
    Mediterranean forests 307–309
    shifting models 312–313
    working landscape 315–316
coppicing 206–208
cork 389
cork oak 110–111, 331–332
Costa Rica
    benefits of silvopastoral systems 16
    biodiversity of windbreaks 220
    cattle pastures
        tree species richness and density 13–14
        chemical composition of indigenous fruits 84

dispersed trees on cattle farms 119–120
fodder banks for cattle 15
future projections 338–341
payment for environmental services concept 335–336
    involvement of farmers 337–338, 341
payment for environmental services mechanisms and strategies 336–337
Council of Europe 321
cover crops 255–256
Crescentia spp. 25
Cuba 15, 221, 236–237, 415–417
cultural landscapes 321, 361–362
culture: benefits of silvopastoralism 360
cut-and-carry systems 26
cutting: impact on heathland pastures 241–243
Cynodon dactylon 170–171
Cytisus spp. 248–250, 269–270

Dactylis glomerata 172–173, 182–183
dairy sludge 192–193
dasometry 185–186
decision-making
    decision support tools (DSTs), computer-based 77
    models and procedures 415–417
Declaration for Silvopastoralism 418
deer 89–90, 306–307
    deer parks 320
degradation: environmental in Latin America 13, 24
degradation, soil *see under* soil
dehesa *see under* Mediterranean regions
Delphi method 363–364
density, tree and shrub *see also* tree-pasture interactions
    definition of optimum level 301
    effects on *Panicum maximum* 177–179
    effects on pasture productivity 103–104, 182–183
    effects on *Robinia pseudoacacia* herbage mass 206–208
    effects on soil characteristics 182–183
    root density 251–254
dependence: on forest resources 300
deserts 346–347
development, rural
    case studies 370–372
    resource diversification 369–370
diet and nutrition *see* nutrition and diet
digestibility
    grasses 140
    improvement in ruminants 85
    oakwood understorey forage 88
    organic matter in dehesa ecosystem 32–33
diversification: of resources 369–370
Domesday Book 400
DPSIR framework 395–396
drought 347
ducks: effect on cocoa yields 62–63
dung: deposition by cattle 239–240

EAA/EAF 97 (European System of Accounts) 326–327
    applied to a dehesa 330–333
easements, conservation 297, 315–316
economics

benefits of silvopastoralism 359, 381–382
Delphi method 363–364
silvopastoralism in USA 75–76
commercial values of silvopastoral systems 295–296
conservation easements and land trusts 297
landowner investment in environmental value 294–295
oak woodlands: costs and values 296–297
elk 89
emetics: to induce aversion to coffee plants 117–118
endangered species 38
environment
economics and policy 75–76
environmental amelioration potential 73–74, 129–131
payment for environmental services
concept 335–336
future projections 338–341
involvement of farmers 337–338, 341
mechanisms and strategies 336–337
Erica ciliaris 269–270
Erica tetralix 241–243, 269–270
Erica umbellata 269–270
erosion *see under* soil
Erythrina spp. 134–135
Eucalyptus 255, 382, 383, 384, 392
fire prevention 409
Europe, Western *see also* Mediterranean regions; individual countries
effects of land-use intensification 19–20
forest fires 380–381
the future of agroforestry 376–379
types of silvopastoral systems 20
European Union
Common Agricultural Policy 19–20, 377–378
EAA/EAF 97 (European System of Accounts) 324–327
applied to a dehesa 330–333
TRANSHOUMANT project 395–396
eutrophication 226
Everglades, Florida 74
Evritania, Greece 60–61
extension agents 337
Eysenhardtia polystachya 108–109

faeces: alkane content 124–126, 127–128
Falkland Islands 406–407
farmers
negative attitudes to trees 53–54
redefinition of role 337–338
resistance to change 16
farms: nutrient loading in farm surface water 272–274
feed: blocks for goats 106–107
fences, live *see* live fences
fermentation, rumen 84–85
fertilization
of acid soils 96, 165–167
biofertilizers 192–193
broiler litter 195–196
dehesa pastures 31
disadvantages 73–74

effects on pasture production and soil characteristics 182–183
impact on heathland pastures 241–243
improved response under tree canopy 276(tab), 277(tab)
sewage sludge 154–155
combined with liming 165–167
fertilizers: from forest matter 390
fibre, nutritional 89(fig)
firebreaks 170–171
fires *see also* burning, controlled
forest fire risk reduction 112–113
forest fires in Europe 380–381, 408–409
prevention 303–305, 383–385, 409
control of combustible plant material 384–385
in reafforested areas 381
role of livestock 384
firewood 14, 389, 390
and coppice stability 36
flock system 60
fodder, leaf and twig 320–321
fodder banks 15, 26
FRAGMENT project (Costa Rica, Nicaragua) 13, 15, 17
France
agroforestry regulations 377
fire prevention 303–305
survey of upland region farms 66–67
fruits 389, 392
berries 355(tab), 356(tab)
chemical composition 84
fuel biomass
effects of grazing 112–113
net firewood return 294(fig)
fungi, soil 192–193

Genista scorpius 114–115
geographical information systems (GIS) 77, 236–237, 362
Germany 397–398
germination: effect of heat 114–115
Gliciridia sepium 134–135
goats
browse impact 304(fig)
on burned shrubland 142–144, 257–258
compatibility with woodland 384
diet composition and preferences 84, 124–126, 127–128, 138(tab)
effects of breed and stocking rate on heath-gorse communities 246–247
effects of breed and stocking rate on live weight 132–133
grazing prevents regeneration in forest stand 365–367
improvement of productivity 106–107
milk production 221(tab)
in mountain zones 58–59
Gorbeia Natural Park, Basque Country 157–158
gorse
grazing by cattle 157–158
grazing by goats 124–126, 132–133, 246–247
grazing by sheep 124–126
vegetation dynamics in burned shrubland 142–144

weight changes of animals grazing burned
shrubland 257–258
grasses
growth with sewage sludge fertilization 154–155
in *Pinus brutia* understorey 145–147
yield increased in alley pasture systems 15
grazing: preferences 124–126, 127–128, 140–141
Great Himalayan National Park 357–358
Greece
assessment of rangeland health 281–282
Pinus brutia understorey 145–147
traditional systems 53–54, 60–61
greenhouse gases 210, 216–217
growth, plant: effect of soil compaction 225
guava 25

health: of rangelands 281–282
heathlands
biodiversity changes in abandoned dehesa 290–292
effect of burning on biodiversity 269–270
Falkland Islands 406–407
goats 124–126, 132–133, 246–247
impact of cutting and fertilization 241–243
mountain grazing by cattle 157–158
sheep 124–126
vegetation dynamics in burned shrubland 142–144
Veluwezoom National Park, The Netherlands 148–149
weight changes of animals grazing burned
shrubland 257–258
herbaceous strata: establishment 96–98
Hi-SAFE model 376
high density systems 26
holm-oak 101–102, 160–161
acorn production in the dehesa ecosystem 174–175, 180–181
interactions with understorey: effects of
management 263–265
light availability in understorey 261–262
soil fertility *vs.* distance from trees 278–280
honey 390
horses 129–131, 148–149, 170–171, 200–201, 384
Hungary 68–69
hunting 38, 392

ibex, Spanish 160–161
Ilex aquifolium 95, 389, 392
income
of Bangladeshi smallholders 42
measurement: comparison of two systems 324–326, 330–333
total social 325
variability 302
India, Great Himalayan National Park 357–358
indices, biodiversity 245
information: dissemination 78, 413–414
insects 256
Integrated Silvopastoral Approaches for the
Management of Ecosystems (GEF-CATIE)
concept and strategies 335–337
future projections 338–341
involvement of farmers 337–338, 341

ipecacuanha syrup: to induce aversion to coffee plants
117–118
Ireland 20, 197–199
Italy 152–153, 170–171, 370

Juniperus oxycedrus 161
Jura Mountains 202–203

Lake Okeechobee, Florida 74
land, value of 296
land trusts 297, 315–316
landowners: investment in environmental value 294–295
landscapes, silvopastoral 321
future 321–322
origins and types 319
use of woody plants 320–321
Latin America
benefits of native tree species 49–50
adoption of silvopastoral systems 16–17
alley pasture systems 15–16
environmental degradation 13, 24
FRAGMENT project 13, 15, 17
grazing within plantations 26–27
improvement of fallows 44–45
integration of trees into ruminant production
systems 83–85
live fences 13
biodiversity of silvopastoral systems 220–221
characterization of tree species in mountain region
351–354
development of silvopastoralism 402–403
dispersed trees on cattle farms 119–120
effect of sheep on coffee quality 204–205
effects of tree and shrub density on *Panicum
maximum* 177–179
Integrated Silvopastoral Approaches for the
Management of Ecosystems (GEF-CATIE)
335–341
models and procedures for decision-making 415–417
prevention of soil erosion 236–237
successful models of silvopastoral systems 221
sustainability of current silvopastoral systems 219–220
legumes *see also Leucaena leucocephala*
brooms 248–250
growth with sewage sludge fertilization 154–155
in improvement of dehesa pastures 32
most important component of Mediterranean swards
37–38
in *Pinus brutia* understorey 145–147
species for Atlantic areas 97–98
tropical tree legume foliages 134–135
Leucaena leucocephala 41–42, 44–45
effects on forage 251–254
palatability for sheep 84
promotes weight gain in cattle 221
seasonal nature of ingestion by cattle 168–169
ley farming 370
liga 389
light
availability in holm-oak dehesa 261–262

effect of shade on nutritive value 172–173
effect of shade on pasture establishment 266–268
liming 96
    combined with sewage sludge 165–167
lithium chloride: to induce aversion to coffee plants 117–118
litter
    broiler 195–196
    leaf: quantification of decomposition 212, 215(tab)
live fences 13, 15(tab), 25–26, 150–151, 351, 352
livestock *see also* individual species
    government subsidies 60–61
    non-ruminant animals 21
loading, nutrient: of surface water 272–274

maize 44–45
Mali: use of live fences 150–151
management
    of arid rangelands 371
    effects on productivity 93, 180–181
    fuel management techniques 304(tab)
    of livestock 384
    of Mediterranean and temperate systems 300–302, 303(fig)
    in tropical mountain regions 353
meat 389
medicinal plants 390, 392
Mediterranean regions
    assessment of rangeland health 281–282
    Australia 255–256
    Californian oak woodlands 313–316
    changes in wild animals and livestock species 38
    classification of silvopastoral systems 39(tab)
    common forest plant species 87(tab)
    dehesa (montado) ecosystem (Spain, Portugal)
        acorn production 101–102, 110–111, 174–175, 180–181
        assessment of the role of grazing 395–396
        changes in biodiversity after abandonment 290–292
        characteristics 30–31
        conservation 33–34
        deer raising and shooting 306–307
        EAA/EAF 97 (European System of Accounts) compared with AAS (Agroforestry Accounts System) 330–333
        fertilization 31
        impact of *Quercus rotundifolia* on the understorey 275–277
        improvement using legumes 32
        income 302
        light availability in holm-oak understorey 261–262
        net protein and organic matter digestibility 32–33
        pasture management 31(fig), 32(tab), 34, 263–265
        products 392
        soil fertility *vs.* distance from trees 278–280
        structure characterization of *Quercus pyrenaica* woodlands 185–186
        types of land use in Portugal 64–65
    digestibility and nutritive value of oakwood understorey forage 88–91
    effects of grazing on fuel biomass 112–113
    floristic stability of pastures 244–245
    forest fires 380–381
    goat production in Portuguese mountain zones 58–59
    goats: improvement of productivity 106–107
    gorse-heathland communities 124–126
    grazing preferences of goats and sheep 124–126
    horse grazing in firebreaks 170–171
    human demands and activities 36–37
    impact of grazing in protected area 285–286
    importance of holm-oak to Spanish ibex 160–161
    land-use dynamics 343–345
    multifunctionality of forests 299, 300(fig)
    non-wood forest products 388–392
    nutritional character of forestry resources 137–138
    Pinus brutia understorey 145–147
    production of *Genista scorpius* seed 114–115
    rural development 370
    swards 37–38
    traditional systems in Greece 53–54, 60–61
    tree and shrub characteristics 37, 391(tab)
    vegetation dynamics of burned shrubland 142–144
    weight changes of animals grazing burned shrubland 257–258
methaemoglobinaemia 73
methane: emitted by sheep 216–217
Mexico
    chemical composition of trees and shrubs 83(tab)
    development of silvopastoralism in Chiapas 402–403
    effect of sheep on coffee quality 204–205
    improvement of fallows 44–45
    palo dulce and tepozan seasonal fodder 108–109
    root density and soil-water relationships 251–254
    sheep in coffee plantations 117–118
    tree species in Tabasco 351–354
microorganisms, soil 192–193, 220(tab), 227
micropropagation 121–123
milk
    fodder banks increase production 15, 26
    goat 221(tab)
    Leucaena and *Sesbania* increase production 41–42
    production results in Latin America 221(tab)
milpa system 44–45
montado (dehesa) *see under* Mediterranean regions
Morocco 300
Morus alba 121–123
mosaic systems 55–57
mountains and mountainous regions *see also* Mediterranean regions, dehesa (montado) ecosystem
    Cantabrian mountains, Spain 55–57
    effects and management of cattle grazing 259–260
    effects of goat breed and stocking rates 246–247
    effects of grazing on fuel biomass 112–113
    effects of tree and shrub density on *Panicum maximum* 177–179
    Evritania, Greece 60–61
    experiments with indigenous cattle breeds 232–234
    farm types, Margeride region, France 66–67

floristic stability of pastures 244–245
goat production 106–107
impact of cutting and fertilization 241–243
impact of grazing in protected area 285–286
oversowing of ski lanes 152–153
palo dulce and tepozan seasonal fodder 108–109
Peneda's Mountain, Portugal 58–59
restoration of wood-pastures 398
transhumance in Great Himalayan National Park 357–358
tree rejuvenation 202–203
tree species in Tabasco, Mexico 351–354
use of gorse and heather by cattle 157–158
Mucuna pruriens 44–45
multi-strata systems 16
mushrooms 356, 390, 392
MUSLE model 349
mycorrhizae, arbuscular 192–193

nematodes: infestation of young animals 221(tab)
Neolithic era 320
Netherlands, The 148–149
networking 403(tab)
New Zealand 20
Nicaragua 13–14
Nigeria 359–360
nomadism 346–347
Northern Ireland 163–163, 404–405
Norway spruce 202–203
nutrient loading: of surface water 272–274
nutrition and diet
estimating diet composition 124–126, 127–128
response of sheep to tropical tree legume foliages 134–135
of ruminants 83–85, 137–138

oak *see Quercus*
Oman 371
orange trees 200–201
organic matter digestibility 32–33
ornamental plants 390, 392

Pakistan: desert nomadism 346–347
palo dulce 108–109
Panama: alley pasture systems 15–16
Panicum maximum 177–179, 200, 251–254
parkland 400–401
peats, acid 406–407
Peneda's Mountain, Portugal 58–59
Penisetum purpureum 134–135
Peru: water supply 371
pests: of plants 175, 180
phosphorus 272–274
phosphorus, loss of 74
Phyllirea latifolia 161
phytocoenotical stability index 245
Picea albis 202–203
pigs 411–412
pine nuts 389
pines *see Pinus* spp.
Pinus spp. 389, 406
and cattle grazing 72–73, 75
fire prevention 409

growth of *P. radiata* with sewage sludge fertilization 154–155
combined with liming 165–167
P. brutia understorey 145–147
plantation productivity 97
uses of pine forests 391–392
Piscida piscipula 44–45
plantations, tree 26–27
ploughing 225
Poland 374–375
pollarding 320, 321
polyethylene glycol (PEG): as food supplement 106–107
poplar: carbon budget 215(fig)
Portugal 58–59, 343–345
montado (dehesa) system *see under* Mediterranean regions
poultry: litter 195–196
price support 75–76
private sector 221, 312
products, non-wood, from forests 299, 355–356, 388–392
projections: of land use 338–341
propagation 352
property rights 150–151
Prosopis spp. 49–50
protection, physical: of trees 302
protein
effect of light levels on content 172–173
microbial 85
net protein of dehesa pastures 32–33
in oakwood understorey forage 89
in Tunisian shrubs 137–138
pruning: effect on acorn production 101–102, 110–111
Pterospartum tridentatum 269–270
Pyrenees 112–113

quality, soil *see under* soil
Quercus
biodiversity changes in abandoned oak dehesa 290–292
bovine/*Quercus* experimental system 197–199
digestibility and nutritive value of oakwood understorey forage 88–91
effect of pruning on acorn production 101–102, 110–111
effects of grazing in dense forests 129–131
effects of grazing in old-growth forests 288–289
effects of management on interactions with understorey 263–265
environmental and commercial value of woodlands 294–297
importance of holm-oak to Spanish ibex 160–161
light availability in holm-oak understorey 261–262
Root Production Method™ 197–199
soil fertility *vs.* distance from trees 278–280
structure characterization of woodlands 185–186
Quercus faginea 129–131
Quercus ilex 101–102, 160–161, 301
acorn production in the dehesa ecosystem 174–175, 180–181
Quercus pyrenaica 185–186
Quercus robur 197–199, 288–289

Quercus rotundifolia 275–277
Quercus suber 110–111, 301, 389

ranchers: values held 314–315
rebollo oak 185–186
regulations: on agroforestry 377–378
rejuvenation, tree 202–203
remote sensing 77, 361
research
    future perspectives 419–420
    results of silvoarable trials 20–21
reservoirs: capacity 348–350
resin, pine 389
respiration, soil 212, 214(tab)
rights, property 150–151
riparian buffers 74
Robinia pseudoacacia 206–208
Root Production Method™ oak 197–199
roots: density 251–254
Rubia Gallega cattle 232–234
rumen
    botanical analysis 160–161
    fermentation kinetics 84–85
Russia: non-wood products 355–356

SAFE (Silvoarable Agroforestry for Europe) 376–379
Salix 406, 407
São Tomé and Principe 62–63, 372
Scotland: ancient wood pastures 361–362
sedimentation: into reservoirs 348–350
seeds
    as commercial product 389–390
    effect of heat on *Genista scorpius* germination 114–115
sequestration, carbon 15–16, 74–75, 210–217, 220
Sesbania acculeata 41–42
sewage sludge: combined with liming 165–167
shade *see under* light
Shannon-Weaver index 245
sheep
    acceptability of foliage 84
    browse impact 304(fig)
    on burned shrubland 142–144, 257–258
    compatibility with woodland 384
    diet composition 124–126, 127–128, 134–135
    effect on coffee quality 204–205
    grazing ski lanes 152–153
    impact of grazing in protected mountain area 285–286
    induction of aversion to coffee plants 117–118
    methane emission 216–217
    population in Poland 375
shelterbelts 68–69
shooting, professional 306
shredding 320, 321
Sierra de Guara National Park, Spain 285–286
Sierra Mágina, Andalusia 244–245
Silvoarable Agroforestry for Europe (SAFE) 376–379
Silvopastoral Agroforestry Toolbox 414
ski lanes: oversowing 152–153
slash and burn systems: improvement of fallows 44–45
sludge
    dairy 192–193
    sewage 154–155, 165–167
society: benefits of silvopastoralism 359, 360(fig)
soil
    acidity 96
    compaction 225
    degradation
        severe global problem 223
        types 224(tab)
    effects of tree density 104(tab)
    erosion 226–227
        prevention strategies 236–237
        scale of the problem 224
            risk factors 33–34
    factors influencing characteristics 182–183
    fertility *vs.* distance from trees 278–280
    leachate carbon 212
    loss of moisture 276
    loss of organic matter 226–227
    microfauna in systems with and without trees 220(tab)
    organic carbon 212, 214(tab), 226
    quality 224
        improvement by conservation 227–228
    respiration 212, 214(tab)
    soil-water relationships 251–254
Solling mountains, Germany 398
Somalia 371
Southeastern Agroforestry Network of Demonstration Sites (SANDS) 78
sowing: oversowing of ski lanes 152–153
Spain
    Atlantic regions *see* Atlantic regions
    beech woods 55–57
    common forest plant species 87(tab)
    dehesa *see under* Mediterranean regions
    Delphi method applied to environmental issues 363–364
    digestibility and nutritive value of oakwood understorey forage 88–91
    effects of forest conservation on reservoir capacity 348–350
    effects of goat breed and stocking rates 246–247
    effects of grazing in old-growth forests 288–289
    European country most affected by forest fires 408–409
    factors influencing pasture production 182–183
    factors influencing soil characteristics 182–183
    heathlands 124–126, 132–133, 246–247
        impact of cutting and fertilization 241–243
    horses in orange plantations 200–201
    importance of holm-oak to Spanish ibex 160–161
    mountain grazing by cattle 157–158
    non-wood forest products 388–392
    oak forests in Navarra 129–131
    Pinus radiata cultivation 97
    production in coniferous and broadleaved forests 409
    sewage sludge fertilization 154–155
Spanish ibex 160–161
specialization, increase in 343–345
species, tree
    effects on pasture productivity 94, 188–190
    effects on soil characteristics 182–183

spruce, Norway 202–203
stability, floristic 244–245
Stellaria media 268
stocking rate
    effect on goat weights 132–133
    effect on vegetation dynamics and biodiversity 246–247
sustainability 404–405
    importance of biodiversity 370
    of silvopastoral systems in Latin America 219–220
swards 37–38
    Agrostis tenuis monoculture 266–268
    on ski lanes 152–153
Switzerland 202–203
    effects and management of Alpine cattle grazing 259–260
    forest grazing by goats 365–367
    history of land use 365

tacotales 338, 339, 340
Taiwan grass 134–135
tannins
    as commercial products 389
    effect on digestibility of woodland forage 89–91
    overcoming effects in goats 106–107
    in Tunisian shrubs 137–138
tenure, land: tree planting rights 150
tepozan 108–109
thinning 301
tillage 228
tobas 346, 347
tourism 370
training 337
trampling: by cattle 239–240
TRANSHOUMANT project (EU) 395–396
transhumant systems 58–59, 60, 346–347, 357–358, 395–396
tree-pasture interactions 188–190 *see also* density, tree and shrub
    in alley systems 15, 76
    effects of tree species, distribution, density and age 94–96
Trifolium brachycalycinum 170–171
truffles 390
Tunisia 106–107, 137–138
Txerrizaleok elkartea (association of farmers) 411

Ukraine 51–52
undergrowth: quality and characteristics 383–384
UNESCO 321
United Kingdom 20
    ancient wood pastures in Scotland 361–362
    animal species used in silvopastoral systems 21
    effects of grazing on biodiversity and tree regeneration 163–163
    English wood-pasture and parkland 400–401
    factors governing adoption of silvopasture 21–22
    policy alignment 21
    technology transfer for farmers 413–414
United States
    C sequestration potential for grazing lands 74–75
    commercial values of silvopastoral systems 295–296
    computerised decision support tools 77
    conservation
        conservation easements and land trusts 297
        easements and land trusts 315–316
        environmental economics and policy 75–76
        history of silvopasture in south-eastern USA 72–73
        information dissemination 77
        landowner investment in environmental value 294–295
        oak woodlands: costs and values 296–297
        Robinia pseudoacacia 206–208
        shifting models 312–313
        values of ranchers 314–315
        working landscape conservation 315–316
universal soil loss equation (USLE) 236–237
Uruguay 49–50

values
    commercial 295–296
    environmental 294–295, 296–297
    land use 296
    private *vs.* public 325
Veluwezoom National Park, The Netherlands 148–149
viability: of acorns 180–181
village system 60
vineyards 255–256, 392

water
    effect of deforestation on supply 371
    reduction of nutrient loading in farm surface water 272–274
    soil-water relationships 251–254
watersheds, forested 348–350
weeds
    control by ducks 62–63
    control by horses 200–201
weevils 180
weight: effect of goat breed and stocking rate 132–133
willow 406, 407
windbreaks 27, 220
women: exploitation of tree resources 320–321
wood-pastures 400–401
woodland: desirable characteristics 382–383
Working Landscapes movement 313

xesteiras 248–250

zebu cattle 168–169